Introductory Food Chemistry

Introductory Food Chemistry

JOHN W. BRADY

Comstock Publishing Associates
a division of
Cornell University Press | Ithaca and London

First published 2013 by Cornell University Press

Printed in China

Library of Congress Cataloging-in-Publication Data

Brady, John W., 1952–
   Introductory food chemistry / John W. Brady.
     p. cm.
   Includes bibliographical references and index.
   ISBN 978-0-8014-5075-4 (cloth : alk. paper)
   1. Food—Analysis.   2. Food—Composition.   I. Title.

TX545.B73 2013
664′.07–dc23

                    2012039748

Cornell University Press strives to use environmentally responsible suppliers and materials to the fullest extent possible in the publishing of its books. Such materials include vegetable-based, low-VOC inks and acid-free papers that are recycled, totally chlorine-free, or partly composed of nonwood fibers. For further information, visit our website at www.cornellpress.cornell.edu.

Cloth printing   10 9 8 7 6 5 4 3 2 1

For Michael, Amanda, and Sam

and Madeline

Contents

Preface

This book was developed from the class notes for the first semester of a two-semester undergraduate course in food chemistry at Cornell University. This course, which is required for food science majors, is taught in the second semester of the junior year and covers the basic general principles of food chemistry, while the second semester, team-taught in the senior year, covers the applications of these principles to specific examples from the various food commodity groups. Typically, the students who take this course have had two semesters of introductory chemistry, introductory physics, calculus, organic chemistry, and biochemistry, as well as food microbiology and food engineering. Many topics from these courses are reviewed in the first three chapters of this book, however; and much of the material in the latter chapters can be understood by people without this background. Because Cornell students enrolled in this course have already taken a laboratory course in analytical chemistry methods applied to food systems, such methods are not discussed extensively here, which allows the material to fit into a single semester. A separate laboratory course in food chemistry accompanies the second semester of this course, using the text *Food Chemistry: A Laboratory Manual*, by D.D. Miller. The material covered in this book is intended in part as preparation for that course. Even with these limitations, the material presented here is somewhat too extensive to be covered in class in just one semester, and in practice, I skip about 10% of the material in my class lectures. Inevitably, difficult decisions were necessary about what material could, and could not be, covered in a single-semester course, and what level was appropriate for an undergraduate course; the choices in part were governed by my own personal interests, as well as by the advice of my colleagues. In general, I have tried to make the discussion somewhat more technical, on the chemical level, than the excellent book *On Food and Cooking: The Science and Lore of the Kitchen*, 2nd Edition, by Harold McGee, and less technical and comprehensive than the equally excellent books *Food Chemistry*, 4th Edition, by H.-D. Belitz, W. Grosch, and P. Schieberle, and *Fennema's Food Chemistry*, 4th Edition, edited by S. Damodaran, K.L. Parkin, and O.R. Fennema, all of which I frequently use as reference sources myself.

I am grateful to many individuals for the assistance they have given me over the years in developing the materials for this course. My colleagues here at Cornell have been very helpful and supportive, including particularly Terry Acree, Carl Batt, David Barbano, Gerald Feigenson, Bruce Ganem, Robert Gravani, Joe Hotchkiss, John Kinsella, Harry Lawless, C.-Y. (Cy) Lee, Betty Lewis, Rui-Hai Liu, Dennis Miller, Steve Mulvaney, Robert Parker, Syed Rizvi, Joe Regenstein, Gavin Sacks, David Wilson , and Robert Zall. Many non-Cornell colleagues, including Seema Bhagwat, Andreas Blennow, Paola Cescutti, Fred Cohen, Valerie Daggett, William Goddard, Douglas Goff, Michael Himmel, Richard Ludescher, Luciano Navarini, Richard Pastor, Sergio Paoletti, Roberto Rizzo, Joseph Sebranek, Fabiana Sussich, and Göran Widmalm, among many others, have been especially generous in sharing information, figures, and data, or in making comments about portions of the manuscript. My friend and mentor Attilio Cesàro has been especially helpful both in his comments and in sharing figures and data. Several of my students and postdoctoral associates, including Peter Hansen, Naomi Kawaji, Gérald Lelong, Adrien Lerbret, Philip Mason, James Matthews, Hitomi Miyamoto, Rebecca Schmidt, Udo Schnupf, Letizia Tavagnacco, and Jakob Wohlert, have contributed figures for the text, and my student Mo Chen was particularly helpful with many last-minute contributions. A number of my former teaching assistants, including Julie Goddard, Meera Iyer, Katherine Kittel, Yizhi Meng, Jose Muñoz, Daina Ringus, David Tisi, Cindy Winoto, and Kelly Wolfe, and as well as the students in my classes over many years, have made numerous suggestions, found many mistakes large and small, and helped me to improve the understandability of the text. Diane Florini provided invaluable assistance in proofreading the manuscript. I thank Hugh Ink, Manager of the Cornell Orchards, and his staff, for spending a summer producing the apple in the Erlenmeyer flask depicted on the cover, and Jason Koski of Cornell University Photography for taking the excellent photograph, and Ted Crane and John Enright for help in technical computer matters. I also am grateful to the College of Agriculture and Life Sciences of Cornell University for generous financial support for the publication of this book. I thank my postdoctoral mentor, Martin Karplus, for first calling to my attention the importance of the chemistry of foods, as well as Giuliana Giuliani, Rich Pastor, and my son-in-law Sam Lipp for many outstanding demonstrations of the culinary arts.

Thanks also are due to Marie-Louise Saboungi, the CRMD/CNRS, and Le Studium, all at the Université d'Orléans, as well as Attilio Cesàro and the Department of Life Sciences at the University of Trieste, for hosting me while parts of the text were being written. Finally, I would like to thank the staff of Cornell University Press, particularly Ange Romeo-Hall, Heidi S. Lovette, Katherine Liu, and Peter Potter, as well as Chris Crochetiere at BW&A, for their patience, hard work, and encourage-ment in putting this book together.

Introductory Food Chemistry

1.
Food Chemistry: An Introduction

WHAT IS FOOD CHEMISTRY?

All foods are made up of chemicals; in fact, foods are simply edible chemicals. Some people might find this statement shocking, since in the popular mind the term "chemicals" conjures up images of synthetic toxins and carcinogens not normally found "naturally" in foods. Many people might insist that they want no chemicals in their food, in the same way that some bartenders might insist that no water is in their beer. Attitudes like these, however, come from misunderstanding or ignorance. Of course, not only foods but all of the material world around us consists of either the chemical elements or of the many kinds of molecules that can be constituted from them. In addition, cooking is, in its essence, chemistry. It is no coincidence that many of the best dishes are prepared by following very specific recipes, because these are actually prescriptions for controlling the many chemical reactions and physical transformations that take place in foods as they are processed. Preparing food, even if that involves just cutting up fresh fruits and vegetables, leads to many chemical reactions that affect the taste, texture, appearance, safety, and nutritional quality of food. For example, the chemical thiopropanal sulfoxide that causes you to cry when you cut up onions does not exist in the onion until a cut is made—the act of cutting initiates the enzymatic reaction that creates this compound by disrupting the cell walls, which also allows it to escape into the air. Given that foods are indeed made up of chemicals, the field of food chemistry, which is the study of the chemical and physical reactions of food molecules, is critical to the broader understanding of foods.

It is not necessary to understand a food to enjoy it, but it is probably necessary to understand something about food to prepare it well, and especially to manufacture a food product that will be tasty, shelf stable, safe, and nutritious. That understanding can be on many levels (such as how to select a ripe melon or sterilize Mason jars for canning), but it also includes the necessity of understanding something about the chemistry and physics of food components in order to know how various reactions and processes will affect the character of a product or a prepared dish.

The food industry has long known the importance of food science, and food chemistry, in the manufacture of quality products. In recent years, professional chefs have also begun to appreciate the value of scientific knowledge in the preparation of fine meals, particularly as the knowledge gained in the laboratory has become more widely available through such popular works as *On Food and Cooking: The Science and Lore of the Kitchen* by Harold McGee. This new approach to cooking is the basis of the postmodern cuisine usually called **molecular gastronomy**, pioneered by the scientists Nicholas Kurti, now deceased, of Oxford University and Hervé This of the Institut National de la Recherche Agronomique (INRA) in Paris. This movement, whose principal exponents include, among many others, the chefs Pierre Gagnaire in France, Ferran Adrià in Spain (who dislikes the term molecular gastronomy and refers to such cooking as deconstructivist cuisine), Heston Blumenthal in England, and Nathan Myhrvold, Wylie Dufresne, and Grant Achatz in the United States, uses the principles of chemistry and physics to develop exciting and innovative new recipes and dishes (*The New York Times*, Nov. 6, 2007). The restaurants where these chefs cook are routinely ranked by such sources as *The Michelin Guide* as among the best in the world, yet these cooks employ ingredients such as xanthan gum and gelling polysaccharides more commonly used in the past for manufactured foods. While these chefs might not agree on the terms used or the approaches taken, possibly all might agree that their common goal is to use a knowledge of ingredients and processes to achieve culinary excellence. This would also make a good statement of the goals of food chemistry in general.

Since food chemistry is one of the major divisions of food science, and we have established what we mean by food, it might be worthwhile to spend a moment reminding ourselves what we mean by "science." Broadly speaking, the term *science* as commonly understood refers to the study of the natural world and the laws that govern it. As such, it refers both to the method of enquiry, the scientific method, and collectively to the body of knowledge about the natural world that has been painstakingly and incrementally accumulated over several centuries. The scientific method is a prescription for learning about nature that starts with a testable (i.e., falsifiable) hypothesis about something, which is then subjected to test through controlled experimentation. The results of such experiments are then tentatively added to the accumulated body of knowledge of nature pending their verification, either by repetition or through subsequent experiments designed to test the logical consequences of the new data. As many such facts are collected, they are woven together into broader unifying general explanations that are termed theories, which are logical frameworks that relate the various observations. Examples of theories include the theory of gravitation, the quantum theory, the theory of relativity, the theory of continental drift and plate tectonics, and the theory of evolution. In this context, the term theory differs somewhat from the meaning it generally has in popular usage, of a new idea or untested hypothesis. Theories of course can be incorrect, such as the Aristotelian model of celestial spheres as an explanation for the motions of the planets. They more often might be found to be incomplete, such as Newtonian mechanics, which was found to be insufficient to explain physics at very small size scales or at very high energies, necessitating the development of the quantum and relativity theories. As unifying explanations of large numbers of observations and facts, however, theories are among the most secure of scientific ideas, not unfounded hypotheses subject to uninformed personal opinion, as is often alleged in the case of the theory of evolution. All scientific theories should be continually reevaluated to ensure that they are indeed consistent with known facts, that they are the best such explanation, and that they are free of any logical inconsistencies, but such questioning must always be based on fact and logic, not arbitrary opinions or beliefs.

The previous paragraph used the term "natural" twice in describing the subject of scientific investigation. Since this term is widely used by the food industry to characterize the contents of their products, it might be worthwhile to briefly consider what this term means. The primary definition of natural means produced by, or a part of, nature and thus governed by the laws of nature. In this context, the opposite of natural is not man-made, but supernatural. Although western religion and philosophy have tended to view mankind as separate from nature rather than a part of it, keep in mind that we are actually animals, and that anything we create or do is subject to the same laws of nature as all of the rest of the natural world. Our cleverness, particularly in the past 150 years, however, has allowed us to manipulate the laws of nature to synthesize or produce things, such as synthetic chemicals or transuranic elements, not encountered in the preindustrial environment. In this context, the term *natural* now has the widespread and probably even more common meaning of "not man-made, not altered from its original state or disguised." This term has no legal, regulatory definition, so that food companies are free to use it in their labels almost as they see fit. The consumer would be well advised to view a label claiming a product is "all-natural" with skepticism; it may simply imply that no miracles were involved in its production!

Because the scientific enterprise is often difficult and is carried out by fallible humans, the possibility of error, large and small, is always present. So the growing body of knowledge of the natural world, which is sometimes referred to as "science," must not be considered as sacrosanct and may well require revision. The information presented in this book comes from this expanding collection of facts concerning food molecules, as it is presently understood. While much of this information is thought to be fairly firmly established, future developments may well prove some parts to be incorrect. There will also be cases where statements are simply incorrect even in the context of current knowledge due to the author's mistakes or limited knowledge. Because the foundation of science is skeptical curiosity, the student would be well advised to keep an open mind about the topics discussed in these pages and to evaluate the experiments reported in the primary literature for her- or himself in cases that may seem questionable.

With respect to the credibility of scientific evidence, perhaps a few words are needed concerning the special cases of toxicological and nutritional data relating to foods. Probably everyone is familiar with the almost daily reports in the popular press about new findings concerning the supposed health benefits of a certain food, or the potential dangers of some chemical in foods or the environment. Often reports with exactly contradictory findings closely follow one another. In judging the meaning of these reports, we should remember that scientific results should properly be presented first in peer-reviewed technical journals, not in the popular press. Reporters generally do not have advanced training in science, and even when they do, they cannot be experts on all of the subjects about which they report. Furthermore, most news stories concern a single recently published technical paper, and not established consensus (which naturally would no longer qualify as "news"). For that matter, even single reports in the technical literature must be confirmed by repetition before they become accepted, and certainly a high level of such confidence-building is needed before responsible scientists and physicians would make recommendations about what people should eat based on the findings. In addition, the fields of toxicology and nutrition are both hampered by various difficulties that frequently make their findings less clear-cut than those of other disciplines. These problems include ethical and practical limitations in designing experiments on human subjects, the ambiguity about applying the results of experiments on animals to humans, or the difficulty of applying "in vitro" (i.e., test tube) experiments to living organisms. Another problem sometimes faced by both fields is that many of their studies are not actually controlled experiments at all, but rather consist of observational data on populations, where there may be many uncontrolled factors affecting results, and which may also be plagued by large statistical uncertainty ("noise"), obscuring any underlying trends. In this respect, they resemble astronomy or paleontology, which rely on observational data, and where experiments are frequently difficult or impossible. For all of these reasons, be cautious about reports in the popular press about what one "should" eat. In this book there will often be references to the findings of such studies, which are beyond the author's area of expertise. All such comments should therefore be considered with the appropriate caution and skepticism.

As an applied science, food chemistry is usually studied in connection with the food manufacture, processing, and retail preparation (restaurant) industries. In this context, it is worthwhile to consider the enormous changes that are likely to overtake these industries in the twenty-first century, and which, in fact, are already transforming the way in which foods are grown, stored, processed, manufactured, traded, and prepared worldwide. Some of the most profound of these potential changes will come about as a result of the "biotechnology revolution." This term might most broadly be understood not just as the enormous potential of direct genetic engineering, but more generally as the explosion of knowledge about basic biology and biochemistry that is allowing us to develop an extraordinarily detailed understanding of how living systems work and to use this knowledge to manipulate biology to our benefit. This development is not only bringing breathtaking progress in medicine, but will, in the foreseeable future, result in great advances in nutrition and in the scientific control of food quality at all stages of its production and preparation. The already-mentioned molecular gastronomy movement is just an early harbinger of what will be possible through the practical application of food science in the near future. Taking all aspects of the biotechnology revolution into account, it may not be too hyperbolic to suggest that the resulting changes for human culture may be as profound as those of the Industrial Revolution, possibly the most earthshaking previous development in human history, as we take control of every aspect of our biology from birth to death, encompassing our growth, health, nutrition, and general happiness, including the contribution to these made by the foods we eat. As a result, the coming years will be an exciting time indeed to be involved in any aspect of biological research and its applications.

A related issue that will be increasingly important in the design of foods is their impact on health. Because of the clear, although poorly understood, connection between food and health, food chemistry can play an important role in public health policy. This may be particularly true with respect to several troubling epidemiological trends in the United States and, to a somewhat lesser extent, worldwide. These include a marked increase in type-II diabetes and a dramatic increase in obesity rates. In 2011, nearly one third of the US population was judged to be obese by the American Heart Association (AHA) standards. The state with the lowest proportion of obese residents in 2011 was Colorado. As an indication of how fast this problem has grown, if Colorado had had the same proportion of obesity in 1995, it would have ranked as the highest. This problem may be primarily cultural; eating less and exercising more would probably nearly eliminate the problem nationally. Nonetheless, given that a large proportion of the food consumed in the nation is either processed or prepared commercially in restaurants, the way in which foods are formulated can have a significant impact on public health. As epidemiology and nutritional science begin to provide more reliable information about the exact connection between specific food components and health, this knowledge must be incorporated into food products, while making every effort to simultaneously maximize satisfaction and availability (that is, affordability).

Another development that will have significant global implications for the food industry in the near term, although not nearly as profound as the biotechnology revolution, is the diminishing place of the United States, where this book was written, in the world economy. Until quite recently the United States was the world's largest economy and dominated both production and consumption on a worldwide basis. As such, consumer preferences and government regulations in the United States dictated the standards for producers and manufacturers worldwide if they hoped to do business in the US market. The explosive development in nations such as China, India, and Brazil and the economic integration of most of the nations of Europe have substantially altered this landscape. The steady growth in the size of the European Union, for example, eclipsed the United States early in the twenty-first century. The European Union now has a population of over 500 million people, mostly in wealthy nations where consumers have substantial buying power, while the US population currently stands at a little over 300 million. Even if the three North American Free Trade Act (NAFTA) nations of the United States, Canada, and Mexico effectively melded into a common market as integrated as that of Europe, it would still be smaller than the European Union. Furthermore, the extraordinary economic rise of several developing nations with very large populations, most notably, of course, China and India, which are becoming the world leaders in manufacture and production, has significantly transformed the world economy. For example, China now produces more steel than the United States, Germany, and Japan combined (*New York Times*, Dec. 21, 2007). Both China and India produce substantially more wheat than does the United States, and collectively the EU production is also much larger than that of the United States. The United States is still the largest producer of maize and soybeans, but the soybean production of Brazil is rapidly rising (at the expense of the Mato Grosso and the Amazon rainforest), and China is now the world's largest market for soy products. China has surpassed Japan as the world's second largest single-nation economy, and is rapidly gaining on the United States. In an integrated and globalized world economy in which North America represents only a part, food producers, processors, and manufacturers will increasingly have to target their products toward global tastes in order to be competitive.

The emergence, in particular, of the European Union as the world's largest single overall market has important implications for foods and food products manufactured in the United States. Because European environmental and product safety standards are in general much stricter than those in the United States, manufacturers of all types of products who wish to do business in the world's largest and wealthiest market must meet those stricter standards. This development means that US government regulatory agencies such as the Food and Drug Administration (FDA), Environmental Protection Agency (EPA), and Department of Agriculture (USDA) may become less important, partially as a result of the political efforts of US companies and trade groups to keep American regulations weak. The ironic result of these efforts is that these same industries will have to observe the more stringent foreign regulations in order to compete in their markets. A shift in regulatory and consumer influence and leadership to Europe might have significant effects in at least temporarily retarding progress in the applications of biotechnology, particularly genetic engineering, because there is strong resistance among European consumers to genetically engineered food products, as is also increasingly true in India. While frequent reference will be made in the following pages to US regulations concerning food safety, it should be kept in mind that the regulations of other nations will increasingly become standards that even the US industry must follow.

A fourth foreseeable development that will increasingly affect the food industry as the current century progresses is the growing environmental crisis facing the world. This crisis has many dimensions, from global warming to mass extinctions to increases in energy costs and shortages in clean drinking water. The collapse of commercial fisheries and the increasing scarcity of pure fresh water may also become serious problems for the food industry. Food manufacturers in the future will have to consider such factors as the "carbon footprint" of their products, the direct energy cost of production, and more general concerns such as their possible impacts on endangered species and ecosystems. Obviously these problems are interrelated and also will be connected with such developments as the biotech revolution. Transgenic crops, for example, offer the prospect of helping to alleviate environmental problems by increasing food production without the need for further land clearance, fertilizer use, or the application of pesticides, but they themselves also represent a potential threat to wild organisms and ecosystems through the possible leakage of genes from the farm and from unforeseen interactions with other organisms. Such environmental considerations will play a role in all industries in the future, including the food industry.

One of the most important environmental problems affecting the food industry in the future will be the increasing cost of energy. The production, transportation, processing, refrigeration, and cooking of foods all require considerable energy inputs. As the world population grows, and as an increasing fraction of the world's people achieve "first world" living standards, the global demand for energy will only increase. At the same time, reserves of nonrenewable energy sources such as coal, natural gas, and oil are finite, so that increasing demand and diminishing supply will inevitably lead to higher prices. The economics of this problem will be further complicated if the cost of carbon dioxide emissions are added to production costs. Thus, efficiency in energy use at every stage of food production and processing will be increasingly important. Particularly in the case of food processing and cooking, efficiency will involve understanding the energetics of the reactions involved, which is an integral part of food chemistry.

Underlying all of our environmental challenges is the fundamental problem of human population and population growth. The current population of the earth is over 7 billion people, and the United Nations estimates that it will exceed 9 billion people by 2050. Doomsayers have been predicting since the time of Malthus that human population growth would outstrip our capacity to produce food, leading to massive famines. Technological developments during the last two centuries have for the most part managed to avert such horrors, but at enormous and tragic cost to the environment in terms of natural ecosystems destroyed by land clearance to open up new farmland. We are nearing the end of our ability to develop new farmlands, even if the destruction of the last remaining wild lands was desirable, and future advances will have to be made primarily through technological progress. This is where food science and technology, including food chemistry, can have a significant beneficial impact. In many developing countries, a large fraction, sometimes even approaching 100%, of agricultural primary production is lost due to problems such as spoilage or pests. While some of this could be averted with traditional infrastructure developments such as roads, railroads, and canals to transport produce to markets and electricity to allow refrigeration, local postharvest processing could also save large amounts and raise total available food supplies more than could substantial increases in primary production. Developing and using technical knowledge to assist in such processing should be a major goal of food science, hand in hand with employing biotechnology, including genetic engineering, to increase primary production with the smallest possible inputs of resources such as water, pesticides, and fertilizer. Failure to do so will almost certainly lead, at some point, to a global tragedy of unprecedented dimensions.

Even if we manage to avoid worldwide famine over the next 40 years, the present human population, without any increase, is unsustainable in terms of the demands that we place on nonrenewable resources such as minerals, hydrocarbon fuels, fresh water aquifers, forests, wildlife, and fisheries. If our descendents are to be able to establish a civilization that is stable and sustainable in the long run, with an enjoyable quality of life that extends beyond having the bare minimum of food necessary to avoid starvation, it will have to be based on a steady-state global population that is only a fraction of its present size. Because nearly all economic systems and plans are based on models of continual growth, making the transition to such a steady-state economy will be difficult. This is particularly true because any successful effort to reduce birth rates implies a future transition period when fewer young, productive workers must support large numbers of older people, many of whom may not be working at all. This problem will be exacerbated if life spans continue to increase, as we would hope they might, resulting in even more old individuals. The necessary increases in productivity might seem almost impossible to achieve. The great efficiency of modern industrial and factory farming, however, as problematic as it may be in its present incarnation, offers the hope that such efficiencies might be achievable, at least in the production of the most basic of all requirements of life (after air and water), while we eventually make the transition to smaller-scale and local food production as the population gradually declines. Food science and technology will play a critical role in making this transition possible.

In general, understanding food chemistry will be useful in addressing all of the challenges and problems facing the food industry in the future, and hopefully, such an understanding will contain the seeds of the solutions to these difficulties. With this in mind, let us turn our attention to a brief review of the basics of chemistry from introductory classes, focusing on those topics that will be most useful in our discussions of food chemistry.

BASIC REVIEW

The many macroscopic properties and attributes of foods with which we are familiar, such as color, taste, odor, texture, and stability, are the result of the properties, interactions and reactions of their component molecules. Changes in these properties can be chemical in nature, such as the browning of the flesh of a cut apple, or physical, such as the melting of a piece of chocolate candy on one's tongue. In all cases, the observed behavior of foods is, like that of all other systems, ultimately governed by the laws of physics and chemistry. The subject of this book will be a survey of what is known about how these laws of chemistry and physics produce the characteristics of foods with which we are familiar on the macroscopic scale, in the kitchen or in the processing plant. We will be interested in the chemical reactions that the component molecules of food undergo during harvest, processing, storage, or cooking, as well as with changes in state or aggregation, such as freezing a fresh vegetable or cut of meat or the creaming of milk.

Food chemistry is largely a subdiscipline of biochemistry or perhaps, more specifically, applied biochemistry. As such, many of the subjects covered in these pages are covered in general introductory biochemistry courses. The selection of topics, however, as well as the emphasis given to various subjects, is different, due to their relevance to foods. For example, nucleic acids are relatively unimportant as foods, but as the medium of information storage, they are of crucial overall importance in biochemistry as a whole. Thus, while nucleic acids, genes, and genetics are a major part of any modern general biochemistry course, they will receive little attention here. Similarly, metabolic pathways and cycles are less important for food chemistry and will not be discussed here, although, again, they are a major part of general biochemistry. Vitamins will be discussed more in terms of their processing stability and interactions than their metabolic roles, which are covered in nutrition courses. Considerable attention will be given to carbohydrates because they are important as functional ingredients in foods, while they get much less attention in biochemistry texts. In lipids, we will focus more on the triglycerides of depot fats, because of their abundance and functionality in foods, than on the phospholipids of membrane bilayers. In proteins, many of those of great importance in foods are hardly covered in general biochemistry courses, and the denatured states of globular proteins will be a significant focus, while biochemistry texts tend to focus more on their folded states. Taste, odor, and color, which are also of lesser importance in general biochemistry, are clearly important in foods, and two chapters are devoted to these subjects alone.

In this chapter we will review some of the concepts most necessary for understanding the behavior of food molecules. This will include a brief discussion of the chemical elements, their properties, and their combination into molecules through chemical bonding. We will then consider how the laws of thermodynamics govern the making and breaking of bonds and the noncovalent interactions of food molecules that determine the characteristic properties and behavior of different foods. In the following two chapters we will review the crucial role of water in all biological systems and also examine the collective behavior that arises in systems when they have structural elements with characteristic sizes intermediate between the molecular level and the macroscopic sizes with which we are familiar in our everyday lives. Many foods have structures in this colloidal, or nanoscale, range. Finally, with this groundwork, we will then proceed to survey the various classes of molecules that are important in foods and their basic reactions.

ATOMS AND MOLECULES

Molecules, including food molecules, are made up of atoms, the basic structural unit of matter. The concept of a smallest, indivisible unit of matter goes back to the ancient Greeks, while the modern atomic theory of matter did not become firmly established in physics until the end of the nineteenth century. Shortly after, however, physicists discovered that atoms are not actually indivisible, but rather are made up of even smaller subatomic particles: protons and neutrons, which are now known to be composed of yet smaller quarks, and electrons. These particles have different masses; protons and neutrons weigh approximately 1836 times as much as electrons. The atoms of the different elements contain different numbers of positively charged protons in their atomic nuclei, with a corresponding equal number of negatively charged electrons in their neutral atomic states. The number of nuclear protons is called the atomic number, and each different atomic number corresponds to atoms of a different element. In addition to protons, atomic nuclei also contain neutrons, particles with approximately the same mass as a proton, but with no electrostatic charge. The protons in a nucleus all have the same charge, but are prevented from flying apart from electrostatic repulsions by the strong nuclear force, which tightly binds protons and neutrons together. The number of neutrons affects the weight of an atom and its stability toward radioactive decay, but does not significantly affect the chemical behavior of the elements, so all atoms with the same number of protons are considered to belong to the same type of element, regardless of how many neutrons they contain. Atoms with the same number of protons but with different numbers of neutrons (and thus different masses) are called isotopes.

Each of the chemical elements possesses its own chemical properties, in terms of its valence, the types of reactions it undergoes, and so on. Ninety-two so-called naturally occurring elements are recognized, with uranium being the heaviest of these, along with a number of man-made transuranic elements. In the middle of the nineteenth century, Dmitri Mendeleev noticed a number of periodicities as a function of atomic number in the chemical behavior of the then-known chemical elements. He arranged the elements into groups according to their chemical similarities in the first periodic table of the elements, the modern form of which is shown in Figure 1.1. It is now known that the reason for these regularities lies in the distribution of electrons in the atomic orbitals of the atoms, an intrinsically quantum mechanical characteristic of the atoms.

The picture of the atom that emerged at the beginning of the twentieth century was like a tiny solar system, with a dense nucleus containing the protons at the center, and with the electrons orbiting around it in stable orbits. The protons and electrons (and later neutrons) themselves were viewed as being tiny spherical particles. From electromagnetic theory, however, such a system should not be stable, but instead the electrons should spiral down to the nucleus, continuously emitting electromagnetic radiation of varying frequencies as their orbits decayed. Various other experimental observations, such as the so-called black body radiation, also could not be reconciled with this picture of the atom. Interpretation of the black body radiation emitted by

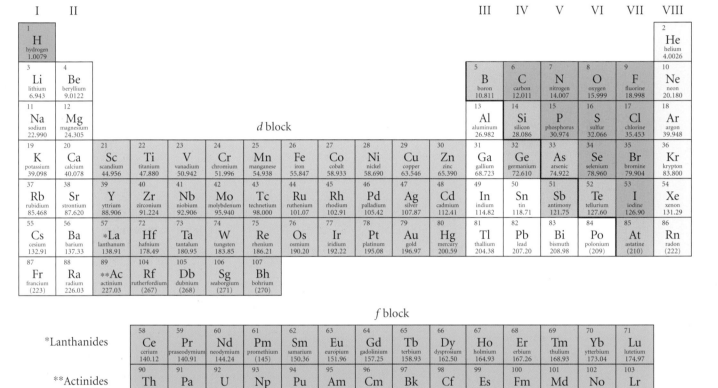

Figure 1.1. The periodic table of the elements. The noble gases are shaded yellow; the other nonmetals are shaded pink, while the semimetals are shaded purple. The *d*-block transition metals are shaded blue, and the *f*-block inner transition elements are shaded green. The average atomic mass of each element is shown in the bottom of each box, and its atomic number is shown in the top left of each box.

a heated body was only possible by assuming that the oscillators that give rise to the radiation could not have just any arbitrary energy, but instead have energies that are integral multiples of a fundamental energy, $\varepsilon = nh\nu$, specified by Planck's constant h and the frequency ν of the oscillator, where n is an integral **quantum number**. The Danish physicist Niels Bohr proposed a model for the atom which "fixed" these theoretical difficulties by hypothesis, but which itself was subsequently found to be inadequate. In this model the electron orbits were assumed to be stable by hypothesis, but that not just any orbit (with its corresponding energy) was allowed. Only certain orbits with specific energies and angular momenta that were integral multiples of $\hbar = h/2\pi$ were allowed. Transitions between two such orbits were accompanied by the absorption or emission of quantized radiation whose energy, $h\nu$, was equal to the energy difference in the two states. The Bohr model, however, still envisioned the electron as a well-defined particle moving in classical, Newtonian-style orbits around the nucleus. The inconsistencies of these early models of the atom with various experimental evidence led to the great crisis of twentieth-century physics that ultimately produced the quantum theory.

In the modern view of atoms, protons and electrons, and indeed even the atoms themselves, are no longer considered to be strictly particles, but rather to have properties of both waves and discrete particles. Their behavior is described by a so-called wave function, whose physical interpretation is that its square determines the probability of finding the "particle" at a given position with a given momentum. This description is also therefore fundamentally probabilistic rather than deterministic, as expressed in the Heisenberg uncertainty principle, which states that the product of the uncertainty in the position Δx and momentum Δp of a particle is equal to Planck's constant h, $\Delta p \cdot \Delta x = h$. This principle also limits the absolute accuracy that can be obtained in any physical measurement. The development of the quantum theory produced a revolution in scientific thinking that extended beyond science to inform much of modern philosophy and culture. One of the great successes of this theory was the development of a modern mathematical description of the chemical bond and molecular structure. Unfortunately, this theory is too complex to be described in detail here, but of course is outlined in all introductory courses in physics and chemistry.

In the quantum theory of atomic structure, the motions of a particle are described probabilistically in terms of a wave function $\psi(x_i)$ of the set of atomic coordinates x_i, by the Schrödinger equation,

$$\hat{H}\psi(x_i) = E\psi(x_i) \tag{1.1}$$

where $\hat{H}$ is an **operator** called the Hamiltonian, a mathematical operation on the wave function $\psi(x_i)$, which produces the total energy E of the particle. In one dimension the Schrödinger equation can be written as

$$-\frac{\hbar^2}{2m}\frac{d^2\psi(x)}{dx^2} + V(x)\psi(x) = E\psi(x) \tag{1.2}$$

where $\hbar$ is $h/2\pi$, m is the particle mass, and $V(x)$ is the potential energy of the particle expressed as a function of the position coordinate x. This wave function has no physical interpretation, but the square of the wave function, $|\psi(x)|^2$, gives the probability of finding the particle at the position x. When applied to real systems such as an atom, Equation (1.2) cannot be solved subject to the applicable boundary conditions unless E is restricted to specific quantized values related to one another by integers. In general, each electron in an atom is described by four quantum numbers: a principal quantum number n, an angular momentum quantum number l, a magnetic quantum number m, and a spin quantum number s. The principal quantum number n can have any positive integral value except 0; the angular momentum quantum number l can assume any integral value from 0 up to $n-1$. The magnetic quantum number m can in turn assume any integral value between $-l$ and $+l$, including zero. The spin quantum number s is the only quantum number with nonintegral values; it can only adopt one of two possible values, either $+\frac{1}{2}$ or $-\frac{1}{2}$. In a one-electron atom such as the hydrogen atom, or single-electron ions of higher number elements, the energy of the sole electron is determined by the principal quantum number n through

$$E = -\frac{2\pi^2 me^4 Z^4}{n^2 h^2} \tag{1.3}$$

where Z is the nuclear charge, and m and e are the mass and charge of the electron. No two electrons in an atom may have exactly the same set of quantum numbers, and as more electrons are added, they successively adopt the unused quantum numbers. The wave functions for these distinct sets of quantum numbers are referred to as the atomic orbitals for the electrons and replace the earlier concept of electron orbits as fixed, stable, circular paths for the electrons to follow around the nucleus.

Molecular bonds between atoms are formed when they approach closely in space and share electrons in common orbitals called molecular orbitals, which result from the linear combination of the atomic orbitals. When the orbitals are combined additively, a lower-energy bonding orbital is produced, and when they are combined subtractively, a higher-energy antibonding orbital results (Figure 1.2).

Figure 1.2. The additive and subtractive linear combination of the 1s atomic orbitals of hydrogen atoms to produce the bonding (σ) and antibonding (σ^*) orbitals of molecular hydrogen H_2.

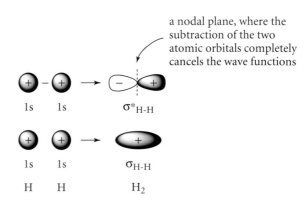

a nodal plane, where the subtraction of the two atomic orbitals completely cancels the wave functions

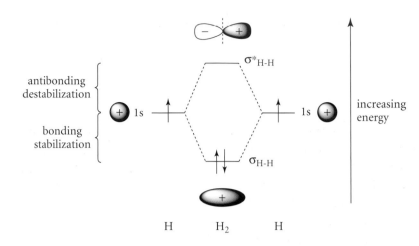

antibonding
destabilization

bonding
stabilization

$\sigma^*_{H\text{-}H}$

1s

1s

increasing
energy

$\sigma_{H\text{-}H}$

H H_2 H

Figure 1.3. The molecular orbital description of H_2, formed from two H atoms, each with a single electron in the lowest-energy atomic 1s orbitals.

The electrons of the individual atoms successively occupy these molecular orbitals, filling the lower-energy ones before the higher-energy orbitals (Figure 1.3). As with atomic orbitals, a transition from a lower-energy to a higher-energy molecular orbital requires the absorption of a quantum of energy equal to the energy difference between the two levels.

Light, or electromagnetic radiation, is also quantized, with a dual wave and particle character. Individual "packets" or quanta of electromagnetic energy are called photons. Promoting one of the bonding electrons in the σ orbital of H_2 to the antibonding σ^* orbital would require the absorption of a photon, or quantum of electromagnetic energy, equal to the energy difference between the two states,

$$\Delta E_{\sigma^*_{H-H}-\sigma_{H-H}} = E_{\sigma^*_{H-H}} - E_{\sigma_{H-H}} \tag{1.4}$$

The frequency v, and associated wavelength $\lambda = c/v$, of this photon are given by the Planck equation,

$$\Delta E_{\sigma^*_{H-H}-\sigma_{H-H}} = hv \tag{1.5}$$

where h is again Planck's constant, and c is the invariant speed of light.

Before we turn our attention to a discussion of the different classes of molecules important in foods, it would be worthwhile to survey a few of the atomic elements that are the primary components of these molecules. Because the majority of food molecules are of course organic, they are largely composed of carbon, hydrogen, oxygen, and nitrogen. Therefore, the chemistry of these elements will be of major significance for understanding food chemistry. Sulfur and phosphorous also are important in biochemistry and thus are significant in foods as well. Various other elements such as the halogens (fluorine, chlorine, bromine, and iodine) and metal ions such as iron, zinc, and manganese have important roles in nutrition and food chemistry, although they are generally present in food in only trace amounts.

Carbon

Carbon, the Group IV element with the atomic number six, containing six protons in its nucleus and six electrons, is the fourth most abundant element in the universe, and due to its special chemical characteristics, is the most important element in biological systems. As such, an entire division of the subject of chemistry, called organic chemistry, is devoted to the study of molecules that contain carbon. In scientific usage, the term "organic" does not refer to how a food is supposedly grown, but rather to whether or not a molecule contains carbon. As such, virtually all foods, with few exceptions, such as table salt, water, and titanium dioxide, used as a white colorant, contain carbon and are thus organic. Secondarily, the term organic refers to the biological world, since all living organisms are primarily composed of carbon.

The six electrons of a neutral carbon atom are arranged in the 1s, 2s, and 2p orbitals, in the configuration $1s^2 2s^2 2p^2$, with only four of the eight electrons needed to complete its outer shell, giving the atom a valence of four, which means that it can make four stable chemical bonds to other atoms. The tetrahedrality of carbon allows it to constitute complex molecules, including polymers, rings, and even three-dimensional space-filling matrices (diamond). Silicon (Si), which is directly below carbon in Group IV, shares this valence, but cannot support the type of chemistry possible for carbon for a number of reasons, including the lower reactivity of silicon, its larger atomic radius, its inability to make stable double bonds and chain compounds, and the insolubility of the oxide of silicon, silicon dioxide, in water.

Elemental carbon is found in three principal forms: amorphous carbon (as in charcoal), graphite, and diamond. A number of nanometer-sized three-dimensional structures called fullerenes, resembling graphite, that have the shapes of geodesic domes, chicken wire-like hollow tubes, or even more exotic geometries have been discovered since the mid-1980s. Carbon atoms can form strong and stable carbon–carbon bonds, allowing the formation of large polymers, such as proteins, carbohydrates, and lipids, which are the primary components of foods. Carbon also makes stable bonds to hydrogen, which are nonetheless higher in energy than bonds to oxygen, so that on oxidation (burning), which replaces all of the carbon–carbon and carbon–hydrogen chemical bonds with bonds to oxygen, considerable energy is released. For example, the burning of the linear hydrocarbon octane (C_8H_{18}), the primary component of gasoline, releases a great deal of stored energy:

$$2C_8H_{18} + 25O_2 \rightarrow 16CO_2 + 18H_2O \quad \Delta H = -1316.6 \text{ kcal/mol} \qquad (1.6)$$

(Note that, in the process, oxygen–oxygen bonds must also be broken, which reduces the total energy yield). This reaction is an example of the most important broad class of chemical reactions, call redox reactions, in which the oxidation numbers of the atoms change. The oxidation number of an atom is the charge that the atom would have if all of its substitutents were removed, along with the electrons shared with those ligands to form their bonds. In a redox reaction, the <u>l</u>oss of <u>e</u>lectrons by an atom is called <u>o</u>xidation (**leo**), and the **g**ain of <u>e</u>lectrons is called <u>r</u>eduction (**ger**) (leo the lion; ger is his roar). In the example in Equation (1.6), the carbon atoms lose electrons to the oxygen atoms, which is why they are said to be oxidized (in this example they also end up bonded to oxygen atoms, but oxidation can also take place without actual oxygen participating in the reaction).

Hydrogen

Hydrogen, with only a single proton in its nucleus, and thus an atomic number of 1, is the most abundant element in the universe. Essentially all of the universe's supply of hydrogen was generated in the immediate aftermath of the Big Bang, and hydrogen is the fuel that powers stars like the Sun as hydrogen atoms are fused together to form helium. Hydrogen is uncommon as a pure element in the biosphere, where it exists as a colorless and odorless diatomic gas, H_2. Hydrogen is abundant in biological molecules, however, and more hydrogen atoms (in number) occur than any other element in biopolymers, although because of its small mass, hydrogen does not make up a significant fraction of the weight of biopolymers. For example, half of all the atoms in the sugar glucose, $C_6H_{12}O_6$, are hydrogen atoms, but they contribute only 6.7% of the mass of the molecule. Hydrogen atoms are often involved in one of the more important types of organic reactions, called reduction reactions, which usually involve either the addition of hydrogen atoms or the loss of oxygen atoms.

Oxygen

Oxygen, the Group VI element with atomic number 8, is the third most abundant element in the universe, after hydrogen and helium. Although oxygen is the most abundant element in the rocks of the Earth's crust, pure elemental oxygen occurs in the atmosphere as a colorless and odorless diatomic gas, O_2, making up 21% of the total atmosphere by volume, and 23% by mass. Oxygen is crucially important for life not only as the central heavy atom of the water molecule but also as one of the principal components of most biological molecules. Oxygen is a constituent of proteins, carbohydrates, and lipids, the three major categories of food molecules, as well as of nucleic acids, the molecules that convey the genetic information of living organisms, and it is a component of many other, smaller food molecules. One of the most important types of reactions in foods is oxidation, which is particularly important as a mode of degradation and loss of quality. Often this involves the formation of new chemical bonds to oxygen, as in the example of the burning of gasoline shown in Equation (1.6). This type of oxidation with its release of energy is the basis of the metabolic use of the energy stored in the chemical bonds in foods.

Nitrogen

Nitrogen, the Group V element with the atomic number seven, containing seven protons in its nucleus and seven electrons, is found under the ambient conditions existing near the Earth's surface as a stable diatomic gas, N_2. Nitrogen gas, which is colorless, odorless, and chemically inert, makes up the largest single fraction, over 75% by weight, of the Earth's atmosphere. The ground state of the nitrogen atom has the electronic configuration $1s^2 2s^2 2p^3$, with five electrons in its outer shell, giving it a standard valence of three, meaning it typically forms three covalent bonds to other atoms. Nitrogen is essential for living organisms because it is a constituent of all proteins and nucleic acids, and thus an important constituent of foods as well. In spite of the fact that nitrogen is so abundant in the biosphere, the majority of it is in the form of atmospheric nitrogen gas, which is generally unreactive due to the strong triple bonds holding the nitrogen atoms together. For this reason, the availability of reactive, higher-energy nitrogen, such as in ammonia or nitrates, is a limiting factor in the growth of plants, which often require that sufficient nitrogen be provided through fertilizers. The conversion of atmospheric nitrogen gas into the higher-energy, reactive form found in ammonia, NH_3, with the consumption of the necessary energy, is called nitrogen fixation. A number of bacteria have evolved enzymatic pathways to carry out the fixation of nitrogen. Plants of the legume (Fabaceae) family, which includes the beans and clovers, are referred to as nitrogen-fixing, but these legumes do not actually carry out the fixation, but rather serve as the hosts for nitrogen-fixing *Rhizobia* bacteria that live symbiotically in their roots. Letting over-used fields "rest" for a season, during which time legumes are allowed to grow, returns reactive nitrogen to the soil to fertilize subsequent plantings of other non-nitrogen-fixing crops. Animals, including humans, obtain their required nitrogen from the proteins and other nitrogen-containing molecules in their diet. Animal wastes are rich in reactive nitrogen, which is why they make good fertilizers.

At the beginning of the twentieth century, the German chemist Fritz Haber developed a method for synthesizing ammonia from nitrogen and hydrogen gases using an iron oxide catalyst, a landmark in industrial chemistry for which he was awarded the Nobel Prize in Chemistry in 1918. Nitrogen is also an important component of many explosives, which before World War I were generally manufactured from materials such as guano (bird or bat droppings) or the sodium nitrate of Chilean caliche. The first large-scale industrial application of the Haber process was during World War I, when Germany made extensive use of the reaction in order to circumvent an Allied blockade, which cut off her access to the raw materials needed to make explosives. The principal use of the Haber process now is to manufacture fertilizers. Stating that the development of this process was one of the most significant advances ever made in industrial chemistry is not an exaggeration. Without the large-scale manufacture of nitrogen-containing fertilizers, perhaps as many as 2.5 billion of the Earth's people would starve to death, due to the inability of slow biological nitrogen fixation to produce sufficient nitrogen to sustain the crops needed to feed the world's current population (Hager 2008).

Phosphorous

Phosphorous is the Group V element directly below nitrogen in the periodic table and accordingly has chemical properties similar to those of nitrogen. It is very reactive and thus is never found in pure form in the environment. Phosphorous is an essential component of nucleic acids (DNA and RNA) as well as the phospholipids of cell membranes (Chapter 7) and is covalently bound to many proteins (Chapter 6). In biological systems phosphorous is usually found bound to oxygen as phosphate, PO_4^{3-}, as in DNA and phospholipids. Phosphates obtained from mined mineral deposits are widely used as fertilizers in industrial agriculture, and as a nonrenewable resource they represent a possible future limit on agricultural production. Morocco is the world's largest exporter of phosphates and may contain as much as 40% of the world's high-quality reserves. Although it is an essential nutrient, a normal diet supplies all the phosphorous that is required, and phosphorous deficiencies usually result either from disease, genetic deficiency, or severe starvation.

Halogens

The halogens are the elements of Group VII, which includes fluorine, chlorine, bromine, iodine, and astatine. The elements of this group all have electronic configurations that lack just one electron to complete their outer shells, and thus all have a conventional valence of one. For this reason, in their pure elemental state they generally occur as diatomics, F_2, Cl_2, Br_2, and I_2 (astatine, At_2, is of little importance in foods). The pure elemental halogens are very reactive, and most can be quite harmful in significant quantities. Chlorine gas was used as a chemical weapon in World War I, and chlorine in the form of sodium hypochlorite, NaOCl, is used as a bleach and disinfectant in water supplies for drinking or swimming, because it is lethal at low concentrations to a wide range of microorganisms. Fluorine is the most electronegative of all of the elements and is very corrosive and reactive; even water will burn under a fluorine atmosphere. All of the halides also form binary compounds with a single hydrogen atom to give diatomic molecules that easily dissociate in water to produce acids, such as hydrochloric acid, HCl. The halogens become progressively less dangerously reactive as their atomic numbers increase, such that iodine, which is the least reactive of the common halogens, is an essential nutrient required in the human diet for the production of thyroid hormones. A deficiency of iodine can lead to the development of goiter and other problems, and intellectual disability in young children. To prevent this deficiency, common table salt is often formulated with supplementary iodine.

Transition Metals

Several of the transition metal elements, the elements in the *d* block (see Figure 1.1), with incomplete *d* subshells, such as manganese, iron, and copper, are necessary in various enzymes and other proteins and are obtained through the diet. For example, iron, the most important of the transition elements and an essential nutrient, is a component of hemoglobin, the oxygen transporting protein in the blood of mammals and many other animals, and in myoglobin, a very similar oxygen storage protein of muscle tissue responsible for the red color of meat (Chapters 5 and 9). Iron in biological systems is usually found as nonneutral ions in the +2 or +3 oxidation states, Fe^{2+} or Fe^{3+}, referred to as ferrous and ferric ions, respectively. Ferric iron in the +3 oxidation state is dangerously reactive for biological molecules, able to generate free radicals, and is a powerful oxidant, responsible for many undesirable reactions in foods. Cobalt, with atomic number 37, the next element in sequence after iron, is an essential component of vitamin B_{12}, where it makes nitrogen–metal covalent bonds to nitrogen atoms in a complex heterocyclic organic molecule called a corrin ring that almost completely surrounds the metal atom.

The metallic elements zinc, cadmium, and mercury are usually treated as transition elements even though they have complete, closed d^{10} shells. Zinc, a component of many proteins, is an essential nutrient supplied in the diet. In excessive amounts, however, zinc is toxic, and in fact, all three of these elements are toxic, with cadmium and mercury being significant environmental pollutants often found as contaminants in foods.

REVIEW OF THERMODYNAMICS

The science of chemistry is primarily, although not exclusively, concerned with the making and breaking of chemical bonds. Secondarily, it is concerned with physical changes in a system such as a food sample at the atomic or molecular level that affect the properties and behavior of the system. In general, understanding the direction of change in foods will require understanding the thermodynamics of the food system. Thermodynamics, as the name implies in Greek, is the study of the flow and exchange of energy, in terms of heat, in physical systems. The laws of thermodynamics govern the direction of spontaneous change in any chemical system, so we will begin by reviewing some aspects of the laws of thermodynamics as they apply to foods. This treatment is not intended to be complete, but rather simply serves as a review of principles that you first studied in your introductory chemistry and physics courses.

Chemical and physical processes generally involve the absorption or evolution of heat. Energy is exchanged between species in chemical systems through chemical reactions and various physical transformations such as melting or boiling, mediated by radiative transfers or by molecular collisions, with the colliding molecules exchanging kinetic and potential energy.

It is useful to recall the definitions of two important measures of energy exchange between systems—heat and work. Work is a quantity of energy transferred across the boundary of a system which can be expressed as collectively and uniformly moving the constituent particles of some body in response to an applied force, and which in thermodynamic terms could be expressed as equivalent to raising a weight to a higher potential energy in a gravitational field. Thus, work is a form of organized motion. In a common example, for a fluid such as a gas or a hydraulic fluid enclosed in a cylinder with a movable piston as one wall, the work done in expanding and moving the piston at a constant pressure would be $P\Delta V$, where P is the pressure and ΔV is the change in volume. Heat is the quantity of energy exchanged across the boundary between two systems at different temperatures in order to bring them into thermal equilibrium. Heat transfer is mediated on the molecular level by molecular collisions, and is fundamentally chaotic in nature, but always flows from regions of higher temperature to regions of lower temperature. On the atomic and molecular level, the temperature of a system is an average measure of the kinetic energy of all of the component molecules. The higher the temperature, the faster the atoms and molecules of the substance move.

The First Law of Thermodynamics

The **First Law of Thermodynamics** states that the energy of an isolated system is constant, which might be thought of as a statistical restatement of the Principle of the Conservation of Energy from Classical Mechanics. Thus, the First Law could be stated as:

a. The energy of an isolated system is constant.
b. When a system changes from one state to another along an adiabatic path (that is, without the exchange of heat), the amount of work done is the same irrespective of the means employed.

The First Law can be used to define a function U, a property of the system, to be called the **internal energy** change in going from state "A" to a new state "B", as,

$$\Delta U = U_B - U_A = q + W \tag{1.7}$$

$$dU = dq + dW \tag{1.8}$$

The sign convention usually adopted for heat transfer and work done is as follows: a positive q is heat added to the system; a negative q is heat flowing out of the system; a positive dW is work done on the system; and a negative dW is work done by the system on the outside world.

The internal energy U is a state function; that is, it depends only on the state of the system, and not how it got that way, so that the internal energy difference between two states A and B can be written as

$$\Delta U = \int_A^B dU = U(B) - U(A) \tag{1.9}$$

The First Law says that ΔU depends only on the initial and final states and not on the path followed from one to the other. The heat evolved and the work done are properties of the paths followed during transformations, and are not properties of the systems themselves, and thus are not state functions. For example, the heat evolved, q, can still be written as the integral of dq over a particular path, but cannot be expressed as the difference of the initial and final states,

$$q = \int_{\text{path}} dq \tag{1.10}$$

but

$$\Delta q \neq q_B - q_A \tag{1.11}$$

The heat exchanged and work done, q and W, may have any number of values, depending on the path followed, but their sum $\Delta U = q + W$ is independent of the path; otherwise one could take a system from state A to B along one path, and return from B to A along another path, and have a net change in energy, violating the First Law, that the energy of a closed system is constant.

The internal energy is undetermined to the extent of an arbitrary additive constant because it is defined experimentally only in terms of a difference between states. However, it may be quoted relative to some arbitrary standard state defined to have $U = 0$.

The relative importance of a particular amount of energy E is determined by its Boltzmann weighted probability, given by

$$P(E) \propto e^{-E/k_B T} \tag{1.12}$$

where T is the temperature (in degrees Kelvin) and k_B is Boltzmann's constant. In a system in thermal equilibrium, $k_B T$ can be thought of as a measure of the average energy likely to be picked up or lost through normal molecular collisions. Large energies will have a low probability, while small energies will have a larger probability. At room temperature, $k_B T$ is approximately 0.6 kcal/mol (2.5 kJ/mol).

Another useful thermodynamic state function called the **enthalpy** H can be defined as

$$H = U + PV \tag{1.13}$$

so that

$$\Delta H = \Delta U + V\Delta P + P\Delta V \tag{1.14}$$

where P is the pressure and V is the volume. Under constant pressure conditions (such as the atmospheric pressure that prevails on laboratory bench tops), ΔP is 0, so ΔH is

$$\Delta H = \Delta U + P\Delta V \tag{1.15}$$

Many condensed phases (liquids and solids) are relatively incompressible, so often for processes in condensed phases $\Delta V \cong 0$, and under these conditions,

$$\Delta H \cong \Delta U. \tag{1.16}$$

For a constant-volume process, $\mathrm{d}V = 0$, and no mechanical work is done, (i.e., $\mathrm{d}W = P\mathrm{d}V = 0$), so any change in internal energy is due to the heat absorbed or lost. If the pressure is kept constant, and only $P\Delta V$ work is done, for some general change between two states A and B,

$$\Delta U = U_B - U_A = q + W = q - P(V_B - V_A) \tag{1.17}$$

$$(U_B + PV_B) - (U_A + PV_A) = q_P \tag{1.18}$$

where q_P is the heat absorbed at constant pressure, so

$$\Delta H = H_B - H_A = q_P \tag{1.19}$$

Thus, the change in enthalpy in going from state A to state B equals the heat absorbed or evolved at constant pressure when no work is done other than $P\Delta V$ work.

Heat Capacity

Transferring heat to or from a system results in a proportional change in temperature,

$$\mathrm{d}q \propto \mathrm{d}T \tag{1.20}$$

as the average energy of the constituent molecules increases, with a proportionality constant C,

$$\mathrm{d}q = C\mathrm{d}T \tag{1.21}$$

where C is called the **heat capacity**. At constant volume,

$$C_V = \left(\frac{\mathrm{d}q}{\mathrm{d}T}\right)_V = \left(\frac{\partial U}{\partial T}\right)_V \quad (\text{if } V = \text{constant}, \mathrm{d}W = 0, \text{so } \mathrm{d}U = \mathrm{d}q) \tag{1.22}$$

At constant pressure,

$$C_P = \left(\frac{dq}{dT}\right)_P \tag{1.23}$$

From Equation (1.14), the differential of the enthalpy can be written as

$$dH = dU + PdV + VdP \tag{1.24}$$

At constant pressure, $VdP = 0$, so the total differential of the internal energy is

$$dU = dq + dW = dq + VdP - PdV = dq - PdV \tag{1.25}$$

Substituting this expression for dU into Equation (1.24) for dH gives

$$dH = dq - PdV + PdV = dq \tag{1.26}$$

Thus, substituting Equation (1.26) into Equation (1.23) gives

$$C_P = \left(\frac{\partial H}{\partial T}\right)_P \tag{1.27}$$

Keep in mind that the heat capacity is a measure of how much energy has to be added to a system to increase its temperature by a certain amount, or removed from the system to cool it. The specific heat for a substance is the heat capacity per unit mass, for example, the amount of heat needed to raise the temperature of one gram of that substance by one degree Celsius. Note that the specific heat itself is a function of temperature. The specific heat of water, defined as the amount of heat required to raise the temperature of 1 gram of pure water at 15 °C and 1 atmosphere pressure by one degree Celsius, is 1 cal/gram/°C.

The amount of heat required to raise the temperature of a substance from temperature T_1 to temperature T_2, can be calculated from the above equation for the heat capacity,

$$\Delta H = C_P \Delta T \tag{1.28}$$

where $\Delta T = T_2 - T_1$, which makes the assumption that C_p is approximately constant over that temperature range.

In a complex food such as a slice of pizza, the different components have different heat capacities, and so will cool at different rates, even though when they come out of the oven all of the components of the slice have the same temperature. The starch and gluten matrix of the crust has a low heat capacity and cools quickly. The tomato sauce, however, which is rich in water, cools slowly, because it has a high heat capacity due to the water (Table 1.1; note that the specific heat of water—its heat capacity on

Table 1.1 Heat capacities at room temperature of selected substances in foods

Substance	Heat capacity, C_p (cal/mol·K)	Specific heat, c (cal/g·K)
tristearin (α form)	395.8	0.444
tristearin (β form)	317.4	0.356
sucrose (solid)	102.0	0.298
limonene	59.6	0.437
vanillin	44.8	0.295
ethanol	27.0	0.586
water	18.0	0.998

Notes: Tristearin is a saturated fat triglyceride such as might be found in beef tallow (Chapter 7; the α and β forms refer to two different crystalline packing arrangements); sucrose is table sugar; limonene is an essential oil from lemons and other citrus fruits (Chapter 8). The specific heat of each substance, c, or the heat capacity on a per weight basis, is shown in the last column; expressing the ability to store heat in this fashion makes it clear how unusual water is in this respect.

a weight basis—is very high; see Chapter 2). The sauce thus can transfer a lot of heat to your tongue or the roof of your mouth without cooling down much. This effect is accentuated in the case of pizza, because the sauce is covered with a layer of melted cheese, which has a specific heat intermediate between that of the crust and sauce. Because the fats in the cheese can serve as an insulation layer (see Table 2.6), they further prevent the sauce and pepperonis from cooling quickly. Such differences in heat capacity also affect the cooking of foods. For example, when roasting a whole turkey, different portions, such as the skeleton and breast meat, will heat at different rates. This is why it is important that meat thermometers inserted into the breast touch only muscle meat tissue, and not the bones, to ensure that all of the meat has reached a uniform temperature of 165°F (73.9°C), to kill all possible *Salmonella* contamination. Otherwise, they may falsely read a higher temperature than has actually been reached by the meat.

Heat Evolved during Physical Changes

Physical changes in systems, such as changes in state like freezing or boiling, will in general lead to the evolution or absorption of heat, as shown in the following examples:

melting	$H_2O(s) \rightarrow H_2O(l)$	$\Delta H_{273.15} = 6.0\,\text{kJ·mol}^{-1}$
vaporization	$H_2O(l) \rightarrow H_2O(g)$	$\Delta H_{373.15} = 40.7\,\text{kJmol}^{-1}$
mixing	$H_2SO_4(l) \rightarrow H_2SO_4$ (aq. soln.)	$\Delta H_n^\circ = -96.2\,\text{kJ·mol}^{-1}$
	$HCl(g) \rightarrow HCl$ (aq. soln.)	$\Delta H_n^\circ = -75.1\,\text{kJ·mol}^{-1}$

As can be seen from these examples, the absorption of heat is required to melt ice to water or to boil liquid water to steam. Conversely, going the other way,

6 kJ/mole of heat is given up when liquid water freezes to ice, which would raise the heat content of the remaining liquid water, thus inhibiting additional freezing until it is conducted away by further cooling.

Thermochemistry: Enthalpies of Formation

Chemical reactions also usually lead to the evolution or absorption of heat; for example, the reaction of a mole of hydrogen gas with one half mole of oxygen gas to produce one mole of water vapor generates 241.75 kJ of energy. An **exothermic** reaction is one that gives up heat to the surroundings, and an **endothermic** reaction absorbs heat from the environment. So, using the water example,

$$\text{exothermic:} \quad H_2 + \tfrac{1}{2}O_2 \rightarrow H_2O(g) \quad \Delta H_{291.15} = -241.75 \text{ kJ}$$

$$\text{endothermic:} \quad H_2O(g) \rightarrow H_2 + \tfrac{1}{2}O_2 \quad \Delta H_{291.15} = 241.75 \text{ kJ}$$

Note that it is necessary to specify the state of the reactants and products, because changes in state also are accompanied by the evolution or absorption of heat. The two reactions in this example, while simply the opposites of one another, are not completely equivalent, because the first reaction, leading to the formation of water from hydrogen and oxygen, proceeds spontaneously (once sufficient initial energy is given to the system to promote the passage of the reactants over the substantial activation barrier), with a net production of energy. Conversely energy must be put into the system to break water into oxygen and hydrogen.

Tables of the heats of reaction are available for many chemical processes. These are generally measured under so-called standard conditions of 1 atm pressure and at the temperature of 298.15 K, or 25 °C. The standard enthalpy of reaction under these conditions is designated as ΔH°_{298}.

Consider the following reactions:

$$C_2H_6 \left(\text{i.e., ethane}\right) + \tfrac{7}{2}O_2 \rightarrow 2CO_2 + 3H_2O(\text{liq}) \quad \Delta H^{\circ}_{298} = -1560.1 \text{ kJ}$$

$$C\left(\text{graphite}\right) + O_2 \rightarrow CO_2 \qquad\qquad\qquad \Delta H^{\circ}_{298} = -393.5 \text{ kJ}$$

$$H_2 + \tfrac{1}{2}O_2 \rightarrow H_2O(\text{liq}) \qquad\qquad\qquad \Delta H^{\circ}_{298} = -285.8 \text{ kJ}$$

The above equations can be combined (added) to give the standard heat of formation of ethane from carbon and hydrogen gas:

$$2CO_2 + 3H_2O\left(\text{liq}\right) \rightarrow C_2H_6 + \tfrac{7}{2}O_2 \qquad\qquad \Delta H^{\circ}_{298} = 1560.1 \text{ kJ}$$

$$2C(\text{graphite}) + 2O_2 \rightarrow 2CO_2 \qquad\qquad \Delta H^{\circ}_{298} = 2(-393.5) = -787.0 \text{ kJ}$$

$$+ \underline{\qquad 3H_2 + \tfrac{3}{2}O_2 \rightarrow 3H_2O\left(\text{liq}\right) \quad \Delta H^{\circ}_{298} = 3(-285.8) = -857.4 \text{ kJ} \qquad}$$

$$2C(\text{graphite}) + 3H_2 \rightarrow C_2H_6 \qquad\qquad\qquad \Delta H^{\circ}_{298} = -84.3 \text{ kJ}$$

As a food example, limonene, the principal component of oil of lemon, is made up entirely of carbon and hydrogen atoms (see Figure 8.25), with the formula $C_{10}H_{16}$. The standard heat of formation of limonene from graphite and hydrogen gas has been measured as:

$$10C(\text{graphite}) + 8H_2 = C_{10}H_{16} \text{ (limonene, liq.)} \quad \Delta H^{\circ}_{298} = -54.4 \text{ kJ}$$

Hess's Law

Because enthalpy is a state function, the enthalpy change for a reaction is independent of the path taken to the final products. This observation, first made by Germain Hess, is called Hess's Law, and means that tabulated thermochemical data is additive, as in the examples shown above. The enthalpy change for an unknown reaction can be found if it can be written as the sum of other reactions for which the enthalpy change is already known, as was done for ethane above. Consider a similar example, the formation of carbon disulfide from phosgene and hydrogen sulfide:

$$
\begin{array}{lll}
& COCl_2 + H_2S \rightarrow 2HCl + COS & \Delta H_{298} = -78.705 \text{ kJ} \\
+ & COS + H_2S \rightarrow H_2O(\text{gas}) + CS_2(\text{liq}) & \Delta H_{298} = 3.420 \text{ kJ} \\
\hline
& COCl_2 + 2H_2S \rightarrow 2HCl + H_2O(\text{gas}) + CS_2(\text{liq}) & \Delta H_{298} = -75.285 \text{ kJ}
\end{array}
$$

The tabulated enthalpies in this example are for the reactions at 298 K (25 °C). To calculate the reaction enthalpy at a different temperature, we can use the heat capacity relation,

$$dH = C_P dT; \quad \text{from Equation (1.28)} \tag{1.28}$$

to compute the change in enthalpy for each of the components in heating or cooling it to the new temperature if the heat capacity is known.

If the number of moles changes, we must take account of this change in the thermodynamic bookkeeping. For example, consider the following reaction of two gases:

$$SO_2 + \tfrac{1}{2}O_2 \rightarrow SO_3 \quad \Delta U_{298} = -97.030 \text{ kJ}$$

What is ΔH for this reaction? It can be calculated from this tabulated value for ΔU_{298} by approximating these chemical species as ideal gases. At constant T, and assuming ideal gas behavior, H and ΔH can be written as,

$$H = U + PV = U + nRT \tag{1.29}$$

$$\Delta H = \Delta U + RT\Delta n \tag{1.30}$$

where $\quad \Delta n = n_{products} - n_{reactants} = 1 - \left(1 + \tfrac{1}{2}\right) = -\tfrac{1}{2}$

so $\quad \Delta H = -97.030 \text{ kJ} - \frac{1}{2} \cdot 8.3143 \text{ } JK^{-1}\text{mol}^{-1} \cdot 298 \text{ } K = -98.269 \text{ kJ mol}^{-1}$

Calorimetry

As we have seen, physical and chemical changes lead to the evolution or absorption of heat, either due to changes in the structures or energetics of the molecules, or their interactions, or from entropic changes (next section). These heat exchanges can be measured in devices called calorimeters. Several types of such devices exist (Figure 1.4). So-called **bomb calorimeters** determine the heat of formation of a molecule from its component elements by measuring the heat evolved as a sample is completely combusted in a well-insulated sample chamber. Reference values for the combustion products can then be used to determine the total heats of formation from the elements in their standard states. A **differential scanning calorimeter** (**DSC**) measures the heat exchange in a sample as a function of temperature, relative to a reference sample, as both samples are heated according to a controlled temperature program. The two samples are subjected to a carefully controlled heating using electrical resistance, with both samples kept at exactly the same temperature. The heat exchange for each sample necessary to keep them at the same temperature is then measured. A frozen solution, for example, might melt as the temperature is increased, absorbing the heat of fusion as the transition occurs. Thus, the heat flow into the sample chamber would be greater than the heat flow into the reference chamber, while the

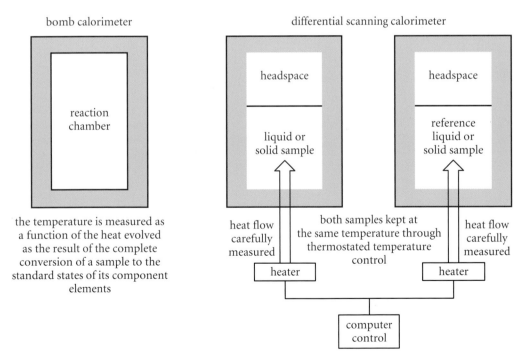

Figure 1.4. Schematic illustrations of calorimeters. On the left, a bomb calorimeter, and on the right, a differential scanning calorimeter.

temperature in the two chambers is kept the same. A similar transition would occur in a sample of chocolate as it is heated past the melting point of cocoa butter. Globular proteins in solution might lose their conformational shape as the solution is heated, in a process called denaturation (Chapter 5), absorbing as heat the energy necessary to overcome the forces that keep them compact at lower temperatures. If the process is subsequently reversed by carefully lowering the temperature, many proteins will spontaneously refold to the original shape, giving up the same amount of heat as was absorbed as they denatured. Similarly, if a previously frozen solution were to be cooled back below its melting point, it would refreeze, giving up the heat of fusion that was previously absorbed when it melted.

Bomb calorimetry measurements are made under constant-volume conditions, so no mechanical work is done, and the heat exchange measured is the change in the internal energy, rather than the enthalpy,

$$q = C_v \Delta T = \Delta U \tag{1.31}$$

Because the volume is constant, the pressure of course is not constant. Variations of this type of calorimetry are the method by which measurements of the number of calories in a particular food are made, by determining the number of calories yielded upon complete combustion of the food sample.

In the differential scanning calorimeter, or DSC, the heat q is exchanged under constant pressure, with the sample able to expand against the headspace, and thus is the enthalpy change. The heat exchanged per unit of time, q/t, to produce a temperature increase ΔT over the same time interval is carefully measured in the DSC experiment. The heat exchanged to produce a measured temperature change ΔT,

$$\frac{q}{t} \bigg/ \frac{\Delta T}{t} = \frac{q}{\Delta T} \tag{1.32}$$

is the heat capacity, from Equation (1.28),

$$C_p = \frac{q_p}{\Delta T} \tag{1.33}$$

A typical DSC measurement would report the heat exchanged as a function of temperature, as in Figure 1.5.

For a given type of transition, such as the melting of a sample of fat or the denaturation of a protein, that takes place over a range of temperatures between T_i and T_f, the total ΔH_{tot} for the transition can be calculated as the integral of the C_p vs T curve over that range,

$$\Delta H_{tot} = \int_{T_i}^{T_f} C_p \mathrm{d}T \tag{1.34}$$

Figure 1.5. A schematic of the type of data yielded from a typical DSC experiment, such as for the denaturation of a protein when it reaches its "melting," or denaturation, temperature T_m.

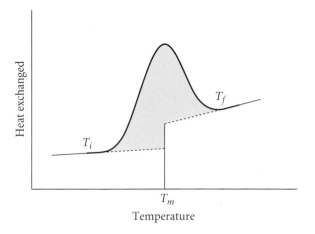

Consider the example of the denaturation of a protein. The enthalpy change at the "melting temperature" T_m (see Chapter 5), $\Delta H(T_m)$, can be calculated from the measured ΔH_{tot} by extrapolation. The melting temperature T_m can be taken to be the temperature at the maximum of the transition curve. The heating curves before and after the transition have different slopes because the native and denatured proteins can be considered to have different heat capacities. As can be seen from Figure 1.5, some of the measured ΔH_{tot} is due to the heating of un-denatured proteins from T_i up to the melting temperature T_m, and some is due to the heating of denatured proteins from T_m up to T_f, so that the enthalpy changes for these two processes must be subtracted from ΔH_{tot} to get $\Delta H(T_m)$,

$$\Delta H(T) = \Delta H_{tot} - \int_{T_i}^{T_m} C_p^{native} dT - \int_{T_m}^{T_f} C_p^{denatured} dT \qquad (1.35)$$

One might wonder why all of the protein molecules do not denature at exactly the denaturation temperature T_m, since all of the protein molecules are the same and thus presumably require the same amount of energy to unfold. The reason is due to the inhomogeniety of the kinetic energy distribution, which fluctuates in time and place. Thus, some of the protein molecules, by random chance, have enough energy to denature transferred to them as the solution is heated before the overall temperature reaches T_m, while others, again due to local, random fluctuations, do not receive enough energy through collisions to denature until the overall temperature has actually exceeded T_m. This situation would be different from the melting of a sample of fat, which would also take place over a range of temperatures. In the case of the fat, the melting range would be much broader and would occur because, unlike the protein denaturation example, where the molecules are indeed essentially all the same, fat samples are highly inhomogeneous, consisting of a range of fatty acids combined in different triglyceride structures (see Chapter 7). For this reason, solid fat is not nearly as regular and crystalline as a sugar crystal, for example. Thus, the fats gradually soften as their triglycerides absorb energy and become more mobile, with locally crystalline regions of different composition melting at slightly different temperatures. Chocolate, which is primarily made up of cocoa butter (which is also a fat), melts over a much narrower temperature range because cocoa butter is much more regular in its structure than most fats (see Chapter 7), although it is still more varied that a protein sample. This feature of cocoa butter gives fine chocolates their property of melting on the tongue, just below body temperature.

The numerous C—C and C—H bonds of the main classes of food molecules (carbohydrates [Chapter 4], proteins [Chapter 5], and lipids [Chapter 7]), contain considerable stored energy, relative to the energies of carbon and hydrogen in C—O and H—O bonds, which are more stable, or lower in energy. Carbon–carbon double bonds, —C=C—, contain more energy than single bonds, and triple bonds are even more unstable, but are uncommon in food molecules. Double bonds, however, occur in numerous types of food molecules, such as in unsaturated fatty acids (Chapter 7) and in many colored compounds (Chapter 9). The energy stored in these various bonds is released when the food molecules are "burned," or oxidized, by metabolic combustion in the mitochondria of cells. This is one of the principal reasons for eating foods, because animals are unable to directly convert the energy of sunlight into a chemically useful form, as plants do through photosynthesis (Equation (4.1)).

Not all of the energy stored in chemical bonds in a food sample, as measured by bomb calorimetry, is actually available to a human or other animal when it is eaten, for a variety of reasons. For example, we do not possess the enzymes necessary to digest some types of food molecules, such as the cellulose in plant cell walls, and even some parts of starch are resistant to digestion (Chapter 4). Some food value is lost to intestinal microflora, some is not efficiently absorbed across the intestinal walls, and some becomes entangled with other molecules in ways that prevent either from being enzymatically deconstructed. In some cases, denatured proteins may also become resistant to both acid hydrolysis in the stomach and enzymatic hydrolysis. In order to determine an estimate of the caloric content of a food that is actually available, nutritionists devised an interesting variation on calorimetry in which an entire small room is essentially converted into a calorimeter. A volunteer subject then stays in this insulated room for an extended period while he/she consumes measured food samples. Their undigested "output", in terms of urine and feces, is carefully measured, and the heat that they give off over the experimental time period is also carefully measured, in terms of the temperature rise in the insulated room over the measurement period. Much of this work was done a long time ago by W. O. Atwater, and there is believed to be considerable uncertainty in this data, which is still used as the basis of calorie counts on food product labels (Merrill and Watt 1973).

It might be worthwhile at this point to mention the units of heat used in such measurements. As a form of energy, the appropriate unit of measurement of heat in SI (Système International) units is the joule, abbreviated J, which is equal to $1\,N{\cdot}m$ (Newton-meter), or $1\,kg{\cdot}m^2/s^2$. A joule is a relatively small amount of energy, and in a chemical context the kilojoule, or 1000 J, abbreviated kJ, is a more useful unit, and is generally reported per amount (in moles) of sample (i.e., kJ/mol). An older, and still more widely used, unit in chemistry is the calorie, defined in various ways, perhaps most commonly as the amount of heat required to heat 1 gram of pure water by 1°C, from 19.5°C to 20.5°C at a constant pressure of 1 atmosphere. Again, this amount of energy is small compared to the typical energies of chemical bonds, which are generally reported in kcal/mol, or 1000 calories per mole of sample. A calorie is equal to 4.186 J, and thus one joule is equal to 0.2389 cal. Nutritionists usually refer to the kilocalorie as a "calorie", or "large calorie", and call the standard calorie a "small calorie". Thus, the "calorie" contents for foods listed on packaging labels in the United States actually refer to kilocalories. European product labels often explicitly label the energy content in kilocalories, or even kilojoules.

It was originally thought that perhaps the direction of spontaneous chemical and physical change would be governed by the evolution of heat; that is, that spontaneous processes would be exothermic and evolve heat as the system moved to a lower internal energy state. There are a number of examples, however, of spontaneous reactions that absorb heat; that is, are endothermic. Clearly something else drives the direction of spontaneous change. That "something else" is entropy, and the Second Law of Thermodynamics.

Entropy and the Second Law of Thermodynamics

The Industrial Revolution, which began around 1830, was largely built upon the development of steam power, and great effort was directed in the nineteenth century toward improving the efficiency of the steam engine. In the course of these efforts, it became clear that there were limitations to what could be achieved in this respect. These studies were primarily carried out by working engineers, and it has been observed that "science owes more to the steam engine than the steam engine owes to science". The German physicist Rudolf Clausius, using the data gathered in this fashion, in mid-century recognized and defined a new property that he called **entropy**, and first formulated the **Second Law of Thermodynamics**. Clausius defined this new property in terms of heat exchanged and work done by a system in a purely macroscopic fashion, that makes no assumptions about the atomic and molecular structure of a system (indeed, in 1850 the atomic theory of matter had not yet been fully established). A few decades later, Ludwig Boltzmann, who played an important role in the eventual acceptance of the atomic theory, redefined entropy in a statistical mechanical sense that is equivalent to the earlier thermal definition. In this approach, entropy can in general be thought of as characterizing the disorder of a system, and the Second Law of Thermodynamics can be thought of as saying that, for any natural process, the disorder of the universe always either increases or stays the same, but never decreases.

The statistical mechanical definition of entropy, S, first given by Boltzmann (and carved on his tombstone), is a measure of the disorder of a system that defines entropy in terms of the number of accessible states W at a given energy, which is the number of different ways to arrange the system at that energy,

$$S = k_B \ln W \tag{1.36}$$

The reason that, on an atomic and molecular level, this relationship defines entropy in terms of the disorder of a system is that the more ways there are to arrange a system (W) at a given energy, the more disordered it will be.

Clausius, however, defined the relative entropy changes for any process of a system thermodynamically, without any reference to its microscopic state, in terms of the heat transferred. This is done with the Second Law of Thermodynamics, one of the most fundamental of all physical laws.

The Second Law of Thermodynamics states that the entropy of an isolated system increases during any natural process, and that no process is possible in which the sole result is the absorption of heat from a reservoir and its conversion into mechanical work; that is, it is impossible to construct a perpetual motion machine. More formally, for a reversible process at a temperature T, the change in entropy is defined as

$$dS = \frac{dq}{T} = \frac{dq_{rev}}{T} \quad \text{and} \quad \Delta S = \int_A^B \frac{dq}{T}. \tag{1.37}$$

Then, for any natural process, $dS \geq \frac{dq}{T}$; in an isolated system $dq = 0$ so $dS \geq 0$. This relation is called the **Clausius Inequality**, and its significance is that it determines the direction of change in any natural process. For any process in an isolated system, then, the entropy, and disorder, must either stay the same or increase. (Natural here means any process that does not involve the supernatural). For a system that is not isolated, the total entropy of the universe must stay the same or increase for any natural process, because the universe represents the ultimate closed system. This of course means that the disorder of the universe is always increasing.

The Second Law is one of the most important of the physical laws and is thought, for example, to be "time's arrow"; that is, to be the reason that we perceive time to "flow" in only one direction (there is no such requirement in basic Newtonian physics). For example, if we watched a movie of a person smoking that was run backwards, we would immediately recognize the unnatural time reversal because the gathering of the dispersed smoke back into the burning cigarette would be seen to be "wrong," since it would clearly and intuitively mean a reduction in entropy.

Free Energy

We can define two additional thermodynamic state functions that include this new property of entropy, the Helmholtz free energy, A, and the Gibbs free energy, G,

$$A = U - TS \quad \textbf{Helmholtz Free Energy} \tag{1.38}$$

$$G = H - TS \quad \textbf{Gibbs Free Energy} \tag{1.39}$$

At constant temperature, the differentials of these two functions are:

$$dA = dU - TdS \qquad (1.40)$$

and

$$dG = dH - TdS \qquad (1.41)$$

These equations can be integrated to give the familiar equations we learned in our introductory chemistry classes governing the change in the Helmholtz or Gibbs free energies for some particular process,

$$\Delta A = \Delta U - T\Delta S \qquad (1.42)$$

and

$$\Delta G = \Delta H - T\Delta S \qquad (1.43)$$

We will make extensive use of the second of these two equations in particular in order to understand many aspects of foods in subsequent chapters, because it governs chemical process under the approximately constant atmospheric pressure conditions of benchtop experiments.

Using the Clausius Inequality, it can be shown that, at constant volume,

$$dA = dq - TdS \leq 0 \quad T, V \text{ constant} \qquad (1.44)$$

Similarly, at constant pressure P, $dH = dq$

So that $\quad dG = dq - TdS \leq 0 \quad T, P \text{ constant} \qquad (1.45)$

Thus, for spontaneous change $\quad dA \leq 0 \quad$ at constant V, T

$$dG \leq 0 \quad \text{at constant } P, T \qquad (1.46)$$

That is to say, the change in these two free energy functions determines the direction of spontaneous change in a chemical system because, for any natural, spontaneous process, the free energy of the system must either stay the same or decrease and, in any practical system, must decrease.

Because chemical lab work is usually done in benchtop glassware open to the approximately constant atmospheric pressure, the Gibbs free energy is the function that most commonly determines the direction of chemical change. In some food processing situations, however, such as in pressure cookers, it is the system volume that is constant rather than the pressure, in which case the Helmholtz free energy is the relevant function.

We should briefly consider what we mean in using the equation $\Delta G = \Delta H - T\Delta S$ to explain changes in food systems such as the folding or denaturation of proteins or the disruption of cell membranes. We will often speak of such processes, when they are not isolated from the rest of the world, as involving the combination of a ΔH term and a $T\Delta S$ term, with their relative magnitudes determining whether the Gibbs free energy change is positive or negative. This is somewhat deceptive because, while the $T\Delta S$ term expresses the contribution from the change in entropy of the system itself, the ΔH term just represents the contribution from the change in the entropy of the surroundings (the rest of the universe), since from our thermodynamic definition of entropy, $dq = TdS$, and $dq = dH$ for a constant-pressure process. Thus, we are still invoking the Clausius inequality, that the universal entropy must increase.

By combining the First and Second Laws, we can write the total derivative of G in another form more useful for many purposes,

$$dU = dq + dW \quad \text{First Law} \tag{1.47}$$

and

$$dS = \frac{dq}{T} \quad \text{Second Law} \tag{1.48}$$

so

$$dU = TdS + dW \tag{1.49}$$

Under most conditions, only PV work is done, so $dW = -PdV$; thus,

$$dU = TdS - PdV \tag{1.50}$$

Because $\quad H = U + PV \tag{1.51}$

$$dH = dU + PdV + VdP \tag{1.52}$$

So, substituting for dU from above,

$$dH = TdS - PdV + PdV + VdP \tag{1.53}$$

$$dH = TdS + VdP \tag{1.54}$$

Now, since $\quad G = H - TS \tag{1.55}$

$$dG = dH - TdS - SdT \tag{1.56}$$

and, substituting for dH,

$$dG = TdS + VdP - TdS - SdT \tag{1.57}$$

so $\quad dG = VdP - SdT$ $\hfill$ (1.58)

At constant temperature ($dT = 0$), the second term drops out, and we are left with

$$dG = VdP \tag{1.59}$$

so $\quad \Delta G = G(P_2) - G(P_1) = \int_{P_1}^{P_2} VdP$ $\hfill$ (1.60)

For liquids and solids, the volume is a very weak function of the pressure, because most solids and liquids are not very compressible, so it can be treated as approximately constant and taken outside the integral, so

$$\Delta G = V\Delta P \tag{1.61}$$

For liquids and solids, unless ΔP is *very* large (P_2 is *very* high) then $V\Delta P$ is very small and easily negligible, so $\Delta G \cong 0$, $G(P_2) \cong G(P_1)$, and G is approximately independent of pressure.

For gases, of course, this will not be true. For the case of an ideal gas,

$$V = \frac{nRT}{P} \tag{1.62}$$

$$\Delta G = \int_{P_1}^{P_2} VdP = nRT \int_{P_1}^{P_2} \frac{dP}{P} \tag{1.63}$$

since $\int \frac{dx}{x} = \ln(x)$, then

$$\Delta G = nRT \ln\left(\frac{P_2}{P_1}\right) \tag{1.64}$$

$$G - G^{ref} = nRT \int_{P^{ref}}^{P} \frac{dP}{P} = nRT \ln\left(\frac{P}{P^{ref}}\right) \tag{1.65}$$

The reference state of choice is the standard state, which for an ideal gas is by convention defined to be the pressure $P°$ of 1 atmosphere,

$$G - G° = nRT \ln\left(\frac{P}{P°}\right) \tag{1.66}$$

Mixtures and Solutions

So far we have considered closed, one-component systems. Most foods, of course, contain more than one type of chemical substance, so we need an expression for the change in free energy of a food with composition. If more than one component is present, and if n_i represents the number of moles of component i, then the Gibbs function, which we defined earlier as a function of P and T, $G = G(P,T)$, is really a function of these variables and of the amount of each component present,

$$G = G(P, T, n_i)\tag{1.67}$$

so,

$$dG = \left(\frac{\partial G}{\partial P}\right)_{T,n_i} dP + \left(\frac{\partial G}{\partial T}\right)_{P,n_i} dT + \sum_i \left(\frac{\partial G}{\partial n_i}\right)_{T,P,n_j} dn_i \tag{1.68}$$

$$dG = VdP - SdT + \sum_i \left(\frac{\partial G}{\partial n_i}\right)_{T,P,n_j} dn_i \tag{1.69}$$

We can now define a new quantity called the **chemical potential** μ_i as the partial molar Gibbs free energy, or the change in G with the composition, expressed as the amount of component i, n_i:

$$\mu_i = \left(\frac{\partial G}{\partial n_i}\right)_{P,T,n_j} \tag{1.70}$$

where n_i is the number of moles of component i. If we incorporate this definition into the total derivative of the Gibbs free energy above, we get

$$dG = VdP - SdT + \sum_i \mu_i dn_i \tag{1.71}$$

the fundamental equation governing any chemical process. The chemical potential of a pure substance is the molar Gibbs function for that substance, G_m. In the case of a single-component system,

$$G = nG_m \tag{1.72}$$

so

$$\mu = \left(\frac{\partial G}{\partial n}\right)_{P,T} = G_m \tag{1.73}$$

and,

$$d\mu = dG_m \qquad (1.74)$$

For a two-component system where we will call the two molecule types "A" and "B," we can write the change in the Gibbs free energy of the system for some process as

$$dG = VdP - SdT + \mu_A dn_A + \mu_B dn_B \qquad (1.75)$$

For a mixing process at constant temperature and pressure, the total Gibbs free energy is given by

$$dG = \mu_A dn_A + \mu_B dn_B \qquad (1.76)$$

For any state function, the total function is given by the sum of the partial molar quantities times the amount of that component present, so for the Gibbs function we have

$$G = n_A \mu_A + n_B \mu_B \qquad (1.77)$$

Consider a system of two ideal gases A and B in a container divided into two compartments by a partition, with only A molecules in one compartment and B molecules in the other, and with the same pressure P in each compartment (Figure 1.6):

The Gibbs free energy for this initial system using Equation 1.66 is:

$$G_i = \sum_i n_i \mu_i = n_A \mu_A + n_B \mu_B \qquad (1.78)$$

$$G_i = n_A \left[\mu_A^\circ + RT \ln\left(\frac{P}{P^\circ}\right) \right] + n_B \left[\mu_B^\circ + RT \ln\left(\frac{P}{P^\circ}\right) \right] \qquad (1.79)$$

If we remove the partition and allow free mixing, in the final state each gas exerts a partial pressure P_i, with the total pressure P being given by $P = P_A + P_B$. The free energy of the final state is

$$G_f = n_A \left[\mu_A^\circ + RT \ln\left(\frac{P_A}{P^\circ}\right) \right] + n_B \left[\mu_B^\circ + RT \ln\left(\frac{P_B}{P^\circ}\right) \right] \qquad (1.80)$$

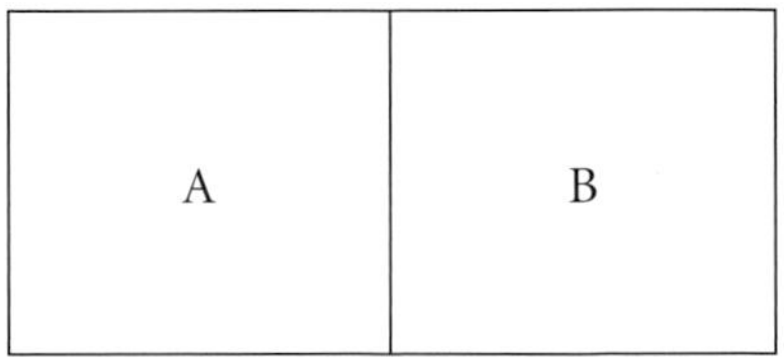

Figure 1.6. A schematic representation of a container divided into two compartments, each initially containing one type of ideal gas, either type A or type B.

The Gibbs free energy change for this mixing, ΔG_{mix} is just the difference between the Gibbs free energies for the final and initial states,

$$\Delta G_{\text{mix}} = G_f - G_i \tag{1.81}$$

Remembering that $\ln a - \ln b = \ln \dfrac{a}{b}$, this gives, since the standard state pressure P^o cancels,

$$\Delta G_{\text{mix}} = n_A RT \ln\left(\frac{P_A}{P}\right) + n_B RT \ln\left(\frac{P_B}{P}\right) \tag{1.82}$$

From Dalton's Law,

$$\frac{P_A}{P} = x_A \quad \text{with} \quad x_A = \frac{n_A}{n} \quad \text{and} \quad n_A + n_B = n \tag{1.83}$$

where x_A is the mole fraction of component A, so

$$\Delta G_{\text{mix}} = n x_A RT \ln(x_A) + n x_B RT \ln(x_B) \tag{1.84}$$

$$\Delta G_{\text{mix}} = nRT(x_A \ln x_A + x_B \ln x_B) \tag{1.85}$$

Because both x_A and x_B are less than 1, their logarithms are negative, so ΔG_{mix} is negative, which means the mixing occurs spontaneously. For an ideal gas, there are no interactions between the atoms or molecules, so the heat evolved upon mixing at constant pressure is zero, and $\Delta H_{\text{mix}} = 0$. Since $\Delta G = \Delta H - T\Delta S$, then

$$\Delta S_{\text{mix}} = -\frac{\Delta G_{\text{mix}}}{T} \tag{1.86}$$

$$\Delta S_{\text{mix}} = -nR[x_A \ln x_A + x_B \ln x_B] \tag{1.87}$$

Because the log terms in brackets are negative, the change in entropy is positive, as we would expect since the mixed system is more disordered than the unmixed one.

The ideal gas state might not seem relevant to real foods, but it is actually a useful model because it allows us a means of estimating the free energy of a liquid as well, such as the liquid water in a solution or food product. When a liquid is in a closed vessel in equilibrium with its vapor, both the liquid and vapor must have the same chemical potential (otherwise more of the phase with the higher chemical potential would be converted into the phase with the lower potential), so the chemical potential of the liquid can actually be written as equal to the expression for the vapor, which is useful because we can adequately approximate the vapor chemical potential in most cases using the ideal gas model as above. In a liquid solution (mixture) the chemical potential can be written in terms of a correction to the chemical potential of the pure solvent $\mu_A^*(l)$ due to the mixing,

$$\mu_A(l) = \mu_A^*(l) + RT \ln\left(\frac{P_A}{P_A^*}\right) \tag{1.88}$$

where P_A^* is the vapor pressure of the pure solvent A at the temperature T and P_A is its vapor pressure in the solution. In a so-called **ideal solution**, the **activity** a_A, defined as $\left(\dfrac{P_A}{P_A^*}\right)$, the partial pressure of component A divided by the equilibrium vapor pressure of the pure component, is equal to its mole fraction x_A,

$$\frac{P_A}{P_A^*} = x_A \tag{1.89}$$

so,

$$P_A = x_A P_A^* \tag{1.90}$$

which is called **Raoult's Law**. Substituting into our equation for the chemical potential gives

$$\mu_A(l) = \mu_A^*(l) + RT \ln(x_A) \tag{1.91}$$

In such an ideal solution the activity a_A of component A is just

$$a_A = \left(\frac{P_A}{P_A^*}\right) = x_A. \tag{1.92}$$

Any nonideality of the solution can be incorporated into an activity coefficient γ_A, so that

$$a_A = \gamma_A x_A \tag{1.93}$$

as $x_A \to 1$ (pure solvent), $\gamma_A \to 1$; also, for an ideal solution, $\gamma_A = 1$.

A slightly less ideal type of solution than one that follows Raoult's Law is one that follows a similar equation at very dilute concentrations,

$$P_B = k_H x_B \tag{1.94}$$

which is called **Henry's Law**.

Henry's Law is useful for describing so-called **ideal dilute solutions**, such as solutions of gases like CO_2 in water, and differs from the Raoult's Law expression only

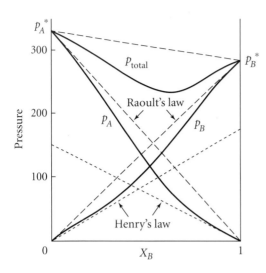

Figure 1.7. A schematic phase diagram for a binary mixture exhibiting negative deviations from ideality. The straight lines represent the Raoult's Law ideal behavior, and the curved lines show the actual partial pressure of each component as a function of composition, expressed as the mole fraction of component B, x_B.

in that the proportionality constant k_H is not equal to the vapor pressure of the pure substance.

Understanding the qualitative interpretation of an ideal solution is important and may not be obvious from the equations and discussion above. Ideality for gases means no interactions between gaseous molecules. Liquids, however, manifestly have strong intermolecular interactions. In ideal solutions, the ideality implies that A : B interactions are the same as A : A and B : B interactions, so that no net tendency for A and B molecules to either preferentially associate with one another or to avoid one another occurs. We will make use of this interpretation when we discuss colloidal and protein properties.

Negative deviations from ideality, such as are illustrated schematically in Figure 1.7, would result from attractive interactions between the two types of molecules in a binary mixture, which reduce the tendency for the molecules of each type to escape into the vapor phase. Similarly, positive deviations would reflect a repulsion between the two types of molecules that would tend to drive them out of the mixture and into the vapor phase. Figure 8.25b displays an example of a binary solution of two food flavorants that deviate strongly from ideal behavior.

The special case of water molecules in solution in a complex food sample is of particular importance in food technology. When the water activity in a food is less than x_A, it is because the interactions of the water molecules with the food are stronger than those with other water molecules in pure water, and the water molecules are in some sense "bound" in the food. What the phrase "bound water" might mean has long been hotly debated. These interactions, however, reduce the availability of the water for other processes such as microbial growth, so measurements of water activity have found operational utility as a guide to the probable stability of a food toward degradative processes (see Chapter 2).

Chemical Equilibrium

The chemical potential determines the ultimate equilibrium that becomes established between reactants and products in chemical reactions. For some general chemical process in which reactants A and B are converted to products C and D,

$$A+B \underset{k_f}{\overset{k_r}{\rightleftharpoons}} C+D \tag{1.95}$$

the reaction will proceed from left to right as long as $\mu_A + \mu_B$ is greater than $\mu_C + \mu_D$. The chemical potentials indicate the direction of change and serve the function implied by their name; at equilibrium,

$$\mu_A + \mu_B = \mu_C + \mu_D \tag{1.96}$$

The standard molar Gibbs free energy change for the reaction is defined as

$$\Delta G_m^\circ = \mu_C^\circ + \mu_D^\circ - \mu_A^\circ - \mu_B^\circ \tag{1.97}$$

If one defines an equilibrium constant K by

$$K = \frac{[C][D]}{[A][B]} \tag{1.98}$$

then

$$\Delta G_m^\circ = -RT \ln K \tag{1.99}$$

For the more general case of an arbitrary stoichiometry

$$aA+bB \underset{k_f}{\overset{k_r}{\rightleftharpoons}} cC+dD \tag{1.100}$$

$$K = \frac{[C]^c[D]^d}{[A]^a[B]^b} \tag{1.101}$$

Remember that the equilibrium constant can also be written in terms of the forward and reverse rate constants,

$$K = \frac{k_f}{k_r} \tag{1.102}$$

which allows the free energy to be expressed in terms of these rate constants.

Consider a simple example, the interconversion of two forms of glucose in water. As we shall see in Chapter 4, there are two stereoisomers of the six-membered ring form of D-glucose, called α and β, which spontaneously interconvert in aqueous solution,

$$\alpha\text{-D-glucopyranose} \underset{k_f}{\overset{k_r}{\rightleftharpoons}} \beta\text{-D-glucopyranose} \tag{1.103}$$

At equilibrium, the relative concentrations of these two forms at room temperature are 63%β/37%α. The equilibrium constant for this reaction is then

$$K = \frac{[\beta\text{-D-glucopyranose}]}{[\alpha\text{-D-glucopyranose}]} = \frac{63}{37} = 1.703 \tag{1.104}$$

and the free energy difference between the two forms is then

$$\Delta G_m^\circ = -RT \ln(1.703) = -0.32 \text{ kcal} \tag{1.105}$$

The negative sign reflects the fact that the reaction favors the β-D-glucopyranose form.

Another example with which the reader is probably already familiar is the self-ionization of water,

$$H_2O + H_2O \rightleftharpoons H_3O^+(aq) + OH^-(aq) \tag{1.106}$$

$$K = \frac{[H_3O^+][OH^-]}{[H_2O]^2} \tag{1.107}$$

Under conditions of normal temperatures and pressures, very few water molecules undergo this autoionization, so that the concentrations of hydronium ions $[H_3O^+]$ and hydroxide ions $[OH^-]$ are low, approximately 10^{-7} each. Since the molar concentration of pure water is approximately constant, ignoring volume changes with temperature and pressure, we can rearrange Equation (1.107) as

$$K_w = K[H_2O]^2 = [H_3O^+][OH^-] \tag{1.108}$$

with K_w equal to 1.008×10^{-14} at 25 °C. The pH, a measure of the acidity of a solution, is defined as the negative logarithm of the hydronium ion concentration,

$$pH = -\log[H_3O^+] \tag{1.109}$$

Since the concentration of both hydronium and hydroxide ions in pure water at room temperature and atmospheric pressure is approximately 1.004×10^{-7}, the pH of a neutral solution is therefore 7.0. A species such as HCl is acidic because it readily dissociates in water to generate additional hydronium ions, because the dissociated protons quickly attach to a nearby water molecule, without generating any corresponding hydroxide ions, thus increasing [H_3O^+] and lowering the pH. NaOH upon dissociation increases the hydroxide concentration, without generating additional hydronium ions, thus raising the pH. Most foods are weakly acidic; only a few are basic. More strongly acidic conditions can inhibit the growth of microorganisms, so several types of weak acids are used as food preservatives (Chapter 10). Several somewhat stronger acids are also used as foods. For example, commercial vinegars in the United States are generally 5% acetic acid, with a pH of ~2.4, and undiluted lemon juice generally has a pH of ~2.1 to 2.2, due to the presence of citric acid in similar concentrations. For comparison, the pH of American coffee is generally ~5.0.

Phases and Phase Changes

A phase is a homogenous state of aggregation of a system that is uniform in composition and physical state throughout. A system that consists of only one phase is said to be homogeneous, and a system that consists of more than one phase is said to be heterogeneous. A phase can be a solution, such as sugar dissolved in water, so long as it is a homogeneous mixture. Substances typically exhibit three main phases, solid, liquid, and gaseous, but the situation can be more complex. In particular, many substances have more than one possible solid form. Solid water (e.g., ice) has as many as 15 different crystal phases, although all but the so-called I_h phase require high pressures or very low temperatures. The fats (triglycerides) found in many foods such as butter and chocolate also can exist in more than one crystalline solid phase, depending on conditions, and a number of globular proteins can crystallize into more than one crystalline lattice (crystal form).

The phase of a substance that exists at any particular temperature at equilibrium is that phase with the lowest chemical potential μ, which at constant pressure means the lowest Gibbs free energy G, with G uniform throughout the phase, or throughout several phases if they are in equilibrium with one another.

Since $G = H - TS$, at constant pressure, the derivative, or slope, of the Gibbs free energy as a function of temperature, $G(T)$, is $-S$,

$$\left(\frac{\partial G}{\partial T}\right)_P = -S \tag{1.110}$$

Using this equation, we can construct a plot like the one in Figure 1.8, indicating which phase or phases will be present at any temperature, because the phase with the lowest free energy will be the form present at equilibrium at that temperature.

From Equation 1.110 we can see that the free energy as a function of temperature is, to a first approximation, a straight line with a slope of $-S$. Each of the three common phases has its own linear relation indicating how its Gibbs free energy changes with temperature. Because crystals are highly regular and ordered, we expect solids to have a lower entropy than liquids, so the slope of the free energy line for the solid is less than that for the liquid, and that for the liquid is very much less than

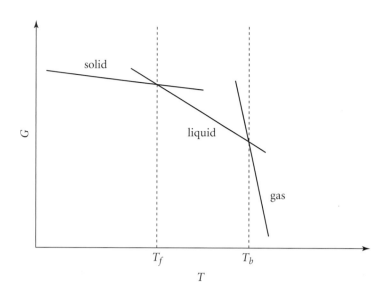

Figure 1.8. The dependence of the free energy of a substance as a function of temperature at constant pressure.

for the vapor, because of the much higher entropy of gases. In other words, for all of the phases, S is positive, and $S_{\text{Solid}} < S_{\text{Liquid}} < S_{\text{Gas}}$. Also, at the melting/freezing point, the solid and liquid are in equilibrium, so $G_{\text{S}}(T_f) = G_{\text{L}}(T_f)$, and at the boiling point the liquid and vapor are in equilibrium, with $G_{\text{L}}(T_b) = G_{\text{G}}(T_b)$. Thus, in Figure 1.8, the freezing/melting point at atmospheric pressure is defined by the intersection of the free energy for the solid and liquid phases as a function of temperature, and the boiling temperature is defined by the temperature at which the liquid and gas functions intersect. For this reason, ice can coexist with liquid water at 0 °C (at atmospheric pressure), and steam can coexist with liquid water at 100 °C.

As heat is added to a solid system, the atoms and molecules of the solid begin to oscillate about their equilibrium positions in the crystal lattice with increasing amplitudes until the melting temperature is reached. At that point, some of the molecules in the solid have enough kinetic energy to escape the confining interaction forces that have held them in place and begin to diffuse freely as a liquid. At the melting/freezing point, the solid and liquid phases of a substance, such as ice cubes floating in an insulated cup of water, have the same free energy and are in equilibrium with one another. The two phases do not have the same entropy or enthalpy, however. As long as both phases are present, the slow, incremental addition of more heat to the system will not raise the temperature from the melting/freezing temperature, but rather will result in the conversion of more of the solid phase into the liquid. The heat required to accomplish this melting, per amount of the solid melted, is called the heat of fusion, or enthalpy of melting, ΔH_f (Table 1.2). Only when enough heat has been absorbed to melt all of the solid will the equilibrium temperature began to increase again.

As a concrete food example of the effects of phase transitions on temperature changes in cooking, consider the making of chocolate fudge, which you may have attempted personally. When making fudge, the temperature of the semiliquid candy

Table 1.2 Melting points, heats of fusion ΔH_f, boiling points, and heats of vaporization ΔH_v for selected food molecules

Substance	Melting point (°C)	ΔH_f (kcal/mol)	Boiling point (°C)	ΔH_v (kcal/mol)
sucrose[1]	186.0	8.40	—	—
β-lactose[2]	225.0	16.2	—	—
vanillin[3]	~82.0	5.5	285.0	13.3
(+)-limonene[4]	−74.4	2.72	176.0	11.8
ethanol	−117.3	1.20	78.5	9.7
water	0.0	1.44	100.0	9.7
cocoa butter[5] (the β_1 crystal form)	28–36	14.5	—	—

Sources: (1) S.L. Shamblin et al., *J. Phys. Chem. B* 103:4113 (1999); (2) N. Drapier-Beche, J. Fanni, and M. Parmentier, *J. Dairy Sci.* 82:2558–2563 (1999); (3) M. Temprado, M.V. Roux, and J.S. Chickos, *J. Therm. Anal. Calorim.* 94:257–262 (2008); (4) H.E. Gallis et al., *J. Chem. Eng. Data* 41(6): 1303–1306 (1996); (5) D.J. Sessa, *J. Sci. Food Agric.* 72:295–298 (1996).

should reach approximately 115 °C for it to set properly. It takes some time to reach this temperature, however, due to the water in the condensed milk used in the recipe. When the mixture reaches 100 °C, further addition of heat goes into the vaporization of the water into steam, so that the overall temperature does not rise above 100 °C until all of the water has boiled off. Once the water is substantially gone, the temperature can again rise until it reaches the target temperature of around 115 °C (~240 °F), the so-called soft-ball stage, as determined with a candy thermometer, when the heating is stopped. This temperature is well below the general caramelization temperature of sucrose (~160 °C; see Chapter 4), so the sugar does not extensively caramelize unless the temperature is allowed to go much higher than the soft-ball stage. This temperature is also well below the melting temperature of sucrose (186 °C, Table 4.7), so sucrose precipitation as large crystals is a potential problem. With the water now gone, the subsequent cooling of the fudge would proceed smoothly past 100 °C without interruption or plateauing. Extensive recrystallization of the sugar, however—an undesirable event that is usually considered a failure of the fudge, because it produces a grainy rather than smooth texture—would give up the heat of fusion of the sucrose lattice, slowing the cooling somewhat.

Note from the table that cocoa butter melts around 36 °C, just below body temperature. Thus, when eaten, it absorbs the corresponding heat of fusion, 14.5 kcal/mol, from the tongue, leading to the pleasing cooling sensation associated with chocolate. Because different substances have different melting and boiling points, these differences can be exploited for practical purposes. It can also be seen from the table that ethanol has a lower boiling point than that of water. This difference, which is of course the basis for the distilling industry, allows these two liquids to be separated by boiling the alcohol at a temperature lower than the boiling point of water.

For some of the substances in Table 1.2, no boiling point is listed. This is because these substances are not volatile, and do not escape into a vapor phase in significant amounts. For example, sucrose molecules in a sugar crystal or a liquid melt interact with one another very strongly, and the hypothetical boiling point would, as a consequence, be very high. Before the molecules could reach a temperature high enough to allow then to escape from the liquid, however, they would break down by pyrolysis, as in caramelization. Similarly, the large triglyceride molecules of cocoa butter are not volatile and cannot escape into the vapor phase by evaporation. However, some short-chain fatty acids like butyric and caproic acids, such as are produced by the action of lipases on triglycerides in the ripening of cheeses, are volatile and contribute strongly to the characteristic aromas of certain cheeses.

The free energy dependences on temperature T illustrated in Figure 1.8 are an approximation based on the exact Equation (1.39), $G = H - TS$. The approximation is that the only temperature dependence in the equation is in the explicit multiplicative term TS. Both H and S, however, can themselves be functions of T; i.e., $H = H(T)$ and $S = S(T)$. Thus the function $G(T)$ might not be strictly linear in T, and

$$\left(\frac{\partial G}{\partial T}\right)_P = \left(\frac{\partial H}{\partial T}\right)_P - T\left(\frac{\partial S}{\partial T}\right)_P - S \tag{1.111}$$

where $\left(\frac{\partial H}{\partial T}\right)_P$ and $\left(\frac{\partial S}{\partial T}\right)_P$ are not necessarily 0. This temperature dependence will be important in Chapter 5 when we discuss why some proteins denature in aqueous solution when they are heated.

Colligative Properties

Several properties of dilute solutions, such as freezing point depression, boiling point elevation, and osmotic equilibrium, depend upon the amount of solute present, but not upon the nature of the solute. These properties are called colligative properties. Let us consider the case of the depression of the freezing point of a solvent by the addition of a solute since this situation exists in all moist foods.

Consider a solution in which the solute is nonvolatile, and so is absent from the vapor phase, where only the solvent gas is present, and which also does not dissolve in the solid solvent. This seemingly special case is actually quite common, because a great many of the solute molecules in foods, such as salt, sugar, and proteins, are nonvolatile, and because, as we shall see in the next chapter, ice, the solid phase of water, freezes out of liquid water as a pure phase because its special structure cannot accommodate solute molecules into its lattice. In such a situation, the chemical potential of the liquid solvent is reduced by $RT \ln x_A$ when the solute B is present, due to the entropy of mixing, which as we have already seen lowers the free energy, because the two substances mixed together are more disordered than the case where both are pure species:

$$\mu_A = \mu_A^* + RT \ln x_A \tag{1.112}$$

Figure 1.9. The effects of a solute in the liquid phase on T_f and T_b.

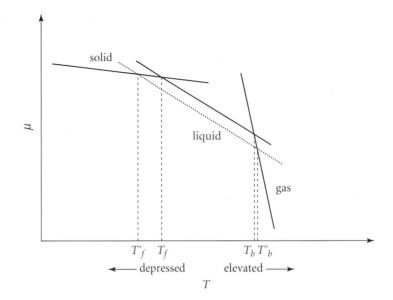

(x_A is a mole fraction < 1, so $\ln x_A$ is negative). The chemical potentials of the solvent vapor and solid phases are unchanged because the solute B is not present in these phases. The dotted line in the graph in Figure 1.9 indicates the chemical potential of the liquid phase, which is lowered by the addition of B. Because it is lower than the line for the pure liquid phase, it intersects the line for its solid phase at a lower T, and that for the gas at a higher T, giving a lowering of the freezing/melting point and a raising of the boiling point.

The lowering of T_f and the raising of T_b are not enthalpic effects because they occur for an ideal solution. As can be seen from Figure 1.9, because of the differences in the slopes for the solid and vapor lines, the magnitude of the freezing point depression will be larger than that of the boiling point elevation. As we have already pointed out, both effects depend only on the amount of the solute present, and not on its chemical nature, so a mole of table sugar would have the same effect as a mole of a protein. In practice, both effects are small, but can have important practical consequences. For example, freezing point depression is why spreading salt on a highway in the winter melts snow and ice, and why adding ethylene glycol to automobile radiators prevents the water in them from freezing and cracking the engine block. In the case of salts, there may also be an additional effect from a large favorable enthalpic contribution resulting from a particularly exothermic heat of solvation. This is true of $CaCl_2$, for example, which also produces a larger entropic contribution, because it dissociates into three kinetically independent species instead of just two in the case of NaCl. Note, however, that this prompt heating effect from the exothermic solvation does not lower the freezing point further and is quickly dissipated on a cold day, while the entropic contribution remains as long as the salt is present.

Freezing Point Depression

Let us now consider how to estimate the magnitude of the freezing point depression that would result from the addition of a given number of moles of a solute B to a liquid solvent A such that the mole fraction of the solvent in the resulting solution is x_A. In such a case, an equilibrium between the solvent in a solution and its solid, crystal phase will exist at some temperature T, different from T^*, the freezing point for pure solvent, when the chemical potentials in the solid and liquid solution phases are equal,

$$\mu_A^*(s) = \mu_A(l) = \mu_A^*(l) + RT \ln x_A \qquad (1.113)$$

$$\ln x_A = \frac{\mu_A^*(s) - \mu_A^*(l)}{RT} = \frac{\Delta G_{f,m}(T)}{RT} \qquad (1.114)$$

$$\ln(1 - x_B) = \frac{\Delta G_{f,m}(T)}{RT} \quad \text{(since } x_A + x_B = 1) \qquad (1.115)$$

where $\Delta G_{f,m}(T)$ is the molar Gibbs free energy of fusion of the pure solvent at temperature T. We want to solve this equation for T in terms of the amount of solute added, expressed as its mole fraction x_B, so we will use a trick to solve it in the form that we want. When $x_B = 0$, the freezing point is T^* and no depression is measured,

$$\ln 1 = 0 = \frac{\Delta G_{f,m}(T^*)}{RT^*} \qquad (1.116)$$

Because this equation is equal to zero, we can subtract it from the previous equation without changing anything, giving

$$\ln(1 - x_B) = \frac{\Delta G_{f,m}(T)}{RT} - \frac{\Delta G_{f,m}(T^*)}{RT^*} \qquad (1.117)$$

Using $\Delta G = \Delta H - T\Delta S$,

$$\ln(1 - x_B) = \left[\frac{\Delta H_{f,m}(T)}{RT} - \frac{\Delta S_{f,m}(T)}{R} \right] - \left[\frac{\Delta H_{f,m}(T^*)}{RT^*} - \frac{\Delta S_{f,m}(T^*)}{R} \right] \qquad (1.118)$$

If we assume that only a small amount of solute is added, then x_B is small, $x_B \ll 1$, and we can make the approximation that $\ln(1 - x_B) = -x_B$, since $\ln(1 - x)$ can be approximated by a Taylor series expansion as

$$\ln(1 - x) = -x - \tfrac{1}{2}x^2 - \ldots = -x \qquad (1.119)$$

where all but the leading term can be ignored, since if x_B is a small fraction, its square will be even smaller. We can also assume that the depression of T_f is small if x_B is small, so we assume that

$$\Delta H(T) \cong \Delta H(T^*) \text{ and } \Delta S(T) \cong \Delta S(T^*) \tag{1.120}$$

Using these substitutions, the ΔS terms cancel, but the ΔH terms do not, because they are divided by different temperatures, so

$$x_B = -\left[\frac{\Delta H_{f,m}}{R}\left(\frac{1}{T} - \frac{1}{T^*}\right)\right] \tag{1.121}$$

$$x_B = -\left[\frac{\Delta H_{f,m}}{R}\left(\frac{T^* - T}{TT^*}\right)\right] \tag{1.122}$$

We have assumed that the freezing point depression δT is small; therefore $TT^* \approx T^{*2}$, and writing $\delta T = T^* - T$,

$$x_B = \frac{\Delta H_{f,m}}{R}\frac{\delta T}{T^{*2}} \tag{1.123}$$

$$\delta T = \frac{RT^{*2}x_B}{\Delta H_{f,m}} \tag{1.124}$$

The nature of the solute does not appear here at all, only its mole fraction.

The mole fraction is the preferred unit of concentration for theoretical work, because it does not depend on volume, pressure, or temperature. Expressing this equation in practical concentration units for lab work is helpful, however, and this is usually done in terms of the molality of the solute m_B, where the molality is defined as the number of moles n_B of solute in 1000 g (1 kg) of solvent. Unlike molarity, which, as a function of volume, is dependent on temperature and pressure, molality is temperature independent.

$$m_B = \frac{n_B}{1 \text{ kg of solvent}} \text{ or } n_B = m_B \cdot (1 \text{ kg of solvent}) \tag{1.125}$$

also,

$$n_A = \frac{1 \text{ kg}}{M_A} \tag{1.126}$$

where M_A is the molecular weight (g/mole) of the solvent. Now, we introduce yet another approximation into the δT equation that is again based on the assumption we already made that x_B is small. Since x_B is small,

$$x_B = \frac{n_B}{n_A + n_B} \approx \frac{n_B}{n_A} \tag{1.127}$$

$$x_B = \frac{n_B}{n_A} = \frac{m_B(1 \text{ kg solvent})}{1 \text{ kg} \Big/ M_A} = m_B M_A \tag{1.128}$$

so

$$\delta T = \frac{RT^{*2}}{\Delta H_{f,m}} m_B M_A = k_f m_B \tag{1.129}$$

where $k_f = \dfrac{RT_f^{*2}}{\Delta H_{f,m}} M_A$ is the **cryoscopic constant** or **freezing point depression constant** and has units of K·kg·mol^{-1} (the units of R are J·K^{-1}·mol^{-1}; for ΔH, J·mol^{-1}; for M, kg·mol^{-1}). For water, k_f is 1.86 K·kg·mol^{-1}.

In the case of boiling point elevation, the equilibrium is between pure vapor and liquid solution with a mole fraction x_B of B present. At the boiling point, the chemical potentials are equal,

$$\mu_A^*(\text{vap}) = \mu_A(\text{liq}) = \mu_A^*(l) + RT \ln x_A \tag{1.130}$$

the same as before.

Parallel arguments then lead to

$$\delta T = \frac{RT^{*2} x_B}{\Delta H_{vap,m}} \tag{1.131}$$

where δT is now $\delta T = T - T^*$, the boiling point elevation. This expression can be written as

$$\delta T = k_b m_B \tag{1.132}$$

The constant k_b is called the **ebullioscopic constant**. Table 1.3 list cryoscopic and ebullioscopic constants for several common liquids.

Figure 1.10 illustrates the freezing point depression of various aqueous solutions of common solutes. The curves for ethanol and ethylene glycol are quite typical for simple alcohols and small, polar molecular solutes. Both molecules are commonly thought of as antifreezes (ethylene glycol is used in automobile radiators, and methanol, which is similar to ethanol, is used as an antifreeze in windshield wiper fluid). At low concentrations the functional dependence appears linear, as predicted by the equation derived above for the limit of small solute concentrations, where the Taylor series expansion is truncated after the leading term. NaCl, which is used on roads in the winter to melt ice, also follows such a typical linear functional dependence but

Table 1.3 Selected cryoscopic and ebullioscopic constants

Solvent	T_m^* (°C)	T_b^* (°C)	k_f (K·kg·mol^{-1})	k_b (K·kg·mol^{-1})
water	0	100	1.86	0.51
acetic acid	17	119	3.90	3.07
camphor	177	204	40	—
benzene	6	80	5.12	2.53
ethanol	−114	78	—	1.23
carbon tetrachloride	−23	77	30	4.95

Figure 1.10. The freezing point depression for water for several common solutes.

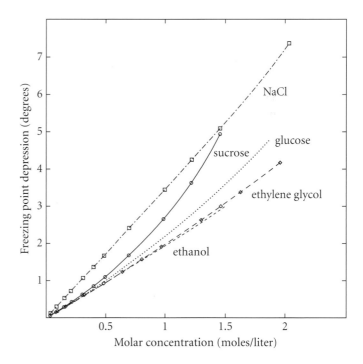

with approximately twice the slope, because each mole of NaCl dissociates in water into two moles of independent ions.

At high concentrations, the two sugars illustrated, however, behave quite differently. Both follow a quadratic dependence, which makes 1.5-molar sucrose as effective as NaCl at the same concentration. Other monosaccharides follow a curve almost indistinguishable from that for glucose, and other disaccharides follow a curve almost indistinguishable from that of sucrose. Apparently the ring structures of the sugars are involved in this behavior, because the curve for the linear sugar alcohol D-mannitol is indistinguishable from that for ethylene glycol, ethanol, and other alcohols. Many cold- and drought-adapted organisms produce excess amounts of sugars such as sucrose and trehalose to control water during periods of freezing temperatures; the reason for their effectiveness is obvious from the figure. There has

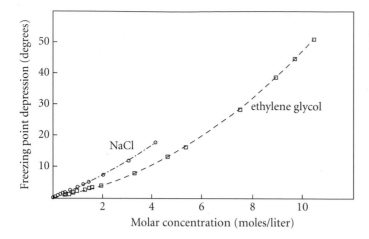

Figure 1.11. The freezing point depression for aqueous solutions of salt and ethylene glycol at high concentrations.

Table 1.4 Freezing point depressions and boiling point elevations for various concentrations of a solute

	M_B (g·mol^{-1})			
	10^3	10^4	10^5	10^6
ΔT_b (in K)	5.19×10^{-3}	5.19×10^{-4}	5.19×10^{-5}	5.19×10^{-6}
ΔT_f (in K)	1.87×10^{-2}	1.87×10^{-3}	1.87×10^{-4}	1.87×10^{-5}

also been interest in the food industry in the possibility of exploiting this behavior of sugar molecules to control freezing in foods.

The derivation of the freezing point depression expression made use of two assumptions, that of low solute concentration and the assumption that $\Delta H(T) \cong \Delta H(T^*)$ and $\Delta S(T) \cong \Delta S(T^*)$. Thus, while all such curves appear linear in the limit of sufficiently small solute concentrations, all are in fact polynomials, and as can be seen in Figure 1.11, both NaCl and ethylene glycol appear quadratic at still higher concentrations.

Colligative properties can be used to estimate molecular weights for an uncharacterized molecular species. For example, freezing point analysis provides the number of moles of solute present in a solution and either the method of preparation or the evaporation of the solution to dryness provides the corresponding mass, thus allowing the determination of the "molecular" weight. Because freezing point depression and boiling point elevations are small, this method can be difficult to apply with any accuracy for colloidal size particles. Table 1.4 illustrates the small magnitudes of ΔT_f and ΔT_b values for 1% aqueous solutions for various molecular weights.

In dairy science, precise measurements of the freezing point depression for raw milk samples are used as a test for adulteration. The market value of milk increases faster with overall weight than it decreases as the fat and protein content decreases, providing an economic incentive to water down milk samples, an illegal practice punishable by both fines and imprisonment. This type of fraud is rare in the United States, but does occur from time to time. While a small amount of natural variation in the freezing point depression of normal milk is found, it is generally around −0.52 °C. If water has been added to milk, it will decrease the concentration of the solutes and thus decrease the magnitude of the freezing point depression. Addition of a saline solution of the appropriate concentration, however, would not be detectable by this method. Devices to measure the freezing points of liquids are called **cryoscopes**, and especially-calibrated versions are available to detect the watering of milk.

Osmosis and Osmotic Pressure

Another familiar colligative phenomenon is osmosis and the development of osmotic pressure. Consider a system like the one illustrated in Figure 1.12. This system consists of a cylinder divided into two regions by a semipermeable partition. In the volume on the right there is pure water, while in the volume on the left there is a solution of solute molecules S dissolved in water. The membrane dividing the cylinder into two has pores that are large enough to allow water molecules to pass through in either direction, but are too small to allow the passage of the much larger solute molecules. The free energy of the region on the left is lowered relative to that of a pure liquid by the favorable (negative) free energy contribution from the entropy of mixing. As a result of this lowering of the free energy due to mixing, there will be a net diffusion of water molecules from right to left in the cylinder. The entropy would also be increased by mixing if the solute particles diffused from the solution compartment into the pure solvent compartment on the right, but they are prevented from doing so by the small size of the holes in the partition. The diffusion of water molecules into the compartment containing solute will continue until this net "flow" is halted by some opposing force (e.g., from the piston in Figure 1.13). That opposing force is the so-called osmotic pressure. As more and more water molecules diffuse into the compartment, the pressure necessarily rises as the density increases, raising the pressure. When this pressure is high enough, it will halt the net diffusion of water

Figure 1.12. A box or cylinder partitioned into two compartments by a semipermeable barrier or membrane. In the compartment on the right is pure liquid water, while in the compartment on the left is an aqueous solution of solute particles, represented by the solid circles, that are too large to pass through the holes in the barrier.

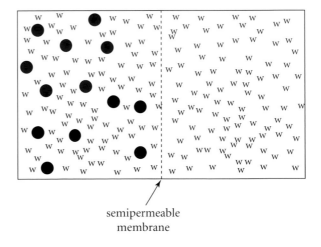

semipermeable
membrane

molecules, and the pressure necessary to stop this "flow" is called the osmotic pressure, usually symbolized by Π.

The chemical potential of the pure water in the compartment on the right is just μ_w^*. At equilibrium, the chemical potential of the solvent water must be the same on both sides of the partition,

$$\mu_w^*(\text{pure liq}, P) = \mu_w(\text{liq solution}) = \mu_w(x_w, P + \Pi) \tag{1.133}$$

where x_w is the mole fraction of the water in the solution on the left and P is the pressure.

$$\mu_w(x_w, P + \Pi) = \mu_w^*(P + \Pi) + RT \ln x_w \quad \text{(Raoult's Law)} \tag{1.134}$$

$$\mu_w^*(P + \Pi) = \mu_w^*(P) + \int_P^{P+\Pi} V_m^* dP \tag{1.135}$$

Substituting both of these equations into Equation (1.133) gives

$$\mu_w^*(P) = \mu_w^*(P) + \int_P^{P+\Pi} V_m^* dP + RT \ln x_w \tag{1.136}$$

or, rearranging,

$$-RT \ln x_w = \int_P^{P+\Pi} V_m^* dP = -RT \ln(1 - x_S) \tag{1.137}$$

where we have used the definition of the mole fraction, $x_w + x_S = 1$ to replace the mole fraction of solvent x_w with $1 - x_S$, where x_S is the mole fraction of the solute. This equation relates the osmotic pressure to the mole fraction of the solute particles. For dilute solutions, $x_S \ll 1$, so we can use the approximation

$$\ln(1 - x_S) = -x_S \tag{1.138}$$

If the molar volume does not change much over the pressure range of interest (which is true here since water is fairly incompressible), then the volume term can be taken outside the integral as a constant, and the integration of dP becomes

$$-RT \ln(1 - x_S) = RT x_S = \Pi V_m^* \tag{1.139}$$

Since $x_S \ll 1$, we can again make the approximation

$$x_S = \frac{n_S}{n_w + n_S} \approx \frac{n_S}{n_w} \tag{1.140}$$

Also, using $n_w V_m^* = V$, the actual volume of the solvent water, V, gives

$$n_s RT = \Pi V \text{ (the van't Hoff equation)} \tag{1.141}$$

This equation relates the amount of dissolved solute or dispersed particles S present in the solution on the left side of the cylinder, as measured by its mole fraction, to the resulting osmotic pressure Π. Notice the similarity of the van't Hoff equation to the $PV = nRT$ form of the ideal gas equation.

Osmosis is an extremely important process in both living systems and in food technology. The osmotic pressure that builds up in living cells from the diffusion of water into the cells, as the result of the many solute molecules in the cytoplasm, is responsible for the turgor of the cells. The cell walls of plants consist of both a matrix of polymers in the outer cell wall (see Figure 4.106) as well as a membrane made up of a phospholipid bilayer like the cell membranes of animal cells (see Chapter 7). These membranes contain pores for both active and passive transport through their barriers, and they can be considered to effectively be semipermeable as in Figure 1.12. The net direction of diffusion can be changed by concentrating solutes on the outside of the membrane. For example, putting large quantities of salt onto plant or animal tissues will dehydrate the cells as the water diffuses out of the cells to the more concentrated salt solution. This process is the reason that drinking seawater is harmful.

Osmosis is also used in the food industry to preserve fruits and meats by removing much of their internal water through a process called osmotic dehydration (Schiraldi 2010). In this process, cut fruit slices are placed in a hypertonic syrup of concentrated sugars. Water has a lower chemical potential in the bath than in the fruit, due to the high concentration in the solution, so water diffuses out of the fruit tissues into the solution, and some sugar also diffuses into the surface tissues of the fruit. Because of the limited permeability of the fruit tissues, however, the sugars do not penetrate far into the fruits. Table sugar (sucrose, Chapter 4) is not popular for this purpose, not only because it adds unnecessary sweetness to the fruits, but also because it easily crystallizes in the surface tissues and gives them a "crunchy" texture. For these reasons, maltose is a better choice for the osmotic bath, because it does not readily crystallize and is not particularly sweet (Table 4.4). Water loss from the fruit tissues is highest during the first two hours and gradually decreases after this period. Osmodehydration is usually accomplished as a batch process, although the diluted syrup can be augmented with additional sugar and reused.

In the laboratory osmosis is exploited in many ways. Like freezing point depression, it can be used to determine molecular weights and is used to measure the binding affinities of two species for one another or as a means of concentrating macromolecular solutions. Food technology exploits osmotic pressure to concentrate liquids such as juices by applying pressure to a solution to force water out through a semipermeable wall or membrane, in a process called **ultrafiltration** or **reverse osmosis**, as is illustrated in Figure 1.13. Osmosis is also used to separate whey proteins from milk and to remove certain undesirable flavor notes from wines. In general, the term "ultrafiltration" is used to refer to processes using semipermeable membranes

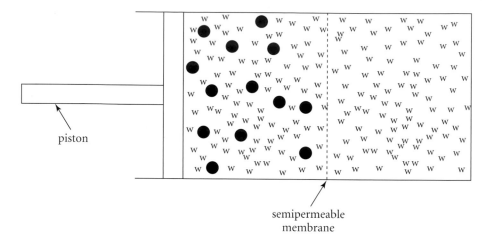

Figure 1.13. The application of a pressure to a solution can exceed the osmotic pressure of the solution and force water back across the semipermeable membrane into the pure water side.

with large pore sizes, which can separate out colloidal-size particles and larger suspended materials from solvent water. The term "reverse osmosis" usually refers to processes in which the pore size of the semipermeable barrier is on the nanometer size scale and can be used to sort out molecules, allowing only water molecules to pass through and retaining molecular and even monatomic ionic solutes. Very large applied pressures are needed in such cases. The advantages of these methods for removing water are both the savings in energy from avoiding having to overcome the high heat of vaporization of water, and also the minimization of damage to the juice from reducing the "cooking" of the product (it remains necessary to briefly heat the product, however, to pasteurize the juice). The fouling or blocking of the pores in such membranes by the dispersed particles is one of the major limitations of this technology in practical applications and requires frequent cleaning.

Reverse osmosis can, in principle, be used to desalinate seawater by allowing pure water to pass through semipermeable membranes while retaining even hydrated sodium and chloride ions. The osmotic pressure of seawater is around 25 atmospheres, the approximate pressure at depths of around 255 meters (~840 ft), so that, if a pipe 255 meters long capped with an appropriate semipermeable membrane was sunk into the ocean, fresh water would accumulate at the bottom. Membranes in such systems need to be quite strong to withstand the applied pressures. Cellulose acetate matrices have been successfully used for this type of application and are able to retain more than 96% of the NaCl in seawater.

In an interesting new approach to desalination using membranes, called forward osmosis, the application of pressure is not necessary. Instead, a high concentration of another volatile solute is added to the right-hand compartment in Figure 1.13. Water then diffuses across the membrane barrier without the application of pressure, because it is diffusing into a region of even higher solute concentration than in the salt water. The volatile solvent, which is selected to have a low boiling point, is then distilled off through the application of gentle heating, leaving behind the pure, desalinated water with a lower expenditure of energy than required in applying high pressures in reverse osmosis.

SUGGESTED READING

The brief review of physical chemistry presented here loosely follows the developments given in several excellent textbooks, which can be consulted for further details.

Atkins, P.W. 1997. *Physical Chemistry*, 6th ed. New York: Freeman.

Cantor, C.R., and P.R. Schimmel. 1980. *Biophysical Chemistry, Parts 1–3*. New York: Freeman.

Castellan, G.W. 1983. *Physical Chemistry*, 3rd ed. Reading, MA: Addison-Wesley.

Moore, W.J. 1972. *Physical Chemistry*, 4th ed. Englewood Cliffs, NJ: Prentice-Hall.

The following general references are also useful.

Atkins, P.W. 1984. *The Second Law*. New York: Scientific American.

Atkins, P. 2003. *Atkins' Molecules*, 2nd ed. Cambridge: Cambridge University Press.

Baianu, I.C., ed. 1992. *Physical Chemistry of Food Processes*. Vol. 1. *Fundamental Aspects*. New York: Van Nostrand Reinhold.

Barham, P. 2001. *The Science of Cooking*. Berlin: Springer.

Belitz, H.-D., W. Grosch, and P. Schieberle. 2009. *Food Chemistry*, 4th revised and extended ed. Berlin: Springer-Verlag.

Coultate, T. 2009. *Food: The Chemistry of Its Components*, 5th ed. Cambridge: RSC Publishing.

Hager, T. 2008. *The Alchemy of Air*. New York: Harmony Books.

McGee, H. 2004. *On Food and Cooking: The Science and Lore of the Kitchen*, 2nd ed. New York: Scribner.

McMurry, J., and R.C. Fay. 1995. *Chemistry*. Englewood Cliffs, NJ: Prentice Hall.

Merrill, A.L., and B.K. Watt. 1973. *Energy Value of Foods ... Basis and Derivation. Agriculture Handbook No. 74*. Washington: Agriculture Research Service, US Dept. of Agriculture.

Myhrvold, N. 2011. *Modernist Cuisine: The Art and Science of Cooking*. Bellevue, WA: Cooking Lab.

Owusu-Apenten, R. 2005. *Introduction to Food Chemistry*. Boca Raton: CRC Press.

Smith-Spangler, C., et al. 2012. "Are organic foods safer or healthier than conventional alternatives? A systematic review." *Ann. Intern. Med.* 157:348–366.

Specter, M. 2009. *Denialism: How Irrational Thinking Hinders Scientific Progress, Harms the Planet, and Threatens Our Lives*. New York: Penguin Press.

This, H. 2006. *Molecular Gastronomy*. Translated by M.B. Debevoise. New York: Columbia University Press.

2.
Water in Foods

PHYSICAL PROPERTIES OF WATER

Water is the most abundant compound on the surface of the Earth and is so important in determining the overall character of the planet that Earth is often referred to as the water planet. Nearly three-quarters of the Earth's surface is covered by water, and most of these ocean depths are less well known than the surface of Mars. The current abundance of water on Earth distinguishes it from most of the other planets and moons of the Solar System, where liquid water is either absent or present in only small amounts. Only Europa, Callisto, and Ganymede, three of the moons of Jupiter, which are covered by frozen seas of ice, and Enceladas, a moon of Saturn, have water in large amounts, although recent discoveries suggest the presence of residual water on Mars and even Earth's Moon.

Water is essential to life; present theories of the origin of life are unable to conceive of life forms that are not dependent on water availability. The recent explorations of the surface of Mars have found the presence of frozen water just below the planet's surface, which may mean that life was possible there in the past. Deep space probes have also found that not only is Europa covered by a sea of frozen water, but that beneath this ice surface there may be liquid water as well, perhaps heated by tidal

Table 2.1 The water content of selected foods

Food	Water content (weight %)
olive oil	~0
raw pork	55–60
raw chicken	74
fish	65–81
beets, broccoli, potatoes	80–90
apples, oranges, grapefruit	85–90
strawberries, tomatoes, green beans, lettuce, cabbage	90–95

Source: O.R. Fennema, in *Food Chemistry*, 2nd ed., O.R. Fennema, ed. (New York: Marcel Dekker, 1985).

stresses from interaction with the giant planet, thus raising the possibility that life could perhaps exist there too.

Not only is water essential for life, it is the most abundant component of living organisms. The human body, for example, is roughly 55% to 60% water by weight (depending on gender). Water is also the most abundant food molecule and is present to some extent in nearly all foods. In many foods it is by far the largest component, making up more than 90% of the weight of some fresh fruits and vegetables. Table 2.1 lists the water content of some representative foods. As might be expected from its overwhelming abundance in foods, the properties of water often dominate the properties of the foods themselves, and understanding the behavior of water is necessary to understanding the foods in which it is found.

Although water is one of the simplest of chemical compounds, the collective behavior of water molecules is quite extraordinary and anomalous. Many of water's physical properties are unusual, as compared with other molecules of similar size. The practical effects of these odd properties are of enormous importance for life in general, and for foods in particular. As examples, note from Table 2.2 that water has a very high melting point and boiling point, as well as very large heats of fusion and vaporization. It also has a high surface tension and a high heat capacity, a large thermal conductivity for a liquid, and a high dielectric constant. Water is also one of the few substances that has its density maximum in the liquid phase (4 °C); that is, solid water (ice) is less dense than liquid water.

Water is the only substance commonly found in abundance in all three phases on the Earth's surface and is the only commonly occurring inorganic liquid. It is also the most widely distributed pure solid (the phase diagram for water is shown in Figure 2.1). Water is sometimes called the universal solvent, because nearly all chemical compounds dissolve in it to some extent. For many of these, called **hydrophobic molecules**, the solubility in water is quite small, however. Although many substances

Table 2.2 Selected properties of water and various other substances

Property	H_2O	NH_3	CH_4	HF	CH_3OH	C_2H_5OH
molecular weight	18.02	17.03	16.04	20.01	32.04	46.07
melting point at 1 atm (K)	273.15	195.45	90.65	190.05	179.25	155.85
boiling point at 1 atm (K)	373.15	239.8	109.15	292.69	338.11	351.65
critical temperature (K)	647.30	132.4	190.85	188.0	512.55	516.15
critical pressure (atm)	218.6	113.1	46.5	66.2	79.2	63.0
heat of fusion (at the mp) at 1 atm (kcal/mol)	1.436		0.224	1.094	0.759	1.200
heat of vaporization (at the bp) at 1 atm (kcal/mol)	9.705	5.576	2.129	6.018	8.979	9.674
density at 293.15 K (g/cm³)	0.998	0.771	0.466 (at 109.2 K)	0.991 (at 292.7 K)	0.791	0.7893
viscosity at 293.15 K (cp)	1.002	—	0.011 (gas)	—	0.597	1.20
surface tension at 293.15 K (against air, N/m)	0.07275	—	—	—	0.02261	0.02405
heat capacity at 298 K (cal/mol·K)	17.98	3.29 (gas)	—	—	19.5	27.0
specific heat (cal/g·K)	0.998	—	—	—	0.600	0.586
entropy of vaporization (cal/K·mol)	26.05	23.18	17.49	6.21	25.10	27.51
thermal conductivity (cal/s·cm·K)	0.001415	—	—	—	0.000483	0.000399
dielectric constant	78.54 (at 25 °C)	25 (at −77.7 °C)	1.70 (at −173 °C)	84 (at 0 °C)	32.63 (at 25 °C)	24.30 (at 25 °C)

Figure 2.1. The phase diagram for water. (Redrawn from data in D. Eisenberg and W. Kauzmann, *The Structure and Properties of Water.* New York: Oxford University Press, 1969.)

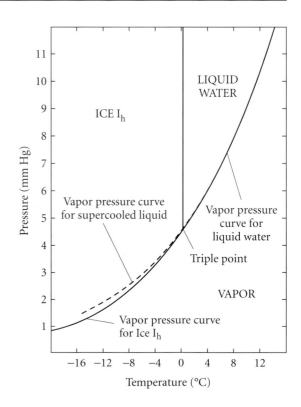

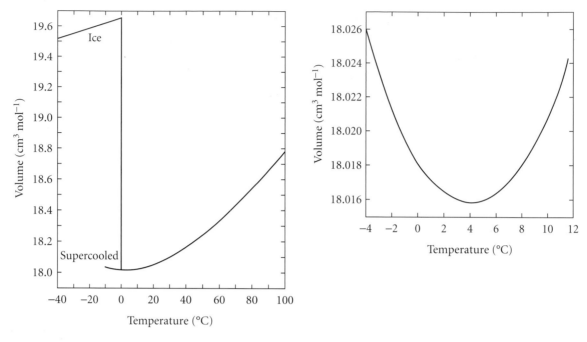

Figure 2.2. The specific volume of water as a function of temperature (remember that the behavior of this quantity is the inverse of that of the density: the more volume a given amount of matter occupies, the less dense it is). Notice that ice has a larger specific volume than water (i.e., is less dense). From the enlargement of the liquid water curve around 0 °C shown on the right, it can be seen that the density maximum and volume minimum occur at ~4 °C. (Redrawn from data given in D. Eisenberg and W. Kauzmann, *The Structure and Properties of Water*. New York: Oxford University Press, 1969.)

dissolve in water to some extent, no instances are known of solutes that make thermodynamically ideal solutions with water; that is, no other molecule is really like water. Water is one of the most reactive of chemicals and can be one of the most corrosive of substances. In spite of this, it is physiologically innocuous.

The consequences of these unusual properties of water can be quite profound. For example, the fact that ice is less dense than water (Figure 2.2), and thus floats, keeps the oceans and lakes from freezing solid. If ice sank, more water would be exposed to the freezing temperatures at the surfaces of lakes and oceans and, as it froze, would also sink. Ultimately much of the water on the planet would end up in the solid state, probably plunging the Earth into a deep and permanent ice age. Under such conditions, it is difficult to imagine how life could survive. The lower density of solid water of course implies that water expands upon freezing, as you almost certainly have noticed from personal observation. Repeated cycles of such freezing and thawing in fissures and cracks in rocks is one of the processes by which much of the Earth's soil has been built up. From a food science point of view, such expansion must be taken into account when designing packaging for frozen food products. Another interesting consequence of the fact that ice is less dense than liquid water is that applying pressure to ice at a constant temperature can cause it to melt. This is because the

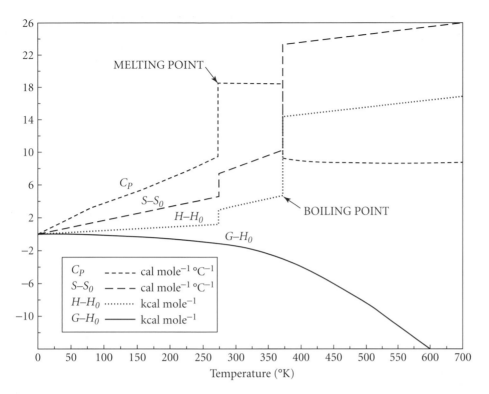

Figure 2.3. The temperature dependence of the free energy, entropy, enthalpy, and heat capacity of water. (Redrawn from data given in D. Eisenberg and W. Kauzmann, *The Structure and Properties of Water*. New York: Oxford University Press, 1969.)

solid/liquid phase boundary actually has a negative slope (i.e., slopes "backwards"), although it is difficult to perceive this in the phase diagram in Figure 2.1, since it concentrates on the region around the triple point, which occurs at a very low pressure. Thus, if one starts from a point close to the solid/liquid boundary, at a temperature just below the freezing point, and applies enough pressure at the same temperature, the vertical line on the phase diagram corresponding to this pressure change will cross the solid/liquid boundary at some point (i.e., will cause the ice to melt without raising the temperature).

The high heat capacity of water also has many important effects. Recall that heat capacity is the molar specific heat, and measures the amount of heat that must be added to or taken away from a substance to change its temperature by a certain amount. The heat capacity is thus a measure of the ability of a substance to store heat, as well as a measure of how much energy must be expended to heat a substance up (or how much must be removed to cool it down). Water's high heat capacity means that a great deal of energy must be added to water in the form of heat to raise its temperature. While both solid and liquid water have high heat capacities, that of liquid water is much greater than that of ice, an unusual situation when compared to other substances (Figure 2.3). Because of this high heat capacity, liquid water can act as a thermal buffer or thermal reservoir. This property explains the climate-moderating effects of oceans and may have been exploited by some dinosaurs to help maintain a stable body temperature; the huge mass of water inside a Brontosaurus (now more properly called "Apatosaurus"), for example, would resist large and rapid changes in its body temperature in response to the daily temperature cycle.

The high heat capacity of water, along with its high melting and boiling temperatures, has significant consequences for the thermal processing of foods. Because a large amount of energy is needed to heat water to its boiling point, and another large expenditure of energy is then needed to vaporize it at the boiling point (because water also has a large heat of vaporization), economics often requires that liquids not be concentrated by boiling off the water. In commercial operations, as in the concentration of juices, dialysis is often used. Removing water from foods is usually accomplished by drying, which takes longer than dialysis or boiling, particularly at low temperatures, but uses far less energy. Very complex thermal gradients may develop in the cooking of certain foods containing many ingredients, because the different water contents of different components (tomatoes versus potatoes, for example) means that their temperatures will respond at different rates to the application of heat.

As can be seen from the plot of thermodynamic properties in Figure 2.3, the heat capacity of liquid water is nearly double that of ice at the same temperature, and much higher than that of steam at the boiling point. The primary means of storing energy in ice is in vibrations of the molecules in the crystalline lattice, and the principal means of storing energy in the vapor is in the translational and rotational energy of the individual water molecules. In all water phases the intramolecular vibrations of the individual water molecules can also store energy. The large heat capacity of the liquid phase is an indication of extensive short-range collective structuring of the water molecules, which requires considerable energy to distort and disrupt, and thus offers a large capacity to store energy when the temperature is increased, as the heat energy goes into disrupting this structure rather than simply raising the temperature (atomic kinetic energy). The heat capacity of liquid water is nearly constant in the range $0\,°C < T < 100\,°C$ because, even as the liquid structure absorbs energy and distorts, it still retains the capacity to distort further as more heat energy is added and more structure is "melted." The entropy continues to rise steadily and almost linearly over this entire range as the disorder grows. Note that the increase in entropy over this interval of temperature where the water is liquid is as great as for the transition from the ordered array of the crystal lattice to the liquid state. The heat capacity of water actually decreases again upon vaporization because the gas can only absorb energy through the molecules translating, rotating, or vibrating faster—there is no further collective liquid structure to "melt."

The viscosity of pure liquid water is a strong function of temperature, decreasing as the temperature increases (Figure 2.4), as is to be expected for liquids. In deep lakes, a temperature gradient as a function of depth generally exists, with the bottom waters always at $4\,°C$, the temperature of maximum density. In winter, the temperature profile increases from $0\,°C$ at the surface to $4\,°C$ below, while in summer, the temperature decreases from a surface temperature of approximately $20\,°C$ in the

Figure 2.4. The decrease in the viscosity of pure liquid water as the temperature increases.

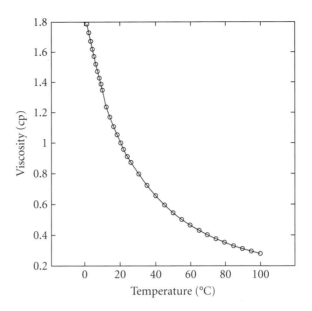

temperate latitudes to again 4 °C at the bottom. Because of the viscosity dependence on temperature, the layers at different temperatures do not mix. In the spring and fall, however, the heating or cooling, respectively, may result in the disappearance of this thermal gradient, with the entire lake having the same temperature, and thus the same viscosity. When this occurs, the waters of the entire lake can easily mix, and a slight wind across the surface might be enough to cause a circulation of the water from the top to the bottom and back, causing a "turnover" of the lake, and bringing sediments and nutrients up from the bottom, leading to an algal "bloom" and a burst of growth and life throughout the lake.

Let us now consider why such a simple molecule as water gives such complex and unusual collective behavior in bulk.

MOLECULAR STRUCTURE AND HYDROGEN BONDING

Although the properties of bulk water are quite unusual and arise from very complex collective behavior, the chemical structure of the individual water molecule is simple. In the valence bond picture of molecular bonding (Figure 2.5), water can be thought of as an sp^3-hybridized oxygen atom with four valence orbitals directed to the vertices of a tetrahedron, two shared with the 1s electrons of the hydrogen atoms and two occupied by the oxygen lone pair electrons.

This model would predict an H—O—H angle of 109.5°, which is somewhat larger than observed. Figure 2.6 shows a van der Waals representation of the actual dimensions of an isolated water molecule: the actual bond angle is 104.5°, with bond lengths of 0.96 Å. This structure can still be reconciled with the qualitative valence bond picture by arguing that the lone pairs occupy more space than a shared bonding pair

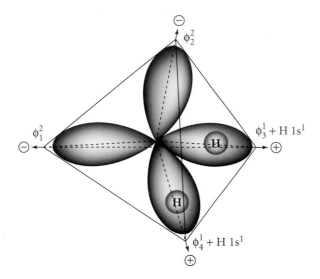

Figure 2.5. The valence bond orbital structure of water.

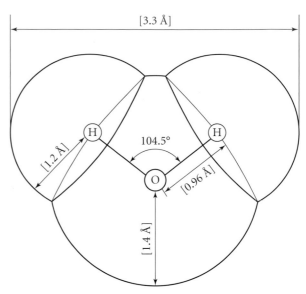

Figure 2.6. The physical dimensions of an isolated (vapor phase) water molecule.

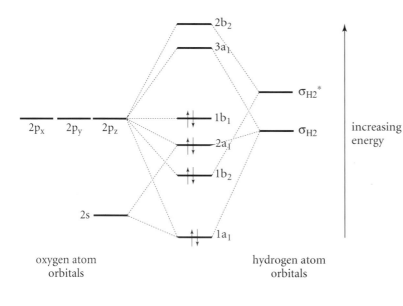

Figure 2.7. The molecular orbital diagram for the water molecule.

and thus crowd the two O—H bonds together, decreasing the bond angle. In the condensed liquid phase and in ice, the close interactions of the water molecules lead to a slight lengthening of the O—H bonds to around 1 Å and an increase in the bond angle.

Figure 2.7 shows the molecular orbital diagram for the orbitals that result from the binding of two hydrogen atoms to a central oxygen atom as in water. The highest occupied molecular orbital is the doubly occupied nonbonding $1b_1$ orbital. This orbital is highly localized on the oxygen atom. Quantum mechanical molecular orbital calculations on water do not find electron density localized in lone pair lobes

Figure 2.8. The boiling points of the hydrides of the Group IV, V, Vi, and VII elements in degrees Kelvin. Note the anomalous behavior of nitrogen, fluorine, and oxygen.

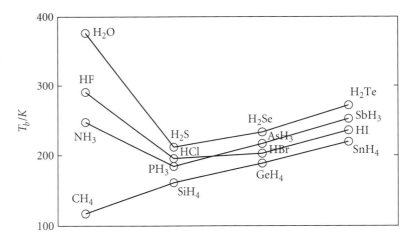

as illustrated in the valence bond view in Figure 2.5, but rather find a more diffuse cloud of high electron density spanning the region where such lone pairs might more conventionally be thought to lie. Nonetheless, we will continue to refer to these notional lone pair positions as a convenient conceptual model for water structure as an approximate tetrahedron.

Hydrogen Bonds

Water is a simple hydride, but its properties are quite different from those of most other hydrides. As already noted, it has a higher heat of fusion, a higher boiling point, a higher heat of vaporization, and a higher surface tension than would be expected by extrapolating from the properties of the other hydrides. The graph in Figure 2.8 illustrates this trend for hydride boiling points. As can be seen, hydrogen fluoride, water, and ammonia all deviate strongly from the trends expected as a function of molecular weight for the other mono-, di-, and trihydrides. Of these three, water deviates most strongly from the trend of the other hydrides of its group in the periodic table. Only the tetrahydride series does not deviate at all in the trend established by the Group IVa elements, with methane, the tetrahydride of carbon, following the expected extrapolated behavior.

The reason for the unusual behavior of water is due to its ability to form **hydrogen bonds** and to the number and geometry of these bonds compared with those of other hydrides. Hydrogen bonds are strong, attractive, noncovalent interactions between certain molecules containing hydrogen atoms covalently bound to electronegative atoms and other functional groups that contain electronegative atoms. It would be difficult to overemphasize the importance of hydrogen bonds in determining not only the properties of water but also, in one way or another, of all of the major types of biopolymers, including proteins, carbohydrates, and lipids, the major chemical components of foods. The double-helical structure of DNA, which is not important as a food component, is also dominated by hydrogen bonding. Because of their

III	IV	V	VI	VII
B 2.0	C 2.5	N 3.0	O 3.5	F 4.0
Al 1.5	Si 1.8	P 2.1	S 2.5	Cl 3.0
Ga 1.6	Ge 1.8	As 2.0	Se 2.4	Br 2.8
In 1.7	Sn 1.8	Sb 1.9	Te 2.1	I 2.5
Tl 1.8	Pb 1.9	Bi 1.9	Po 2.0	At 2.1

Figure 2.9. Electronegativities of the Group 3A (IIIa) through Group 7A (VIIa) elements.

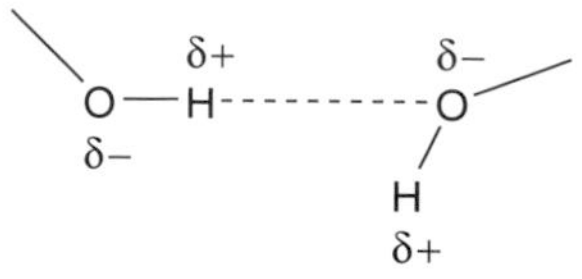

Figure 2.10. The hydrogen bond represented as an electrostatic interaction between the partial charges of the O–H bonds resulting from the high electronegativity of oxygen.

importance throughout food science, the properties of hydrogen bonds are worth reviewing.

Hydrogen bonds are possible whenever a hydrogen atom is chemically bonded to an **electronegative atom** (Figure 2.9 lists the electronegativities of the Group III through Group VII elements). The oxygen atom of water is highly electronegative and draws excess electronic density toward it, thus partially polarizing the bond. This results in a partial positive charge on the hydrogen atoms and a partial negative charge on the oxygen (Figure 2.10). This polarization allows the hydrogen atoms to have a strong electrostatic interaction with the partial negative charge of the oxygen atom of another water molecule. Hydrogen bonds are possible only when hydrogen atoms are chemically bound to strongly electronegative atoms. The most electronegative atoms are F, O, N, and Cl, in decreasing order. S and Br are also weakly electronegative. Carbon is not strongly electronegative, and aliphatic protons bound to carbon atoms do not participate in hydrogen bonds under normal circumstances.

The importance of hydrogen bonds arises from their tendency to impose structure on systems in which they occur. This structuring ability comes from three properties of hydrogen bonds: their great strength, their short range, and their angular dependence. Hydrogen bonds typically have interaction energies in the range 2 to 6 kcal/mol (~8 to 25 kJ/mol). While this is only a small fraction of the energies of covalent chemical bonds, which generally have energies in the range 50 to 150 kcal/mol, it is very large compared with the energy of the most typical of noncovalent interactions, van der Waals attractions, which are usually in the range 0.01 to 0.3 kcal/mol. Significantly, the interaction energy of water or biopolymer hydrogen bonds is an order of magnitude larger than k_BT, which is 0.6 kcal/mol at room temperature. This quantity, which appears throughout statistical mechanics as the basic "yardstick" of energy, due to its occurrence in the Boltzmann weighting factor (see Equation

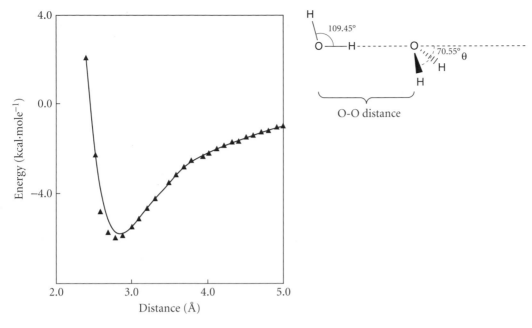

Figure 2.11. The energy of a water–water hydrogen bond, as determined from quantum mechanical calculations, as a function of the oxygen–oxygen atom distance between the two molecules. (Data taken from J.M. Goodfellow et al., in *Water and Aqueous Solutions*, G.W. Neilson and J.E. Enderby, eds. Bristol: Adam Hilger, 1986.)

(1.12)), is a measure of the importance of a particular amount of energy compared to the energy of thermal fluctuations. An energy ten times larger than $k_B T$ is unlikely to be picked up by a molecule in a typical molecular collision, so hydrogen bonds are too strong to be easily disrupted by normal thermal jostling. Hydrogen bonds are also relatively short range, having their strongest interaction energy over a distance of only a few Ångstroms. Figure 2.11 sketches out the distance dependence for a theoretical model of the hydrogen bond energy (approximated as the total interaction energy between two water molecules). As can be seen, the magnitude of the interaction energy decreases from −6.0 kcal/mol at around 2.7–2.8 Å to less than −2.0 kcal/mol at 4.0 Å and around −1.0 kcal/mol at 5.0 Å.

The hydrogen bond is also angle dependent, preferring a linear —O—H···O direction with the proton pointed directly at one of the lone pairs of the oxygen atom. For angles above 270° or below 90°, the hydrogen bond energy rises sharply, as can be seen for the theoretical model illustrated in Figure 2.12, again based on the interaction energy of two water molecules. Within that range the energy is more permissive, allowing the angle to distort rather easily within these bounds. This observation is also confirmed experimentally by surveying X-ray diffraction studies of crystals, where the hydrogen bond distances cluster tightly around 2.8 Å, while a wider range of angles are observed, although the distribution of angles still displays a preference for linear interactions, with angles clustered within the allowed range.

These three characteristics of hydrogen bonds—their relatively great strength, their short range, and their angular dependence—mean that they can impose

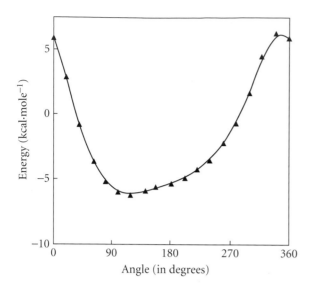

Figure 2.12. The calculated dependence of the hydrogen bond energy between two water molecules as a function of the angle made by the molecular plane with the bisector axis between the two molecules (see Figure 2.11 for an illustration of this angle). (Data taken from J.M. Goodfellow et al., in *Water and Aqueous Solutions*, G.W. Neilson and J.E. Enderby, eds. Bristol: Adam Hilger, 1986.)

considerable structuring on systems that contain hydrogen-bonding groups, as they order themselves to accommodate these bonds. Such groups are common in proteins and carbohydrates and are responsible for the considerable structuring in their conformations. Water, of course, has a high density of hydrogen bonding groups per weight and per molecule, and thus its collective structuring is also dominated by the requirement to maximize its hydrogen bonding.

Types of Hydrogen Bonds in Biological and Food Molecules

Hydrogen bonding groups are found in many types of food molecules. These include the hydroxyl-to-hydroxyl hydrogen bonds found in water, but also those found in hydrogen bonds between sugar molecules and between the amino acid side chains of serine or tyrosine and water. Water molecules can also form hydrogen bonds with the peptide functional groups of protein backbones. The carbonyl (—C=O) moiety of a peptide group can only make hydrogen bonds as an acceptor through the two lone pair groups of the oxygen atom, while the amide (—N—H) moiety can only make a single hydrogen bond as a donor. These two types of groups can also make a strong mixed hydrogen bond to one another, and such peptide hydrogen bonds are an essential feature of alpha helices and beta sheets, the secondary structural elements that are the basic building blocks of protein conformational organization (see Chapter 5). Water molecules can also have strong interactions that resemble hydrogen bonds with the charged groups of proteins, as donors with the oxygen atoms of the carboxylic acid groups of the aspartic and glutamic acid residues and the C-termini of proteins, and as acceptors with the amino group of lysine and the N-termini.

The first four types of interactions listed in Figure 2.13 are conventional hydrogen bonds with energies generally in the range ~4 to 10 kcal/mol, with most having energies of ~5 to 6 kcal/mol (around 10 times $k_B T$). This type of hydrogen bond is

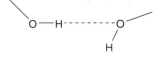

water–water, sugar–sugar, or sugar–water hydrogen bonds, and Thr, Ser, or Tyr amino acid side chains to water, sugars, or one another

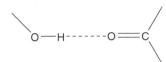

water–carbonyl hydrogen bonds, or Thr, Ser, or Tyr side chains hydrogen bonded to backbone carbonyl groups of proteins, or the ester oxygen atoms of lipids

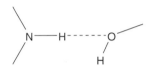

peptide backbone hydrogen bonds, as in alpha helices and beta sheets

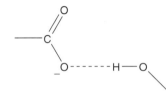

backbone amide groups hydrogen bonded to water, sugars, or Thr, Ser, and Tyr side chains

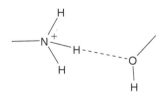

Asp or Glu side chains, or the C-terminus, to water or a hydroxyl group

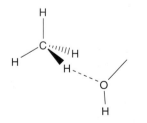

Lys or Arg side chains, or the N-terminus residues of proteins, to water or a hydroxyl group

methyl or methylene groups, of an amino acid side chain (e.g., Ala, Ile, Leu, or even Lys), or a lipid, to a water (this type of interaction is **not** usually thought of as a hydrogen bond, except by a few authors, e.g., Jeffrey and Saenger 1997)

Figure 2.13. The types of hydrogen bonds typically found in biological and food molecules.

common in biological molecules. The next two involve charged groups interacting with a neutral polar group and are referred to as strong hydrogen bonds, or occasionally as "ionic hydrogen bonds," because their energies range from ~15 to 40 kcal/mol. The upper limit of this range is within a factor of two of covalent bond strengths. The last type of interaction shown in Figure 2.13, a —C—H···O interaction, is not generally thought of as a hydrogen bond, although in recent years some authors have begun to call these interactions hydrogen bonds (Jeffrey and Saenger 1997). The electronegativity of carbon is greater than that of hydrogen, so the C—H bond could in principle serve as a hydrogen bond donor, but because carbon is only weakly electronegative, it does not significantly polarize the C—H bond, and energies of these types of interactions are at most only around ~0.4 to 0.8 kcal/mol. These energies are not much greater than general van der Waals interactions and are about the same magnitude as $k_B T$, so these interactions are not bound at room temperature. Accordingly, we will **not** refer to this type of interactions as a hydrogen bond, because they do not have sufficient energy to impose significant structural order at room temperature on systems containing them and cannot compete for partners with more conventional hydrogen bond pairs.

Hydrogen Bonds in Water: Ice and Liquid Water

The strongest conventional hydrogen bonds are formed by HF, because fluorine is the most electronegative element. Notice from the trends in the hydride boiling points, however, that HF is not as anomalous in its behavior as is water. This is because of the special symmetry of the water molecule, where the oxygen atom has two protons and two lone pairs. In HF, the fluorine atom has three lone pairs but only one proton; similarly, in NH_3, there are three protons and only one lone pair. This imbalance in the number of hydrogen bonds that can be formed as donors and acceptors leads to extensive frustration in the ability of these molecules to make hydrogen bonds and prevents them from forming regular space-filling three-dimensional networks. Each water molecule, however, can form two hydrogen bonds as a donor and two as an acceptor, with each of these four hydrogen bonds directed toward one of the four vertices of a regular tetrahedron, as shown in Figure 2.14, so that none interferes with the other. This parity allows water molecules to form regular space-filling networks of hydrogen bonds where each molecule makes four hydrogen bonds to neighbors. This larger number of hydrogen bonds per molecule accounts for the anomalous properties of water.

The structure of ice, shown in Figures 2.15 and 2.16, is a tetrahedral array of these hydrogen-bonded molecules, each making four very regular hydrogen bonds to neighboring molecules. These hydrogen bonds constitute an extended space-filling three-dimensional network with hexagonal symmetry, as can be seen by looking down the so-called c axis of the crystal in Figure 2.16. For this reason, this form of ice, the only one that is stable at atmospheric pressure and ambient temperatures, is referred to as hexagonal ice, symbolized as I_h. This basic symmetry arises from

Figure 2.14. An illustration of a central water molecule, making four hydrogen bonds to quasi-tetrahedrally arrayed partners, as in an ice I_h lattice.

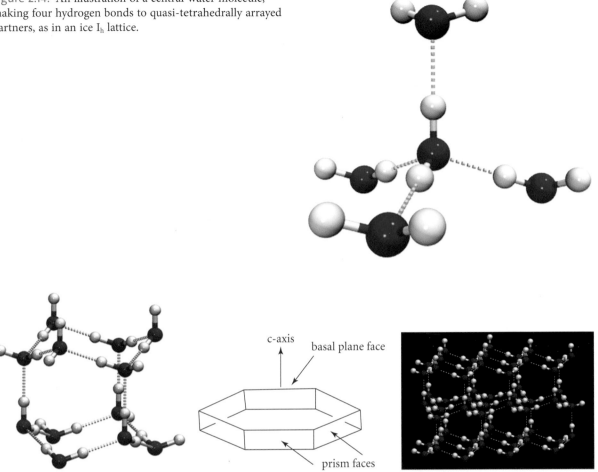

c-axis   basal plane face

prism faces

Figure 2.15. On the left is an illustration of a portion of the ice lattice, where each water molecule is in exactly the same geometric arrangement as the central water molecule shown in Figure 2.14. On the right is an illustration of an extended portion of the ice I_h lattice. Note that, in the figure on the left, six water molecules form a cycle linked by successive hydrogen bonds with the resulting hexagonal ring having the shape of a chair or chaise lounge. As can be seen in the figure on the right, such puckered six-molecule cycles join to form extended puckered surfaces called the basal planes. The crystal surfaces perpendicular to these planes are called prism faces.

hexagonal rings of six water molecules each hydrogen bonded to the next, with a puckered, chair-like geometry for the array of oxygen atoms of these water molecules resembling that of the carbon atoms in cyclohexane. These six-molecule hexagonal cycles hydrogen bond to others to constitute puckered average planes in the lattice like accordion folds, as can be seen in Figure 2.15. Hydrogen bonds perpendicular to these average planes are part of other six-molecule cycles that have an alternate, boat-like geometry, as can also be seen in Figure 2.15. Other crystal forms of water are possible at very high pressures (and for some, lower temperatures), leading to more than a dozen other types of ice, designated with Roman numerals II–XV, but because of the extreme conditions needed to form these crystal packings, they have no

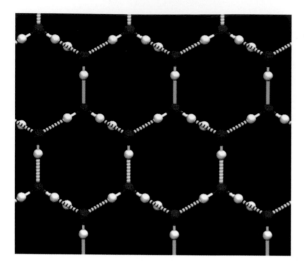 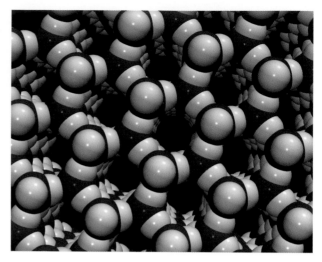

Figure 2.16. On the left, the structure of ice, as seen from "above," looking down the "c" axis in Figure 2.15. On the right, a space-filling view of ice looking down the "c" axis, showing the large open spaces in the lattice, which are nevertheless too small to accommodate a guest species.

relevance for foods. As can be seen, the I_h lattice is open, accounting for its low density. It cannot accommodate substitution, apart from a few scattered fluoride or ammonia ions, because no other molecules have the ability to make four such hydrogen bonds. In addition, even though the lattice is open and thus relatively low in density, the spaces between the water molecules are too small to accommodate any other type of molecule. As a result, when water freezes, it necessarily does so as a pure solid phase.

The inability of hexagonal ice I_h to accommodate solutes has important consequences for freezing in foods and other biological systems. As the water of a food is frozen, all of the substances dissolved in the liquid phase must remain in that phase as more and more of the water molecules are locked up in the solid lattice, which means that this liquid solution becomes progressively more concentrated as the freezing progresses. This process can be seen in ordinary ice cubes made in home freezers. The liquid water put into the ice trays contains salts and dissolved gases such as oxygen and nitrogen. The cubes freeze from the outside in, trapping these dissolved substances on the inside. As the last of the liquid water freezes, all of these gases are forced out of solution as small bubbles trapped inside the ice cube, which gives the interior its whitish appearance.

When ice is heated, the individual water molecules become more mobile as they oscillate about their lattice positions. As the melting temperature of 0 °C is reached, the energy of these oscillations is sufficient for the molecules to escape their lattice positions and begin "squeezing" past adjacent lattice neighbors, and the system becomes liquid. Although the heat of fusion of water is large, it is actually only enough energy to break about 15% of those hydrogen bonds present in the solid phase. The liquid phase must therefore also be extensively hydrogen bonded. In fact, very few hydrogen bonds are actually broken upon melting, because not enough

Figure 2.17. Contours of high neighboring water molecule density around a central molecule in liquid water as calculated from molecular dynamics computer simulations. (P.E. Mason and J.W. Brady, *J. Phys. Chem. B.* 111:5669, 2007.)

thermal energy is present to do so, but most are distorted, being bent and stretched or compressed as the molecules move past one another. Nevertheless, each water molecule still attempts to make approximately four hydrogen bonds as before, two as a donor and two as an acceptor, but with the number of bonds either going up or down temporarily, depending on the instantaneous and chaotic local geometry of its neighbors. On average, however, the coordination number goes up as more neighbors crowd around when the molecules squeeze past one another, and as the open ice lattice breaks down, which is the reason that liquid water is denser than ice. Thus, even though we normally think of liquids as being relatively random and disordered, a considerable amount of residual average tetrahedral structure remains in liquid water, with the behavior of the liquid still dominated by these numerous, if distorted, hydrogen bonds.

This average local structuring can be seen from computer simulations of liquid water, such as is shown in Figure 2.17, which displays in red the areas of high probability of finding another water molecule around any given water molecule. These four areas are still arrayed around the central molecule in a roughly tetrahedral fashion, with two clouds of high density representing the average positions occupied by the water molecules that are making hydrogen bonds as acceptors to the two protons, and two more diffuse clouds representing the positions of water molecules making hydrogen bonds as donors to the central molecule's lone pairs. While in ice the four nearest neighbors would occupy precisely located positions in the lattice, the diffuse clouds seen in this figure show that in the liquid they are less restricted, but still on

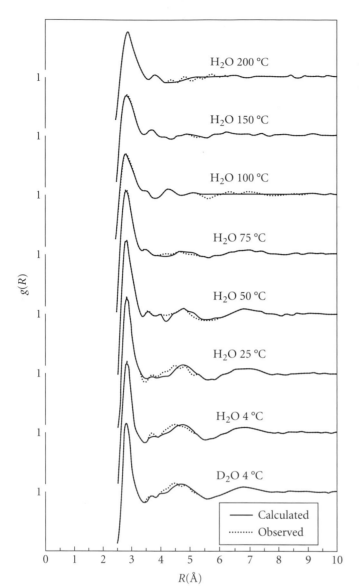

Figure 2.18. Experimental water oxygen radial distribution functions $g_{OO}(r)$ as determined by X-ray diffraction at various temperatures. Note that, at the boiling point, the first peak is broad and less intense, like that for the nonpolar group shown in Figure 2.24. (Reproduced from Narten et al. 1967. Disc. Faraday Soc. 43:97–107 with permission of The Royal Society of Chemistry.)

average try to maintain four hydrogen bonds, since not enough kinetic energy is available to disrupt them completely.

Neutron and X-ray diffraction experiments on liquid water confirm this picture of the average structure of the fluid. Figure 2.18 displays radial distribution functions $g_{OO}(r)$ for the distribution of other water molecules around any given central water molecule extracted from such scattering experiments. These curves essentially give the radially averaged probability of finding a water molecule a given distance from the reference molecule (such as those constituting the red clouds seen in Figure 2.17). The sharp, strong peak around 2.8 Å corresponds to those water molecules hydrogen bonded to the reference molecule. This sharp, narrow first peak is charac-

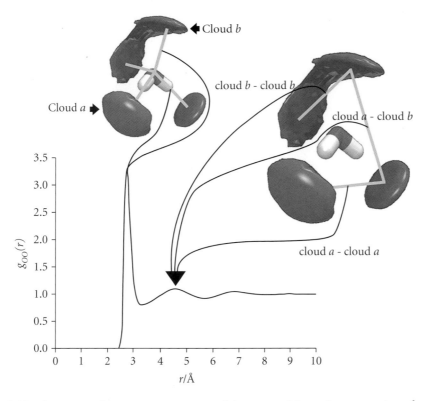

Figure 2.19. Shown is a schematic representation of the origin of the peaks at 2.8 and 4.5 Å in the water radial distribution function $g_{OO}(r)$ using data calculated from a molecular dynamics (MD) computer simulation as an example. The dark gray clouds represent regions of space, relative to the illustrated central water molecule, in which there is a high probability of finding another water molecule; in fact, these clouds enclose the likely locations of the hydrogen-bonded neighbors of this central molecule. These hydrogen-bonded first neighbors are responsible for the sharp first peak in the radially averaged probability distribution function $g_{OO}(r)$; correlations between these hydrogen-bonded first neighbors are responsible for the second peak in this radial probability distribution function. (Reprinted with permission from P.E. Mason and J.W. Brady. 2007. *J. Phys. Chem. B.* 111:5669–5679; copyright 2007, American Chemical Society.)

teristic of the strong localization in space of hydrogen-bonded pairs. Integration of this probability over a spherical volume out to the first minimum reveals on average 4.4 such hydrogen-bonded neighbors (somewhat more than in ice). Figure 2.19 shows schematically using the results of computer modeling the origins of the first two peaks in the $g_{OO}(r)$ in terms of the interactions involved.

Notice that in the series of diffraction curves at different temperatures shown in Figure 2.18, as the water approaches its boiling point at 100 °C, the first peak corresponding to the hydrogen-bonded first neighbors becomes less intense and much broader, and the peak moves to a slightly longer distance. This is characteristic of the close packing of nonpolar spheres seen for liquids such as argon (at very cold temperatures) or pseudo-spherical molecules like methane (which is roughly spherical because the four protons barely stick out of the van der Waals radius of the central carbon atom). This broadening is due to the increased mobility of the water molecules with increasing temperature, allowing them to rotate more freely, decreasing the angular structuring, as the spinning water molecules appear more spherical to one another, and slightly increasing the average inter-oxygen distance as the hydrogen bonds are weakened.

The liquid-state hydrogen bonds give water its high surface tension, because spreading a layer of water, to increase its surface, means pulling against these strong interactions. The local tetrahedral structure that they impart is responsible for the large heat capacity of liquid water, as absorbed heat goes into stretching and bending the bonds further. Because this network is deformable (although for an energy price), the heat capacity remains large and relatively constant over a large temperature range. The residual attractions of even these substantially distorted hydrogen bonds at the boiling point are the reason for the high enthalpy of vaporization, since all remaining hydrogen bonding must be overcome for a molecule to escape into the vapor phase.

AQUEOUS SOLUTIONS

As we have seen, considerable hydrogen bonding occurs in liquid water, and this hydrogen bonding dominates the behavior of water and solutes in aqueous solutions as well. Polar solutes, such as ions and molecules containing hydrogen-bonding atoms and/or groups, will in general be quite soluble in water. The primary mode of interactions between these solutes and the solvent will be by hydrogen bonds or hydrogen bond-like interactions (charge-dipole interactions, for example). We have already seen the experimentally determined radial distribution function for pure water, which is essentially the probability of finding another water molecule at a given radial distance from some reference water molecule's oxygen atom. In pure water, which is dominated by water–water hydrogen bonding, this radial distribution function exhibits a strong and very sharp first peak at lower temperatures (Figure 2.18), characteristic of hydrogen bonds, with an integral out to the first minimum that corresponds to somewhat more than four nearest neighbors. A similar radial distribution would be expected for water molecules hydrogen bonded to polar solute functional groups, such as sugar hydroxyl groups or the carbonyl and amide groups of polypeptides (Figure 2.13). Figure 2.20 shows such a distribution for water molecules around a carbonyl oxygen atom of an alanine dipeptide. This distribution displays a sharp first peak as in the case of pure water, centered around the optimal hydrogen bond distance of 2.75 Å, with a deep first minimum, but the integral of this probability function out to this first minimum gives only two first neighbors, because the carbonyl oxygen atom, which lacks a proton, can only make hydrogen bonds as an acceptor with its two lone pairs. The hydroxyl groups of serine, glucose, or ethanol might be expected to make up to three such hydrogen bonds, two as an acceptor to the lone pairs of the oxygen, and one as a donor involving the hydroxyl proton.

When small nonpolar solutes such as methane are in aqueous solution, it is still necessary for them to be adjacent to water. They cannot hydrogen bond to the solvent,

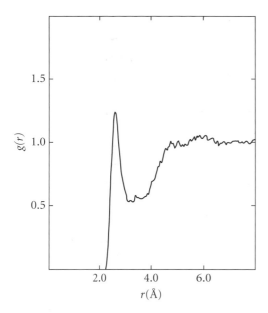

Figure 2.20. The $g(r)$ for water molecules around a hydrogen-bonding carbonyl oxygen atom of a dipeptide, as calculated from a computer simulation. J. W. Brady and M. Karplus, unpublished results.

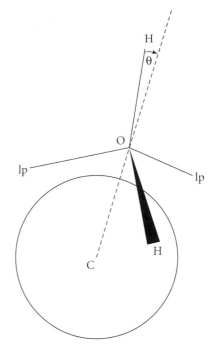

Figure 2.21. A schematic representation of a water molecule straddling a nonpolar solute like methane or an atom of neon.

however, so the water must reorganize so as to straddle the solute and make all of its hydrogen bonds with other water molecules, without pointing either a proton or lone pair at the nonpolar solute (which would result in the loss of a hydrogen bond, and thus require approximately 5 or 6 kcal/mol of energy, or around 10 times $k_B T$ at room temperature). This reorganization reduces the rotational freedom of the water molecules, lowering the entropy (Stillinger 1980). This reduction in rotational options can be seen by looking at the angles describing the orientations of these water molecules. While this orientational distribution is not readily measured experimentally, it can be calculated from computer simulations.

The angular probability distribution function for water molecules around the nonpolar methyl side-chain group of an alanine dipeptide in water, as calculated from computer simulations, is shown in Figure 2.22. This distribution shows the probability of observing an angle θ (indicated in the drawing in Figure 2.21) defining the orientation of the water molecules. This distribution exhibits a maximum at $\cos(\theta) = 1$, corresponding to an angle of 0°, meaning that a proton or lone pair is pointing directly away from the nonpolar group. The broad peak approximately centered around the tetrahedral angle at $\cos(\theta) = -0.33$ results from the nearly tetrahedral structure of water, because the other protons or lone pairs must be making approximately the tetrahedral angle if one proton or lone pair is pointing directly away. Notice particularly the very low probability of a proton or lone pair pointing directly

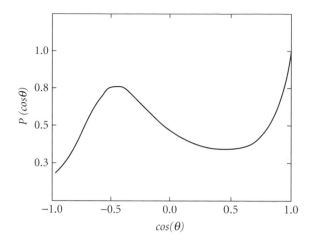

Figure 2.22. $P(\cos(\theta))$ versus $\cos(\theta)$, the probability of observing the cosine of the angle θ as a function of θ. The $\cos(\theta)$ is used to correct for the cosine law distortion. This distribution exhibits a maximum at $\cos(\theta)$ of 1, corresponding to an angle of 0°, meaning that a proton or lone pair is pointing directly away from the nonpolar group; the broad peak at the tetrahedral angle of $\cos(\theta) = -0.33$ results from the quasi-tetrahedral structure of water, because the other proton or lone pair must be making the tetrahedral angle if the first proton or lone pair is pointing directly away.

at the methyl group ($\cos(\theta) = -1$), because this would involve the loss of a hydrogen bond. A hydrogen-bonding solute would exhibit the opposite behavior from that illustrated in Figure 2.22; that is, one of the water hydrogen atoms would be pointed directly at an electronegative atom in the solute, or one of the water molecule's lone pairs would be directly pointed at a solute polar hydrogen atom (one chemically bound to an electronegative atom).

This type of straddling of the solute molecules is only possible because of the small size of methane, with its high radius of curvature. Not only would methane hydrate in this fashion, but so would the methyl group side chain of the amino acid alanine (Chapter 5), as would each of the methylene (CH_2) groups of the hydrophobic chains of lipids (Chapter 7). At low temperatures (~4 °C) and high pressures, such as at the bottom of the ocean, this type of arrangement locks into a regular cage-like crystalline lattice called a clathrate hydrate (Figure 2.23). The faces of this type of cage have pentagonal symmetry, with five water molecules hydrogen bonded in a cycle, rather than the six-molecule hexagonal structure of I_h ice. Each pentagonal face shares two molecules with each adjoining face to completely enclose the guest methane in a dodecahedral cage of twenty water molecules. These gas clathrate hydrates can be stable at subzero temperatures even at atmospheric pressure, but melt at higher temperatures, releasing both the frozen water and the encaged methane. Vast amounts of methane, which is at least 10 times worse than CO_2 as a greenhouse gas, are trapped in this way in the Arctic permafrost. The melting of this permafrost, as a consequence of global warming, releases the sequestered methane and further accelerates that warming, threatening a runaway heating of the planet.

Non-hydrogen-bonding solutes, such as methane or the methyl and methylene groups of proteins, lipids, and sugars, would also be expected to exhibit a radial distribution profile for their solvating water molecules that is different from that seen for hydrogen bonding as in Figures 2.18 and 2.20. Water molecules structured around a hydrophobic group as in Figure 2.21 would not be drawn into the close separations of hydrogen-bonding groups. Figure 2.24 shows the radial distribution function for water molecules around the methyl side chain of an alanine dipeptide in water, as

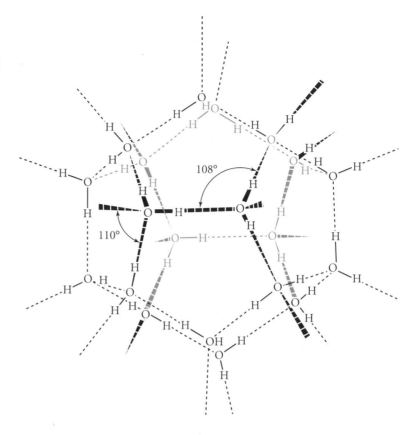

Figure 2.23. The pentagonal dodecahedral clathrate hydrate structure of water molecules, such as would form around a methane molecule, which could be accommodated inside the cavity enclosed by this structure.

calculated from computer simulations. As can be seen, this distribution is very different from that of hydrogen-bonded water molecules, with a first peak that is lower in amplitude and with its maximum at a longer distance, at 3.6 Å, and with more than twelve nearest neighbors. The broad range of this peak is an indication that these water molecules are not strongly localized by the methyl group, even though it is imposing considerable orientational structure on the solvent as discussed above.

The structuring induced in the solvent by this orientational reorganization around small nonpolar solutes reduces the number of possible arrangements and rotational freedom, thus lowering the entropy, which means that its contributions to the free energy change for the solvation process is unfavorable (positive). The heat capacity increases, however, as new structure is imposed on those water molecules adjacent to the nonpolar solute, which provides more modes in which to store energy. Ions and hydrogen-bonding solutes also impose structure on the solvent, but in these cases a large gain in enthalpy (which is thus more negative) that is due to the strong interaction energy of the hydrogen bonds compensates for the loss in entropy. The process of solvation of a non-hydrogen-bonding methane molecule from the gas phase and the transfer of this molecule from solution in a nonpolar solvent to aqueous solvent is illustrated schematically as a thermodynamic cycle in Figure 2.25. Although the solvation of gaseous methane in either solvent is favored by the entropy of mixing, both processes are accompanied by a large loss of translational freedom for the gas

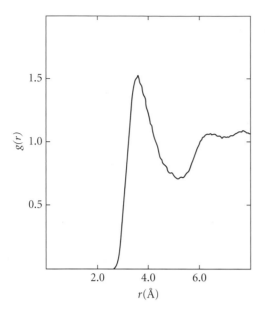

Figure 2.24. The *g(r)* for water molecules around the non-hydrogen-bonding methyl carbon atom of the side chain of an alanine dipeptide. J. W. Brady and M. Karplus, unpublished results.

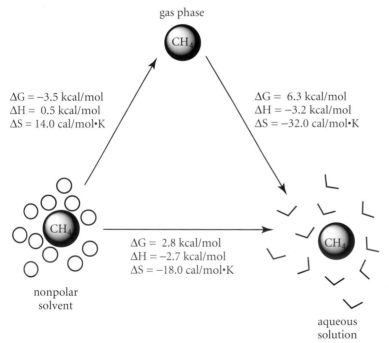

Figure 2.25. A thermodynamic cycle representing the solvation of methane in water and a nonpolar solvent and its transfer from the nonpolar solvent to water. (Redrawn from T.E. Creighton, *Proteins: Structures and Molecular Properties*. New York: Freeman, 1984; using data given in W.P. Jencks, *Catalysis in Chemistry and Enzymology*. New York: McGraw-Hill, 1969.)

molecule, due to the caging effect of its neighboring solvent molecules crowding around, so that the total entropy change for both processes is unfavorable (that is, negative). The additional structure imposed on the water makes the entropy change much worse for the aqueous solution, so that the entropy of transfer from nonpolar to aqueous solution is unfavorable. Perhaps unexpectedly, the enthalpy term for the transfer from nonpolar to aqueous solution is favorable, although not enough to cancel the entropic cost. The enthalpic gain comes mostly from the more regular hydrogen bonds made between water molecules in the quasi-clathrate-like first solvation shell around the methane, where the imposed structure suppresses the disruption of these bonds.

If two of these nonpolar solutes should come together, they exclude water from their interface, thereby freeing it from these configurational restrictions. Thus, their association in aqueous solution would be favored by free energy, driven by this increase in entropy, and would occur spontaneously. As more and more solute molecules came together, more water would be freed to restructure as it preferred, leading to a spontaneous segregation of the solute and solvent into two separate phases (Raschke and Levitt 2005). Calorimetrically, the gain in entropy would dominate the aggregation. This is the reason that oils and other nonpolar substances are not soluble in water, because similar structuring would occur for the successive methylene groups in fatty acid chains in lipids as well.

The situation is somewhat different for more extensive hydrophobic surfaces. Once the dimensions of the hydrophobic surface exceed approximately 1 nm, the orientations of the first layer of adjacent water molecules changes (Lee, McCammon, and Rossky 1984; Chandler 2005). Above this size scale, water molecules are not able to approach the solute closely while simultaneously making a full complement of four hydrogen bonds to other water molecules, "looking past" the solute as in the clathrate structure. Thus, if the water molecules approached with the same orientation as in Figure 2.21, they would be sacrificing three hydrogen bonds, as illustrated in Figure 2.26. In this case they will behave in the opposite fashion as in Figure 2.21, pointing one proton or lone pair directly at the surface, because in that case at least it is losing only one hydrogen bond instead of three, as shown in Figure 2.26, II, perhaps forming an ice-like layer (as in Figure 2.15). In such a situation, the association of two such nonpolar surfaces might be enthalpy driven instead of entropy driven, because the water molecules liberated by the approach of two such surfaces would recover those sacrificed hydrogen bonds once freed of contact with the surface. A similar ordering

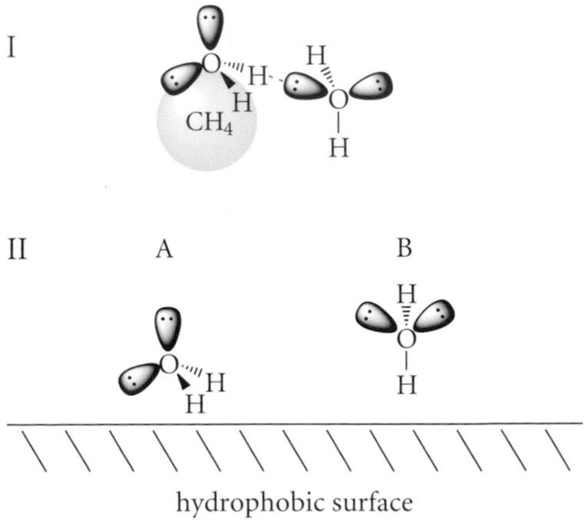

Figure 2.26. A schematic illustration of (I) how water molecules would orient around a nonpolar molecule like methane; and (II); how water molecules might orient relative to a hydrophobic surface. A molecule that was oriented as in (A) would sacrifice three hydrogen bonds to other water molecules, while a molecule oriented as in (B) would lose only one water-water hydrogen bond.

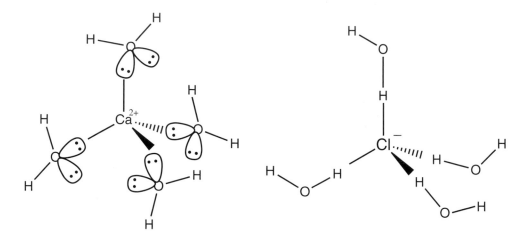

Figure 2.27. Schematic representations of the solvation of a cation (Ca^{+2}) and an anion (Cl^-). Only four neighbors are shown for chloride, while the actual hydration number is probably between 6 and 7.

would be observed for the last layer of water in contact with air, either at the top of a container such as a cooking pot or beaker, or at the surface of a bubble in a foam.

The interaction of monoatomic ions with water imposes even greater structuring on the water molecules because of the enormous strength of the charge–dipole interactions between the ions and the water. The strength of these interactions causes the water molecules to crowd in closely around the ion until their mutual steric repulsions prevent any closer approach or the approach of other water molecules. These interactions are not generally thought of as standard hydrogen-bonding interactions. The strength of these interactions results in very short ion–water distances, and the tight structuring of the first-neighbor water molecules leads to an even greater lowering of the system entropy than does the hydration of nonpolar solutes. Most ions nonetheless dissolve in water readily because this entropic cost is strongly outweighed by the enthalpic gain from binding to the water molecules along with the entropy gain as the ions are liberated from the regular crystal lattice and mix with the solvent. The illustrations in Figure 2.27 show schematically how water molecules might interact with ions as they are structured by them in aqueous solution. Because cations are smaller than anions, and thus have a higher surface charge density, these ions tend to impose more structuring upon solvent molecules than anions, and among the cations the small doubly charged ions such as Ca^{2+} have the largest effects. The great strength of the binding of water molecules to ions causes solvated ions to significantly increase the surface tension of aqueous solutions. The charge valence of only −1 and the large van der Waals radius of anions like chloride (see Table 2.4) mean that the number of water molecules that crowd around them is much larger than around Ca^{2+}; the coordination number of solvated chloride can be as high as 6 or 7.

A typical food molecule, such as a protein or sugar, contains several types of chemical functionalities in close proximity. For example, a single lysine residue in

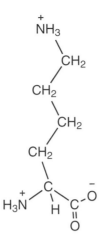

Figure 2.28. The structure of the amino acid lysine. While the terminal amino group of the molecule is very polar, it is connected to the standard acid and amino groups through four nonpolar, hydrophobic methylene groups.

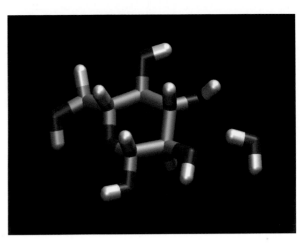

Figure 2.29. An illustration of a water molecule making two rather poor and strained bifurcated hydrogen bonds to two adjacent hydroxyl groups of a typical sugar molecule, β-D-glucopyranose (dextrose, or blood sugar; see Chapter 4). (P.E. Mason et al., *J. Phys. Chem. B.* 109:13104, 2005.)

a protein (Figure 2.28; see also Chapter 5) will have a charged amino group on the end of its side chain, an extended hydrophobic surface consisting of the methylene ($-CH_2-$) groups of carbon atoms β through δ of the side chain, and the neutral but polar hydrogen bonding N—H and C=O groups of the backbone. Each of these functional groups will have its own specific hydration requirements and will attempt to structure the adjacent water molecules accordingly, but with different strengths. The interactions of these differing hydration requirements mean that such a solute will impose a complex structuring pattern on its "first-shell" water neighbors (often called the monolayer water in food science).

In the lysine example, the molecule is flexible, because rotations can take place around the single bonds that connect the methylene groups of the side chain. In more rigid solutes, such as the rings of sugar molecules, conformational changes may be more difficult, and the monolayer or first shell solvent molecules may be strongly localized in their interactions with the functional groups of the solutes. Figure 2.29 shows an example of a water molecule in the first solvation shell of a sugar molecule, D-glucose in its pyranose ring conformation (see Chapter 4), as calculated from a computer simulation. Of course, in aqueous solution, there are a number of other water molecules which are not shown jostling to interact with the sugar molecule too. In general, each of the hydroxyl groups of the sugar could make as many as three hydrogen bonds to water, one as a donor and two as an acceptor to the oxygen atom's lone pairs. The architecture of the sugar ring crowds these hydroxyl groups so close together, however, that there is not enough room for three hydrogen-bonding water molecules around each group, because those hydrogen bonded to one group would crowd and clash with those of the neighboring groups. The bifurcated or straddled

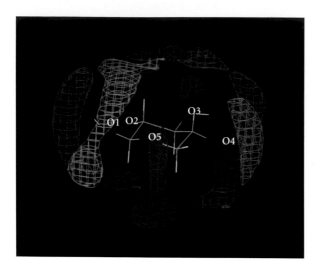

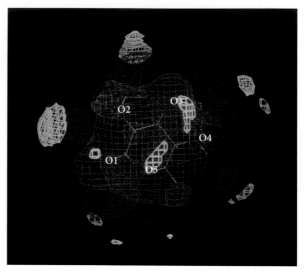

Figure 2.30. Contours of water density around β-ᴅ-xylopyranose as calculated from MD simulations. The blue density clouds contain water molecules that on average make two hydrogen bonds to the xylose solute. The purple cloud contains water molecules that on average make only one hydrogen bond to the sugar, and the red clouds enclose water molecules that on average make no hydrogen bonds to the sugar. The solvent density contours represented by the surfaces enclose regions of space where the water density is on average twice as high as in the bulk of the water, far from the solute. (R.K. Schmidt, M. Karplus, and J.W. Brady, *J. Am. Chem. Soc.* 118:541, 1996.)

Figure 2.31. A view of the water density "clouds" for the same xylose, but seen from "above" the ring. Here, the blue mesh represents the van der Waals surface of the sugar, from which the water molecules are excluded; the yellow clouds are regions of the solution where the water density is actually below the average density of the solution (15% lower); and the red contours are again regions twice as dense as the bulk.

hydrogen-bonding geometry seen in Figure 2.29 is the result, as a single water molecule attempts to hydrogen bond to two adjacent hydroxyl groups, making poor, nonlinear bonds to both. The various compromises that the water molecules make as they crowd in to occupy specific regions around a complex food solute like this sugar produce very specific localized hydration patterns in space. As an example, Figures 2.30 and 2.31 illustrate the structuring imposed on water by the sugar ᴅ-xylose (see Chapter 4), as calculated from computer simulations. The coloring scheme in the figures helps illustrate the complexity of the interactions of water with the solute. As can be seen, some water molecules occupy regions where they are close to the sugar molecule but make no hydrogen bonds to it because they are hydrating nonpolar groups like the CH_2 and CH surfaces, while other water molecules even make more than one hydrogen bond to the solute. As will be discussed in Chapter 4, sugar molecules tend to interact strongly with water and significantly affect the water-binding properties of the foods that contain them. As a result, both Nature, in living systems, and food technologists in designed foods make use of the water-binding properties of sugars to control water.

Figure 2.32. An illustration of a cubic sample of a solution of guanidinium chloride in water, as calculated from a molecular dynamics computer simulation. The chloride ions are indicated as green spheres, while the guanidinium ions (illustrated separately on the right) are indicated with either standard atomic coloring (aqua for carbon, blue for nitrogen, and white for hydrogen atoms), or are colored either red or purple, depending on whether the particular guanidinium ion is complexed with other guanidinium ions or with chloride ions. (P.E. Mason et al., *J. Am. Chem. Soc.* 126:11462, 2004.)

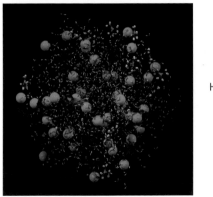

guanidinium

Solutions with Multiple Solutes: The Hofmeister Series

Actual solutions in typical foods are generally much more complex than just sugar in water and often contain many solutes, such as proteins, sugars, and salts, all interacting with each other and affecting each other's behaviors, such as solubility and mobility, as well as the overall properties of the solution. For example, some of the solutes may actually bind together or aggregate, while others may avoid one another. Figure 2.32 illustrates such complexity for a rather simple solution of guanidinium chloride in water, a three-component solution. As might be expected, the chloride ions (Cl^-) tend to avoid one another. Because they tend to fully hydrate, there is also little tendency for them to remain bound to their oppositely charged guanidinium counterions (Gdm^+), as they would be in the anhydrous salt crystal (GdmCl). Perhaps somewhat surprisingly, there is a significant tendency for the like-charged guanidinium ions to associate together in a stacked fashion, as can be seen in the upper left and lower right. All of these ions bind to adjacent water molecules. In the case of the guanidinium ions, each of the N—H functionalities can make a rather regular hydrogen bond as a donor to a water molecule, restricting the water's translational freedom, and collectively these interactions increase the viscosity of the solution as well as raising the surface tension.

The presence of ions in aqueous solutions of proteins can affect both the solubility of the proteins and their stability toward denaturation, a process by which globular proteins lose their specific "folded" shapes (discussed in greater detail in Chapter 5 on proteins) when heated or exposed to denaturants. Some ion co-solutes cause proteins to denature more easily, while others stabilize them against denaturation. Furthermore, some ions tend to cause proteins to precipitate or crystallize out of solution (called "salting out"), while others have little effect on their solubility or even increase their solubility.

Ions can often act independently of one another in affecting the properties of macromolecules in solution, and can be ranked in a series in terms of the magnitude of the effects they produce. Such a ranking is called a Hofmeister Series, from the

Table 2.3 The Hofmeister Series

Cation Series

Pauling radius*	0.99 Å		0.65 Å		0.60 Å		0.95 Å		1.33 Å		
cation	Ca^{2+}	>	Mg^{2+}	>	Li^+	>	Na^+	>	K^+	>	NH_4^+

conformation stabilizing	$\rightarrow$	most denaturing
salting out	$\rightarrow$	salting in
increases hydrophobic interactions the most	$\rightarrow$	increases hydrophobic interactions the least
increases water surface tension the most	$\rightarrow$	increases water surface tension the least

Anion Series

radius*	1.81 Å	2.16 Å

$SO_4^{2-} > HPO_4^{2-} > CH_3COO^- > citrate > tartrate > Cl^- > NO_3^- > ClO_3^- > I^- > ClO_4^- > SCN^-$

Neutral Molecular Species Series

sucrose, glucose, glycerol	>	urea, guanidinium (guanidine hydrochloride), 2-chloroethanol, methoxyethanol
conformation stabilizing	$\rightarrow$	denaturing

*L. Pauling, *The Nature of the Chemical Bond*, 3rd ed. (Ithaca, NY: Cornell University Press, 1960).

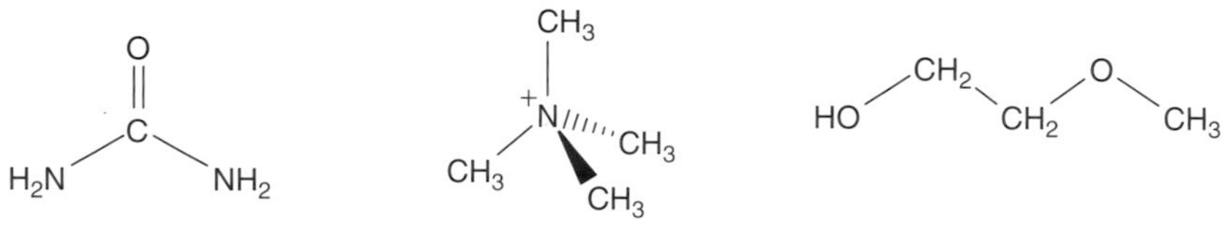

Figure 2.33. The structures of some chaotrope molecules.

initial studies of Franz Hofmeister over 100 years ago on the use of salts to precipitate proteins from solution.

Those ions that stabilize protein conformations are sometimes called kosmotropes. This class includes Mg^{2+}, Ca^{2+}, Li^+, SO_4^{2-}, and HPO_4^{2-}. Molecular species that destabilize protein conformations, such as urea, guanidinium, tetramethylammonium, SCN^-, ClO_4^-, I^-, and HCO_3^- (Figure 2.33) are called chaotropes.

The explanation of the Hofmeister ordering of the monatomic cations might be understood in terms of the atomic radii, listed above each ion in Table 2.3, and their effects upon surface tension. As we shall see in the next chapter, creating any surface between two phases is accompanied by an unfavorable free energy change, and in fact the surface tension is actually defined in terms of the work required to increase the surface area. Ions in an aqueous solution increase the surface tension of the water. This can be thought of as being due to the ions making stronger charge–dipole interactions to the water molecules than the hydrogen bonds that another water molecule

Table 2.4 Effect of surface charge density on the hydration energy of ions

Ion	Pauling radius (Å)	Surface charge density ($q/Å^2$)	Hydration energy (heat of hydration in kcal/mol)*
K^+	1.33	0.045	86.1
Ca^{2+}	0.99	0.162	398.8
Al^{3+}	0.50	0.955	1141.0
Cl^-	1.81	0.024	81.3
I^-	2.16	0.017	64.1

Note: The surface charge density, defined as the charge valency ($v = \pm1$, +2, or +3) divided by the surface area of the van der Waals contact surface ($4\pi r^2$), determines the strength of the interaction of the ions with water molecules; note that choride and iodide, which are the largest of the ions, have a very low surface charge density and thus do not interact as strongly with an individual water molecule as another water molecule does.
*Excess enthalpy of hydration for the transfer of the ion from the gas phase to water; H.L. Friedman and C.V. Krishnan, "Thermodynamics of Ionic Hydration," in *Water: A Comprehensive Treatise*, Volume 3, F. Franks, ed. (New York: Plenum, 1973).

could make to it. As the ions in the series become smaller, their surface charge density increases; this effect is particularly strong for Mg^{2+}, which not only has the smallest radius, but a charge valence of +2. As the surface charge density in the series increases, the strength of the binding of the ions to water goes up, giving a corresponding increase in the surface tension of the water. This increase in surface tension might in turn increase the unfavorable free energy of unfolding, because unfolding increases the surface area of the interface between the protein and water as the protein becomes less globular and more uncoiled, thus stabilizing the protein toward denaturation. At the same time, as the free energy of the solution increases, the solubility of the protein decreases, and the tendency to precipitate ("salting out") increases.

Explaining the ordering of the anionic series is more difficult than for the monoatomic cations, particularly because most of the members of this series are molecular ions rather than simple spheres. Surface tension effects will still play an important role for these anions. Consider the two sulfur-containing anions sulfate (SO_4^{2-}) and thiocyanate (SCN^-), on opposite ends of the Hofmeister ranking. The sulfate ion has a −2 charge valency and thus a relatively high charge density on each of its four oxygen atoms. Solutions of sulfate thus have a relatively high surface tension due to the strong hydrogen bond interactions with water, stabilizing the proteins toward increasing their surface area by unfolding, but by raising the free energy of the solution, making it more likely that the proteins will precipitate out of the solution altogether. The surface tension argument also works for polar but nonionic solutes such as glucose and glycerol, which are conformation stabilizing and promote salting out. Strongly hydrating species such as these are referred to as **osmolytes**. For the denaturing, "salting in" molecular species such as urea, the explanation of their denaturant power is that these molecules bind directly to the protein. Since urea is hydrophobic, it can associate with other hydrophobic functional groups, such as hydrophobic side

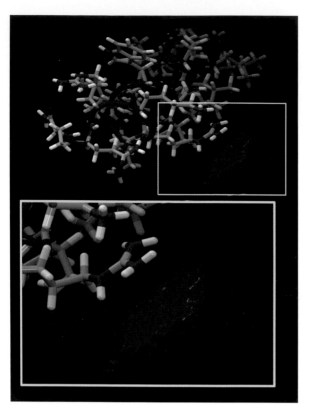

Figure 2.34. The density of the nitrogen atoms of Gdm⁺ around the guanidinium group of the arginine 22 side chain of the small alpha helical bee venom protein melittin, as calculated from computer simulations. The alpha helix is shown as a red ribbon. The high density of Gdm⁺ stacking against the Arg side chain is clear. The contour shown is 11 times the bulk number density of these nuclei. (P.E. Mason et al., *Biophys. J.* 93:L04, 2007.)

chains of the amino acids in proteins. By complexing with these groups, their non-polar surfaces are shielded from interacting with water even if the protein unfolds in a way that would otherwise expose them to direct interactions with solvent. Surprisingly, this is also the explanation for the denaturing and salting-in effects of guanidinium, even though it is an ion with a +1 charge. The reason is that the flat surface of the guanidinium ion is so extensive that the charge density, delocalized over the entire surface, is very low and thus does not permit the "top" and "bottom," dominated by the central carbon atom, to make hydrogen bonds with water, giving these flat surfaces an almost hydrophobic character even though the molecules are charged. Unlike guanidinium, however, urea can directly hydrogen bond to the $C{=}O$ and $N{-}H$ groups of the protein backbone, disrupting secondary structures (Chapter 5), and this effect may be as important in its denaturant power (Bennion and Daggett 2003; Lim, Rösgen, and Englander 2009; Raschke, Tsai, and Levitt 2001).

Figure 2.34 illustrates guanidinium ion density next to an arginine side chain of a model helical "protein," the bee venom polypeptide melittin, as calculated from computer simulations. The side chain of arginine (Arg) contains a guanidinium group that is almost identical to the free guanidinium ion (Gdm⁺, see Chapter 5), and these free Gdm⁺ ions stack against the side chain groups just as they associate with one another in solution, thus reducing the exposed hydrophobic surface area of each. Guanidinium ions will similarly stack against purely hydrophobic surfaces such as the side chains of phenylalanine or tryptophan (Chapter 5).

Figure 2.35. The density of the atoms of water (red, O atoms; white, H) around the guanidinium group of the Arg22 side chain of melittin as calculated from computer simulations. As in the previous figure, a superimposed red ribbon indicates the overall path of the helix. (P.E. Mason et al., *Biophys. J.* 93:L04, 2007.)

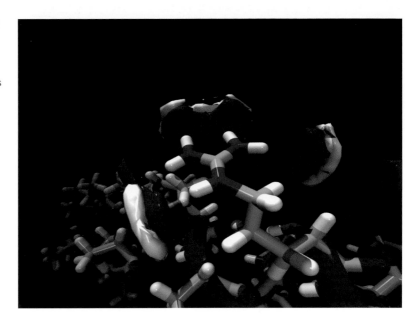

When a guanidinium ion stacks against a hydrophobic group on the surface of a protein, it is still able to make hydrogen bonds through its amine ($-NH_2$) groups to water molecules. The same is true for the guanidinium groups of the side chains of the amino acid arginine such as in Figure 2.34 when they are complexed with a free guanidinium ion from a salt such as GdmCl. This can be seen in Figure 2.35, which shows contours of water atom densities around an arginine amino acid's guanidinium side chain group. As can be seen, water molecules are making fairly linear hydrogen bonds in the plane of the guanidinium group with the NHs serving as donors to their oxygen atoms.

Thus, in a guanidinium chloride solution of a protein such as melittin, there is a tendency for the guanidinium ions to associate with hydrophobic groups in the protein, as illustrated in Figure 2.36. In the case of a folded globular protein, which by virtue of being folded has segregated many of its hydrophobic groups in its interior away from contact with water, such binding of guanidinium to these hydrophobic groups when the protein is unfolded would make this state more favorable and thus induce denaturation.

Solutions of Gases in Liquids: Carbonated Beverages

Carbonated beverages are aqueous solutions of carbon dioxide. Figure 2.37 displays the solubility of carbon dioxide in water as a function of temperature. Notice that, unlike most solutes such as sugar, table salt, or vanillin (see Figure 8.23), the solubility of gases in water decreases as the temperature increases. This is true in general for all gases dissolved in liquids. The reason is due to differences in the solvation entropy contributions for the different types of solutes.

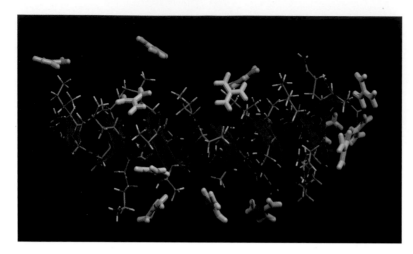

Figure 2.36. A view taken from a computer simulation of an aqueous solution of melittin and guanidinium chloride, illustrating the instantaneous location of the guanidinium ions complexed with the protein. Note particularly the guanidinium ions that can be seen edge-on in the upper left, which are paired with nonpolar methylene groups. The helical backbone is indicated by a red ribbon cartoon, and the guanidinium cations are all shown in yellow. The chloride counterions as well as the water molecules are omitted for clarity. (P.E. Mason et al., *Biophys. J.* 93:L04, 2007.)

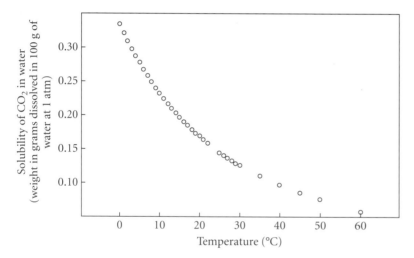

Figure 2.37. The solubility of carbon dioxide in water as a function of temperature at 1 atm.

Whenever one substance dissolves in another, there is a favorable contribution to the free energy that comes from the entropy of mixing (see Chapter 1). Greater disorder results when two substances are randomly mixed together than when they are both segregated into pure phases. This lowering of free energy comes from any random mixing, regardless of the character of the solutes and solvents. Of course, other contributions to the total free energy come from the specific character of the molecules of the solvent and solute, and contributions which arise from the physical state of the solute before dissolution. When a crystalline solid dissolves in a liquid, the entropy increases significantly, due to the entropy of mixing and also because the solute molecules are going from the very ordered state of the crystal lattice to the more freely diffusing disorder of a liquid. The contribution to the free energy is $-T\Delta S_{dis}$, where the ΔS_{dis} term is positive, so that the contribution is favorable due to the negative sign. The situation is different, however, when the solute is a gas. The entropy of mixing is still favorable, but the entropy contribution from the change of

Figure 2.38. The solubility CO_2 of as a function of pressure at 18 °C.

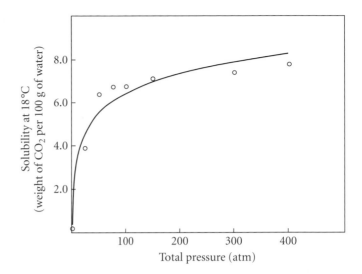

state of the solute is unfavorable, because the entropy of a gas is very large. In liquid solution the solute has much less translational freedom, due to the "caging" effect of the surrounding solvent molecules restricting the solute molecule's motion as it crashes into them. As a result, the total ΔS term is negative, so the $-T\Delta S$ term is positive (unfavorable), and the higher the temperature, the more unfavorable it will be and the less soluble the gas will be in the liquid. Thus, to dissolve a solid in a liquid, raising the temperature increases the solubility, while when dissolving a gas, raising the temperature decreases solubility. This is why CO_2 boils out of a carbonated soft drink taken from a refrigerator as it heats up, but sugar dissolves more readily in hot tea than in iced tea.

The solubility of CO_2 in water increases sharply with pressure up to about 50 atm, beyond which the increase in solubility is very much lower; the functional dependence is approximately logarithmic, as can be seen from the graph of solubility as a function of pressure shown in Figure 2.38. The solubility of CO_2 in water is also affected by the presence of other solutes in the solution. As is illustrated in Figure 2.39, salt decreases the solubility of CO_2 as the concentration of salt is increased. The ordering of the effect for the two cations on chlorine in this example follows their Hofmeister ranking.

Dissolved CO_2 exists in an ionic equilibrium between the neutral molecule and bicarbonate ion, with a pK_a of 6.352 at 25 °C.

$$CO_2 + H_2O \rightleftarrows H^+ + HCO_3^- \tag{2.1}$$

The bicarbonate ion is also in equilibrium with carbonic acid, with a true pK_a for carbonic acid of 3.88,

$$H_2CO_3 \rightleftarrows H^+ + HCO_3^- \tag{2.2}$$

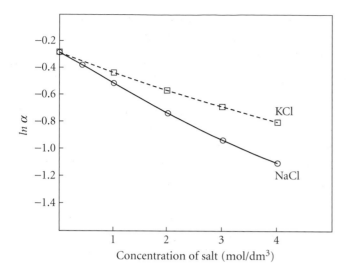

Figure 2.39. The solubility of CO_2 in water as a function of added salt concentration. The parameter α is the Bunsen coefficient, which is the volume of gas, corrected to a partial pressure of 1 atm at 273 K, dissolved in one volume of solution.

so that carbonated water solutions are mildly acidic. A saturated solution of CO_2 has a pH of around 3, while a typical carbonated soda has a pH of around 2.5 to 3.5, which is also partly due to the presence of added phosphoric or citric acids (Chin et al. 1995).

Calcium ions can precipitate carbon dioxide from solution through the formation of calcium carbonate, $CaCO_3$.

$$CO_2 + Ca^{2+} + H_2O \rightleftarrows CaCO_3 + 2H^+ \tag{2.3}$$

This molecule is the principal chemical constituent of the shells of most marine organisms such as mollusks (e.g., clams, oysters), barnacles, and corals, as well as the main chemical constituent of the eggshells of birds. Dissolved calcium carbonate is also one of the most common components of so-called **hard water**, which refers to water with a high concentration of dissolved mineral salts of calcium and magnesium. The other main contributors to water hardness are calcium sulfate and magnesium sulfate and carbonate. The world's aquatic systems exist in a complex equilibrium with the carbon dioxide of the atmosphere that has vast global implications. Carbon dioxide in the atmosphere dissolves in oceans, lakes, and streams, lowering the pH in the process. Calcium ions weathered from rocks of the Earth's crust dissolve in rain and ground water and are eventually carried to the oceans, where they precipitate the dissolved carbonate ions, leading to the formation of limestone on the ocean floors (if subsequently buried and subjected to intense heat and pressures, limestones can be transformed by recrystallization into marble). Because limestone is basically composed of calcium carbonate, it can be dissolved by acid, which is not only a test for limestone, but is also the explanation for the widespread occurrence of caverns, caves, potholes, and sinkholes in limestone layers in the Earth's crust.

Evaporation of water containing dissolved calcium carbonate leaves a white, relatively insoluble, and hard-to-remove residue of crystalline calcium carbonate. This scale can be seen, for example, in the bottoms of coffee pots, teakettles, and food processing plant vats, and even on the rotating trays in home microwave ovens. In areas with particularly hard water, dripping taps can become coated with this scale as water evaporates away. Water hardness can be a significant problem in food processing. The buildup of scale in equipment can clog valves and other small openings, interfere with heat transfer, and even provide substrates where bacteria can escape cleaning. In some cases, it can even directly affect food quality, because, as we will see in Chapter 4, calcium ions can lead the pectin in the intercellular matrix of plant tissues to form stiff gels. This can sometimes be beneficial, because it helps preserve the firmness of vegetable tissues that would otherwise be softened by heat treatment, but in other cases it can be undesirable. Excessive calcium in peas, for example, can lead to unnatural toughness. Water can be softened by the addition of hydrated lime (calcium hydroxide, $Ca(OH)_2$, also called "slaked" lime), which precipitates the magnesium and calcium salts of bicarbonate. Salts of sulfate and chloride are harder to deal with and must be removed using ion exchange resins. This type of hardness is referred to as permanent hardness, while water containing only carbonate salts is referred to as exhibiting temporary hardness. In kitchen applications, calcium carbonate scales are difficult to remove, but they can be dissolved and washed away with scrubbing, using weak acid solutions such as vinegar (acetic acid) or lemon juice (citric acid).

The rapid increase in the concentration of carbon dioxide in the Earth's atmosphere as the result of human activities has been partially slowed and buffered by dissolution in the oceans and the formation of limestone through precipitation with calcium, but the rate of atmospheric increase may be leading to an increase in the acidity of the oceans (i.e., a lowering of the pH). The growing ocean acidity can dissolve the calcium carbonate of the shells of marine organisms. This process may eventually lead to widespread deaths among such species and may also slow or even halt the growth of coral reefs and cause the rapid erosion of coral atolls and islands.

WATER ACTIVITY AND MOBILITY

Water is required by all living organisms, and it greatly facilitates many types of chemical reactions as a solvent, plasticizer, or catalyst. Thus, its abundance in foods represents a potential storage problem because it can promote many different types of degradative processes. For this reason the control of water is essential in food preservation. The oldest food preservation technology, simple drying, which dates back to the Ice Age, deals with this problem by just removing the water, and this approach is still widely used today. Freezing also acts by limiting the availability of water by locking it up in ice crystals. The problem with these approaches is that water is so integral to the structure and properties of biological materials and tissues that its removal often leads to irreversible changes in the food (think of how beef jerky differs from fresh meat or how a thawed frozen strawberry differs from a fresh one). Nevertheless, the need to control the availability of water in foods has led to considerable interest in their **water activity**. Recall that the activity is defined as

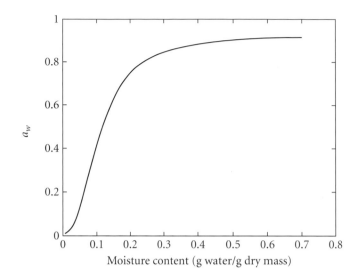

Figure 2.40. Water activity as a function of moisture content at constant temperature.

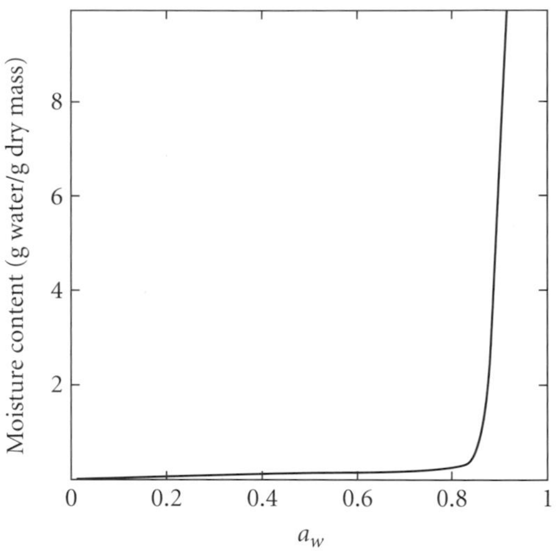

Figure 2.41. A moisture sorption isotherm (moisture content as a function of water activity).

$$a_{\mathrm{W}} = (P/P^o) \tag{2.4}$$

where P is the partial pressure of water over the food sample at equilibrium and P^o is the equilibrium vapor pressure of pure water at the same temperature. The water activity of a food is important because it is a measure of the free energy of the water in the food (Equation 2.5), since at equilibrium, the free energy of the water remaining in the food must be equal to that in the vapor (otherwise, more water would either evaporate or condense).

$$\mu_{\mathrm{w}}(\text{in food sample}) = \mu_{\mathrm{w}}^{*}(\text{liq}) + RT \ln a_{\mathrm{w}} \tag{2.5}$$

The less the water of a food system is affected by the other components of the food, the more it will be able to behave like pure water, which has a partial pressure of P^o, and thus the more closely its activity will approach 1. The activity can be plotted as a function of the moisture content, as shown in Figure 2.40, with the function asymptotically approaching unity as the concentration approaches pure water.

In food science applications, the inverse function is usually plotted; that is, moisture content as a function of water activity, as in Figure 2.41, with the curve diverging (i.e., the amount of water increasing without limit relative to the other components of the food) as the activity approaches 1. These curves are referred to as **moisture sorption isotherms** because they measure the tendency of the water to be absorbed by the food product at a constant temperature as a function of overall water content.

Focusing only on the very low moisture content portion of the curve gives the familiar type of plot shown in Figure 2.42. The measurement of the water activity as a function of moisture content has proven to be of some practical utility because

Figure 2.42. An enlargement of part of Figure 2.41.

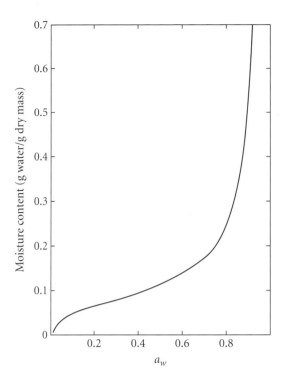

under operational conditions a number of processes have been shown to be correlated with the water activity, as shown in Figure 2.43. For most of these degradative processes, lowering the measured water activity reduces the rate of the process, thus increasing shelf life.

Because the interactions of the water with the various other components of food systems tend to reduce its partial pressure, and thus its activity, the effect is as if the water is being held in the food, and thus the older literature frequently discussed water activity in terms of "water binding" and "bound water" molecules. Water molecules in the food sample will have different interaction energies with the different functional groups of the food molecules. T.P. Labuza of the University of Minnesota pioneered the description of the relationship between water activity and various types of degradative processes in foods and broadly divided the range of activities into three zones, as shown in Figure 2.43 (Labuza 1971). For example, those water molecules that are not directly adjacent to any part of a food molecule such as a protein, lipid, or carbohydrate will be the most easily removed, because the molecules with which they are most directly interacting are all other water molecules. Nevertheless, their tendency to escape might still be affected, even if weakly, so that their vapor pressure relative to that for pure water would be reduced (by the entropy of mixing, for example). They are removed most easily as the food is dried; therefore their removal corresponds to the Zone 3 region in Figure 2.43. Also contributing to this Zone 3 might be water molecules in direct contact with the food molecules but interacting with hydrophobic groups such as the long hydrophobic chains of fats without hydrogen bonds. Those water molecules that are hydrogen bonded to food molecules have

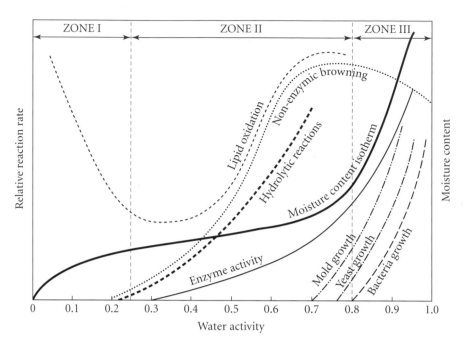

Figure 2.43. A schematic illustration of how various degradative process depend upon water activity. (Adapted from data in O.R. Fennema, *Food Chemistry*, 1st ed. New York: Marcel Dekker, 1976.)

much stronger interaction energies and are thus harder to remove, corresponding to Zone 2 in the figure. Those water molecules that are interacting with salts and charged groups in the food are more strongly bound still and would presumably correspond to water activities generally below 0.4 or 0.3. Finally, water molecules entrapped in a matrix of the food, such as water molecules folded up inside globular proteins, would be the hardest to remove as the food is completely dried and the activity approaches zero.

In this context, the effect of varying the temperature on the measured water activity can be predicted. At any given water content, higher temperatures will result in higher energy for individual water molecules on average, a greater tendency for molecules to escape, and thus a higher activity. This allows one to predict the direction of displacement of the curves as the temperature is increased in Figure 2.44. Similarly, adding salt will also affect the sorption isotherm. Because the ions of the salt hydrate so strongly, as we have seen, their presence will reduce the overall average tendency for water molecules to escape from the solution in the food. Thus, for any given moisture content, the partial pressure will be reduced, so the curve will be displaced to the left. This, of course, is why salting is an effective method of preserving food, such as meat, and its use for this purpose is very ancient.

An important feature of many of the measured sorption isotherms is that they exhibit **hysteresis** (a mathematical term meaning a situation where a function exhibits two possible values for one value of the independent variable), as shown in Figure 2.45. That is, a different activity is recorded for a given moisture content depending on whether the food sample is being dried out or rehydrated. Many explanations have been advanced for this phenomenon based on assumptions about the microstructure of foods, because microcapillarity and oddly shaped cavities can give rise to hysteresis

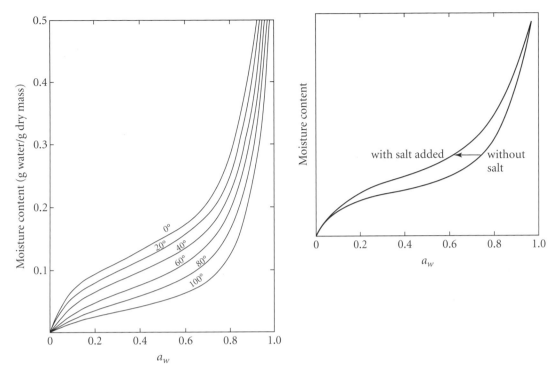

Figure 2.44. On the left is an illustration of the effect of increasing temperature on the moisture sorption isotherm for a generalized food product. An increase in temperature makes it easier for a water molecule to escape from a food at any given moisture content and thus displaces the curve to the right. On the right is an illustration of the general effect of adding salt; added salt will "bind" to the water, reducing its tendency to escape and thus displacing the curve to the left for any given moisture content.

Figure 2.45. Hysteresis in water activity curves for adsorption and desorption taking place at the same temperature.

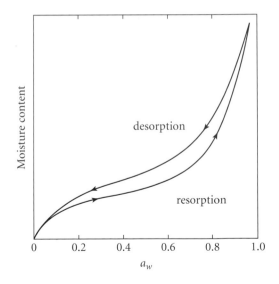

even in systems like ceramics. These models will be discussed in more detail in Chapter 3. It is interesting that in food systems, however, it has been shown that, after one or two cycles of dehydration and rehydration, the hysteresis collapses and disappears (Rizvi and Benado, 1984a,b). This suggests that what is going on is primarily irreversible structural changes in food systems. This is not particularly surprising given the extremely important role of water in food systems. As the water is removed, proteins denature, carbohydrates may tend to crystallize, and membranes and lipid bilayers disrupt. Many of these changes are irreversible (think of the consequences of the loss of cell membranes for tissues) and thus would not be repaired by subsequent rehydration or repeated on subsequent cycles of dehydration and rehydration.

Water activity is measured by placing a sample of the food product in a sealed dessicator and allowing the water vapor to come to equilibrium between the sample and the enclosed head space. This equilibration often takes weeks to months, making such measurements difficult in practical, time-critical situations on industrial production lines. A more fundamental problem in using a_W to judge food stability must be considered, however. Water activity is an equilibrium property, and is not always a good measure of food product stability in the highly nonequilibrium conditions of normal food storage. Most foods are not thermodynamically stable, and their shelf life is a question of kinetics rather than thermodynamics. The real issue in the preservation of foods by freezing, drying, or any other means is the availability of the remaining moisture, which is in part a function of mobility. It is possible for a food product to be judged stable, based on its water activity using something like Figure 2.43 as a guide, and for it still to degrade rapidly as a result of inadequate control of the mobility of its molecular components. In recent years there has been much interest in the food industry in preserving foods as polymer glasses, based on the kinetics of water mobility, in which diffusive motions are so slow that degradation is halted on practical (shelf-life) timescales.

Food Glasses

The solid state is one of the three principal states of matter with which most people are directly familiar. Like liquids, solids are condensed states, which means that the constituent atoms and molecules are closely packed in some fashion. Unlike liquids, however, which are condensed fluids, solids have a static structure on practical timescales. Solids resist deforming in response to an applied external force, due to the strong internal cohesive forces that lock the molecules into stable positions with respect to one another, preventing the rapid reorganization that takes place in liquids.

When all of the molecules or atoms in a solid occupy a regular array of spatial positions, they constitute a crystal. We have already considered the high symmetry of ice, the lowest-energy crystal state of water. Water has more than a dozen other crystalline phases, all characterized by different ways of packing the water molecules into regular lattices. As we have seen, these other types of ice require high pressures

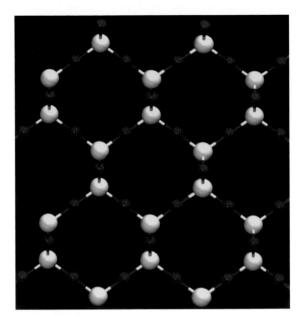 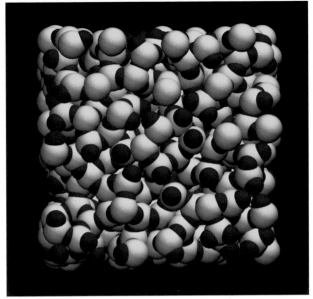

Figure 2.46. (a) On the left is an illustration of an ideal crystalline lattice of β-cristobalite, a silica dioxide polymorph of quartz; (b) on the right is a silica dioxide glass, the principal constituent of the glass in window panes and drinking glasses, where the SiO_2 molecules are randomly disordered as in a liquid, but are locked into this random arrangement by a lack of sufficient energy to overcome these interactions and diffuse about. (Note that these two figures are not drawn on the same scale, as can be seen from the smaller size of the SiO_2 molecules in the figure on the right.)

or very low temperatures and are not important for food science. Other types of substances, such as many minerals, have more than one stable crystalline phase under ambient conditions, although they may have been formed under much different conditions of temperature and pressure. We will see in Chapter 7 that the triacylglycerols (triglycerides) of food fats can crystallize into several different crystalline packing patterns of different energies. One of these lattices can be formed at a higher temperature, where it is the most stable phase, and to then be cooled to a lower temperature where it is not the most stable phase, but where it is trapped in its thermodynamically less favorable state because the individual molecules do not have enough energy to surmount the energy barriers that are preventing them from rearranging into the most favored state at the new temperature.

Our intuitive definition of a crystalline solid thus requires that the atoms are strongly localized, at least on a short-term time scale (i.e., are not moving around), and that they occupy a very regular array in a crystalline lattice (Figure 2.46a). This state is then different from a liquid both dynamically, where we expect the individual molecules to be mobile and to diffuse about throughout the liquid, and structurally, because on the molecular level we expect the arrangement of molecules in a liquid to be chaotic and disorganized, with no regular lattice or long-range structuring (Figure 2.46b). In a glass, however, the liquid state has been cooled down so quickly, compared to the time required for crystals to form, that the kinetic energy of diffusive motion was drained out of the system before it could crystallize, resulting in a solid-like substance, in the sense that it does not flow and is relatively resistant to deformation, but with no long-range ordering on the atomic and molecular level. In other words, a glass is structured chaotically like a liquid on the molecular scale, but the individual molecules have so little kinetic energy that they cannot diffuse past one another and are trapped in their liquid-like structure.

In a food, the glassy state is a shelf-stable way to store a food, because the lack of mobility in the food means that degradative processes cannot take place—not only is microbial growth halted, but even many chemical reactions are slowed or stopped, because they require that the reactants diffuse together in order to react. Many examples of food glasses can be identified: the continuous matrix of unfrozen water and dissolved sugar and proteins between the ice crystals and fat globules of ice cream is in a glassy state, as are the starch molecules of dry pasta. The latter example is interesting because the pasta is a glass at room temperature. The starch granules of the pasta are gelatinized when the pasta is initially prepared (see Chapter 4), releasing amylose and amylopectin polymers to form a random gel, which solidifies as it is dried, locking the starch molecules into a glass-like chaotic matrix without water. In this dry state the pasta is stable for very long times. When the pasta is cooked by boiling in water, it is rehydrated, plasticizing the matrix by replacing starch-starch hydrogen bonds with similar hydrogen bonds to water, and restoring its flexibility.

In many products, particularly those stored at normal temperatures, the small water content remaining in the stabilized product is not actually fully immobilized; only the larger, generally polymeric, components are truly in a glassy state. Thus, the dynamics of such systems can often be successfully described using a sophisticated theory of polymer kinetics called the WLF, or Williams-Landel-Ferry, theory. The application of this theory to food, pioneered by L. Slade and H. Levine, has proven to be quite useful in understanding the general properties of water in foods (Slade and Levine, 1991). Unfortunately, the measurement of polymer kinetics is even more difficult in an operational sense than a_W and, thus, is of still less direct practical use in many production line situations. It is impossible to adequately discuss this topic here, but a good introduction is given in Chapter 2 of the textbook *Food Chemistry* by O. Fennema (3rd ed., Dekker, New York, 1996). Conceptually, however, the idea of a glass transition in foods is quite useful, and we will invoke the concept repeatedly, as, for example, when we consider the staling of bread (see Figure 5.21 and the accompanying discussion).

Let us consider a simple example. Figure 2.47 shows the phase diagram for sugar (sucrose) in water (that is, sugar water, or Popsicles, for example). The phase diagram maps the state of the solution (liquid or solid) as a function of temperature and concentration. The left side of this graph represents a pure water sample with no sucrose, and the right side represents pure sucrose. Pure water melts at 0 °C at atmospheric pressure, and pure sucrose melts at 186 °C. Mixing the two together lowers the melting point through the freezing point depression effect from the entropy of mixing, as we have already discussed. Consider as a specific example a solution of ~15% sucrose in water initially at a temperature near 20 °C, indicated by "A" on the plot. If placed in a freezer and cooled, its temperature will initially fall without any

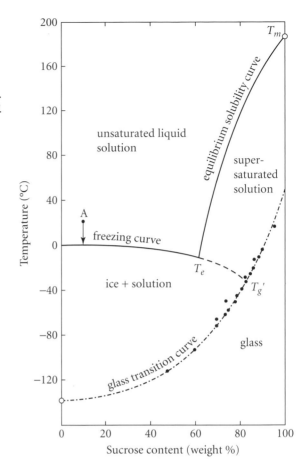

Figure 2.47. The phase diagram for sugar (sucrose) in water, as in a Popsicle, for example. Because this diagram contains rate information (the glass transition curve is defined in terms of kinetics, not thermodynamics), this is not solely a phase diagram, but is a useful summary of the behavior that can be expected from the system in realistic nonequilibrium processes. (Figure redrawn from data in F. Franks, R.H.M. Hatley, and S.F. Mathias, *Pharm. Technol. Int.* 3:24–34, 1991.)

composition change until the equilibrium freezing curve is reached, at a temperature a little below 0°C. At this point, ice begins to form, leaving the solution and thus increasing the sucrose concentration, so the system moves to the right along the equilibrium freezing curve to higher sucrose concentrations until it reaches the eutectic point T_e. At this point sucrose should also freeze out, because the equilibrium solubility curve for sucrose intersects the freezing curve for water solutions at this point. In practice, however, due to the slow rate of sucrose crystallization, when the cooling is done in real time, the system will probably continue along the extrapolation of the equilibrium line until so little free water, which serves as a plasticizer, is present per sugar molecule that the system becomes locked into a glassy state. The T_g curve, the glass transition curve, is generally taken to be the locus of points where the viscosity is 10^{15} P (i.e., 10^{14} N·s·m^{-2}), and the glass transition temperature T_g', which is usually reported in the food literature, is then the point at which the extrapolation of the freezing curve of the water crosses the glass transition curve. This temperature is therefore the lowest at which the supersaturated solution can exist in the liquid state. The T_g values for several food carbohydrates in the pure state are given in Table 2.5. Note that trehalose has the highest T_g of the simple sugars.

Table 2.5 Glass transition temperatures for some carbohydrates found in foods

Carbohydrate	T_g (°C)
ribose	−10
sorbitol	−4
fructose	10
glucose	37
sucrose	68
cellobiose	77
maltose	92
trehalose	125

Source: A. Cesàro and F. Sussich, in *Bread Staling*, P. Chinachoti and Y. Vodovotz, eds. (Boca Raton, FL: CRC Press, 2001). *Note*: See Chapter 4 for the structures of these molecules.

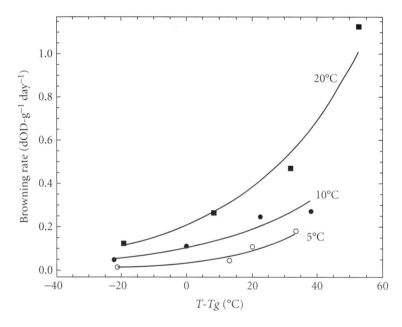

Figure 2.48. The rate of Maillard browning is a strong function of how high the temperature is above the glass transition temperature T_g. This figure illustrates the rate of nonenzymatic browning in a simple maltodextrin/L-lysine/D-xylose system as a function of $T–T_g$. In the experiment, the product temperature was kept constant at three different values (5, 10, and 20 °C), and the T_g was varied by changing the water content, which then changed $T–T_g$. dOD refers to the change in the optical density at a wavelength of 420 nm, considered to be typical for the organic molecules responsible for the brown color of foods. (Figure redrawn from data given in Y.H. Roos and M.-J. Himberg, *J. Agric. Food Chem.* 42:893, 1994.)

The determination of glass transition temperatures is notoriously difficult. For example, Table 2.5 lists the T_g for sucrose as 68 °C, but numerous reported experimental values span a 20° range between 50° and 70 °C (Cesàro and Sussich 2001). Determining T_g for nonhomogeneous polymers such as starch or gluten is even more difficult. In general, the higher the temperature of a food product is above its glass transition temperature, the more rapidly various reactions proceed, including reactions that degrade food quality. For example, Figure 2.48 shows the rate of Maillard browning (Chapter 4) as a function of temperature. In a low-moisture food system, the glass transition temperature is a function of the plasticizing water content. As can be seen from Figure 2.49, wheat gluten (Chapter 5) with a moisture content of 16%

Figure 2.49. The glass transition temperature is a strong function of moisture content. As the amount of water in wheat gluten increases, plasticizing the polymer motions, the T_g falls. (Figure redrawn from data given in R.D. Hosney, K. Zeleznak, and C.S. Lai, *Cereal Chem.* 63:285, 1986.)

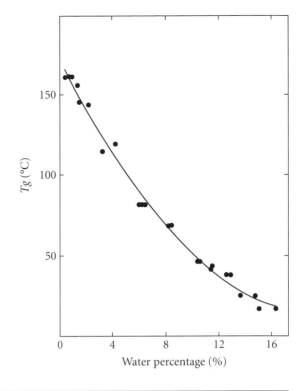

Figure 2.50. The glass transition temperature T_g for native starch (black dots) and pregelatinized starch (open dots; see Chapter 4) as a function of moisture content. (K.J. Zeleznak and R.C. Hoseney, *Cereal Chem.* 64:121–124, 1987.)

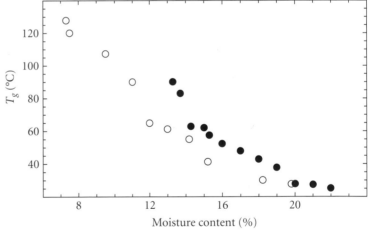

is in a stable glassy state when the temperature is below around 20 °C; that is, room temperature. The water molecules make it easier for the polymer molecules in the gluten to move around and slide past one another. Thus, increasing the amount of water present will lower the glass transition temperature—that is, the food must be cooled to a lower temperature before microscopic motions are frozen out. Figure 2.50 shows the dependence of the glass transition temperature of starch as a function of temperature. Both starch and gluten are major components of bread, and the dependence of the glass transition temperature for bread on moisture content is shown in Chapter 5 (Figure 5.21).

Freezing and Freeze-Drying

The most effective method for preserving the quality of most types of fresh foods is freezing. Freezing locks up the water of the food tissues in ice and makes it unavailable for degradative processes such as microbial growth or chemical reactions. In general, freezing does not kill microorganisms or prevent chemical reactions but simply slows growth and reaction rates. To be effective, very low temperatures must be achieved and maintained. In commercial food processing, foods are flash frozen by supercooling to temperatures of −18 to −20 °C or colder in order to produce large numbers of small ice crystals, rather than the few large crystals that would develop by slow freezing (see next section). In general, freezing works best in maintaining quality for intermediate- and lower-moisture tissues like meat, rather than for products that contain very high amounts of water like some fruits (Table 2.1). Freezing can potentially maintain edible conditions for long periods. Many wooly mammoths and other Ice Age mammals have been found in the permafrost of Siberia and Alaska after being frozen solid for as much as 30,000 years, although their flesh is no longer palatable or even edible (Kurtén 1986).

Freeze-drying, or **lyophilization**, is a preservation technique in which a food is dehydrated by first freezing the product, after which the water is removed by sublimation; that is, by direct evaporation from the solid (ice) phase. The basic approach is to try to put as much of the food sample as possible into the glassy state before the removal of the water, in order to minimize the changes that take place in the food microstructure as a result of the dehydration. Figure 2.51 illustrates an example of the type of structural changes that can occur upon freezing in a food product. The principal goal of freeze-drying is to preserve mesoscale structural organization, such as cell walls and organelles, so it is most often applied in the food industry to products containing ingredients with such cellular organization, such as fruits, vegetables, and

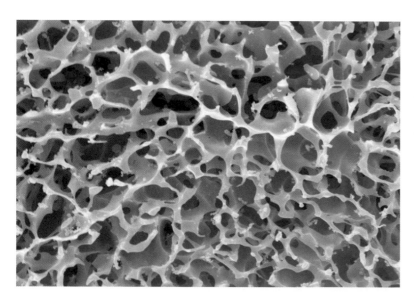

Figure 2.51. An electron micrograph of the microstructure of a freeze-dried alginate gel (the formation of ice crystals during the freezing has compressed many of the polymer strands into lamaella, which remain after the water is sublimed away). (Figure courtesy of Attilio Cesàro, University of Trieste.)

meats. This differs somewhat from the application of lyophilization in the pharmaceutical industry, where a major goal may be to prevent protein or peptide conformational changes. Of course, since freezing the water in a tissue is itself a type of dehydration in situ, the goal of preserving cellular organization and microstructures cannot be completely met, and freeze-drying, like ordinary freezing, works better with some products than with others. Freeze-dried food products are almost always intended to be consumed after being re-hydrated, so the technique is most often applied in situations in which the freeze-dried product re-hydrates successfully to a state that closely resembles the original product. While the goal of lyophilization is preserving product quality through maintaining original structure, other techniques such as simple freezing or canning may often work about as well for less cost or effort, so freeze-drying is usually used for products where refrigeration is unavailable or where weight is a factor, as in freeze-dried meals used in camping and backpacking. Freeze-drying is also most valuable when the structures being maintained are heat sensitive and are likely to suffer from being heated to remove water.

In commercial food freeze-drying the food is flash frozen to temperatures as low as −46 °C (−50 °F) to quickly freeze as much bulk water as possible and to immobilize microstructures and the remaining water in a glassy state. The product is then placed under a high vacuum, so that the frozen water is below the triple point on its phase diagram (see Figure 2.1). If the temperature is then raised while maintaining the low pressure, the water can sublimate, which is to go directly from ice to the vapor phase without passing through the liquid state. In lyophilization, often as much as 98% of the original moisture is removed, which also reduces the original weight by around 90%. After this has been achieved, the product can be warmed in a second heating phase to room temperature. Placed in a vacuum-sealed container, the product can then be stored for long periods. The growth of microorganisms is significantly inhibited, as are most other types of degradative chemical processes. Freeze-dried foods can be prepared for eating by rehydrating, usually by adding boiling water or briefly boiling the product in water. The removal of the crystals of I_h ice by sublimation tends to produce a porous dessicated product with voids where the ice crystals were located, which facilitates the permeation of water back into the food when it is rehydrated. Apart from backpacking meals and emergency food products, coffee for more conventional use is also prepared by freeze-drying because of its brief shelf life.

Ice Nucleation and Growth

The freezing point of water (273.16 K = 0 °C) is the temperature at which liquid water and ice coexist in equilibrium at atmospheric pressure. Lowering the temperature of a sample of water below this point shifts the equilibrium such that the solid form has the lower free energy, so thermodynamics dictates that the entire sample will freeze. How long it will take for the water to freeze, however, a question of kinetics, is not determined by thermodynamics, and in practice it is possible to undercool liquids, including water, below their freezing points, because the crystallization of a solid

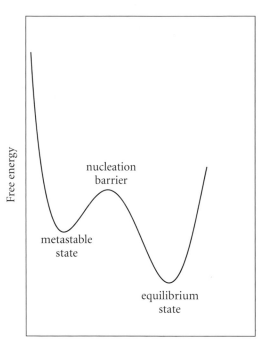

Figure 2.52. A schematic representation of the free energy barrier inhibiting nucleation.

phase requires a nucleus or template of crystal-like structure for the molecules to "condense" upon as the new solid phase grows in the liquid. This nucleation process is an example of a more general phenomenon involved in any phase transition in which the new phase must grow out of the older one. Other examples include the superheating of water, as can happen when heating a cup of water in a microwave oven, where under certain conditions the water can be heated above 100 °C without boiling, or in the preparation of supersaturated solutions by cooling a concentrated solution below the solubility limit of the solute. In each of these cases the system is thermodynamically unstable, but temporarily trapped while it waits for a small nucleus of the new phase to arise by spontaneous structural fluctuation. The reason that these thermodynamically unstable states can exist is that the initial formation of these nuclei by spontaneous fluctuations is unlikely, which is another way of saying that the free energy for their initial formation is unfavorable, even though the overall free energy change for the complete transformation of the unstable phase to the more stable one is, of course, favorable (Figure 2.52).

The origin of this barrier can be thought of in thermodynamic terms as arising from the formation of a surface between the nascent ice nucleus and the surrounding liquid water. The formation of a surface always results in a positive (unfavorable) change in free energy, with the amount of surface energy being proportional to the surface area, which increases as the square of the nucleus radius, $4\pi r^2 \sigma$, where σ is the surface free energy per unit area. In these same terms, one can also assume that for each molecule transferred from the bulk liquid to the interior of the growing solid nucleus, a favorable free energy increment ΔG_b that is proportional to the volume of

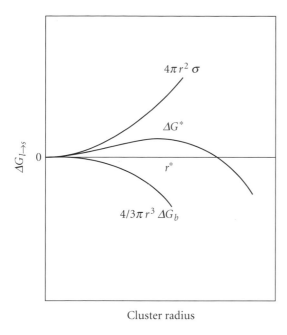

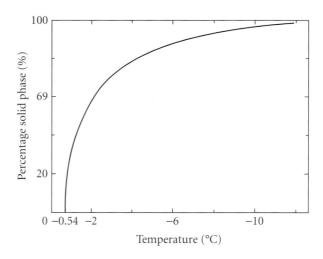

Figure 2.54. Freeze concentration in a 1.54 M NaCl solution (called an isotonic saline solution, because it has the same solute concentration as blood). (Figure adapted and redrawn from F. Franks, *Biophysics and Biochemistry at Low Temperatures*. Cambridge: Cambridge University Press, 1985.)

Figure 2.53. A schematic illustration of the origin of the nucleation free energy barrier.

the nucleus occurs, which increases as the cube of the radius, $4/3\pi r^3$. For very small nuclei, the unfavorable surface area term which is a function of r^2 increases faster than the favorable volume term which is a function of r^3 decreases, producing the small barrier to growth seen in Figure 2.53. Once a cluster larger than r^* is formed, transferring a molecule from the liquid to this solid particle is favorable, so all of the remaining molecules freeze out by condensing onto it, leading to the freezing of the entire phase. For water, the critical radius r^*, or critical cluster size, corresponds to an ice cluster of around 175 to 200 molecules. In other words, almost 200 water molecules must come together more or less simultaneously in the correct structural arrangement (that is, an ice I_h lattice) before the further growth of this cluster is spontaneous and favorable. This process is called **homogeneous nucleation**.

The fact that the ice lattice cannot accommodate solutes, and thus freezes out as pure water, means that the process of freezing results in a progressive concentration of the remaining liquid solution (Figure 2.54). This behavior has a number of consequences for food processing, such as freezing point depression, protein denaturation, and pH changes.

In frozen foods the size, shape, and overall morphology of the ice crystals formed during freezing are important. This is most true in foods that are meant to be consumed while still frozen, such as frozen desserts, because these factors will determine the overall texture and "mouthfeel" of the product. The nature of actual crystals of water is complex, however, and depends on many factors. Everyone has heard the old saw that no two snowflakes are exactly alike, although they share many overall features, including a hexagonal symmetry and extensive branching. **Dendrites**

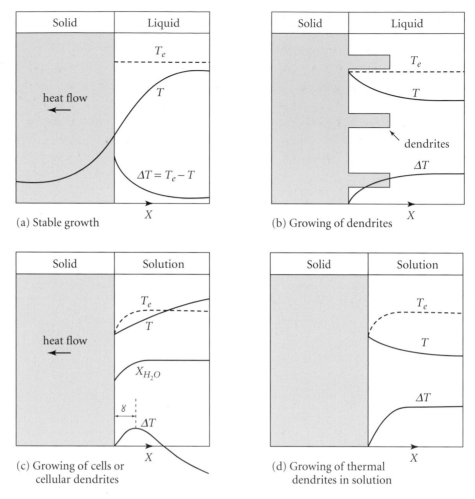

Figure 2.55. Freezing pattern for different cooling regimens. (a) Slow freezing in a liquid that is warmer than ice; (b) dendritic growth in an undercooled liquid; (c and d) the same situations as above but for liquid solutions with freezing point depression. (Adapted and redrawn from A. Calvelo, "Recent studies on meat freezing," in *Developments in Meat Science*, R. Lawrie, ed. London: Applied Science Publishers, 1981.)

(branched crystals) are characteristic of "runaway" freezing in undercooled water, while hexagonal cells are more likely to form when the temperature is closer to freezing. Typically, freezing in food samples occurs through an arrangement like that illustrated in (a) in Figure 2.55. When placed in a freezer, the exterior of the sample will be colder than the interior, through contact with the chilled air, so the heat flow will be out through the exterior layer of the sample as illustrated. This arrangement will lead to slow stable growth of the ice in a uniform front, because if any part of the advancing ice layer by chance should start to extend further out into the unfrozen liquid than the rest, it would be extending into a region of higher temperature, as indicated by the temperature profile, where the degree of undercooling is less, so the rate of freezing will slow down until the rest of the ice face catches up, because the rate of freezing is proportional to the degree of undercooling.

Dendritic growth is characteristic of growth in an undercooled medium, as on the surface of a pond or in a puddle. Once an ice nucleus forms in the liquid phase, if any part of the small ice nucleus grows faster than the rest, it will extend out into water that is colder (because the release of the latent heat of fusion upon freezing will raise the temperature at the crystal face to the freezing point, so that the temperature falls as one moves away from the crystal out into the undercooled liquid). This situation is illustrated in part b of Figure 2.55. As a result of extending out into colder liquid, this small protuberance will then tend to grow faster by the freezing of more liquid, which will have the effect of extending it even further out into the liquid away from the bulk of the growing crystal face, leading to runaway growth. Because the process would be repeated if any protuberances developed on this extension, branching is favored, which leads to the dendrites. The situation is further complicated if solutes are dissolved in the aqueous phase, as will usually be the case in a food, since freezing will lead to a freeze concentration at the crystal face, because the solute is excluded from the ice, resulting in freezing point depression (parts c and d in Figure 2.55).

Table 2.6 Approximate thermal conductivities of several substances

Substance	Thermal conductivity ($W/m \cdot K$*; all values are approximate)
copper	390.0
ice	2.0
water	0.6
fats	0.2
air	0.2
dry wood	0.06

Note: Fats are good insulators because of their low thermal conductivities, as well as their relatively low heat capacities. Water also has a relatively low thermal conductivity, but is not a good insulator because of its high heat capacity—it will draw heat out of an adjacent body, without conducting it on efficiently. The thermal conductivity of ice is more than three times higher than that of liquid water, while its heat capacity at the melting point is less than half that of water (Figure 2.3). Copper has a high thermal conductivity because of the free electrons in its conduction band, which is why it is effective on the bottoms of pots and pans, where it distributes heat evenly.
Wood has a very low thermal conductivity, which is why cooking spoons made of wood can be comfortably held even when immersed in very hot foods or liquids, unlike metal spoons.
*Watts per meter per degree Kelvin.

A number of practical consequences for food systems present themselves due to the differences in the properties of water and ice. One example is the difference in freezing and thawing rates for foods. The heat capacity of water is much greater than that of ice, and more importantly for freezing and thawing, the thermal conductivity of ice is much greater than that of liquid water (Table 2.6). This means that a given amount of heat can be transmitted through a sample of ice more rapidly than it is transmitted through an equal amount of water. In the freezing of a food, the sample is typically placed in a freezer unit where the air surrounding the food is colder than the sample itself, so it begins to freeze from the outside in, and as additional heat is drawn out to also freeze the interior, this heat must then pass through an outer layer of ice. This thermal exchange is relatively rapid, however, because of the high thermal conductivity of ice. When the food sample is taken out of the freezer to thaw, however, the surrounding air is now warmer than the frozen food, and it also begins to thaw from the outside in. In this case the heat being added to the sample must pass through an outer layer of liquid water which has a lower thermal conductivity and which absorbs more of the energy because of its high heat capacity. As a result, the thawing process is much slower than the original freezing (which illustrates the very great advantage of microwave heating technology).

SUGGESTED READING

Chandler, D. 2005. "Interfaces and the driving force of hydrophobic assembly." *Nature* 437:640–647.

Damodaran, S., K.L. Parkin, and O.R. Fennema, eds. 2007. *Fennema's Food Chemistry*, 4th ed. New York: Dekker.

Eisenberg, D., and W. Kauzmann. 1969. *The Structure and Properties of Water*. New York: Oxford University Press.

Franks, F. 1985. *Biophysics and Biochemistry at Low Temperatures*. Cambridge: Cambridge University Press.

Jeffrey, G.A., and W. Saenger. 1997. *An Introduction to Hydrogen Bonding*. Oxford: Oxford University Press.

Raschke, T.M., and M. Levitt. 2005. "Nonpolar solutes enhance water structure within hydration shells while reducing interactions between them." *Proc. Natl. Acad. Sci. USA* 102:6777–6782.

Slade, L., and H. Levine. 1991. "Beyond water activity: Recent advances based on an alternative approach to the assessment of food quality and safety." *Critical Reviews in Food Science and Nutrition* 30:115–360.

Stillinger, F.H. 1980. "Water revisited." *Science* 209:451–457.

3.
Dispersed Systems: Food Colloids

COLLOIDS AND TYPES OF DISPERSIONS

Many food systems consist of components having complex structures on a number of levels, from the molecular to the macroscopic (visible). Often the properties of these foods are dependent on the nature of this structure and cannot be understood in terms of the chemical properties of the molecular components alone. In addition, many of these food systems are not in an equilibrium state and thus cannot be fully described using the equilibrium thermodynamic properties of the components. Dispersions exhibit both of these features and are common in both unprocessed and manufactured foods. A **dispersion** consists of discrete particles scattered in a continuous medium. The term "particle" need not connote a solid in this context; if two liquids are mutually insoluble, then the "particles" might be small droplets of one liquid dispersed in the other liquid, a system referred to as an **emulsion**. Table 3.1 lists the various types of dispersions that are possible; because gases freely mix, to disperse one in the other is not possible except on the molecular level.

A dispersion is generally considered to be a **colloid** if the discrete components of the system have sizes in the range between $10\,\text{Å}$ ($1\,\text{nm}$) and $1\,\mu$ ($10^{-6}\,\text{m}$). These limits are chosen somewhat arbitrarily, and vary depending on the tastes of the

Table 3.1 Types of dispersions

Dispersion type	Dispersed phase	Continuous phase	Example in foods
fog, mist, aerosol	liquid	gas	—
smoke, aerosol, haze	solid	gas	smoke?
foam	gas	liquid	beer, meringue, whipped cream
emulsion	liquid	liquid	salad dressing, butter, mayonnaise (o/w)*, margarine (w/o)*
sol, colloidal solution, gel, suspension	solid	liquid	skim milk
solid foam	gas	solid	foam candy
solid emulsion	liquid	solid	margarine(w/o)*, butter, ice cream (o/w)*
alloy	solid	solid	—

*(o/w): oil in water emulsion; (w/o): water in oil emulsion.

user; often the lower limit is taken to be larger, around 10 nm, and the upper limit is also often taken to be much larger, even up to 1 mm. The choice is not completely without a logical basis however; ~10 to 100 Å is taken as the lower limit so that the size scale of the dispersed particle will be substantially larger than that of the molecules of the continuous phase, to justify considering this bath as continuous rather than discrete. Further, the upper limit of 1μ is selected because particles below this limit are sufficiently small as to still be substantially affected by Brownian motion.

Named for the nineteenth-century Scottish botanist Robert Brown, **Brownian motion** refers to the random jiggling motion of very small particles when observed through a microscope. Brown noted that when he looked through a microscope at pollen grains suspended in water, they were constantly jiggling, no matter how carefully he tried to isolate his apparatus and its base from sources of outside vibrations and disturbances. The theory that describes this phenomenon, developed by Albert Einstein and Marian Smoluchowski, among others, explains the effect as resulting from instantaneous imbalances in the forces acting on small particles from opposite directions (Figure 3.1). These forces result from interactions with the molecules of the phase in which the particles are suspended. In colloidal science, we would like to treat this phase as continuous, but of course, it is in fact also made up of discrete molecules. The approximation of treating it as continuous, and thus qualitatively different from the suspended particles, requires that the colloidal particles be much, much larger in dimension than the molecules of the "continuous" phase. If the dispersed particle is very large, say large enough to be seen easily with the naked eye, then the number of solvent molecules hitting it from left and right, up and down, forward and behind, are truly enormous, on the order of fractions of a mole (6.02×10^{23}). As a result, any instantaneous differences in any direction are so small, as the

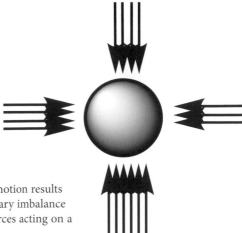

Figure 3.1. Brownian motion results when there is a temporary imbalance in the instantaneous forces acting on a particle.

result of averaging over so many individual collisions, that no net effect is noticeable on the enormous inertia of the particle. As the particle gets smaller, however, the instantaneous averaging over much smaller numbers of collisions produces more significant differences in the opposing directions, compared to the decreasing weight (and inertia) of the particles, so that below a certain size level, the motion produced in response to these imbalances becomes perceptible.

The reason that the upper limit in size for particles to be considered colloidal is related to Brownian motion is that these random forces help to keep the discrete particles dispersed. In any dispersion on the Earth's surface, the system is subjected to the Earth's gravitational field, which exerts a downward force on all components of the system. Except in the unlikely situation that the continuous and dispersed phases have the exact same densities, this force tends to sort out the system, so that particles heavier than equal volumes of the continuous phase will tend to settle out, while those lighter than the continuous phase will cream out to the top. Gravitational forces are weak, however, and when the particles are smaller than ~1 μ, they can effectively be balanced by the random forces resulting from Brownian motion which thus greatly extend the time required for the system to separate out under the gravitational forces.

In general, colloidal dispersions are thermodynamically unstable and thus will tend to break down, given enough time. For example, in an emulsion, if the dispersed droplets consist of the less-dense liquid, these droplets will not only tend to encounter one another as the result of collisions and merge to form larger droplets, but in a gravitational field such as exists on the Earth's surface, will tend to rise to the top of the system, eventually separating out completely as a less-dense continuous phase on top of the heavier continuous liquid phase. Because of the small size of colloidal particles, they have a large surface-to-volume ratio, and thus their properties tend to be dominated by surface effects.

As already noted, the definition given for colloids is somewhat arbitrary and thus can frustrate some attempts to generalize about such systems. For example, the

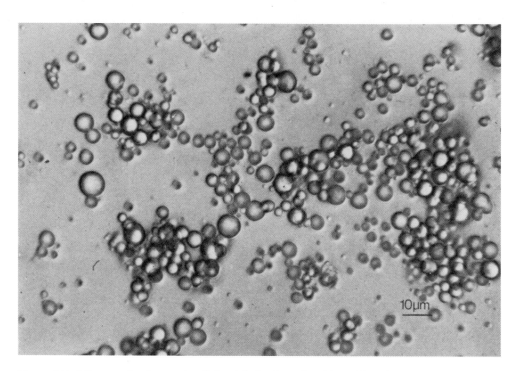

Figure 3.2. Micrograph of natural milk fat globules flocculated by cold-agglutination. Notice that these individual globules have aggregated but not coalesced, so there has been no reduction in the overall surface area, and the globules could in principal be dispersed again by agitation or other expenditure of energy. (Reproduced from H. Mulder and P. Walstra, *The Milk Fat Globule*. Farnham Royal, Bucks, England: Commonwealth Agricultural Bureaux, 1974.)

lower size limit includes many proteins and polysaccharides, many of which form true, stable solutions, randomly but uniformly mixing on the molecular scale, and thus are not thermodynamically unstable. These types of dispersions are sometimes referred to as "**lyophilic**" or "**hydrophilic**" colloids. Particularly in foods, the upper size limit is also vague. For example, in raw milk, most fat globules are larger in diameter than 1μ (see Figure 3.2), while many are also below this arbitrary limit.

The following definitions are useful in studying food colloids:

emulsion: a colloidal dispersion of discrete droplets of one liquid in a continuous phase of another liquid.

foam: a colloidal dispersion of gas bubbles in a continuous liquid phase.

aerosol: a colloidal dispersion of either solid particles or liquid droplets in a gas.

lyophilic colloids: colloidal systems in which the dispersed phase has a high affinity for the continuous phase and forms a true solution. If the continuous phase is water, this type of dispersion is termed "hydrophilic."

lyophobic colloid: colloidal dispersion in which the dispersed phase has little or no affinity for the continuous phase. If the continuous phase is water, this type of system is referred to as "hydrophobic."

flocculation: the clumping together of small particles into clusters, called **flocs**, without the fusing of the individual particles into larger particles. In flocculation, no reduction of surface area occurs, although some of the surface sites may be blocked by touching other particles.

coalescence: the process whereby two or more small particles fuse together to form a single larger particle. The most important feature of this process is that the surface area is reduced. In coalescence, all evidence of the smaller particle is lost.

coagulation: an aggregation in which the interparticle distances are on the order of atomic dimensions. The forces between coagulated particles are strong, and coagulation is often irreversible.

peptization: the redispersing of coagulated particles, in those cases in which it is possible.

gel: a continuous phase of interconnected particles and/or macromolecules intermingled with a continuous liquid phase, usually having a soft, elastic consistency, but exhibiting elastic solid-like behavior; that is, in response to a small stress, it deforms recoverably.

emulsifying agent or emulsifier: a material that can be added to an emulsion, which prevents flocculation of the emulsion particles and thus makes the emulsion more stable.

monodisperse colloid: a colloidal dispersion in which all of the particles are the same size and shape.

polydisperse colloid: a colloidal dispersion in which a distribution of different particle sizes or shapes occurs.

Gels

Because of their importance in foods, gels deserve special mention in the context of colloidal science. Gels are "jelly-like" solid systems, generally colloidal in one component, that deform elastically and display solid-like properties, even though they consist primarily of water or some other liquid. Usually they consist of an extensively entangled or cross-linked polymer network forming an open framework or matrix, with a continuous liquid phase interpenetrating and occupying the spaces between the polymer molecules. The polymer framework might be considered a second, continuous solid phase, because it extends throughout the gel, but of course it is not really a "phase" in the true sense. Such systems are generally treated as colloidal due to the large size of the polymer molecules. Generally, the polymers constituting the lattice are present in a low overall concentration, often as little as a few percent by weight, but their interactions with the liquid phase are sufficient to prevent it from draining out or undergoing any sort of bulk flow. Nuclear magnetic resonance (NMR) studies of such gels find that the water molecules exhibit relaxation behavior like that of water in a liquid solution rather than a solid lattice. Not all polymers can form such lattices; generally, gelling polymers have regions of their sequence that closely associate with other chains in **juncture zones**, but these sequences alternate with other parts of the chain that cannot or do not associate. Without this alternating character, the chains would self-associate to the point of precipitating out as a crystal, or more probably, as an amorphous solid, but the nonassociating sequences allow the chains to wander off in other, random directions, setting up the framework of the gel. The juncture zones can consist of regions without branching side chains, as in the example on the top in Figure 3.3, or regions bound together by ions strongly interacting with charged residues on both chains, as shown on the bottom in the figure, such as occurs in alginate and pectin (Chapter 4). A number of other food polymers, and particularly polysaccharides, can form gels. These include methyl cellulose, agar, gellan, locust bean gum, and konjac. In addition, the protein collagen of animal connective tissue is widely used in foods to make gels, as are denatured soy proteins, as in tofu.

Emulsions

Dispersions are very common in foods, and emulsions play a particularly prominent role. None of these systems is thermodynamically stable, but shelf life stability is a particular problem in the case of liquid emulsions due to the fluidity of their continuous phases. The random Brownian motions of the dispersed particles in this continuous fluid medium allows them to encounter each other more rapidly, promoting the formation of flocs. In the case of liquid-in-liquid emulsions, the fluidity of the dispersed droplets also promotes their coalescence into larger droplets once they encounter one another. For this reason, food emulsions frequently require that various emulsifiers be added to stabilize the system. These additives can function in various ways, such as by increasing the viscosity of the continuous medium to slow the motions of the dispersed droplets, by adsorbing onto the interfaces between the phases to stabilize them thermodynamically, or by acting as a coating to mechanically inhibit coalescence once dispersed droplets encounter one another.

Many food emulsions have very complex structures on the mesoscopic (nanometer) scale. Milk is a prime example, as illustrated in Figure 3.4, with soluble whey proteins in the nanometer size range, casein "micelles" in the 0.01- to 0.3-μ size range, and dispersed fat globules ranging in size from 0.1μ up to ~9μ. Other examples include ice cream, which is a multicomponent frozen foam; mayonnaise, an oil in water (o/w) emulsion made by using the emulsifier lecithin in egg yolks to stabilize the dispersion of oil in the water of either vinegar or lemon juice; and Béarnaise sauce, which is a similar tarragon-and-shallot-flavored oil-in-water emulsion of clarified butter in vinegar, again stabilized by the lecithin of egg yolks. A signal feature of all of these systems is that, because they contain a large number of small discrete particles dispersed in a continuous phase of a different character, their behavior is dominated by the large surface area between these two phases, because each particle has a surface of one type in contact with the continuous phase of the opposite type.

Figure 3.3. Illustrations of polymers interacting through limited juncture zone regions to form an open lattice for a gel network.

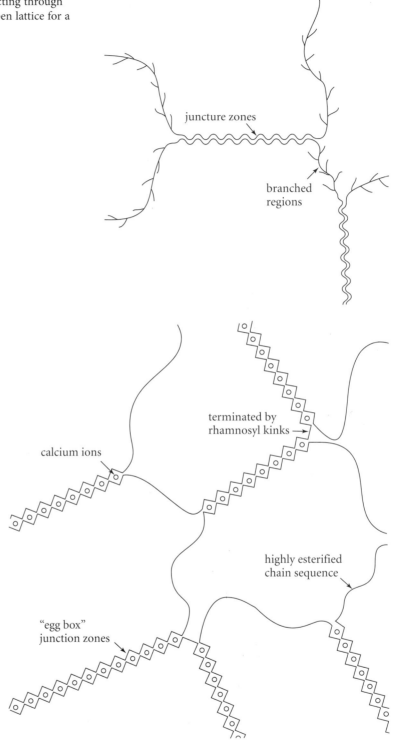

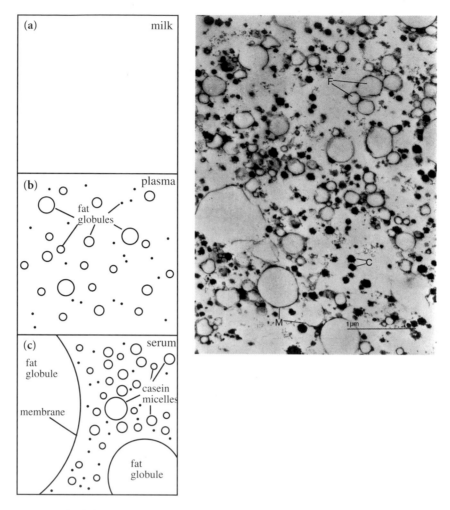

Figure 3.4. On the left is a schematic representation of milk at three different size scales, to illustrate the relative size of the components. (a) is magnified × 5; (b) magnified x 500; (c) magnified × 50,000. On the right is an actual electron micrograph of partially homogenized milk; F points to examples of undisrupted fat globules, M shows examples of fat globule membranes, and C indicates examples of so-called casein micelles. The partially obscured distance scale at the lower right indicates 1μ. (Figure on the left is redrawn after a figure in P. Walstra and R. Jenness, *Dairy Chemistry and Physics*. New York: Wiley-Interscience, 1984, and the figure on the right is reproduced with permission from H. Mulder and P. Walstra, *The Milk Fat Globule*, Farnham Royal, Bucks, England: Commonwealth Agricultural Bureaux, 1974.)

THE IMPORTANCE OF SURFACE AREA

For a small particle, as the size of the particle decreases, the surface of the particle becomes more and more important, so surface properties and phenomena become more dominant. This can be illustrated by looking at the surface area per unit volume for spherical particles. For a sphere, the surface area A_s is given by

Figure 3.5. In a small particle or droplet, nearly all of the constituent molecules are on or near the surface, while in a larger one, many molecules in the interior are less affected by the surface.

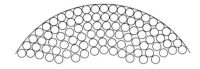

$$A_s = 4\pi r^2 \qquad\qquad (3.1)$$

and the volume is given by

$$V_s = 4/3\pi r^3 \qquad\qquad (3.2)$$

so that the surface area per unit volume is

$$A_s/V_s = 4\pi r^2/(4/3\pi r^3) = 3/r \qquad\qquad (3.3)$$

As can be seen from this formula, as the particle gets smaller, the surface area per unit volume increases, diverging to ∞ as r goes to zero. In small particles, all of the molecules may be surface molecules or significantly perturbed by the presence of the surface, while in large particles, most of the component molecules are deep in the interior and experience significant interactions only with like molecules (Figure 3.5).

It can easily be shown with simple geometric arguments that, when two spherical droplets of the same size fuse into a larger droplet, their overall collective surface area decreases by just under 21%. Because the unfavorable thermodynamic free energy of dispersed systems can be expressed as an unfavorable surface free energy term, decreasing the surface area (as through coalescence) decreases this unfavorable free energy, driving the emulsion to spontaneously separate out.

This increasing importance of the surfaces of particles as the size of the particles decreases is responsible for the danger posed by dust in facilities such as grain silos. Those who are from the Midwest are probably aware that on occasions these facilities explode as if they had been dynamited. The reason for these tragedies is that they are frequently full of dust particles that are composed mostly of flammable carbohydrates (starch). The dust abrades from the grain as the silos are filled, and these small particles contain large amounts of energy per mass (because, as we will see in the next chapter, starch is the principal means of long-term chemical energy storage in plants). For a very small particle, a large fraction of these starch molecules are on the surface and exposed to oxygen, so that a spark could surmount the activation barrier for oxidation of these carbohydrates, releasing the energy of combustion and thus promoting the burning of nearby dust particles and so on in a classic chain reaction fashion. Given this crucial importance of surfaces for very small particles, the thermodynamics of surfaces is clearly important in determining the properties of colloidal dispersions. With such small particles dispersed in a continuous medium, there is an enormous amount of total surface area between the two phases. Accordingly we will briefly review the thermodynamics of interfaces in order to understand the behavior of colloidal dispersions.

THE THERMODYNAMICS OF SURFACES

The molecules of a liquid experience a cohesive, attractive force with one another that keeps them associated together and overcomes their tendency to fly off as vapor molecules with the associated gain in translation entropy. In the bulk of such a condensed liquid phase the molecules are surrounded by these attractive interactions in every direction, while those on the surface of the liquid experience such attractive interactions only from the direction of the bulk, looking "down" into the liquid interior, because no neighbors occur in the direction of the vapor phase (Figure 3.6). This imbalance of attractive forces tends to draw the surface molecules down into the interior of the bulk phase, thus tending to contract the volume of the liquid and in particular the surface area. The unbalanced forces increase the interior pressure, and the contraction stops when the increasing pressure favoring expansion balances the attractive forces favoring contraction. This process produces a surface tension, defined as the resulting force acting at right angles to the surface and directed inward.

This qualitative picture of surface tension applies not just for a liquid boundary with a vacuum/vapor, but also for any two phases in contact. For example, if two immiscible liquids are in contact, the molecules of each phase experience stronger or more favorable interactions with their own kind of molecule than with those of the other phase and are thus drawn inward; otherwise the two liquids would freely

vacuum/vapor

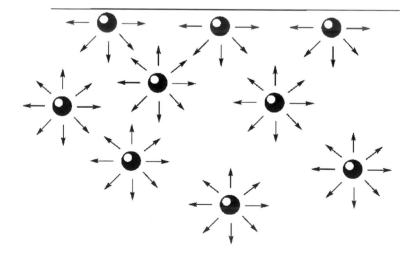

Figure 3.6. At a vacuum/liquid interface, the molecules in the top layer of the liquid do not feel the cohesive binding forces of liquids above them, where there are few or no molecules, so they are drawn down into the bulk, creating a surface tension and tending to contract the volume.

Table 3.2 Surface tension γ_0 against air and interfacial tension γ_i against water for some liquids at 20 °C in N·m^{-1}

Liquid	γ_0	γ_i
water	7.28×10^{-2}	—
benzene	2.89×10^{-2}	3.50×10^{-2}
ethanol	2.23×10^{-2}	—
n-hexane	1.84×10^{-2}	5.11×10^{-2}
mercury	48.5×10^{-2}	37.5×10^{-2}

mix and form a solution. Thus, the surface of one phase in contact with another is associated with an unfavorable free energy term arising from the tendency of the molecules of each phase to prefer to be drawn into the interior of that phase. The surface tension γ can be defined in terms of the work involved in increasing the interfacial surface area, A_s, between two such phases,

$$dw = \gamma dA_s \qquad (3.4)$$

From this definition, we see that increasing the surface area between two phases requires work and is an unfavorable process.

At constant temperature and pressure, γ can be written in terms of the Gibbs free energy,

$$\gamma = \left(\frac{\partial G}{\partial A_s} \right)_{T,P} \qquad (3.5)$$

The units of γ can be seen from these equations to be $J/m^2 = N \cdot m/m^2 = Nm^{-1}$. Table 3.2 gives some sample interfacial tensions.

The surface tension is thus defined thermodynamically as the change in the Gibbs free energy per incremental change in surface area. Because increasing the surface area between two phases increases the free energy, a system will spontaneously tend toward the state with the smallest interfacial area. The higher the surface or interfacial tension, the stronger this tendency will be, and the harder it will be to increase the interfacial surface area.

As we saw in Chapter 2, water has a particularly high surface tension against air (Table 2.2). The high surface tension of pure liquid water results from the strong attraction of water molecules for one another as a result of their numerous hydrogen bonds. This cohesiveness is weakened as the temperature is increased, because the additional kinetic energy allows the molecules to stretch against these hydrogen bonds. Thus, the surface tension of water against air is a function of temperature and decreases as the temperature rises (Figure 3.7).

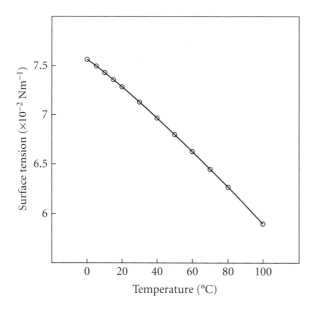

Figure 3.7. The surface tension of pure liquid water against air as a function of temperature.

The "Mentos Eruption"

In recent years many people have become familiar with the amusing result of putting Mentos candies into a bottle of carbonated soda, usually referred to as the "Mentos eruption," "Mentos explosion," or "Mentos geyser," which has been widely demonstrated on the Internet, late-night television, and the cable TV show *MythBusters*. This simple reaction illustrates in a dramatic fashion several of the principles that we have discussed thus far. Mentos are a popular Dutch mint candy, marketed worldwide, consisting of small oblate spheroids a little less that 2 cm in diameter. When these candy pieces are dropped into a 2-liter bottle of carbonated soda such as Diet Coke or Diet Pepsi, the soda will quickly "erupt" out of the bottle in a "geyser," ranging from a meter to several meters in height. There seems to be some confusion about the origins of this effect, and several explanations have been put forward, in some cases mutually contradicting one another, and not all aspects of the process seem to have been adequately explained. In general terms, however, this phenomenon is an example of rapid nucleation of carbon dioxide gas from a supersaturated solution.

Commercial soda drinks are of course supersaturated solutions of carbon dioxide, as we saw in Chapter 2. Usually diet drinks are chosen for Mentos geysers, in part, because the mess is easier to clean up afterward (however, the eruption is sufficiently messy that it definitely should not be tried indoors, regardless of the soda type, but rather on a roomy patch of dirt or grass outside, so that no cleanup is required). Diet sodas are also chosen, however, because they produce the most impressive eruptions. Regular sugar-containing sodas will also produce an eruption, although generally not as high as that produced by a diet drink. Colas are not required, as other carbonated soft drinks will also erupt. Thus, the phosphoric acid of colas is not directly a factor. An eruption with simple carbonated water, as in club soda, can be produced, but the height reached by the geyser is considerably less than for either type of cola sodas. Even the Mentos brand candies are not specifically required, because an eruption can also be produced with other similar products, such as Tums tablets, for example, or even with conventional laboratory boiling chips.

When the Mentos candies are dropped into the soda, they immediately fall to the bottom because of their greater density, which is an important part of the process. Once on the bottom of the bottle, the candies quickly produce an enormous number of bubbles of carbon dioxide. The seemingly smooth surfaces of the candies are rough on the microscopic level and contain many small holes, pits, and cracks, which are initially filled with air, and which do not fill with water when they are immersed. Thus, they serve as nucleation sites for the CO_2 in the supersaturated cola to come out of solution as a gas, forming bubbles that rapidly grow and cream off, leaving the crevice as it was, ready to again generate another bubble by the same process. Because a large number of these holes and crevices exist, a great many bubbles are quickly generated on the candies, propelling the liquid above them out of the narrow opening as they rush to the top of the container. Boiling chips and other matte-finished candies also contain numerous surface pits and cracks like Mentos and thus can also cause eruptions. Candies with a smooth surface, however, such as would be produced by confectioner's glaze, do not produce much of a reaction.

It has been suggested that the gum arabic (Chapter 4) in Mentos plays a part in the eruption by modifying the surface tension, but this is probably not a significant factor. It does touch on an important point, however, which is the role of surface tension. As we have just seen, surface tension can be defined as the work required to increase the surface area between two phases, as in a bubble interface between gas and liquid. The higher the surface tension, the harder it will be to generate bubbles. This is why diet sodas produce a more energetic eruption than a sugared soda. Sugar (sucrose) is very soluble and hydrophilic, due to its many polar hydroxyl groups, and thus increases the surface tension of water as it dissolves. This is the reason that sugar is a conformation stabilizing and salting-out osmolyte in the Hofmeister ranking (Chapter 2). The artificial sweeteners used in diet sodas are hundreds of times sweeter than sugar (see Chapter 8), so far less of them are necessary in diet drinks to generate the same level of sweetness. More importantly, however, aspartame, which is the most common of such sweeteners, has both hydrophobic and hydrophilic functional groups (see Chapter 5), because it has a hydrophobic phenylalanine side chain along with a carboxylic acid group. It thus is surface active, and tends to accumulate at surfaces such as the interface between the aqueous soda and air bubbles. As a result, the surface tension of diet drinks is lower (see below), which is why diet drinks foam more than regular sodas even when poured into a glass—their lower surface tension makes it easier to nucleate bubbles with their unfavorable surface free energy.

As might be expected, the best treatment of this phenomenon in popular sources was on the Discovery Channel's *MythBusters* series, which devoted an episode to this subject. In that show, the Mythbusters determined by experimentation that three of the ingredients in diet colas—aspartame, potassium benzoate, and caffeine—contributed to the explosive bubbling, while the phosphoric acid had no effect. In addition to aspartame, both benzoate, used as a preservative (Chapter 10), and caffeine, used as a flavorant and a stimulant (Chapter 9), are surface active, and serve to further reduce the surface tension of the sodas, promoting bubbling.

Curved Surfaces, the Kelvin Equation, and Ostwald Ripening

Because of the unfavorable free energy term associated with the presence of a surface, a system will tend to try to contract to minimize its surface area, giving rise to curved surfaces; this process will cease when the increasing internal pressure balances the gain in free energy through further surface reduction, giving rise to the **Laplace equation** for the pressure difference across the curved surface of a sphere,

$$P_{in} - P_{out} = \frac{2\gamma}{r} \tag{3.6}$$

or, for an ellipsoidal droplet with radii r_1 and r_2

$$P_{in} - P_{out} = \gamma \left(\frac{1}{r_1} + \frac{1}{r_2} \right) \tag{3.7}$$

As a result of this pressure difference across curved surfaces, spherical particles of different sizes are not equally stable. It can be shown that for such spherical droplets,

$$RT \ln \frac{P}{P^\star} = \frac{2\gamma V_m}{r} = \frac{2\gamma M}{\rho r} \tag{3.8}$$

where P is the vapor pressure, $P^\star$ is the vapor pressure of the pure liquid with an uncurved surface, V_m is the molar volume, M is the molar mass, ρ is the mass density, and γ is the surface tension.
Thus,

$$P = P^\star \exp\left(\frac{2\gamma V_m}{RTr} \right) \tag{3.9}$$

This equation, called the **Kelvin (or Thompson) Equation**, after William Thompson, who became Lord Kelvin, shows that the vapor pressure of a small droplet of liquid is higher than that of a larger one, giving it a greater tendency to evaporate. Similar arguments apply for the solubility of a small solid particle in a liquid compared with a larger one. For example, if one had a beaker with small sugar (sucrose) crystals of different sizes scattered on the bottom, covered with a saturated solution of sucrose in water, the system would not remain unchanged with time. Even though the total amount of crystallized sugar would remain the same, because the liquid solution is

Table 3.3 The effect of polysaccharides in controlling ice crystal growth in ice cream

	Mean ice crystal diameter (µm)	
Time (weeks)	Unstabilized ice cream	Ice cream stabilized with 0.15% locust bean gum and 0.02% carrageenan
0	43.3	35.4
3	61.9	57.0
24	113.7	95.4

Source: H.D. Goff, K.B. Caldwell, and D.W. Stanley, J. Dairy Sci. 76:1268–1277 (1993).

saturated and cannot hold any more sugar, over time the greater tendency for individual sugar molecules to escape from the smaller crystals and to precipitate onto the larger ones would lead to the gradual disappearance of the small crystals and the growth in size of the large crystals as a result of the different Kelvin equation behavior of the different-size crystals.

One of the most important manifestations of the Kelvin equation effect in food systems is in the tendency of small ice crystals in ice cream to decrease in size with time, and for large ones to grow over time, leading to the development of large ice crystals that have a major impact on the product quality, making the product grainy or "sandy" in texture. Lactose crystals can also grow in size over time and impact the texture. This effect is the direct consequence of the Kelvin equation, because the water molecules that make up the small crystals have a greater tendency to escape and migrate to the larger ones. To control these processes, most commercial ice creams contain polysaccharides such as carrageenan, tara gum, or alginate to reduce mobility in the glassy matrix between the air bubbles of the foam (see Table 3.3).

This same process is the underlying phenomenon behind so-called **Ostwald ripening**, a source of instability in liquid emulsions. This process does not readily occur in oil in water (o/w) food emulsions, because the triacylglycerol oils are not sufficiently soluble in water to migrate from the small droplets to the larger ones. Such ripening is seen in dispersions of essential (citrus) oils in water due to the greater solubility of these non-triglyceride oils in water, and it is a source of instability for water in oil (w/o) emulsions.

Antifreeze Proteins and Ice Nucleating Proteins

Because freezing conditions are common in temperate and arctic regions, organisms living in these latitudes have had to adapt strategies to deal with the damaging effects that could result from the possibility that their constitutional water might freeze. Warm-blooded organisms (endotherms) avoid this potentially lethal event by maintaining an internal temperature significantly above freezing. Plants, however, and many cold-blooded animals do not have this option and have adopted a number of

Figure 3.8. Left, a cartoon representation (see Chapter 5) of the conformation of the type-III antifreeze protein from ocean pout as determined by X-ray diffraction from the protein crystal. Right, a van der Waals representation of this protein, showing the relatively flat binding surface on the bottom of the figure. Oxygen atoms are colored red, nitrogen atoms blue, sulfur atoms yellow, and carbon atoms are colored metallic blue-green. Hydrogen atoms are omitted. (F.D. Sonnichsen et al., *Structure* 4:1325–1337, 1996; A.A. Antson et al., *J. Mol. Biol.* 305:875, 2001.)

other strategies to deal with freezing temperatures. Many have evolved antifreeze compounds, either to prevent their internal water from freezing, or to protect vital internal structures, such as globular proteins and membrane bilayers, from being disrupted by freezing. The disaccharide trehalose, a nonreducing disaccharide (Chapter 4), has significant cryoprotective activity and is produced by a number of organisms evolved to be able to resist freezing or drought. Many other organisms have evolved a series of antifreeze proteins that are able to inhibit the formation or growth of ice crystals. There has been considerable interest in the food industry in the use of such molecules to control water in food products. Unilever has already developed an antifreeze protein from the ocean pout, *Macrozoarces americanus*, a type of cold-water fish, for use in preventing the Ostwald ripening-driven growth of ice crystals in frozen dessert products, where the maintenance of small ice crystal size is necessary for retaining a creamy texture. The conformation of this protein, as determined by X-ray crystallography, is shown in Figure 3.8.

The pout antifreeze protein may function by preferentially binding to the fastest-growing prism faces (Figure 2.15) of ice crystals. The relatively flat and mainly hydrophobic surface seen in the bottom of Figure 3.8 is believed to be the binding surface of the protein. By preventing the ice from growing underneath the bound protein, the protein molecules also induce a curvature to the growing uncovered surface between the adsorbed molecules. From the Kelvin equation (Equation 3.9), this curvature between the adsorbed proteins also tends to slow growth, by making escape across the curved surface more likely than further accretion, because water molecules

liquid water

decreased growth rate across
curved surface—faster water
escape; slower addition

ice surface

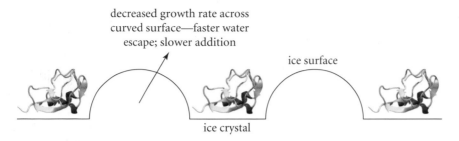

ice crystal

Figure 3.9. A schematic illustration of the mechanism by which antifreeze proteins inhibit ice crystal growth.

already in the ice face are more likely to escape from the curved ice surface than free molecules in the liquid are to add to the surface (See Figure 3.9).

By using genetic engineering techniques, scientists at Unilever have inserted the gene that codes for this protein into the genome for Baker's yeast, which allows the production of the protein in large amounts. This protein ingredient is listed on their frozen dessert product labels as "ice structuring protein." As a naturally occurring protein, this ice structuring protein has GRAS (generally recognized as safe; see Chapter 10) status, but due to the manner by which it is produced, this protein would be considered a genetically modified organism (GMO) food (see Chapter 5). Many other antifreeze proteins have been identified from other species, and other possible applications exist for their use in frozen foods and in cryobiology in general.

ADSORPTION AT LIQUID SURFACES

Surface-active molecules, that is, those that tend to adsorb onto or concentrate at surfaces, are called **surfactants**. The adsorption of a surface-active substance onto the interface between two liquid phases is governed by the so-called **Gibbs adsorption isotherm**. This equation expresses the surface tension γ as a function of the surface excess concentration Γ of the surfactant, defined as

$$\Gamma = \frac{n^s}{A_s} \tag{3.10}$$

where n^s is the excess amount of a solute adsorbed on the surface, relative to the amount in a comparable volume in the bulk phase, far away from the surface, and A_s is the surface area. If the bulk concentration of the solute is c, then the Gibbs adsorption isotherm for the solute at the interface is given by

$$\Gamma = -\frac{1}{RT}\left(\frac{\partial \gamma}{\partial \ln c}\right)_{T,P} \tag{3.11}$$

$$dy = -RT\Gamma d(\ln c) \tag{3.12}$$

The surface excess can be a positive or negative quantity. A positive surface excess means that the solute is a surfactant, because it has a tendency to accumulate in the interfacial region, giving an "excess" concentration over what occurs in the bulk. A negative surface excess would indicate a solute with a high affinity for one of the bulk phases, so that it is drawn down into the bulk in order to maximize its interactions with the molecules of that bulk phase, thus decreasing its concentration in the interfacial region. In doing so, it increases the surface tension. This is the type of behavior that would be expected in water for ions (electrolytes) and highly polar molecules such as sugars. As we can see from the Gibbs sorption isotherm, a positive surface excess concentration lowers the surface tension, and thus the free energy (from Equation 3.5), so a surfactant molecule that accumulates on a surface tends to stabilize the interface because it lowers the surface tension. On the other hand, a solute that is strongly drawn down into the bulk, giving a negative surface excess, raises the surface tension and thus the free energy. A zero surface excess concentration would have no effect on the surface tension. Figure 3.10a illustrates schematically how these

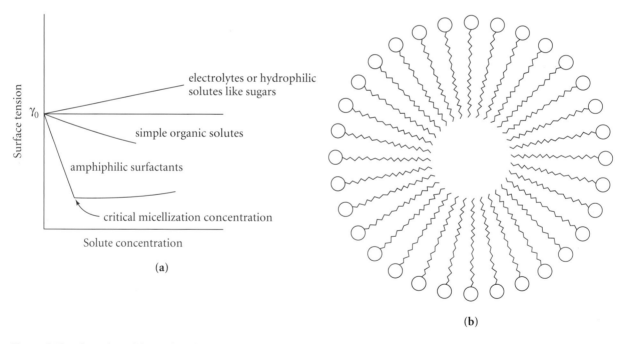

(a)

(b)

Figure 3.10. Electrolytes (charged molecules) and strongly polar molecules are drawn down into the bulk of an aqueous solution, increasing the surface tension, and the more of these types of solute that are added, the more γ increases. Barely soluble, mostly hydrophobic species tend to aggregate at the surface of the liquid solution, lowering the surface tension according to Equation (3.12). Amphiphilic surfactants adsorb on the interface, with their polar portions in the aqueous phase and their hydrophobic portions in the nonpolar phase (or in the air/vacuum in the case of a liquid/vapor interface). The surface tension decreases with increasing concentrations of both types of solutes. When the concentration reaches a characteristic concentration called the critical micellization concentration, surfactants reorganize into micelles such as is shown schematically on the right, and with their polar groups now pointing out, behave like a polar species, being drawn down into the bulk and increasing the surface tension again as their concentration is further increased.

different types of solutes would affect the surface tension of a solution as a function of concentration.

Amphiphilic molecules such as phospholipids or fatty acids, which have a polar, water-soluble group attached to a nonpolar, hydrophobic group or hydrocarbon chain are drawn to surfaces because, at the interface between an aqueous and hydrophobic phase, they can put their hydrophilic functionalities into the polar water phase and their hydrophobic functional groups into the oil phase. In doing so, they lower the surface tension and thus the unfavorable free energy associated with the creation of the surface, thus stabilizing the interface thermodynamically. As a result, the surface tension falls steeply with increasing total concentration of the surfactant, as illustrated in Figure 3.10a.

When the increasing surfactant concentration reaches a threshold value of surface coverage, called the critical micellization concentration, sufficient amphiphilic molecules are present to support the formation of micelles, such as illustrated schematically in Figure 3.10b. In these approximately spherical, nanometer-scale aggregates, the surfactant molecules self-organize into compact structures with their polar head groups pointing out into the aqueous phase and their hydrophobic tails segregated in the interior, away from contact with the water. Once most of the interface is covered with amphiphilic surfactant molecules and the excess are tied up in micelles, they are no longer driven to migrate to the water/oil interface, and adding more surfactant molecules just promotes the formation of more micelles. Since these colloidal-size aggregates have all-polar surfaces, they now act like a single polar solute molecule and are drawn down into the interior of the aqueous phase, actually increasing the surface tension again as the concentration is further increased. Thus, as can be seen in Figure 3.10a, γ then increases with the addition of more surfactant.

The Gibbs sorption isotherm is most useful in describing the sorption of species onto the interface between two liquids. This model is not less valid in other cases, but the specification and measurement of surface tensions when one phase is a solid presents difficult problems due to the much slower kinetics. For this reason, other sorption isotherms are more convenient for discussing sorption onto interfaces where one phase, particularly the dispersed particle, is a solid. In describing this type of situation, such as the condensation of water onto a solid surface, other models are more useful for describing the water sorption isotherms like those discussed in the previous chapter. Two theoretical models for the moisture sorption isotherm in particular are widely used in food chemistry. These are the Langmuir Adsorption Isotherm and the BET sorption isotherm.

ADSORPTION AT SOLID SURFACES

Because of the difficulty in applying the Gibbs adsorption isotherm to solid surfaces, we must seek another description of adsorption onto solids. If we have an adsorbent solid surface in contact with a solution of adsorbate, a specific amount of the solute will adsorb onto the surface under any given set of conditions. If the total surface area is known and the amount of solute lost from solution can be measured, then the amount adsorbed as a function of the equilibrium concentration of the solution at a given temperature is called an adsorption isotherm, as with the moisture sorption isotherms already discussed in Chapter 2. One of the easiest such isotherms to understand is the Langmuir Isotherm, which assumes that the surfaces of the solid particles contain a fixed number of adsorption sites, all of which are equivalent, with the same adsorption energy for the solute molecules. The adsorption process could be written as an equilibrium between the adsorption of new particles onto the surface and the desorption of some of those particles which have already adsorbed,

$$A + S_{\text{urface sites}} \underset{k_{\text{a}}}{\overset{k_{\text{d}}}{\rightleftharpoons}} A \cdot S \qquad\qquad (3.13)$$

where the equilibrium constant is $K_{\text{eq}} = \dfrac{k_{\text{a}}}{k_{\text{d}}}$.

If the fraction of the adsorption sites that are occupied is called θ, then the fraction that is unoccupied is $(1 - \theta)$. The desorption of particles from the surface is a first-order process, which means that the desorption rate R_{d} is proportional to the number of adsorbed molecules,

$$R_{\text{d}} = k_{\text{d}} \theta N \qquad\qquad (3.14)$$

where N is the total number of sites. The adsorption process is second order, because it involves two species—the adsorbing solute molecule colliding with the site on the surface. The concentration of adsorbing molecules is [A], and the concentration of sites available is the total number of sites N times the fraction which are unoccupied, $(1 - \theta)$. The rate of this second-order condensation process R_{a} is then given by

$$R_{\text{a}} = k_{\text{a}} [A] (1 - \theta) N \qquad\qquad (3.15)$$

At equilibrium for any given temperature these rates balance, so the two separate equations can be set equal to each other,

$$k_{\text{d}} \theta N = k_{\text{a}} [A] (1 - \theta) N \qquad\qquad (3.16)$$

Solving for θ,

$$\theta = \frac{k_{\text{a}} [A]}{k_{\text{d}} + k_{\text{a}} [A]} = \frac{K_{\text{eq}} [A]}{1 + K_{\text{eq}} [A]} \qquad\qquad (3.17)$$

This equation describes the extent of adsorption as a function of concentration at a particular temperature and is called the **Langmuir Adsorption Isotherm** (named after Irving Langmuir, Nobel laureate and founder of the field of surface science). Its essential qualitative feature is that the available sites saturate asymptotically as a function of concentration, but that the adsorption stops when monolayer coverage is

Figure 3.11. A schematic representation of a Langmuir Adsorption Isotherm. θ is the fraction of occupied surface sites, and [A] is the concentration of adsorbant.

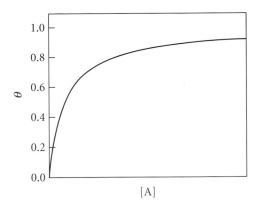

achieved (Figure 3.11). Note that for infinite dilution, when there are very few solute molecules to adsorb, so that their concentration [A]→0; then [A] $K_{eq} \ll 1$, so

$$\theta = K_{eq}[A] \qquad (3.18)$$

At very high concentrations, K_{eq} [A] $\gg 1$, so

$$\theta = 1$$

The Langmuir isotherm assumes a fixed number of adsorption sites, all of which are equivalent, and that the sorption process ends when these have all been filled. The amount of adsorbing molecules needed to fill all of these sites is called a monolayer coverage. A more realistic description of the adsorption process, particularly for the case of water molecules condensing from the vapor onto a solid surface, has been developed that assumes that adsorption can continue after monolayer coverage is achieved, but with subsequent layers of water interacting with other water molecules in the monolayer rather than directly interacting with the surface. The energy of interaction, or adsorption energy ε, is assumed to be different from that for the first-layer waters and also different from the energy of vaporization of pure water, ε_v. In this model, there are also no fixed adsorption sites, as is assumed in the Langmuir model. This multilayer adsorption isotherm, called the BET Isotherm, after its originators, S. Brunauer, P.H. Emmett, and E. Teller (the last of these is the controversial Edward Teller, "father" of the hydrogen bomb, principal exponent of the so-called Star Wars program of the Reagan administration, and model for the "Dr. Strangelove" character in the Stanley Kubrick movie of the same name), can be written as

$$\frac{1}{V}\frac{a}{(1-a)} = \frac{a(c-1)}{cV_m} + \frac{1}{cV_m} \qquad (3.19)$$

$$\frac{V}{V_m} = \frac{ac}{(1-a)(ac-a+1)} \qquad (3.20)$$

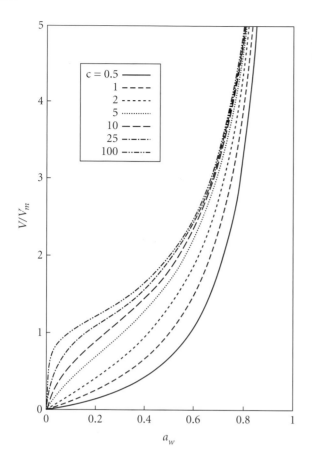

Figure 3.12. Plots of the BET isotherm equation as a function of the activity $a = P/P_0$ for several values of the adjustable constant c.

$$\frac{V}{V_m} = \frac{ac}{a^2(1-c)+a(c-2)+1} \tag{3.21}$$

where a is the water activity $\dfrac{P}{P_0}$, c is a constant, V is the volume of water adsorbed, and V_m is the volume of water which would constitute a monolayer coverage. The parameter c is related to the binding energy to the particle surface, relative to the vaporization energy, $\varepsilon{-}\varepsilon_v$, and thus also describes the temperature dependence of the sorption.

As can be seen from the plot of the BET isotherm equation as a function of the water activity in Figure 3.12, for larger values of the constant c, the isotherm has a functional form that very closely approximates that observed for the experimentally determined moisture sorption isotherms for food products (see Figure 2.42). Thus, for an extensively hydrated food sample, the food molecules may be covered by many layers of water, so that the first water molecules removed during dehydration are primarily making interactions with other water molecules that are similar to those that they might make in a pure water sample, while layers closer to the food surface are progressively more affected by the nature of the food molecules and their specific functional groups and hydration characteristics.

HYSTERESIS IN MOISTURE SORPTION ISOTHERMS

The occurrence of hysteresis in moisture sorption isotherms has been explained using a number of models that make use of the Kelvin equation, which states that vapor pressure across a curved surface varies with the radius of curvature of that surface. These models also make assumptions about the microstructure of the food sample, since they require that the food contain microcapillaries or pores of specific geometries. The development of hysteresis in these models is due to making the assumption that different radii of curvature characterize the desorption and adsorption processes.

In these models, the Kelvin equation for the vapor pressure during desorption contains a minus sign because the desorbing water molecules are assumed to be escaping across the curved surface of a meniscus for the water filling the pore, with the radius of curvature being measured on the outside of the meniscus, rather than on the inside of a drop as before,

$$RT \ln \frac{P}{P^*} = -\frac{2\gamma V_m}{r} = -\frac{2\gamma M}{\rho r} \tag{3.22}$$

$$P = P^* \exp\left(\frac{-2\gamma V_m}{RTr}\right) \tag{3.23}$$

The hysteresis in sorption isotherms always has adsorption taking place at higher pressures than desorption, which would seem to imply that condensation during the adsorption process is taking place in a larger pore than the evaporation during the desorption process. This could be explained using the top two illustrations of Figure 3.13. Desorption could be thought to take place by emptying the capillary through evaporation across the meniscus of radius R, which in the upper diagram on the right is illustrated for the case where this meniscus has a radius equal to that of the capillary, and in the lower diagram a case is shown in which that radius is much larger. In these cases, the vapor pressure will be given by Equation 3.23, with the appropriate value of the radius.

The adsorption could be thought of as taking place as the deposition of a layer of moisture around the surface of the capillary, as in the upper left illustration of Figure 3.13, with a thickness t, but where this layer does not coalesce with a meniscus until the capillary is nearly filled. This unfilled inner part of the capillary could be approximated by a long ellipsoid with a length (semimajor axis) of r_{cyl}, and a radius of $(r - t)$. In that case, the Laplace equation for an ellipsoid applies

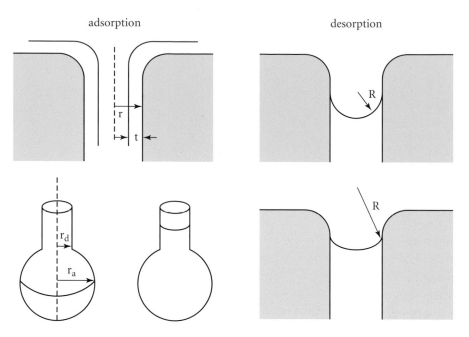

Figure 3.13. The models used to explain hysteresis in moisture sorption isotherms assume that the desorption and adsorption processes are characterized by curved surfaces with different radii. In one model, evaporation is assumed to be an emptying of a water-filled pore across a curved meniscus of radius R, as in the illustrations on the right. Adsorption, however, is then modeled as taking place by deposition onto the surface of an empty pore, coating it with a layer of water that gradually thickens until it finally fills it in completely. In this model, the adsorption is characterized by the radius of the empty part of the pore, which is the radius of the pore minus the thickness of the water layer. In an alternate model, illustrated in the lower left, the pore is envisioned as having different radii due to broadening out like a round-bottom flask. In that model the adsorption is viewed as pooling in the bottom of the pore, recreating a meniscus. (Adapted with permission from *Principles of Colloid and Surface Chemistry* by P.C. Hiemenz, New York: Marcell Dekker, 1986.)

$$P_{in} - P_{out} = -\gamma\left(\frac{1}{r_1} + \frac{1}{r_2}\right) \tag{3.24}$$

giving a Kelvin equation of

$$\ln\left(\frac{P}{P^o}\right) = \frac{-\gamma V_m}{RT}\left(\frac{1}{r_1} + \frac{1}{r_2}\right) \tag{3.25}$$

For the cylinder, one of these radii is very long and thus effectively infinite (i.e., $r_{cyl} = \infty$), and the other is $(r - t)$, so for adsorption:

$$\left(\frac{1}{r-t} + \frac{1}{r_{cyl}}\right) = \frac{-\gamma V_m}{RT(r-t)} = \frac{-\gamma V_m}{RTr} \quad if\ t << r \tag{3.26}$$

When water is desorbing, however, it is escaping across an intact meniscus, and the conventional Kelvin equation (3.9) applies. Thus, for desorption:

$$\ln\left(\frac{P}{P^\circ}\right) = \frac{-2\gamma V_m}{RTr} \tag{3.27}$$

$$\frac{1}{2}\ln\left(\frac{P}{P^\circ}\right)_d = \ln\left(\frac{P}{P^\circ}\right)_a \tag{3.28}$$

so,

$$\left(\frac{P}{P^\circ}\right)_d^{1/2} = \left(\frac{P}{P^\circ}\right)_a \tag{3.29}$$

Because P/P° is less than 1 for both adsorption and desorption, $P_d < P_a$.

Another model assumes that the food sample contains numerous small pores with a peculiar design, somewhat like a round-bottomed flask such as illustrated in the lower left of Figure 3.13, with a narrow neck of radius r_d, and a much larger well with a maximum radius r_a. When the food sample is being dried, water is initially removed from the narrow neck with r_d, while when the dried food is being rehydrated, water initially goes into the well, with its much larger radius r_a.

While it is unlikely that either of these simple geometries fully characterizes the nanoscale structure of common foods, it is likely that they contain many cavities and pores resulting from the jumbled packing of polymer chains like starch, cellulose, and denatured proteins. It should also be remembered that in many foods, the hysteresis collapses when the drying/rehydration cycle is repeated (Rizvi and Benado, 1984a,b; see Chapter 2), indicating that irreversible changes are taking place in the mesoscopic-scale structure of the food upon drying.

PRODUCING EMULSIONS

Emulsions are manufactured by breaking up large volumes of liquid to make smaller ones (droplets), as in a colloid mill. This process requires the expenditure of energy to overcome the surface tension of the droplets, and more energy is required as the droplets become smaller. Emulsification is most commonly accomplished by tearing the droplets apart with large shear forces. Lowering the surface tension between the two liquid phases reduces the energy needed to disrupt the large droplets. In the two homogenizers illustrated on the left and in the center of Figure 3.14, the feed stock mixture of liquids is forced under high pressures through the narrow valves of the homogenizer, and the shear stresses set up from friction with the walls as a result of the high flow velocities tear the streams apart, creating droplets. In the colloid mill illustrated on the right in Figure 3.10, the feedstock is forced between a rapidly spinning rotor and the stationary vessel wall, again setting up the necessary shear forces.

Emulsifiers are surface-active molecules (surfactants) that adsorb at liquid interfaces, reducing the surface tension and thus promoting the stability and formation of emulsions. A variety of molecules are used for this purpose in the manufacture of

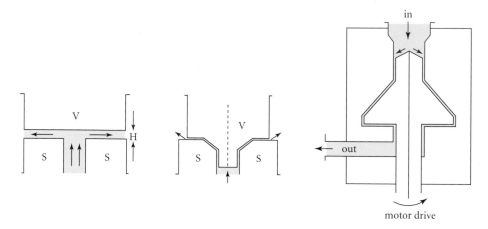

Figure 3.14. Left and middle, schematic diagrams for two types of homogenizers; on the right, a diagram of a colloid mill. (Figure adapted and redrawn from E. Dickinson and G. Stainsby *Colloids in Foods*. London: Applied Science Publishers, 1982.)

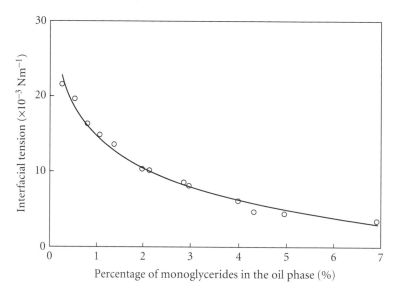

Figure 3.15. The interfacial tension between water and cottonseed oil as a function of the concentration of monoglycerides added to the oil as an emulsifier. (Data redrawn from Kuhrt et al., *J. Am. Oil Chem. Soc.* 27:310, 1950.)

food emulsions, including triglycerides and other lipids, various proteins, and various synthetic molecules (such as the so-called "tweens" and "spans"). Figure 3.15 shows the effect of monoglycerides, one of the most common food emulsifiers, as a function of the bulk concentration, on the interfacial tension between water and cottonseed oil.

Emulsifiers adsorb at interfaces because they are amphiphilic molecules that have hydrophobic functional groups that prefer to be in the oil phase side of an interface, and hydrophilic functional groups that are favorably hydrated by the aqueous side of an oil/water interface. In general terms, when the physical sizes of these two components are very different, this difference can be used to predict which phase will be the continuous phase and which phase will be the discrete phase. Broadly speaking,

Figure 3.16. When a large surfactant molecule is more soluble in one phase than another, it tends to promote that phase as the continuous phase, because more of the bulky part of that molecule can be packed on the convex side of the curved surface than on the concave part (Dickinson and Stainsby 1982).

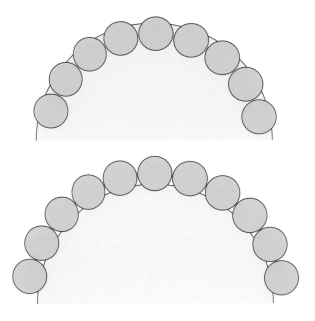

the phase in which the largest part of the molecule is soluble will be the continuous phase, because a curved surface that is convex with respect to the closely packed surfactants can accommodate more of these molecules than one that is concave, so the surfactants will tend to place their smaller portion on the concave side of the interface and their larger portion on the convex side. This concept is illustrated schematically in Figure 3.16, where the surfactants are represented as spheres. In the top case the molecules are preferentially adsorbed on the concave inside of the curved surface, while in the lower panel they are preferentially adsorbed on the convex outside of the same curved surface. In this schematic example, the same surface arc can accommodate 9 of the spheres on the inner concave side but can place 11 on the outer convex surface. The packing on the outer surface would thus be favored if the surfactant concentration is high, and that phase would then tend to be the continuous phase as a result.

HLB Numbers

Empirical scales have been developed to rate the various amphiphilic emulsifying agents in terms of their relative lipophilic or hydrophilic characteristics, and hence their tendency to promote either o/w or w/o emulsions, which is based on assigning an HLB (hydrophile–lipophile balance) number to each type of molecule that is approximately the ratio of the weight percentages of hydrophilic and hydrophobic groups in the molecule. The least hydrophilic surfactants are assigned the smallest HLB values in this system. An approximate empirical equation for calculating such HLB numbers has been developed based on the functional groups making up the molecular structure of the surfactant (Griffin 1954; Davies 1957). It is

Table 3.4 Contributions made by various functional groups to HLB numbers

Group	n_H	Group	n_H
—SO$_4$Na	38.7	—COOH	2.1
—COOK	21.1	—OH (free)	1.9
—COONa	19.1	—O—	1.3
tertiary amine	9.4	—OH (sorbitan)	0.5
ester (sorbitan)	6.8	—(CH$_2$—CH$_2$—O)—	0.33
ester (free)	2.4	—(CH$_2$—CH$_2$—CH$_2$—O)—	−0.15[a]

[a] A negative sign indicates that this group is actually lipophilic (hydrophobic) (Davies, 1957).

Table 3.5 Behavior expected from emulsifiers as a function of HLB number

HLB range		HLB range	Dispersability
3–6	w/o	1–4	none
7–9	wetting agents	3–6	poor dispersabilibty
8–15	o/w	6–8	unstable milky dispersion
13–15	detergent	8–10	stable milky dispersion
15–18	solubilizer	10–13	translucent dispersion/solutions
		13+	clear solution

$$n_{HLB} = 7 + \sum_i n_H(i) - \sum_j n_L(j) \tag{3.30}$$

where the summation i is over all of the hydrophilic segments and the summation j is over all of the lipophilic (hydrophobic) segments. Each segment (distinct chemical group) will then contribute some characteristic value n to the total HLB number. All hydrophobic hydrocarbon segments were considered to make the same contribution, $n_L = 0.475$ (that is, all groups such as —CH—, —CH=, —CH$_2$—, —CH$_3$). The contributions from hydrophilic groups vary with the group, and are given in Table 3.4.

Emulsifiers with HLB values in the range 3 to 6 promote w/o emulsions, because they are predominantly hydrophobic and are most adsorbed on the oil side, which by the arguments we gave before tends to make the oil phase the continuous phase. Emulsifiers with HLB values in the range 8 to 15 or above promote o/w emulsions (see Table 3.5). A surfactant with an HLB number of 7 is considered to be "neutral."

Table 3.6 lists HLB numbers for some commonly-used food emulsifiers. Often a blend of two or more emulsifiers gives a more stable emulsion than any one emulsifier at the same HLB value would. When mixing emulsifiers, approximate algebraic additivity holds in computing resulting HLB values. For example, 20% sorbitan

Table 3.6 HLB numbers for some commonly used food emulsifiers

Emulsifier	HLB value	FAO/WHO ADI (mg/kg body wt)
potassium oleate	20.0	—
polyoxyethylene sorbitan monolaurite ("Tween 20")	16.7	0–25
polyoxyethylene sorbitan monostearate ("Tween 60")	14.9	0–25
gum acacia	11.9	—
methylcellulose	10.5	—
gelatin	9.8	—
sorbitan monopalmitate ("Span 40")	6.7	0–25
commercial lecithins	4 to 9	—
sorbitan monooleate ("Span 80")	4.3	—
mixture of mono-and diglycerides (61–69% total mono-)	3.5	—
sorbitan tristearate ("Span 65")	2.1	0–25
oleic acid	1.0	—

Source: Dickinson and Stainsby (1982).

tristearate (HLB 2.1) plus 80% polyoxyethylene sorbitan monostearate (HLB 14.9) would give

$$(0.2)\,2.1 + (0.8)\,14.9 = 12.3 \Rightarrow \text{o/w emulsion}$$

One of the more widely used synthetic food emulsifiers is **polysorbate 80**, or polyoxyethylene sorbitan monooleate (Figure 3.17), marketed by Uniqema/ICI as Tween 80. It is often used in ice cream to prevent the milk proteins from binding to fat droplets. Polysorbate 80 is synthesized from the monounsaturated fatty acid oleic acid and the sugar alcohol sorbitol. As can be seen from its structure, it can serve as a surfactant even though it is not ionic (charged), because the esterified end has many oxygen atoms capable of hydrogen bonding to water molecules, while the fatty acid tail is completely hydrophobic. Figure 3.18 illustrates schematically how tweens and spans stabilize oil droplets in emulsions and prevent coalescence.

The most widely used industrial emulsifier is **lecithin** (Figure 3.19), a natural product. Commercial lecithin is a mixture of phospholipids from cell membranes and is primarily obtained from soybean oil. Egg lecithin is also used in many foods like mayonnaise and cakes but is not widely used as a separate commercial ingredient. With a charged head group on one end (on the left in Figure 3.19) and two long hydrophobic tails, this molecule is a classic surfactant and functions as a detergent-like emulsifier. This amphiphilic character is often schematically illustrated with a cartoon such as is shown on the right in the figure, where the circle represents the hydrophilic head group, and the zigzag lines represent the hydrophobic fatty acid tails. Phospholipids will be discussed in more detail in Chapter 7.

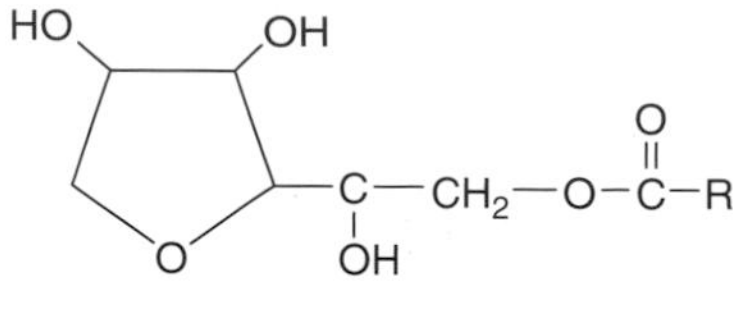

A "Span"
Note: *R* is a nonpolar chain or group

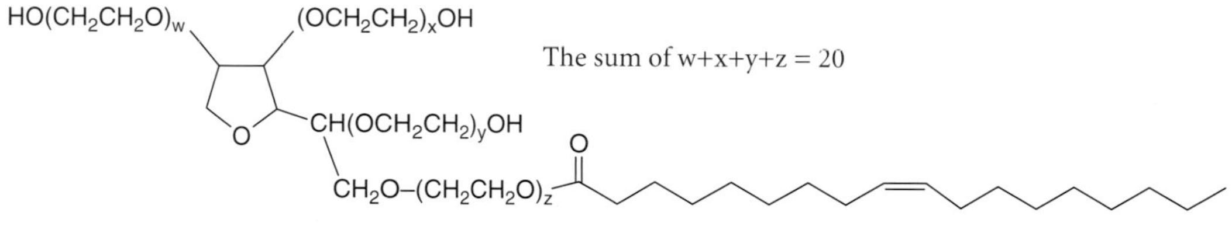

The sum of w+x+y+z = 20

Polysorbate 80 ("Tween 80"),
a polyoxyethylene sorbitan ester

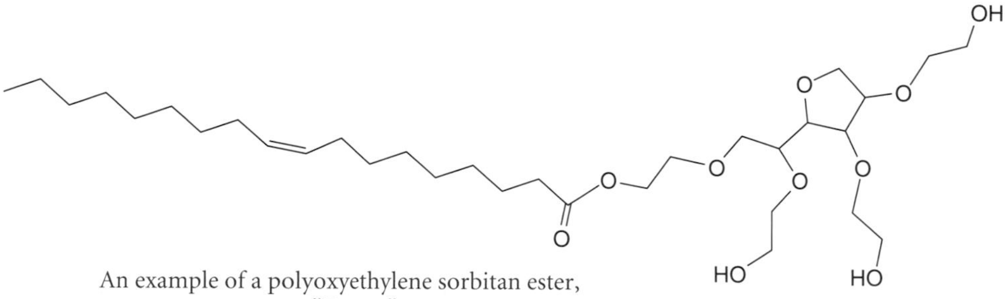

An example of a polyoxyethylene sorbitan ester,
or "Tween"

Figure 3.17. The structures of some food emulsifiers.

EMULSION STABILITY

Solvation Forces

In solutions, solvent molecules assume a specific spatial distribution around the solute, which is referred to as the solution structure. This can be as nonspecific as the packing of spheres at a planar surface or as anisotropic and complicated as the structuring we have already seen of water organizing around a sugar molecule. The further one gets from the solute molecule, the less structured the solvent is with respect to the position of the solute. The presence of these structured layers can have a significant effect on the interactions of colloidal particles dispersed in a liquid, as is illustrated schematically in Figure 3.20. As the surfaces of two such particles approach one another, the space available to be occupied by the solvent decreases, so that more and more of it is squeezed out. As can be seen from the diagram, at distances equal

Figure 3.18. A schematic illustration of how "tweens" and "spans" adsorb at the interfaces between oil droplets and water in an o/w emulsion and serve to stabilize the emulsion. The reason for the choice of the name "tweens" becomes clear from this diagram. (Figure adapted and redrawn from E. Dickinson and G. Stainsby, *Colloids in Foods*. London: Applied Science Publishers, 1982.)

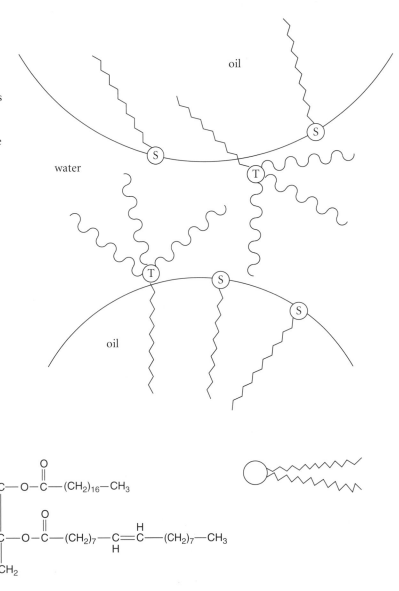

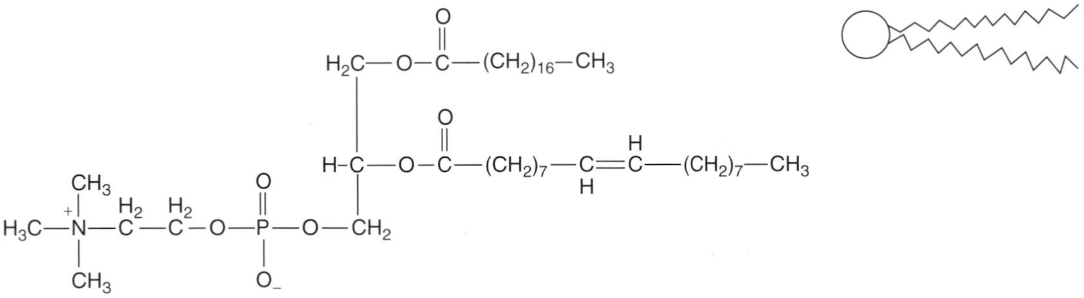

1-Stearoyl-2-oleoyl-3-phosphatidylcholine

Figure 3.19. On the left, the structure of lecithin, often symbolically represented as on the right.

to roughly integral multiples of the solvent molecule dimensions, the force needed to further reduce the distance between the solute particles increases dramatically as an entire solvent layer is squeezed out. Following its removal, the density between the surfaces is anomalously low, so pressure from the other side of the particles drives them together until they once again are approximately an integral number of solvent diameters apart. This behavior in principle leads to an oscillating interparticle force,

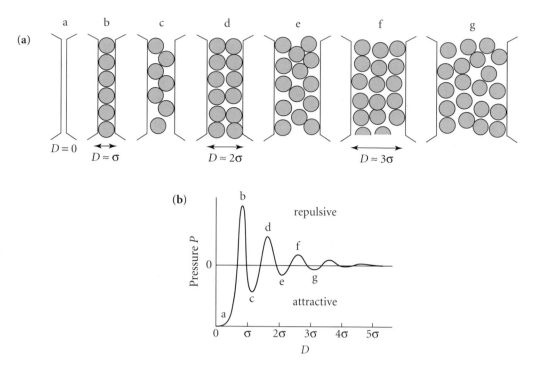

Figure 3.20. As two colloidal particles approach one another to molecular separation distances, the discrete nature of the molecules of the continuous phase begins to affect their approach. As shown in the top panel, eventually only a few layers of solvent separate the surfaces, and for each further approach, another layer of solvent must be squeezed out. Once the particles have approached just slightly closer that nR, however, where n is a small integer and R is the solvent diameter, there is more space than needed by the remaining solvent molecules, so the surfaces are drawn together until they are separated by $(n–1)R$, where it becomes even harder to squeeze out yet another layer of solvent, and so on, leading to an oscillating average force between them, as shown schematically in the bottom panel, expressed as a pressure because this effect is normally measured as the osmotic pressure needed to force the particles further together by dialysis . (Figure adapted and redrawn from J.N. Israelachvili, *Intermolecular and Surface Forces*, 3rd ed. London: Academic Press, 1992, with permission from Elsevier.)

with the magnitude of the oscillations increasing as the distance gets smaller. In practice, the surfaces of the dispersed droplets or particles are not rigid, and neither are the solvent molecules nor their overall organization. The result of this flexibility is to smear out the distance distributions, so that the hydration or solvation forces observed by experiment actually increase exponentially with decreasing distance, stabilizing the dispersion by keeping the particles apart.

DLVO Theory

In addition to hydration forces, colloidal dispersions can be stabilized toward flocculation by electrostatic repulsions if the surfaces of the dispersed particles are charged. This can happen in several ways and is actually very common. Particle

Figure 3.21. Large anions like chloride have a tendency to adsorb onto surfaces, and in some cases, as in proteins and phospholipids, the surface may carry a permanent charge. In either case, the presence of negative charge on the surface makes it more probable that cations will be in the solution layer immediately adjacent to the surface, and less likely that anions will be there, leading to a "double layer of charge" (top), with the bottom panel illustrating schematically how the population of both types of ions would then vary in the solution as a function of distance from the surface. (Figure adapted and redrawn from J.N. Israelachvili, *Intermolecular and Surface Forces*, 2nd ed. London: Academic Press, 1992.)

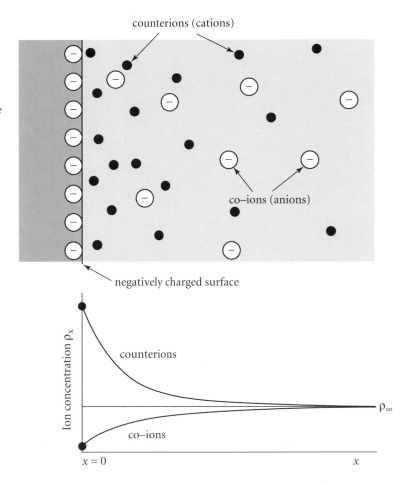

surfaces could become charged by the ionization of surface functional groups, such as the loss of a proton by a surface carboxylic acid group —COOH to give —COO⁻. Even when a particle or droplet surface is neutral and not easily ionizable, ions may be adsorbed on the interface, particularly if they are large, with a low surface charge density, as with Cl⁻. Highly charged ions like small cations, on the other hand, will be drawn down into the bulk aqueous phase, away from the surface, leading to a depletion of these ions near the surface. This depletion and enrichment with oppositely charged ions of the solution layers near a surface is referred to as the **electrical double layer** (Figure 3.21).

The distribution of counterions as a function of distance from the particle surface is governed by the Boltzmann distribution, which when combined with the Poisson equation for the relationship between charge density and electric potential gives the Poisson-Boltzmann equation connecting the electric potential, electric field, and counterion density as a function of distance from the surface. From this it is possible to calculate the pressure resisting the approach of two particles due to the repulsions arising from their electrical double layers.

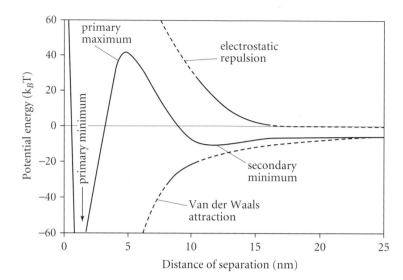

Figure 3.22. DLVO theory balances the interparticle repulsion from the electrical double layers on colloidal particles with their van der Waals attractions, mediated by the van der Waals interactions with the molecules of the intervening continuous medium. Depending on the details of both terms, this can lead to a deep primary minimum at close separations and a shallow secondary minimum at larger separations, separated by a substantial barrier imparting stability to the dispersion. (Figure adapted and redrawn from G. Narsimhan, "Emulsions," in *Physical Chemistry of Foods*, H.G. Schwartzberg and R.W. Hartel, eds. New York: Dekker, 1992.)

The repulsive interactions arising from the electrical double layers decrease with increasing distance as illustrated in Figure 3.22. These repulsions are countered by an attractive force arising from the van der Waals attractions of the atoms of the two particles for one another, also illustrated in Figure 3.22. This mathematical analysis is referred to as DLVO theory, after B.V. Derjaguin, L. Landau, E.J.W. Verwey, and J. Overbeek, the Russian and Dutch scientists who first developed it.

Depending on the relative magnitudes of the attractive and repulsive force components, the overall potential energy between the two particles as they approach can be all repulsive or can exhibit two minima at different distances: a deep primary minimum at close distance and a shallow secondary minimum at much greater distance, due to the more rapid falloff with distance of the van der Waals term. A barrier often called the primary maximum separates the two and thus provides kinetic stability to the system, preventing it from flocculating by approaching to distances close enough to realize the energy gain of being in the primary minimum. A food technologist can do little to change the van der Waals attractions between dispersed particles, but the electrical contribution is a function of environmental conditions such as the ionic strength and dielectric constant of the medium, which can be manipulated. The range of the electrical forces is expressed through a quantity called the electrical double layer thickness κ, which is a measure of the extent of the region in which the ionic concentrations are perturbed by the surface sorption of ions. For a symmetrical 1:1 electrolyte of valency v and bulk concentration n_o, κ is

$$\kappa^2 = \frac{4e^2 v^2 n_o}{\varepsilon_D \varepsilon_o k_B T} \tag{3.31}$$

The double layer thickness is actually $1/\kappa$, so this equation implies that as the dielectric constant ε_D is decreased, as when alcohol is added to water, the double layer thickness is decreased, destabilizing the dispersion. This equation would also predict that increasing the salt concentration or adding a salt of higher valency v (i.e., ± 2 or 3 rather than 1) to the continuous aqueous phase of an o/w dispersion would screen the electrostatic interactions more effectively at the same bulk concentration and thus destabilize the dispersion.

DLVO theory has been successful in explaining the behavior of some types of dispersions such as paints, and when it was first introduced, it was thought that it would also be useful in understanding the stability of food dispersions. For example, Equation (3.31) would seem to explain why adding alcohol to milk causes the casein micelles to curdle. The alcohol would lower the effective average dielectric constant ε_D of the serum. From the equation, this would increase κ^2, which would thus decrease the double layer thickness $1/\kappa$, allowing aggregation. Equation (3.31) would also predict, however, that adding salt would increase κ^2 as well, and thus also lead to coagulation, but in fact adding salt to milk has little effect on its stability toward curd formation. The reason for this failure is that in general, DLVO forces are too short range to be dominant in most food systems, where steric and hydration forces are more important. The stability of dispersions of casein micelles, which will be discussed in more detail in Chapter 5, is generally believed to be governed by steric stabilization forces, which we will now briefly consider.

Steric Stabilization

Often in food formulations, emulsifying polymers are added that impart kinetic stability to the dispersion by adsorbing onto the surface of the dispersed particles. As two such particles approach one another, their adsorbed coatings of polymers interact and impart stability to both flocculation and coalescence by several mechanisms. These processes are collectively referred to as "**steric stabilization.**" Such adsorbed polymers may stick out like tails, be closely adsorbed, usually called a train, or loop out into the solution, as is shown in Figure 3.23. Often a single molecule may do all three.

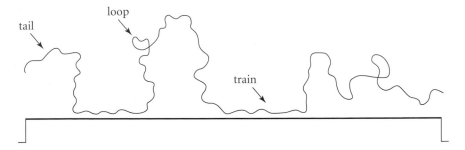

Figure 3.23. In a food dispersion, dissolved proteins or other polymers may adsorb onto the surface of the particles. These polymers may stick out like tails, be closely adsorbed, usually called a train, or loop out into the solution. Often a single molecule may do all three. (Figure adapted and redrawn from E. Dickinson and G. Stainsby, *Colloids in Foods*. London: Applied Science Publishers, 1982.)

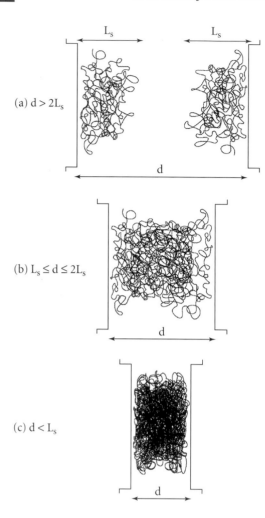

(a) $d > 2L_s$

(b) $L_s \leq d \leq 2L_s$

(c) $d < L_s$

Figure 3.24. Steric interactions from polymers adsorbed onto the surface of colloidal particles. (Figure adapted and redrawn from G. Narsimhan, "Emulsions," in *Physical Chemistry of Foods*, H.G. Schwartzberg and R.W. Hartel, eds. New York: Dekker, 1992.)

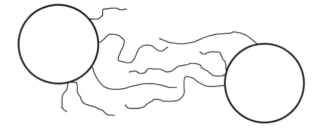

Figure 3.25. As two colloidal particles with adsorbed polymers of their surfaces approach one another, their interdigitation reduces the entropy of the chains, leading to a free energy term favoring repulsion.

Figure 3.24 illustrates how adsorbed polymers interact as two such coated particles approach one another. As the polymer coats begin to directly interpenetrate, they repel one another through two different processes. One is an entropic effect resulting from the restrictions on the configurational freedom of the chains by clashes with the other chains. The other repulsive force arises from the diffusion of water molecules into the region between the particles as a result of the increase in the local concentration of the monomer units of the polymer chains. Together these effects produce an exponential repulsive force as a function of separation distance.

Figure 3.25 shows in greater detail how the adsorbed polymer chains on the surface of two colloidal particles may interact as the approaching particles close to distances shorter than the sum of the lengths of their adsorbed polymers. The situation depicted in this illustration is thought to be representative of the structure of the casein submicelles of milk (see Chapter 5), where the arms sticking away from the submicelle are κ-casein chains. At short distances, the arms of the adsorbed polymers begin to interdigitate, as illustrated in Figure 3.25. In doing so, they begin to restrict one another's conformational freedom, reducing the number of conformational states and thus lowering the entropy of the chains, thereby increasing the free energy as the particles approach, producing a force keeping them apart. Another entropic contribution also arises as a result of this interdigitation. As the local concentration of the monomer units that make up the polymers increases as they mix together in this way, water molecules will diffuse into the space between them by osmosis, thus increasing the local osmotic pressure and again contributing a force driving the particles apart. Complex formal theoretical developments (Dickinson and Stainsby 1982) predict that both of these contributions will decrease exponentially with distance and, of course, disappear when the particles are sufficiently far apart that the sorbed chains no longer interact.

Finally, as the colloidal particles touch, a thick layer of adsorbed polymer chains can impart simple mechanical resistance to coalescence, preventing the fluid oil of the two droplets from actually touching, as in Figure 3.24.

Effects of Dissolved Polymers on Emulsion Stability

Large dissolved polymers, such as proteins in the aqueous phase of an oil-in-water emulsion, can also affect the overall force between the dispersed particles.

Figure 3.26 shows how the presence of dissolved polymers can affect the force between two dispersed colloidal particles as a function of the distance between them. When the colloidal particles are far apart, as in the top illustration, the dispersed polymers have no net effect on the forces acting between the particles. At intermediate distances, as the colloidal particles approach, they must "squeeze out" these polymers, along with their hydration shells, which can increase the force between the two particles. When the particles have approached so closely that the large polymer molecules can no longer fit in the remaining space between them, osmosis draws more water out of this region of pure water, drawing the colloidal particles even closer together.

The overall attraction of the two particles at very close separations, as the result of water being drawn out of the pool of pure water between them by osmosis, is sometimes referred to as **depletion flocculation**. The barrier to closer approach at intermediate distances due to the force needed to squeeze the polymers out is sometimes referred to as **"depletion stabilization"** (Marenduzzo et al. 2006).

Sedimentation and Creaming

If a density difference between the two liquid components of an emulsion is present, then the dispersed droplets will either sediment to the bottom or cream out, depending on the relative densities of the two phases (Figure 3.27). As the dispersed droplets encounter one another and clump together to form flocs, these larger flocs rise more quickly, sweeping out those single droplets that they encounter and overtake on the

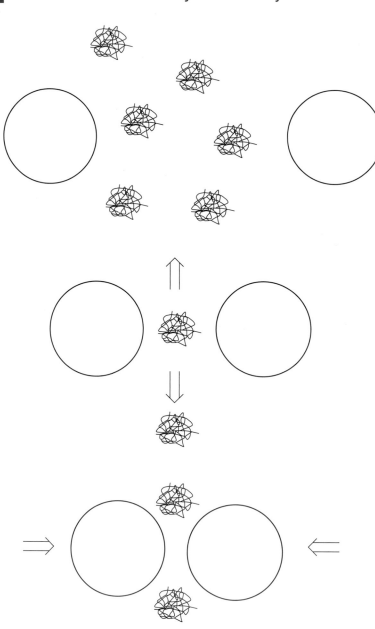

Figure 3.26. The effects of dissolved polymers on the tendency of colloidal particles to flocculate or aggregate at different distances. (Figure redrawn from G. Narsimhan, "Emulsions," in *Physical Chemistry of Foods*, H.G. Schwartzberg and R.W. Hartel, eds. New York: Dekker, 1992.)

way, which leads to a more rapid creaming rate. If the component droplets of the flocs coalesce to form larger droplets, the creaming or sedimentation rate is further enhanced, ultimately leading to a complete phase separation, with a flat interface, which is the minimum possible total surface area between the two phases. Disruption of the emulsion droplets to make smaller droplets, as in homogenization, stabilizes the emulsion by slowing down the rate of breaking (Figure 3.27).

Figure 3.27. Schematic diagram of various processes involved in emulsion instability in a gravitational field.

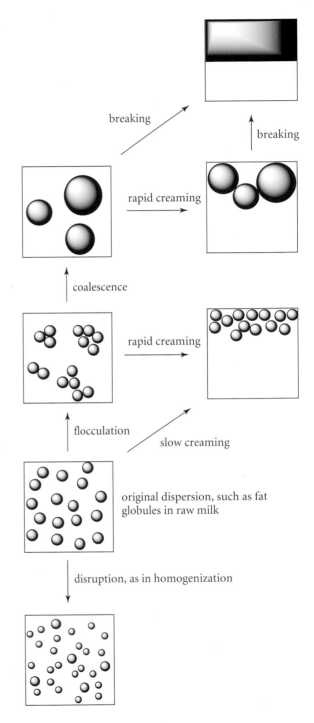

FOAMS

> The world is the ever-changing foam
> that floats on the surface of a sea of silence.
> —Rabindranath Tagore, *Fireflies*

Foams are a dispersion of a gas in a liquid. If the continuous phase subsequently hardens, the dispersion, which is substantially stabilized kinetically, is known as a solid foam. Well-known food examples include ice cream, whipped cream, meringue, the foam on beer or carbonated drinks, and foam candies. In many food examples the foam is a dispersion of air or CO_2 in water, in which case the low density of the gas molecules in the gas bubble means that no hydrogen bond partners are available for the water molecules at the air/water interface, so the gas bubble behaves in some ways as if it was a hydrophobic fluid. In nearly all foams of practical interest to us the air bubbles in the foams are much larger (mm size range) than the upper size limit that we specified in our definition of a colloidal system. These bubbles rapidly aggregate by creaming under the action of gravity. The system is still considered colloidal, however, because the thin liquid film layer between the aggregated bubbles is usually in the colloidal (micron) size range. In this foam layer, the liquid continues to drain out of the thin layers separating the bubbles, resisted by the cohesion of the liquid molecules. When enough liquid drains out, the bubbles break, and the gas therein joins with the adjacent bubbles or with the atmosphere, leading to the collapse of the foam.

Foams are formed either by mechanical means, such as by vigorous whipping or by directing a high velocity gas stream through a liquid, where the shear forces will set up turbulences that will break the stream up into small bubbles, or by nucleation of a gas bubble in situ. This second process can be the result of chemical reactions, as in the release of CO_2 from chemical leavening agents such as sodium bicarbonate ($NaHCO_3$) in weak acid, or can result from bubbles nucleating from a supersaturated solution, as when the overpressure is released from a bottle of a carbonated soft drink. The nucleation seen when this soda is poured into a glass is excess CO_2 coming out of solution and going into microscopic air pockets already present in the glass at scratches and imperfections in the surface. Ceramic boiling chips work the same way because they are riddled with microscopic pores that are filled with gas rather than liquid, due to the failure of the high-surface-tension liquid to wick into these small pores. As more and more CO_2 comes out of solution, the bubbles grow and eventually cream away, leaving the microscopic crevices still filled with gas to serve as a growth site for the next bubble, which is why you see streams of bubbles originating from the same points in your glass.

Some of the solid foams mentioned before, such as ice cream, are not really foams in the colloidal sense because they have been stabilized while the bubbles are rather isolated in the continuous phase, so that the intervening liquid layers are not colloidal in size. The bubbles in meringues and bread dough are other examples. As bread is baked, the normal thermal expansion of the gas causes the void to expand dramatically, eventually releasing the gas, while the protein and polysaccharide matrix of the dough that was surrounding the dispersed bubble becomes fixed by dehydration.

Foams cannot form from pure liquids because the fluid drains from the interfaces between bubbles too rapidly. Formation of a stable foam requires that surfactant molecules be dissolved in the liquid phase, which lower the surface tension of the interface. Good emulsifying agents that adsorb at the gas/liquid interface and inhibit coalescence help stabilize a foam. If the tail of the surfactant that extends into the liquid phase strongly binds and localizes the solvent molecules, it will significantly increase the viscosity of the fluid in the liquid boundary layers, further inhibiting drainage and stabilizing the foam. In food formulations, several proteins, including the whey protein β-lactoglobulin (when denatured) and the egg white albumin proteins ovoglobulins G2 and G3, have been found to be very effective at stabilizing foams. Presumably when these proteins denature they adsorb onto the interface with their hydrophobic regions sticking out into the air bubble, which acts like a hydrophobic surface, and with the hydrophilic portions of their chains sticking out into the water, binding the water and raising the local viscosity. How well a particular protein functions in this capacity might depend more on the distribution of residues in the denatured chain than on the surface features of the folded globular chain. Some denatured proteins can form quite durable films on bubble surfaces if they become cross-linked either by strong noncovalent entanglement or by chemical cross-linking, perhaps by intermolecular disulfide linkages between different chains. Such cross-linking might explain the effectiveness of β-lactoglobulin, which contains one buried free cysteine residue in each chain, which could cross-link if the globular proteins denature on the bubble surface (such disulfide bonds will be discussed further in Chapter 5). Foams can also be stabilized by adding thickening agents to the aqueous phase, such as polysaccharides, which can significantly increase the viscosity of the resulting solution.

After a dispersion of bubbles has been formed, the bubbles rapidly cream to the top of the fluid under the force of gravity, creating a thick foam layer. A major source of instability of foams in this layer, and even while they are being formed by whipping, is Ostwald ripening. As we saw previously in our discussion of the Kelvin equation, small bubbles have a higher vapor pressure than larger ones, leading to a tendency for small bubbles to shrink further in size as gas molecules escape and join larger ones, which thus increase in size. This process is particularly favored in foams in water because many gases are soluble in water. Ostwald ripening leads to a narrowing of the size distribution as all of the bubbles quickly become fairly similar in size. When two bubbles come in contact, the increase in stress in the adjacent water molecules leads to a flattening of the interface. Where three bubbles meet, a small prism of water will form between the highly curved boundaries which connect the newly flat interfaces between the bubbles. As the bubbles accumulate and acquire new neighbors, they assume polyhedral rather than round geometries.

A simple experiment that you can do in your kitchen the next time you are washing dishes illustrates some of the behavior of foams. Dish soap is made up of surfactant lipid or phospholipid molecules which adsorb onto the lamellar water interfaces between the air bubbles which are created by vigorous scrubbing or by the vortices resulting from the water rushing turbulently out of the tap. Depending on the type of dish soap being used, this foam can be stable for quite a while and eventually breaks as much from evaporation of the water molecules as from mechanical drainage. If you add a few drops of rubbing alcohol (isopropyl alcohol) to the continuous aqueous phase in the lamellae between the bubbles, simply by sprinkling it over the foam, you would observe a rapid collapse of the foam. This collapse is caused by the alcohol diffusing into the aqueous layer and changing the solvent quality of the liquid, making the surfactants more soluble in the fluid and thus less surface active, destabilizing the foam.

SUGGESTED READING

Dickinson, E., and G. Stainsby. 1982. *Colloids in Foods*. London: Applied Science Publishers.

Friberg, S., ed. 1976. *Food Emulsions*. New York: Marcel Dekker.

Hiemenz, P.C. 1986. *Principles of Colloid and Surface Chemistry*, 2nd ed. New York: Dekker.

Israelachvili, J.N. 1992. *Intermolecular and Surface Forces*, 2nd ed. London: Academic Press.

Marenduzzo, D., K. Finan, and P.R. Cook. 2006. "The depletion attraction: An underappreciated force driving cellular organization." *J. Cell Biol.* 175:681–686.

Mulder, H., and P. Walstra. 1974. *The Milk Fat Globule*. Farnham Royal, Bucks, England: Commonwealth Agricultural Bureaux.

Schwartzberg, H.G., and R.W. Hartel, eds. 1992. *Physical Chemistry of Foods*. New York: Dekker.

Walstra, P. 2003. *Physical Chemistry of Foods*. New York: Marcel Dekker.

Walstra, P., and R. Jenness. 1984. *Dairy Chemistry and Physics*. New York: Wiley-Interscience.

4.
Carbohydrates

CARBOHYDRATES IN FOODS

Carbohydrates are the most abundant class of biological molecules; in fact, there is more by weight of just one type of carbohydrate, cellulose, than of all other biomolecules combined. Cellulose is so abundant because it is the major component of plant cell walls. Other carbohydrates, such as the chitin of insect exoskeletons, also have structural roles. As the end product of photosynthesis, carbohydrates are even more important for energy storage and use, which is of course the reason that they are eaten as food. However, carbohydrates have many other roles in manufactured foods, such as imparting flavor, controlling water, and determining texture, among others. Carbohydrates occur most commonly as polysaccharides, but are also widely found in foods as monomers and dimers called sugars. Tables 4.1 and 4.2 give the sugar contents of various foods, including fruits and vegetables.

Table 4.1 Sugar (sucrose) content of various processed foods

Food	Sugar content %
Coke	9
cracker	12
ice cream	18
ready-to-eat cereals (dry)	1–50
orange juice	10
ketchup	29
cake (dry mix)	36
nondairy creamer (dry)	65
Jell-O (dry)	83

Source: R.L. Whistler and J.R. Daniel in *Food Chemistry*, 2nd ed., O.R. Fennema, ed. (New York: Marcel Dekker, 1985).

Table 4.2 Free sugar content of fruits and vegetables (% fresh weight basis)

Fruit or vegetable	D-Glucose	D-Fructose	Sucrose
apple	1.2	6.0	3.8
grape	6.9	7.8	2.3
peach	0.9	1.2	6.9
pear	1.0	6.8	1.6
cherry	6.5	7.4	0.2
strawberry	2.1	2.4	1.0
beet	0.2	0.2	6.1
broccoli	0.7	0.7	0.4
carrot	0.9	0.9	4.2
cucumber	0.9	0.9	0.1
onion	2.1	1.1	0.9
spinach	0.1	0.04	0.1
sweet corn	0.3	0.3	3.0
sweet potato	0.3	0.3	3.4
tomato	1.1	1.3	0.01
snap beans	1.1	1.2	0.3
peas	0.3	0.2	5.3

Source: R.L. Whistler and J.R. Daniel in *Food Chemistry*, 2nd ed., O.R. Fennema, ed. (New York: Marcel Dekker, 1985).

CARBOHYDRATE STRUCTURES

Carbohydrates are polyhydroxy aldehydes and ketones, many of which have the general formula $C_x(H_2O)_y$,

$$HOCH_2-(CHOH)_n-CHO \quad \text{and} \quad HOCH_2-(CHOH)_{n-1}-\overset{\displaystyle O}{\overset{\displaystyle \|}{C}}-CH_2OH$$

aldoses   **ketoses**

Both types of molecules contain a $-\overset{\displaystyle H}{\underset{\displaystyle OH}{C}}-\overset{\displaystyle O}{\overset{\displaystyle \|}{C}}-$ group, called a **saccharose group**.

Because these molecules contain only one such moiety, they are referred to as **monosaccharides** (Figure 4.1). In the systematic nomenclature for sugars, the suffix "ose" indicates a monosaccharide sugar aldehyde. To distinguish them from the aldoses, the ketose sugars are named with the suffix "ulose" (e.g., levulose is another name for D-fructose). The term *sugar* most commonly means sucrose (table sugar), but is also widely used to mean any of the low-molecular-weight carbohydrates, such as glucose, galactose, and maltose. The term *saccharide* is almost synonymous with the broader meaning of sugar, as is the term *glycose*.

Figure 4.1. Fischer projection formulas for the prototypical aldose sugar D-glucose and for the prototypical ketose D-fructose.

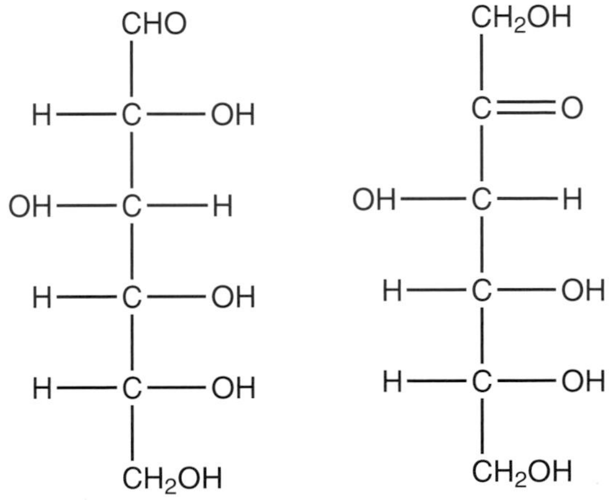

D-glucose (+)
positive rotation of
plane polarized light

D-fructose (-)
negative rotation of
plane polarized light

An important feature of the sugar molecules is the large number of asymmetric carbon atoms that they contain; the simple monosaccharides are all essentially stereoisomers of one another that nevertheless have distinct biological properties. Because of the large number of hydroxyl groups in sugars, hydrogen bonding is an important feature of their interactions with other molecules, and in particular is important in their interactions with water. Many sugars bind considerable amounts of water as a result of their extensive hydrogen bonding capacity, and most are very soluble in water (see Table 4.7). Glucose and its polymers are important energy storage molecules because their C—C and C—H bonds contain considerable chemical energy that can be released upon oxidation. Because of their large number of hydroxyl groups, the sugars are already partially oxidized, reducing their energy content relative to that of the fatty acid chains of lipids, which are mostly oily hydrocarbons, but imparting their water solubility and allowing them to serve as circulating energy molecules in the blood.

Perhaps the most typical monosaccharide sugar is D-glucose, also called blood sugar in medicine and **dextrose** in the food industry, which is produced as the endpoint of photosynthesis in the chloroplasts of green plants using light energy captured by chlorophyll,

$$6CO_2 \text{ (g)} + 6H_2O \text{ (l)} \xrightarrow{\;hv\;} C_6H_{12}O_6 \text{ (s)} + 6O_2 \text{ (g)} \quad \Delta H° = 2816 \text{ kJ}$$

Open-chain hydroxy aldehydes are highly reactive. They readily enolize in alkaline solution (see below) to reduce the charge of ions such as Cu^{++} (cupric) or $Fe(CN)_6^{3-}$ (ferric, Fe^{3+}) and so are called "**reducing sugars**." In the process, the aldehyde group is oxidized to an acid.

$$\underset{\text{blue colored}}{R\text{-}\overset{\overset{\displaystyle H}{|}}{C}=O \;+\; Cu^{2+}SO_4^{2-}} \xrightarrow{\;OH^-\;} RCOOH \;+\; \underset{\text{reddish or orange precipitate}}{Cu^+ \;\; (\text{i.e., } CuO \text{ or } Cu_2O \downarrow)}$$

Ketose sugars like fructose are also reducing sugars, even though they do not contain an aldehyde group, because under basic conditions they also enolize to produce a reactive enediol ion (discussed below). Thus, if the aldehyde or ketone group of a carbohydrate is free, it is called a reducing sugar. The reagent copper sulfate $(CuSO_4)$, when used as a test for the presence of reducing sugars, is called **Benedict's reagent** or Benedict's solution. This test is useful in food science because reducing sugars can lead to Maillard browning when proteins are present. Quantitating the amount of reducing equivalents in a polysaccharide sample is useful in characterizing the structure of the polymer in terms of length.

Many of the common food carbohydrates were discovered long ago and given trivial names, often indicating the primary natural source. These terms convey no structural information and are not consistent with the systematic nomenclature, but are too well-established to be abandoned. Some examples are:

dextrose	=	glucose
fructose		fruit sugar
levulose		levorotatory sugar = fructose
maltose		malt sugar
lactose		milk sugar
xylose		wood sugar
cellulose		from cellular membranes of plants

The commonly occurring monosaccharides contain varying numbers of carbon atoms, with five- and six-carbon sugars being the most common in food systems. These are referred to as **pentoses** and **hexoses**, respectively. A number of different types of sugars containing the same number of carbon atoms occur, and these are all structural or configurational isomers of one another. Within a given class of sugar (aldose or ketose), each stereoisomer for each carbon number represents a molecule with distinct physical and biological properties and is thus given a distinct individual name. Only a handful of these sugars are commercially important, although some of the rare sugars are being studied as possible low-calorie sweeteners or products for diabetics. A few of the more uncommon monosaccharides or their derivatives occur in polysaccharides widely used as ingredients in the food industry.

All of the aldoses can in principle be synthesized from glyceraldehyde (Figure 4.2) by the successive application of the **Kiliani-Fischer synthesis**. In a sense, glyceraldehyde can be thought of as the simplest aldose carbohydrate, a three-carbon saccharide containing one saccharose group and one asymmetric carbon atom. Two stereoisomers of this molecule are possible, one of which rotates plane-polarized light in the positive direction (+), and one of which rotates it in the (−) direction, designated R or S. For historical reasons, these are also designated D and L, with the two isomers being nonsuperimposable mirror images of one another. (Recall that when using the Fischer projection–type formulas in the bottom panel of Figure 4.2 that the drawing cannot be rotated without changing the stereochemistry being represented, as can be appreciated from the perspective drawing in the panel directly above the Fischer projection.)

The Kiliani-Fischer synthesis, which involves adding cyanide in the presence of acid, increases the length of the carbon backbone of the saccharide by adding one carbon to the reactive aldehyde end of the sugar (Figure 4.3). Note that, while this process can be repeated to add additional carbon atoms, none of these transformations affects the stereochemical configuration at the already existing asymmetric carbon atoms. Because of this, the stereochemistry of the highest-numbered asymmetric carbon atom can always be related back to that of the original glyceraldehyde molecule at the beginning of the chain of syntheses. For this reason, all aldose sugars are named D or L depending on whether the stereochemistry of this highest-numbered asymmetric carbon atom is the same as that of D- or L-glyceraldehyde.

The Kiliani-Fischer synthesis is never actually used in the practical production of simple sugar molecules, which in food science are obtained from natural sources.

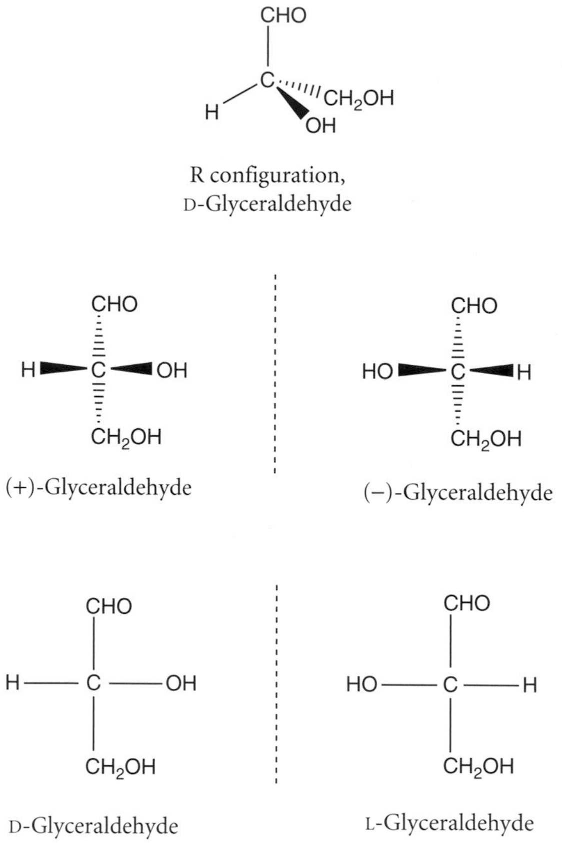

Figure 4.2. The stereochemistry of glyceraldehyde, the simplest aldose carbohydrate.

R configuration,
D-Glyceraldehyde

(+)-Glyceraldehyde

(−)-Glyceraldehyde

D-Glyceraldehyde

L-Glyceraldehyde

Rather, it is introduced here as a way of illustrating the conceptual relationship of the different sugar molecules to one another and as a way of explaining the approach to the nomenclature of the simple saccharides. In the IUPAC system of naming organic molecules, the aldehyde group carbon would be assigned the highest priority and be labeled as carbon number 1, even though after each successive step of the Kiliani-Fischer synthesis this aldehyde carbon is the new carbon from the cyanide reagent, as could be proven through isotope labeling.

The D-Aldoses

The D-series of aldose sugars, derived from D-glyceraldehyde is shown in Figure 4.4. Notice that all of these molecules have the same configuration at the highest-numbered asymmetric carbon atom. An exactly analogous series is derived from the L-glyceraldehyde molecule. Each L-sugar is the mirror image of the corresponding D-sugar. Note that switching the configuration of only the highest-numbered asymmetric carbon produces an L sugar, but not the L isomer of the original sugar; that conversion requires changing the configuration of *all* of the chiral carbon atoms to give the mirror image compound.

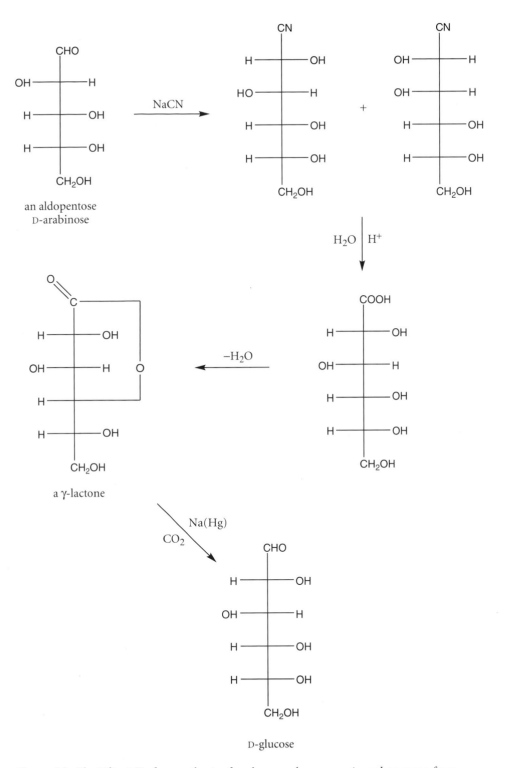

Figure 4.3. The Kiliani-Fischer synthesis of D-glucose, a hexose, or six-carbon sugar, from D-arabinose, a pentose. Because the first step produces a racemic mixture of isomers at the new C2 position, D-glucose would not be the only product, and the difficult separation of the isomers would be necessary to yield only D-glucose.

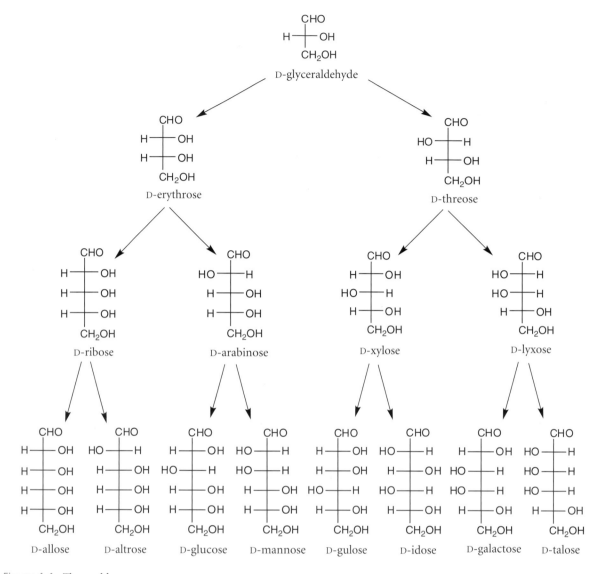

Figure 4.4. The D-aldose sugars.

D-Allose, D-altrose, D-idose, and D-talose are rare in foods, and only D-glucose and D-galactose are at all common. D-Glucose is widely distributed in foods, but D-galactose is found principally as one of the two monosaccharides that make up milk sugar (lactose), which is a dimer, or disaccharide consisting of D-glucose and D-galactose. Remember that within each carbon-number class, the aldose sugars are all stereoisomers of one another. As we have already noted, they nonetheless have differing biological, and even physical, properties. For example, the hexose sugar D-mannose is 2.8 times more soluble in water than D-glucose and 56 times more soluble than D-galactose.

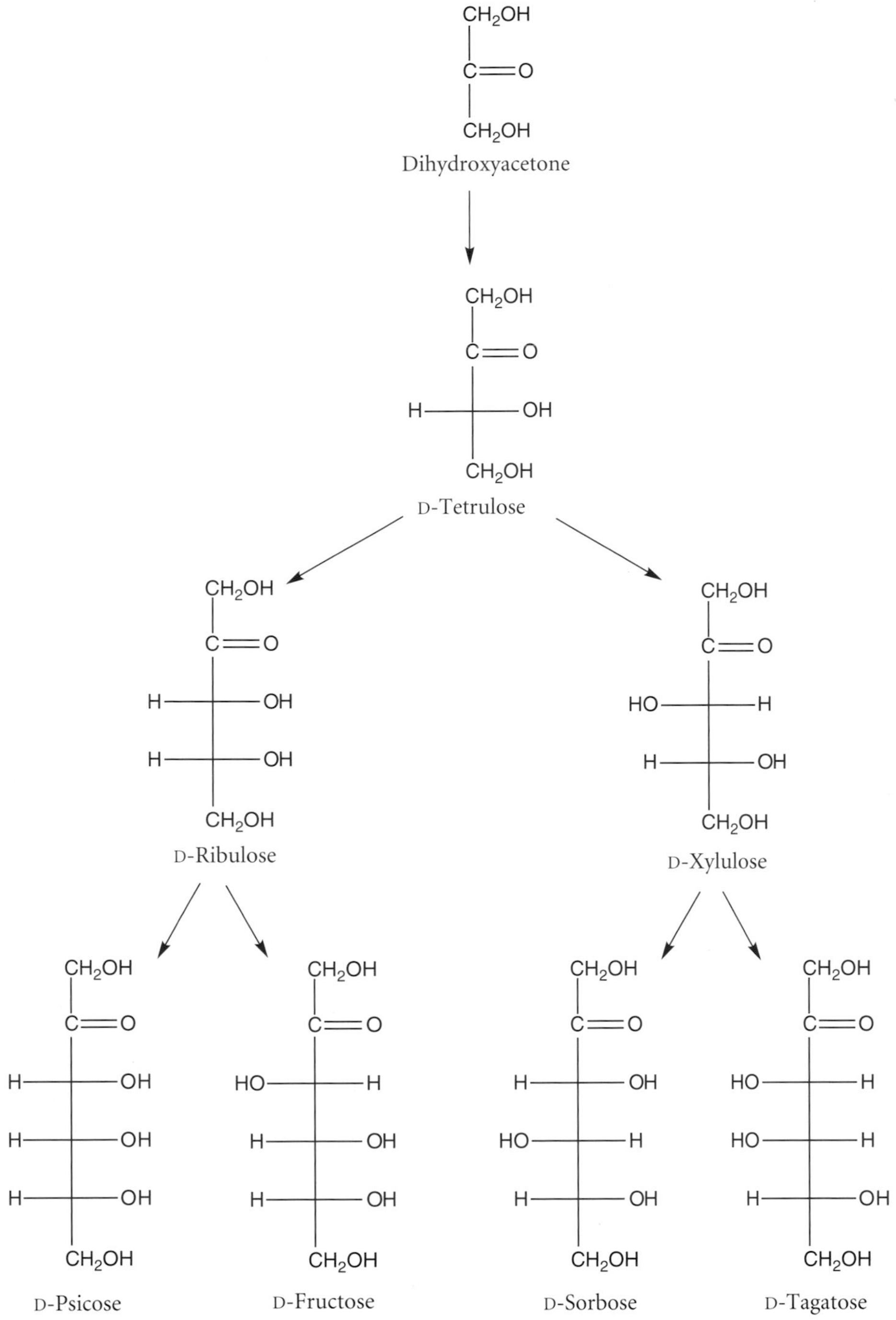

Figure 4.5. The D-ketose sugars.

The D-Ketoses

The D-ketoses (Figure 4.5) can be thought of as related to dihydroxyacetone in a manner analogous to the way the D-aldoses are related to D-glyceraldehyde. Note that dihydroxyacetone, the three-carbon parent molecule, a structural isomer of glyceraldehyde ($C_3H_6O_3$), does not contain an asymmetric carbon atom. Therefore, in the ketose series, the molecules are designated D or L with reference to the tetrulose. Because the ketoses contain one fewer asymmetric carbon atom, correspondingly fewer stereoisomers exist.

Systematic Nomenclature

The systematic names of the aldoses are formed by adding tetrose, pentose, or hexose to the trivial prefixes given in Figure 4.6, separated by a hyphen. Thus, D-arabinose is more properly "D-arabino-pentose" and D-glucose is "D-gluco-hexose." These systematic names are almost never used in the food chemistry literature and can even cause confusion. Their principal value is in allowing precise specification of the structures of synthetic sugar derivatives that have little food relevance.

In naming the 2-ketoses (glyculoses), the "X" in Figure 4.6 is $\mathrm{-\!\!\overset{\displaystyle O}{\overset{\|}{C}}\!\!-CH_2OH}$, instead of CHO, so that D-fructose, is D-*arabino*-hexulose.

$$
\begin{array}{c}
CH_2OH \\
| \\
C=O \\
HO\!\!-\!\!|\!\!-\!\!H \\
H\!\!-\!\!|\!\!-\!\!OH \\
H\!\!-\!\!|\!\!-\!\!OH \\
| \\
CH_2OH
\end{array}
$$

Sugars that differ in configuration at a single asymmetric carbon other than C1 are called **epimers** (as we shall soon see, while the C1 carbon is achiral in the open chain form, it is chiral in the more common ring structures). For example, D-mannose is the C2 epimer of D-glucose, and D-galactose is its C4 epimer (Figure 4.7). Note that Figure 4.7 uses a shorthand form of the Fischer projection representation where hydroxyl groups are represented by a line on the right or left of each chiral carbon, while protons are omitted.

Configurational isomers are referred to collectively as **stereoisomers**. If two stereoisomers are related as an object and its mirror image, they are called **enantiomers**. Stereoisomers not related in this way are called **diastereomers**.

In Figure 4.8, A and B are enantiomers—mirror images, while A and C are diastereomers. For the sugars, the D and L prefixes refer to the asymmetric carbon atom furthest removed from the carbonyl atom. L-Aldoses and ketoses are mirror images of their D counterparts, as illustrated in Figure 4.9.

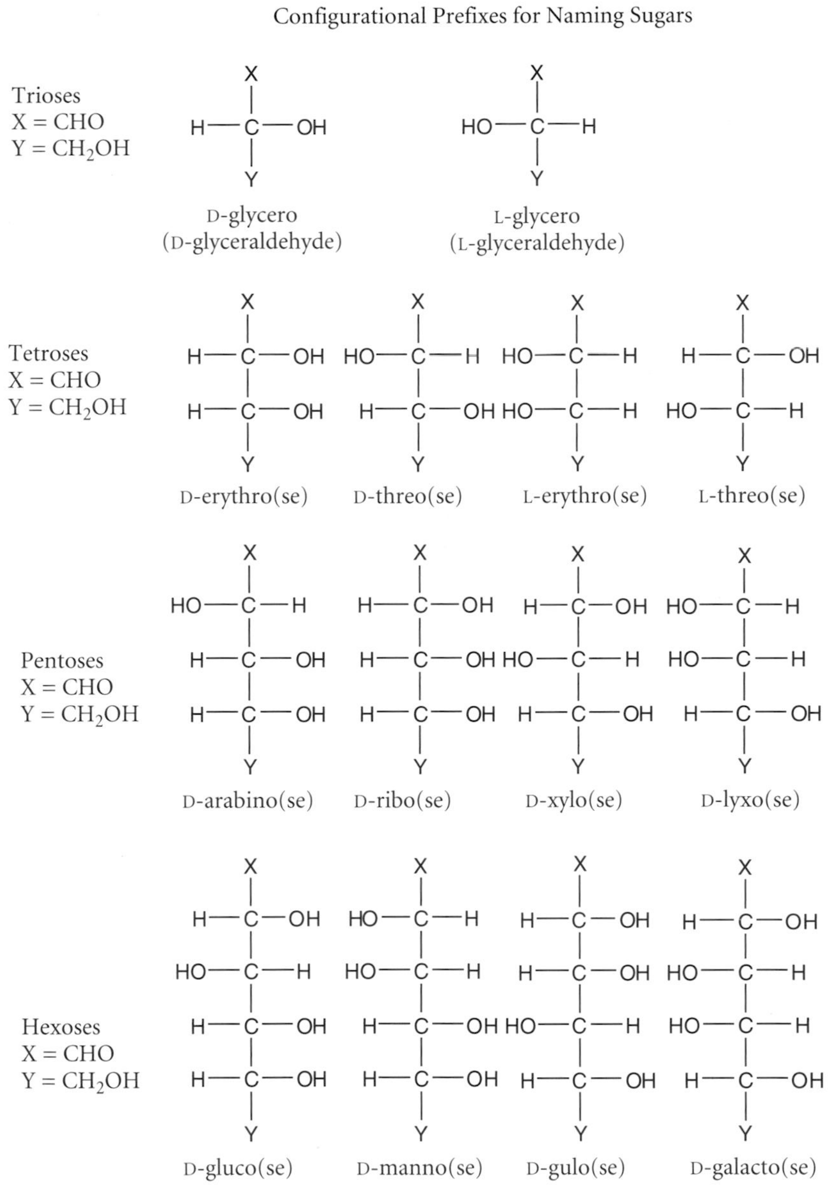

Figure 4.6. The prefixes used in naming the aldoses in the systematic nomenclature.

Configurational Prefixes for Naming Sugars

Trioses
X = CHO
Y = CH$_2$OH

Tetroses
X = CHO
Y = CH$_2$OH

Pentoses
X = CHO
Y = CH$_2$OH

Hexoses
X = CHO
Y = CH$_2$OH

Carbohydrate Rings

The open-chain aldehyde and ketone forms of carbohydrates are highly reactive. The aldehyde group will readily add water or simple alcohols to form **hemiacetals** (an acetal would have a second —OR′ group instead of an —OH), as in Figure 4.10. For example, they can add water to produce a hydrate (Figure 4.11),

In most cases, the hemiacetals are not particularly stable, and the equilibrium lies far to the left. Sugars undergo this kind of reaction with alcohols; the aldehyde group

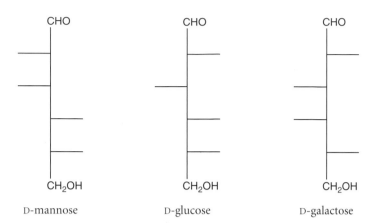

Figure 4.7. D-Mannose and D-galactose, the C2 and C4 epimers of D-glucose.

D-mannose   D-glucose   D-galactose

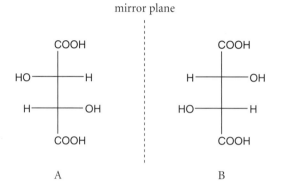

Figure 4.8. The three forms of tartaric acid.

A   B   C

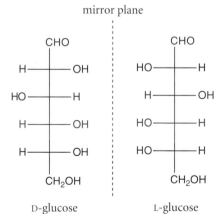

Figure 4.9. The mirror-image relationship between D and L sugars.

D-glucose   L-glucose

can also fold around and can react with the alcohol groups on carbon atoms #4 or 5 of its own chain to make a ring, which unlike most hemiacetals, is more stable than the aldehyde form (Figure 4.12). Ketone sugars can also undergo such cyclization.

If the aldehyde reacts with the C4 hydroxyl group to form a five-membered ring, it is called a **furanose**, because of the vague similarity to the structure of furan (Figure 4.13). If it reacts with the C5 hydroxyl group to form a six-membered ring, it is called

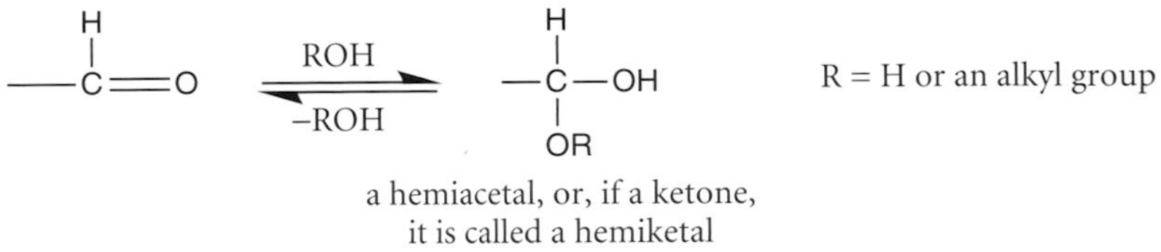

Figure 4.10. The reaction of aldehydes with alcohols to form hemiacetals.

a hemiacetal, or, if a ketone, it is called a hemiketal

R = H or an alkyl group

Figure 4.11. The combination of an aldehyde with water to form a hemihydrate.

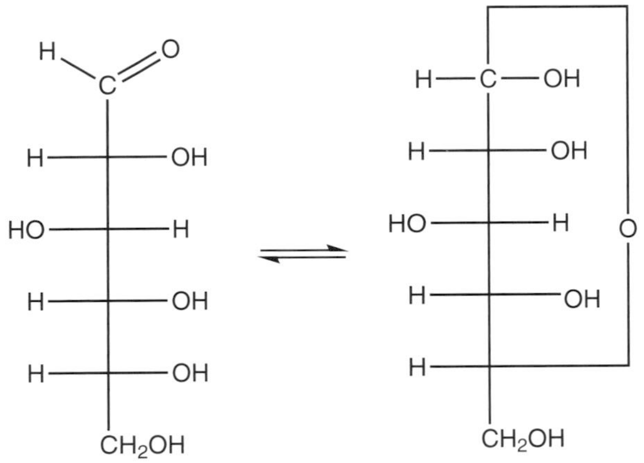

Figure 4.13. On the left, furan, and on the right, pyran, used as templates for naming sugar rings.

Figure 4.12. Fischer projection representations of the equilibrium between open chain and cyclic D-glucose.

a **pyranose**, after the resemblance to pyran (note, however, that the sugar rings do not contain any double bonds):

The cyclization equilibria lie far to the right, but because the reaction is reversible, the ring forms are still considered to be reducing sugars, since the free carbonyl group is occasionally regenerated and is thus temporarily available to reduce ions.

The closure of the ring generates a new asymmetric carbon at C1, which is chemically different from all of the other carbons in the ring, because it is bonded to two highly electronegative oxygen atoms instead of one. This carbon is called the **anomeric carbon**, and additional rules of nomenclature are necessary to specify the two possible configurations of the new OH group at this carbon. The two forms are given the designations α and β. The α and β forms are really epimers, but because of the special chemical characteristics of the hemiacetals, which are not aldehyde alcohols, they are referred to as **anomers**. In the Fischer projection formulas, if the anomeric hydroxyl is on the same side as the D/L descriptor hydroxyl, it is called the α-anomer, and β if it is on the opposite side.

The Fischer projection formulas convey little sense of actual spatial relationships in the cyclic sugar structures, so alternate representational conventions were developed to depict sugar rings. The first of these was the still widely used Haworth projection, which represents the sugar ring as a planar hexagon viewed at an oblique angle. The principal advantage of these formulas is that they are relatively easy to draw quickly and consistently. In the Haworth perspective formulas, the rule for specifying anomeric configurations is the following: orient D-series sugar rings so that, when looked at from above, the numbering of the carbon atoms increases in the clockwise direction. By convention these rings are usually drawn tilted slightly down with the C2 and C3 carbon atoms in front and with the ring oxygen at the two o'clock position. From this perspective, the α-anomer has the C1 hydroxyl group below the ring, and the β anomer has it above the ring. For this orientation, the pyranose ring forms of D sugars will have the primary alcohol substituent on the C5 atom above the ring, while L sugars will have it below the plane of the ring. In the complete nomenclature for rings, the "ose" suffix is dropped, and replaced by the ring type identifier, either furanose or pyranose, so that the complete name for the molecule in Figure 4.14 is **α-D-glucopyranose**.

The principal drawback of the Haworth projection formulas is that they still substantially distort the true spatial arrangement of the substituent groups, due to the simplified representation of the sugar geometry. Sugar rings are not hexagons, and the sp^3-hybridized carbon atoms do not have C—C—C angles of 120°, and H—C—OH angles are clearly not 180°, as implied in the drawings. The most useful representations are conformational drawings, which represent the rings as a chaise-lounge-type chair (or boat or twist boat) as is done for cyclohexane. These projections are attempts to depict the true spatial relationships in the molecule in a realistic, three-dimensional manner (Figure 4.15). Their principal drawback is that they are harder to draw correctly.

To convert from the Haworth formulas to conformational formulas, orient the conformational perspective ring such that the ring oxygen is in the same position as in the Haworth formula. Then, those substituents that are up in the Haworth formula will be up in the conformational formula, and those that are down in the Haworth drawing will be down in the perspective drawing.

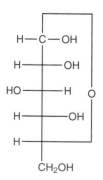
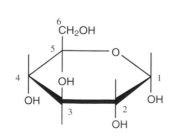

Figure 4.14. On the left is the Fischer projection formula for α-D-glucopyranose, and on the right is the Haworth formula for the same molecule.

Figure 4.15. Left, conformational formula representation for α-D-glucopyranose; right, computer generated "licorice" bond representation of the same molecule.

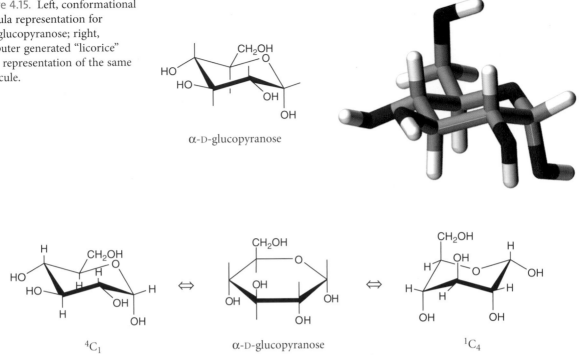

α-D-glucopyranose

$^4C_1$ α-D-glucopyranose $^1C_4$

Figure 4.16. There are two possible nonequivalent chair conformations for α-D-glucopyranose, shown on the left and right, as well as many boat and twist-boat conformations. The Haworth formula, however, shown in the center, cannot represent these conformational differences.

In the perspective formulas, two distinct types of substituent positions should be noticed, those that are in so-called axial positions, perpendicular to the approximate average ring plane, and equatorial positions, which are approximately in the average plane of the ring, pointing only slightly up or slightly down. When one makes this distinction, it can be seen that there are actually two different ring chair conformations possible, as illustrated in Figure 4.16, which can be distinguished with the conformational perspective formulas, but which both are represented by the same Haworth drawing (the names of these forms are explained below). In this book, both conformational and Haworth representations will be used.

The five-membered furanose rings are represented with Haworth formulas by a planar pentagon, generally drawn with the ring oxygen position in the back (Figure 4.17).

As was the case with the Haworth representation of the pyranose rings, this formula distorts the true spatial relationships in the sugar molecule, but in the case of the furanose rings, this distortion is not as great, because these rings tend to be more flat than the six-membered rings. As with the pyranose ring formulas, the α-D-anomers have the anomeric hydroxyl group below the plane of the ring, and the β anomers have it above the ring plane; the D-sugars again have the highest-numbered exocyclic hydroxymethyl group up, and L sugars have this group in the down position.

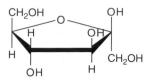

β-D-fructofuranose

Figure 4.17. The Haworth representation of the furanose ring form of D-fructose.

Figure 4.18. A conformational representation of a pyranose ring in a boat conformation.

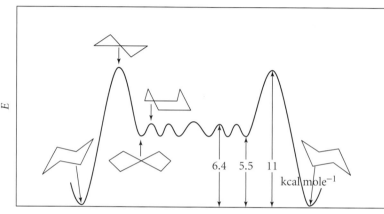

E

6.4   5.5   11

kcal mole^{-1}

Conformation coordinate

Figure 4.19. Schematic conformational energy profile for cyclohexane. (Figure adapted and redrawn from J.F. Stoddart, *Stereochemistry of Carbohydrates*. New York: Wiley-Interscience, 1971.)

Other conformations for pyranose rings are also possible besides the two chair forms. As with cyclohexane, the six-membered sugar rings can also adopt various "boat" conformations, as shown in Figure 4.18. Actually, perfect boats of this type are less stable, and the rings are usually somewhat twisted; such conformations are referred to as twist-boat conformations.

The different ring conformations do not all have the same energy, and generally one form is sufficiently lower in energy than all of the others so as to be overwhelmingly predominant. The energy differences are not so great that the conformations cannot interconvert, however, and these conformational changes can have practical consequences in food systems. One case where such ring conformational changes could have effects on physical properties is in polymer conformations, which will be discussed later. In general, chair conformations are usually lower in energy than boats, as is illustrated schematically in Figure 4.19. This drawing shows the change in energy of cyclohexane as it moves along some generalized reaction coordinate that converts it from a chair to a high-energy transition state, then to a twist-boat, and on to a boat conformation, which can also be seen to be a less preferred form relative to the twist conformations. In the symmetric cyclohexane, several equivalent twist-boat conformations and several equivalent boat conformers occur, as well as two

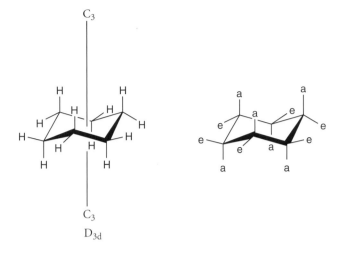

Figure 4.20. The conformational structure of cyclohexane, showing the axial and equatorial substituents.

C_3

C_3

D_{3d}

Figure 4.21. The equatorial substituents of one chair form of a pyranose ring are converted into the axial substituents of the opposite chair conformation.

equivalent chairs of equal energy. In the asymmetric sugars, these forms will all be different and will all have different energies.

In cyclohexane (Figure 4.20), the axial and equatorial positions (labeled *a* and *e* on the right in the figure) are equivalent, but this is not the case in the asymmetric sugar rings.

The interconversion of one of the two possible pyranose ring conformers to the other, for example, converts all of the axial substituents to equatorial positions, and all of those that were equatorial to axial positions, as is illustrated in Figure 4.21.

Because the various possible ring conformations can have different effects on the physical and chemical behavior of the carbohydrates, a system for specifying the ring conformation in the systematic nomenclature is necessary.

The rules for specifying pyranose ring conformations are the following: Chair conformations are designated with a C, boat conformations are designated with a B, and twist-boat conformations are indicated with an S (because they are also called skew-boats, to distinguish them from the twist conformations of five atom rings, described below, which had already been assigned the designation "T"). The ring atoms are numbered with the oxygen as zero, and the anomeric carbon as number 1, etc., and the ring is oriented such that the carbon atom numbers increase in the clockwise direction when viewed from above (Figure 4.22). A reference plane is chosen containing four of the ring atoms, and those ring atoms that lie above this plane are then written as a superscript to the left of the letter designating the conformation type, while those ring atoms lying below the plane are written as a subscript

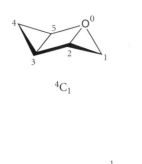

$^4C_1$

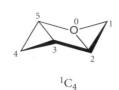

$^1C_4$

$^{2,5}B$

Figure 4.22. Examples of pyranose ring conformational designations.

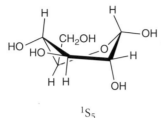

$^1S_5$ $^1S_5$

Figure 4.23. A skew-boat conformation for α-D-glucopyranose.

to the right. For the boat conformations there will only be one possible plane that contains four ring atoms, while for the chair and twist-boat conformers, there is an ambiguity, because there is more than one possible plane of this type. In those cases, the plane is chosen such that the lowest-numbered carbon atom is displaced from the ring. This selection of reference plane can be particularly difficult to see for skew boat conformations, especially without using physical models. Figure 4.23 illustrates an example of a skew boat conformation for α-D-glucopyranose. The reference plane is the one containing the C2, C3, C4, and O atoms, because the other possible reference plane would be C5, O, C1, C3, but it must be excluded since the rule says that, of the two, the one that does not have the lowest-numbered carbon atom in it should be chosen. In this example, the anomeric (C1) carbon, the lowest-numbered carbon atom, is above the plane of the ring, and the C5 carbon is below the plane of the ring, so the conformation is $^1S_5$.

Five-membered furanose rings also can adopt a number of different conformations, although in general only two overall conformer types exist; envelope forms in which one atom is displaced up or down from the plane occupied by the other four, or twist forms in which three consecutive ring atoms lie in a plane, while one of the other two atoms is displaced above the ring and one is displaced below the ring. The rules of nomenclature for these five-membered rings are similar to those for the pyranose rings, with the symbols T and E representing the twist and envelope forms, respectively, and the "up" and "down" directions again defined such that the ring atom numbering is increasing in the clockwise direction when viewed from above. The reference plane in the envelope case is obvious, and in the twist forms is the one containing the three consecutive coplanar ring atoms. As in the pyranose cases, atoms displaced above the ring are written as superscripts to the left, and those displaced below the ring are written as subscripts to the right (Figure 4.24).

Examples of the application of these nomenclature rules to two different envelope conformations are given in Figure 4.25.

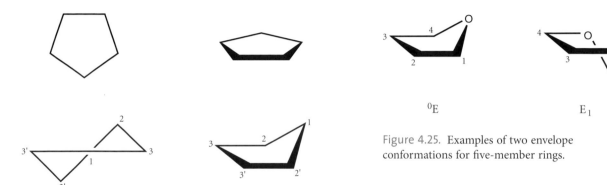

Figure 4.25. Examples of two envelope conformations for five-member rings.

Figure 4.24. Five-member rings have essentially two nonplanar conformations: the envelope, shown on the right, and a twist form, shown on the left.

The Relationship between D- and L-Sugars

The L-sugars are the mirror images of the D forms. In the Fischer projection formulas, converting from one to the other involves switching the direction (right/left) of each hydroxyl in the formula. In the Haworth projection formulas, one swaps the up/down direction of all of the substituent groups, including the exocyclic carboxymethyl (C6) group. For a Haworth formula, in the standard orientation, the highest-numbered carbon atom outside the ring projects upward in the D series and downward in the L-series. Although the conformational perspective drawings convey the most realistic representation of sugar structures, in the food science literature the Haworth formulas are probably still more common, due to the ease and speed with which they can be drawn. For this reason, familiarity with both types of representations and how to convert from one to the other is important (Figure 4.26).

Naturally occurring L-series sugars are uncommon and thus are rarely found in typical foods. The majority of the enzymes that transport, hydrolyze, or process carbohydrates will not recognize L-sugars (that is, they will not bind to them). This characteristic can sometimes be exploited in the design of low-calorie foods, because a carbohydrate fraction composed of L-sugars might be noncaloric. Since some of the L-sugars are sweet, they could potentially be used as noncaloric sweeteners. An important limitation on such uses would be their natural scarcity, which could make their use uneconomical.

Tautomerization

Even though their aldehyde groups are not free, the monosaccharides are still referred to as reducing sugars because they exist in equilibrium with a small amount of their open-chain forms, as well as with all of the other possible ring forms for that molecule

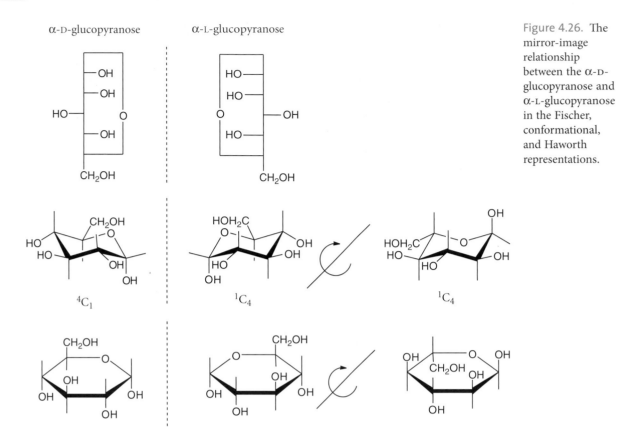

Figure 4.26. The mirror-image relationship between the α-D-glucopyranose and α-L-glucopyranose in the Fischer, conformational, and Haworth representations.

(see Figure 4.27). All of these interconverting forms are called **tautomers**, and the dynamic equilibrium is referred to as the **tautomeric equilibrium**. In the crystalline form, each of these species is stable and remains pure, locked into its ring form by the crystal. In solution, raising the temperature and increasing or decreasing the pH increases the rate of attainment of equilibrium, because the reaction is acid and base catalyzed, although raising the pH beyond a certain point prevents equilibrium because it induces enolization (discussed below with reactions under basic conditions).

The attainment of this tautomeric equilibrium is often referred to as **mutarotation**, because it is accompanied by an overall change in the optical activity of the solution from that characterizing one pure form of the sugar in its crystal state to that characteristic of the equilibrium mixture of all of the tautomers. For example, it is possible to obtain pure crystals of α-D-glucopyranose, which have a rotation of the sodium D line at 20 °C, $[\alpha]_D^{20}$ of +112.2°, and also of β-D-glucopyranose, which have an $[\alpha]_D^{20}$ of +18.7°. Pure crystals of each are possible because neither sugar anomer fits properly into the regular lattice of the other's crystal. When either sugar anomer is dissolved in water, however, the tautomeric equilibrium begins to become established as the rings open up and reclose, producing all of the possible species. The

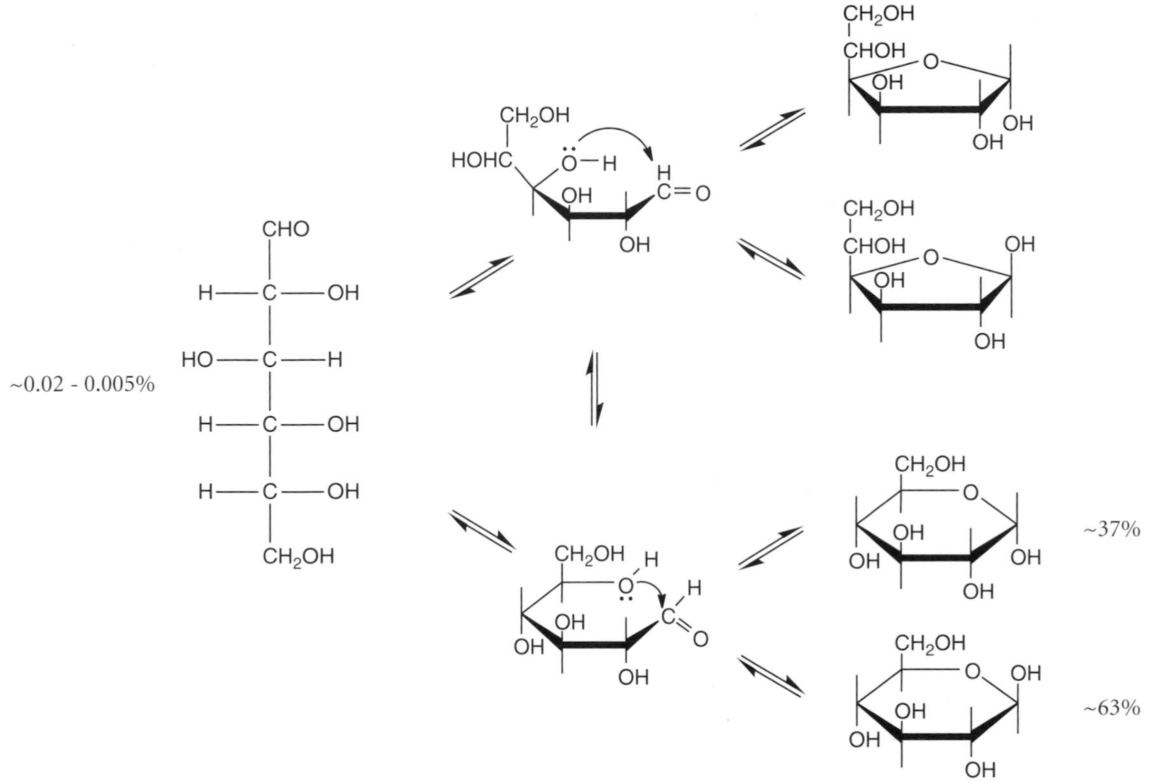

Figure 4.27. The tautomeric equilibrium in D-glucose.

equilibrium amount of each species is determined by its relative free energy, which is a combination of internal energetics and of the energy associated with its interaction with the solvent water. As the molecules of the dissolved crystal begin to interconvert, more and more of the other forms, which have different optical rotation values, will accumulate. The overall optical rotation, which is the average of that of all of the species present, will also change, which is the reason that the process is called mutarotation. Figure 4.28 illustrates this process for D-glucose. In the case of glucose, where only two tautomers are present to any appreciable extent, and where those two tautomers are the starting forms, it is possible to use the final equilibrium $[\alpha]_D^{20}$ to determine the approximate equilibrium amounts of each. Note that the attainment of this equilibrium is relatively slow, taking several hours. More accurate concentration determinations, which also measure the small amounts present of the other species, have been made by NMR experiments (Angyal 1969; Table 4.3).

For D-glucose, only the two pyranose rings occur in any appreciable amount; the furanose rings and the open chain are present in only vanishingly small amounts at any given time, due to their higher free energies. Even after equilibrium is attained, however, individual rings are continually opening and closing, thus occasionally regenerating the reactive free aldehyde form. For this reason the solution is still able

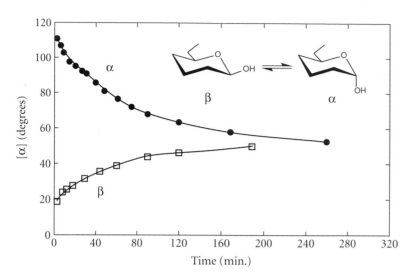

Figure 4.28. The time development of the tautomeric/anomeric equilibrium in D-glucose as followed by measuring the optical rotation of the sodium D line. (Adapted and redrawn from R.S. Shallenberger, *Advanced Sugar Chemistry*. Westport, CT: AVI Publishing Co. Inc., 1982.)

Table 4.3 Tautomeric distributions (as percentage) for various sugars in aqueous solution at 20 °C, as measured using NMR

Sugar	α-Pyranose	β-Pyranose	α-Furanose	β-Furanose
D-glucose	36	64	—	—
D-mannose	67	33	—	—
D-galactose	32	64	1	3
D-fructose	—	76	4	20
D-xylose	35	65	—	—
D-talose	40	29	20	11
D-idose	31	37	16	16
D-ribose	20	56	6	18

Source: Data taken from R.S. Shallenberger, *Advanced Sugar Chemistry* (Westport, CT: AVI Publishing, 1982).

to reduce ionic charges, and the sugar is referred to as a reducing sugar. Nonreducing sugars, such as the disaccharides sucrose and trehalose, which will be discussed later, do not undergo mutarotation and do not have alternate tautomeric forms, because their anomeric and structural isomers do not interconvert in solution. The mutarotation reaction can have important consequences in food systems, particularly because it allows the so-called Maillard browning reactions, which require the free aldehyde form of reducing sugars, to take place. The tautomeric equilibrium is also temperature dependent. In fructose, this can affect the sweetness in solution, because at elevated temperatures the equilibrium shifts to favor the furanose forms, which are less sweet, at the expense of the sweeter β-D-fructopyranose form favored at 20 °C (Tables 4.3 and 4.4). Thus, soft drinks and other foods sweetened with fructose or high-fructose corn syrups will be less sweet when warm than when cold.

H$\diagdown$C$\diagup\!\!^{O}$

$\xrightarrow{\text{NaBH}_4}$ CH$_2$OH

CH$_2$OH

CH$_2$OH

an alditol, glucitol in this case,
which has the trivial name sorbitol

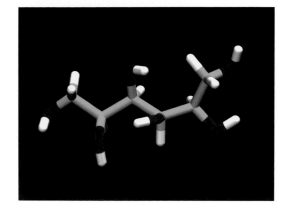

Figure 4.29. The reduction of D-glucose to an alditol, D-sorbitol, shown in a conformational drawing on the right.

Sugar Derivatives

Alditols

The aldehyde groups of sugars can be reduced to hydroxyl groups, producing sugar alcohols, called **alditols**, by treatment with sodium borohydride under basic conditions (Figure 4.29). The systematic names for sugar alcohols are formed by replacing the "-ose" or "-ulose" endings of the parent sugar with an "-itol" suffix, so that the sugar alcohol of D-glucose is called "D-glucitol." Several alditols, including xylitol, sorbitol, and mannitol, are used in foods as low-calorie sweeteners, to control water activity in intermediate-moisture foods, or to inhibit crystallization. Humans can oxidize alditols back to monosaccharides during digestion, but only glucose is involved in insulin-dependent metabolism, and sugar alcohols like xylitol and mannitol are tolerated by diabetics. All of the sugar alcohols are poorly absorbed and so do not yield as many actual metabolized calories as sucrose and do not raise insulin levels as much as sugar. It is not true, as is sometimes claimed, that glucitol has no effect on blood sugar levels. Note that because the alditols lack a reactive carbonyl group, they are no longer capable of reducing ionic charges and thus do not cause Maillard browning. None of the alditols promotes tooth decay. Sugar alcohols are also natural products; sorbitol, for example, is abundant in prunes, and its indigestibility may contribute to their laxative effect.

Because they do not possess a reactive aldehyde group, sugar alcohols do not form ring structures. Instead, they remain as fairly extended open-chain structures as shown in Figure 4.29. They are quite stable under most food processing conditions, including elevated temperatures and both acidic and basic pHs.

Sorbitol is widely used as a lower-calorie sweetener and humectant. It has only about two thirds as many calories as sucrose, but is also only 60% as sweet as sugar. Mannitol is used as a powdery dust to coat chewing gum because it can keep the gum from absorbing water and becoming sticky (see Table 4.6). **Maltitol**, shown in

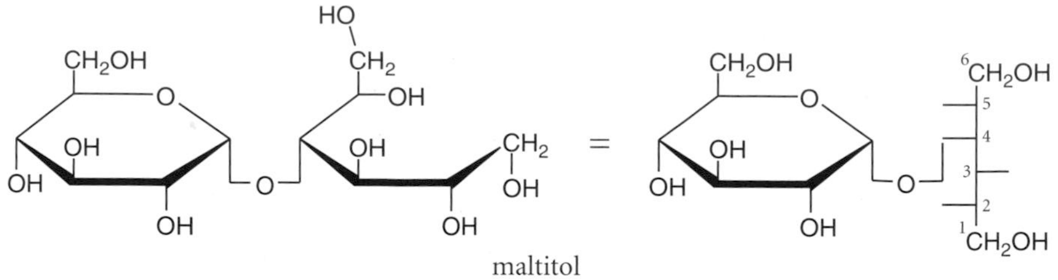

Figure 4.30. A Haworth formula representation of maltitol (left) and a hybrid Haworth/Fischer representation on the right.

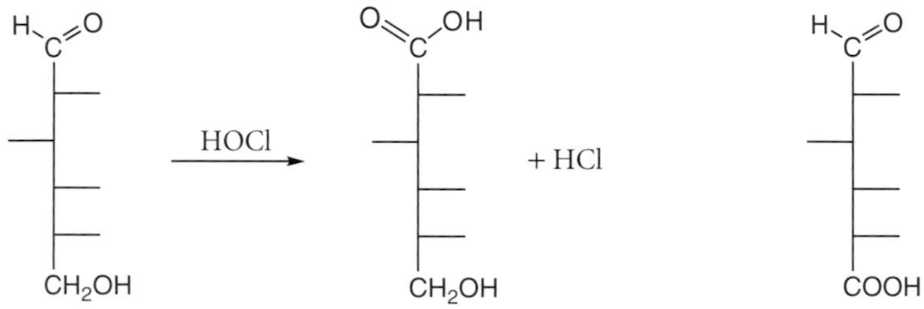

Figure 4.31. Fischer representations of aldonic and uronic acids.

an aldonic acid,
D-gluconic acid

an uronic acid,
D-glucuronic acid

Figure 4.30, resembles sugar in taste and behaves somewhat like sucrose in terms of bulk, mouthfeel, and interaction with other ingredients such as water. Note from the illustration that the glycosyl group of this dimer is a conventional glucopyranose ring, while the other unit is of course an open chain sorbitol. In the combined Fischer/Haworth representation shown on the right in Figure 4.30, the sorbitol has been illustrated "upside down" compared to the traditional orientation, to facilitate the depiction of the linkage, but to avoid confusion the atomic numbering is also indicated. (Remember that a Fischer projection cannot be rotated by 90° without changing the stereochemistry, but a 180° rotation preserves the configuration.)

Aldonic and Uronic Acids

When treated with a weak acid such as hypochlorous acid, HOCl, the aldehyde groups of sugars can easily be oxidized to carboxylic acids, giving **aldonic acids** (Figure 4.31), which have the general chemical formula $C_6H_{12}O_7$. These are named with the "-onic acid" suffix. If only the primary alcohol is oxidized, one has an **uronic acid**, such as the D-mannuronic and L-guluronic acids of the alginates.

The laboratory synthesis of uronic acids is more difficult than that of aldonic acids because the reactive aldehyde group must be protected, but their artificial synthesis is not necessary for foods because they are not used as additives. Uronic acids are found naturally in food polysaccharides, most importantly in alginate and pectin, to be discussed more fully later, where they are synthesized enzymatically. The basic

Figure 4.32. Haworth representations of the three uronic acids commonly found in food polysaccharides; all three are shown in their unprotonated pyranose forms.

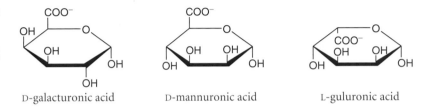

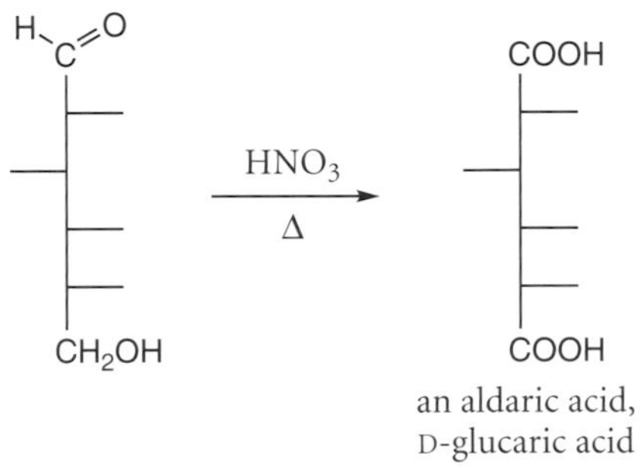

D-galacturonic acid   D-mannuronic acid   L-guluronic acid

Figure 4.33. The oxidation of an aldose sugar to give an aldaric acid.

an aldaric acid, D-glucaric acid

uronic acid repeat unit of pectin, D-galacturonic acid, and the monomer subunits of alginate, D-mannuronic acid and L-guluronic acid, are shown in Figure 4.32. Gluconic acid is found in rotting fruit as the result of the action of glucose oxidase enzymes from yeasts. As a sugar acid, it contributes both sour and sweet notes to wines made from grapes that begin to rot on the vine under oxidizing conditions.

Aldaric Acids

With a stronger acid such as nitric acid, one can oxidize both ends of the sugar molecule, giving **aldaric acids** (Figure 4.33). In the systematic nomenclature these are named with an "-aric" suffix.

Sugar Amides

Several sugar amides are important in polysaccharides from natural sources and thus can be found in foods (Figure 4.34). These modified sugars include **N-acetylglucosamine**, abbreviated as GlcNAc or NAG, which is the amide of D-glucose substituted with acetic acid at the 2-carbon; **N-acetylmuramic acid**, abbreviated as NAM; and **N-acetylneuraminic acid**, abbreviated as NANA or Neu5Ac. N-Acetylmuramic acid is the ether of N-acetylglucosamine with lactic acid linked at the 3-carbon position. N-Acetylglucosamine is the monomeric repeat unit of chitin, the structural polysaccharide from which the exoskeletons of insects and crustaceans like crabs and lobsters are built. N-Acetylglucosamine and N-acetylmuramic acid together form the monomer repeat units of the alternating copolymer polysaccharides of bacterial cell walls and are the principal substrate of

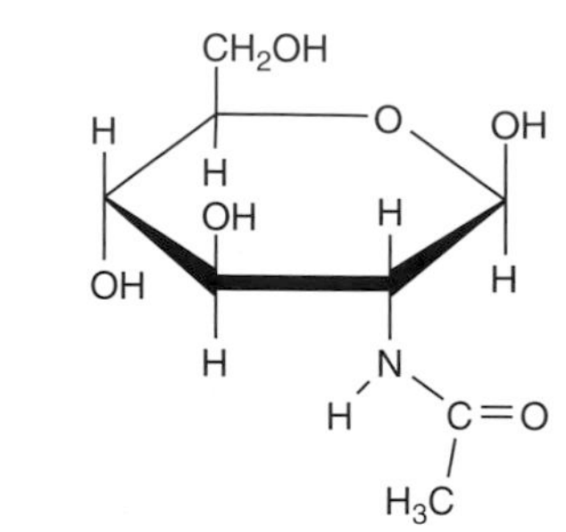

N-Acetyl-D-glucosamine

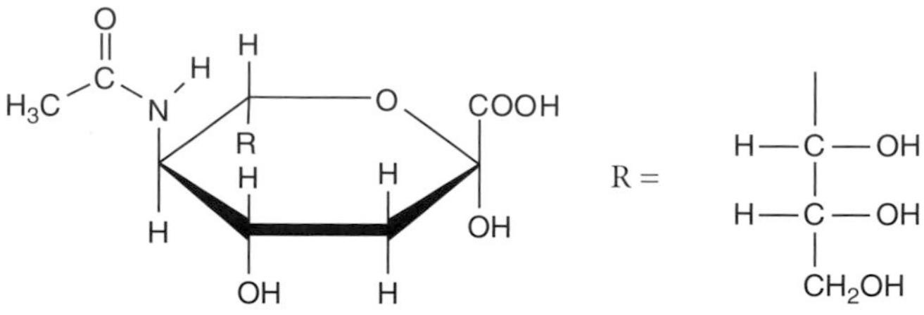

N-Acetylneuraminic acid (NANA)

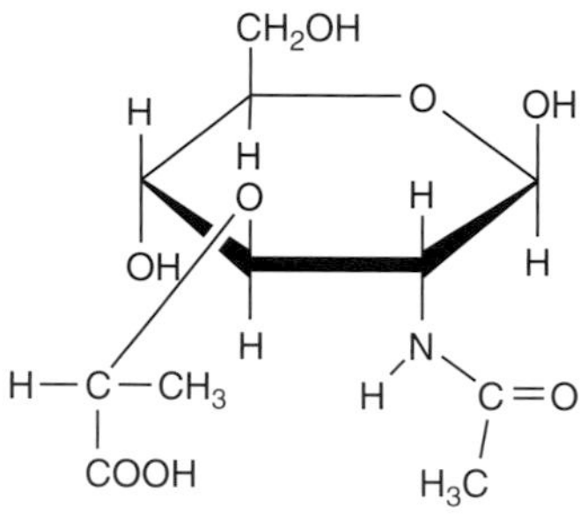

N-Acetylmuramic acid (NAM)

Figure 4.34. The common sugar amides NAG, NANA, and NAM.

the defensive enzyme lysozyme (see Figure 5.50), which kills bacteria by hydrolyzing the glycosidic linkages their cell walls.

Deoxy Sugars

Deoxy sugars have had one of their hydroxyl groups replaced by a hydrogen atom (Figure 4.35). Important biological examples are L-rhamnose, L-fucose, and 2-deoxy-D-ribose. L-rhamnose is an important component of pectins (see below) and other polysaccharides. Such sugars are uncommon in foods. L-Fucose is found in many cell-surface oligosaccharides and plays a role in molecular recognition. 2-Deoxy-D-ribose of course is an integral component of DNA.

Inositols

Inositols are cyclic polyols, six-carbon ring compounds that can be considered as polyol derivatives of cyclohexane, or as analogs of pyranose sugars without a ring oxygen atom (Figure 4.36). Nine possible inositols occur, with different stereochemistries for the various hydroxyl groups; four of these are illustrated in Figure 4.36. The most common is *myo*-inositol, with one axial hydroxyl group. *Scyllo*-inositol is

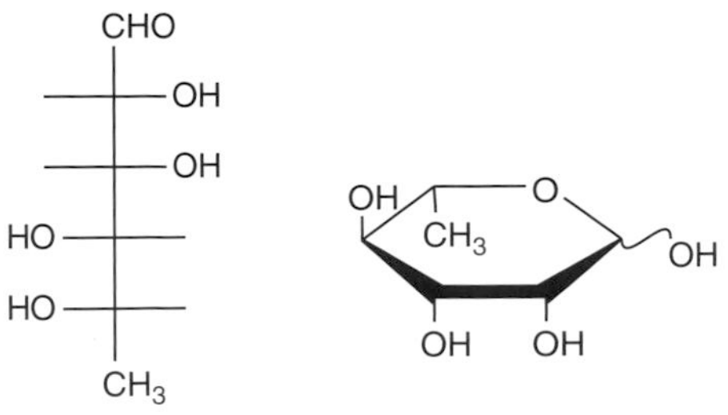

L-rhamnose
(6-deoxy-L-mannose)

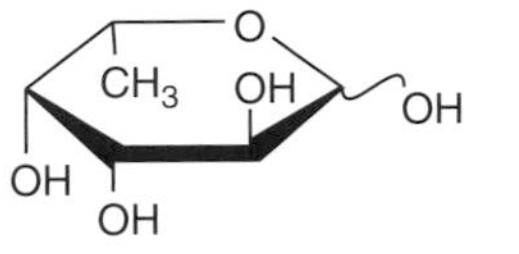

L-fucose
(6-deoxy-L-galactose)

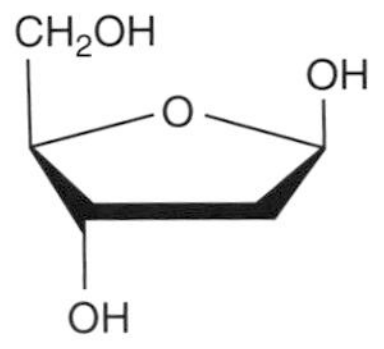

2-deoxy-D-ribose

Figure 4.35. Examples of deoxy sugars commonly found in biological systems.

Figure 4.36. Four of the nine possible inositol structures.

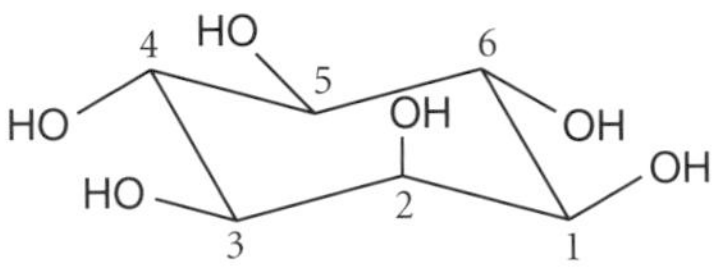

myo-inositol

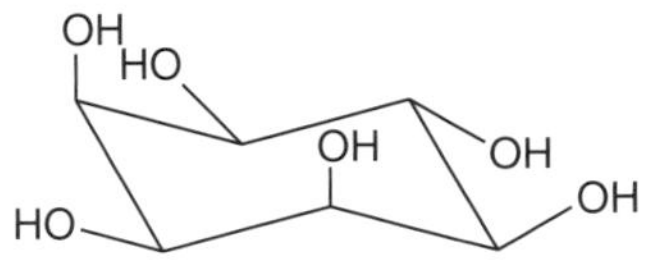

epi-inositol

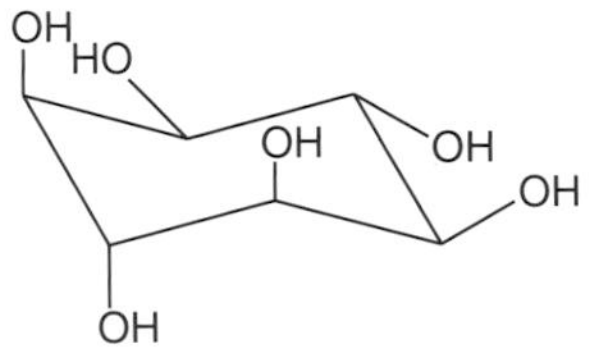

muco-inositol

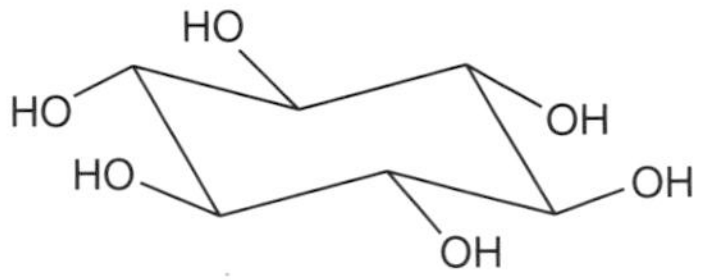

scyllo-inositol

the variant in which all of the hydroxyl groups are equatorial, as in the lowest-energy conformation of β-D-glucopyranose. Theoretical studies have found that the *scyllo*-inositol form is the lowest-energy form (Liang et al. 1994). **Myo-inositol** is found in many foods and is part of such important molecules as the phospholipid phosphatidylinositol (see Chapter 7) and phosphatidylinositol phosphate. Inositols are involved in signal transfer in intracellular signaling. *Myo*-inositol is synthesized in humans from glucose-6-phosphate, although it was once thought to be an essential nutrient and was referred to as **vitamin B$_8$**.

REACTIONS OF CARBOHYDRATES

Acid-Catalyzed Reactions

Mutarotation

Mutarotation is catalyzed by small amounts of acid or base, so the addition of weak acid to a sugar solution will produce all of its tautomeric forms in their appropriate ratios.

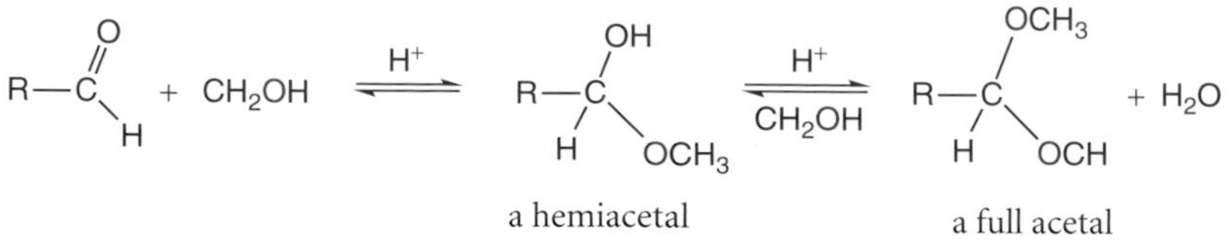

Figure 4.37. The acid-catalyzed condensation of an aldehyde with an alcohol to form a hemiacetal. These can further condense with another alcohol to form a full acetal.

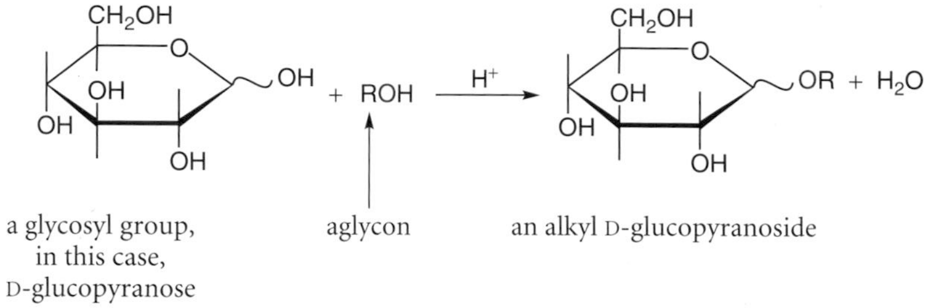

Figure 4.38. The formation of a glycoside by the condensation of the reactive anomeric hydroxyl group of a hemiacetal sugar ring with another alcohol. The red color-coding indicates that the resulting linking oxygen atom was donated by the aglycon alcohol, while the sugar's anomeric hydroxyl oxygen atom was lost to form a water molecule. The ~OH indicates that both anomers are present in their equilibrium concentrations.

Glycosidic Condensation: The Formation of Glycosides

As already noted, aldehydes can react with alcohols to form hemiacetals and full acetals (Figure 4.37). This reaction is the basis of the cyclization of the aldose sugars, where the aldehyde is reacting with one of its own alcohol groups to form a ring. As we have also seen, the formation of these rings is reversible, and in the presence of acid and another alcohol, the sugar can form a full acetal through combination with the second alcohol before it cyclizes. This second addition is accompanied by the elimination of the hydroxyl group of the intermediate hemiacetal as a water molecule.

The products of these condensation reactions with other alcohols are called **glycosides**. When a water molecule is eliminated between the anomeric hydroxyl group of one sugar, called the glycose, and the hydroxyl group of another alcohol, called the **aglycon**, **O-glycosides** are formed (Figure 4.38). Formally, the sugar radical that would remain after removal of the anomeric hydroxyl group is called the **glycosyl group**. These products are full acetals with mixed substituents (mixed acetals).

The addition of another alcohol in acid solution to a sugar in a ring structure proceeds with the ring first opening to regenerate the reactive open-chain aldehyde, followed by the addition of the aglycon to form the glycoside, and then the reaction again with one of its own hydroxyl groups (in the glucose example in Figure 4.38, at the C5 position) to generate the full acetal; that is, the pyranose ring structure with a glycosidic bond to the aglycon. Once the solution pH is neutralized, the new bond, called a **glycosidic linkage**, is stable, and subsequent ring opening and tautomeric equilibration are frozen out. Notice that, in Figure 4.38, the anomeric configuration

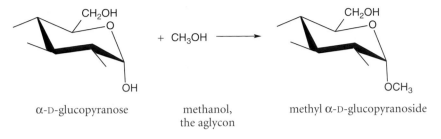

α-D-glucopyranose   methanol,   methyl α-D-glucopyranoside
the aglycon

Figure 4.39. The condensation of D-glucopyranose with methanol to form methyl α-D-glucopyranoside. Although the glycosyl group is shown as the α anomer, in fact both anomers would be present, and both the α and β anomeric configurations of the glucopyranoside would be formed.

is not specified, but is replaced by a wavy line. This is done to indicate that in the original sugar, the anomeric/tautomeric equilibrium guarantees that both anomers will be present, and thus in the product glycoside, both α and β stereochemistries will also be present (Figure 4.39). Most food glycosides are synthesized enzymatically, with a specific anomeric configuration that is stable and retained through many processing operations, but would be lost again if heated in acid, because the glycosidic bond formation is reversed.

Sugars are themselves alcohols, and when the alcohol with which the sugar is combining happens to be another sugar group, a **disaccharide** is produced. These disaccharides are referred to as **O-glycosyl-glycoses**. The linkages can be specified in systematic names by using -(1→x)-, where 1 and x are the numbers of the carbon atoms of the two sugars involved in the glycosidic linkage (or -(2→x)- in the case of ketose glycosyl groups). If water is eliminated between the anomeric hydroxyl groups of two monosaccharides, the resulting disaccharide is **nonreducing**, such as sucrose and trehalose, because there is no longer the possibility of either ring opening to give the reactive carbonyl group capable of reducing ions. If the disaccharide still contains a reducing group, it may undergo another such reaction, to form a trisaccharide, and so on. The number of sugar rings in such a chain is referred to as the **degree of polymerization**, or **DP**. Those polymers with a DP in the range from two to about ten are referred to as **oligosaccharides**. Food oligosaccharides and polysaccharides are biologically synthesized by enzymes and again have specific anomeric configurations, which then characterize the geometry of the glycosidic linkage, and thus must be specified in the nomenclature designating the linkage. Thus, for example, in **lactose** (milk sugar), the disaccharide formed from D-galactopyranose linked to D-glucopyranose via an equatorial (β) glycosidic linkage between the anomeric C1 group of the galactose and the C4 hydroxyl groups of the glucose (Figure 4.40), the linkage is specified as β-(1→4).

In a disaccharide like lactose, the ring form (pyranoid) and linkage configuration of the galactose residue are fixed by the formation of the glycosidic bond, but the glucose aglycon group still has a free anomeric carbon, and thus can still undergo ring opening to regenerate the free aldehyde, and in fact all of the other tautomeric

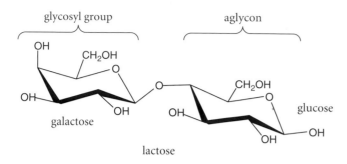

glycosyl group   aglycon

galactose

lactose

glucose

Figure 4.40. When the alcohol group with which a sugar glycosyl condenses is in fact part of another sugar molecule, the resulting glycoside is a disaccharide, such as this disaccharide, lactose, which is a glycoside of D-galactose (the glycosyl group) and D-glucose (in this case, the aglycon).

forms for that sugar. Thus, lactose in solution will be a mixture of α- and β-lactose, referring to the anomeric configuration of the glucose half of the molecule, with proportions that are similar to those for free glucose. In addition, lactose is thus still a reducing sugar, because it occasionally regenerates the free aldehyde capable of reducing the charge of metal ions, but with only one half the reducing equivalents per weight as a molecule of galactose and a molecule of glucose would have if both were free sugars.

Glycosidic condensations are easily reversible under acidic conditions, so the product glycoside must be removed as it is produced to give good yields. Under neutral and weakly basic conditions, however, O-glycosides are stable, and most food glycosides are also stable under mildly acidic conditions. In addition to oligosaccharides and polysaccharides, glycosides occur in foods in a variety of other types of molecules, such as the anthocyanins of fruits and wines. Many are bitter tasting, like naringin in grapefruits, especially if the aglycon is larger than a methyl group, although some are extremely sweet, such as stevioside from the sweet herb of Paraguay, osladin, and glycyrrhizzic acid (glycyrrhizin) from licorice root (see below).

In the natural polymerization of monosaccharides, water is eliminated between the anomeric hydroxyl of one molecule and one of the several possible hydroxyl positions of the other monomer. Disaccharides and higher oligosaccharides can be formed in small amounts by heating glucose or sucrose syrups in the presence of acid catalyst. When starch syrups are manufactured by acid-catalyzed hydrolysis, oligosaccharides are formed that are not structural units of the starch molecules, but rather are mixtures of all possible random combinations of α and β oligosaccharides of D-glucose, with the disaccharides isomaltose and the bitter-tasting gentiobiose (with α- and β-(1→6) linkages, respectively) predominating. These recombination products are called **reversion products**.

Glycyrrhizin and Stevioside

There are a number of interesting complex glycosides from plant sources that are important in foods. These include the anthocyanin pigments of many flowers and fruits (to be discussed in Chapter 9) and other food polyphenolics, and the natural plant sweeteners **glycyrrhizin** and **stevia**. Glycyrrhizin is the molecule responsible for the flavor of licorice from the root of the Eurasian legume *Glycyrrhiza glabra*. It

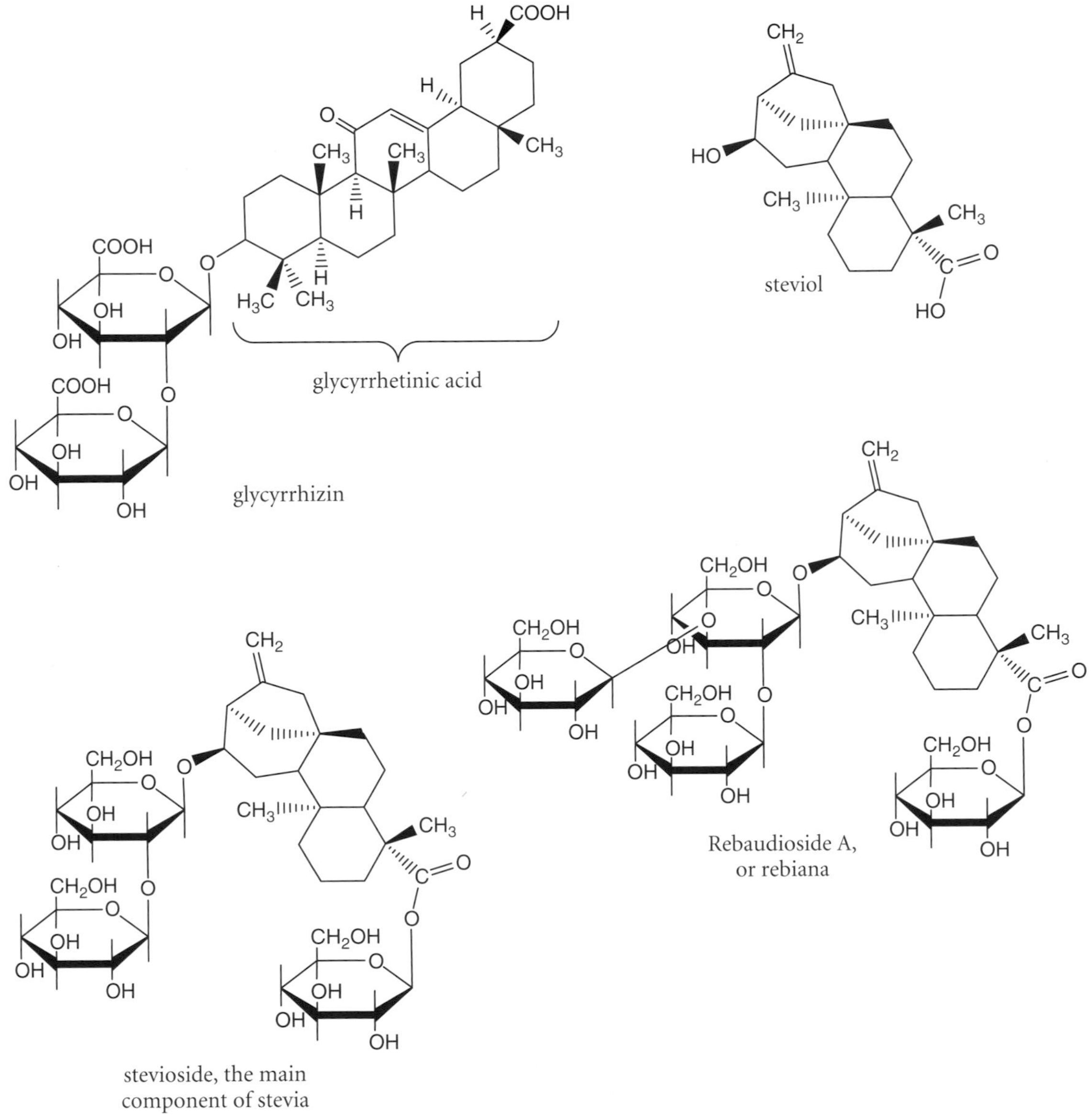

Figure 4.41. The structures of the food glycosides glycyrrhizin, stevioside, and rebiana.

is a glycosidic complex of the aglycon glycyrrhetinic acid β-linked to a (1→2)-linked disaccharide of glucuronic acid (Figure 4.41). Stevia is extracted from the so-called sweet leaf plant, *Stevia rebaudiana,* from Paraguay. Used in Japan in a variety of products such as chewing gums and beverages, it was only recently approved for use in the United States. Stevia contains a number of complex glycosides called steviol glycosides. The best known of these glycosides is stevioside, which is a double glycoside of the aglycon steviol, bound to a single glucose on one end and a β-(1→2)-linked disaccharide of glucose on the other end. It is water-soluble and very sweet

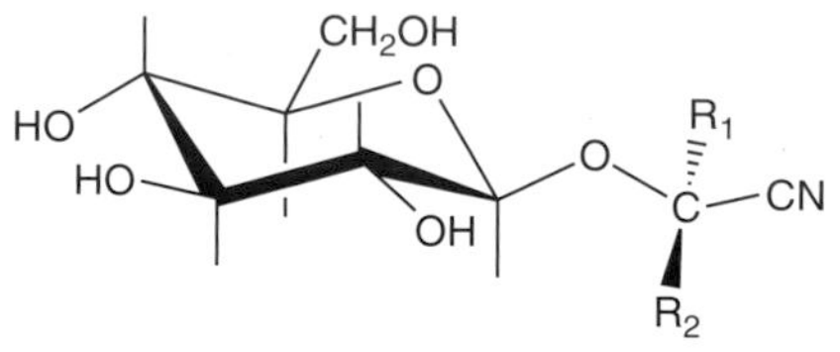

Figure 4.42. The structure of β-D-glucosic cyanogenic glycosides such as linamarin.

cyanogenic glycoside

but has a slightly bitter-menthol aftertaste. Another glycoside found in stevia, Rebaudioside A (or rebiana, also colloquially called "reb A"), is identical to stevioside except that the disaccharide has an additional glucose residue linked β-(1→3) to the reducing end of the disaccharide (Figure 4.41).

Both glycyrrhizin and stevioside are quite sweet tasting. In the United States glycyrrhizin has GRAS (generally recognized as safe) status as a flavor additive, but not as a sweetener. Both are used as sweeteners in Japan, although the licorice flavor of glycyrrhizin limits its utility for this purpose. Rebiana is sold in the United States as a tabletop sweetener.

Cyanogenic Glycosides

The seeds of the legumes (beans) contain a number of antinutritional or toxic substances, including protease inhibitors that interfere with the digestion of proteins, and various lectins that bind to glycoproteins on the cell walls of the intestines, interfering with the adsorption of nutrients (Chapter 5). In addition, several types of beans, as well as the seeds of other types of plants, also contain cyanogenic glycosides, which under certain conditions produce potentially fatal quantities of cyanide when eaten. These molecules generally have structures of the type shown in Figure 4.42.

Several of the most important of these are **linamarin** from lima beans and cassava (where R_1 and R_2 are both methyl groups), dhurrin from sorghum (R_1 = phenol, R_2 = H), and **amygdalin** from apricot pits. Amygdalin, or laetrile, was widely claimed to have various anticancer properties, but these have not been demonstrated in scientific tests. Amygdalin is a glycoside of gentiobiose, the β-(1→6)-linked disaccharide of glucose (Figure 4.43). Linamarin in cassava is a significant limitation on the use of cassava as a food, which must be extensively processed in order to make it safe to eat. Depending on the level of linamarin present in the cassava root, this processing can consist of simply cooking, or grinding and soaking in water to remove the soluble linamarin.

When foods containing cyanogenic glycosides are eaten, the plant cell structures are broken down, allowing β-glycosidase enzymes that hydrolyze glycosidic bonds to react with these glycosides, producing glucose and hydroxynitrile (Figure 4.44).

Hydroxynitrile lyase, another plant enzyme, acts in turn on the hydroxynitrile to produce a free aldehyde or ketone and hydrogen cyanide (Figure 4.45).

Cyanide is extremely toxic, acting to shut down mitochondrial respiration by irreversibly complexing with the ferric ion of cytochrome c oxidase. A lethal dose of cyanide can be as little as 0.5 mg per kilogram of body weight (see Table 11.1), or

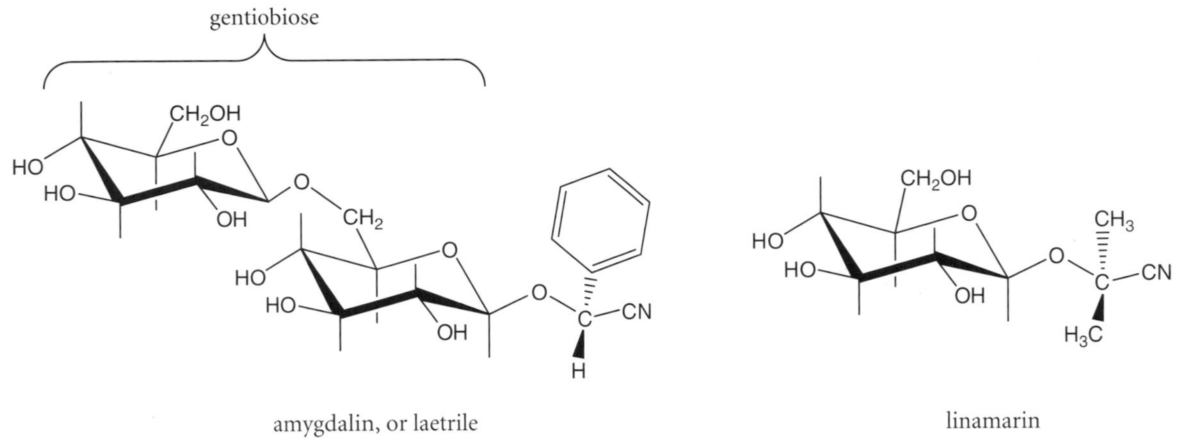

Figure 4.43. The structures of the cyanogenic glycosides amygdalin and linamarin.

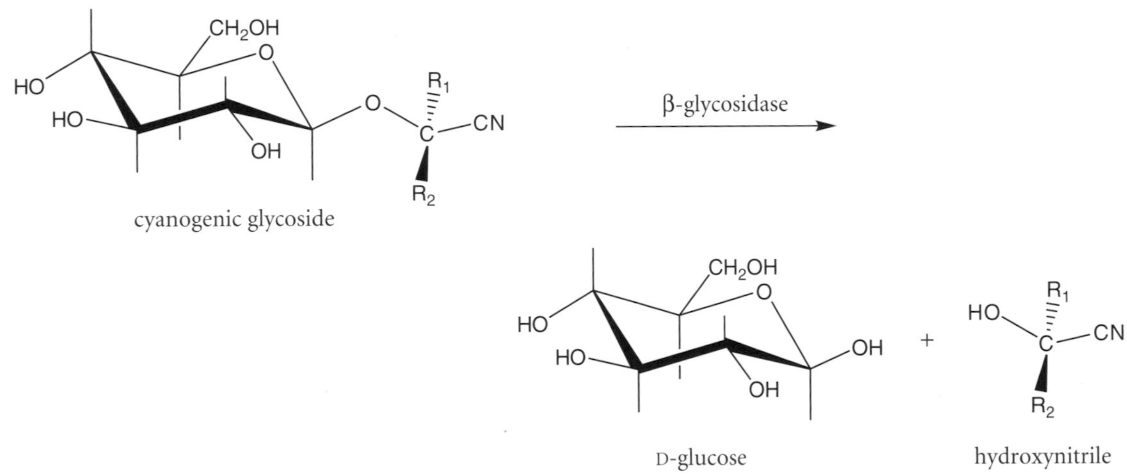

Figure 4.44. The production of hydroxynitrile from a cyanogenic glycoside by the action of a β-glycosidase enzyme.

Figure 4.45. The production of cyanide from hydroxynitrile by the action of the enzyme hydroxynitrile lyase.

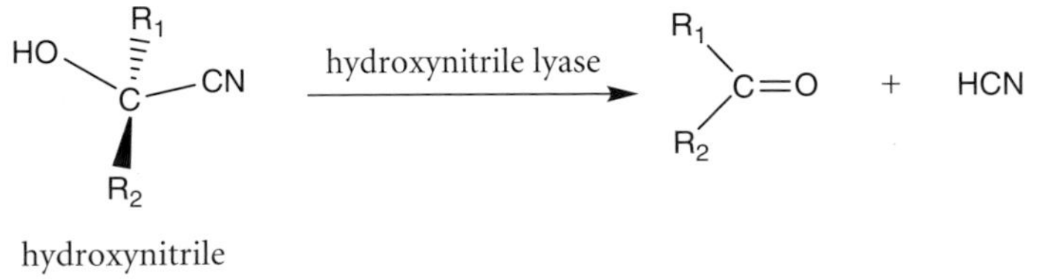

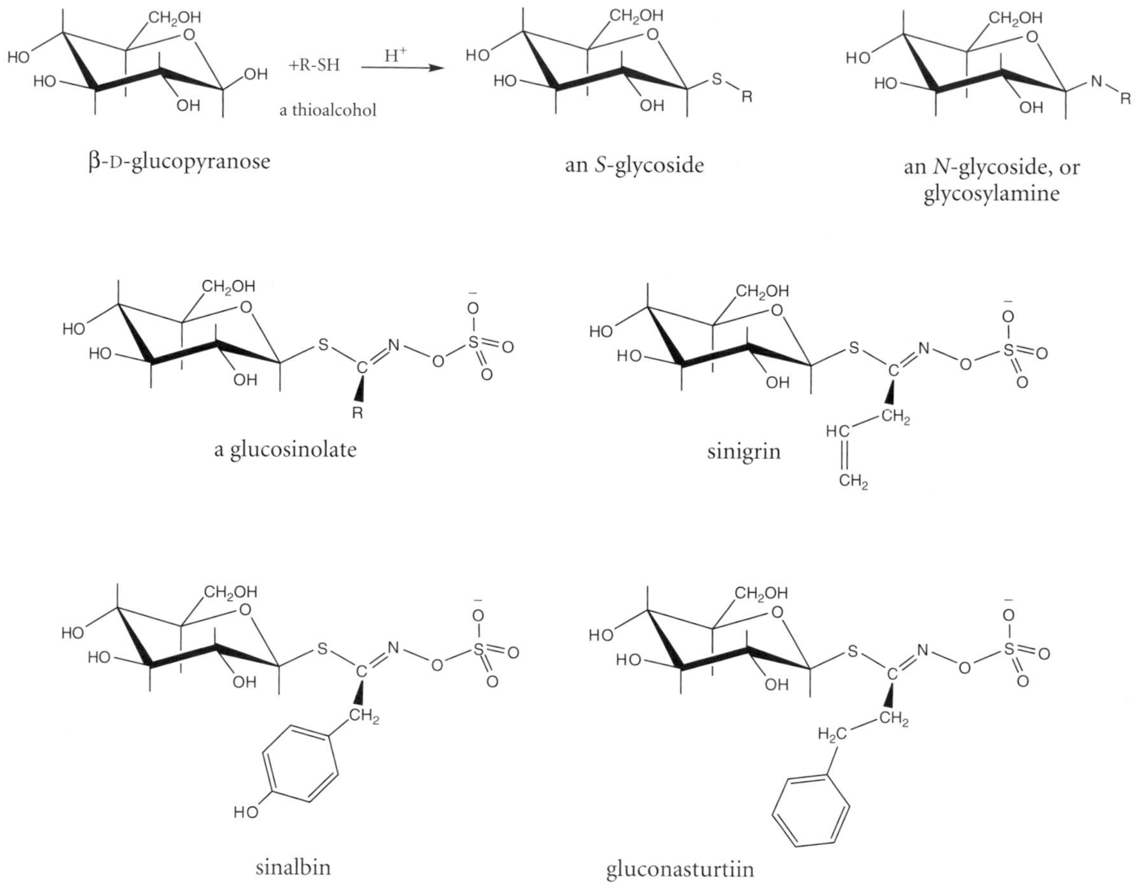

Figure 4.46. The formation of S-glycosides under acidic conditions, and the structures of several important food S-glycosides, as well as the generic structure of N-glycosides (glycosylamines).

about 27 mg for a 120-lb (55 kg) individual. It is possible to receive a lethal dose of cyanide from eating only a few bitter almonds (apricot pits).

S-Glycosides: Mustard, Horseradish, and Wasabi

Reducing sugars can also be combined with thioalcohols, molecules with an —S—H functionality, to form glycosides in which the linking atom is a sulfur instead of an oxygen (Figure 4.46). Such molecules are referred to as S-glycosides. There are also many examples of N-glycosides, sugars linked to aglycons through a nitrogen atom, but according to IUPAC rules they are referred to as glycosylamines instead of N-glycosides. Unlike the O-glycosides, N-glycosides can undergo mutarotational ring opening. They form whenever reducing sugars are present with free amines such as the amino groups of proteins, and the formation of glycosylamines is the first step of Maillard browning reactions that will be discussed in greater detail below.

Several examples of S-glycosides of glucose, called glucosinolates, are found in plants from the *Brassicaceae* family that are used as foods, including brussels sprouts,

Figure 4.47. Allyl isothiocyanate.

allyl isothiocyanate

broccoli, cabbages, mustards, horseradish, and wasabi. These molecules generally have sharp or bitter flavors and serve a protective function by discouraging herbivores from eating their tissues. They are probably the principal reason that many of these plant foods are unpopular with children. When the cellular structure of these plants is disrupted, enzymes called thioglucosidases are freed to act on their glucosinolate substrates and cleave the sugar–sulfur bond, producing isothiocyanates and free glucose. In wasabi, black mustard, and horseradish the main glucosinolate is **sinigrin** (Figure 4.46), so that the product of thioglucosidase cleavage is the more volatile **allyl isothiocyanate** (Figure 4.47), which contributes strongly to the pungent taste of these condiments and is largely responsible for their aroma. Relatively pure allyl isothiocyanate, called **oil of mustard**, can be obtained from ground mustard seeds by heating and distillation. These plants contain other glucosinolates as well, including sinalbin from the seeds of white mustard and the related gluconasturtiin, which is abundant in horseradishes.

Common Food Disaccharides

A number of disaccharides are important in foods. Several of the most significant of these are shown in Figure 4.48. As we have already seen, the milk sugar lactose is the β-(1→4)-linked disaccharide of D-galactose and D-glucose (Figure 4.40). **Cellobiose** is the β-(1→4)-linked disaccharide of D-glucose, and is the fundamental repeat unit of cellulose. It differs from lactose only in the configuration of the C4 hydroxyl group of the galactose ring. **Maltose** is the α-(1→4)-linked disaccharide of D-glucose, and is the basic repeat unit of starch (amylose). It differs from cellobiose only in the configuration at the C1 carbon atom of the nonreducing ring. Cellobiose, maltose, and lactose are all reducing disaccharides, because the anomeric carbon of one of their rings is unsubstituted and the ring can reopen, regenerating the reactive aldehyde group. **Sucrose**, table sugar, is the α-(1→2)-linked disaccharide of D-glucose and D-fructose. **α,α-trehalose** is the α-(1→1)-linked disaccharide of two D-glucose molecules where both of the anomeric groups are axial. Two other types of trehalose are possible: β,β-trehalose, where both anomeric groups are equatorial, and α,β-trehalose, where there is one axial and one equatorial anomeric linkage. The α,α-trehalose disaccharide is an important cryoprotectant molecule found in drought- and cold-adapted organisms, and its special hydration properties are potentially useful for controlling water mobility in frozen foods. Both sucrose and trehalose are nonreducing disaccharides, because the glycosidic linkage involves the anomeric group of both monomer rings, leaving no unsubstituted anomeric group. Their rings do not reopen, and they have no free aldehyde or ketone form. In the systematic nomenclature nonreducing disaccharides are named as pyranosides.

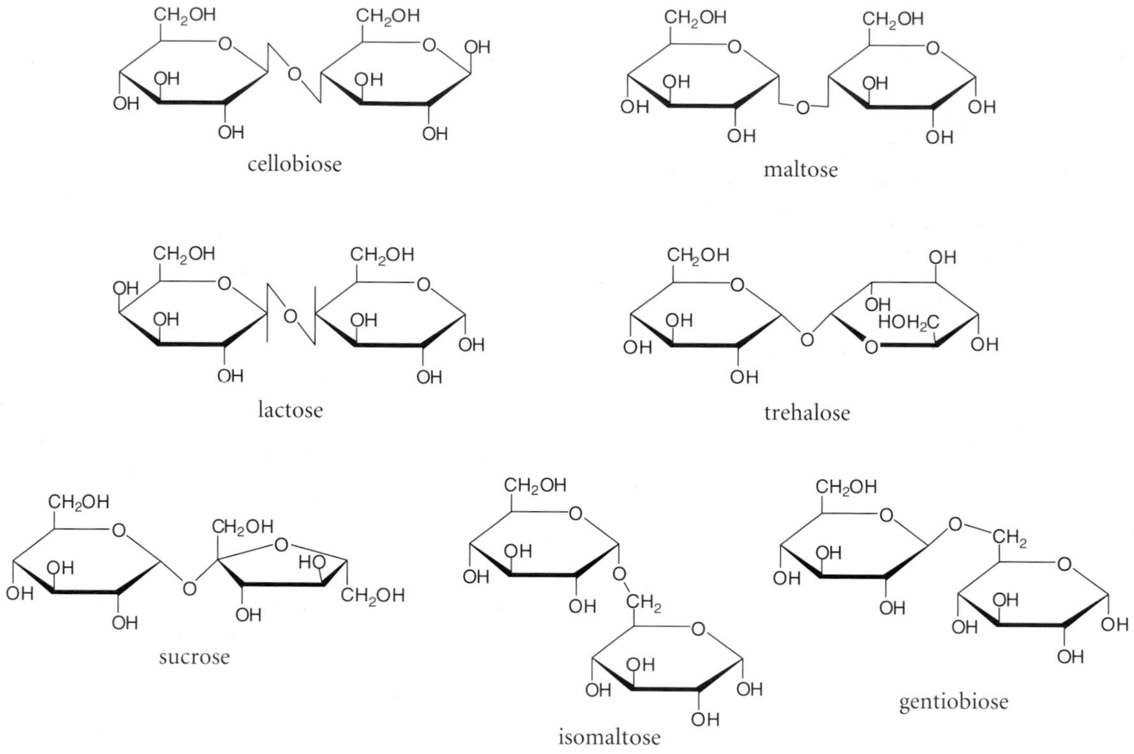

Figure 4.48. Haworth formula structures for seven of the most common disaccharides found in foods.

As with the monosaccharide rings, Haworth projections give a poor and distorted picture of the actual spatial relationships of the various functional groups on the two rings in disaccharides. Figure 4.49 presents a conformational drawing of the crystal structure of the disaccharide maltose. Note, for example, the hydrogen bond between the O2 hydroxyl group of the nonreducing ring and the O3′ hydroxyl group of the ring on the reducing end of the disaccharide. It is difficult indeed to understand such a hydrogen bond using the Haworth structure alone (as in Figure 4.48).

Note also that the two sugar rings can change their relative orientations through rotations about the two single bonds of the glycosidic linkage, C1—O1 and O1—C4′. Not all conformations about these bonds will have the same energy, due to such factors as steric clashes and interactions like the O2—O3′ hydrogen bond. Since maltose is the basic repeat unit of the polysaccharide amylose, the preferred conformations of the maltose disaccharide will help determine the shape of the starch polysaccharide as well.

Sucrose

Unquestionably the most important disaccharide for foods and cooking is sucrose (Figure 4.49). Sucrose is produced commercially in larger amounts than almost any other pure chemical substance and is one of the most important of all food

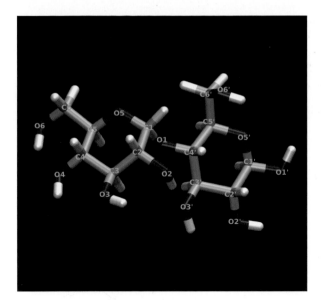 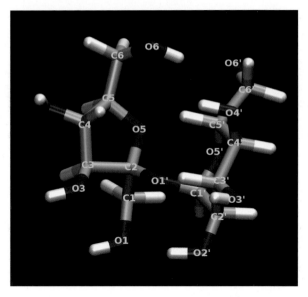

Figure 4.49. A computer-generated perspective representation of the conformation of maltose, the basic disaccharide repeat unit of amylose from starch, on the left, and the non reducing disaccharide sucrose, or table sugar, on the right.

commodities. The demand for sugar has played a major role in Western history for hundreds of years. It was one of the driving forces in New World colonialism, in the introduction of the trans-Atlantic slave trade, in the imperial wars of the eighteenth century, and even in the American annexation of Hawaii and the long-standing estrangement from and embargo of Cuba. Even today, one of the most important agricultural products of the European Union is sugar beets for sucrose production, and sugarcane is one of the most important commodities produced in many Caribbean and Latin American nations. Sugarcane produces more calories per hectare than any other crop except oil palm.

Sucrose, of course, has been used for centuries as a sweetener because of its strong, clean, sweet taste, and it is used as the standard against which all other sweeteners are measured. As a nonreducing disaccharide, sucrose does not directly promote Maillard browning, which is often important for the stability of food products (see below). It is easily used in cooking and baking, and replacing its range of properties when substituting other molecules is often difficult. Sucrose can be problematic in the diet, however. It is readily hydrolyzed by oral bacteria, promoting the development of dental cavities, and contributes a large number of calories without any other food value, and thus is widely disparaged as "empty calories." When metabolized, sucrose yields 3.94 kcal/mol in energy. Because the double anomeric linkage is more labile than other types of glycosidic linkages, sucrose is readily hydrolyzed at moderate temperatures under acidic conditions. Thus, even though sucrose is nonreducing, it can still sometimes be a problem with regard to Maillard browning.

Sugar (sucrose) is refined from sugarcane juice or sugar beet juice by boiling to concentrate and partially crystallize the sugar. The sugarcane stalks are crushed and pressed to express the juice, which contains many other substances besides sucrose, and which has a greenish color. This juice is clarified by heating with lime (calcium oxide, CaO) to precipitate out many of these impurities. The resulting liquor is boiled in a centrifuge to concentrate the sucrose and rapidly separate off the sugar crystals in the centrifugal field. The thick, dark brown-colored liquor remaining in an intermediate stage in this refining after the sucrose that crystallizes from the boiled cane juice has been carried off is called **molasses** (actually, at this stage, "first molasses"). It is approximately 60% sucrose (on a dry weight basis) and 40% other components, including inorganic salts, raffinose, kestose, organic acids, and amino acids, principally glutamic acid. The presence of significant amounts of raffinose in beet juice can interfere with sucrose crystallization, due to the similarity of raffinose to sucrose, which can thus reduce the yield. Molasses is brown due to the presence of these various organic and inorganic impurities, including dehydration and fragmentation products resulting from partial caramelization during the boiling. Molasses can also be made from sugar beet juice, but beet molasses is too bitter for human consumption. As a final step in the refinement of white sugar, the sugar crystals are redissolved, and the solution is filtered through activated charcoal to decolorize it by adsorption of the chromophores onto the charcoal. The sucrose is then recrystallized under controlled conditions to give relatively uniform crystals that are 99.8% pure. While sugar-processing plants can be messy and foul smelling, the processing of sugar involves only simple steps of cooking, crystallization, redissolution, and filtering, and the only added reagent is lime. Because the charcoal used in the filtration step is sometimes produced from animal bones, many vegans will not eat refined sugar.

Brown sugar is refined sucrose that contains molasses, from which it derives its color, either as a residue of the refinement or, more commonly, added to pure sucrose after refinement. Manufacturers prefer to add the molasses back after the refinement stage for the white sugar because under those circumstances they can carefully control the amount of molasses in the product. Because beet juice molasses is so bitter, brown sugar made from beets actually uses added molasses from cane sugar. Brown sugar is moister, and thus clumps more, than refined white sugar because molasses is hygroscopic and contains water.

Sugar of course is the principal ingredient in a variety of candies and confections, and some, such as rock candy and cotton candy, are almost entirely sucrose. Sucrose melts at 186 °C (see Table 4.7), and in the manufacture of cotton candy, sugar is melted in the center of a rotating drum, mixed with food coloring, and then spun out by centrifugal force through many small pinholes in the center spindle. As it exits from the holes the liquid stream is cooled by the air and resolidifies in long, thin strands.

A number of foods, and particularly beverages, contain high concentrations of sucrose, including fruits and vegetables, soft drinks, juices, and wines. A measure called the "**degrees Brix**" is widely used in industry, particularly for juices and wines, to characterize the sucrose content. The degrees Brix, abbreviated as °Bx, is defined as the mass ratio of sucrose in a solution to the total mass of the solution (i.e., g sucrose/100 g solution). Thus, a 15 °Bx solution is 15% sucrose, or 15 g of sucrose and 85 g of water. The degrees Brix can be measured using a device called a saccharimeter, which measures the specific gravity of a liquid solution, or with a refractometer, which is calibrated to the degree to which sucrose in solution affects the refractive index of the water in the solution. In winemaking, the degrees Brix measurement can be used to evaluate when grapes are ready to be harvested.

Food Oligosaccharides

A number of oligosaccharides are found in foods in small quantities, but only a few occur widely enough to be of significance. Two of the most important of these oligosaccharides, the trisaccharide **raffinose** and the tetrasaccharide **stachyose** (Figure 4.50), are derivatives of sucrose. In raffinose, a molecule of D-galactopyranose is linked α-(1→6) to the exocyclic primary alcohol of the glucose residue in sucrose, and in stachyose, a second D-galactopyranose molecule is linked α-(1→6) to raffinose.

Figure 4.50. Haworth representations of the structures of raffinose and stachyose.

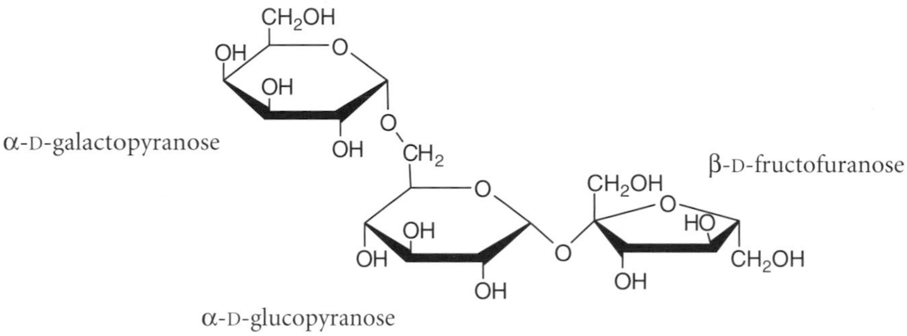

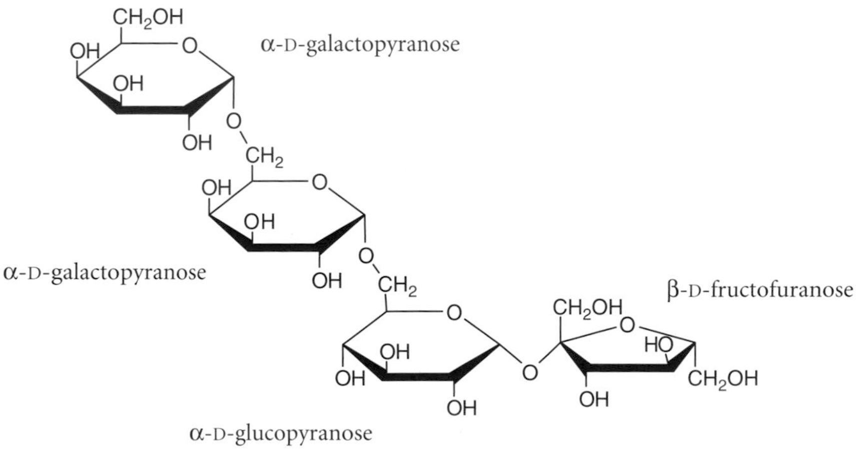

Raffinose occurs in small amounts in sugar beets, and its presence reduces the recovery of sucrose from beet juice by interfering with the sucrose crystallization. To prevent this from happening, in the industrial production of sugar from beets, enzymes called **α-galactosidases** are used to cleave off the galactose sugar to give the monosaccharide and sucrose. Stachyose is found in legumes such as lentils, garden beans, and soybeans, in quantities ranging from 1 to 4% of the dry mass by weight. This and other oligosaccharides are hydrolyzed by microorganisms in the intestinal flora of humans to produce gases such as H_2, CO_2 and CH_4, which causes flatulence. A commercial, food-grade α-galactosidase preparation called Beano is marketed by GlaxoSmithKline, which can be eaten at the same time as beans containing raffinose and stachyose are consumed, thus converting the oligosaccharides to galactose and sucrose and preventing the development of flatus.

Another class of plant storage carbohydrates consists of the **fructans**, which are oligomers or polymers of fructose, with a sucrose molecule on the aglycon end. The prototypical molecule of this class is **inulin** (Figure 4.51), which is a short polysaccharide of β-(2→1)-linked fructose units found in many plants of the composite family. Typically inulins are 20 to 30 fructose monomers in length. The plant of this category most used as a food is the Jerusalem artichoke (*Helianthus tuberosus*), which is a type of North American sunflower, unrelated to actual artichokes, with an edible tuberous root. Inulin is also extracted for commercial use from chicory root. The most important examples of fructans in foods are the kestoses, including **6-kestose**, where the additional fructose is linked β-(2→6) to the C6 primary alcohol of the fructose (see Figure 4.51), and 1-kestose, where the additional fructose is linked β-(2→1) to the C1 primary alcohol group as in inulin. These trisaccharide fructans are found in several foods, including sugar beets, onions, and honey. Fructans such as the kestoses are indigestible for humans, and inulin might be considered to be dietary fiber. Inulin has been used as a starch substitute for diabetics and has been suggested as a source of fructose for sweeteners, but has not proved economical for this purpose.

Various so-called malto-oligomers, also called maltodextrins, are also found in small amounts in certain foods. The most common are maltotriose and maltotetraose, which are found in particular in the hydrolysis of amylose from starch to produce corn syrups (Figure 4.52). These oligosaccharides are small fragments of amylose, which consist of D-glucopyranose sugars linked together by α-(1→4) linkages (see the discussion of starch below).

Internal Glycosidic Condensation: Levoglucosan

Certain sugars can undergo an internal glycosidic-type condensation. An example important in foods is β-D-glucose. In the pyranose form this molecule is usually in the $^4C_1$ chair arrangement, but occasionally flips over to the $^1C_4$ conformation, especially at elevated temperatures. In the $^1C_4$ geometry, the exocyclic hydroxymethyl group O6 is close to the C1 hydroxyl group. These two groups can undergo an internal glycosidic-type condensation with the elimination of a water molecule to form a new O—C bond in a molecule called 1,6-anhydro-β-D-glucopyranose. In food science this molecule is widely referred to by the trivial name **levoglucosan** (Figure 4.53).

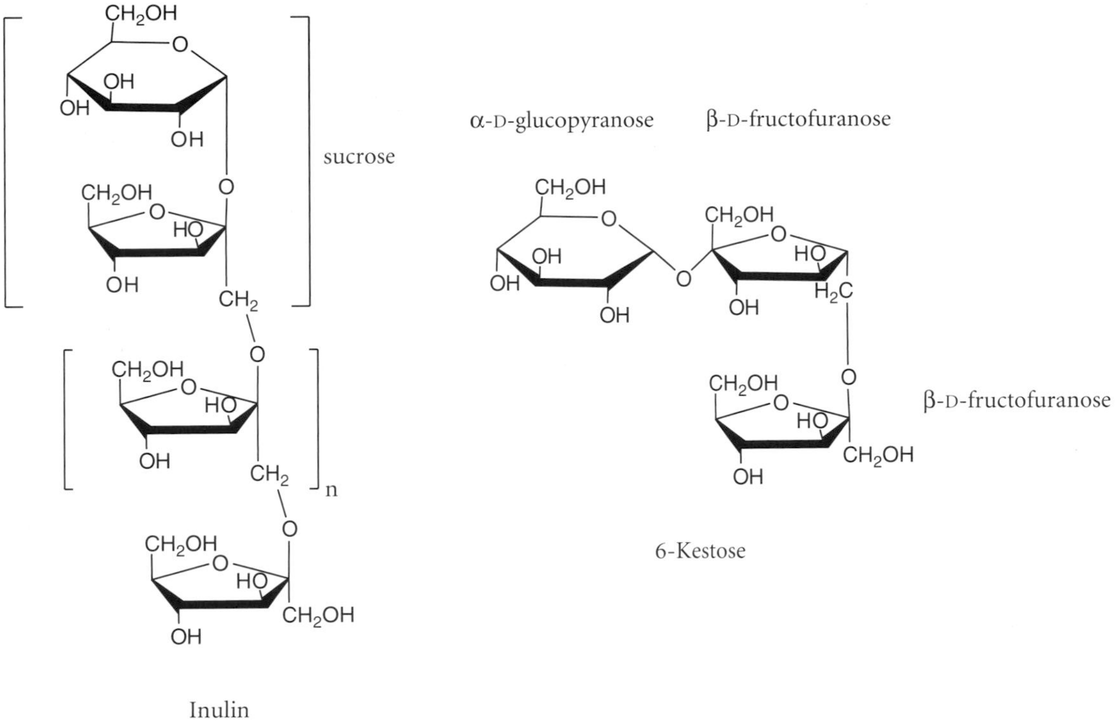

α-D-glucopyranose   β-D-fructofuranose

β-D-fructofuranose

6-Kestose

sucrose

Inulin

Figure 4.51. The Haworth representations of inulin and 6-kestose.

Figure 4.52. A Haworth representation of a short, four-residue malto-oligomer fragment of amylose.

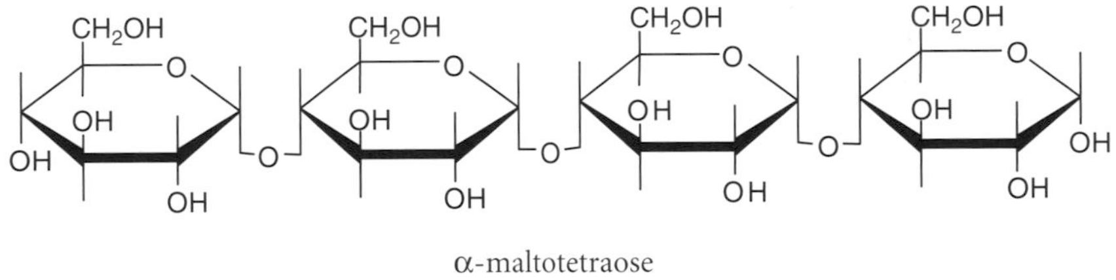

α-maltotetraose

Figure 4.53. When β-D-glucopyranose changes from the $^4C_1$ to the $^1C_4$ chair conformation, the O6 and O1 hydroxyls are sufficiently close as to allow an internal glycosidic condensation, generating a bicyclic molecule called levoglucosan.

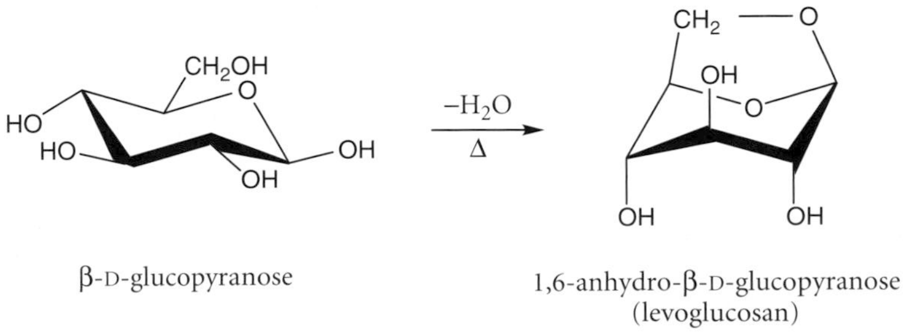

β-D-glucopyranose

$-H_2O$
Δ

1,6-anhydro-β-D-glucopyranose
(levoglucosan)

Molecules of this type are called **glycosans**. The reaction occurs upon heating, which promotes the conformational transition necessary to bring the two hydroxyl groups into close proximity. This strained bicyclic structure is formed, along with the furanose analog, 1,6-anhydro-β-D-glucofuranose, upon the pyrolysis (destructive distillation) of D-glucose, starch, or cellulose, all of which occur commonly in foods. Small amounts can be formed during the heating of glucose syrups above 100 °C during candy making. Because this compound has a slightly bitter and astringent taste, its accumulation in large amounts is undesirable, but it does contribute one of the flavor notes associated with caramel.

This reaction is not possible for the α-anomer, because even in the $^1C_4$ conformation, α-D-glucopyranose has its anomeric hydroxyl group oriented away from the exocyclic group (Figure 4.54).

Acid Hydrolysis

The addition of water across glycosidic bonds is called **hydrolysis**. Hydrolysis of glycosidic linkages is the reverse of the glycosidic condensation reaction and regenerates the original sugar and aglycon alcohol (Figure 4.55). This reaction breaks

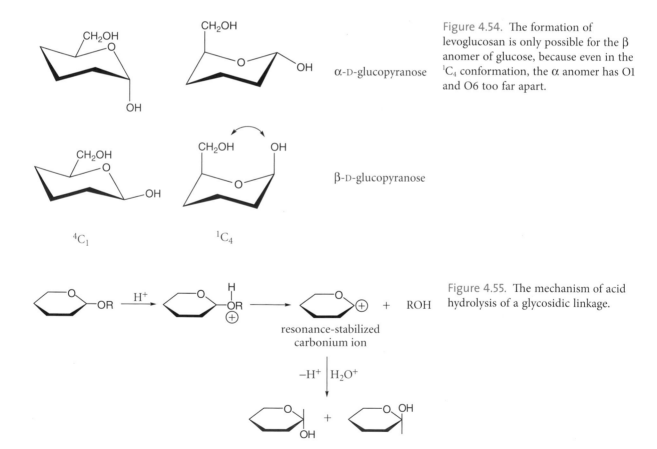

α-D-glucopyranose

Figure 4.54. The formation of levoglucosan is only possible for the β anomer of glucose, because even in the $^1C_4$ conformation, the α anomer has O1 and O6 too far apart.

β-D-glucopyranose

$^4C_1$ $^1C_4$

resonance-stabilized carbonium ion

Figure 4.55. The mechanism of acid hydrolysis of a glycosidic linkage.

disaccharides into the component monosaccharides and polysaccharides into smaller polymers and ultimately into component sugars. Thermodynamics favors hydrolysis over condensation, and thus hydrolysis releases energy. Hydrolysis of glycosidic linkages can proceed directly by acid attack at elevated temperatures; increasing the temperature significantly increases the rate of this hydrolysis, which also goes faster for α-D-anomers than for β-D-anomers.

It is commercially feasible to hydrolyze cheap corn starch to glucose, and then to isomerize some of the glucose to fructose, producing a low-cost sweetener solution that has captured a significant share of the sucrose market. These are referred to as **high-fructose corn syrups**. This can be done by several methods, including by direct acid conversion—starch is cooked at 140 to 160 °C for ~20 minutes in the presence of HCl, after which it is neutralized with soda ash and purified. When prepared in this way, troublesome bitter-tasting reversion products are formed by glycosidic condensation. The degree of conversion of corn syrups is measured in terms of **dextrose equivalents** (DE), the percentage of reducing sugars in corn syrup, calculated as dextrose (glucose), on a dry weight basis. These hydrolysis reactions are pursued to some particular DE $\cong$ 60.

Enzymatic Hydrolysis

Hydrolysis is also the basic step in the degradation and digestion of polysaccharides in foods. In digestion, hydrolysis is primarily catalyzed enzymatically, although it also occurs nonenzymatically by acid hydrolysis in the acidic environment of the stomach, to a limited extent. Enzymes that catalyze the hydrolysis of glycosidic bonds are called **glycosidases**.

The only polysaccharides that humans can digest are starch (which consists of two polymers, amylose and amylopectin), glycogen, and maltodextrins, along with the disaccharides maltose, sucrose, and lactose. (Dextrins are short oligosaccharides of glucose formed by the partial degradation of amylose; in maltodextrins, all of the linkages are α-(1→4)). The human body cannot assimilate polysaccharides or even disaccharides; they must be hydrolyzed to monosaccharides for transport into and through the blood. The hydrolysis of starch and glycogen begins in the mouth by the action of salivary **α-amylase** (Figure 4.56). The α-amylases, also secreted by the pancreas into the duodenum, are enzymes that hydrolyze α-(1→4) linkages scattered throughout the amylose chain to yield dextrins and ultimately maltose. Because these enzymes can hydrolyze bonds anywhere within the chain sequence, they are called **endo-glycosidases,** endoglycanases, or endoglucanases. The hydrolysis proceeds with retention of configuration at the anomeric carbon, which means that the resulting anomeric configuration is α. As the activity of α-amylase proceeds, it generates more and more short fragments that are progressively shorter, until some single glucose molecules are produced. If one chews a piece of bread for a long time without swallowing it, this process will begin to generate enough free D-glucose that the bread will begin to taste sweet.

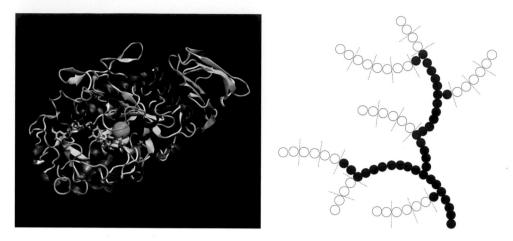

Figure 4.56. On the left is a cartoon diagram of the conformational structure of human pancreatic α-amylase (Qin et al. 2011). The salivary enzyme is similar. The alpha helices are colored purple, and beta sheet segments are shown in yellow. The active site is located in a deep cleft visible in the lower left part of the structure in this view, where an inhibitor called acarviostatin, which resembles amylose, is bound into the groove that would be occupied by substrate. This enzyme is a metalloenzyme that requires a bound calcium ion, shown as an orange sphere, to maintain its active conformation. A chloride ion, shown in green, is also required for activity. On the right is a schematic illustration of the digestion of amylopectin by β-amylase. Amylopectin is a polysaccharide of D-glucose, with a linear backbone of sugars linked α-(1→4) and with branching side chains linked α-(1→6). The β-amylase processively cleaves off maltose disaccharide units, indicated by the dashed lines, one at a time from the nonreducing end, but cannot cleave the α-(1→6) or proceed past it. Each sugar is schematically represented as a circle, with those making up the limit dextrin indicated in black.

A second type of enzyme, **β-amylase**, can also hydrolyze starch. This type of enzyme is produced by higher plants, and some types of bacteria, but not by humans and other mammals. Any β-amylase found in the human lower digestive tract is produced by the intestinal flora. β-Amylases are **exo-glycosidases**, or exoglycanases, cleaving off maltose disaccharides processively from the nonreducing end of the polysaccharide chain. The original α-linkage is inverted, producing β-maltose (4-O-(α-D-glucopyranosyl)-β-D-glucopyranose). Both types of amylases will also attack the branched starch polymer amylopectin, but because neither can hydrolyze (1→6) linkages or proceed past them, the degradation eventually stops. Because it is an endoglycosidase, α-amylase can hydrolize parts of amylopectin that β-amylase cannot, since while neither can attack the branching linkage, the α-amylase can jump past them to make further endo cleavages if a significant unbranched sequence exists where the α-amylase would not be inhibited by steric crowding. The polysaccharide that remains after amylopectin is exhaustively digested by amylases is referred to as a **limit dextrin** (Figure 4.56). This indigestible core of amylopectin is a component of what is known as **resistant starch**, starch that is not digested by humans. Highly branched amylopectin, called waxy starch, can be especially resistant to digestion. Retrograded amylose is also difficult to digest and contributes to resistant starch. All of these topics will be discussed below in the section on starch.

In ripening fruit, β-amylases hydrolyze starch to give free glucose, contributing to the sweet taste of the fully ripe fruits. Amylase is also present in the yolks of eggs. Because these amylases are very stable, when making cream fillings for pies using eggs, the cream must be heated above the boiling temperature to denature the enzymes, or their hydrolysis of the starch from the flour in the mixture will cause the filling to gradually liquefy.

Maltose is hydrolyzed by **maltase** in the intestine into D-glucose. Sucrose is hydrolyzed in the intestine by **sucrase**, also called **invertase**, into D-glucose, which is rapidly absorbed, and D-fructose, which is more slowly absorbed. Lactose is hydrolyzed in infants and many adults by intestinal **lactase** into D-glucose and D-galactose. Until the development of animal husbandry and dairying, our Pleistocene ancestors had no need to produce lactase beyond early childhood. It appears that the genetic capacity to continue expressing the lactase gene beyond infancy evolved separately and quite recently among cultures that practice some form of dairying, and is thus unevenly distributed in modern populations. As a result lactose intolerance is widespread, with certain populations having larger proportions of intolerant individuals than others (Beja-Pereira et al. 2003).

Polysaccharides other than starch and glycogen, such as celluloses, hemicelluloses, and pectin, are not hydrolyzed by human gastrointestinal enzymes and proceed on to the large intestine basically intact, where they apparently play a useful physiological role by providing bulk to aid peristaltic action. Thus, the majority of foods like lettuce, spinach, and celery cannot be digested and yield few calories. This group of indigestible polysaccharides is called **dietary fiber**. Grazing animals like cows or horses also produce no enzymes capable of hydrolyzing cellulose, but are able to partially utilize this most abundant of carbohydrates by indirectly taking advantage of the **cellulases** generated by microorganisms in their digestive tracts. Limited digestion of this type can also sometimes take place in the human gut, making some portion of the calories in cellulose available.

The disaccharide sucrose (table sugar), which does not have a free anomeric carbon and thus is not a reducing sugar, consists of equal amounts of the two monomer units D-glucose and D-fructose. Because it is not a reducing sugar, sucrose does not undergo mutarotation. In the presence of invertase it is easily hydrolyzed to its component monosaccharides. This process is called **inversion**, which explains the origin of the name of the enzyme, because it results in the inversion of the sign of the solution's optical rotation of the sodium D line from dextro- to levo-rotary. This occurs because the three sugars have the following optical activities:

for sucrose: $[\alpha]_D^{20} = +66.5°$

for D-glucose: $[\alpha]_D^{20} = +52.5°$

for D-fructose: $+\dfrac{[\alpha]_D^{20} = -92.0°}{\Delta = -39.5°}$

Because the hydrolysis of sucrose produces D-glucose and D-fructose in equimolar amounts, the optical rotation of this solution is just the sum of the individual optical rotations for the two sugars, or −39.5°. Because the optical rotation for D-fructose is so negative, this sum is negative rather than positive, and it thus is opposite in sign, or inverted, from that of sucrose.

This equal mixture of D-glucose and D-fructose is called **invert sugar**. Because enzymes such as invertase are expensive, invert sugar can be prepared commercially via acid hydrolysis using citric acid. Invert sugar has several advantages over sucrose when used as a food ingredient. It is sweeter than sucrose and has a considerably higher water binding capacity. It also is not susceptible to crystallization as sucrose is. One of its principal limitations is that, as a reducing sugar, it is susceptible to Maillard browning (see below), while the nonreducing sucrose is not. The enzyme invertase, produced commercially from yeast, is also sometimes used as a food additive in confectioneries to produce in situ creamy or liquid fondants as fillings for the interiors of chocolates. The hydrolysis of sucrose prevents crystallization and causes the fondant to partly liquefy over time.

Honey somewhat resembles invert sugar, in that it also has high and nearly equal proportions of glucose and fructose, because bees produce invertase to hydrolyze the sucrose in nectar. It differs from invert sugar not only in that honey contains small amounts of other sugars such as maltose, as well as many other minor components, but also in that it contains more fructose than glucose. Nonetheless, invert sugar has also been referred to as artificial honey in the past. The water content of honey is only about 17%, so the sugar solution is very concentrated. As a result, its water activity is low, only about 0.75. Thus, it is relatively stable toward microbial growth (see Figure 2.43), but it can absorb atmospheric moisture, increasing moisture content and allowing the limited growth of certain yeasts. Honey contains small amounts of amino acids as well as other proteins (mostly enzymes like invertase), and so it can be susceptible to limited amounts of Maillard browning at elevated temperatures (next sections), particularly because this type of reaction is promoted by low-moisture environments. One of the enzymes in honey is glucose oxidase, which catalyzes the oxidation of glucose to gluconic acid (Figure 4.31), with the production of hydrogen peroxide as a by-product,

$$C_6H_{12}O_6 \text{ (glucose)} + O_2 + H_2O \xrightarrow[\text{glucose oxidase}]{} C_6H_{12}O_7 \text{ (gluconic acid)} + H_2O_2$$

Hydrogen peroxide inhibits the growth of bacteria, and its presence in small amounts in honey gives it a mild preservative property, as well as contributing to the flavor, as does the gluconic acid, which also lowers the pH, along with other acids, to around 4.

Production of Corn Syrups and High-Fructose Corn Syrups

Corn syrups are concentrated aqueous solutions of malto-oligomers produced from the partial hydrolysis of starch, used in foods as a thickener and to control water, particularly in confectionary products. **High-fructose corn syrup (HFCS)** is a liquid syrup made from the free glucose monomers of hydrolyzed corn syrup by enzymatic isomerization of the glucose to fructose using the enzyme xylose/glucose isomerase. It is widely used as a sweetener and as a humectant because of its water retaining (binding) capacity. Generally, as the name implies, the source of HFCS is corn, but it can in fact be starch from other sources such as potatoes. There are several methods for producing corn syrups.

1. *Direct Acid Conversion.* Starch is cooked at 140 to 160 °C for 15–20 minutes in the presence of ~0.12% HCl until the desired DE is reached. The mixture is then neutralized with soda ash and purified. This method is simple and cheap, but numerous reversion products are formed during this procedure, some of which are bitter tasting.

2. *Acid-Enzyme Conversion.* Direct acid conversion, as above, is used until the DE is $\cong$ 40 to 50%. The mixture is then neutralized and treated with glycosidases, usually α-amylase. The enzyme is stopped by thermal deactivation around a DE of ~60. **High-maltose corn syrups** are prepared by allowing the acid conversion to proceed only until the DE is about 20, followed by treatment with β-amylase, which produces maltose as a product.

3. *Enzyme–Enzyme Conversion.* Gelatinized corn starch (discussed below) is treated first with α-amylase, followed by treatment with glucoamylase, leading to near complete conversion to D-glucose.

4. *High-Fructose Corn Syrup.* For the production of HFCS, the corn syrup is then treated with **glucose isomerase**, which converts glucose to fructose and thus produces a mixture of D-glucose and D-fructose. The reaction does not lead to a complete conversion of glucose to fructose, but rather yields only a little more than 40% conversion, because these two molecules have almost the same free energies, with glucose being a little lower in energy, so that thermodynamics does not favor complete conversion (see Chapter 6). In industrial production of HFCS, the percentage of fructose in the syrup is increased using liquid chromatography to as high as 90%, which is marketed as HFCS 90. This is then mixed with 100% glucose corn syrup to give whatever proportions desired. HFCS 55 (55% fructose, 45% glucose) is the form most commonly used to sweeten soft drinks.

HFCSs as Sweeteners

Although sugar (sucrose) is relatively cheap in the United States, its price is higher than would be dictated by market forces alone due to a variety of circumstances, such as the decades-old US embargo of Cuban products and import restrictions designed to protect domestic production. As a result, using HFCS to sweeten products such as soft drinks is less expensive than using sucrose (remember that when billions of an item are sold per year, even a small fraction of a cent decrease in production cost per unit can lead to large profits). Beginning in the 1970s, most soft drinks in the United States were reformulated to partially or completely replace sucrose with HFCS. High-fructose corn syrups are also used to sweeten baked products, candies, juices, soups, tomato sauces, and ketchups. In 1970 the per capita consumption of HFCS in the United States was just 0.23 kg per year, but increased two orders of magnitude to 28.4 kg by 1997 (Putnam and Allshouse 1999). It has not escaped the attention of a number of workers that this reformulation correlates with the apparent beginnings of the modern epidemics of diabetes and obesity, and it has been suggested that this correlation is not necessarily coincidental. At first this suggestion might seem counterintuitive, because fructose is sweeter on a per weight basis than sucrose, in principle allowing a reduction in calories for an equivalent level of sweetness. Physiological arguments for thinking that fructose might be responsible at least in part for some of these problems should, however, be considered. The metabolism of fructose in humans differs somewhat from that of glucose. When glucose is consumed, its metabolism in the liver first involves conversion to glucose 6-phosphate, followed by that molecule's conversion to fructose 6-phosphate by the enzyme phosphoglucoisomerase, and this is in turn converted to fructose 1,6-bisphosphate by the enzyme 6-phosphofructokinase. This last step is inhibited by citrate and ATP and limits the production of circulating lactate. Fructose metabolism escapes this regulating step. The uptake of fructose by the liver, where it is directly metabolized into two triose phosphate molecules, is not inhibited at the level of phosphofructokinase as glucose is, so that direct consumption of fructose produces a higher level of blood lactate than does glucose. While small intakes of fructose are likely to be helpful in glycemic control in type 2 diabetes, larger doses of fructose, bypassing the phosphofructokinase bottleneck, result in the unregulated production of acetyl-CoA and increased lipid (fat) production. Fructose also has a low glycemic index, so its consumption does not stimulate the production of insulin for control of energy homeostasis, thus providing a sense of satiety. Therefore, one may be tempted to eat (or drink) again, leading to the consumption of more calories.

On the other hand, a recent study was unable to find any statistical evidence for a decreased sense of satiety when drinking soft drinks sweetened with HFCS (Monsivais et al. 2007). Furthermore, the actual difference between sodas sweetened with HFCS and sucrose is small. As noted, the most typical sweetener for such soft drinks is HFCS 55, which is 55% fructose and 45% glucose, which is fairly similar to the 50–50 composition of sucrose. Sucrose is essentially completely hydrolyzed by intestinal sucrase/invertase (in fact, because the pH of typical colas can be as low as 2.5 to 3.5, significant acid hydrolysis of sucrose can take place in the bottle, before being consumed, if it is stored at warm temperatures). Furthermore, there is little indication that either monosaccharide is not completely adsorbed across the intestinal lumen. As of this writing, this question remains unresolved, with some workers arguing that there is little difference between consuming HFCS or an equivalent amount of sucrose (Elliot et al. 2002), while others have suggested that as a food additive fructose should be considered as toxic (Lustig et al. 2012)!

Reactions under Basic Conditions

Enolization

While sugars are fairly stable under acidic conditions, basic conditions can lead to extensive changes and even the destruction of sugars. One of the most important reactions of simple sugars in base is **enolization**, in the so-called Lobry de Bruyn–Alberda van Ekenstein reaction. Under alkaline conditions, or if sugar solutions are heated above 100°C, the open-chain aldehyde forms undergo enolization, the formation of an enediol ion. This is an acid-base catalyzed reaction, but base is much more efficient in promoting it. Under basic conditions, the weakly acidic proton on the C2 carbon atom, α to the carbonyl group, is abstracted, resulting in loss of chirality at that carbon (Figure 4.57).

This step is easily reversible, which leads to epimerization at the C2 position because the proton can go back on by attacking from either side. As can be seen from Figure 4.57, the enediol ion has several resonance forms, so that the abstracted proton can also be replaced by another proton binding to the C1 carbon, leading to the isomerization of the sugar to a ketose. Thus, in the example shown, adding base to D-glucose leads to the rapid formation of D-mannose and D-fructose, in addition to some regeneration of the original D-glucose. This same set of products would be expected from the treatment of D-mannose with base, and among those resulting from such treatment of D-fructose as well. The extent of this type of reaction is dependent upon concentration, temperature, time, and the natures of the specific sugar and cation.

The reaction can continue on in either direction from the carbonyl group of D-fructose, because both the C1 and C3 protons are "α" to the ketose carbonyl group (Figure 4.58).

Enolization can also proceed, although much less efficiently, by acid catalysis. The carbonyl group is first protonated by the acid, then the OH^- of water or the anion of the acid acquires the expelled proton (Figure 4.59).

Dehydration and Thermal Degradation

At higher temperatures, acidic or basic reagents can cause elimination of a hydroxyl group from the position β to the carbonyl group, ultimately leading to the formation of the yellow-colored pigment **hydroxymethylfurfural** (HMF) and various brown-colored polymers of this molecule and its derivatives. The acid mechanism is shown in Figure 4.60.

Alkaline conditions are relatively uncommon in foods, since most foods are acidic rather than alkaline. However, egg white is alkaline, and foods treated with alkali, such as hominy and pretzels, may be weakly basic. Maple syrup may also become alkaline during open-kettle boiling. Baking soda is of course alkaline, and slightly alkaline conditions have been used to conserve the green color of chlorophyll in processed vegetables.

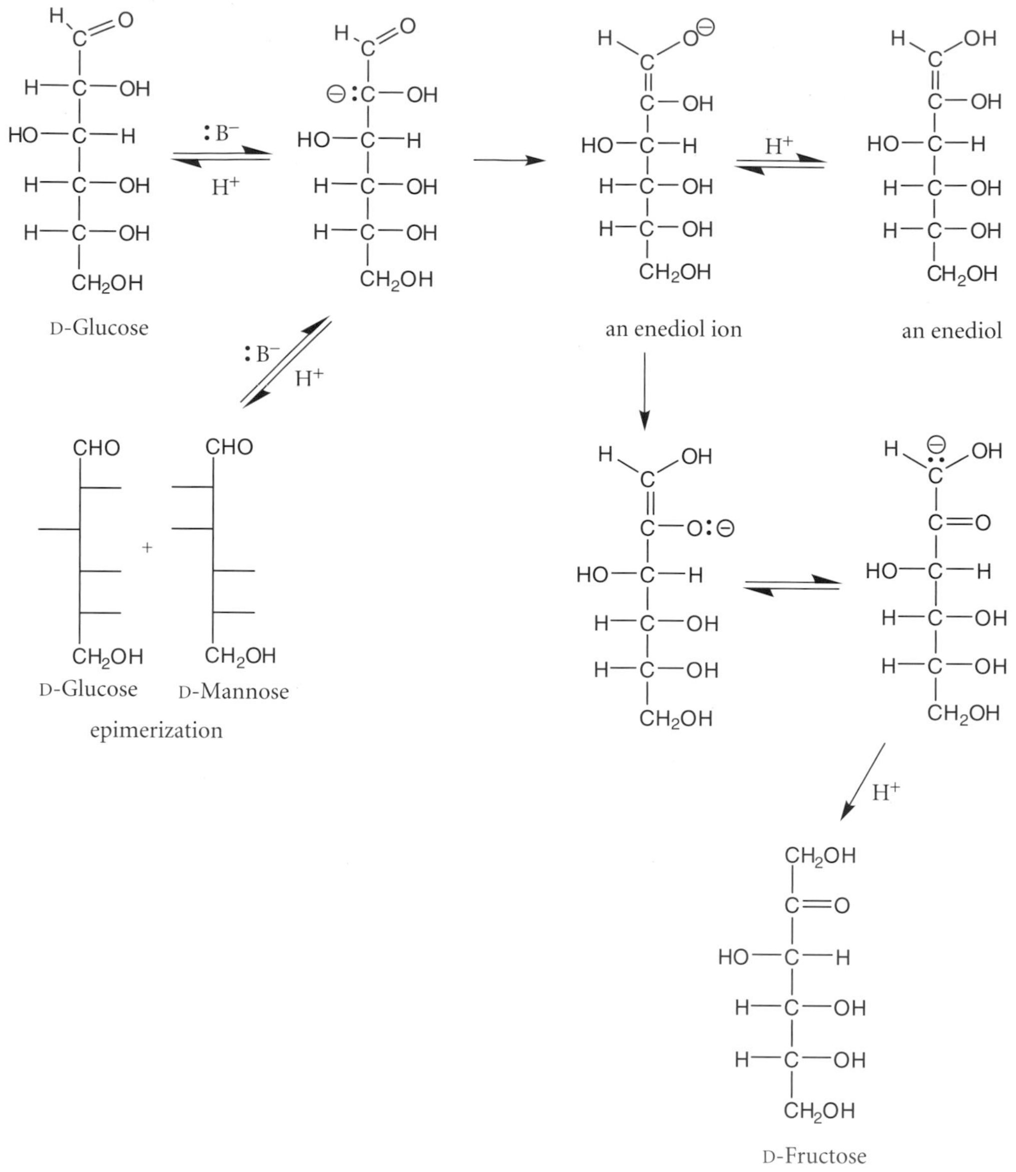

Figure 4.57. The treatment of a reducing sugar with base abstracts the aliphatic proton in the alpha position relative to the carbonyl group, resulting in epimerization and the formation of an enediol ion, which can isomerize to produce a ketose.

In alkaline media, dehydration of carbohydrates by β-elimination may also occur through base catalysis (Figure 4.61). In acid, enolization is slow, and dehydration reactions are rapid and unaffected by air oxidation, with few fragmentation products. In weakly alkaline media, enolization is rapid, dehydration reactions are slower than enolization, air oxidation changes the product composition, and many fragmentation products are formed, such as those shown in Figure 4.61. The various dehydration

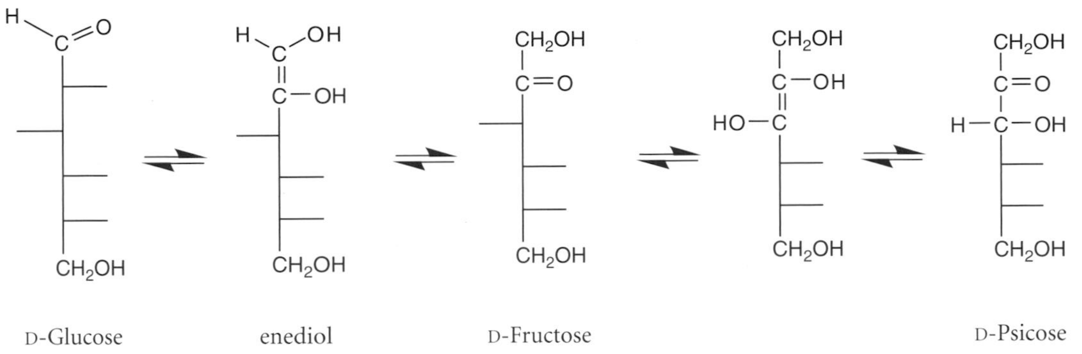

Figure 4.58. Ketose sugars can also undergo enolization.

Figure 4.59. Enolization can also result from treatment with acid, although less efficiently.

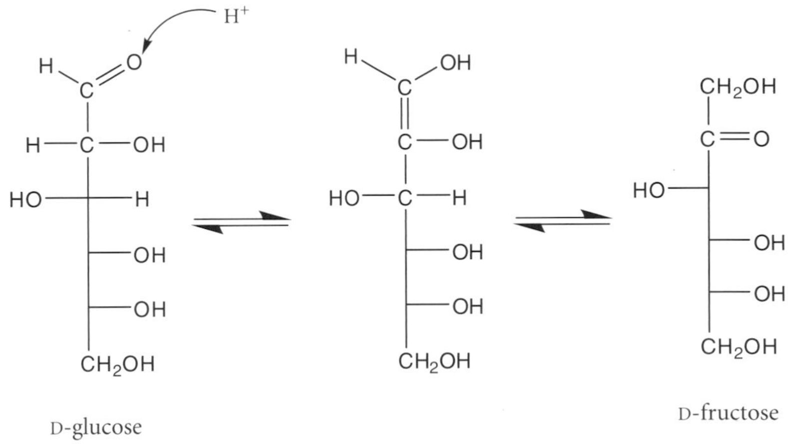

and fragmentation products resulting from enolization are a major contributor to the processes of nonenzymatic browning.

Nonoxidative Browning

Two major types of browning in foods occur, **oxidative** and **nonoxidative browning**. Enzymatic oxidative browning involves the enzyme-catalyzed reaction between O_2 and a phenolic substrate. The enzymes that catalyze these reactions are called polyphenol oxidases (PPO), and are found only in plants. Enzymatic browning, which is discussed in more detail in Chapter 6, is the reason that damaged fruits and vegetables turn brown. This type of browning does not involve carbohydrates. The nonenzymatic reactions directly involve carbohydrates and are principally of two types. Both of these reactions are quite complex. One is called caramelization and results from heating sugars to high temperatures. Although many types of reactions

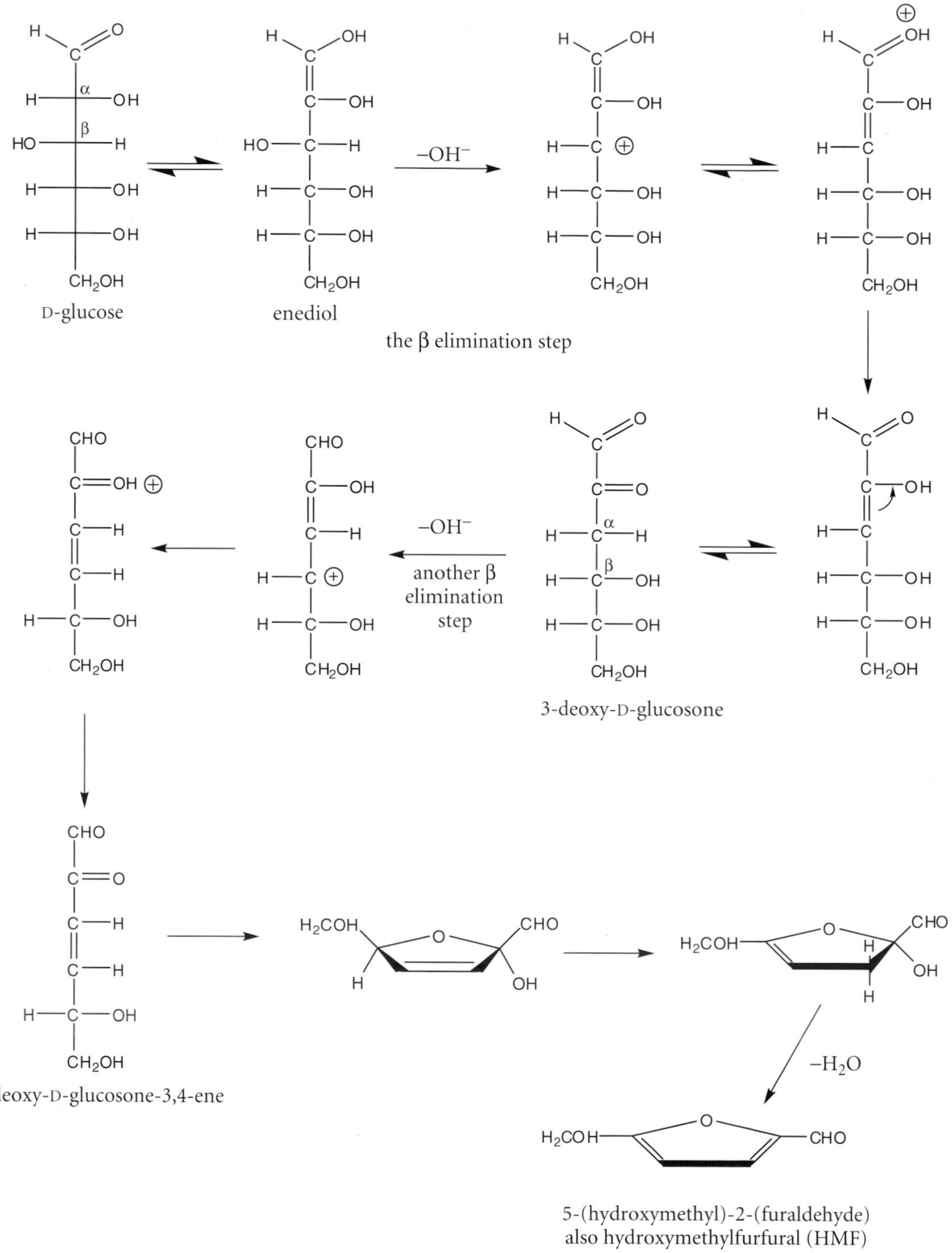

Figure 4.60. The thermal dehydration of a reducing sugar to produce hydroxymethyl furfural.

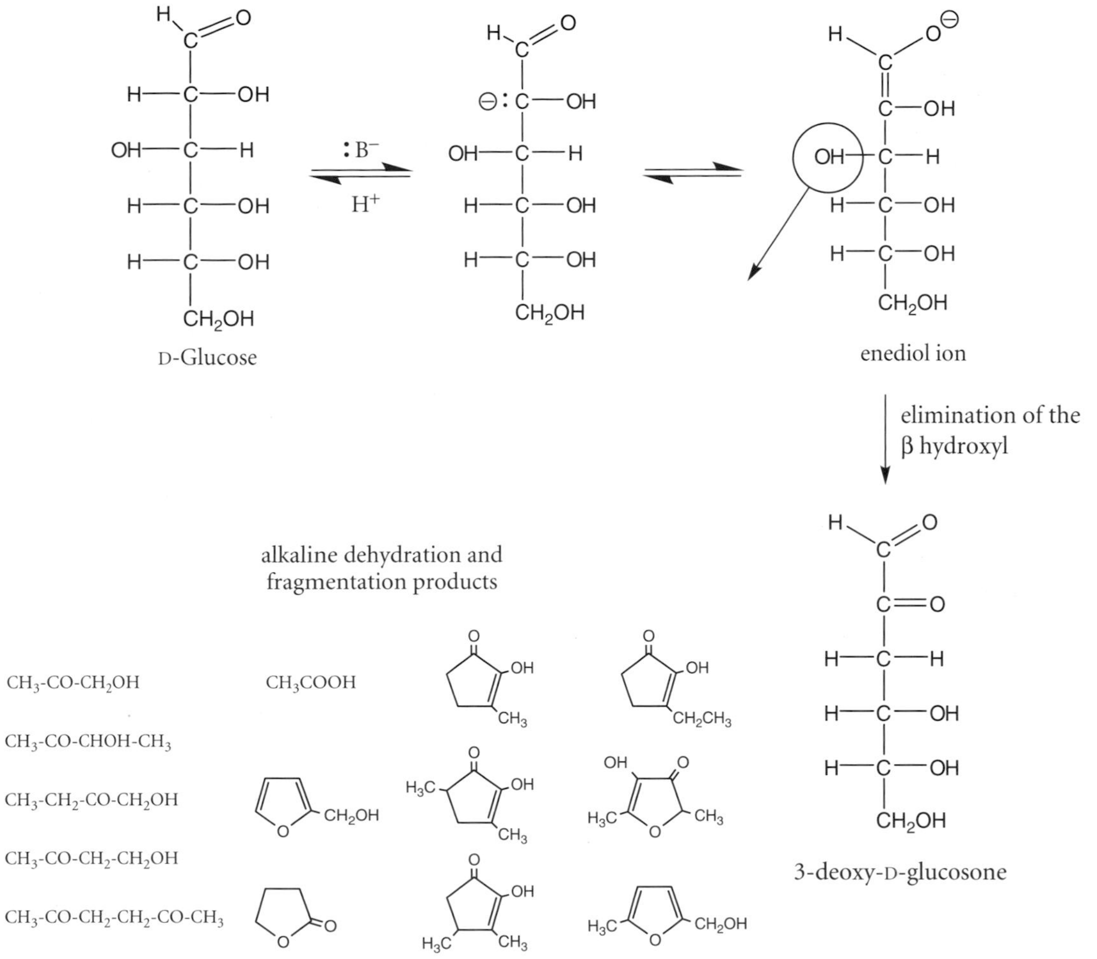

Figure 4.61. Examples of dehydration products formed through β-elimination. (Redrawn with permission after a figure from J.E. Hadge and E.M. Osman in *Principles of Food Science, Part I, Food Chemistry*, O.R. Fennema, ed. New York: Marcel Dekker, 1976.)

contribute to caramelization, a number of these involve the decomposition of sugars. The other type of browning, called Maillard browning, is a reaction between reducing sugars and the free amines of proteins and amino acids. Both of these nonenzymatic reactions are important in cooking and food processing, and both can be desirable in some situations and unwanted in others, so that understanding and controlling both is important.

Caramelization

Direct heating of carbohydrates produces a complex set of reactions called **caramelization**, which is assisted by addition of small amounts of acid and salts. While many different types of reactions are taking place, the most characteristic lead to the pyrolytic decomposition of sugars. These reactions include:

1. inversion of sucrose to D-glucose and D-fructose
2. anomeric and tautomeric equilibration
3. intermolecular condensation to produce randomly-linked di- and tri-saccharides (reversion products)
4. intramolecular condensation (i.e., formation of glycosans)
5. isomerization of aldoses and ketoses (e.g., epimerization, through enolization)
6. enolization and dehydration reactions, producing double bonds in sugar rings, HMF, and various decomposition products.

For most simple sugars, temperatures of 160°C are required for caramelization. Sucrose is mostly used for caramelization to make colors and flavors. There are three main kinds of caramel:

1. Acid-free caramel made with ammonium bisulfate catalyst to produce color for cola drinks.
2. Brewer's color for beer is made by heating sucrose solution with ammonium ion.
3. Baker's color, produced by direct pyrolysis of sucrose to give a burnt sugar color.

The distribution of flavor and aroma compounds in a particular caramelization process depends on the details of the recipe and processing. Caramelization also occurs when foods containing high concentrations of sugars, like many vegetables, most typically onions, are cooked over high heat. The flavor of caramel is complex and in addition to sweetness, from residual un-degraded sugar, also contains bitter notes (from reversion products, levoglucosan, and dehydration products, for example), and even sour notes. The longer the sugar is cooked, the more of the original sugar will be pyrolyzed, and the darker and less sweet the resulting caramel will be. Because the oxidation of sugars is exothermic, adding extra energy to the cooking process, careful temperature control in caramelization is important.

The fragmentation and dehydration reactions in caramelization produce hundreds of molecules (Figures 4.61. 4.62, and various polymers of these molecules, particularly pyyrole), a number of which absorb visible wavelengths of light, giving rise to the brown caramel color. Others contribute to the characteristic caramel flavors and aromas. Among those not shown are acetaldehyde, ethanol, ethanedial, pyruvaldehyde, diacetyl (Figure 8.20), furan (Figure 4.13), and furan derivatives, including acetylformoin and 4-hydroxy-2,5-dimethyl-3-furanone. Maltol (Figure 4.62) has a distinctive aroma of caramel, often identified as cotton candy by subjects in odor test panels, and is also often described as the aroma of cedar, pine, or other conifers. Maltol not only contributes to the taste and aroma of caramel, but also to the aroma of roasted malt (from which it gets its name) and baked bread. In each case, this is the result of sugar caramelization. The US FDA permits its use as a food additive for flavor enhancement. The related ethyl maltol and isomaltol are also characteristic of baked breads and caramelized vegetables, and 4-hydroxy-2,5-dimethyl-3-furanone

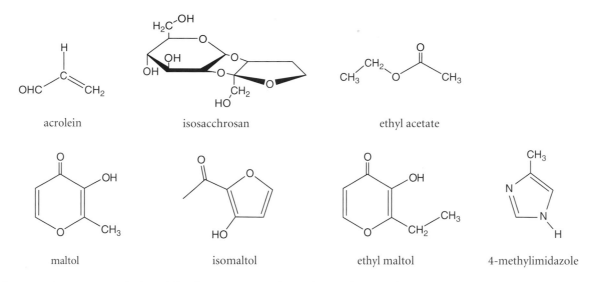

acrolein   isosacchrosan   ethyl acetate

maltol   isomaltol   ethyl maltol   4-methylimidazole

Figure 4.62. Structures of a few of the characteristic products of caramelization of sugars.

makes an important contribution to the flavor of caramel. Both maltol and ethyl maltol are able to chelate ferric iron and other metal ions (Chapter 10).

One common product of caramelization, 4-methylimidazole, which is also formed during Maillard browning, has been implicated as a toxin and possible carcinogen at high levels. This molecule, which is soluble in both water and alcohol, is the side chain of the amino acid histidine (see Figure 5.3) and is present in very small amounts in commercial caramel color. As a result of the concerns about its safety, some soft drink manufacturers have modified the processes used in the production of their caramel color in order to reduce the levels of this compound. As a product of both types of nonoxidative browning reactions, 4-methylimidazole is found in low levels in a number of cooked and roasted foods and also results naturally from some fermentation processes.

Maillard Reaction

When reducing sugars are heated in solution in the presence of amines, the sugar reacts with the amines to form a glycosylamine. Although this reaction is reversible, it initiates a series of additional reactions that can lead to the formation of a wide range of flavor, odor, and color molecules, including HMF (Figure 4.60) and other dark-colored compounds, in a process referred to as **Maillard browning**, named after the French chemist and physician Louis Camille Maillard, who first studied these reactions. Maillard reactions between simple sugars and proteins are common in foods, because several of the amino acids that constitute proteins, most notably lysine and proline (see Figure 2.28, and Chapter 5), contain free amine functionalities. The reactions are used to produce colors and flavors in a number of food products, including fudge, caramels, and milk chocolate, and also take place in the production of soy sauce, the baking of bread, and the roasting of meats. Although in many cases these reactions are desirable, they also have drawbacks that might make it desirable to inhibit the reactions. One drawback is that the dark product pigments might not be wanted in all cases. In addition, the reactions involve the destruction of the essential amino acid lysine through the reaction with its ε-amino group. Also, some of the Maillard products may be mutagenic or otherwise harmful in large quantities.

The most common sugars involved in Maillard reactions in foods are the monosaccharides glucose and fructose and the disaccharides maltose and lactose. Sucrose does not undergo this reaction, because it is not a reducing sugar, unless it is first hydrolyzed to produce free glucose and fructose. Maillard reactions proceed more readily for primary amines like the amino group of lysine than for secondary amines like proline (Figure 5.3). These two amino acids are primarily the ones involved in Maillard browning, although the reaction also takes place with free tryptophan and arginine. As with most reactions, the rate is significantly increased by raising the temperature, and it is favored by low water activity and higher pHs. Raising the temperature by 10 °C approximately doubles the browning rate (Figure 2.48).

Maillard browning begins with the nucleophilic attack by an amine on the carbonyl group of a sugar to form a glycosylamine (see Figure 4.63). Thus, the reducing sugar must be in its open-chain, free aldehyde or ketone tautomer to react. The formation of the glycosylamine is a reversible process, and can be reversed to regenerate the reactants. The glycosylamine can also quickly undergo an internal rearrangement to form an aminoenol that can further rearrange to produce an aminoketose, a 1-amino-2-keto sugar commonly referred to as an **Amadori compound**. The Amadori products exist in equilibrium not only with the aminoenol intermediate but also with an enediol form; both of these can undergo further reactions to generate additional products. Alternatively, the enediol intermediate can rearrange to produce either 1-deoxy-2,3-hexodiulose, with the loss of the amine group, or 1-amino-1,4-dideoxy-2,3-hexodiulose, with retention of the amino group. Subsequent rearrangements lead to the formation of ethyl reductones and ultimately to the production of melanoidin pigments. The aminoenol intermediate, on the other hand, can undergo several rearrangements that produce the corresponding 3-deoxy-2-hexulose. This molecule exists in equilibrium with a number of heterocyclic forms, and one product of this pathway is hydroxymethylfuraldehyde, and again, ultimately, an array of melanoidin pigments, which are the source of the brown color.

Cereal grains, such as the wheat used in making bread and the malt barley used in beer, contain significant amounts of the amino acid proline. Thus, even though this secondary amine reacts less readily in Maillard browning, the Maillard products of proline are particularly important for the flavors and odors of foods made from grains. Many of the products of the Maillard reaction of proline are volatile nitrogen-containing heterocyclic compounds. Tryptophan in protein or polypeptide chains is not significantly reactive in Maillard browning, but free tryptophan does undergo the reaction. Since there is free tryptophan in soy sauce, complex polycyclic products from Maillard reactions of tryptophan contribute to the flavor and color of soy sauces. Arginine, even when bound in a polypeptide or protein chain, can undergo, like

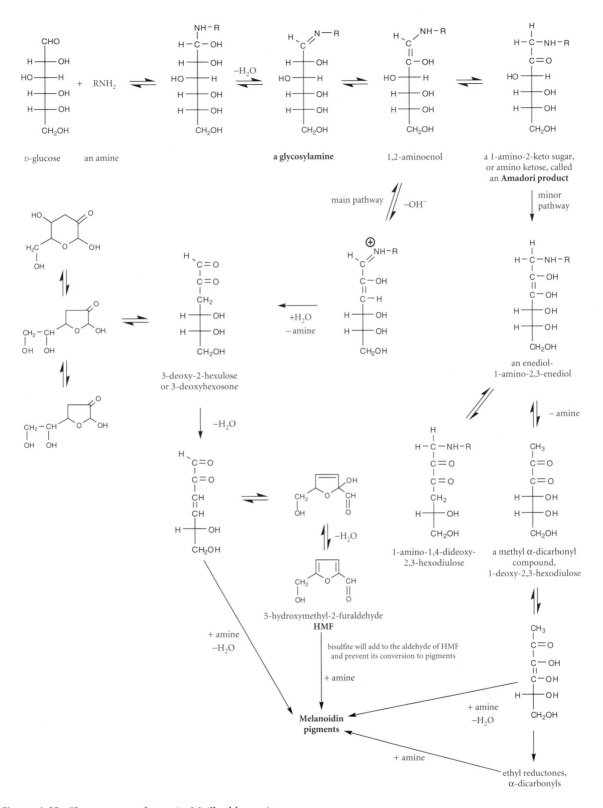

Figure 4.63. The sequence of steps in Maillard browning.

lysine, significant modification due to Maillard browning. Because arginine is not an essential amino acid, however, its loss through this route is not as nutritionally significant as that of lysine. Cysteine cannot directly undergo Maillard browning, but releases hydrogen sulfide upon heating that can react with other Maillard products (see below). Free ammonia, which of course is a primary amine, can also undergo Maillard browning reactions with sugars. There is little free ammonia in unspoiled foods, but ammonia is used to manufacture caramel coloring for use in cola drinks. The other amino acid side chains do not readily undergo Maillard reactions when they are part of protein chains because they do not contain free amines. All free amino acids in solution of course contain an amine group and, thus, could undergo Maillard browning under such conditions.

The melanoidin compounds that ultimately form as the products of the Maillard reaction have not been adequately characterized. They are thought to be a diverse and complex group of macromolecules. Several possible models for their structure have been advanced. One model treats them as polymers of the furans and pyrroles that result from the breakdown of hexose and pentose sugars (Figure 4.64). These

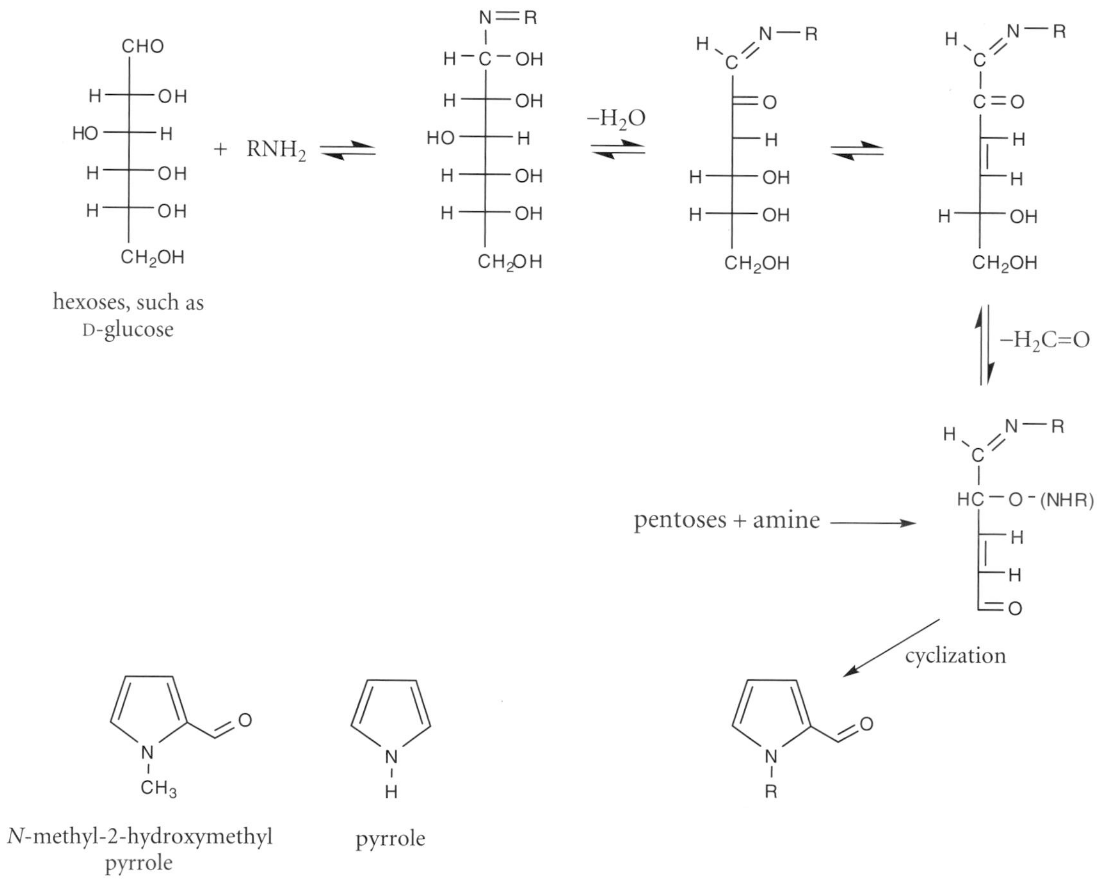

Figure 4.64. A mechanism for pyrrole derivative formation in Maillard browning.

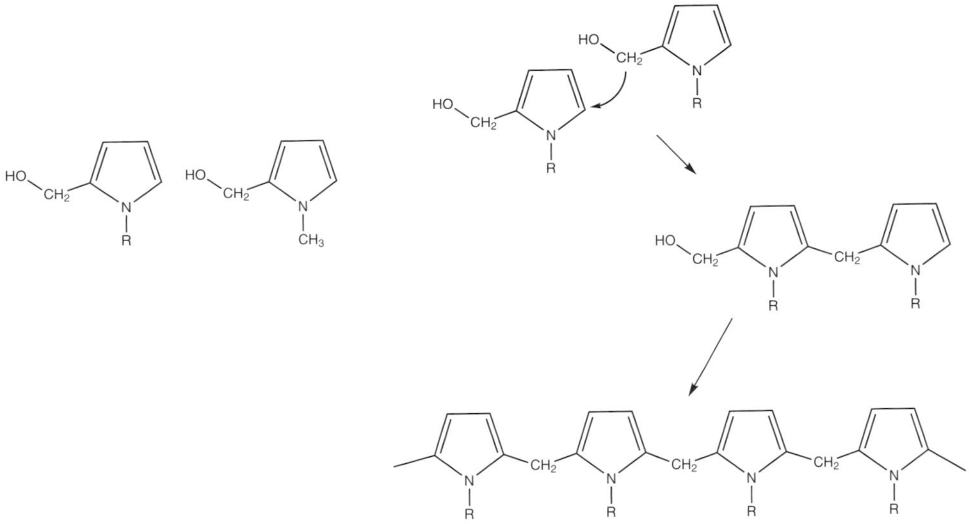

Figure 4.65. In one possible model for the structure of melanoidins, they result from polymerization of pyrrole and furan breakdown products of both pentose and hexose sugars. (R. Tressl et al., *J. Agric. Food Chem.* 46:104–110, 1998.)

molecules polymerize with similar molecules, as shown in Figure 4.65, to form an extensive macromolecular chain. Pyrrole derivatives, unsubstituted pyrroles, and Strecker aldehydes (next section) could be incorporated into a complex, irregular, and possibly branched, macromolecular backbone such as is shown in Figure 4.66. Another set of complex molecules that have been identified in melanoidin pigments are adducts of 4-[1-formyl-2-(2-furyl)ethenyl]-5-(2-furyl)-2-[(E)-(2-furyl) methylidene]-2,3-dihydro-α-amino-3-oxo-1H-pyrrole derivatives, which have a brown-orange color, to the ε-amino groups of lysine side chains of polypeptides (Figure 4.67).

Maillard browning is the reaction responsible for the development of the brown color of bread as it is baked, and of toast. This complex reaction generates hundreds of aroma, taste, and color molecules. A large number of Maillard products are responsible for the aromas of toasted and roasted foods, such as roasted meats, roasted coffee, baked goods, toast, rice cakes, popcorn, and malted barley. Many of these aroma molecules are volatile heterocyclic ring compounds, such as those shown in Figure 4.68. Hundreds of derivatives of these basic ring structures are found in the aromas of baking and roasting. Many other types of heterocyclic amine compounds result from Maillard reactions of proline, which starts out as a heterocyclic ring (see Figure 5.3). The thiazole ring compounds result from the heating of cysteine, which releases hydrogen sulfide, which can react with Maillard intermediates.

Two of the most important of the aroma products of Maillard browning in breads and baked goods are 2-acetyl-1-pyrroline and 6-acetyl-2,3,4,5-tetrahydropyridine, shown in Figure 4.69. Both of these molecules are thought to form specifically from the Maillard reaction of the amino acid proline. The molecule 6-acetyl-2,3,4,5-tetrahydropyridine exists in an equilibrium with 6-acetyl-1,2,3,4-tetrahydropyridine, with the latter being favored.

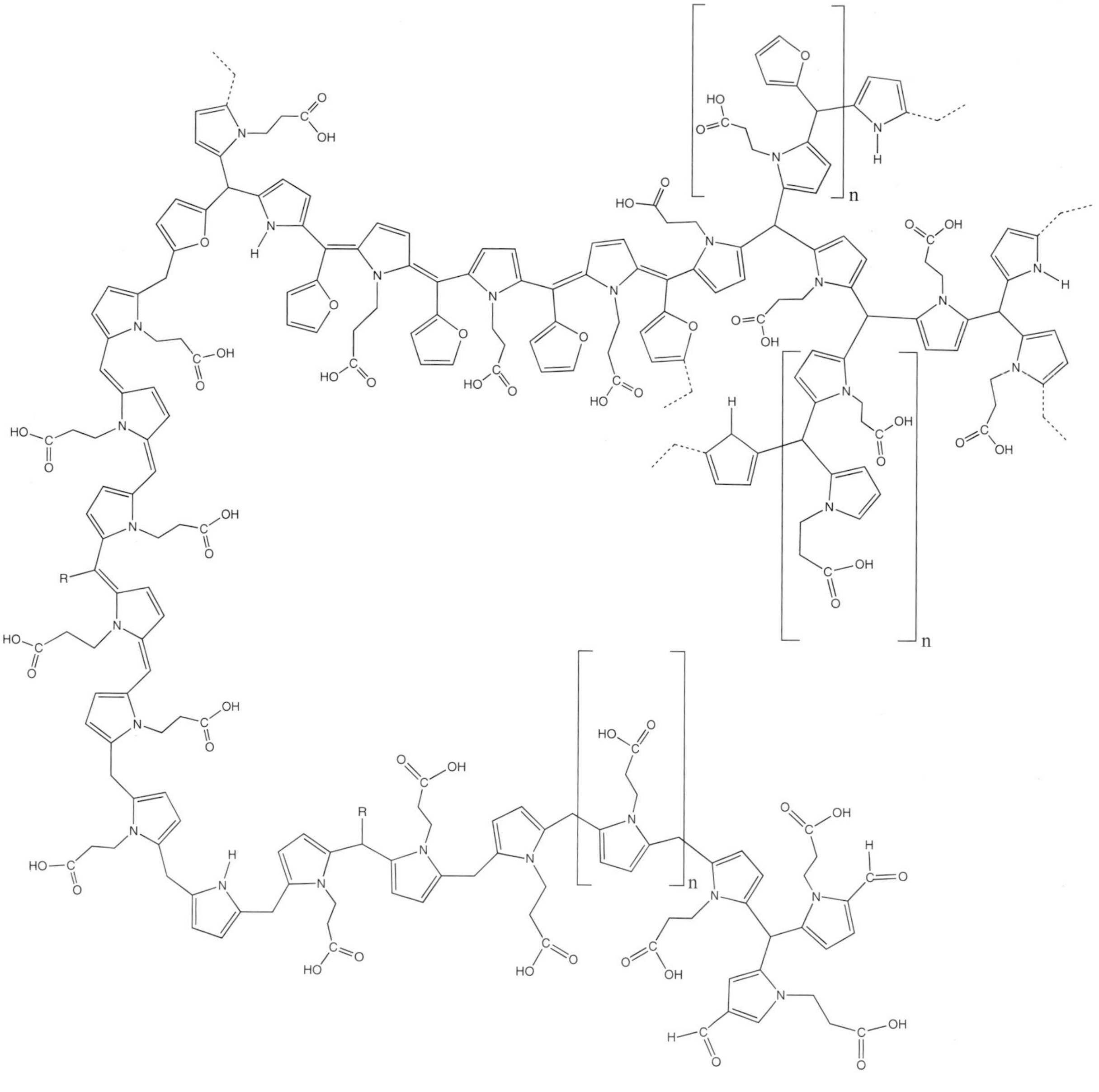

Figure 4.66. A possible structure for some of melanoidins that could be formed by the polymerization of substituted pyrroles in the Maillard reaction. (R. Tressl et al., *J. Agric. Food Chem.* 46:1765–1776, 1998.)

If not desirable, Maillard browning can be inhibited in several ways. The first and most obvious of these is to avoid reducing sugars in formulating products where this type of browning could be expected. Even when using the most common nonreducing sugar, sucrose, this is not altogether possible, however, because the glycosidic bond in sucrose can be hydrolyzed under acid conditions such as exist in many food products. Using sugar alcohols (alditols) will eliminate this problem. Low pHs inhibit Maillard browning (while increasing sucrose hydrolysis), because when the ε-amino

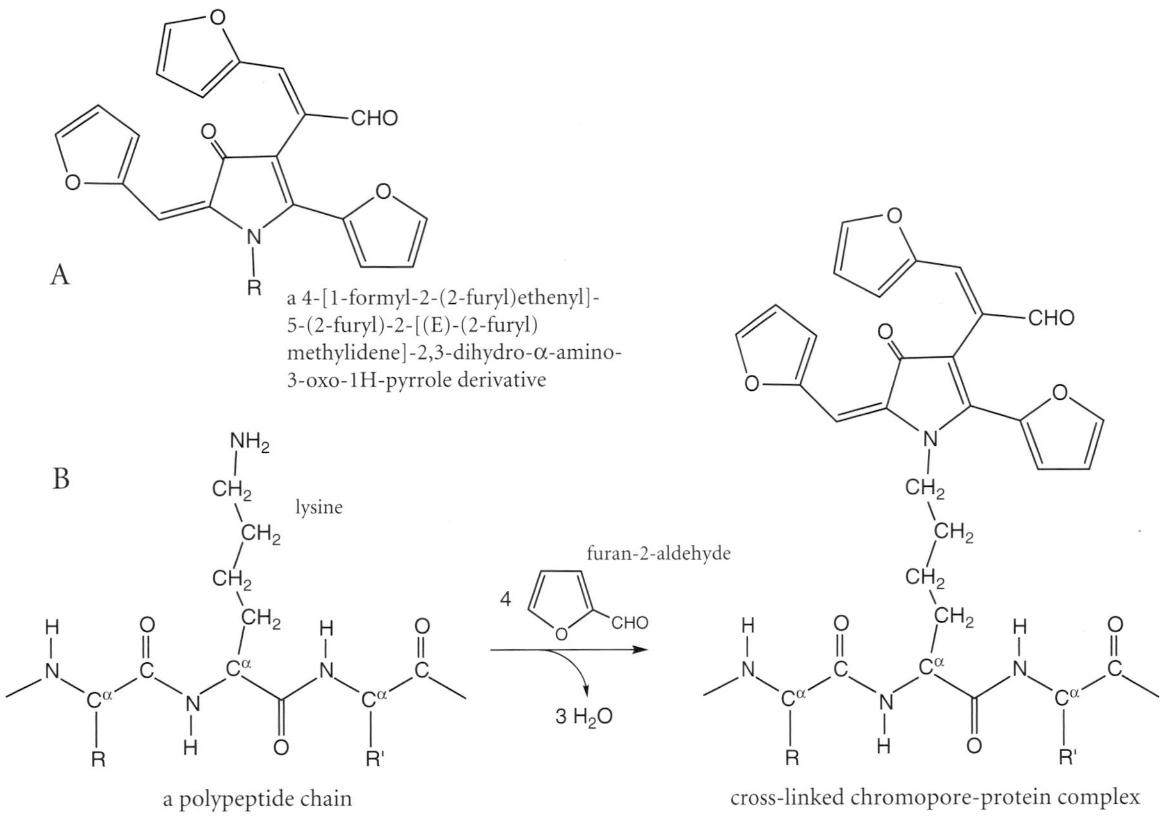

Figure 4.67. A proposed structure for melanoidins based on a model of a derivitized protein chain in which lysine residues are covalently linked to complex pyrrole derivatives (upper left). (T. Hofmann, *Z. Lebensm. Unters Forsch. A* 206:251–258, 1998.)

Figure 4.68. Examples of some of the parent heterocyclic ring compounds formed by Maillard reactions making up the aromas of baked and roasted foods.

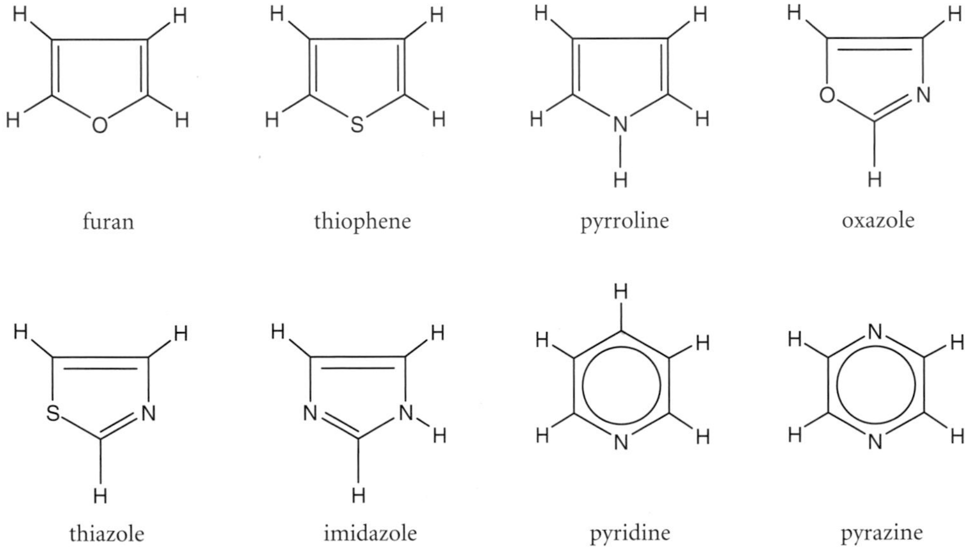

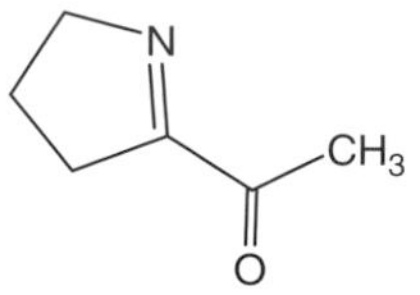

2-acetyl-1-pyrroline

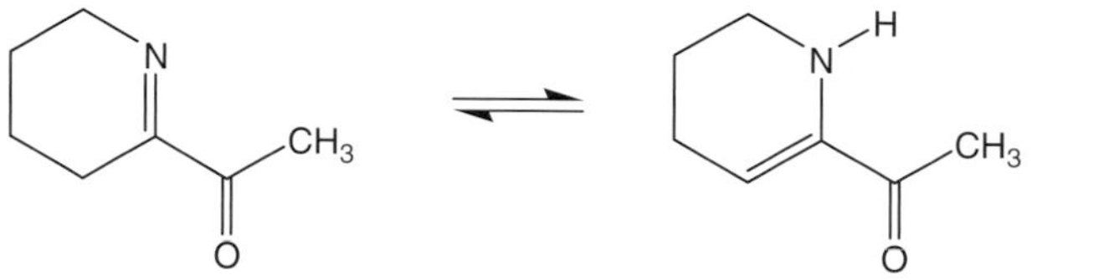

Figure 4.69. The structures of some of the heterocyclic Maillard products contributing to the odors of baked and toasted breads and roasted nuts.

6-acetyl-2,3,4,5-tetrahydropyridine   6-acetyl-1,2,3,4-tetrahydropyridine

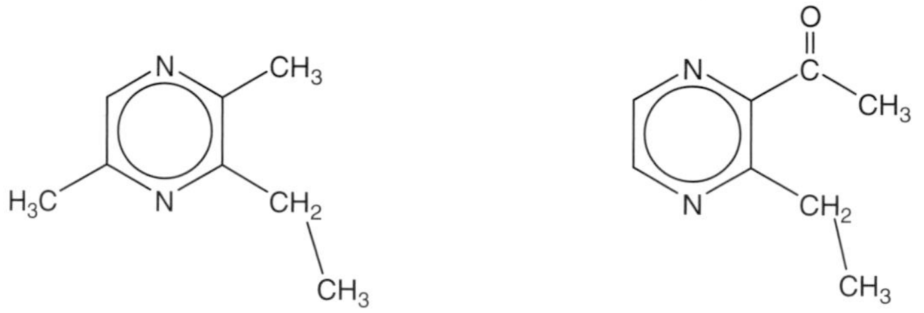

substituted pyrazine derivatives contributing
to the aromas of roasted nuts

group of lysine is protonated the sugar aldehyde groups cannot bind to it. Sulfites can inhibit the final development of brown pigments but do not prevent the loss of lysine (a sulfite is a salt or ester of sulfurous acid, H_2SO_3). Dilution with water also inhibits Maillard browning, because it proceeds most efficiently at low water activities, as in dried or powdered products.

Because the Maillard reaction involves the destruction of amino acids, it has potential nutritional implications. Lysine, the main amino acid susceptible to Maillard browning, is an essential amino acid, and thus its loss reduces the nutritional quality of protein that has undergone Maillard reactions. In experiments, animals fed on protein that had been subjected to Maillard browning showed reduced growth rates due to this nutritional loss. Most Americans consume a large excess of protein in their diets, so that this loss is not always significant in modern diets, but could be more important for those who do not overeat. Lysine deficiency can be a significant problem in preparing powdered milk products, especially in infant formulas, where the nutritional content is crucially important. Historically Maillard browning has been a particular difficulty in manufacturing condensed and powdered milk because there is abundant lactose present in milk, and brown-colored milk is unpopular with consumers. Clearly, care must be taken to control lysine loss in the production of powdered milk for infant formulas. (Ledl and Schleicher 1990).

The Strecker Reaction

The Strecker degradation is a reaction that occurs as a consequence of the Maillard reaction. It is a reaction between α-dicarbonyl compounds and amino acids, in which the amino acid is deaminated and decarboxylated. As we saw in the mechanism outlined in the previous section, the Maillard reaction produces dicarbonyl compounds by at least two different pathways. In an associated reaction, the amino ketones produced by the Strecker reaction can combine to form pyrazine derivatives (Figure 4.70). The reaction occurs in foods with free amino acids when heated to high temperatures as in caramelization.

Strecker aldehydes are usually volatile, with distinctive odors contributing to the aromas of foods that contain them. Pyrazines are also volatile and impart burnt odor notes to aromas and are important in the flavor of foods heated over 100 °C.

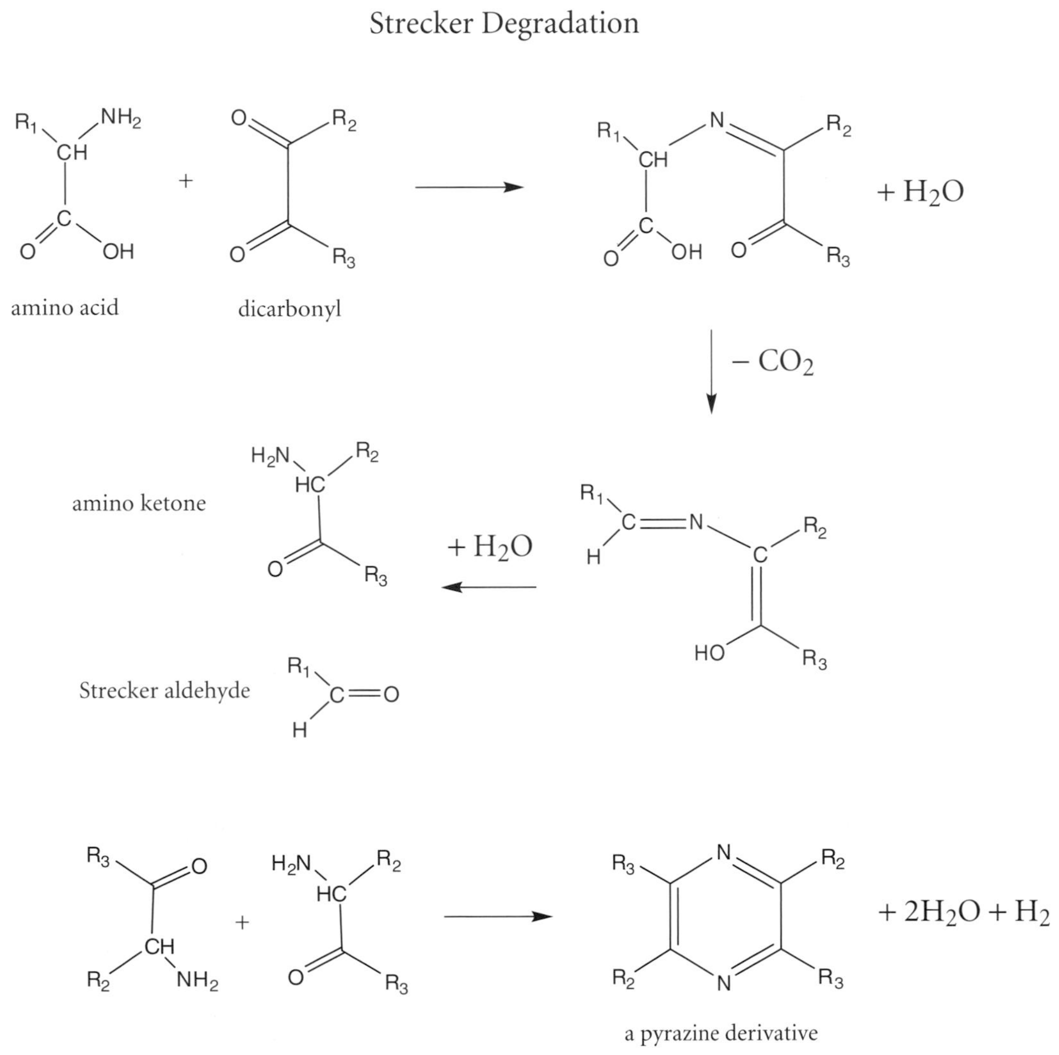

Strecker Degradation

Figure 4.70. The Strecker reaction.

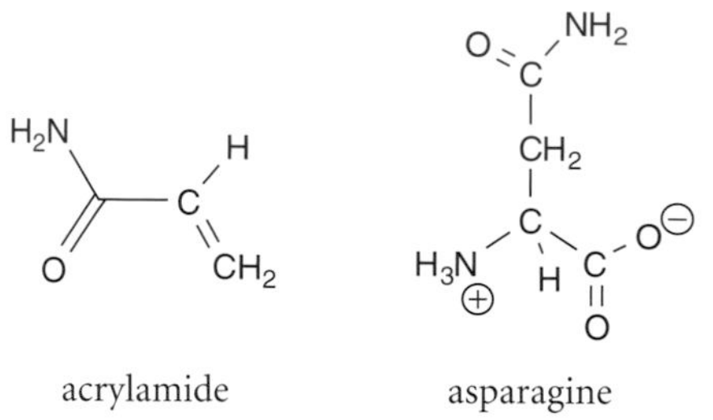

Figure 4.71. The structures of acrylamide and asparagine.

acrylamide

asparagine

Acrylamide

Beginning in 2000, several reports appeared that low levels of the potentially cancer-causing chemical acrylamide (Figure 4.71) had been detected in a number of common foods, particularly potato products such as potato chips and French fries, and even in the crust of baked breads (Tareke et al. 2000; Rosen and Hellenas 2002; Becalski et al. 2003; Lineback 2012). The list of products in which acrylamide has been found has since grown quite lengthy and includes such items as dried banana chips and roasted asparagus. The levels found are not always trivial; they are often more than two orders of magnitude higher than the levels allowed in drinking water by the US Environmental Protection Agency (EPA) or recommended by the World Health Organization (WHO). The health consequences of the acrylamide in these foods is unclear; bread, of course, has been consumed in large amounts by billions of people since almost the dawn of the agricultural age.

The acrylamide found in these products is not present in the raw ingredients and is not present if those ingredients are prepared by other methods, such as boiling, for example. Although the exact mechanism of the formation of this acrylamide is not completely understood, it seems that it is formed by reaction of reducing sugars with the amino acid asparagine (Becalski et al. 2004). Most workers believe that the acrylamide is formed through the Maillard reaction, as outlined in the mechanism shown in Figure 4.72.

It has also been shown that acrylamide can be formed with 2-deoxyglucose, which cannot undergo the Maillard reaction, suggesting that it may be formed by a different mechanism, involving a ring-opening step (Figure 4.73; Ganem 2002).

Like Maillard reactions, the formation of acrylamide is inhibited by excess water, as in boiling, which is why it is not present in boiled potatoes. Becalski et al. (2004) suggested that the levels of acrylamide in French fries could be reduced by selecting potatoes with low sugar and asparagine contents.

FUNCTIONAL PROPERTIES

Sweetness

One of the most useful properties of sugars in general, and sucrose in particular, is their sweet taste. Not all sugars are equally sweet, however, and some are not

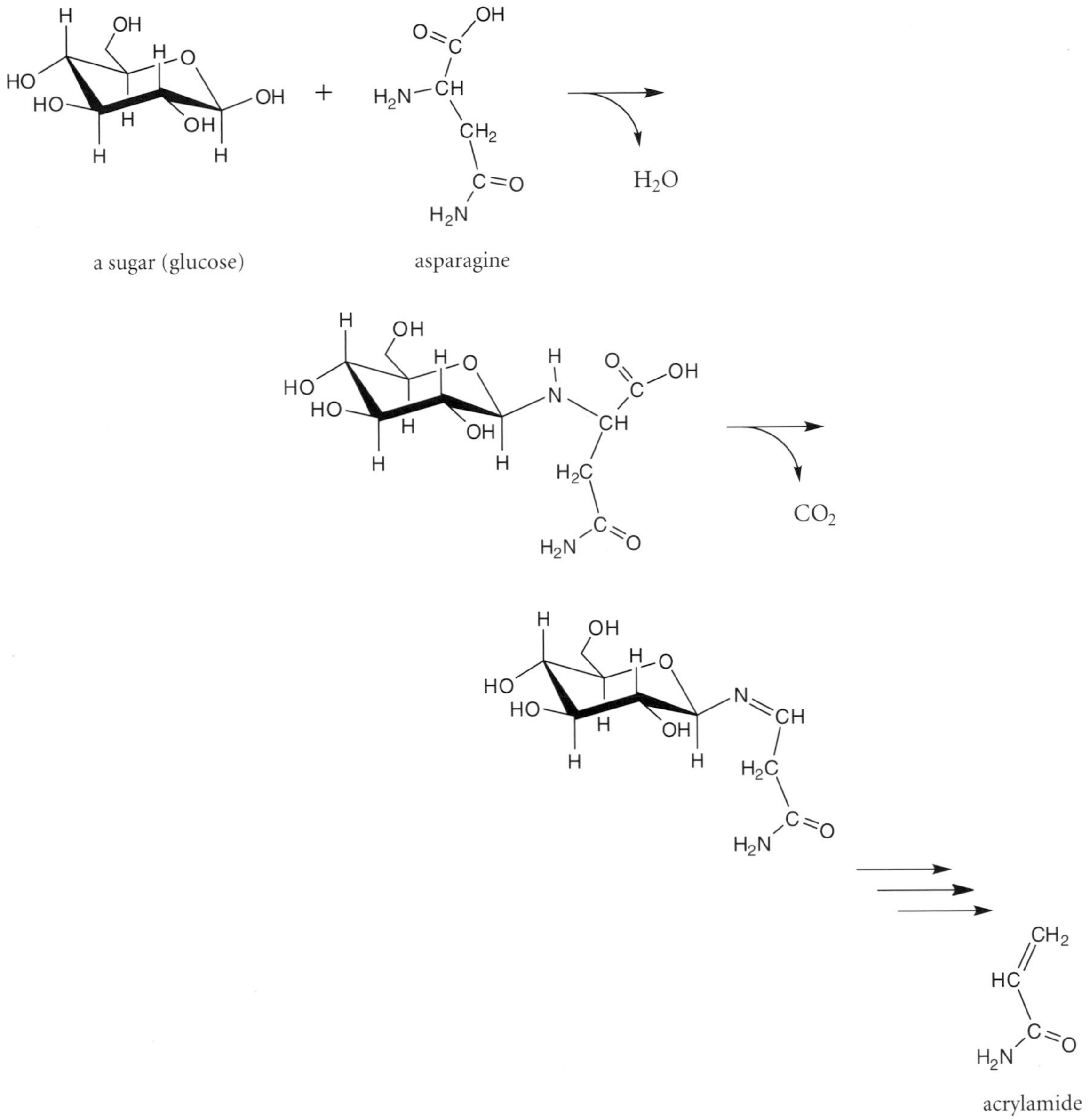

Figure 4.72. The formation of acrylamide from asparagine and glucose.

sweet at all. Tables 4.4 and 4.5 compare the relative sweetness on a per weight basis of various sugars and sugar alcohols to that of sucrose, as determined by taste panels.

Other molecules besides sugars can also be sweet; noncaloric artificial sweeteners such as saccharin and aspartame, for example, are quite sweet. Ethylene glycol (antifreeze), while very toxic, is also very sweet, as are several related small alcohols, including glycerol. Aspartame (Nutrasweet) is actually a dipeptide, and several other

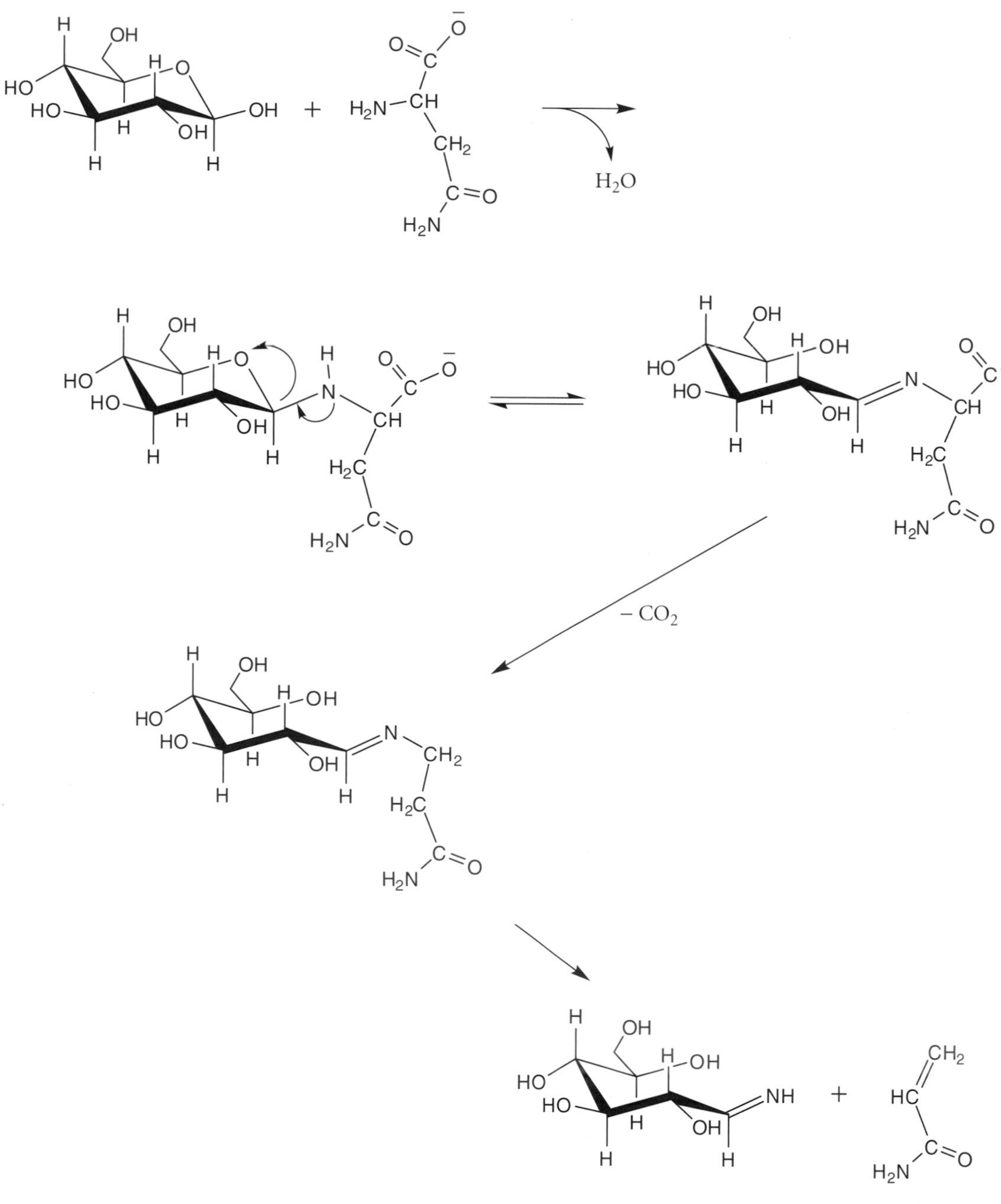

acrylamide

Figure 4.73. A proposed mechanism for the formation of acrylamide from asparagine and glucose via a ring opening that does not involve a Maillard reaction. (B. Ganem, *Chem. Eng. News* 80:6–9, 2002.)

Table 4.4 Relative sweetness (RS) of selected sugars (w/w%), with sucrose = 100

Sugar	Solution RS	Crystalline RS
β-D-fructose	100–175	180
sucrose	100	100
α-D-glucose	40–79	74
β-D-glucose	<α-anomer	82
α-D-galactose	27	32
β-D-galactose	—	21
α-D-mannose	59	32
β-D-mannose	bitter	bitter
α-D-lactose	16–38	16
β-D-lactose	48	32
β-D-maltose	46–52	—
raffinose	23	1
stachyose	—	10

Source: R.L. Whistler and J.R. Daniel in *Food Chemistry*, 2nd ed., O.R. Fennema, ed. (New York: Marcel Dekker, 1985).

Table 4.5 Relative sweetness of selected sugar alcohols (alditols), with sucrose = 100

Sugar alcohol	Relative sweetness (RS)
xylitol	90
sorbitol	63
galactitol	58
maltitol	68
lactitol	35

Source: R.L. Whistler and J.R. Daniel in *Food Chemistry*, 2nd ed., O.R. Fennema, ed. (New York: Marcel Dekker, 1985).

peptides and proteins are also sweet. In fact, the sweetest molecules known, on a weight basis, are the proteins monellin and thaumatin.

Although sucrose is considered to have the "cleanest" and most "natural" sweet taste, the growth in demand for low-calorie diet foods has increased the need for alternate sweeteners. Because of the ongoing search for new sweet-flavored molecules, it would be desirable to be able to develop an understanding of why some substances are sweet and others are not.

Unfortunately, the complete mechanism underlying the perception of sweetness is not yet understood and will surely be found to be complex, involving not only chemical detection but also neurological and psychological factors. The initial step of sweetness perception involves the interaction of a sweet molecule, such as a sugar, with the receptor protein T1R3 on the tongue (Chapter 8). The three-dimensional conformation of this protein has not yet been definitively determined, although the structure of a related homologous protein is known (Figure 8.1). Most likely this interaction will involve some type of complementary binding of the sugar to the functional groups of the protein, perhaps in a manner similar to the binding of glucose to yeast hexokinase, illustrated in Figure 4.74, where specific groups in the sugar match up with specific groups in the protein binding site.

Because the receptor protein has not yet been fully characterized, efforts have been made to deduce some of the important features of this interaction by surveying all of the known sweet molecules with the objective of determining what structural features they have in common. This approach is often used in pharmaceutical design when a series of partially useful drugs is available, but where there is no known model for how they work. This approach was applied to the study of sweetness in the late 1960s by R. Shallenberger, T. Acree, and C.-Y. Lee, in one of the first uses of the method. They developed the **AH/B theory** of sweetness by noting that all of the known sweet molecules had hydrogen bonding and hydrophobic groups in certain relative positions (Shallenberger and Acree 1969; Shallenberger, Acree, and Lee 1969). In particular, they possessed a hydrogen bond donor (AH), an acceptor (B), and a hydrophobic group (γ) in a triangular arrangement at specific distances (Figures 4.75 and 4.76), presumably in order to match up with a complementary arrangement of functional groups in the receptor protein.

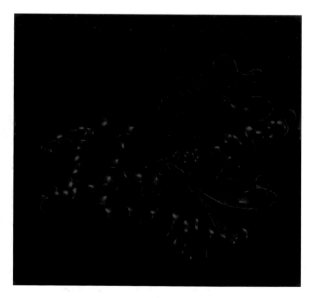

Figure 4.74. A cartoon representation of the protein yeast hexokinase with a glucose molecule bound in its active site.

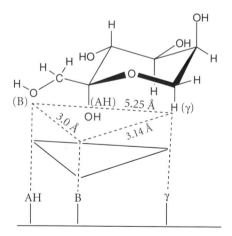

Figure 4.75. AH/B model for the sweetness of fructose. (Adapted from R.C. Lindsay, "Flavors," in *Food Chemistry*, 2nd ed., O.R. Fennema, ed. New York: Dekker, 1985.)

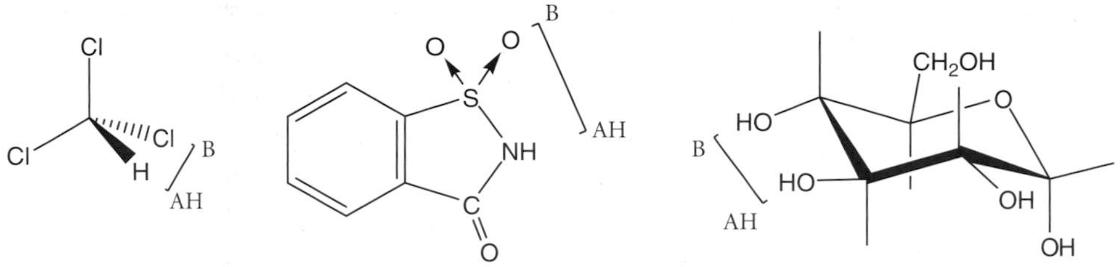

Figure 4.76. Examples of sweet molecules, identifying possible AH/B groups. (Adapted from R.C. Lindsay, "Flavors," in *Food Chemistry*, 2nd ed., O.R. Fennema, ed. New York: Dekker, 1985.)

Table 4.6 Percentage of water absorbed by carbohydrates from moist air

Carbohydrate	Relative humidity (RH) at 20°C		
	60%, 1 hr	60%, 9 days	100%, 25 days
D-glucose, crystalline	0.07	0.07	14.5
D-fructose, crystalline	0.28	0.63	73.4[a]
sucrose, crystalline	0.04	0.03	18.4[a]
invert sugar, commercial	0.19	5.1	76.6[a]
invert sugar, pure	0.16	3.0	74.0[a]
maltose, commercial, anhydrous	0.80	7.0	18.4
β-maltose hydrate, pure	5.05	5.1	—
β-lactose, anhydrous	0.54	1.2	1.4
α-lactose hydrate, pure	5.05	5.1	—
raffinose, anhydrous	0.74	12.9	15.9
cornstarch	1.0	13.0	24.4
cellulose	0.9	5.4	12.6
D-mannitol	0.06	0.05	0.42

Source: Data taken from J.E. Hodge and E. Osman, "Carbohydrates," in *Principles of Food Science. Part I. Food Chemistry*, O.R. Fennema, ed. (New York: Marcel Dekker, 1976).
[a]Water uptake still increasing at 25 days.

Water Binding and Solubility

Because of their hydrogen-bonding capacity, most sugars are extremely hygroscopic, which is an important factor in their use in specific foods (Table 4.6). Exceptions to this general trend are maltose and lactose, which are not strongly hygroscopic. The sugar alcohol D-mannitol is especially notable in this respect, adsorbing almost no water at all. Because of their limited uptake of moisture, these sugars are particularly useful in bakery toppings, coffee whiteners, and instantly reconstitutable powders or granules that must not become sticky. Hygroscopic sugars such as invert sugar and glucose syrup help retain moisture in baked goods, plastic candies, and fillings that should not become brittle. If the sugar crystallizes, water is given up and the plasticizing effect is lost. As can be seen from Table 4.6, different dry sugars absorb different amounts of water on an equal weight basis, and do so at different rates.

Sugars also differ in their solubility in water, depending on various factors, including how favorably their functional groups interact with the water molecules through hydrogen bonding. But another factor equally important in determining how soluble a particular sugar is in water is the lattice energy of the crystal. If the crystalline solid of a given sugar has a particularly favorable energy due to ideal packing, then it will have a lower tendency to go into solution than another crystal with a higher-energy crystal lattice due to less favorable crystal packing. Such lattice energy differences can arise from topological differences in the structures of the sugar molecules. Table 4.7 gives the solubilities for several sugars at selected temperatures. Fructose is the most soluble of the simple sugars. As can be seen, galactose is the least soluble of the monosaccharides listed. Cellobiose, the repeat unit of the polysaccharide cellulose, discussed later, is also fairly insoluble, which is perhaps to be expected in light of the fact that cellulose is totally insoluble in water. Lactose, which is similar to cellobiose, is also sparingly soluble, which is a significant problem in the reconstitution of powdered milk. Sugar alcohols like sorbitol and xylitol are generally more soluble than their parent saccharides.

Table 4.7 Solubilities of selected sugars, in terms of mole fraction of the sugar at saturation

Sugar	Mole fraction at saturation			Crystal melting temperature (°C)
	20 °C	25 °C	30 °C	
xylose	0.118	0.130	0.141	157
arabinose	0.074	0.082	0.095	156 (α and β)
glucose	0.080	0.094	0.114	150 (β); 146 (α)
fructose	0.441	—	0.449	103 (β)
mannose	0.222	0.259	—	132
galactose	0.004	0.043	0.050	170
sucrose	0.095	—	0.101	186
lactose	0.009	—	0.013	225 (β); 216 (α)
cellobiose	0.009	0.008	0.009	225

Sources: Data taken from Gray et al., *Appl. Biochem. & Biotech.* 105–108:179–193 (2003); H.M. Pancoast and W.R. Junk, *Handbook of Sugars* (Westport, CT: AVI Pub. Co., 1980); *CRC Handbook of Chemistry and Physics*, 54th ed. (Cleveland, OH: CRC Press, 1973); N. Drapier-Beche, J. Fanni, and M. Parmentier, *J. Dairy Sci.* 82:2558–2563 (1999).

POLYSACCHARIDES

Polysaccharides are widely exploited in the food industry for their functional properties, including their ability to modify viscosity, form gels, stabilize emulsions, and control water, among other uses. These polysaccharides include not only those likely to be found in foods in large amounts anyway, such as starch and cellulose, but also various gum additives. The **gums** are water-soluble materials used primarily to increase the viscosity of solutions; the term **hydrocolloid** is almost synonymous with gums. Most gums are polysaccharides, and in our usage we will apply this term only to polysaccharides.

Naturally occurring polysaccharides can belong to one of several types. Some like amylose or cellulose are homopolymers; that is, they consist entirely of one monomer type:

Homopolymers: −A−A−A−A−A−A−A−A−A−A−

These polymers can be linear like amylose and cellulose, or branched like amylopectin. Heteropolymers can be of several varieties, as shown in Figure 4.77.

Polysaccharides can exist in a number of conformations. Many, such as amylose, occur as helices, with different pitches and numbers of monomers per turn. Others can exist as extended, relatively flat ribbons, such as is the case for cellulose. Still others may exist as irregular flexible coils, as occurs when there are (1→6)-linkages, or even as double helices as in starch.

Hydrated polysaccharides can significantly affect the viscosity of the aqueous dispersing medium. As an extended polysaccharide tumbles in solution, it frequently makes contact with other polymer molecules, sweeping out a roughly spherical volume as they rotate, as in part (a) of Figure 4.78. Highly-branched polysaccharides of the same molecular weight sweep out a smaller volume and thus come in contact less often, so they do not increase the viscosity as much (part c of Figure 4.78). If linear polymers coil, it reduces their effect (as in part b of Figure 4.78). Short branches (part d of Figure 4.78) can stiffen the chain by inhibiting folding and coiling, and thus increase the effect on viscosity. The branches on polysaccharides generally inhibit interchain association and thus the formation of gels.

Figures 4.79–4.81 illustrate how effective various hydrocolloids (gums) can be at significantly increasing solution viscosity even at low concentrations. This property

Figure 4.77. Schematic illustrations of possible polysaccharide structural motifs.

−A−B−A−B−A−B−A−B−A− copolymer; ex: carrageenan

−A−A−A−A−A−B−B−B−B− block copolymer, ex: alginate

−A−A−A−A−A−A−A−A−A−
  | | | | |
  B   B   B B   B branched copolymer; ex: guar

−B−B−A−B−A−B−A−A−A−B− random copolymer

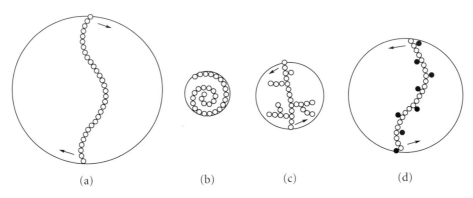

Figure 4.78. A schematic illustration of the relationship between polysaccharide molecular architecture and its effect on solution viscosity. Each illustrated chain has the same number of monomers, represented as small spheres. (a) A stiff extended chain sweeps out a large radius as it tumbles in solution, interacting with many water molecules and often entangling with other chains, and thus producing a large increase in viscosity. (b) A chain that folds into a compact shape, interacting mostly with itself and with little of the solvent, and rarely colliding with other polymer molecules, has only a small effect on viscosity. (c) Branching increases stiffness but reduces the radius swept out by tumbling on a per weight basis. (d) A small amount of branching prevents folding, thus promoting solvent and interchain interactions, without significantly shortening the chain on a per weight basis. (Redrawn with permission after a figure from J.E. Hodge and E.M. Osman, "Carbohydrates," in *Principles of Food Science. Part I. Food Chemistry*, O.R. Fennema, ed. New York: Marcel Dekker, 1976.)

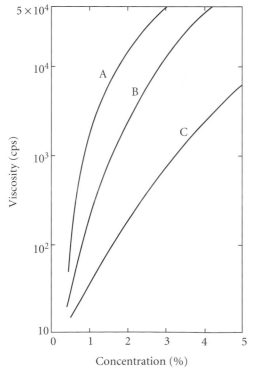

Figure 4.79. The viscosity of aqueous alginate solutions for three types of alginate with (a) high, (b) medium, and (c) low viscosity. Note that for all three, small increases in concentration can increase solution viscosity by orders of magnitude. (Redrawn from data given in R.L. Whistler, *Industrial Gums*. New York: Academic Press, 1973.)

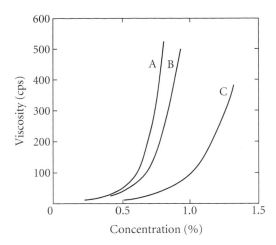

Figure 4.80. The viscosity of mixed aqueous carrageenan and alginate solutions as a function of total polysaccharide concentration for alginates from (A) *Eucheuma spinosum*, (C) *Chondrus crispus*, and (B) both A and C in the ratio 2:1. (Redrawn from data given in R.L. Whistler, *Industrial Gums*. New York: Academic Press, 1973.)

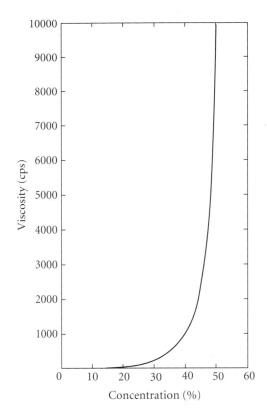

Figure 4.81. Viscosity of an aqueous solution of gum arabic as a function of the concentration of the polymer. (Redrawn from data given in R.L. Whistler, *Industrial Gums*. New York: Academic Press, 1973.)

is what makes the gums so useful in product formulation, giving them their ability to determine mouthfeel, slow diffusion, and inhibit starch retrogradation.

Most polysaccharide gums produce significant increases in viscosity at low polymer concentrations, but as can be seen in Figure 4.81, gum arabic is different in its behavior. Solutions of this polysaccharide do not exhibit significant increases in viscosity until the concentration is above 40% (compare with Figures 4.79 and 4.80). This difference in behavior is largely due to the fact that gum arabic is so highly branched, as in diagram (c) of Figure 4.78.

Homopolymers

Starch

The major energy reserve of most plants is **starch**, which is most abundant in seeds, roots, and tubers, and this substance is also the most important energy source in the

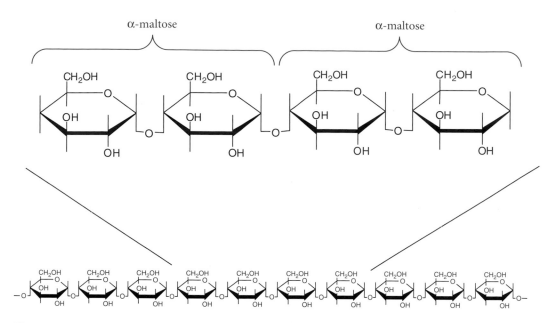

Figure 4.82. Haworth projection representation of amylose.

human diet on a worldwide basis. Starch is usually consumed without being isolated from the source plant material, but refined starch is also important in food processing. Common commercial starches are corn, wheat, rice, tapioca, potato, and sago.

Starch consists of two different but closely related molecules, **amylose** and **amylopectin**, both of which are large homopolymers of D-glucose. Amylose is the mostly linear (that is, unbranched) polymer consisting of α-(1→4)-linked glucose units, with DPs in the many hundreds or thousands (Figure 4.82). Although it is overwhelmingly linear, some amylose chains have been found to have very limited branching, usually with a single glucose unit bound by an α-(1→6)-linkage to the main chain.

In crystals, crystalline powders, and drawn fibers, amylose exists as a regular helix. A number of helices are possible in principle, with different pitches (distance advanced per monomer residue) and numbers of residues per turn. These helices will not be of equal energy, and in general, only a few of the possible helices are likely to be stable. As it happens, only one such helix is actually stable for pure amylose. Different crystal forms are called polymorphs, and substances that organize in different forms are called polymorphic. Starch has been observed in at least three different polymorphs, called A, B, and C, using X-ray diffraction from powders or ordered fibers. These different forms could be the result of different helical conformations for the polymers, or from different types of packing of the helices. In fact, the latter is the case, and generally native starch molecules exhibit only one type of regular helical conformation. Experiments on the starch from different food sources have shown that often the starch from a particular food plant is found in only one of the three common polymorphic forms, and thus the different food starches can be classified as either A, B, or C type. The starch from cereal sources (such as wheat and corn) are of the A type, while starches from tubers such as potatoes and yams are generally of the B type. C type starch is less common, but is found in foods like peas, beans, and bananas.

Unfortunately, powder and fiber diffraction studies cannot give the same atomic-level structural resolution seen in single-crystal X-ray diffraction experiments such as are often done for proteins (the yeast hexokinase structure shown in Figure 4.74 was determined by this method). Single-crystal diffraction experiments were not possible for starch or dextrins until quite recently because naturally occurring amylose contains molecules of many different lengths, which thus cannot pack into a regular crystal. For this reason, the actual conformations of the different solid-state forms of starch remained controversial. Although the conformational structures of the different starches cannot be determined exactly from fiber diffraction, the data from such experiments, combined with computer modeling, has produced consensus structures for both the A and B type starches (Imberty and Perez 1988; Takahashi et al. 2004). These crystalline forms are thus believed to consist of left-handed double helices of two amylose chains intertwined together (Figure 4.83). Both chains have six glucose residues per turn of the helix. The conformations of the double helices in the A and B forms are almost the same, and they differ primarily in the way that the double helices pack together in the crystals. The C type of starch was subsequently found not to be a distinct conformational structure or packing pattern, but rather a mixture of the A and B types of amylose.

Another helical conformation of amylose whose structure has been well-established is the so-called V structure. This conformation is the familiar one that forms when starch is treated with iodine, as in the common high school chemistry experiment. The appearance of a dark blue color in that test is an indication of the presence of starch. In the V structure, the iodide ions occupy the central, hollow core of a low-pitch single helix, with the sugar rings almost lined up in the direction of the helix (see Figure 4.83). The iodine in these complexes is thought to be in the form of triiodide, I_3^-, or pentaiodide, I_5^-. Triiodide in solution is yellow-brown, but the blue color of the complex apparently arises because the presence of the sugar atoms perturbs the electronic structure of the I_3^- ion, shifting the energy level splittings to the blue. The complexes are fairly strong, with an enthalpy change of about −71 kJ per mole of bound I (Cesàro 1986). The interior of the V helix is hydrophobic and can also accommodate single-chain nonpolar species like fatty acids or monoglycerides (without a color change).

Amylose is insoluble in cold water, but moderately soluble at higher temperatures. What form the amylose strands take in aqueous solution is not known exactly, but it is generally thought that large stretches of the chain remain relatively helical, interrupted by nonhelical regions of varying lengths. Amylose readily precipitates out of solution into a solid phase when cooled. Figure 4.84 schematically represents a random-coil-type conformation for amylose in the upper left, which in precipitating out at lower temperatures, could gradually assume a helical conformation when one turn, with its stabilizing hydrogen bonds, randomly forms and serves as a "nucleation site," promoting the formation of additional helical turns. Precipitation would then

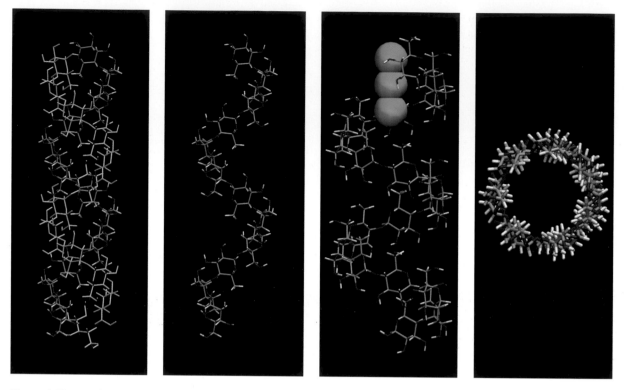

Figure 4.83. Amylose occurs as a double helix (left, with one helix colored yellow for clarity, while the other uses conventional atom coloring, with red for oxygen atoms, blue for carbon, and white for hydrogen), with each of the two strands having the same helical conformation, shown for a single chain on the middle-left. The individual helices each have six glucose monomers per complete turn. V-type starch, shown on the middle-right and right, is a single chain that has a much lower pitch, and thus a larger central cavity, as can be seen in the end-on view on the far right, which can be occupied by iodide ions. Their presence, indicated by a linear triiodide molecule, I_3^-, in the middle right figure, stabilizes the conformation of the single chain into the V-helix, which would not otherwise be stable in this low-pitch conformation.

involve intertwining and regular stacking of many helices, but would not produce crystalline order, and overall the precipitate would be amorphous.

The other major molecular fraction of starch, amylopectin (Figure 4.85), is highly branched. It consists of a linear amylose-like backbone chain with numerous (1→6) branches, which also contain branches, and so on. This branched, "bushy" topology is essentially the same structure as **glycogen**, the energy-storage molecule in animals, with the only difference being the number of branches; glycogen has a larger number of shorter branches than amylopectin. Amylopectin can be suspended in water but is not soluble, whereas the more highly branched glycogen is soluble, since the numerous branches prevent close association. The branching arms of amylopectin are generally coiled together into double helices like amylose (Figure 4.83) in native starch, with the helices disrupted by the branching points.

Amylopectin molecules are very much larger than amylose chains, with DPs in the tens of thousands and molecular weights in the range 10^7–10^8, making them

Figure 4.84. A schematic representation of the transition of amylose from a random coil at high temperature to a helix at lower temperature. (Adapted and redrawn from D.A. Rees, *Polysaccharide Shapes*. New York: John Wiley & Sons, 1977.)

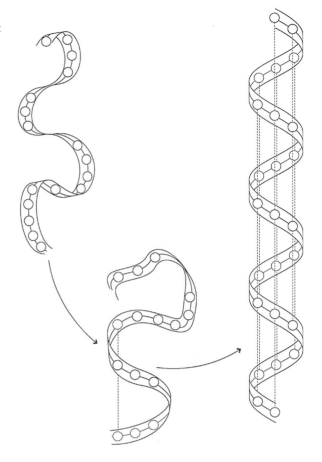

among the largest of biological molecules. Each amylopectin molecule has a single reducing end and consists of two types of chains of different lengths, called A and B, with the larger backbone B chains generally containing around 40 glucose monomers. Each B chain generally has one other B chain branching α-(1→6) from it, but with several of the shorter A chains, with lengths of around 15 monomers, branching from them. The A chains are not themselves branched.

Starch is always produced by plants in small packets called **granules** (Figure 4.86). The starch granules produced by different plants are distinctive and can be used to identify the source (Table 4.8). Potato starch has the largest granules of the commercial starches, and rice the smallest. The large range in sizes seen in wheat starch granules in Table 4.8 is because there are actually two types of wheat starch granules: larger so-called lenticular granules with sizes in the range 15 to 35 μm and much smaller polyhedral granules with diameters less than 10 μm (Oates 2001).

All of the granules have a cleft called the **hilum**, either a small spot or the intersection of two lines, which is the nucleation point around which the granule developed. When illuminated by polarized light, starch granules are birefringent, exhibiting

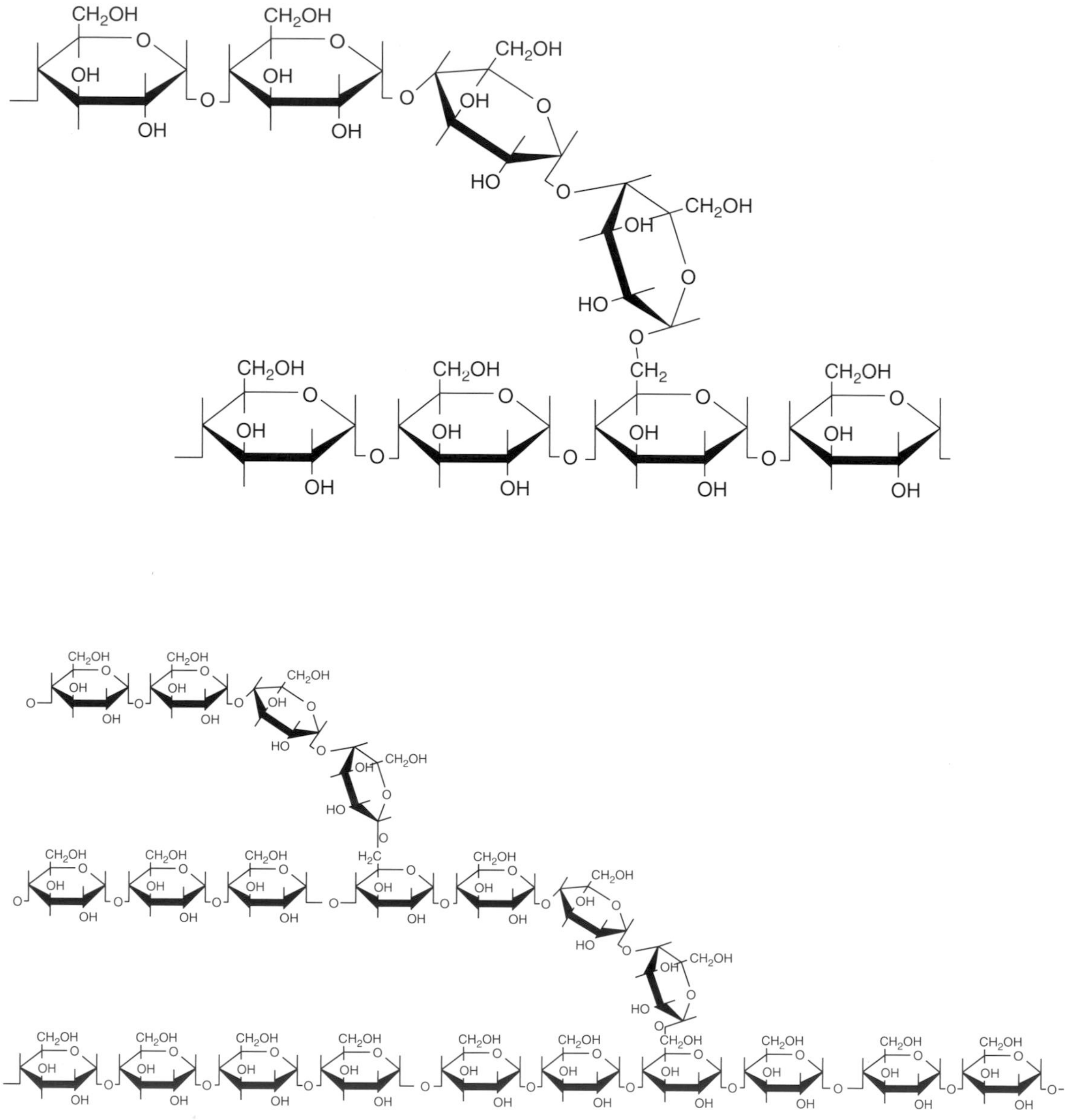

Figure 4.85. Haworth projection representation of amylopectin.

a Maltese cross centered at the hilum, indicating spherical crystallinity (Figure 4.87); that is, the starch molecules radiate out from the center (hilum), but with a great deal of order in their packing that primarily results from the intertwining of the amylose and amylopectin arms into double helices and the orderly packing of adjacent coils. Normally starch is about 25% amylose, but plant varieties producing granules containing as much as 85% amylose have been developed by selective breeding. Other starches, called **waxy starches**, such as waxy corn, waxy beans, and waxy or glutinous

Figure 4.86. Top: a granule of corn starch (left), and seen in cross section (right) under ordinary light; Bottom: a granule of potato starch (left), and seen in cross section (right). The granules are colorless (and appear white to the naked eye), but are illuminated with orange light here to emphasize their structure. (Figure courtesy of Andreas Blennow, University of Copenhagen.)

Table 4.8 Characteristics of starch granules

| Source | Diameter | | Gelatinization temperature (°C) |
	Range (μm)	Mean (μm)	
corn	4–26	15	61–72
white potato	15–100	33	62–68
sweet potato	15–55	25–50	82–83
tapioca	6–35	20	59–70
wheat	2–38	20–22	53–64
rice	3–9	5	65–73

Source: R.L. Whistler and J.R. Daniel in *Food Chemistry*, 2nd ed., O.R. Fennema, ed. (New York: Marcel Dekker, 1985).

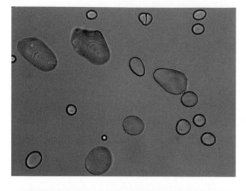
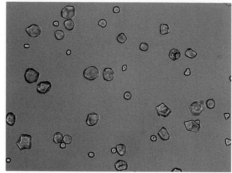

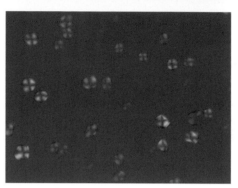

Figure 4.87.
Micrographs of potato (left) and corn starch (right) granules seen under ordinary light (top), and on the bottom, plane-polarized light. (Figure courtesy of Andreas Blennow, University of Copenhagen.)

rice, contain only amylopectin. The crystalline character of the starch granules may come primarily from the amylopectin fraction, with the various branching A chains forming parallel, hydrogen-bonded double-helical coils between the disrupting branching points. Amylopectin molecules, particularly from potatoes, may contain very small amounts of phosphate groups esterified to some of their hydroxyl groups (Blennow et al. 2002).

Granules that are 100% amylose are not possible, because the density of such granules would have to decrease radially as the linear amylose chain ends become further and further apart as the granules grow. In actual granules, the amylopectin molecules fill in these spaces by A chain branching at regular intervals so that the branches occupy the spaces as the parent chains diverge. This branching occurs when the ends of the chains have diverged enough to allow the branching enzymes sufficient access to add a branch point, which geometrically would occur over the entire granule surface at roughly the same rate, so that the branches tend to be somewhat in register (Figure 4.88). This alternation of closely packed chains with radially decreasing density, interrupted at intervals by branching, gives the appearance of growth rings in the granules when they are seen in cross section under a microscope (Figure 4.86).

Starch granules are not soluble in water. The water content of cereal starches is normally around 12 to 14%. The granules can absorb cold water in excess of this amount, up to a maximum of about 30%, depending on the starch source. This water

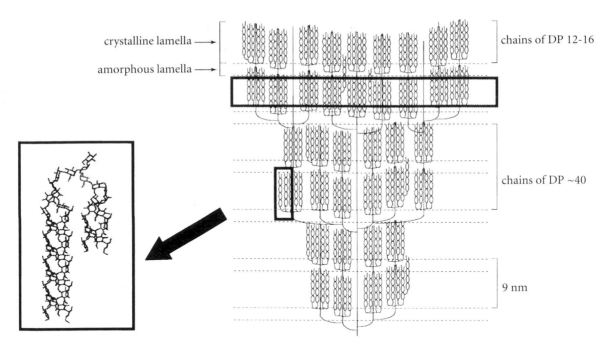

crystalline lamella ⟶

amorphous lamella ⟶

chains of DP 12-16

chains of DP ~40

9 nm

Figure 4.88. A schematic illustration of the branching in amylopectin, alternating with helical layers.

adsorption does not significantly disrupt the packing structure of the granules, and there is only a small increase in the granule diameter, from ~10 to 30%. Upon heating in water, however, the behavior is different. The heating provides the energy to disrupt some of the crystalline-like packing of the amylopectin strands, allowing much greater swelling as much larger quantities of water are absorbed to take the place of the lost interchain hydrogen bonds. Much of the ordered structure remains intact, however, unless the starch is violently agitated to mechanically disrupt the granules. As this process is going on, some of the amylose molecules slip out of the granule and into solution. The resulting starch paste contains swollen but still discrete blocks of starch, the remnants of the granules, suspended in a solution of free molecules. This process, illustrated schematically in Figure 4.89, is referred to as **gelatinization**. As the gelatinization proceeds and the ordered packing of the strands is lost, the birefringence disappears. The temperature at which birefringence first disappears is called the **gelatinization temperature** or gelatinization point, which is actually usually a narrow temperature range. This temperature will vary with the source of the starch (Table 4.8) and can thus sometimes be used for identification purposes. Gelatinization is *not* a reversible process. The molecular structure of the resulting gel or paste might be represented schematically by Figure 4.89, where the drawing on the left represents the semicrystalline packing of the amylose and amylopectin strands in the granule and the drawing on the right represents the entangled polymers of the paste.

A slurry of starch granules in cold water has low viscosity, but as the slurry is heated and the granules swell, they absorb much of the water around them, and the

Figure 4.89. A schematic illustration of the transition from native starch to the gelatinized state. (Figure adapted and redrawn from P.C. Paul and H.H. Palmer, *Food Theory and Applications.* New York: John Wiley & Sons, 1972; reproduced therein from K.H. Meyer and P. Bernfeld, *Helv. Chim. Acta* 23:890, 1940.)

viscosity increases, producing a thick paste primarily as the result of the swollen granules bumping into one another. The native starch granules do not have a surface membrane, but as the linear amylose molecules diffuse out of the granule during heating, some may adhere on the surface to form a membrane-like surface. Stirring can break the swollen granules, with a large decrease in viscosity. As the amylose molecules continue to diffuse out, they eventually leave a collapsed, wrinkled granule, like a deflated balloon. If the thickened hot mixture of starch components is further stirred as it is cooled, a gel will form. This will not be the case, however, when only amylopectin is present, as in waxy starches, apparently because the branches inhibit the necessary degree of interchain association.

Because the free amylose is not very soluble at low temperatures, it precipitates out as a relatively regular solid (actual perfect crystals are precluded by the irregular mix of chain lengths), most rapidly near 0 °C, a process that is called **retrogradation**. This name was chosen originally because it was thought that the amylose was returning to its crystalline state before the gelatinization, but it should be emphasized that this is very little like the actual original state in the granule. Retrogradation can also occur during frozen storage of processed foods, and usually negatively affects product quality. For frozen products like gravies or pie crusts, many of the gross changes resulting from amylose retrogradation can be avoided by using waxy starches that are almost all amylopectin. Amylopectin can also undergo more limited retrogradation, but this primarily consists of the A chain branches retwining as double helices and hydrogen-bonding together in a more regular fashion that more closely resembles the original state of the amylopectin than is the case for retrograded amylose. The retrogradation of amylopectin can be reversed by heating, but that of amylose is not as easily reversed by ordinary heating, making it all the more important that it be prevented in the first place in starchy foods stored at low temperatures. The crystalline regions of retrograded starch are not susceptible to enzymatic hydrolysis, lowering the digestibility of the starch. The formation of crystalline regions during retrogradation results in some shrinkage, and the exclusion of water from the ordered regions, a process called **syneresis**. At the macroscopic level, this can appear as though water is "bleeding" out of the product, possibly even forming small pools visible to the naked eye.

Staling in baked products like bread or cookies is also the result of starch retrogradation, particularly of the amylopectin fraction, which occurs more gradually

Figure 4.90. Amylographs illustrating the time course of starch gelatinization for several starch concentrations. Note that the temperature was varied over the course of the gelatinization process as indicated on the top. (Figure adapted and redrawn from data given in R.J. Smith, "Characterization and analysis of starches," in *Starch: Chemistry and Technology*, vol. 2, R.L. Whistler and E.F. Paschall, eds. New York: Academic Press, 1967.)

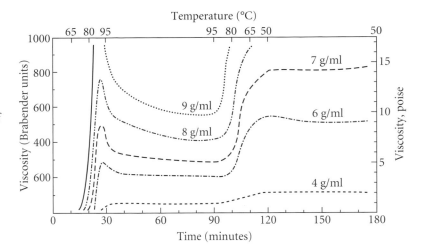

than that for the amylose part. As staling proceeds, the amylose molecules stale quickly, re-entangling as moisture is lost, while the staling resulting from the amylopectin occurs more slowly. Staling can be partially reversed by heating with moisture, with the heat disrupting the orderly packing and allowing water to replace starch-starch hydrogen bonds. Baked products with higher fat contents stale less readily, and lipids and other surfactants can be used as antistaling agents in baked foods. Some that are used are glyceryl monopalmitate and other monoglycerides, and sodium stearyl-2-lactylate. Carboxymethyl cellulose is also sometimes used to inhibit staling. These molecules function by entangling with the starch molecules and interfering with the regular packing of the starch chains.

The course of the gelatinization process can be followed by monitoring the viscosity of the system as a function of time in a viscometer, and the plots produced by these devices are sometimes referred to as amylographs. An example is shown in Figure 4.90, with the temperature indicated along the top. From such a plot one can easily see the onset of the gelatinization as the amylose molecules escape and as the swollen granules collide, followed by a collapse in viscosity as the system is stirred, and finally the setting of a firm gel as the temperature is lowered.

Various factors can affect the course of the gelatinization. As already noted, waxy starches will not exhibit the final "setting" stage as they do not gel. The presence of other ingredients in the system can slow down the swelling and collapse of the starch granules. Various hydrophilic species, such as ions or free sugars, can bind the available water and make it less available for penetrating the granules and for reorganizing around the amylose and amylopectin chains in the gelation phase. Figure 4.91 displays the effect on gelatinization of various concentrations of sucrose on a 5% corn-starch slurry. Other sugars also affect gelatinization, with the degree of the effect depending upon the hydration characteristics of each sugar, as can be seen from Figure 4.92.

Lowering the pH a small amount can slow the gelatinization slightly, but lowering it significantly causes a collapse of the gel as acid hydrolysis destroys the polymer

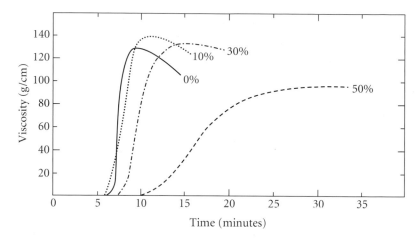

Figure 4.91. The effects upon starch gelatinization of various concentrations of added sucrose. (Figure adapted and redrawn from data given in M.L. Bean and E.M. Osman, *Food Res.* 24:665, 1959.)

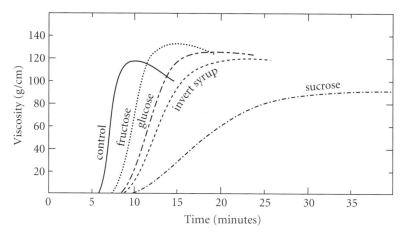

Figure 4.92. The effects at equal weight concentrations of various sugars on the gelatinization of starch. (Figure adapted and redrawn from data given in M.L. Bean and E.M. Osman, *Food Res.* 24:665, 1959.)

chains (Figure 4.93). Extremely basic conditions can also accelerate the rate of viscosity increase, but can also cause a marked decrease in viscosity as time progresses.

Cyclodextrins

Cyclodextrins are short, cyclic amylosic oligomers, where the amylose chains form a covalently linked ring head-to-tail, generally with 6, 7, or 8 glucose monomers in the ring. As in amylose, all of the linkages are α-(1→4). Rings with 6 glucose molecules are called α-cyclodextrin; those with 7, β-cyclodextrin (see Figure 4.94), and those with 8, γ-cyclodextrin. These molecules are natural products of the bacterial degradation of starch and can be prepared industrially using bacterial enzymes. The most commonly occurring natural cyclodextrin is the seven-glucose β-cyclodextrin ring.

The cyclodextrins have the shape of truncated cones, with the radius of one entrance to the "donut hole" being larger than the other. As can be seen from Figure 4.94, this structure creates a pocket inside the ring that is lined with the aliphatic carbon and hydrogen functionalities, and the relatively less polar ether oxygen atoms

Figure 4.93. The effects of pH on starch gelatinization as determined using viscosity measurements. (Figure adapted and redrawn from data given by E. Osman in *Food Theory and Applications*, P.C. Paul and H.H. Palmer, eds. New York: Wiley, 1972.)

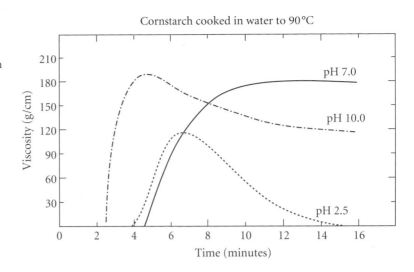

Figure 4.94. The structure of β-cyclodextrin, in a so-called liquorice representation on the left, and in a van der Waals representation on the right.

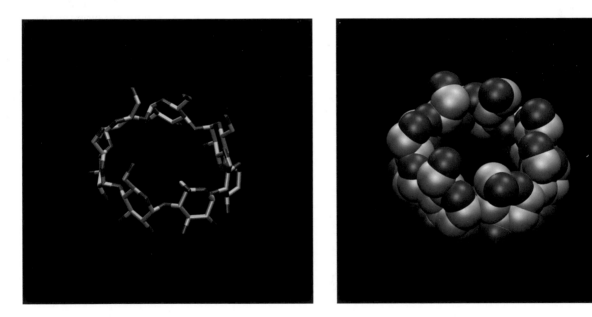

of the glucose rings and the glycosidic linkages. This inner pocket is therefore relatively hydrophobic, while the outer surfaces of these rings are hydrophilic due to the exposure of the hydroxyl groups. Thus, small hydrophobic molecules, such as vanillin, limonene, or carvone (see Chapter 8), will form 1:1 complexes with the cyclodextrins by binding inside the donut hole in an aqueous environment. It has been shown that such a complex with vanillin not only increases the solubility of the vanillin in water (Figure 8.23), but also protects the vanillin from oxidation to vanillinc acid (Karathanos et al. 2007).

Table 4.9 Free energies, enthalpies, and entropies for the association of various ligands in 1:1 complexes with β-cyclodextrin

Ligand	ΔG (kJ/mol)	ΔH (kJ/mol)	ΔS (J/mol·K)
pyridine	−12.6	−10.5	8.4
indole	−44.4	−3.3	138
phenol	−23.8	−7.5	54
benzoic acid	−17.1	−40.2	−75
p-nitrophenol	−12.1	−30.6	−62.3
perchloric acid	−9.2	−31.4	−71.2
acetic acid	−21.8	−5.0	54
L-phenylalanine	−23.4	−4.6	62.8
L-tyrosine	−16.7	−4.2	42
L-tryptophan	−8.4	−7.5	4
vanillin[a]	−3.47	−6.99	−11.8

Source: E.A. Lewis and L.D. Hansen, *J. Chem. Soc. Perkin Trans.* II: 2081–2085 (1973).
[a]V.T. Karathanos et al., *Food Chem.* 191: 652–658 (2007).

As can be seen from Table 4.9, nonpolar species like indole and phenol have strong binding affinities with β-cyclodextrin, with large favorable entropy changes. The story is more complicated than simple nonpolar binding, however, and hydrophobic association is not the only mode of binding of small molecules to cyclodextrins. Several species have negative entropy changes upon binding, and even vanillin has a small unfavorable entropy change. For these species, of course, the strength of the binding enthalpy outweighs the unfavorable entropy change and gives a negative overall Gibbs free energy change. Even polar species like acetic acid (vinegar) will bind to cyclodextrins (see Table 4.9). Such ligands may be stabilized in cyclodextrin complexes by dipole–dipole interactions or hydrogen bonding.

The natural cyclodextrins, in spite of their surface polarity, also have only limited solubility in water, apparently because they readily crystallize due to the many good intermolecular hydrogen bonds that can be formed when the molecules are stacked together in crystals. Their solubility can be significantly increased by substituting functional groups onto the hydroxyl groups to disrupt the possibility of such hydrogen bonding in the crystal state. The most common derivatives are methyl and hydroxypropyl substitutions.

The amylosic rings of cyclodextrins are not hydrolyzed by human α-amylases, making these molecules indigestible. Natural β- and γ-cyclodextrins have GRAS status in the United States and are used in foods to incorporate water-insoluble molecules like vitamins or flavors such as vanillin, limonene, or eugenol (Chapter 8) into aqueous-based foods. β-Cyclodextrin is also used as the active ingredient in room deodorizing or odor-eliminating products, where it functions by binding volatile hydrophobic odor molecules.

Chemically Modified Starches

The range of functional properties exhibited by starches can be expanded and controlled by various forms of chemical modification. These process can include shortening the polymer chains by partial hydrolysis, changing the ability of the chains to closely associate as in retrogradation via derivatization of hydroxyl groups, or cross-linking of different chains.

Acid-Modified Starches are starches that have been partially hydrolyzed by treatment with acid. Typically the starch granules are treated with HCl or H_2SO_4 at ~25 to 55 °C for times ranging from 6 to 24 hours. Amorphous regions of the granule are most affected, and the amylopectin is more extensively hydrolyzed than is the amylose. These starches have a decreased hot-paste viscosity, but gel firmly on cooling, and are often used in candy manufacture.

Starches can also undergo chemical modification or substitution at their free hydroxyl groups to produce **substituted starches**. This is often characterized by the degree of substitution, DS, ranging from 0 (no modification) to 3 (all 3 free hydroxyl groups per monomer modified). Hydroxyethyl starch ethers are produced by reacting starch with ethylene oxide at 50 °C (Figure 4.95). Usually low DSs (0.05–0.1) are used in foods, but these changes significantly alter the physical properties. The substitutions inhibit the orderly packing of chains in crystalline arrangements, and disrupt interchain hydrogen bonding, and thus prevent retrogradation. Starch phosphate monoesters are prepared by treating starch with acid salts of ortho-, pyro-, or tri-polyphosphate at high temperatures (50–60 °C for 1 hour). DSs of ~0.25 are usually achieved. These modified starches have excellent resistance to retrogradation and are used extensively in frozen and refrigerated foods.

Cross-Linked Starches are produced by reaction in aqueous solution with phosphorylchloride, or by treating dry starch with trimetaphosphate (Figure 4.96). The cross-links are phosphate groups esterified to two different hydroxyl groups, each on a different polymer chain, thus serving to covalently attach the two chains.

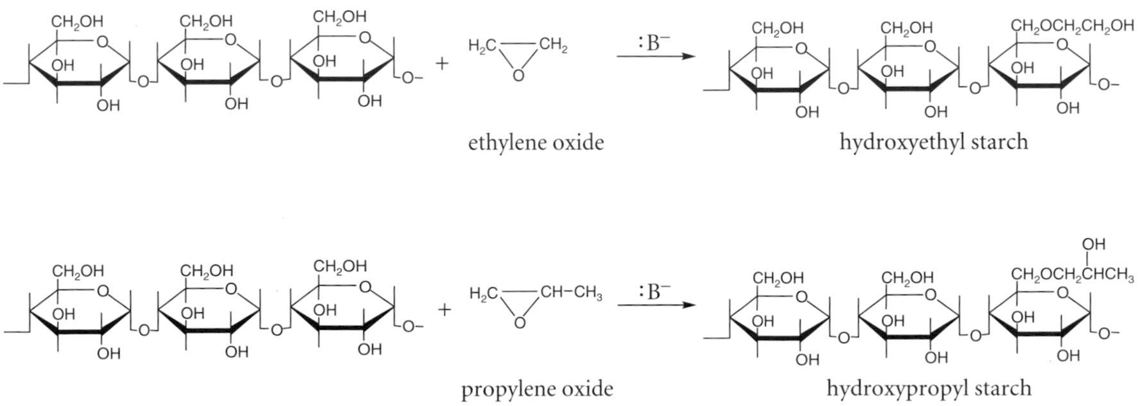

Figure 4.95. The modification of starch with ethylene and propylene oxide.

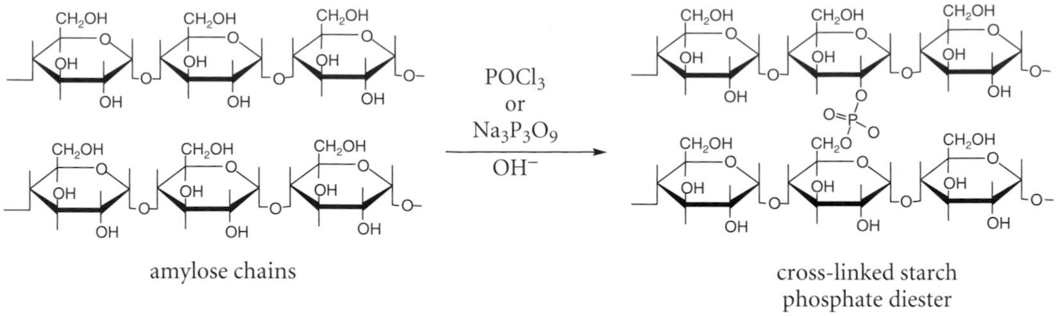

Figure 4.96. The cross-linking of starch as a phosphate diester.

amylose chains

cross-linked starch phosphate diester

Table 4.10 Properties of various corn starches

Type	Amylose/amylopectin ratio	Gelatinization temperature range (°C)	Distinguishing properties
normal	1:3	62–72	poor freeze-thaw stability
waxy	0:1	63–72	minimal retrogradation
high-amylose	3:2–4:1	66–92	granules less birefringent than normal starches
acid-modified	variable	69–79	decreased hot-paste viscosity compared with unmodified
hydroxyethylated	variable	58–68 (DS 0.04)	increased paste clarity, reduced retrogradation
phosphate monoesters	variable	56–66	reduced gelatinization temperature, reduced retrogradation
cross-linked	variable	higher than unmodified; depends on degree of cross-linking	reduced peak viscosity, increased paste stability
acetylated	variable	55–65	good paste clarity and stability

Source: R.L. Whistler and J.R. Daniel in *Food Chemistry*, 2nd ed., O.R. Fennema, ed. (New York: Marcel Dekker, 1985).

The principal types of commercial starches and starch derivatives and their distinguishing properties are summarized in Table 4.10.

Cellulose

The most abundant organic molecule is the polysaccharide cellulose, a linear β-(1→4)-linked polymer of D-glucose; in fact, there is more by weight of this one molecule in the biosphere than of all other biomolecules combined. This polysaccharide is the primary structural component of plant cell walls, which accounts for its widespread abundance. It is the main molecular constituent of cotton, wood, and paper, and thus is the basis for several major industries. Cellulose does not generally occur as a single

Figure 4.97. The solubility of cellooligomers in room temperature water as a function of oligomer DP. Cellobiose, the dimer, is also quite insoluble relative to other dissacharides (see Table 4.7) D.W. Wilson, personal communication of unpublished data.

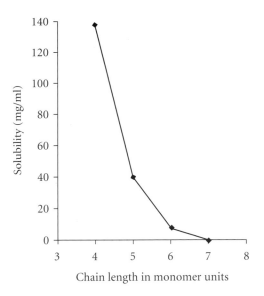

chain, but rather is synthesized in close proximity to many other chains, which organize into microfibrils as the fundamental structural unit; these microfibrils in turn are organized into larger fibrils, generally associated on their surfaces with lignin and hemicellulose (an irregular mixed polysaccharide containing a number of different sugars, but with xylose as the most abundant component), and these fibrils form the structural scaffolding for plant cell walls.

Cellulose oligomers longer than cellohexaose (six glucose units) are almost completely insoluble in water (see Figure 4.97), although slightly longer chains (DP 7 and 8) can be slightly solubilized in very cold (undercooled) water. The glucan chain length (degree of polymerization) of natural cellulose varies from about 2,000 to more than 15,000 glucose residues. In part because of its insolubility, cellulose is resistant to degradation, and it can persist in the environment for long times. It is also stable under normal food processing and cooking operations.

The disaccharide repeat unit of cellulose is cellobiose (Figures 4.48 and 4.98). In Figure 4.98, the two glucose rings are shown in their standard Haworth orientation, with both ring oxygen atoms in the two o'clock position, and with their C6 exocyclic groups on the same side. In cellulose, however, the values of the angles ϕ and ψ, specifying rotations about the C1—O1 and O1—C4′ single bonds, are such that each successive ring is "flipped" by 180° with respect to the previous one, so that these exocyclic groups alternately point to the right and the left, and the equatorial–equatorial β-(1→4)-linkage produces a relatively flat, ribbon-like conformation, as shown in Figure 4.99.

Because all of the hydroxyl groups are equatorial, all of the axial positions in cellulose are occupied by nonpolar (and non-hydrogen-bonding) protons, which means that the sides of the cellulose chain are polar and hydrogen bonding, while the tops and bottoms are hydrophobic (Figure 4.100). The chains can then stack together in

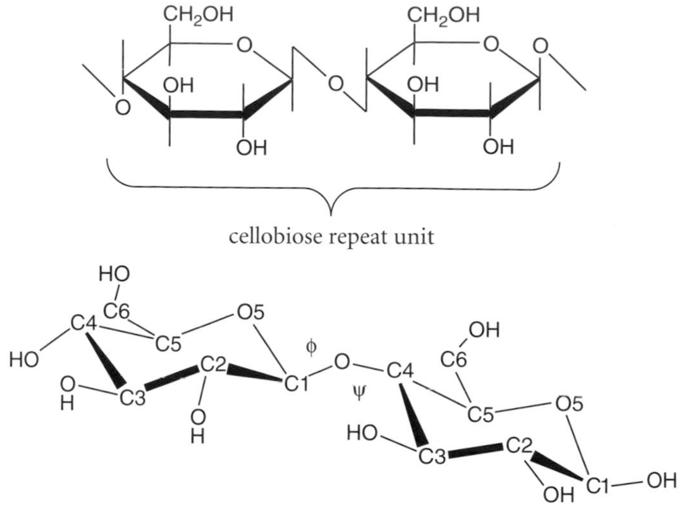

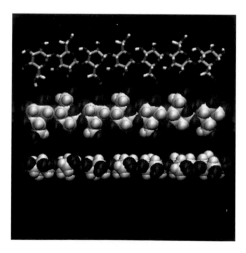

Figure 4.98. The cellobiose repeat unit of cellulose, shown both in the Haworth and conformational representations.

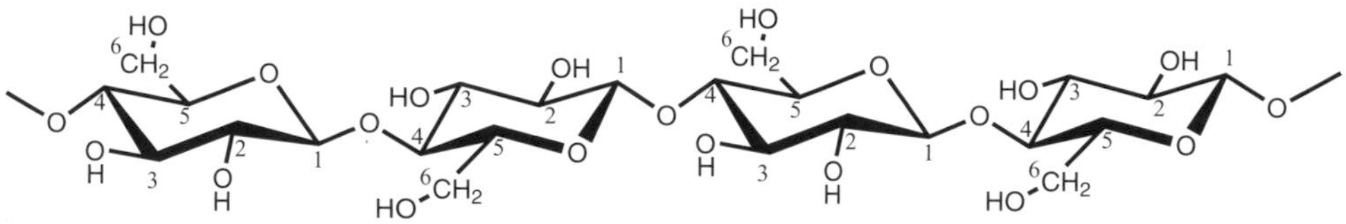

Figure 4.99. A short, four residue segment of a cellulose chain, illustrating the relatively flat, ribbon-like conformation, with the exocyclic primary alcohol groups alternately extending out from opposite sides of the chain.

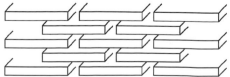

Figure 4.100. Upper: the cellulose Iβ crystal structure; in the middle and bottom van der Waals drawings, the polar, hydrogen-bonding atoms are colored red, while the non-hydrogen-bonding and hydrophobic aliphatic carbon and hydrogen atoms are colored white. Lower: the aliphatic, nonpolar and non-hydrogen-bonding tops and bottoms of the glucose rings allow the chains to pack against each other, with a staggered stacking as illustrated schematically. (Adapted and redrawn from D.A. Rees, *Polysaccharide Shapes*. New York: John Wiley & Sons, 1977.)

a regular crystalline packing, as illustrated in Figure 4.100, matching up hydrophobic faces as well as allowing hydrogen bonds between chains, and between successive residues along the same chain. This hydrophobic component to the association of cellulose chains is the reason that short cellooligomers can be made more soluble at low temperatures; the $-T\Delta S$ term favoring association becomes less important as T decreases, a typical signature of hydrophobic association.

The individual fibrils of cellulose exhibit significant crystalline order, with at least two different packing arrangements, called Iα and Iβ, occurring in most fibrils. The fibers are not completely crystalline, however, but also have significant regions of amorphous structure mixed with the crystalline regions in the same fiber. The crystal packing gives the fibers great strength, but the amorphous regions are necessary to decrease the rigidity and allow the plant cells to respond to environmental stress, such as a plant's swaying in the wind or changes in cell turgor.

As many as seven different crystal forms for cellulose have been proposed, but only two are common in foods, the Iα and Iβ forms (Figures 4.101 and 4.102). These two crystal structures are similar, with the differences primarily involving only crystal packing, including their hydrogen bonding patterns between adjacent cellulose chains, and the register between chains in different layers. The conformations of the individual chains are nearly identical. The Iβ crystal form is somewhat lower in

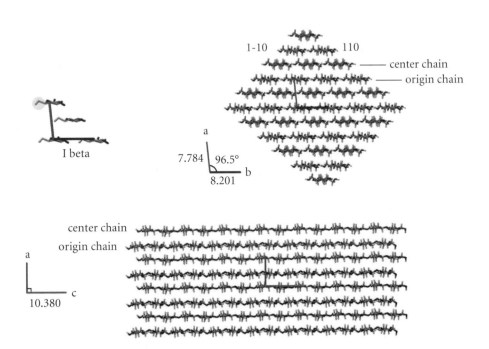

Figure 4.101. End-on and side views of the packing of cellulose chains in the Iβ crystal form (Y. Nishiyama et al., *J. Am. Chem. Soc.* 124:9074, 2002). Note how the aliphatic, non-hydrogen-bonding, nonpolar tops and bottoms of the glucose rings pack against each other in this structure. This type of staggered stacking is illustrated schematically in Figure 4.100. (Adapted and redrawn from D.A. Rees, *Polysaccharide Shapes*. New York: John Wiley & Sons, 1977.)

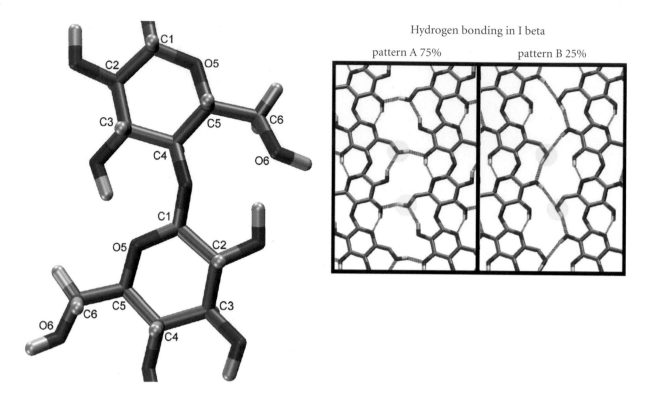

Figure 4.102. Left: the crystal structure of cellulose Iβ, as determined by X-ray fiber diffraction. Note that in this structure the chains form internal hydrogen bonds from the O3 hydroxyl group of one glucose monomer to the O5 ring oxygen of the preceding residue, and another hydrogen bond between the O2 hydroxyl group and the O6 hydroxyl group of the next residue. Right: the hydrogen bonding pattern in this Iβ crystal structure, illustrating the hydrogen bonds between adjacent chains within the same layer of the crystal. Hydrogen bonds involving the O6 hydroxyl group are shown in blue, and those involving the O3 hydroxyl group are shown in red. The doubly hydrogen-bonded hydroxyl groups of the exocyclic C6 carbon atoms are highlighted in yellow. (Y. Nishiyama et al., *J. Am. Chem. Soc.* 124:9074, 2002; J.F. Matthews et al., *Carbohydr. Res.* 341:138, 2006.)

energy than the Iα form, which will irreversibly convert to the Iβ form if heated above 260 °C. (see Figure 4.103). In both of these crystal forms, the chains are parallel to one another (that is, with the vectors from their nonreducing to reducing ends aligned in the same direction). Also, in both of these crystal forms, the flat, straight, ribbon-like conformation of the chains is stabilized by two hydrogen bonds between successive residues: one from the hydroxyl group of the C3 carbon of the residue on the reducing end side of each glycosidic linkage to the O5 ring oxygen atom of the preceding residue, and the other from the C6 hydroxyl group to the O2 hydroxyl group of the preceding residue (see Figure 4.102). These hydrogen bonds stabilize the extended conformation, because any significant rotations around ϕ and ψ would tend to disrupt these bonds. The chains are then stacked together in the crystals like bricks or lumber (see Figure 4.100), with additional hydrogen bonds between chains on either side in the same plane (see Figure 4.102). No hydrogen bonds occur between the stacked layers in the reported crystal structures (Figure 4.101), however, and the planes are presumably held together by hydrophobic forces, because the "tops" and "bottoms" of the glucose residues consist entirely of hydrophobic axial protons (Bergenstråhle et al. 2010). This model structure would then superficially resemble the stacking of protein layers in silk (see Figure 5.14). However, cellulose does not exhibit the flexibility of silk that results from such a topology.

It should be noted that the reported X-ray "crystal" structures of cellulose are not based on single crystals like those used to determine protein conformations (Chapter 5) but are actually based on the crystalline regions of partially disordered fibers. Recent computer simulations have found that this reported "crystal" structure is not energetically-stable with any of the standard models (Matthews et al. 2011, 2012), which suggests that small revisions to this picture may be required.

Cellulose II, used in the textile industry, is a lower-energy, stable form that results from treating cellulose fibers under tension with alkali, a process called **mercerization**. In cellulose II adjacent chains have traditionally been believed to run in antiparallel directions to one another, and this assumption was built into the recent analyses of the neutron and X-ray diffraction of cellulose fibers (Langan, Nishiyama, and Chanzy 1999, 2001). This model, however, has been called into question by chemically labeling the reducing ends of all of the chains in a mercerized cellulose sample, which in one experiment resulted in labeling only at one end of each fiber, as would be expected if all the fiber chains are parallel (Maurer and Fengel 1992). Another such experiment found a mixture of fibers labeled at only one end and fibers labeled at both ends, as would be expected for antiparallel chains (Kim et al. 2006). Again, further research would seem warranted. The cellulose III crystal forms result from treating either cellulose I or II with small nitrogen-containing compounds like ammonia or ethylenediamine (Figure 4.103).

For humans, cellulose and cellulose derivatives are dietary fiber, because we have no enzymes capable of hydrolyzing the linkages in these polymers. Even herbivores lack cellulases to digest plant material and must rely on cellulases produced by their intestinal microflora to digest their food. Ruminants like cows have multiple stomach chambers that function as reactors for the growth of cellulolytic microorganisms that

Figure 4.103. The relationships between the crystal polymorphs of cellulose.

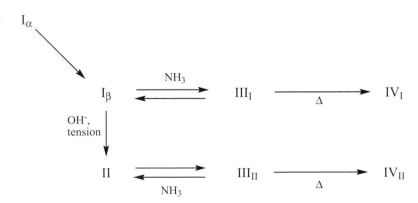

produce cellulases. The equids, however, such as horses and burros, have single-pass digestive systems that are less effective at hydrolyzing cellulose, because their microflora has less time to act on the cellulosic material. Among animals, only certain species of termites are able to produce cellulases and thus to directly process cellulose as a food. In some cases, microorganisms in the flora of the human digestive tract can hydrolyze some of the cellulose in nominally low-calorie foods like lettuce, so that metabolizable glucose is produced and absorbed across the intestinal wall, contributing dietary calories.

A variety of functionally useful modified cellulose chains can be prepared by alkylating the hydroxyl groups (Figure 4.104). Treating cellulose with methyl chloride, chloroacetic acid, or propylene oxide in a strongly alkaline solution causes some of the hydroxyl groups to become substituted with methyl, carboxymethyl, or hydroxypropyl groups.

Although the presence of substitutions, in principle, makes the chains more hydrophobic, because one is taking a hydrogen-bonding hydroxyl group and placing a non-hydrogen-bonding, hydrophobic group on it, the substitution actually makes the cellulose chains soluble in water. The explanation is that the methyl, carboxymethyl, or isopropyl groups disrupt the microcrystalline packing arrangement of the solid phase. In aqueous solution, these substituted celluloses are thixotropic; that is, they exhibit shear thinning, meaning that the viscosity of their solutions decreases with applied shear as the polysaccharides align with the flow of the liquid. In methylcellulose the viscosity also decreases initially with increasing temperature, but at higher temperatures exhibits an unusual behavior by forming thermoreversible gels as the temperature is raised. The gelation temperature depends on concentration, but

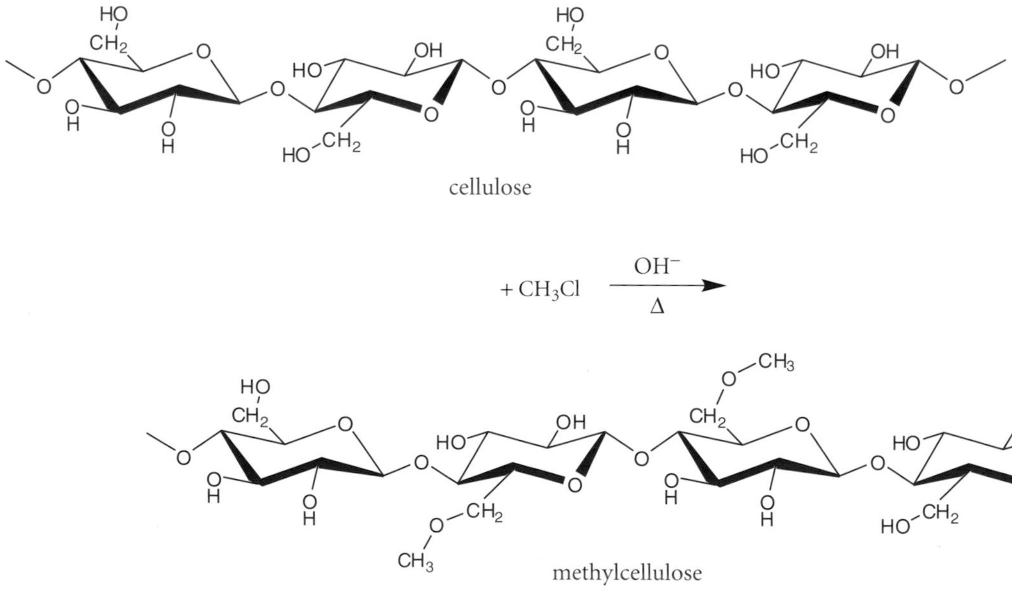

Figure 4.104. The derivatization reactions of cellulose.

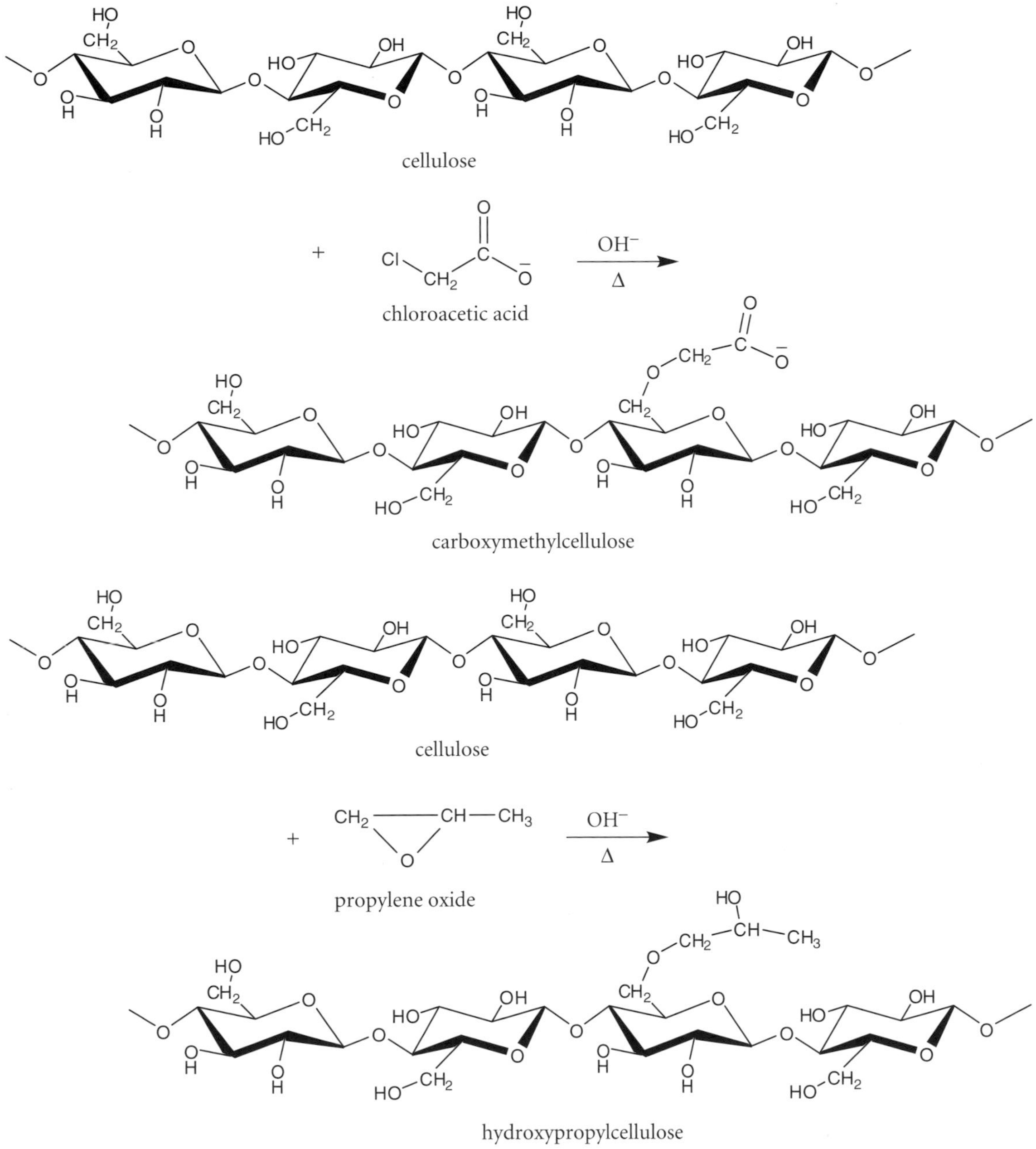

cellulose

chloroacetic acid

carboxymethylcellulose

cellulose

propylene oxide

hydroxypropylcellulose

Figure 4.104. *Continued*

solutions of methylcellulose that have less than a 5% concentration in the polysaccharide reversibly gel around 42°C.

Conventional thermoreversible gels melt at elevated temperatures and then gel again when the temperature is lowered below a critical value. Because they actually gel at higher temperatures and return to the solution state when the temperature is again lowered, modified celluloses are quite unusual. The gelling temperature is a function of the degree of substitution, DS. This odd property is useful in fried food manufacture, such as doughnuts or fried meats, because they can be used to create a barrier to oil absorption by the product at cooking temperatures, and retard the loss of water during the cooking, and improve the adhesion of batter to the product. Substituted celluloses are also used in baking with gluten-poor starches to improve batter consistency and decrease crumbliness and in conventional baked products to inhibit staling. Carboxymethyl cellulose is referred to as **cellulose gum** in the food trade (see Table 10.1).

Chitin and Chitosan

Chitin is a β-(1→4)-linked homopolymer of N-acetylglucosamine residues (Figure 4.105). Structurally it resembles cellulose, but with each of the C2 positions of the glucose rings acetylated. Chitin is one of the principal components of the exoskeletons of arthropods (insects) and crustaceans (e.g., crabs, lobsters, shrimp), where it forms a hard matrix with proteins and calcium carbonate. It thus also resembles cellulose functionally, playing a structural role in these animal exoskeletons similar to that of cellulose in plant cell walls. Like cellulose, native chitin can pack into quasi-crystalline microfibrils with two different arrangements, called α and β, where the chains in the more abundant α form are packed in an alternating antiparallel manner, while they are all parallel in the much less abundant β form.

Chitosan (Figure 4.105) is prepared commercially from chitin by treating it with NaOH at elevated temperatures to randomly deacetylate the N-acetylglucosamine residues. Various fungi also produce enzymes called chitin deacetylases that catalyze the deacetylation reaction. The degree of deacetylation in commercial chitosans ranges from 60 to 100%. Like the amino groups of proteins, the amino groups of chitosan are protonated (charged) at neutral pH, making the polysaccharide water soluble. Like cellulose, chitosan cannot be digested by humans and in the diet constitutes dietary fiber. Chitosan can bind to fats and may complex with them in foods either as they are prepared or as they are digested.

A number of dubious health claims have been made for chitosan. The most common of these is that as a dietary supplement it can block the adsorption of fat,

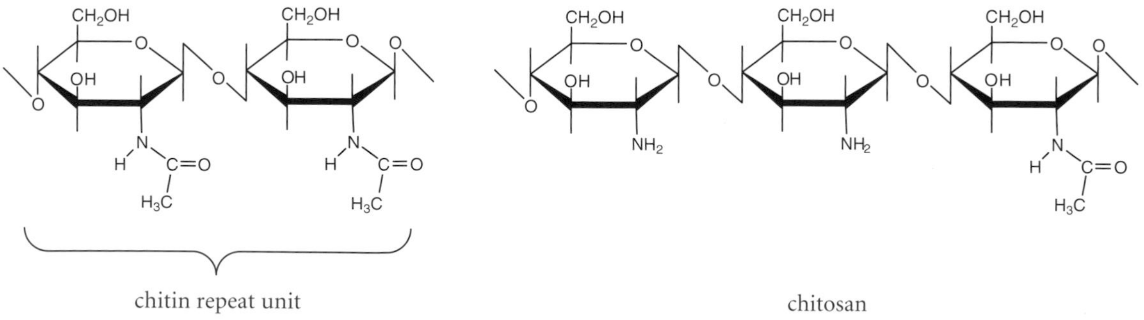

Figure 4.105. Haworth formulas for chitin and chitosan.

leading to weight loss. It also has been reported that it can lower serum cholesterol. Although based on the ability of chitosan to bind to fats, these claims do not yet appear to be supported by reliable clinical studies and should be treated with skepticism until further information is available.

Heteropolymers

There are a number of heteropolymers that are important as gums in the food industry. Most are obtained from plant sources, although some commercial gums are also produced by bacteria.

Hemicellulose

The three main polymeric components of plant cell walls are cellulose, lignin, and hemicellulose, of which cellulose is the most abundant. (Figure 4.106 shows a schematic model for the structure of such a cell wall.) Hemicellulose, which is not structurally related to cellulose, is a random heteropolymer primarily composed of the sugar monomers xylose, arabinose, mannose, galactose, mannuronic acid, galacturonic acid, and rhamnose (see Figure 4.118), with xylose being the most abundant. The composition and structures of hemicelluloses can vary with the plant source; some hemicelluloses can be largely or almost entirely xylans. Most hemicelluloses are insoluble fibers that surround the cellulose microfibrils and fibrils in plant cell walls. They cannot be digested by humans and thus represent dietary fiber in foods. While they are indigestible for humans, hemicelluloses are easily hydrolyzed by acid treatment because they do not pack tightly into crystalline arrays like cellulose due to their flexibility and heterogeneity.

Figure 4.106. A schematic illustration of one model for the structure of plant cell walls. In the secondary wall, large, partially crystalline cellulose fibrils are embedded in a matrix of lignin and hemicellulose. Some pectin (pp. 262–265) is also intertwined in this matrix, while the majority of the lignin is found in the middle lamella between adjacent cells. Cellulose, hemicellulose, and lignin are also the primary constituents of the outer, primary cell wall. The inner plasma membrane is a phospholipid bilayer as in Figure 7.12 (p. 414). (Figure reproduced from Achyuthan et al. 2010. Molecules 15:8641–8688.)

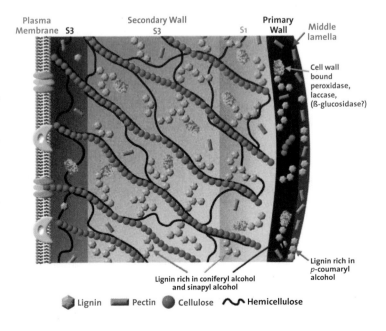

Seaweed Polysaccharides

A number of polysaccharides useful in the food industry are isolated from marine plants commonly referred to collectively as seaweeds. The most important of these polymers are agar, widely used as a culture medium in microbiology experiments; alginates; and the carrageenans.

Alginates

Alginates are water-soluble block copolymers of uronic acids extracted from the cell walls of the brown alga commonly known as giant kelp. In the extraction, the macerated seaweed material is first demineralized by treatment with dilute acid, to replace any calcium ions, and is then further ground in the presence of an alkaline salt, to neutralize the acid and remove the soluble polysaccharide from cellulosic and other insoluble material. It is then dried and collected as the salt of the corresponding alkaline cation used in the neutralization. Sodium alginate is the most commonly used in industrial applications.

The alginates are unbranched copolymers of β-D-mannuronic acid and α-L-guluronic acid. These two types of residues occur primarily in blocks of one type or the other, although regions also occur where the two sugars alternate in sequence. The composition of the alginates is variable, although it is often approximately a 3:2 ratio (mannuronic acid/guluronic acid). Both types of sugars are linked through (1→4)-linkages, but the geometries of the glycosidic linkages for these two different types of monomers result in different conformations for homopolymers of the two residues. As can be seen in Figure 4.107, poly-D-mannuronic acid, which is linked by two equatorial bonds, forms a flat ribbon similar to cellulose, while the axial–axial linkage of poly-L-guluronic acid has a highly buckled conformation, which creates pockets along the chain lined with polar hydroxyl groups and the negatively charged acid group. Alginate is soluble in cold water but gels strongly in the presence of

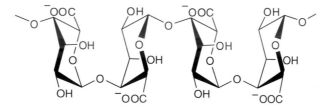

a poly (L-guluronic acid) segment of alginate

a poly-(D-mannuronic acid) segment of alginate

Figure 4.107. The structures of poly-L-guluronic acid and poly-D-mannuronic acid.

calcium ions due to the presence of these highly buckled L-guluronic acid sequences. It can be dissolved in cold milk by using a calcium complexing agent.

In the presence of Ca²⁺ ions two of these buckled sequences on two different chains can associate, with the cation canceling the repulsions of the like-charged acid groups and cementing these two sequences together to form a crystalline juncture zone where a "hole" with the dimensions of the calcium ion is occupied by these ions. The poly-D-mannuronic acid homopolymeric chains do not have a marked tendency to associate in solution under these conditions and thus do not precipitate out in the presence of calcium ions. The association of the poly-L-guluronic acid homopolymers by this mechanism is referred to as the "egg-box" association mechanism due to the resemblance of the calcium ions in these cavities to eggs in a cardboard carton, as illustrated schematically in Figure 4.108. In the alginate copolymers, in the presence of calcium ions, the regions of poly-L-guluronic acid can associate through egg-box-like juncture zones, while the poly-D-mannuronic acid regions can still go off in different, essentially random directions, allowing the formation of an open gel rather than a precipitate as in the poly-L-guluronic acid case.

Recent studies have questioned some of the aspects of this egg-box model for alginate juscture zones (similar egg-box juncture zones are also thought to occur in low-methoxy pectin gels; see below). Using enzymes to produce strictly alternating copolymers of guluronic and mannuronic acids, Donati et al. proposed that blocks of such alternating sequence also occur in natural alginates and could serve as gel

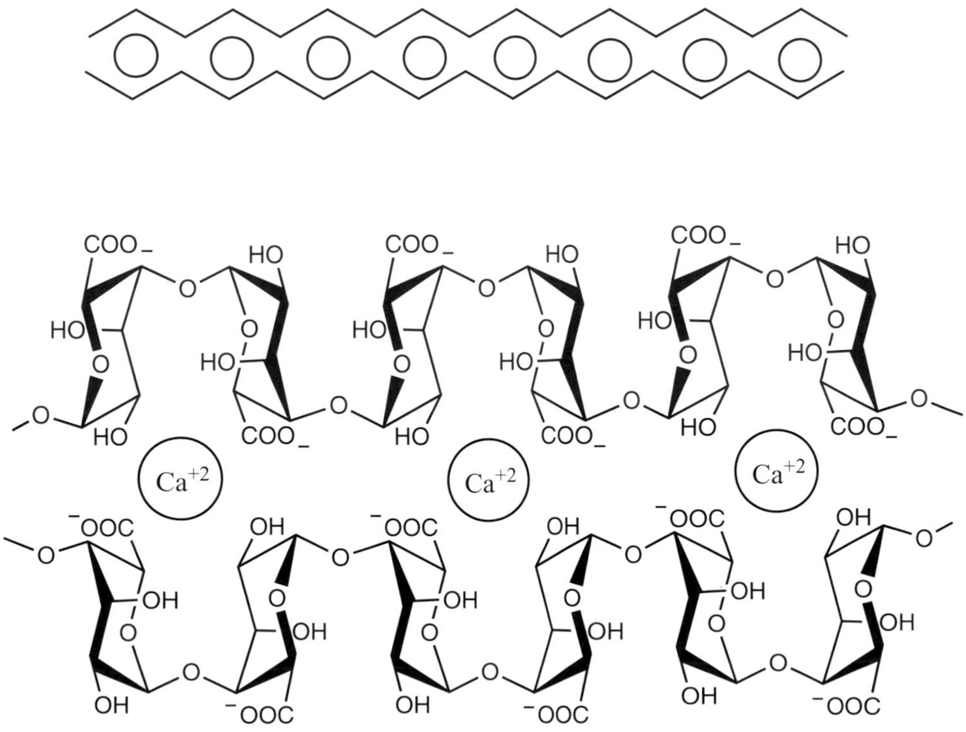

Figure 4.108. An illustration of the proposed egg-box model for the juncture zones in poly-L-guluronic acid, showing a schematic representation on the top and a conformational molecular representation on the bottom.

juncture zones (Donati et al. 2005). Computational energy studies find that the egg-box juncture zone may be only metastable (Braccini et al. 1999; Braccini and Pérez 2001), and x-ray diffraction studies suggest the occurrence of other types of juncture zones as well (Li et al. 2007), which led these latter workers to propose that the egg-box juncture zones form from fast gelation but that slow gelation leads to more stable 3/1 helical juncture zones.

The ability to gel is one of the reasons for the wide utility of the alginates, which are used in a number of gel products such as puddings and dessert gels. It is also useful as a thickener and stabilizer due to its water binding capacity and is added to baked products, chocolate, salad dressings, and ice creams. In ice cream, alginates control the mobility of water, thus preventing the growth of ice crystals when it is stored at higher temperatures.

A common science fair demonstration is to use sodium alginate solutions to create instant, ersatz "Gummi worms" (real Gummi candies are made with gelatin; see Chapter 5). To accomplish this, a concentrated solution of sodium alginate is poured into a glass or beaker containing a concentrated solution of calcium chloride. The calcium ions cause the stream of alginate solution to gel on the outside as it hits the $CaCl_2$ solution, with some ions then permeating into the interior to cause the entire stream to gel, creating a worm-shaped, rubbery tube. Small alginate spheres can be created in the same fashion by allowing single drops of the alginate solution to fall into the $CaCl_2$ solution. This method has been used in various molecular gastronomy recipes to create flavored artificial "caviars" and liquid-filled spheres, and alginate has been popular more generally in molecular gastronomy for a variety of purposes.

Propylene glycol alginate is a derivative of alginate in which some of the carboxylic acid groups are esterified with propylene glycol. In the United States, it has GRAS status (Generally Recognized As Safe; see Chapter 10). It is nongelling due to the esterification of the acid groups, which disrupts potential juncture zones. It is used as a thickener and stabilizer in products like salad dressings, mustards, and ice cream.

Carrageenans

The **carrageenans** are a family of polysaccharides extracted from marine red algae (Irish moss, *Chondrus crispus*) of the North Atlantic, which are of considerable practical interest because of their ability to form thermoreversible gels, and because of their effects on solution viscosity. The basic structure of these polysaccharides is a regular, linear alternating copolymer sequence of galactose derivatives with alternating β-(1→4) and α-(1→3) linkages (Figure 4.109). Many of the monomer units are sulfated at their C2, C4, or C6 hydroxyl groups. The sequence of the carrageenans consists of alternating O3-linked sulfated β-D-galactopyranosyl units and either O4-linked sulfated α-D-galactopyranosyl or 3,6-anhydro-α-D-galactopyranosyl

Figure 4.109. Haworth and conformational formulas for the carrageenans.

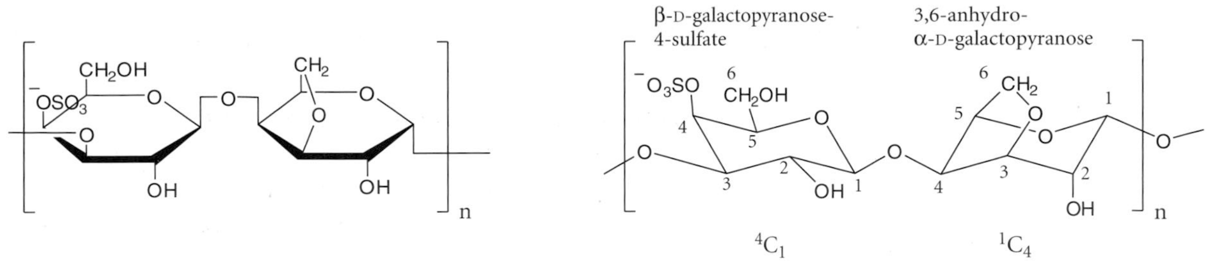

β-D-galactopyranose-4-sulfate
3,6-anhydro-α-D-galactopyranose
$^4C_1$
$^1C_4$

Haworth representation

conformational formula

basic repeat unit of κ-carrageenan

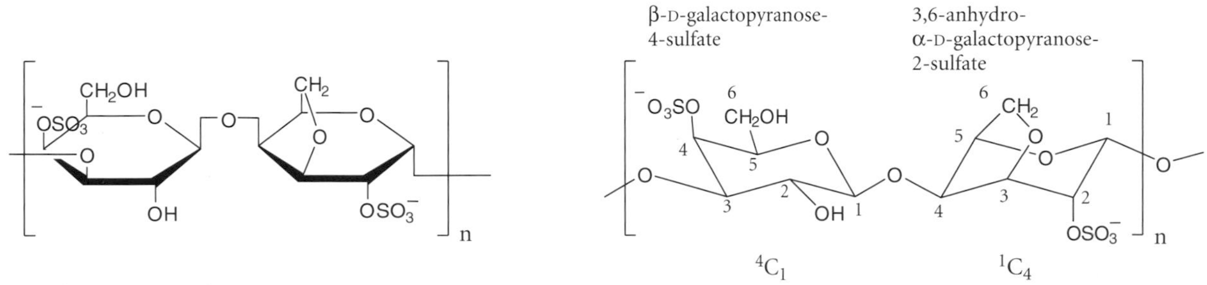

β-D-galactopyranose-4-sulfate
3,6-anhydro-α-D-galactopyranose-2-sulfate
$^4C_1$
$^1C_4$

Haworth representation

conformational formula

basic repeat unit of ι-carrageenan

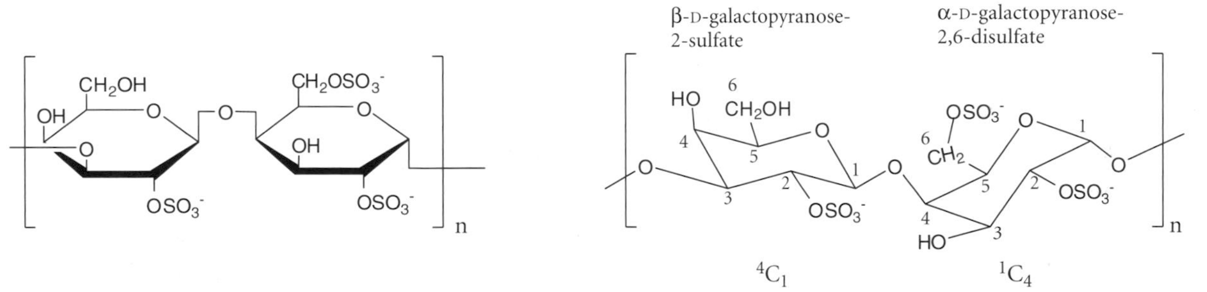

β-D-galactopyranose-2-sulfate
α-D-galactopyranose-2,6-disulfate
$^4C_1$
$^1C_4$

Haworth representation

conformational formula

basic repeat unit of λ-carrageenan

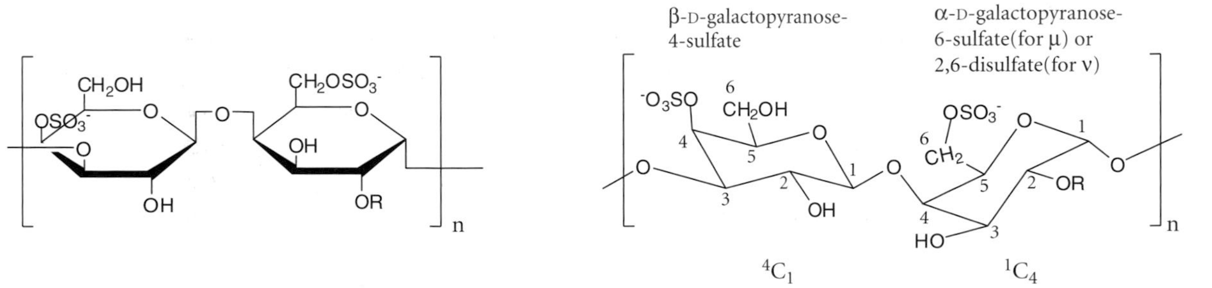

β-D-galactopyranose-4-sulfate
α-D-galactopyranose-6-sulfate(for μ) or 2,6-disulfate(for ν)
$^4C_1$
$^1C_4$

Haworth representation

conformational formula

basic repeat unit of μ and ν-carrageenan

μ: R = H

ν: R = SO_3^-

Table 4.11 Summary of carrageenan types

Type	Residue types and ring conformation	Linkage types	Number of sulfates	Gelling behavior	Remarks
ι	β-ᴅ-galactopyranose-4-sulfate, $^4C_1$ 3,6–anhydro-α-d-galactopyranose-2-sulfate, 1C4	β-(1→4) α-(1→3)	2	gels with Ca^{+2}	does not synerese
κ	β-ᴅ-galactopyranose-4-sulfate, $^4C_1$ 3,6–anhydro-α-ᴅ-galactopyranose,$^1C_4$	β-(1→4) α-(1→3)	1	gels with K^+	rigid gels; synereses
λ	β-ᴅ-galactopyranose-2-sulfate, $^4C_1$ α-ᴅ-galactopyranose-2,6-disulfate, $^4C_1$	β-(1→4) α-(1→3)	3	does not gel	
μ	β-ᴅ-galactopyranose-4-sulfate, $^4C_1$ α-ᴅ-galactopyranose-6-sulfate, $^4C_1$	β-(1→4) α-(1→3)	2	does not gel	
ν	β-ᴅ-galactopyranose-4-sulfate, $^4C_1$ α-ᴅ-galactopyranose-2,6-disulfate, $^4C_1$	β-(1→4) α-(1→3)	3	does not gel	

groups. The latter of these is a bicyclic sugar similar to levoglucosan, in which a water molecule has been eliminated between the O6 and O3 groups to form a second ring. At least five different polysaccharides belong to this class of molecules, being generally designated as ι, κ, λ, μ, and ν (Table 4.11). The members of this family are characterized by the sulfate content of their repeating units. For example, the κ- and ι-carrageenans have one and two sulfate groups on their repeating units, respectively. The unsulfated polymer, which does not occur naturally in food, is called β-carrageenan. In ι- and κ-carrageenan, the β-ᴅ-galactopyranose rings are in the normal $^4C_1$ conformation, but the 3,6-anhydro-α-ᴅ-galactopyranose rings are in the generally less stable $^1C_4$ conformation needed to bring the C3 and C6 hydroxyl groups close enough together to form the second cycle. In λ, μ, and ν-carrageenan, all the monomer rings are in the normal $^4C_1$ conformation.

In the presence of cations, including K^+, Rb^+, Cs^+, and NH_4^+, κ and ι-carrageenan form strong and thermoreversible gels even at low concentrations. Not all carrageenans form gels; the degree of sulfation affects this property. The highly sulfated λ-carrageenans do not gel; they are used in the food and pharmaceutical industries as thickeners. Carrageenans are widely used in dairy products to stabilize dispersed milk proteins, suspend cocoa powder in milk, and to control texture in cheese and ice cream (for example, to inhibit the development of "sandiness" through sugar and ice crystal growth; see Table 3.3). They are also used to produce thin films for edible coatings.

There has been considerable debate about the conformation of the carrageenans in food systems, because the experimental determination of this conformation has proven to be extremely difficult. The individual polymer chains are almost certainly a regular helix. It has long been believed that the chains exist as pairs of polymer strands woven together as a double helix (Figure 4.110). A variety of recent evidence indicates that carrageenan chains may actually exist as single helix strands, but this

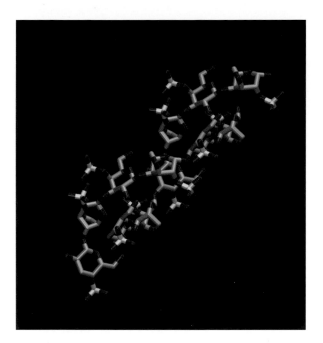 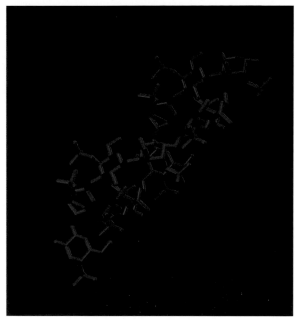

Figure 4.110. The proposed double helical conformation of ι-carrageenan. On the left, atomic coloring is used, while in the right panel all atoms in each of the two chains are colored the same to illustrate the double-helical coiling. (S. Arnott et al., *J. Mol. Biol.* 90:253–267, 1974.)

controversial question remains unresolved. Because of this uncertainty, it is not yet understood exactly how the presence of cations induces gelation, but they must somehow complex with the charged sulfate groups on different chains, thus reducing their tendency to fly apart as a result of the repulsions of their like-charged groups.

Carrageenans cannot be hydrolyzed by human glycosidases, so these polysaccharides can be considered to be dietary fiber. The name carrageenan has been widely reported to be derived from either a county, town or village called "Carragheen," in Ireland, a country where carrageenan has been harvested from the sea for at least 600 years. There apparently is no such place in Ireland, however, and it seems more likely that the name derives from the Irish Gaelic words for "little rock," possibly a reference to the submerged rocks on which the algae grows. As seaweed extracts with centuries of traditional use, the carrageenans are natural products and have GRAS status in the United States, but there have been studies suggesting the possibility that high consumption levels might present a cancer risk (Tobacman 2001; Bhattacharyya et al. 2008).

Agar, Agarose, and Agaropectin

Agar is a gelatinous material derived by hot water extraction from the cell membranes of various marine red algae seaweeds of the class Rhodophyceae, including species from the genera *Pterocladia*, *Gelidium*, and *Gracilaria*. It is widely used as a growth medium for bacterial cultures in microbiological research, and as a stabilizer

Figure 4.111. The basic repeating unit of agarose.

β-D-galactopyranose

3,6-anhydro-α-L-
galactopyranose

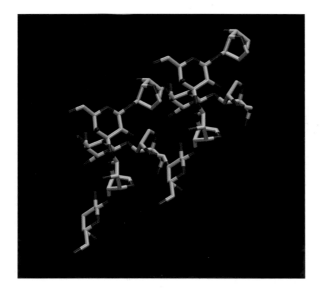 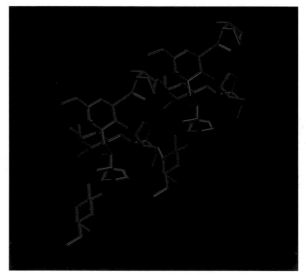

Figure 4.112. The proposed double helical conformational structure of the agarose fraction of agar. On the left, atomic coloring is used, while in the right panel all atoms in each of the two chains are colored the same to illustrate the double-helical coiling. (S. Arnott et al., *J. Mol. Biol.* 90:269–284, 1974.)

and thickener in a number of food products including jellies and certain diet products. Agar is a complex substance primarily composed of two polysaccharides, agarose and agaropectin. These polysaccharides are similar to the carrageenans because they are principally made up of β-D-galactopyranose and its derivatives linked through alternating β-(1→4) and α-(1→3) linkages, as in carrageenan, but in the agar polysaccharides half of the residues are 3,6-anhydro-α-L-galactopyranose rather than the similar 3,6-anhydro-α-D-galactopyranose of κ-carrageenan (Figures 4.111 and 4.112). Also like the carrageenans, a small number of the galactose residues are sulfated. The agarose polysaccharide fraction of agar has sulfate groups on about one tenth of the galactopyranose residues. The agaropectin has a much higher proportion of sulfates and also has pyruvic acid side chains that interfere with close chain associations, making agaropectin nongelling, while agarose does gel. Agar is used commercially as a vegetarian substitute for gelatin (Chapter 5) in foods.

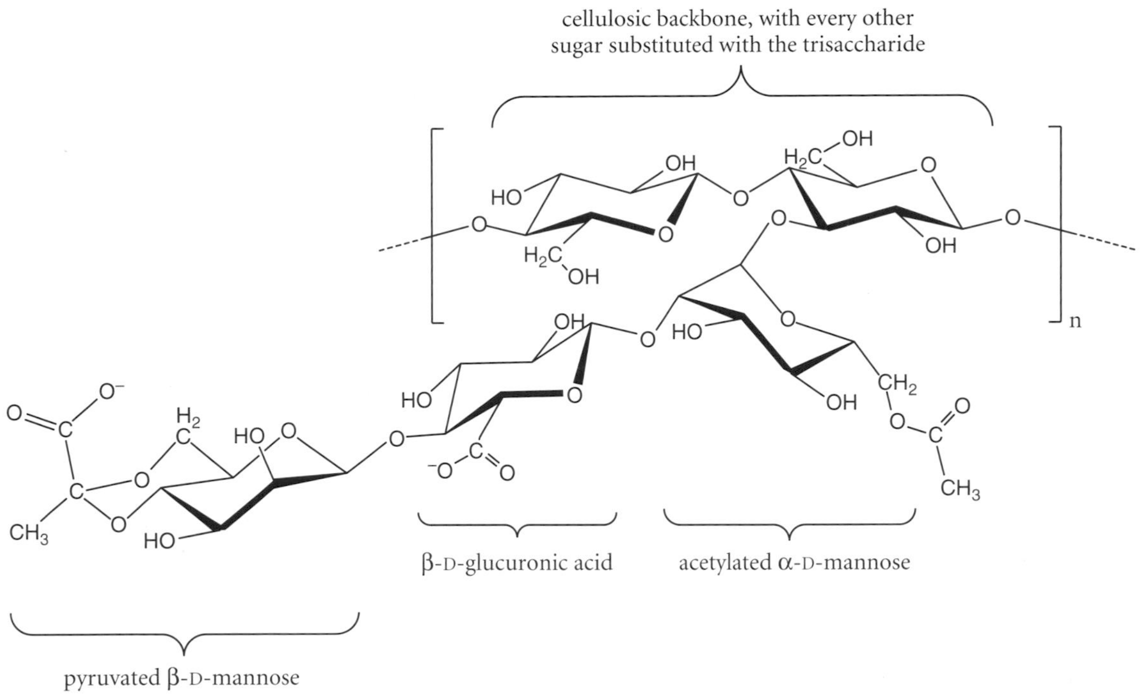

Figure 4.113. The repeating unit structure of xanthan.

Bacterial Polysaccharides

Xanthan

Xanthan gum is a heteroglycan obtained from the bacterium *Xanthomonas campestris*. Xanthan is a branched polysaccharide, with the backbone being the same as cellulose; that is, β-(1→4)-linked glucopyranosyl units (Figure 4.113). In xanthan every other glucose ring is branched at the O3 position, with a β-D-mannopyranosyl-(1→4)-β-D-glucouronopyranosyl-(1→2)-6-O-acetyl-β-D-mannopyranose trisaccharide attached. In approximately half of these branches the terminal mannose rings are modified by having a pyruvic acid attached in a cyclic 4,6 acetal form. Xanthan is insoluble in organic solvents and is isolated from the bacterial batch fermentation broth by precipitation using isopropyl alcohol, dried, and then ground to specific particle sizes. Because of the substitutions on every other C3 group, the cellulose-like backbone cannot form half of the stabilizing O3—O5 hydrogen bonds that stabilize the conformation of cellulose itself. In this powdered state, xanthan gum assumes a regular fivefold helical conformation, with the trisaccharide side chains folded back against the backbone.

Xanthan has several special properties that make it particularly valuable in food formulations. Perhaps the most important of these is that, like methyl cellulose, xanthan gum exhibits **shear thinning** (Figure 4.114). When a shear stress is applied to a xanthan solution, the viscosity decreases sharply, but when this shear stress is

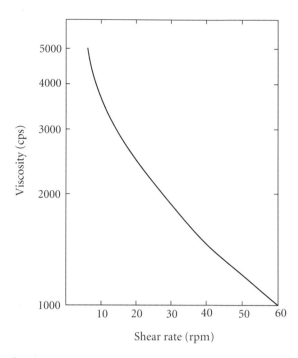

Figure 4.114. The viscosity of xanthan solutions exhibit shear thinning, as the viscosity decreases with shear rate. (Redrawn from data given in P.C. Paul and H.H. Palmer, *Food Theory and Applications*. New York: John Wiley & Sons, 1972.)

removed, the viscosity quickly goes up again. As a result, it is excellent for maintaining suspensions at low apparent viscosity. This viscosity drop arises because, as the shear increases, the polysaccharide fibers tend to align with the direction of the fluid flow, decreasing their tendency to collide and entangle with one another. Thus, using xanthan, a commercial salad dressing with large particles of dispersed spices, diced vegetables, or grated cheese, for example, can remain relatively stable while sitting on the table but flow readily when poured and not produce a stiff or rubbery mouth feel when chewed.

Xanthan exhibits significant water-binding capacity, is soluble in both hot and cold water, and produces a high solution viscosity at a low polysaccharide concentration. Unlike many other gums, the viscosity of xanthan solutions changes little with temperature in the range from 0 °C to 100 °C. This property makes it useful in stabilizing products such as chocolate syrups or salad dressings that must remain pourable after refrigeration, and for gravies, which must not thin too much at high temperature nor thicken too much upon cooling. The high acid stability means that xanthan is useful for salad dressings, which can have very low pHs, in the range 3.0 to 3.5, because it can stabilize these suspensions without being degraded by acid hydrolysis of the glycosidic linkages. Xanthan is often used in combination with other gums. Combining xanthan with guar gum gives very high viscosities in solution, and when combined with propylene glycol alginate the mix produces heat-reversible gels. Xanthan mixed with either guar or propylene glycol alginate is widely used as an ice cream stabilizer.

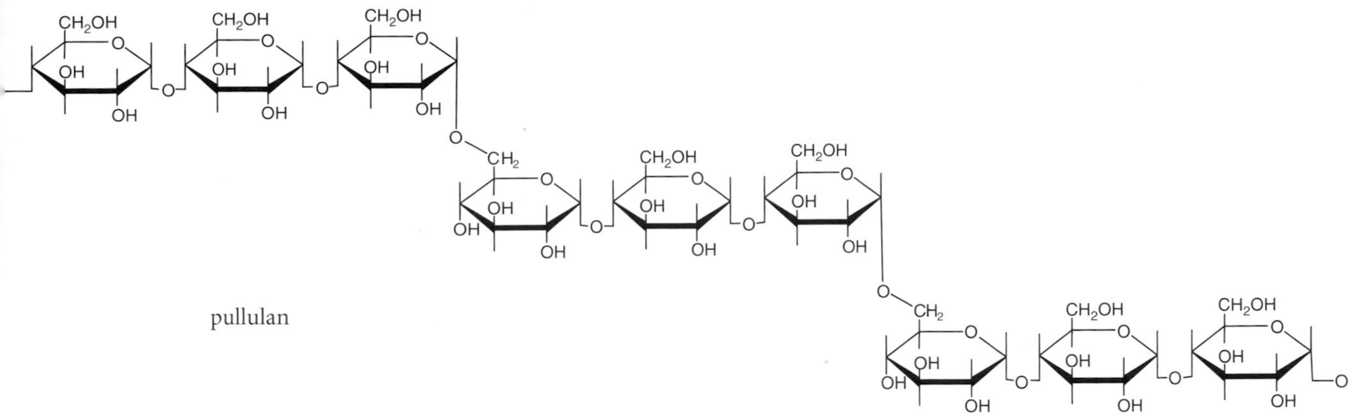

Figure 4.115. The structure of pullulan in Haworth representation.

Pullulan

Pullulan is a linear extracellular bacterial polysaccharide produced from starch by *Aureobasidium pullulans*. It consists of linear maltotriose units (that is, three D-glucose units successively linked together through α-(1→4) glycosidic bonds), with these trisaccharides linked together through α-(1→6) linkages (Figure 4.115). Pullulan also contains small amounts of maltotetraose, making up from 1 to 7% of the total residues of the polysaccharide. The α-(1→4) and α-(1→6) linkages are present approximately in the ratios 2:1, except for the deviations resulting from the presence of maltotetraose sequences.

Pullulan chains are water soluble, very flexible, and have no regular helical conformational structure. The individual pullulan chains have a high affinity for one another and become entangled due to their flexibility, forming useful edible films that have a number of practical applications. They are used as coatings for candies and other food products, and in the manufacture of capsules in pharmaceutical applications.

Gellan

Several other bacterial polysaccharides are finding uses in the food industry, including wellan, rhamsan, and gellan. **Gellan** is an acidic polysaccharide isolated from bacterial cultures of *Pseudomonas elodea*. This complex, water-soluble polysaccharide is used in microbiological research as a culture medium because, like agar, it can form firm gels. Gellan gels are stable up to 120 °C. As a food additive it is used as a thickener and stabilizer. The chemical structure of gellan has been determined and found to consist of a repeating tetrasaccharide sequence, [→3)-β-D-glucose-(1→4-β-D-glucuronic acid-(1→4-β-D-glucose-(1→4-α-L-rhamnose-(1→]$_n$, with the first D-glucose unit being acetylated on the O6 position in 50% of such residues and with an L-glyceric acid esterified to the O2 hydroxyl group of the same residue (Figure 4.116).

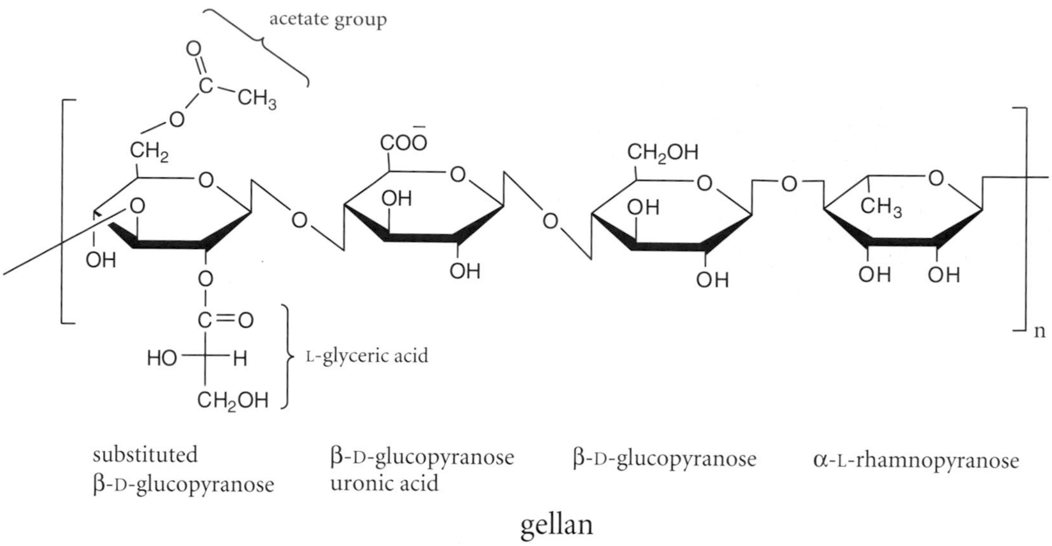

acetate group

substituted β-D-glucopyranose   β-D-glucopyranose uronic acid   β-D-glucopyranose   α-L-rhamnopyranose

L-glyceric acid

gellan

Figure 4.116. The structure of gellan. (From M.-S. Kuo and A.J. Mort, *Carbohydr. Res.* 156:173–187, 1986.)

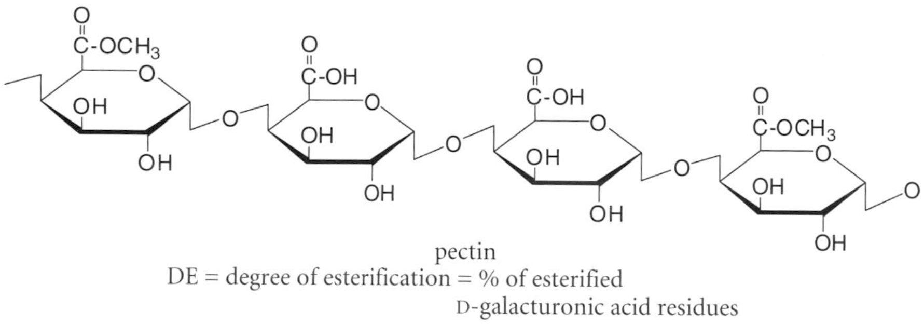

pectin
DE = degree of esterification = % of esterified
D-galacturonic acid residues

Figure 4.117. The structure of pectin in Haworth projection representation.

Plant Polysaccharides

Pectins

Pectins are complex polysaccharides derived from the soft tissues of plants. They are found in large amounts in a variety of fruit sources, including citrus peels and apple pomace. Pectins are found in the middle lamella between plant cells and help to bind the cells together. These gums consist of a backbone of $\alpha(1 \rightarrow 4)$-linked D-galacturonic acid monomers (Figure 4.117), perhaps with occasional short "side chains" consisting of single L-arabinose or D-galactose sugars. Many of the backbone D-galacturonic acid units have their acid groups esterified with a methyl group, and the extent of this modification is indicated by the Degree of Esterification (DE) or Degree of Methylation (DM), which can range up to 75%. Up to 10% of the sugar residues in pectin can be rhamnose (Figure 4.118) , and there may be small amounts of other

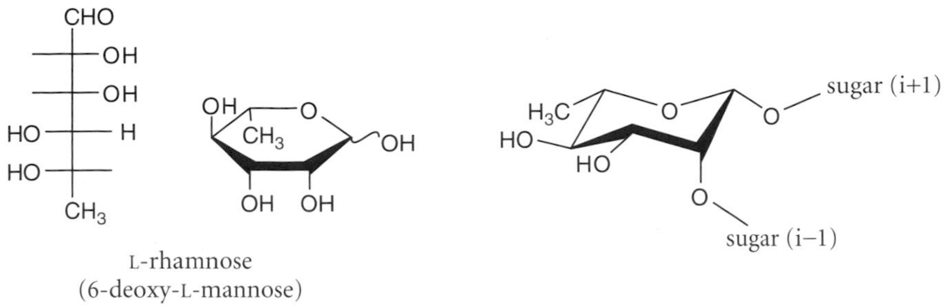

Figure 4.118. The structure of L-rhamnose in the open-chain Fischer projection, left, and the Haworth representation of the pyranose ring form, center. On the right is an illustration of the way this pyranose ring is linked through $\beta(1{\to}2)$ and $\beta(1{\to}4)$ linkages to preceding and succeeding galacturonic acid residues in pectin chains.

sugars including xylose and fucose. At random points in the backbone sequence the galacturonic acid sugars may be replaced by a β-$(1{\to}2)$-linked L-rhamnose sugar unit, which introduces a significant "kink" in the chain.

The esterified groups are randomly distributed along the chain, with occasional clumps of esterified groups and with regular, unmodified bare regions. Similarly, the arabinose and galactose "branches" or "hairs" can be irregularly distributed along the chain, as are the occasional rhamnose units of the backbone.

Because the C4 hydroxyl group of galacturonic acid is axial when the ring is in the typical, lowest-energy $^4C_1$ conformation, the pectin backbone linkage is the same type of axial–axial linkage found in the poly-L-guluronic acid sequences of the alginates discussed above, and produces the same sort of highly buckled chain found in those polysaccharides. In the presence of calcium ions these sequences can associate in egg-box junction zones. Because of their nonuniform distribution in the polysaccharide chains, these associating segments allow the pectins to form gels (Figure 4.119). As with the alginates, the ions are necessary to counteract the electrostatic repulsions of the acid groups. The polysaccharide forms the irregular matrix of a gel, rather than a regular crystal, because the methyl ester groups and particularly the side chain sugars prevent the portions of the chains that contain them from packing together closely in a regular fashion. The reason that methylation disrupts this association is clear, because without the charged acid group, the methyl ester cannot coordinate to a calcium ion in the egg-box sequence. The unusual $\beta(1{\to}2)$ linkage at the rhamnosyl groups also introduces a significant disruption in the association necessary for the regular egg-box juncture zones.

Pectin polymers with high DEs have few uninterrupted unesterified sequences, and thus cannot form significant egg-box interaction zones, and therefore do not gel in the presence of calcium ions. If they have a high molecular weight, however, these polymers can also form gels under acid conditions in the presence of high concentrations of sucrose. Under these conditions the chains are thought to form helices with juncture zones between separate molecules consisting of regions where the chains are intertwined in double helices. The acid conditions (pH < 3.2) are necessary

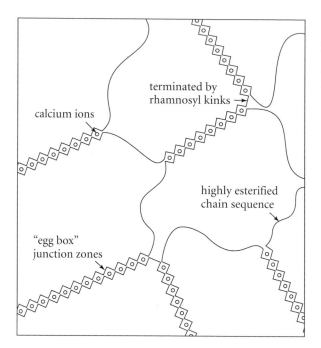

Figure 4.119. Proposed model for the formation of pectin gels. (Figure adapted and redrawn from D.A. Rees et al., "Shapes and interactions of carbohydrate chains," in *The Polysaccharides*, vol. 1, G.O. Aspinall, ed. New York: Academic Press, 1982.)

because, if the carboxylic acid groups of the chains are charged rather than neutral (protonated), they would strongly repel one another. The sucrose concentrations needed are very high—between 55 and 63% sucrose by weight. Sucrose solutions of this concentration even without polysaccharides are very viscous syrups; the viscosity of a 52% sucrose–water solution is 19 times that of pure water at 20°C, and that of a 64% sucrose–water solution is 120 times that of pure water (see Figure 4.120). Considering this high viscosity, the overall role of the polysaccharide in the gel is to provide a lattice framework to entrap the sucrose molecules, localizing them in space. Their strong interactions with the water then produce most of the immobilization of the water. Good gels can be formed with very low polysaccharide concentrations (<1%).

Pectin is present in almost all plant material, but is most common in various fruits, including apples and orange peels, from which it is extracted commercially as a byproduct of juice production. Pectins can be extracted from macerated plant tissues by treatment with hot dilute hydrochloric acid or ammonium oxalate ($(NH_4^+)_2$ $OOC—COO^-$). As a food additive, pectin has GRAS status and is widely used commercially as a gelling agent, a thickener, or as a suspension agent. Pectins are used in a variety of products, including jams and jellies, candies, beverages, and as a stabilizer for emulsions. For industrial purposes pectins are classified as **low-methoxyl, LM**, if the DE is less than 50%, or **high-methoxyl, HM**, if greater than 50%. LM pectin gels are thermoreversible, melting at higher temperatures and resetting at low temperatures. HM pectins, which, as we just saw, require acid conditions and a high level of soluble solids such as free sugar (sucrose) to gel, produce stiffer gels. HM pectins

Figure 4.120. The viscosity of aqueous sucrose solutions as a function of sugar concentration.

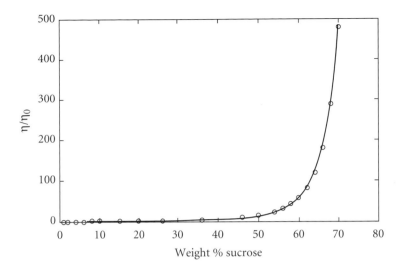

with DEs in the range 60 to 75% are widely used in the manufacture of jams and jellies, often using citric acid to lower the pH, neutralizing the interchain repulsions. The HM pectins are also classified according to how fast they gel, and are described as being slow-, medium-, or rapid-set pectins.

The in situ chemistry of pectin is also important in various vegetable foods. For example, $CaCl_2$ is often added to canned tomatoes to keep them firm because the calcium promotes the gelling of the pectins in and between the cell walls, thereby increasing or preserving stiffness even after cooking. In cooking dried beans and lentils, the cooking time is dependent upon the state of the pectin in the beans. Under acid conditions, as with added vinegar or lemon juice, the acid groups of the pectins are protonated. Thus, even in the absence of calcium ions, the polysaccharide chains remain entangled, preserving the cell wall structure of the beans and prolonging the time needed to cook them compared to pure neutral water. Under basic conditions (if one added baking soda, for example), all of the acid groups are deprotonated, and thus without calcium, their like charges tend to repel one another; this helps to "unravel" the cellular structure of the beans and shortening the cooking time. This overall picture is complicated somewhat, because when heating pectins under acid conditions, some acid hydrolysis of the glycosidic linkages takes place, weakening the tissue strength.

In the human diet pectin constitutes dietary fiber, because humans possess no enzymes capable of hydrolyzing it. Pectin is however broken down by bacteria in the large intestines. Because of its abundance in plant foods, pectin has always been a part of the human diet as we evolved from exclusively plant-eating primate ancestors. There is some evidence that pectin in the diet can lower serum cholesterol, perhaps by increasing viscosity in the intestines and interfering with absorption of cholesterol from foods.

Seed Gums

The Galactomannans

Galactomannans are water-soluble polysaccharides extracted from the endosperms of the seeds of various legumes. One of these gums, **guar**, or guaran, used commercially as a nongelling thickener since the 1950s, is extracted from the endosperm of the seeds of the legume *Cyamopsis tetragonoloba*. This annual plant is native to India and Pakistan, but is now cultivated in the Southern United States as well. **Locust-bean gum**, also called **carob seed gum**, which has been known since Antiquity, is another galactomannan that is extracted from the seeds of the Mediterranean locust, or carob tree, *Ceratonia siliqua*, originally native to Syria and Asia Minor but now cultivated all along the Mediterranean coast. Both of these polysaccharides are soluble in water, but not in organic solvents, so they are collected by precipitation with isopropyl alcohol from the seed extracts. These gums are branched copolymers consisting of D-galactose and D-mannose. Guar has a fairly regular, branched copolymer structure (Figure 4.121), where the backbone is a regular repeat of β-(1→4)-linked D-mannose monomers, with D-galactose side chains linked α-(1→6) to every other mannose of the backbone (Figure 4.122). Because of its extensive branching, guar gum can produce some of the highest viscosities per weight of any of the common commercial gums. Its effect on viscosity is shear rate dependent, decreasing significantly as the rate of shear increases (shear thinning). The extensive regular branching also prevents guar from forming gels. Guar gum is used in ice cream and salad dressings as a thickener and stabilizer.

Locust-bean gum has a similar copolymer structure with a mannose backbone and galactose side chains, but with only one galactose branch for four backbone residues. In locust-bean gum, the galactose side chains or "hairs" are not uniformly distributed, but are clustered in blocks of "hairy" chain sequence, separated by "smooth," or unsubstituted regions. This difference in structure from guar is the

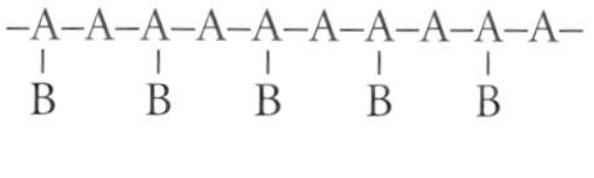

Figure 4.121. The sequence of guar gum; A represents the D-mannopyranose residues, and B represents the D-galactopyranose "branches."

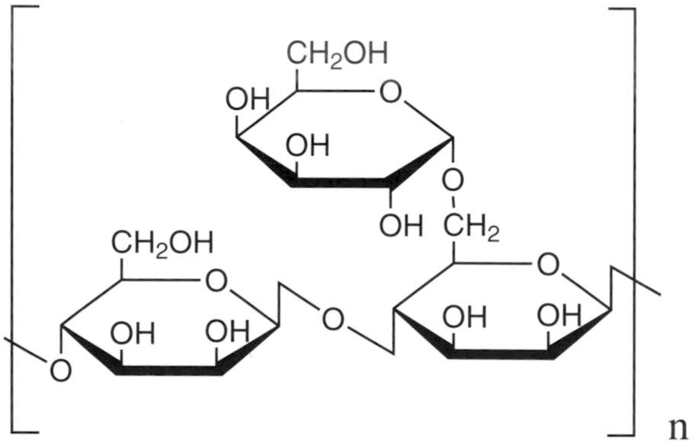

Figure 4.122. A Haworth representation of the basic repeat unit of guar gum.

Figure 4.123. The proposed model for the formation of locust-bean gum gels. (Figure adapted and redrawn from Dea et al., *Carbohydr. Res.* 57:249–272, 1977.)

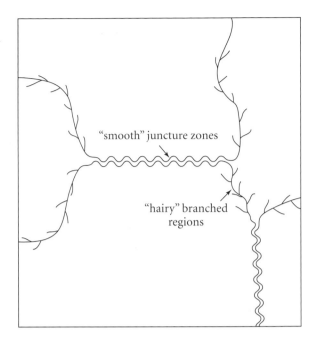

reason that carob gum can form gels; apparently the smooth regions closely associate in juncture zones between different chains, while the galactose "hairs" prevent the branched regions from closely associating and force the chains off in different directions, establishing the open matrix of the gel (Figure 4.123). Locust-bean, or carob, gum will not gel by itself, but will do so readily in combination with xanthan gum. These gels are stable over a wide range of pH, because the locust-bean gum is nonionic.

Locust-bean gum is used as an emulsion stabilizer, as a thickener, and as a gelling agent in a variety of processed foods, including dairy products (cheese, ice cream), sauces, canned foods, and processed meats.

A third galactomannan that is used commercially is **tara gum**, which is produced from the seeds of the South American tara tree (*Cesalpina spinosa*). This polysaccharide has the same basic covalent structure as guar and carob gums, with a mannose backbone and galactose side chains, but with one galactose per every three mannose residues, unlike the 1:4 ratio of locust-bean gum or the 1:2 ratio of guar. As in locust-bean gum, the distribution of galactose units is in random blocks. Tara also does not gel by itself, due to the branching, but like locust bean gum will gel when combined with xanthan gum, hence its use as a thickening agent and stabilizer in foods, particularly in ice cream. In aqueous solution it is less viscous than guar gum at the same concentration but more viscous than locust bean gum, consistent with the model discussed above for the relationship between viscosity and the degree of branching in polysaccharides of the same molecular weight. Because it is the most highly branched per unit weight of these three, the guar gum polysaccharide is the least flexible and thus the most extended, leading to the highest viscosity.

One other galactomannan, **fenugreek gum** from the seeds of the fenugreek plant, *Trigonella foenum graecum*, has received considerable recent attention from the food industry as a possible thickening agent in foods. Fenugreek gum has a 1:1 ratio of galactose to mannose; that is, with a galactose branch on every backbone mannose residue (Song et al. 1989; Brummer et al. 2003; Wu et al. 2009). As such, like guar, it is nongelling and exhibits an even higher viscosity per weight used than does guar.

Tamarind Seed Gum

Not all seed gums are galactomannans. An interesting example of one that is not is the gum produced from the seeds of the tamarind tree. This tree, *Tamarindus indica*, native to Africa and Southern Asia, has been introduced in tropical areas around the world. This tree produces large edible seed pods that are used for a variety of food purposes. The green fruit is used as a spice or seasoning in a number of tropical cuisines, and to make popular beverages in Latin America and Egypt, and the pulp of these seed pods is an important food product. The tamarind seeds available as a by-product of pulp production are dehulled and crushed to produce tamarind kernel powder, from which a soluble polysaccharide, usually called **tamarind seed gum**, is extracted. This polysaccharide is a xyloglucan, with a cellulose-like backbone of β-(1→4)-linked D-glucopyranose monomers with α-(1→6)-linked D-xylopyranose branching side chains (Gidley et al. 1991). Tamarind seed gum is actually a galactoxyloglucan, because it also contains galactose residues β-(1→2)-linked to some of the xylose side chain branches, with the ratios of the three sugars being variously reported to be in the range 4:2:1 to 3:2:1 to 3:2.25:1 to 2.25:1.25:1 (glucose:xylose:galactose). The structure is thought to be of the type illustrated in Figure 4.124. Tamarind seed gum may also contain L-arabinofuranose residues linked (1→6) to the glucose backbone, but Gidley et al. assert that any arabinose in tamarind gum is an artifact that results from the preparation procedure. Tamarind seed gum is used in the food industry as a stabilizer or thickener in such products as ice cream and as a gelling agent. It behaves somewhat like pectin, gelling in the presence of high concentrations of sucrose (>50%) and can be used to replace pectin in the manufacture of jellies. Because the cellulosic backbone is indigestible, tamarind seed gum constitutes dietary fiber for humans.

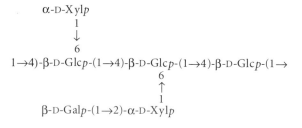

Figure 4.124. The proposed structure for tamarind seed gum.

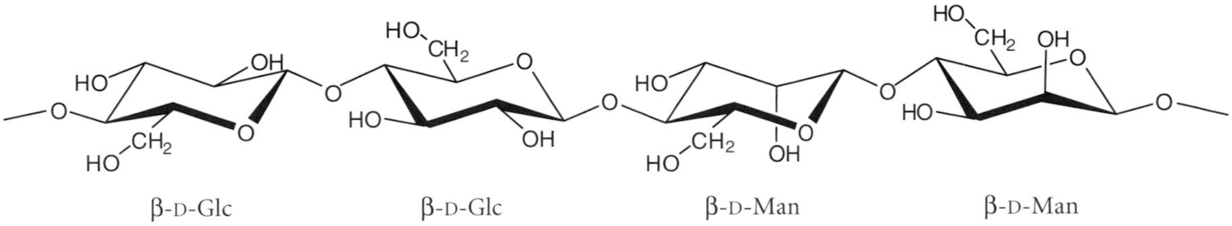

β-D-Glc   β-D-Glc   β-D-Man   β-D-Man

konjac

Figure 4.125. The covalent structure of konjac.

Glucomannans

Glucomannans are neutral hydrocolloidal polysaccharides composed primarily of glucose and mannose. Unlike the galactomannans, where the backbone is composed only of mannose residues, with the galactose residues occurring as branching side chains, the glucomannans have both sugar types occurring in the main chain backbone, joined by β-(1→4) glycosidic linkages (Figure 4.125). They are found widely in the plant kingdom and may have both storage and structural functions. One such glucomannan, **konjac**, has been used as a food ingredient since ancient times in East Asia, and more recently has begun to be used in the food industry around the world. Konjac, called *konnyaku* in Japan, is the flour obtained from the tuber-like corm of the perennial plant *Amorphophallus konjac*, commonly called the Voodoo Lily or Devil's tongue in English, native to Eastern Asia and widely cultivated in Japan, China, and Korea. In Japan *konnyaku* flour has been traditionally used to make gels, jellies, and candies and as one of the ingredients in the hot pot dish *oden*. It is also made into many types of noodles, such as *shirataki* and *ito-konnyaku*. These are popular with dieters because such noodles have no bioavailable calories, since humans cannot digest konjac.

Konjac flour contains a number of components, including starch and protein, but by far the largest fraction by weight is made up of the β-(1→4)-linked konjac glucomannan. This polysaccharide has found many uses in food processing as a thickening and gelling agent and a source of dietary fiber. It is increasingly used to form films, particularly in combination with other polysaccharides like xanthan. Konjac has been characterized structurally (Millane and Hendrixson 1994; Cescutti et al. 2002; Katsuraya et al. 2003); it consists of a linear glucose and mannose backbone with variable sequence consisting of individual glucose and mannose units along with various cello- and manno-oligomer sequences. The ratio of mannose to glucose monomer residues has been found to be 1.6:1.0 (Davè and McCarthy 1977). A small amount of branching also occurs in konjac. It was initially thought that the branching residues were galactose and occurred at the C3 positions of both the glucose and mannose residues, but Katsuraya et al. reported only glucose and mannose branches and at the C6 carbon of the glucose units as a β-(1→6)-linkage. The chains are also lightly acetylated, which increases the solubility of the polysaccharide, with acetyl groups occurring in general on about one tenth or fewer of the backbone residues.

Konjac exhibits good stability toward acid conditions, down to pHs as low as 3.3. It does not gel in water at room temperature because of the interference to close association from the acetyl groups. Under differing conditions, konjac can form either thermoreversible or thermo-irreversible gels. Thermo-irreversible gels are produced when the polysaccharide is heated to 85 °C under alkaline conditions. Thermoreversible gels can be formed at ambient temperatures in combination with xanthan gum, as in the case of locust-bean gum.

The use of konjac, officially known as E425 konjac in the European Union, in a type of jelly confectionery called jelly minicups was banned in the European Union in 2003 because of the possibility that they could lead to death by choking in children and the elderly, and the US FDA has also warned about the possible danger from choking posed by these products. The problem lies with the product size, shape, and texture, and not with any intrinsic danger from konjac.

Exudate Gums

Various trees, mostly found in arid or semiarid tropical regions, exude protective resins when injured to seal off living tissue and protect it from desiccation. These **exudates** dry into hard nodules that can be gathered by hand for marketing. Because the plants that produce these resins are found in some of the oldest centers of civilization (North Africa, the Middle East, and India), their useful properties have long been known. Perhaps the best-known of these, **gum arabic**, was used to secure the wrappings of mummies in ancient Egypt more than 4,500 years ago.

Gum arabic, produced by the *Acacia senegal* tree, is a mixture of polysaccharides and glycoproteins. Its principal component is a complex and highly branched polysaccharide made up of several different monosaccharides. The backbone consists of β-(1→3)-linked D-galactopyranosyl residues with many branching side chains that are themselves further branched. The principal sugars in these branches are L-arabinose, L-rhamnose, D-galactose, and D-glucuronic acid. The varying composition of gum arabic as a function of source and year, in terms of the proportions of polysaccharides and glycoproteins, and particularly in the structures of the polysaccharides, makes the functional behavior of gum arabic somewhat unpredictable. In general, however, it is soluble in water and readily dissolves. It is used as a stabilizer and emulsifier in food manufacture, and as a glue or paste in other applications, and as a binder in paints.

As we have already seen (Figure 4.81), Table 4.12 again demonstrates that gum arabic behaves differently from other polysaccharide gums, because it does not significantly increase the viscosity of aqueous solutions until high concentrations (~50%) are reached. Even at this high level the viscosity is less than that produced by most of the other industrial gums at 3% concentration. This makes gum arabic useful in applications where high viscosities are not desired, as in beverages, where it is used to stabilize foams in beer and as an emulsifier in soft drinks. It is most widely used in candy manufacture to control or prevent sugar crystallization and to emulsify other components.

The *Acacia senegal* tree, as its scientific name implies, is found from Senegal and Mauritania on the Atlantic coast of West Africa to the Horn of Africa in the East, and from Egypt to South Africa. The saps from the trees in much of this wide range are less useful for food applications, and industrial gum arabic is primarily produced from countries in the Sahel, the arid southern margin of the Sahara Desert that stretches from Sudan to Mauritania. The largest producer of gum arabic is Sudan. Political instability in Sudan and several other main centers of production (Somalia, Eritrea, Chad) has in the past led to shortages of gum arabic and price fluctuations. Gum arabic is also sometimes listed on food labels as "gum acacia," or just "acacia."

Tragacanth is collected from the *Astragalus* bush found in the nations of the northern Middle East. It consists of two principal polysaccharide fractions, one of which is the insoluble arabinogalactan bassorin, while the other fraction is the soluble, highly branched polysaccharide **tragacanthic acid** (Figure 4.126). Like pectin, tragacanthic acid has a backbone of polygalacturonic acid residues linked α-(1→4), with a number of branching side chains. These branches consist of

Figure 4.126. The structure of tragacanthic acid.

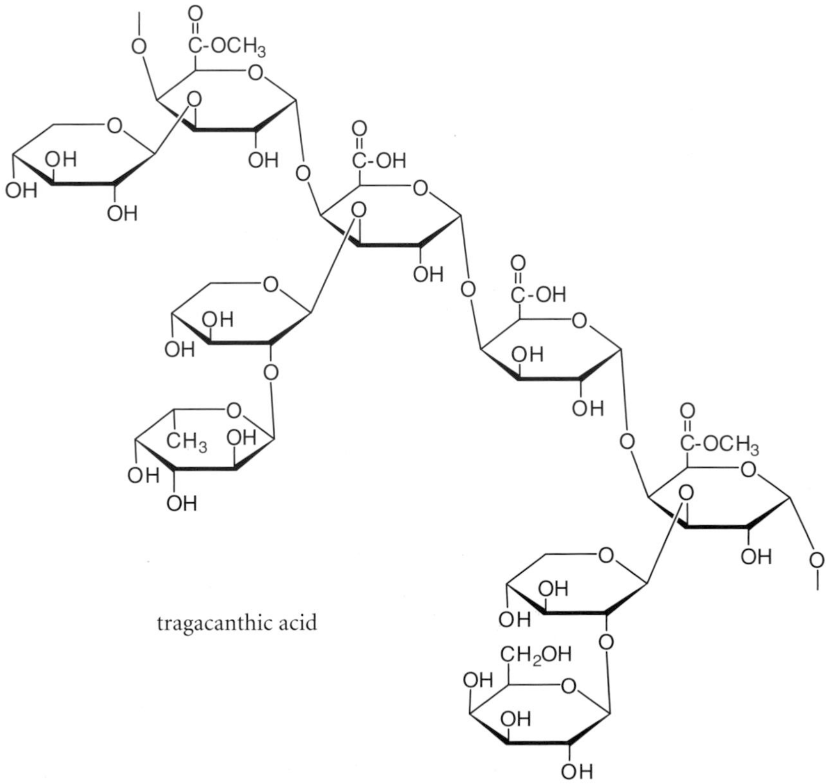

tragacanthic acid

Table 4.12 Viscosities of solutions of various hydrocolloidal polysaccharides as a function of concentration, in cps

Concentration (%)	Gum arabic	Tragacanth	Carrageenan	Sodium alginate	Methyl cellulose	Locust-bean gum	Guaran gum
1	—	54	57	214	38.9	58.5	3,025
2	—	906	397	3,760	512	1,114.3	25,060
3	—	10,605	4,411	29,400	,850	8,260	111,150
4	—	44,275	25,356	—	12,750	39,660	302,500
5	7.3	111,000	51,425	—	67,575	121,000	510,000
6	—	183,500	—	—	—	—	—
10	16.5	—	—	—	—	—	—
20	40.5	—	—	—	—	—	—
30	200.0	—	—	—	—	—	—
40	936.3	—	—	—	—	—	—
50	4,162.5	—	—	—	—	—	—

Source: H.-D. Belitz, W. Grosch, and P. Schieberle, Food Chemistry, 4th rev. and ext. ed. (Berlin: Springer-Verlag, 2009).

xylopyranose monomers linked β-(1→3) to the galacturonic acid rings, with about two-thirds of these further substituted with either L-fucopyranose linked α-(1→2) or D-galactopyranose linked β-(1→2) to the xylose rings. This exudate gives one of the highest viscosities per weight of any of the gums (Table 4.12). It is used as a stabilizer and thickener in emulsion products such as salad dressings or ketchup.

Our brief survey of the most important polysaccharides used in foods has only scratched the surface of the range of functionalities that can be exploited with these ingredients. As we have seen, many of these polymers can serve more than one functional role in a food product, and conversely, several might have similar properties and uses. The selection of the appropriate polysaccharide for a particular product depends on many factors, not excluding cost and possible consumer resistance to ingredients whose names on the labels sound undesirable to the uninformed, even in the case of natural products. In an actual food product, the gums will be interacting with a number of other components, such as other polysaccharides, proteins, lipids, and a range of smaller molecules, which may significantly affect their functionality and produce collective behavior that could not be predicted a priori from the known properties of the individual components. Unfortunately, this often means that the best choices of polymers and their proportions must be determined by trial and error, although many food companies have internal, proprietary, empirically determined data about which formulations have worked for their types of products in the past. Table 4.13 summarizes the properties of some of the gums most commonly used in foods.

Table 4.13 Properties of the common food gums

Source type	Polysaccharide	Type	Monomers	Source	Solubility	Gelling behavior	Functional uses
seaweed polysaccahrides	alginates	block copolymer	L-guluronic acid, D-mannuronic acid	brown algae (giant kelp)	soluble in cold water in the absence of Ca^{+2}	gels in the presence of Ca^{+2}	thickener, gelling agent
	ι-carrageenan	alternating sulfated copolymer	β-D-galactopyranose-4-sulfate, 3,6-anhydro-α-D-galactopyranose-2-sulfate	red algae (Irish moss)	soluble in hot water	gels in the presence of Ca^{+2}	thickener, gelling agent, gels don't synerese
	κ-carrageenan	alternating sulfated copolymer	β-D-galactopyranose-4-sulfate, 3,6-anhydro-α-D-galactopyranose	red algae (Irish moss)	soluble in hot water	gels in the presence of K^+	thickener, gelling agent; rigid gels but synereses
	λ, μ, or ν-carrageenans	alternating sulfated copolymer	various sulfated β-D-galactopyranose and α-D-galactopyranose	red algae (Irish moss)	soluble in cold water	none of these gel	thickener
	agarose	alternating copolymer, lightly sulfated	β-D-galactopyranose, 3,6-anhydro-α-L-galactopyranose	various red algae seaweeds	insoluble in cold water	gels	growth medium for bacterial cultures; jellies; diet products
	agaropectin	alternating copolymer, highly sulfated	β-D-galactopyranose, 3,6-anhydro-α-L-galactopyranose, pyruvic acid branches	various red algae seaweeds	insoluble in cold water	doesn't gel	thickener

Table 4.13 Properties of the common food gums *Continued*

Source type	Polysaccharide	Type	Monomers	Source	Solubility	Gelling behavior	Functional uses
bacterial polysaccharides	xanthan gum	branched polymer with cellulosic backbone	β-D-glucopyranose, D-guluronic acid, α-or β-D-mannopyranose	*Xanthomonas campestris*	soluble in cold water	doesn't gel	thickener
	pullulan gum	flexible homopolymer	α-D-glucopyranose	*Aureobasidium pullulans*	soluble in water	gels	forms edible films, coatings
	gellan	complex but unbranched copolymer	β-D-glucopyranose, β-D-glucopyranose uronic acid, α-L-rhamnopyranose	*Pseudomonas elodea*	soluble in water with heating	gels with cations	thickener and stabilizer; forms strong gels at low concentrations
pectins	HM pectin	complex, randomly branched and methylated polymer and limited copolymer	α-D-galactouronic acid with a high degree of methyl esterification, L-rhamnose, with some sidechains of L-arabinose or D-galactose	apples, citrus, many other fruits	soluble in water	gels under acid conditions in presence of sucrose	gelling agent in jams, jellies
	LM pectin	complex, randomly branched and methylated polymer and limited copolymer	α-D-galactouronic acid with a low degree of methyl esterification, L-rhamnose, with some sidechains of L-arabinose or D-galactose	apples, citrus, many other fruits	poor solubility in presence of Ca^{+2}	gels in the presence of Ca^{+2}	jams, jellies

category	name	structure	composition	source	solubility	gelling	application
seed gums	guar gum	branched galactomannan	D-mannose, D-galactose, in a 2:1 ratio	seeds of a legume	very soluble with very high viscosity	doesn't gel	thickener in ice creams, salad dressings
	locust bean gum	branched galactomannan	D-mannose, D-galactose, in a 4:1 ratio	seeds of the carob tree	very soluble with high viscosity	gels in the presence of xanthan	gelling agent, thickener in ice creams
	tara gum	branched galactomannan	D-mannose, D-galactose, in a 3:1 ratio	seeds of the tara tree	soluble, less viscous than guar	gels in the presence of xanthan	stabilizer, thickener in ice creams
	fenugreek gum	branched galactomannan	D-mannose, D-galactose, in a 1:1 ratio	seeds of the fenugreek plant	very soluble with very high viscosity	doesn't gel	thickener, stabilizer
	tamarind seed gum	highly branched galactoxyloglucan	β-D-glucopyranose, α-D-xylopyranose, β-D-galactopyranose	seeds of the tamarind tree	soluble in water	gels in presence of sucrose	stabilizer, thickener in ice creams
tuber polysaccharide	konjac gum	linear glucomannan with limited branching and acetylation	D-glucose, D-mannose	voodoo lily, *Amorphophallus konjac*	soluble at low temperatures	gels at high temperatures under alkaline conditions; forms a thermoreversible gel with xanthan	diet noodles, gels, jellies, and candies
exudate gums	gum arabic	complex branched polymer	D-galactose, L-arabinose, L-rhamnose, D-glucuronic acid	*Acacia senegal*	soluble; lower viscosity	doesn't gel	stabilizer, emulsifier
	tragacanth	highly complex branched, methylated polymer	D-galactouronic acid, D-xylose, L-fucose, D-galactose	*Astragalus* bush	soluble; very high viscosity	doesn't gel	stabilizer, thickener

SUGGESTED READING

Chinachoti, P., and Y. Vodovotz, eds. 2001. *Bread Staling*. Boca Raton, FL: CRC Press.

Lineback, D. 2012. "Acrylamide in Foods: A Review of the Science and Future Considerations." *Annual Review of Food Science and Technology* 3:15–35.

Lustig, R.H., L.A. Schmidt, and C.D. Brindis. 2012 "The Toxic Truth about Sugar," *Nature* 482:27–29.

Rees, D.A. 1977 *Polysaccharide Shapes*. London: Chapman and Hall.

Robyt, J.F. 1998. *Essentials of Carbohydrate Chemistry*. New York: Springer.

Stephen, A.M., ed. 1995. *Food Polysaccharides*. New York: Marcel Dekker.

Stick, R.V. 2001. *Carbohydrates: The Sweet Molecules of Life*. San Diego, CA: Academic Press.

Stoddart, J.F. 1971. *Stereochemistry of Carbohydrates*. New York: Wiley-Interscience.

Wrolstad, R.E. 2012. *Food Carbohydrate Chemistry*. Chichester: Wiley-Blackwell.

5.
Proteins

PROTEINS IN FOODS

Proteins constitute one of the most important categories of molecules found in living organisms, and thus also one of the most important classes of molecules in foods. These macromolecules are polymers of amino acids and play many essential roles in living systems. They are coded for by the information in the sequences of our genes, which control our very makeup. Protein catalysts, called **enzymes**, mediate the majority of the reactions in living organisms. The immune system, which allows animals to ward off infections, functions using protein antibodies. Many hormones are also proteins or shorter amino acid polymers usually referred to as polypeptides, or often, just peptides. Proteins are also vital as structural constituents, such as the collagen of tendons, skin, and ligaments; the keratin of hair, nails, hooves, and scales; and the myosin of muscles. The proteins myosin and actin and the other proteins associated with them that constitute muscle tissue are much more important than simple structural constituents, because their ability to reversibly contract allows animals to move.

Muscle tissues are the principal component of meats used in foods and are a major source of protein in the human diet. Because of the ubiquity of proteins in living tissues, protein is also available in many other types of foods, and many plant food sources can provide all the protein needed in the diet. Protein is necessary in foods because humans cannot synthesize all of the amino acids needed to make our own proteins. Those amino acids that humans cannot synthesize are called essential amino acids. As such, they are limiting for growth and development.

Although a major role of protein in foods is nutritional, they can also play functional roles as well, in the sense of determining the properties of the food. One of the most important examples of this is the gluten in wheat doughs, which is responsible for the remarkable properties of breads and other baked goods and pastas that use wheat flour. Of course, the chemistry of muscle tissue actin and myosin and connective tissue collagen substantially determines the properties of meat. Enzymes are sufficiently important in food quality and processing that they merit a separate chapter of their own and will be primarily discussed in Chapter 6. Even more than in the case of the polysaccharides, the conformational structures of proteins determine their overall properties. In turn, proteins exhibit a much wider array of conformational structures than polysaccharides, because they have much more diverse sequences and because, in general, the sequence details determine the conformation. In this chapter we will first discuss the general properties of amino acids and proteins and how they determine protein conformations, and will then survey some examples of specific proteins that are important in food preparation and processing, as well as the more important reactions of proteins in foods.

POLYPEPTIDE CHAINS

Proteins are **polypeptides**, long polymers of L-**amino acids** linked together with the loss of a water molecule by the formation of so-called **peptide bonds** (Figures 5.1 and 5.2). Amino acids have, as their name implies, both a carboxylic acid group and an amine group bound to a central chiral carbon atom. These chiral carbons, called alpha carbons (C^{α}), are also bound to an aliphatic proton and a fourth functional group, generally referred to as a **side chain** because in the polypeptide polymers they stick out to the side of the main path of the polymer. The peptide bonds link together the acid and amine groups of two amino acids (Figure 5.1). The repeating backbone of these polymers is then an alternating sequence of peptide bonds joining together the asymmetric alpha carbons. With the exception of prolines, which will be discussed in greater detail below, the only sequence variation comes in the identity of the "R" groups attached to these asymmetric carbons, with each amino acid having a different functional group as the so-called side chain. Twenty commonly occurring amino acids are found in living organisms (Figure 5.3), although many other types of amino acids that do not occur in biology are in principle possible. Some of the naturally occurring amino acids have polar side chains, some have nonpolar or hydrophobic side chains, and several are charged. For one, glycine, the side chain is a proton, in which case the C^{α} carbon is no longer chiral. Proteins are linear polymers; that is, no

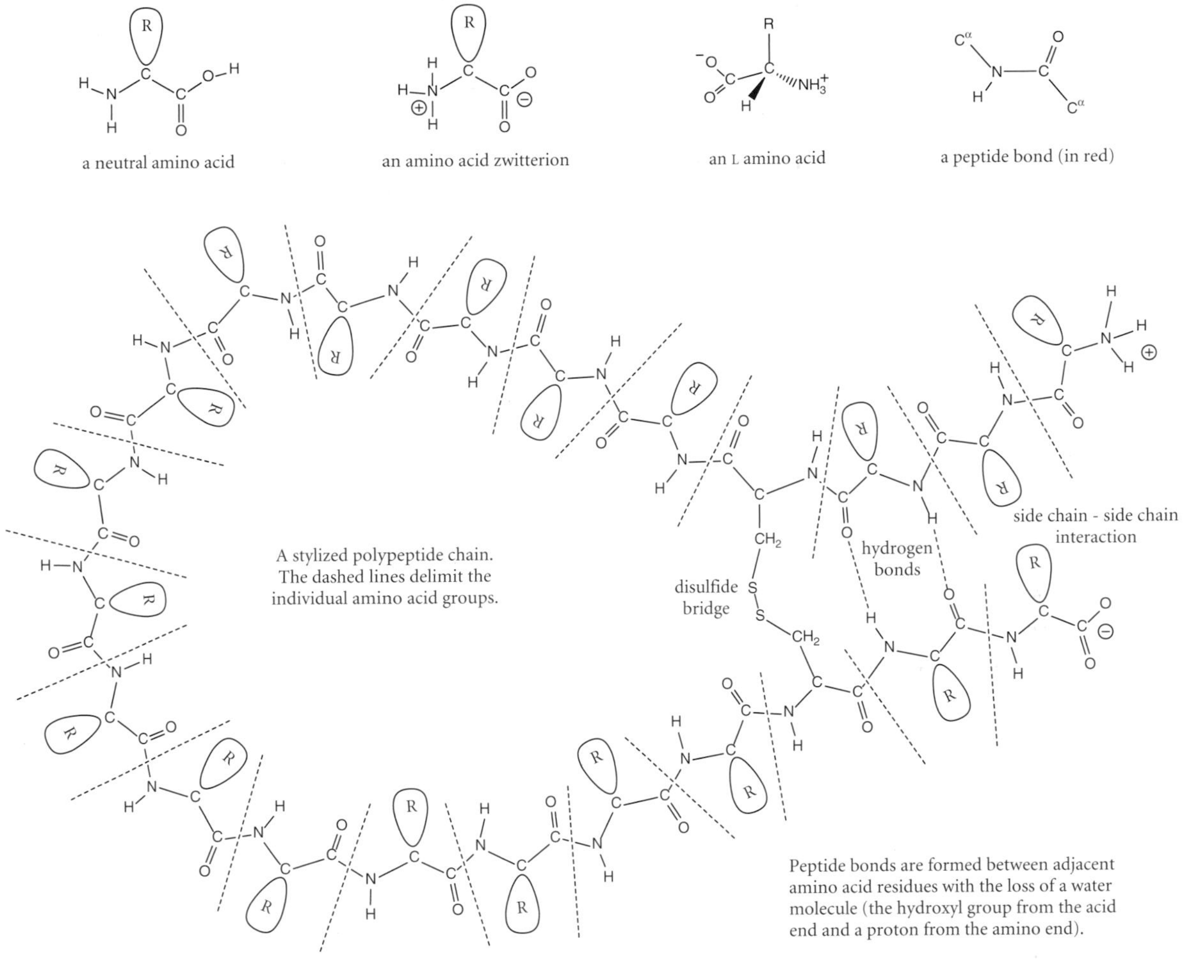

Figure 5.1. A schematic illustration of a polypeptide chain made up of amino acids. On the upper left is a generalized neutral amino acid, while the middle left shows a generalized amino acid zwitterion. The upper middle right shows the stereochemical structure of L-amino acids, while the upper right-most illustration shows a peptide bond in red.

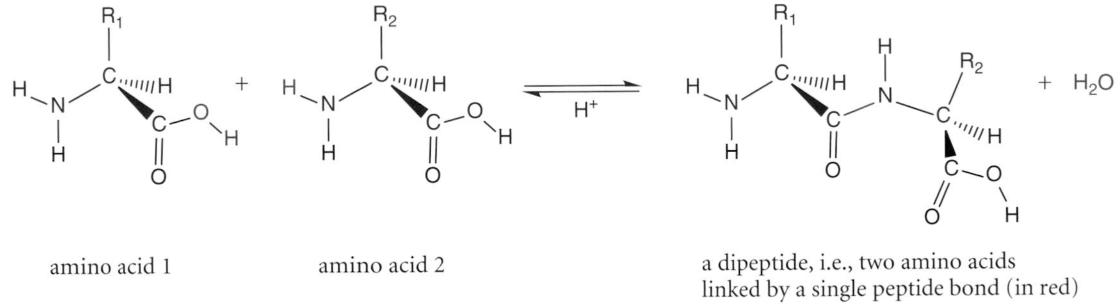

Figure 5.2. The formation of a peptide from two amino acids, with the elimination of a water molecule. The reverse reaction, the disruption of the peptide by the addition of a water molecule, is called hydrolysis.

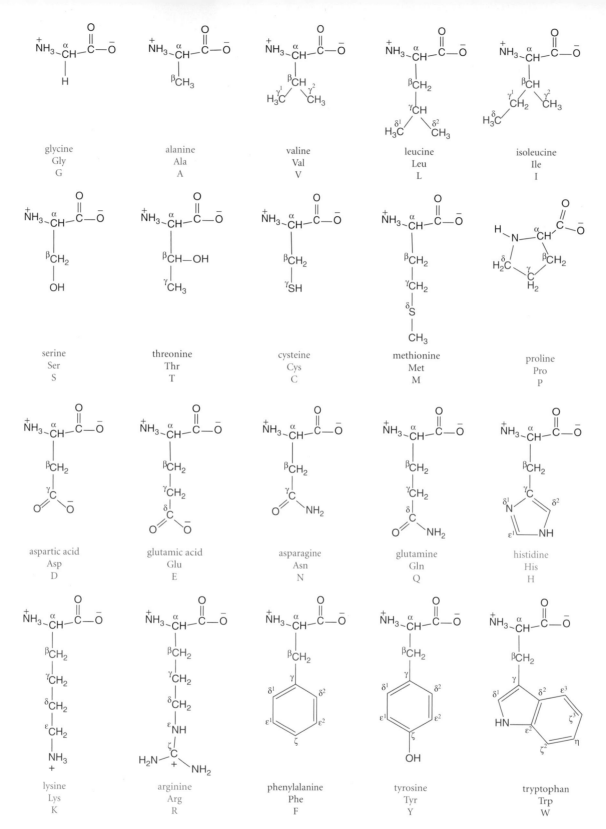

Figure 5.3. The 20 commonly occurring amino acids, with their 3-letter and 1-letter designations. The side chains are shown in red. Note that Gly, Ala, Val, Leu, Ile, and Phe are hydrophobic, while Lys and Arg are positively charged at pH 7, and Glu and Asp are negatively charged. Ser, Thr, Asn, and Gln, while uncharged, are polar due to their ability to participate in side chain hydrogen bonds. Some, like Tyr and His, have mixed character. Pro is unique in that its side chain is covalently bonded to the nitrogen atom of the backbone peptide group, producing a five-atom ring. The labels of essential amino acids are shown in red; those of nonessential amino acids in blue, and those of semi-essential amino acids are shown in green (see text).

chain branching occurs as can happen in polysaccharides, but they can be chemically cross-linked by **disulfide bonds**, where the sulfhydryl —S—H groups of two appropriately placed cysteines are oxidized to give —S—S— bonds, as illustrated in Figure 5.1. In living organisms on Earth, these chiral amino acids always have the "ʟ" configuration. It is common practice to refer to amino acids by the three-letter abbreviations given in Figure 5.3; in specifying the detailed sequence of amino acids in a protein, biochemists generally use the one-letter abbreviations also shown in the figure.

The linking of amino acids to form peptide bonds occurs with the elimination of a water molecule, and the chain can be broken apart again to give free amino acids by adding water across the peptide bond under acid conditions, a process that is thus called hydrolysis (Figure 5.2). Living organisms contain a number of enzymes that catalyze the hydrolysis of peptide bonds. This reaction can also be accomplished non-enzymatically by heating a polypeptide in an acidic solution. The formation of the peptide bond is endoenergetic; that is, the dipeptide is higher in energy than the two component amino acids, and hydrolysis of the bond releases energy.

The amino acids in protein and polypeptide chains are conventionally numbered sequentially from their N-terminus to their C-terminus. They are synthesized one residue at a time, in a process called translation, at the ribosomes in cells, as shown in the micrograph in Figure 5.4, using the information encoded in the genes. The sequence in which the amino acids occur in a protein chain is referred to as the

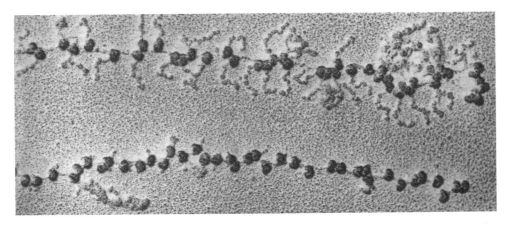

Figure 5.4. An electronmicrograph, magnified 140,000 times, showing ribosomes (small black balls) moving stepwise along a single thin strand of messenger RNA. Only one RNA strand is shown; the top and bottom segments are connected by a loop off the left edge of the figure. New protein chains, which have not yet assumed their final conformation, can be seen extending out from the ribosomes, progressively longer as one moves down the RNA chain. (Reprinted by permission from Macmillan Publishers Ltd.: *Nature* 344:585, copyright 1990.)

primary structure of the polymer. Proteins of a given type, such as the hemoglobin molecules of the blood, may vary from individual to individual of the same species due to the accumulated effects of random mutations, which are passed on to the individual's progeny. Within a given species, these variations in each individual's proteins, resulting from corresponding differences in each individual's genes, are small. The same type of protein may occur in many species (as does hemoglobin, for example), performing the same functions, due to the common ancestry of the species sharing the protein, and in such cases there will likely be many more sequence differences between the molecules of the same type from different species, but with many similarities in sequence in the portions of the chain required for the proper functioning of the molecule. The more distantly related two species are, reflecting a longer time since they diverged from a common ancestor, the greater the number of random amino acid substitutions that may be expected in the nonessential portions of the sequences of the proteins. Those residues that are shared by the molecules from different species are referred to as conserved and are often important to the proper functioning of the molecule, because a mutation at that position would probably impair the survivability of the organism and be eliminated by natural selection.

During the course of evolution, a protein may, as the result of cumulative mutations directed by natural selection, develop a new function (while probably losing its original function), resulting in a new type of protein that nevertheless shares a significant sequence similarity with the original ancestral molecule, and thus also with the later versions of that protein as well (although they also would have undergone changes over time). In such a case the two protein types are said to be **homologous**.

AMINO ACIDS IN FOODS

Proteins are important nutritionally as a source for the amino acids needed to build our own body's proteins. Some amino acids can be synthesized endogenously, depending on species, while others cannot and thus must be obtained from food. Those amino acids that cannot be synthesized in our bodies are termed **essential.** The essential amino acids for humans are valine, leucine, isoleucine, phenylalanine, tryptophan, methionine, and threonine. Lysine and arginine are semi-essential, meaning that a dietary source is needed for proper nutrition, and histidine is essential for infants. Because a dietary source for these amino acids is needed, they are typically considered essential by nutritionists as well. The **nonessential amino acids** are glycine, alanine, proline, serine, cysteine, tyrosine, asparagine, glutamine, aspartic acid, and glutamic acid.

In addition to the 20 common naturally occurring amino acids, a few modified versions occur with some frequency in foods. These modified amino acids are produced by posttranslational enzymatic modification. Two such residues important in collagen (Table 5.3) are γ-hydroxyproline and δ-hydroxylysine (Figure 5.5).

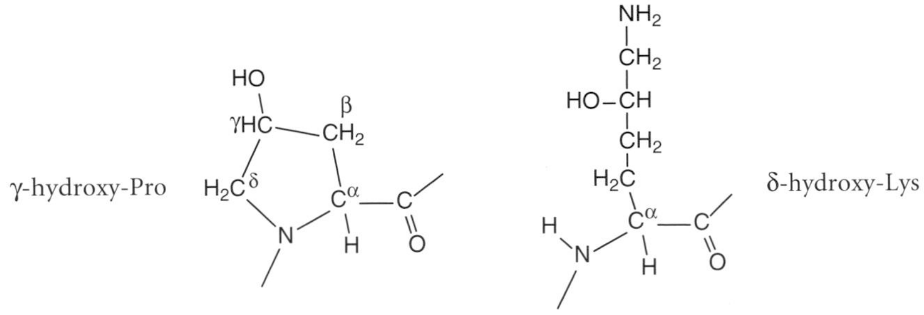

Figure 5.5. The structures of
γ-hydroxyproline and
δ-hydroxylysine.

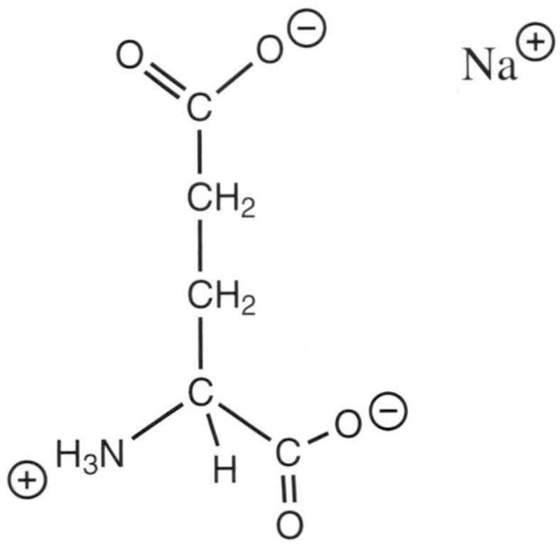

γ-hydroxy-Pro   δ-hydroxy-Lys

Figure 5.6. The structure of monosodium glutamate, MSG.

Amino Acid Configurations and Taste: MSG

Although the amino acids in natural proteins are all L-amino acids, except for glycine, it is possible to prepare the D-amino acids synthetically. These isomers are chemically equivalent to the L-series but have some differences in biological properties. The most important example of these differences are in taste; the D-amino acids are generally sweet (in many cases intensely so) or tasteless, while a number of the L-amino acids are bitter-flavored. In particular, among the L-series, Arg, His, Ile, Leu, Phe, Trp, and Tyr are all bitter tasting, while Ala, Ser, Thr, and Gly are sweet. In general, bitterness is often correlated with an extended region of hydrophobicity, which also explains why some amino acids that consist of both polar and hydrophobic functional groups or regions (for example, Lys) can exhibit both sweet and bitter flavor notes.

As would be expected for an acid, Asp is sour tasting. L-glutamic acid, Glu, however, is something of a special case. The sodium salt of glutamic acid, **monosodium glutamate** (**MSG**) (Figure 5.6), is important in its own right as a food ingredient. Because it is a standard amino acid, it is present in nearly all protein foods, although some foods contain more than others; glutamic acid is disproportionately abundant in some meats, cheeses, and tomatoes. MSG is widely used as a flavor enhancer because it is able to make foods taste more savory. It has been used as a separate flavor ingredient in Asian foods for many decades, following its initial production for this purpose in Japan in the early years of the twentieth century. The taste of MSG, described as "meaty" and called "*umami*" in Japanese, was identified as the first addition to the four basic tastes (sweetness, sourness, saltiness, bitterness; see Chapter 8) in the last years of the twentieth century, when scientists were able to identify the protein receptor responsible for the flavor perception of MSG (Chaudhari et al. 2000; Nelson et al. 2002). This protein belongs to a class of glutamate receptor proteins, called metabotropic glutamate receptor proteins (mGluRs), which specifically recognize and bind to glutamate. These proteins are involved in neural transmission, and in this context glutamate functions as an excitatory neurotransmitter. The protein of this class that is responsible for the perception of umami taste is called "taste-mGluR4." Some people apparently do not tolerate added MSG well, and it has long been argued that it is responsible for the so-called Chinese Restaurant Syndrome (Chapter 11). As a naturally occurring component of almost all proteins, glutamate has GRAS (generally recognized as safe) status, but under FDA regulations, if it is specifically added to a food as an ingredient, it must be identified as monosodium glutamate in the ingredient list. Recently the gene for the sweet taste receptor protein was discovered and identified as being homologous to glutamate binding proteins (Bachmanov et al. 2001; Kitiagwa et al. 2001; Li et al. 2001, 2002; Temussi 2006). Although the structure of this protein has not yet been determined, its homology with a known glutamate receptor protein allows its conformation to be approximated. It has a receptor site for glutamate that is separate from the receptor site for binding sweet-tasting molecules, which could explain how glutamate can modify the perception of other tastes.

Aspartame

Some of the sweetest substances known are globular proteins, such as the protein thaumatin (see Figure 5.32), but in general short peptides are bitter tasting. This bitterness generally does not depend upon sequence and probably arises because short polypeptides cannot fold up into a compact conformation that places their hydrophobic functionalities on the interior. A notable exception is **aspartame**, a methyl ester derivative of the dipeptide of aspartic acid (Asp) and phenylalanine (Phe); that is, L-aspartyl-L-phenylalanine methyl ester (Figure 5.7), which is intensely sweet even though one of its two residues, Phe, has a large hydrophobic side chain.

Aspartame is approximately 200 times as sweet as sucrose and is now widely used as an artificial sweetener (see Chapter 8). Note that it is not a noncaloric sweetener, because it can be hydrolyzed to amino acids during digestion, but its intensity allows much less to be used to achieve a similar level of sweetness, thus making it a low-calorie sweetener.

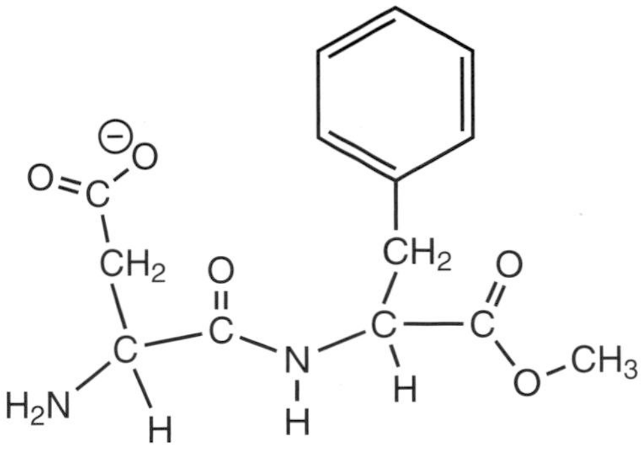

Figure 5.7. The structure of the dipeptide sweetener aspartame.

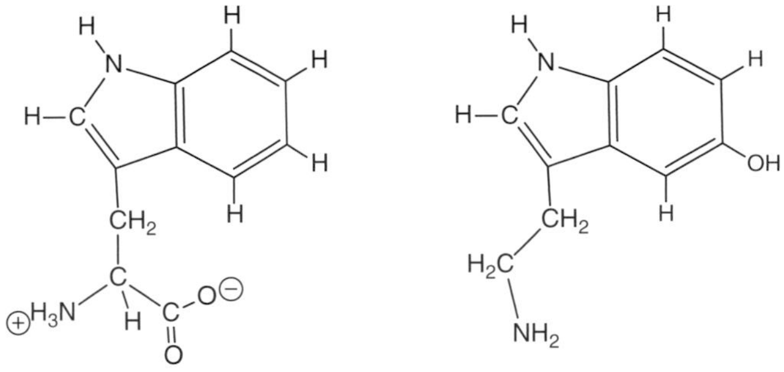

tryptophan   serotonin

Figure 5.8. The structures of the amino acid tryptophan and the related compound serotonin.

Tryptophan and Serotonin

Serotonin is an important neurotransmitter molecule involved in the regulation of emotions such as anger and aggression as well as mood in general. It also plays a part in regulating the neurological control of blood pressure, appetite, and sleep. The chemistry of serotonin in the body is extremely complex, but low levels of serotonin in the brain can lead to depression, while elevated levels are thought to promote drowsiness and sleep. Serotonin is synthesized in the body from the essential amino acid L-tryptophan through the action of two enzymes, tryptophan hydroxylase, which adds a hydroxyl group to the 5-position of the indole group, and aromatic L-amino acid decarboxylase that strips away the carboxylic acid group of the amino acid (Figure 5.8).

Because of the role of tryptophan in serotonin synthesis, it is widely believed by the general public that foods that are high in protein, and particularly those that are high in proteins rich in tryptophan, such as milk and turkey, can induce sleepiness when consumed in large quantities (such as just before the football game on Thanksgiving day!). Whether this is actually true remains unclear, but it may be an urban myth. As with most questions concerning serotonin, this issue is complicated by other factors, such as the plasma concentrations of the other large neutral amino acids (such as phenylalanine and tyrosine) that compete for transport across the blood–brain barrier. Studies have also shown that the consumption of large amounts of carbohydrates can also play an important part (Wurtman et al. 2003), because carbohydrates stimulate the release of insulin, which in turn facilitates the transport of tryptophan across the blood–brain barrier. Finally, turkey and milk are not that much richer in tryptophan than many other foods. In general, however, drowsiness follows the consumption of a large high-protein meal in many animals, particularly predators such as lions and tigers that consume protein-rich meats.

PROTEIN CONFORMATIONS

Protein chains in general do not have random geometries, but rather adopt a specific three-dimensional shape or conformation. In some cases such as collagen and myosin, these specific conformations take the form of extended helices, but more commonly under physiological conditions proteins adopt a compact globular or "folded" shape called the tertiary structure that is a specific function of their sequence. In almost all cases this conformation is important to the biological function of the proteins, such that losing this shape results in a loss of function. Many folded proteins lose their native conformation under nonphysiological conditions such as elevated temperatures, high concentrations of certain solutes including urea, guanidinium, or thiocynate, or nonphysiological pHs. This process is referred to as **denaturation**. In foods (particularly manufactured foods), the role of a protein might be different from that designed by Nature for it to play in a living system, and often the functional properties of a protein in a food may be more governed by its denatured state than by its native conformation. For this reason, specifying and understanding both the folded and unfolded conformations is important.

Folded proteins assume their native, or physiological conformation spontaneously either during or immediately after synthesis at the ribosome (Figure 5.4). This spontaneous folding requires no energy input and implies that the folded state is the lowest free energy state for the polypeptide chain under physiological conditions. In 1957 Christian Anfinsen demonstrated that a high concentration of urea with mercaptoethanol denatures the enzyme ribonuclease A, as monitored by a loss of enzymatic activity, but that when the urea is removed from the solution by dialysis under oxygen, the proteins spontaneously refolded to their native state, recovering their original enzymatic activity. This landmark experiment, for which Anfinsen was awarded the Nobel Prize, proved that this protein is in its global minimum free energy state and is not locked into a higher-energy but kinetically metastable conformation (Anfinsen et al. 1961; Anfinsen 1973). It has subsequently been found that most proteins are in their global minimum free energy state. Thus, no folding template or input of energy is required for normal protein folding. It has been established that the first step in the folding process is a rapid hydrophobic collapse, which results from the chain twisting and coiling in such a way as to remove as much of its hydrophobic surface area as possible from direct contact with water molecules by sequestering nonpolar side chains as much as is topologically possible in the interior, maximizing their contact with each other. This condition of being globular necessarily also buries sections of the backbone with their polar and hydrogen bonding C=O and N—H functionalities in the interior, away from water. To satisfy the hydrogen bonding requirements of these sections of polypeptide backbone, most of the buried segments of the chain organize themselves into relatively regular secondary structures such as alpha helices or beta sheets, which will be discussed below, that satisfy the hydrogen bonding requirements, so that protein–protein hydrogen bonds replace hydrogen bonds to water.

Some protein conformations are remarkably stable, but for the majority, under physiological conditions of temperature and pressure, the free energy difference between the native and denatured states is generally in the range of 5 kcal/mol, which is roughly equivalent to the energy of only one water–water hydrogen bond (Chapter 2). Since an energy of this magnitude is roughly $8k_BT$ at body temperature, proteins are generally quite stable in living organisms. The low denaturation energy for most proteins probably has an evolutionary utility, because in a living cell it is necessary to continually recycle proteins by disassembling those that have become nonfunctional through random posttranslational modifications to reuse their amino acids to synthesize new functional proteins. The low denaturation energy, however, means that under the frequently harsh conditions of industrial food processing and cooking, many proteins may denature.

The free energy change for the folding of a complex molecule can be thought of as being made up of a number of different contributions. For example, as a protein folds, a favorable contribution to the free energy comes from removing a large part of the hydrophobic surface area from contact with the water, as just discussed. There may also be a favorable enthalpy contribution resulting from this process because the intramolecular hydrogen bonds found in protein secondary structures are often more regular, due to the cooperativity of making a number of correlated bonds in helices and sheets, than those that may be formed to water molecules. There will be an unfavorable contribution, however, from the loss of configurational entropy for the chain itself as it goes from the many possible conformations of the unfolded coil to the much more highly constrained conformation of the folded state (see below).

Although they are extremely large, proteins are simply molecules, and like smaller molecules, many proteins can be induced under the right conditions to form solid-state crystals. This can be achieved by lowering the temperature, adding a nucleation surface, or by the addition of cosolutes that raise the free energy of the solution and thus favor salting out, as discussed in Chapter 2 in connection with the Hofmeister series. Care must be taken, of course, to avoid denaturing the protein, which leads to an amorphous solid precipitate rather than crystals. Because proteins are so large, and their surfaces topologically complex, crystals of proteins can be difficult to nucleate, and may grow slowly. This can be understood by considering that for a new protein molecule to be added to the growing face of a nucleated crystal, it must randomly arrive at an empty lattice position on the crystal surface in the right position and orientation, and must also somehow dehydrate—that is, remove most of the water molecules that both cover the surface of the protein and fill the vacant lattice site. Proteins are now routinely crystallized by biochemists for the purpose of studying their conformations using X-ray diffraction. Before protein crystallization had been achieved, it was not clear that proteins could be isolated in pure form, because in the early twentieth century it was not clear whether they were actually simply ordinary molecules. In 1946, James Sumner shared the Nobel Prize with John Northrup for the first crystallization of the enzymes urease and pepsin, which represented the first isolation of enzymes in pure form and demonstrated that they were subject to ordinary chemical laws.

By exploiting the wave character of electromagnetic radiation (which in the quantum mechanical description has both wave and particle character), it has proven possible to determine the structure and conformations of thousands of proteins at atomic resolutions. In a crystalline lattice, the high symmetry places the atoms in regular arrays. These arrays of atoms can diffract radiation of the appropriate wavelength, producing both constructive and destructive interference patterns that are a function of the positions of the atoms in the diffracting array, which means their positions in the molecules as well. For molecules, the lattice spacings are on the order of nanometers, so the appropriate wavelengths are in the X-ray region (see Figure 9.4). Although extraordinarily complex, it is possible to deconvolute this diffraction information to give the complete structure of the protein. Only the positions of the hydrogen atoms of the proteins are not determined by this method, due to the weak scattering cross section of hydrogen atoms for X-ray radiation. Even the positions of the hydrogen atoms, however, can be determined by exploiting the wave character of neutrons. By bombarding the crystals with narrow beams of neutrons and analyzing the resulting diffraction patterns, complete conformational structures of proteins can be determined. The US government, through grants from the National Institutes of Health, supports a freely available, on-line repository for the coordinates of proteins called the Protein Data Bank (PDB), located at *www.rcsb.org/pdb/home/home.do*. It contains literally tens of thousands of protein structures determined by various methods, such as crystallography or two-dimensional nuclear magnetic resonance (NMR).

PROTEIN FOLDING

The overall "folding" of a protein is characterized by specifying the conformation of the backbone chain, which is simplified by a special feature of the peptide group. The peptide bond has a partial double bond character due to the delocalization of the carbonyl lone pair electrons into the C—N bond. This partial double bond character inhibits rotations about the peptide bond, forcing all of the atoms illustrated in Figure 5.9 that are connected by white lines to lie approximately in the same plane. Most peptide groups in proteins are in the *trans* conformation illustrated in the left part of Figure 5.9; only rarely do they isomerize to the *cis* conformation with the carbonyl and amide groups on the same side, which is also much higher in energy. As a result of this planarity of the peptide bonds, the path that the backbone follows through space can be described by specifying the relative orientations of successive peptide planes about the asymmetric carbon atoms, which can be characterized by measuring the angles ϕ and ψ illustrated in Figure 5.9.

The angles ϕ and ψ that specify the local conformations of polypeptide chains are defined using successive four-atom sequences in the backbone of the polymer chain. The angle ϕ is defined using C_{i-1}—N_i—C_i^α—C_i, and the angle ψ is defined using N_i—C_i^α—C_i—N_{i+1}. These angles are dihedral angles; a dihedral angle is defined as the angle between two planes. A mathematical plane can be defined by three points in

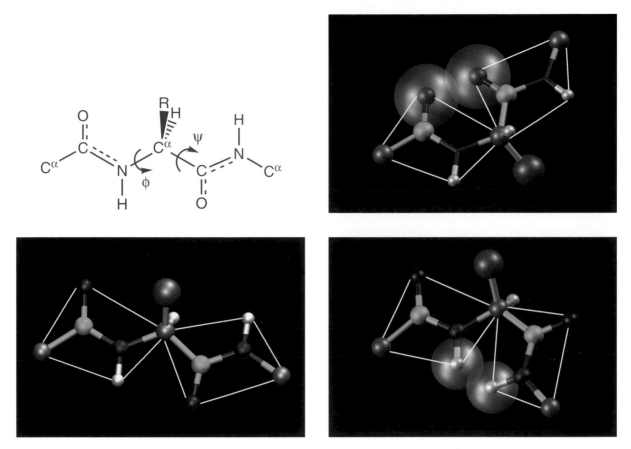

Figure 5.9. The problem of describing the conformation of a polypeptide chain can be greatly simplified by treating the six atoms in each peptide bond, shown linked by the rectangular white lines, as lying in a common plane. Then, specifying the local conformation around the alpha carbon atom of each residue reduces to specifying the angles ϕ and ψ (upper left panel) which describe the orientation of the two successive planes relative to each other. In these figures, the peptide carbonyl carbon atoms are green, their oxygen atoms are red, the amide nitrogen atoms are blue, the backbone alpha carbon atoms are gold, and hydrogen atoms are white. The side chains are indicated generically by purple spheres. In the two examples on the right, steric clashes are indicated generically by gray spheres (not drawn to the scale of the respective atomic radii).

Cartesian space, so the two planes defining the ϕ angle are between the C_{i-1}—N_i—C_i^α plane and the N_i—C_i^α—C_i plane (see Figure 5.9), and the planes defining the ψ angles are defined as N_i—C_i^α—C_i and C_i^α—C_i—N_{i+1}. Since, if we ignore for the present the conformations of the side chains, the local conformation of the polypeptide chain about C_i^α can be described with just these two angles, this conformation could be thought of as a single point in a two-dimensional, (ϕ,ψ) coordinate space (Figure 5.10). Any other point in this two-dimensional coordinate system, specifying different ϕ and ψ angles, would correspond to another conformation.

Not all conformations of the backbone are allowed, because some, such as those illustrated on the right in Figure 5.9, lead to unacceptable steric clashes between atoms of the backbone. In fact, only limited ranges of these angles do not produce

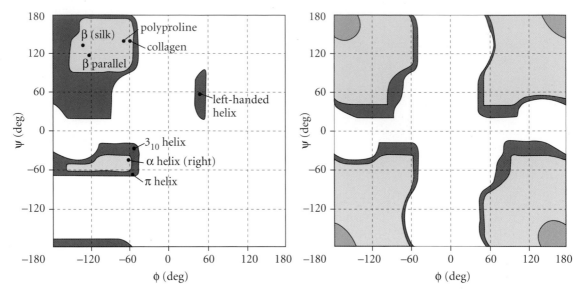

Figure 5.10. Ramachandran conformational energy maps for alanine (left) and glycine (right). The light gray areas are conformations that do not produce steric clashes for the dipeptide when using average atomic radii determined by X-ray diffraction from a series of crystal structures of small molecules containing C, O, N, and H atoms in similar covalent situations. The darker gray regions are those conformations that do not produce steric clashes when using the smallest values for these radii from the same series of crystal structures.

such steric clashes; the conformational maps in Figure 5.10, referred to as **Ramachandran maps**, delineate the possible combinations of ϕ and ψ that do not produce such steric overlaps. Such maps were first prepared by G.N. Ramachandran and coworkers using physical models of the peptide group made of balls and sticks scaled to the appropriate relative dimensions.

Ramachandran Maps

The map on the left in Figure 5.10 outlines those regions in (ϕ,ψ) space that do not produce backbone steric clashes for the case where the side chain R is the methyl group of alanine. The areas shown in white are completely disallowed by the steric overlaps of atoms that result for the conformation corresponding to that (ϕ,ψ) angle pair. The light gray regions were calculated using the average atomic radii for the various atoms as observed in many crystal structures of small organic molecules. The much larger, darker areas, plus the new region corresponding to the left-handed alpha helix, result from using the smallest radial values observed in small organic crystals. As can be seen, more than three-quarters of the potential conformations are disallowed by atomic overlaps, with only three allowed regions. As the labels indicate, these three regions incorporate all of the principal types of protein secondary structures such as the alpha helix and beta sheet conformations. Glycine residues have more open maps, with larger allowed areas, because their smaller side chains (a single

Figure 5.11. The Ramachandran conformational energy map for alanine calculated from an empirical energy function that approximates the way in which the energy of a model alanyl residue in a polypeptide chain changes with conformation. The energy contours are lines connecting all conformers with the same energy, indicated by numbers, in kcal/mol above the global minimum (that is, the very lowest point). The dots represent low points or minima in their respective energy "valleys," and the white areas represent regions of extremely high repulsive energy, as in Figure 5.10. (Figure reprinted with permission from S.S. Zimmerman et al., 1977, *Macromolecules* 10(1): 1–9. Copyright 1977 American Chemical Society.)

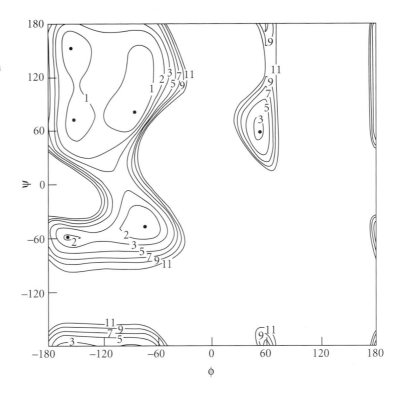

proton) produce fewer atomic overlaps (Figure 5.10, right). Side chains branched at the β carbon, like valine, have greater steric clashes, and thus more restricted allowed conformational regions than alanyl-like residues.

Using theoretical models, it is possible to make contour maps of the peptide energy as a function of these conformational angles, as is shown in Figure 5.11. These maps attempt to capture all of the principal contributions to the total conformational energy of the peptide group, such as bond stretching or compression, bond angle distortions, hindered rotations about bonds, and van der Waals and electrostatic interactions (Gibson and Scheraga 1966; Zimmerman et al. 1977; McCammon and Harvey 1987; Brooks, Karplus, and Pettitt 1988). The Ramachandran maps shown in Figure 5.10 could be thought of as resulting from an extremely simple energy description, one in which the energy is infinity if the conformation results in atomic overlaps, and zero otherwise. When all of the torsion angles from protein structures known from X-ray diffraction are plotted on such a field, they are seen to mostly fall into the allowed regions. Those that do not lie in the three allowed regions of the alanine map are usually glycine residues.

Secondary Structures

The regular repeat of backbone torsion angles (φ,ψ) produces extended structural motifs such as helices. Several of these structures are important for protein

Table 5.1 Torsion angles and characteristic features of the principal secondary structural elements

	Torsion angle (in degrees)			Residues per turn	Translation per residue (Å)
	ϕ	ψ	ω		
antiparallel β-sheet	−139	135	−178	2.0	3.4
parallel β-sheet	−119	113	180	2.0	3.2
right-handed α-helix	−57	−47	180	3.6	1.5
polyproline I	−83	158	0	3.33	1.9
polyproline II	−78	149	180	3.0	3.12
polyglycine II	−80	150	180	3.0	3.1

The angle ω defines the peptide torsion itself; a value of 180° is a perfectly planar *trans* arrangement, and 0° represents a perfectly planar *cis* configuration.

conformations. The most well-known of these are the so-called alpha helices and beta sheets. These secondary structures are important in globular proteins because they allow the hydrogen-bonding CO and NH functionalities of the peptide group to be buried in the center of the protein, away from water, without needing to give up their hydrogen bonds, because the secondary structures satisfy the hydrogen bonding requirements by making internal hydrogen bonds (Table 5.1).

Alpha Helices

In an **alpha helix**, each successive residue of the helix has the same set of backbone torsional angles ϕ and ψ, with the numerical values of these angles being approximately (−57°, −47°). This repeated conformation results in an extended helical coil with a rise of 1.5 Å per residue and a nonintegral repeat of 3.6 residues per complete turn of the helix, 5.41 Å in length. In an extended sequence of alpha helix the structure is stabilized by a series of hydrogen bonds along the axis of the helix between backbone carbonyl and amide N—H groups, as shown in Figure 5.12. These hydrogen bonds form between the carbonyl C=O group of residue i and the N—H group of residue (i + 4). The alpha helix is the most common structural element in proteins. Some globular proteins, such as myoglobin, are mostly alpha helix. Most fibrous proteins, including muscle fiber myosin and the keratin of hair and wool, are also predominantly alpha helical, although these proteins are extended helices rather than globular. The hydrogen bonds of the alpha helix account for the flexibility and extendibility of wool fibers; stretching the fiber pulls against the alpha helical hydrogen bonds, which snap back when the stress is released.

The chain can also coil in a left-hand spiral, producing a left-handed helix with (ϕ,ψ) values of approximately (57°,47°). This type of helix puts the side chains in close contact with the backbone, and so is rarely observed in proteins and only for residues with small side chains like glycine.

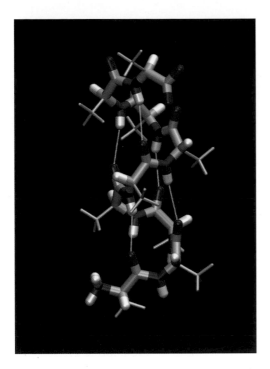

Figure 5.12. A "liquorice"-type representation of a polyalanine chain in an alpha helical conformation. The methyl side chains are shown in gold, and the i to i+4 hydrogen bonds between C = O and N–H groups are shown as dashed lines.

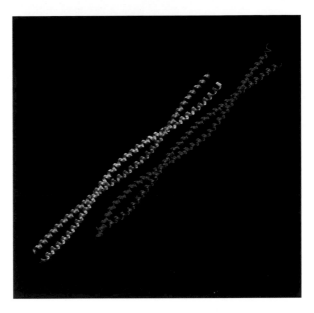

Figure 5.13. Backbone traces of the coiled-coil structure of the β-myosin II dimer conformation as determined from X-ray crystallography. (W. Blankenfeldt et al., *Proc. Natl. Acad. Sci. USA* 103:17713–17717, 2006.)

Figure 5.13 illustrates the backbone path for several strands of myosin II, a so-called coiled-coil or superhelix of two extended alpha helical strands wound around one another. Myosin is the principal protein component of the muscle fiber of meat and, as such, is one of the most important of food proteins. Myosin II is a dimer that has a characteristic heptid sequence, where residues *a* and *d* in the repeating seven-residue pattern are small hydrophobic residues, with a valine typically at position "*a*" and a leucine at position "*d*." In the coiled-coil the ϕ and ψ values deviate slightly from those of a perfectly straight alpha helix, with only 3.5 residues per turn. The coiled-coil superhelix has a repeat of around 140 Å, and this coiling means that the 5.4 Å repeat of the individual component alpha helices is no longer perfect. This spacing places the small hydrophobic side chains of the two peptide chains in contact along a hydrophobic "seam" and provides hydrophobic stabilization for their pairing in an aqueous environment.

The regular repeat of any combination of dihedral angles would generate a helix, so in principle many different types of helices other than the alpha helix are possible for proteins, but they are not observed because they are energetically unfavorable. One particular variant of the alpha helix is called the 3_{10} helix because it has three residues per turn and a stabilizing hydrogen bond that incorporates 10 atoms in the distorted ring enclosed by the hydrogen bond along the helix axis, with (ϕ, ψ) values for each residue of $(-49°, -26°)$. This type of helix is more extended (stretched) than an alpha helix, and its carbonyl groups point out and slightly away from the helix axis, so that its hydrogen bonds are less linear. As a result, the 3_{10} helix is less stable than an alpha helix, and it is not found in proteins as an independent extended structure. However, it is sometimes observed as a single turn at the ends of alpha helices. In some of the protein conformational structures illustrated in this and other chapters, short turns of 3_{10} helix are indicated in dark blue (see, for example, Figures 4.56, 5.25, 5.50, and 6.19).

Beta Sheets

Another type of secondary structure widely found in proteins is the **beta sheet**. Actually two types of beta sheets are known (Figure 5.14). In these structures the polypeptide chains are almost completely extended, with the hydrogen bonds being

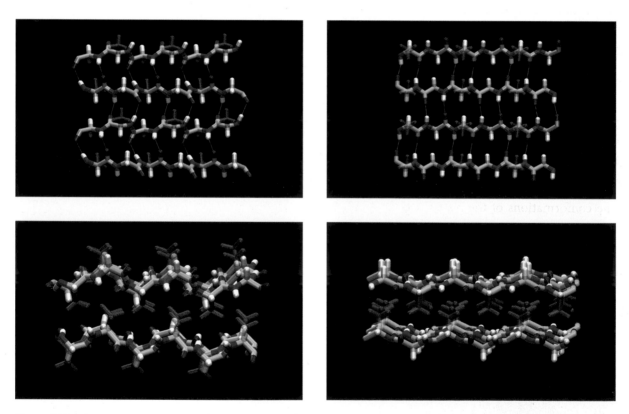

Figure 5.14. The structure of a parallel β sheet (on the left) and an antiparallel β sheet (on the right) for polypeptides with the sequence —(Ala—Gly)$_n$— of silk. The hydrogen bonds between the chains are shown as dashed yellow lines, and the Ala side chains are shown in gray. The lower panels illustrate how two of these sheets would associate together; the illustration on the lower right is the actual structure of silk, with alternating layers of alanine or glycine residues stacked against each other.

formed between two adjacent strands of the polypeptide, rather than along the immediate sequence as in helices. Because two strands are involved, they can be either parallel or antiparallel. The dihedral angles required to produce these two types of structures are somewhat different, but both are sterically allowed. The antiparallel beta sheet results when a chain reverses its direction and then runs back in the opposite direction alongside itself, or when two more distant sequences of a chain are adjacent but running in opposite directions, with (ϕ,ψ) values of $(-139°,135°)$. The parallel beta sheet has (ϕ,ψ) values of $(-119°,113°)$.

Silk is made up of polypeptide chains with many repeats of the sequence —(Gly—Ser—Gly—Ala—Gly—Ala)$_n$— in a beta sheet conformation, stacked with the alanines against alanines and the glycines against glycines. The chains are extensively hydrogen bonded within the layers of the sheet but make no hydrogen bonds between layers. Silk is strong and flexible, resistant to tearing but with a low extendibility. Stretching the silk in one direction strains against the almost fully extended covalent structure, and thus yields very little, while in the other two dimensions stresses pull against the weak van der Waals attractions and intermediate strength hydrogen bonds, allowing flexibility. In globular proteins beta sheets are much less regular, and often exhibit a right-handed twist (see, for example, Figure 5.25).

Reverse Turns

For a protein to be globular, the extended helices and sheets must reverse direction occasionally to produce a compact ball. The loop regions that accomplish these turns are often in a special conformation called a **reverse turn**, in which four successive residues have specific (ϕ,ψ) values that reverse the direction of the chain but that are not the same for each of the linkages. Most of these reverse turns are stabilized by a hydrogen bond across the turn as shown in Figure 5.15.

Nearly a dozen different types of reverse turns have been described in terms of the conformations of the residues in their sequence. Table 5.2 lists the ideal (ϕ,ψ)

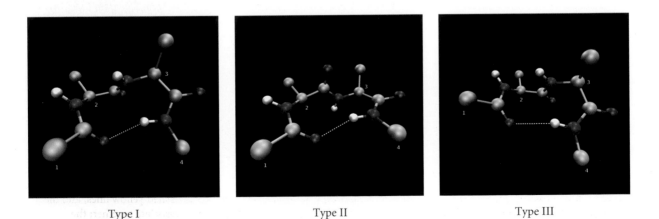

Type I   Type II   Type III

Figure 5.15. Representations of three types of standard reverse turns.

Table 5.2 Conformations of the second and third residues in several ideal reverse turns

Bend type	(ϕ,ψ) for residue 2 of the turn (in degrees)	(ϕ,ψ) for residue 3 of the turn (in degrees)	Sequence restrictions
I	(−60, −30)	(−90, 0)	residue #3 ≠ Pro
II	(−60, 120)	(80, 0)	residue #3 must be Gly
III	(−60, −30)	(−60, −30)	none

angle values for three of the most common types in terms of the conformations required for their second and third residues. In type I turns, any amino acid can occupy any of the four positions, except that residue 3 cannot be a proline. The type II turn requires a glycine residue in position 3, while the type III turn has no restrictions on sequence. Each of these three types of turns has a mirror image, designated I′, II′, and III′, in which the signs for (ϕ,ψ) are reversed; the sequence restrictions for these turns are different, however.

Prolines

Prolines are unique among the amino acids in that their aliphatic side chains loop around and covalently bond to the nitrogen atom of the backbone peptide group to give an imino acid. Because of this, no amide hydrogen atom occurs in the peptide group to participate in the hydrogen bonds of secondary structures (α helices and β sheets).

The rigidity of this pyrrolidine ring fixes the backbone torsional angle ϕ at approximately −60° to −75° (Figure 5.16), so the conformational freedom of proline residues in a chain lies only in the ψ angle. Figure 5.17 shows how the energy of a proline residue changes with ψ; as can be seen, there are two minima, around 145° and −55°. Chains of pure polyproline can adopt one of two regular left-handed helical conformations, called polyproline II, with (ϕ,ψ) values of approximately (−75°,150°), and another conformation called polyproline I where the peptide bonds are in the *cis* arrangement, with (ϕ,ψ) values of approximately (−75°,160°).

The unique structure of the proline residue also affects the conformational possibilities for the residue that precedes it in the polypeptide sequence because of steric clashes with the bulky ring where there is usually only a proton. The Ramachandran map for an alanyl-type residue preceding a proline is shown in Figure 5.18; as can be seen, these steric clashes completely "wipe out" the alpha helix low energy region of the map. This means that prolines cause a termination of alpha helices, because they force the residue preceding them out of the alpha helical conformation. (Note, however, that the prolines themselves can adopt the alpha helical conformation, since one of their two possible conformations is (−60°, −55°), approximately the alpha helical conformation, so they can be the first residue of an alpha helix.) Prolines are important in determining the functionality of several types of food

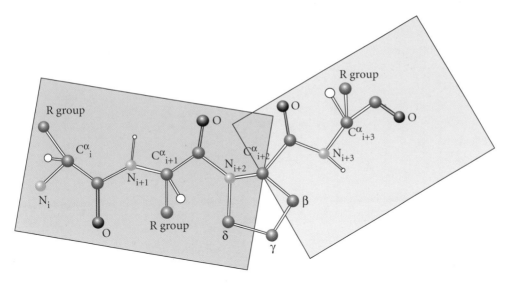

Figure 5.16. The conformational restrictions imposed by a proline in a polypeptide.

Figure 5.17. A plot of the conformational energy of proline as a function of the peptide torsional angle ψ. As can be seen, only two values of ψ produce a low conformational energy. (Figure redrawn from data given in C.R. Cantor and P.R. Schimmel, *Biophysical Chemistry. Part I. The Conformation of Biological Macromolecules*. San Francisco: Freeman, 1980.)

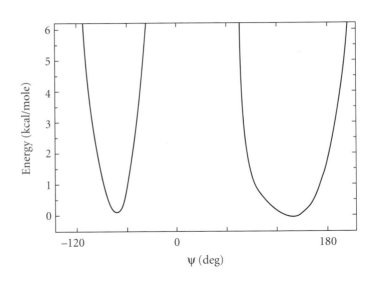

proteins, including collagens, glutens, and caseins, which we will consider in greater detail below.

Collagen

Collagen is the major protein component of connective tissue (e.g., tendons, skin, ligaments, blood vessels). It is the most abundant protein in higher animals and makes up about one-third of the total protein content of mammals, but a smaller proportion of total fish protein. The collagens are a family of around a dozen different types of nonglobular polypeptides about 1000 residues in length characterized by repetitive sequences of the type (—Gly—X—Y—)$_n$, with every third residue a glycine.

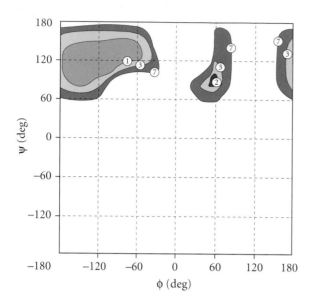

Figure 5.18. The Ramachandran conformational energy map for an alanyl-like residue preceding a proline in a polypeptide chain. The energy contour values are indicated in kcal/mol above the minimum in circles superimposed on the lines. (Figure adapted and redrawn from C.R. Cantor and P.R. Schimmel, *Biophysical Chemistry. Part I. The Conformation of Biological Macromolecules*. San Francisco: Freeman, 1980.)

Many of the residues X and Y are prolines. Many of the proline residues at Y are hydroxylated by posttranslational modification by the enzyme proline hydroxylase to form hydroxyproline (abbreviated Hyp), so that collagen contains many Gly—X—Pro and Gly—X—Hyp sequences. Lysine is also relatively abundant in collagen, and lysines in the Y position are also frequently hydroxylated to δ-hydroxylysine by the enzyme lysine hydroxylase (these hydroxylated residues Hyp and Hyl are found only in the Y position). As would be expected from the sequence, glycine and proline each constitute about a third of the total residues of collagen. Collagen contains little of the nonessential amino acid cysteine and usually none of the essential amino acid tryptophan (see Table 5.3), so it is a poor source of dietary protein.

Individual collagen strands coil into an extended left-handed helix similar to that of polyproline, and three of these helices then twist together as a right-handed triple helical bundle called **tropocollagen**, as illustrated in Figure 5.19. The (ϕ,ψ) values of the peptide linkages are similar to those for polyproline (because so many of the residues are proline), but are very different from those of the alpha helix. (As we have already seen, proline is not compatible with the alpha helix structure except as the first residue of the helix.) This overall structure was almost simultaneously deduced in 1955 from X-ray fiber diffraction and modeling studies by two different groups, those of Alexander Rich and Francis Crick (who had earlier shared in determining the structure of DNA) in Britain and G.N. Ramachandran and G. Kartha in India.

The opposite handedness of the coiling of the individual strands and of the coiled-coil gives the tropocollagen great tensile strength, because this opposite coiling prevents the coiled-coil from unraveling under tension, or an individual strand from being pulled out, much like a rope wound from smaller fibers or a steel bridge cable. The triple helix structure of collagen is the reason that every third residue must be a glycine. In the superhelix, the residues in these third positions are in such close

Table 5.3 Amino acid composition in weight percentage of a typical collagen sample

Amino acid	Weight %[a]
glycine	26–31
proline	15–18
hydroxyproline	13–15
total prolines	28–33
glutamic acid	11–12
lysine	4–5
hydroxylysine	8–12
total lysines	12–17
alanine	8–11
arginine	8–9
aspartic acid	6–7
leucine	3–4
serine	3–4
tryptophan and cysteine	little or none

The composition varies depending on the source of the material and somewhat on the extraction process.
[a]w/w; that is, grams amino acid per 100 g protein.

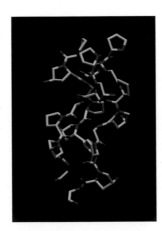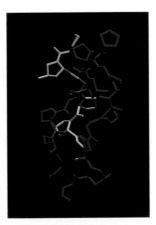

Figure 5.19. Illustration of a short segment of the triple helical structure of three strands of collagen with the repeat sequence –Gly–Pro–Pro–. Each of the three strands is individually a helix, and the three helices are intertwined to form the tropocollagen triple helix.

contact with the other chains that there is not room for a side chain other than the single proton of Gly. Mutations at the glycine positions disrupt the helical packing and thus affect the functional properties. In humans, mutations that disrupt the helical structure of collagen fibrils cause various collagen diseases, some of which are fatal.

The three chains of the superhelix are held together by hydrogen bonds between their backbone C=O and N—H groups. The hydroxyl groups of hydroxyproline residues in different chains can also form interchain hydrogen bonds, stabilizing the triple helix. In addition to interchain hydrogen bonds, the collagen supercoil is further stabilized when the individual chains also become covalently cross-linked, primarily between lysine side chains, which reduces their solubility. As an animal ages, the lysine residues of collagen can be converted by an enzyme called lysyl oxidase into an aldehyde called allysine, as shown in Figure 5.20. Two of these allysine residues on different chains can then cross-link to form allysine aldol. This allysine aldol can further react with a histidine side chain on a third strand of the tropocollagen to form a three-way cross-linking between all three chains. This cross-link can even further combine with yet another lysine on one of the three chains to form a four-way cross-linked complex. These cross-links were initially an impediment to determining the structure of collagen, until it was realized that they are not present in the connective tissues of very young animals. The number of these cross-links increases with age, except in fish, which is why meat from older animals is tougher.

In general the three chains of the triple helix of collagen are not necessarily the same. In type I collagen, the most common type, two of the chains, called the α1(I) chains, are the same, while the third, called α2(I), has a different sequence. Type I is the form of collagen found in skin, bones, and tendons. In type II and type III collagen, however, all three chains have the same sequence. Type II, composed of α1(II) chains, is the primary collagen component of cartilage, while type III, composed of α1(III) chains, is the principal constituent of blood vessels, and therefore is not as important as a food, although present in all meat tissues. Collagen from different sources will in general have different sequences.

Gelatin

Collagen is useful as a functional ingredient in foods, because of its ability to form gels, and provides an economic value to what otherwise would be a waste material. It is the principal component of commercial gelatins such as Jell-O. When collagen fibers are heated in water the triple helix unravels as the strands denature and become disordered. When cooled, parts of the chains will become entangled as portions of the individual chains once again twist into short regions of the triple helix, providing the juncture zone of a gel, while the disordered portions of the chains wandering off in random directions constitute the framework of the gel. The temperature at which the strands denature, or melt, depends upon the collagen type and the sequence. In general, the proline and hydroxyproline residues reduce the potential flexibility of the individual chains, and the melting temperature of collagen increases as the content of these two imino acid residues increases. The melting temperature is also sensitive to the ratio of prolines to hydroxyprolines; fewer hydroxyprolines leads to a lower melting temperature.

The conversion of collagen to gelatin can be observed in the ordinary cooking of meats, which will exhibit the development of gelatinous deposits when cooled.

Figure 5.20. The process of cross-linking of individual collagen strands in tropocollagen. The initial formation of allysine from a lysine side chain is catalyzed by the enzyme lysyl oxidase, with two such aldehydes subsequently undergoing an aldol condensation to cross-link the chains. (Voet and Voet, 1995.)

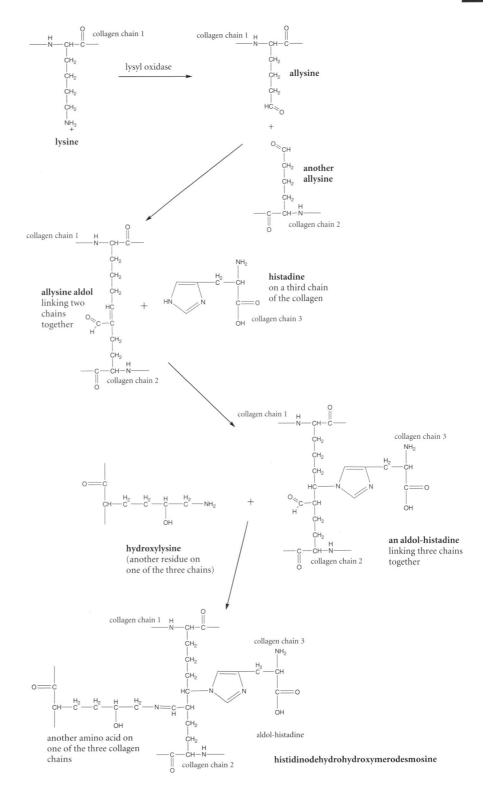

Tougher low-quality meats such as stew beef contain a high proportion of collagen, but as this is cooked for long periods of time, more of this collagen is denatured, destroying more of the fiber strength and tenderizing the meat.

Disulfide Bonds

The thiol or sulfhydryl group of cysteine is weakly acidic and reactive. The cellular cytoplasm surrounding the ribosomal sites of protein synthesis is a reducing environment, but proteins that are excreted outside of the cellular membrane are exposed to oxidizing conditions. In the presence of atmospheric oxygen, when the S—H thioalcohol groups of two cysteine residues are close together, they can be readily oxidized to form a cross-linking (—S—S—) **disulfide bond**.

$$R_1{-}S{-}H + H{-}S{-}R_2 \rightarrow R_1{-}S{-}S{-}R_2 \tag{5.1}$$

Such disulfide cross-links can stabilize the native, folded conformation of a protein toward denaturation—not by making the folded state more stable per se, but rather by making the unfolded state less favorable. The disulfide cross-links constrain the unfolded state by requiring that the two cysteine residues remain within one S—S bond length of one another, reducing the number of possible unfolded conformations relative to the situation with two reduced —S—H groups. This reduced randomness of the unfolded states caused by the cross-linking thus lowers the entropy of the manifold of unfolded states and reduces the favorable contribution to unfolding coming from the change in chain configurational entropy. Small increases in the thermal stability of a protein (leading to a slightly higher melting temperature) can sometimes be achieved through genetic engineering methods by changing two residues in van der Waals contact in the native conformation into cysteine residues, thus introducing an additional disulfide cross-link into the chain under oxidizing conditions.

The presence of free cysteines can lead to instability in food processing. Food proteins that are denatured by heat or the presence of denaturants can also make intermolecular cross-links with free cysteines of other proteins, thus preventing their refolding and leading to changes in functional properties or precipitation. Disulfide cross-linking and the rapid exchange of these linkages are thought to be crucial to the functional properties of gluten in wheat-based doughs, permitting elasticity while the exchange reactions allow deformability.

Gluten

Wheat is one of the most important staple crops in the world, and has been almost since the beginning of agriculture. It is the third most widely planted cereal grain, after maize and rice, and hundreds of millions of people eat diets centered around the breads, pastas, and other products made from wheat. Wheat was first domesticated from wild grasses of the genus *Triticum* in the Middle East during the Neolithic Period following the end of the Ice Age. Wheat grains are made up of approximately

12% protein and over 70% starch; wheat flour ranges from 8 to 18% protein and ~80% starch. The protein content of wheat is much higher than that of maize and rice. Although wheat flour is less than 20% protein, the special viscoelastic properties of dough are due to the protein fraction rather than to the starch. The proteins of wheat are unique in their properties and are the reason that wheat flour can be used to create such products as leavened bread and baked goods. The principal protein fraction of wheat flour is called **gluten**. Gluten is found only in wheat and a few related grasses such as barley and rye, although other flours such as corn meal contain unrelated proteins sometimes called gluten. Gluten is responsible for the elasticity, viscosity, extensibility, and cohesiveness of wheat dough, and thus for all of the remarkable functional properties of wheat as a food (see the review by Wieser 2007).

Gluten is actually a mixture of two types of water-insoluble proteins, **glutenin** and **gliadin**, that are left after wheat dough is washed in water to remove starch granules and water-soluble fractions. The gliadins are soluble in aqueous alcohol, while the glutenins are also insoluble even in alcohol. Both fractions contain a number of closely related similar proteins, all of which have high contents of proline and glutamine. The gliadin proteins are relatively smaller molecules with molecular weights generally between 28,000 and 55,000, while the glutenin proteins consist of a fraction with molecular weights in the range 32,000 to 45,000 and a fraction of larger molecules with molecular weights in the range 67,000 to 125,000. These glutenin proteins are aggregated together through intermolecular disulfide bonds into much larger multiprotein covalent complexes with molecular weights ranging from 500,000 up to 10,000,000. Both of these proteins have very low contents of charged amino acids and very high contents of glutamine and proline. Neither of these types of proteins have well-defined tertiary conformations, perhaps due in part to the high proline content. The glutenins are cohesive and elastic and are principally responsible for the elastic properties of dough that permit it to be kneaded and rolled, for example. Gliadins have little elasticity and are not as cohesive as glutenins, but function as a plasticizer for the glutenins in the dough. Gluten proteins contain cysteine, but at only about 2% of the amino acid content, it makes up a less than statistically random fraction (~5%; i.e., 1/20 of the total). Nonetheless, the cysteine content is extremely important to the functional properties of wheat dough. Most of the Cys residues exist in an oxidized form as disulfide bonds within glutenin subunits or as intermolecular disulfides between subunits, linking the chains together.

Careful analysis has determined that there are many variants of the gliadin proteins. The gliadin fraction of gluten was originally divided into four groups of monomeric proteins, called α, β, γ, and ω gliadins, based on their mobility when separated by gel electrophoresis. Later electrophoretic studies established that the α and β gliadins actually are only one group now called α/β-type gliadins. High-pressure liquid chromatography (HPLC) can divide the gliadins into over 100 different molecule types, which are now grouped into a different set of four types, called ω5, ω1, α/β, and γ gliadins. Most of these proteins consist of many repeating short segments of peptides with the same or similar sequences. The combined Gln and Pro content of these proteins ranges from 50 to over 75%. They also contain a significant proportion of phenylalanine, ranging from 5% in the α/β- and γ-gliadins to 9% in the ω-gliadins. With proline contents ranging from 16 to 26%, the gliadins are unlikely to have globular conformations, but they have been reported to have β-turns in their N-terminal regions and alpha helices and beta sheets in their C-terminal regions. The α/β-gliadins have six cysteines in their C-terminal domains, and the γ-gliadins have eight, with these groups involved in three and four intramolecular disulfide bonds, respectively. The gliadins form a sticky, viscous liquid when mixed with water and impart extensibility to dough.

Like the gliadins the glutenins are relatively rich in glutamine and proline residues in repetitive segments with sequences such as Gln—Gln—Gln—Pro—Pro—Phe—Ser. The C-terminal regions are similar to those of the α/β- and γ-gliadins. The repetitive sequences with large proline compositions are thought to lead to a series of repeated β-turns, producing a coiled, open, helical-like structure resembling a weak spring. Such a springlike structure would contribute significantly to the elasticity of gluten doughs. Extending these coils would decrease their entropy, favoring recoil, in a manner similar to the role of entropy in the elasticity of rubber. As already noted, one of the most important features of the glutenin proteins in determining their functional behavior in doughs is that they are cross-linked by their disulfide bonds into much larger covalent complexes. With an upper bound in molecular weight approaching 10 million, these molecules are among the largest of all proteins. Although the native glutenins are insoluble in alcohol, if their disulfide linkages are reduced, their monomeric units become as soluble in aqueous alcohol as the gliadins. The cysteines of the glutenins are involved in both intramolecular and intermolecular disulfide bonds under the normal oxidizing environment of the home and bakery.

The disulfide bonds of gluten are crucial to the functional characteristics of its proteins, and in particular their rheological behavior, which allows them to be kneaded, shaped, and rolled. Disulfide cross-linking and the rapid exchange of these linkages are thought to contribute to elasticity, with the exchange reactions allowing extensibility and deformability. The importance of these disulfide bonds to the quality of dough is demonstrated by the effects of reducing agents, which significantly weaken dough, while oxidizing agents strengthen dough. Dough quality is also affected by thiol terminators, small thioalcohol-containing molecules that can form disulfide-like bonds with free cysteine thioalcohol groups, thereby making them unavailable for protein cross-linking. Dough is also weakened by urea, a protein denaturant that has been shown to weaken or break protein intramolecular hydrogen bonds by directly hydrogen bonding to their backbone C=O and N—H groups.

In the moist dough, the gluten proteins form an essentially continuous network entangling the starch granules. In leavened bread, CO_2 gas is generated, either through the reactions of chemical leavening agents or as the result of fermentation from added yeast. The yeast also produce α– and β–amylases, which carry out limited hydrolysis of the outer ends of the polymers in the ungelatinized starch granules, producing small amounts of free reducing sugars. As CO_2 is produced, a small amount escapes, but the majority is trapped in small bubbles completely encased by gluten films. When the bread is baked, these bubbles of course begin to expand dramatically as the gas pressure rises, stretching against the enclosing, and extensible, gluten film like a balloon being inflated. At the same time, as the dough is heated, starch granules in the interior of the loaf, which is called the **crumb**, begin to absorb water and gelatinize. The amylases from the yeast remain active as the dough is heated and continue to hydrolyze the starch until they denature at around the gelatinization temperature. The escaping amylose polymers entangle with the gluten and further contribute to the structural matrix of the bread, with the remaining swollen granules, as well as the expanding gas bubbles, trapped in this mixed polymer network. On the surface of the loaf, there is far less moisture, since the water in this region can readily escape from the dough by evaporation, so that starch granules in this outermost layer, which becomes the **crust**, do not gelatinize extensively, and when bread crust is examined under a microscope, relatively intact starch granules are observed. The low moisture, elevated temperature, and reducing sugars produced by the amylases, however, promote Maillard browning, and the development of the brown color and flavor of the crust, and particularly the generation of the aroma compounds characteristic of freshly baked bread. As they expand, the growing gas bubbles merge and lose their spherical shape, compressing the gluten and polysaccharide films on their surfaces into thin lamellae. Eventually these bubbles rupture, allowing the gases and much of the remaining moisture to escape, and creating the familiar open network of connected pores of bread. As more and more moisture is lost, the gluten and polysaccharides "set" into a fixed, but still flexible, high-temperature matrix as a result of the loss of much of the plasticizing water molecules. Figure 5.21 shows the temperature dependence of the glass transition temperature for white bread as a function of water content. The water content of baked bread is around 35% (approximately 44% for the crumb, and around 20% for the crust; the values vary with time after baking), so at room temperature fresh white bread is above its glass transition temperature. This is why it "spoils" from mold growth after a few days, and undergoes staling, which is primarily due to retrogradation of starch, which is a partial recrystallization facilitated by the mobility allowed by the plasticizing water.

Figure 5.21. The glass transition temperature of white bread as a function of water content. (A. Cesàro and F. Sussich, 2001.)

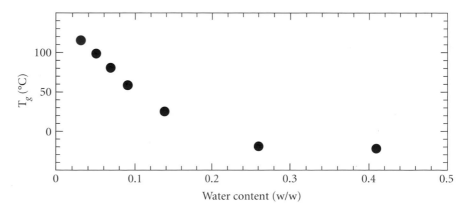

Solid or semisolid fats, called shortenings, are added to doughs to be used for such things as pastries or pie crusts to produce light, flaky, and moist textures in the baked products. These fats work by binding to the gluten proteins and interfering with the formation of the protein–protein cross-linked networks responsible for the extensibility of the dough, thus reducing the amount that the dough rises, or "shortening" it, when it is cooked and preventing the formation of stiff matrices of proteins and gelatinized, retrograded starch. These shortenings can consist of semisolid vegetable oils, lard, butter, or margarine (see Chapter 7).

The gluten isolated from doughs by washing away the starch and other water-soluble components is an important commercial product. It can be added back to doughs to give a higher gluten percentage, for a higher-rising dough. Products such as pet foods add gluten to increase their protein content. Gluten is sometimes used as a food additive as a stabilizer, but the frequency of gluten allergies (see below) limits such uses. Because of the firm, elastic, and somewhat fibrous texture of cooked gluten, it is used in artificial meat substitutes in vegetarian diets. Gluten is widely eaten in Asia, particularly in vegetarian Buddhist diets, and in macrobiotic foods in the United States, where it is referred to as **seitan**, a word of Japanese origin. In Japan itself, however, gluten is usually called *fu*, and in Chinese cuisines, it is called *mian jin*.

The gluten proteins of wheat products are responsible for celiac disease, one of the more widespread food allergies. Individuals who suffer from this allergic disease are unable to tolerate any gluten-containing foods. Eating any significant amounts of gluten causes serious gastrointestinal distress and poor absorption of other nutrients for these individuals, who constitute an important market for substitutes for gluten in baked products. Unfortunately, this is of course difficult, because no other proteins have quite the same properties, although progress is being made.

FOLDING OF GLOBULAR PROTEINS

As we have already seen, many proteins have a compact globular shape somewhat like wadding up a string of pearls in one's hand. Unlike the "balled-up" pearls, however, in the native state proteins generally have a specific and unique arrangement that they assume spontaneously after they are synthesized at the ribosomes, and are stable in this conformation under biological conditions. While this folded native state represents the lowest free energy conformation for most proteins, the possibility remains that some proteins are kinetically trapped in a higher-energy conformation separated from the lowest energy state by a high barrier. This situation exists for some proteins such as insulin and botulinum toxin that are enzymatically modified after they fold (see below). It has been found that cells contain proteins called molecular chaperones that help newly synthesized protein chains to fold properly to their native states, but do not themselves determine the folding pattern, because the proteins are able to fold up in vitro without these chaperones. The chaperones are necessary to prevent hydrophobic aggregation

Figure 5.22. The thermal denaturation curve for the enzyme ribonuclease. (Redrawn from data given in A. Ginsburg and W.R. Carroll, *Biochemistry* 4:2159, 1965.)

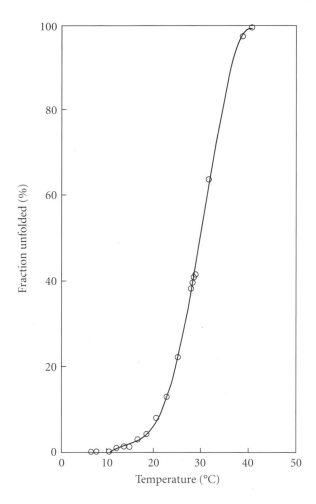

and precipitation of the unfolded chains, which would keep them from folding properly.

As we have seen, globular proteins can lose their folded native conformations when the ambient conditions, such as temperature or pH, are changed in such a way that the folded state is no longer favored. Denaturation of the native state is usually a highly cooperative process that takes place over a narrow range of conditions of pH, temperature, or denaturant concentration. An example of the thermal denaturation of the protein ribonuclease is shown in Figure 5.22. Such unfolding is generally associated with an absorption of energy, and the melting temperature and enthalpy change are generally determined by differential scanning calorimetry (DSC; see Chapter 1). In general, a fraction of unfolded protein equal to 0.5 does not mean that the proteins are half unfolded, but rather that half of the proteins are still in their native conformation and half are completely denatured.

Proteins generally have a biological function, which is the role that they play in promoting the evolutionary fitness and survival of the species that produced them.

Table 5.4 Denaturation temperatures T_m and heats of unfolding for selected food proteins. See also Table 5.8.

Protein	T_m (°C)	ΔH_{unf} (kcal/mol)
collagen (halibut)[1]	18	1.1
collagen (sheep skin)[1]	37	1.4
myosin[1]	53	0.45
chymotrypsin[1]	57	164
bovine β-lactoglobulin[3]	70	55
myoglobin[1]	72	134
lysozyme[1]	78	140
glycinin[1]	80	—
ovalbumin[2]	80	112
papain[1]	84	212
bovine pancreatic trypsin inhibitor	103	70

For the fibrous proteins collagen and myosin, the reported temperatures are the uncoiling temperatures, given per mole of residues.

[1] W. Pfeil, in *Thermodynamic Data for Biochemistry and Biotechnology*, H.-J. Hinz, ed., (Berlin: Springer-Verlag, 1986).

[2] S. Photchanachai et al., *Biosci. Biotechnol. Biochem.* 66: 1635–1640 (2002).

[3] J.N. de Witt and G.A.M. Swinkels, *Biochim. Biophys. Acta* 624: 40–50 (1980).

Often food proteins are denatured during the course of processing operations, with the loss of their biological function. Their functionality, which is the design role that they play in the food, such as stabilizing a foam, or forming an edible film, may actually depend upon their denaturation. In the case of certain endogenous, degradative enzymes (see Chapter 6), denaturing these proteins may be essential to maintaining a food's stability. Thus, food chemists are often at least as interested in the denatured states of proteins as in their folded native state.

Thermal denaturation as in cooking or processing is probably the most important cause of protein denaturation in foods. Table 5.4 lists the denaturation temperatures of several important food proteins and their enthalpy change upon unfolding, as determined by DSC. An example with which everyone is familiar is the setting of egg whites as they are heated (Table 5.8). Surfaces can also be a significant source of conformational instability for proteins. The interface between an aqueous phase and air, glass, or even the stainless steel of cooking pots, acts like an extended hydrophobic surface. Many proteins tend to absorb onto such surfaces, often with some degree of denaturation, as the hydrophobic portions of the polypeptide rearrange so as to be in contact with the "hydrophobic" air or glass surfaces. Many types of processing involve high shear from whipping, homogenization, or shaking, which can lead to extensive protein denaturation through the formation of bubbles, with their large surface area, and the subsequent sorption of globular proteins onto this new large surface area. These sorbed proteins can substantially stabilize the drainage of liquid from the surface of these bubbles, thus leading to the formation of stable foams. Under proper concentration and shear rate conditions, whey proteins can be induced to adsorb with denaturation in large amounts onto air bubbles in an aqueous dispersion, with the resulting "foam" behaving much like an emulsion. This similarity in mouthfeel to emulsion behavior allows all-protein fat replacements to be manufactured from what is otherwise a waste byproduct of cheese making.

The Chain Configurational Entropy Contribution to the Free Energy of Folding

When a polypeptide chain "folds" up from a random coil to a specific conformation in a protein tertiary structure (or for that matter into a regular secondary structure such as a beta sheet or alpha helix), there is a considerable decrease in the configurational entropy of the chain, because in the random state it has many possible conformations, while in the regular structure it has only one conformation and is thus in a much more ordered, lower entropy state. This decrease in configurational entropy of the chain makes an unfavorable contribution to the overall free energy of folding for the polymer, which must be overcome by other favorable contributions if the chain is to spontaneously fold. These other contributions could include such things as a possible lower enthalpy due to improved hydrogen bonds or salt bridges, or an increase in the entropy of the solvent as the protein folds, counterbalancing the decrease in the entropy of the chain. One can estimate how much free energy must be gained from these other contributions by making a simple estimate of the energy cost of folding due to the loss of configurational freedom of the chain. This can be done using the Boltzmann equation for the definition of entropy,

$$S = k_{\mathrm{B}} \ln W \tag{5.2}$$

or on a molar basis,

$$S = R \ln W \tag{5.3}$$

where W is the number of accessible states of equal energy. Consider the case for the folding to a helix. The change in entropy for the folding process can then be written as the difference in the configurational entropy of the folded and unfolded chains,

$$\Delta S = S_{\mathrm{H}} - S_{\mathrm{C}} = R \ln W_{\mathrm{H}} - R \ln W_{\mathrm{C}} \tag{5.4}$$

$$= = R \ln \frac{W_{\mathrm{H}}}{W_{\mathrm{C}}} \tag{5.5}$$

where W_H is the number of states for the helix and W_C is the number of states for the random coil. If the helical region is 10% of the total allowed conformational space on the Ramachandran map (that is, if there are approximately 9 other allowed denatured conformers for each residue besides the folded one), then $W_H = 0.1\ W_C$. Thus, for a 100-residue polymer, the number of possible conformations would be 10^{100} in the random state but just $1^{100} = 1$ for the helix. Thus,

$$\Delta S_{chain} = R \ln \frac{W_H}{W_C} = R \ln \left(\frac{1}{10}\right)^{100} = 100 R \ln (0.1) \tag{5.6}$$
$$= -0.46 \text{ kcal/K·mol}$$

Thus the configurational free energy change at 37 °C (approximately body temperature) is

$$\Delta G_{chain} = \Delta H_{chain} - T\Delta S_{chain} = \Delta H_{chain} + 143 \text{ kcal/mol at 37 °C} \tag{5.7}$$

This is 1.43 kcal/mol per residue. For the chain to fold to the helical conformation, the other contributions to the total system free energy, such as the enthalpy terms and the solvent entropic change, must counterbalance this amount,

$$\Delta G_{total} = \Delta H_{chain} + \Delta H_{solvent} + \Delta H_{chain-solvent\ interaction} - T\Delta S_{chain} - T\Delta S_{solvent} \tag{5.8}$$

It was initially thought that a denatured protein was a random coil. That is, that the conformations of each individual peptide linkage adopted one of the several allowed low-energy conformations with roughly equal probability and made frequent transitions between these allowed conformations with roughly random probability, producing a coil with no specific or long-lasting structure, coiling, uncoiling, and recoiling like a wet spaghetti noodle in boiling water. In this picture, the conformation at any given instant is one of a vast number of possibilities, with the large family of such possible conformations representing a highly disordered, and thus high entropy, state with correspondingly low free energy. In contrast, as we have just seen, the folded native state would represent a low entropy situation, with only one conformation or a relatively few closely related conformations. A variety of experiments in recent years have shown that, in many cases, denatured proteins do not go to extended random coils, but rather they have simply lost the essential features of their tertiary conformation that give them their activity. Often they remain in a collapsed globular condition called a molten globule, which would reduce the entropic cost of folding from the value estimated above.

Because the ϕ angle of a proline residue is essentially fixed at roughly $-60°$, a proline residue has only two possible conformations, at around $(-60°, 145°)$ and $(-60°, -55°)$. As a result, a section of polypeptide chain containing a proline residue has many fewer possible conformations than one containing any other residue in that position. The reduction in conformational possibilities is even more significant because the residue preceding the proline cannot adopt the alpha helical conformation and thus also has fewer conformational options. As we have just seen, a denatured, random-coil polypeptide has a greater configurational entropy than the folded native state, neglecting the entropy arising from fluctuations and oscillations about that primary folded state, giving the unfolded state a free energy advantage over the folded state. Reducing the conformational options of the unfolded chain decreases this free energy advantage, and thus has the effect of stabilizing the folded state toward denaturation. Therefore, an appropriately placed amino acid substitution that replaces one of the other 19 standard amino acids with a proline residue, whether it results from a natural mutation event or is the result of site-directed mutagenesis through genetic engineering, could have the result of stabilizing the mutant protein toward thermal denaturation. The residue must be properly placed in the folded protein, because the presence of a proline can disrupt secondary structural motifs such as alpha helices and, in the wrong place, as in the middle of such a helix, could prevent the protein from folding up properly at all. Nonetheless, experiments placing proline residues in strategically selected positions, such as loops or reverse turns, have succeeded in raising denaturation ("melting") temperatures, but usually by only a few degrees. By similar arguments, replacing almost any residue in a polypeptide sequence by a glycine can in principle destabilize the folded state, because the Ramachandran map is so much more permissive for a glycine residue (Figure 5.10), resulting in a huge increase in the number of possible conformational options for the unfolded chain.

The alert reader at this point might be confused as to why a protein should denature when the temperature is raised. If the folding is driven by a net entropy increase (i.e., a positive ΔS), due to the release of structured water molecules when hydrophobic groups are buried inside the folded state, then raising the temperature should make the folded state more stable, not less, since

$$\Delta G = \Delta H - T\Delta S \tag{5.9}$$

From this argument, the higher the temperature, the more stable the folded state would be (that is, the more negative ΔG would be). The reason this is not so is that neither the ΔH nor the ΔS terms are independent of temperature. For example, the magnitude of ΔS decreases as the temperature is raised, because at high temperatures the water molecules have so much energy that they can fluctuate considerably even when adjacent to hydrophobic groups and thus suffer less rotational restriction relative to water molecules in the bulk. In addition, the structure of the bulk itself is less constrained by the residual hydrogen bonds at high temperatures (see for example Figure 2.18). The magnitude of the ΔH term also decreases, but not as fast as the $T\Delta S$ term, so that as the temperature is raised, the difference between the two contributions to ΔG decreases; the denaturation temperature is the temperature at which $\Delta H = T\Delta S$.

TERTIARY STRUCTURE OF GLOBULAR PROTEINS

Although several of the most important food proteins have extended structures such as the fibrous collagen and myosin II, and others such as the caseins and gluten may have no fixed conformation, probably the majority of food protein types have globular shapes on the molecular level. Understanding their functional properties in general requires understanding their conformational structure. Fortunately, great progress has been made in recent years in characterizing many important food proteins, and their structures are often freely available from the Protein Data Bank. Advances in computer graphics have also made it easy to visualize these structures and to use their information to understand how they function.

As an example, in Figure 5.23 is a so-called ribbon diagram of the small (58 amino acid residue) globular protein bovine pancreatic trypsin inhibitor (bpti), which controls the activity of the protease trypsin. The ribbon traces the backbone path by connecting the positions of the C^α carbon atoms. This is one of the smallest proteins that adopts a compact, globular conformation. In spite of its small size, this protein contains a short alpha helix, a segment of twisted antiparallel beta sheet, at least one reverse turn, and three disulfide bonds. It was also one of the first proteins whose structure was determined to high resolution by X-ray crystallography. For these and other reasons it was for a while one of the most studied of all proteins. Four water molecules enclosed inside the protein fold are also indicated. These molecules are an integral part of the folded structure and cannot be removed without denaturation.

Various conventions have been developed for characterizing the principal structural features of complex protein structures using simplified caricatures or "cartoons," such as illustrated in Figs. 5.24 and 5.25. These schematic drawings use ribbons to

Figure 5.23. The backbone conformation of the small 58-residue regulatory protein bovine pancreatic trypsin inhibitor, indicated by a green virtual bond connecting the alpha carbons of each successive residue. The sulfur atoms of the cysteine residues involved in this protein's three disulfide bonds are shown as yellow spheres. Four water molecules that are trapped in the interior of the folded protein are also shown.

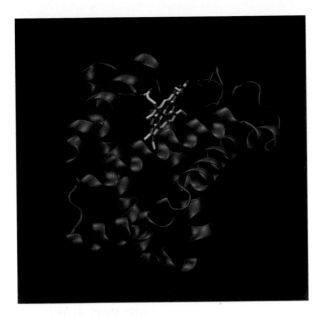

Figure 5.24. Schematic drawing of myoglobin, an almost all α-helical protein (F. Yang and G.N. Phillips, *J. Mol. Biol.* 256:762–774, 1996). The heme group is also indicated.

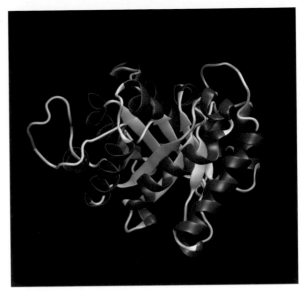

Figure 5.25. Schematic drawing of the backbone structure of triose phosphate isomerase, an α-β barrel protein. (P.J. Artymiuk et al., 1998 incomplete.)

represent secondary structural elements; a coil is used for alpha helices and arrows are used for beta sheet strands; the arrowheads then allow easy visual determination of whether two adjacent strands are parallel or antiparallel. Figure 5.23 is also a ribbon-type diagram, but where straight tubes are used to connect the α carbon atoms of each amino acid, with a change of direction marking the boundaries between the successive residues. In Figures 5.24 and 5.25, regions of the chain without a well-defined secondary structure are shown as narrow tubes approximately following the path of the chain in space. Several computer programs are now available that can prepare such cartoons automatically from the crystallographic coordinates for proteins.

MILK PROTEINS

Whey Proteins

As we saw in Chapter 3, milk is a complicated system with a number of components in the colloidal size range dispersed in an aqueous medium. As shown in Figure 3.2, on one level, milk is a dispersion of spherical fat globules in the micron size range in a fluid phase called the plasma. As summarized in Figure 3.27, the principal source of instability of this dispersion is due to the creaming that results from the difference in densities of the fat globules and the continuous plasma. Seen at a higher level of magnification, however (Figure 3.4), the plasma, which appears continuous on the micron size scale, is also a colloidal system of casein micelles dispersed in the continuous medium of the serum. As we will see in the next section, the casein micelles are composed of a family of nonglobular proteins defined by their aggregation behavior rather than by structural homology. As with the plasma, however, examination of the serum on a smaller size scale reveals that it is also not entirely continuous, but rather consists of a thermodynamically stable aqueous solution of several different types of globular proteins, including β-lactoglobulin, α-lactalbumin, serum albumin, and cryoglobulins. Also dissolved in the serum is the milk sugar, lactose (Figure 4.48).

In cheese making, milk is treated with bacterial cultures, usually *Lactococci* or *Lactobacilli*, which ferment the lactose of the serum into lactic acid, and in the process lower the pH. This lowering of pH results in a certain amount of aggregation of the casein micelles into loose clumps called curds. In most fresh cheeses, this acid clumping is the only aggregation that occurs, but in the majority of cheeses the enzyme chymosin, from rennet obtained from the fourth stomachs of calves (also called rennet; see Chapter 6), is used to further aggregate the casein particles, producing a firmer gel-like matrix. The fat globules are entrapped in the matrix of the aggregated caseins. The remaining fluid, which is drained away, is the whey, which is essentially the same as the serum, with the exception that for the rennetted cheeses it contains a peptide cleaved by proteolysis from one of the casein proteins. The soluble proteins in this fluid, such as β-lactoglobulin and α-lactalbumin, are called whey proteins. We will first survey the properties of the principal globular whey proteins and will then examine the casein proteins and how they are organized and how they change during cheese making.

β-Lactoglobulin: An Example of a Typical Globular Food Protein

β-Lactoglobulin (blg) is the most abundant **whey protein**. In many ways, it is a typical globular protein, with 162 amino acid residues and a molecular weight of 18,362 kilodaltons. Its secondary structure consists primarily of eight strands of antiparallel beta sheets with one extended alpha helix and three other short helical segments (Figure 5.26). Several genetic variants, differing slightly in amino acid sequence and called A through D, occur in the milk of the common dairy cattle breeds. The A form of blg molecules tends to aggregate as a function of pH. Its isoelectric point is at 5.4; under acid conditions, it exists in an equilibrium between monomers and dimers. At pHs below 3.5, it is primarily found as a monomer, but as the pH increases above 3.5 it tends to associate into a noncovalent dimer, and as the pH increases further in the range between 3.7 and 5.4, these dimers undergo further aggregation to octamers. Above 5.4, these octamers again dissociate into dimers, and at pHs above about 7.5 the dimers again dissociate into monomers. Only the A variant of blg undergoes these aggregations, and not the B, C, and D variants.

Blg contains five cysteine residues, with two intramolecular disulfide bonds and one free cysteine sulfhydryl group. One of these disulfide bonds is between Cys residues 66 and 160, while the other is between residues Cys 106 and 119. The nearby

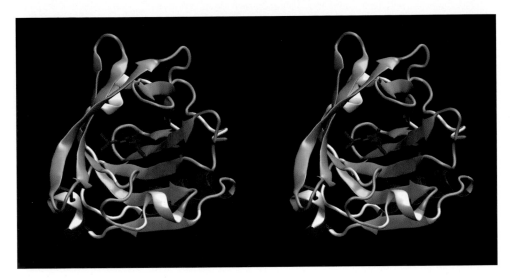

Figure 5.26. A stereo representation of the backbone folding pattern of the whey protein β-lactoglobulin. The alpha helices are colored purple, and the beta sheet strands are colored gold. Reverse turns are indicated in turquoise, and 3_{10} helices are shown in white. A retinol ligand molecule is also shown as a red stick figure bound into the inside of the binding pocket. This figure is a so-called wall-eyed stereo pair; if the observer looks at the left image with the left eye and the right image with the right eye, while merging the two images and focusing, the image will appear three dimensional.

residue 121 is also a cysteine. The free cysteine leads to intermolecular cross-linking when the protein is denatured. Blg is moderately thermostable, denaturing irreversibly between 70 and 90 °C.

The function of blg in milk is unknown, and it is possible that it has no specific function beyond providing essential amino acids. It is not present in the milk of humans and other primates. The protein is highly homologous to retinol-binding protein, a transport protein which complexes with vitamin A in the liver and carries it through the aqueous blood plasma, where vitamin A is insoluble and subject to oxidation, and delivers it to cells where it is needed. Both proteins are members of a general class of small transport proteins called lipocalins. These proteins have a so-called **β-barrel** conformation, with the antiparallel β-sheet strands twisted into the shape of a calyx (somewhat like an ice cream cone), with an inner pocket lined with nonpolar functional groups so that it has a high affinity for hydrophobic ligands such as retinol (Figure 5.26). It has been demonstrated that blg binds vitamin A, but it is not clear that this protein would be necessary to transport retinol in milk, because the vitamin could easily reside in the fat globules. It is interesting in this context, however, that blg is resistant to acid hydrolysis and survives the stomach intact (Papiz et al. 1986).

Whatever its biological function, however, it should be remembered that in many cases the functional uses of a protein in a manufactured food may have little to do with its function in vivo. This is the case with blg, which has found a number of applications in the food industry (and a good thing too, because it is a significant waste byproduct of cheese making). One use of this protein is as a carrier for hydrophobic molecules in aqueous formulations. The protein can also be used in its denatured form as a protein emulsifier to help form or stabilize manufactured emulsions. Because it tends to denature at surfaces, and to subsequently undergo cross-linking by the formation of intermolecular disulfide bonds under oxidizing conditions, blg can quite effectively stabilize foams. This same tendency toward cross-linking allows whey protein concentrates to be used to form thin-film edible coatings.

Other Whey Proteins

The second most abundant whey protein, **α-lactalbumin**, which makes up less than 4% of the protein content of milk, is a sequence and structural homologue of lysozyme. It has 123 amino acids and a molecular weight of 14,200. Like lysozyme, the native conformation of α-lactalbumin is primarily made up of alpha helices, with two short strands of antiparallel beta sheet, organized into two lobes separated by a deep cleft. In mammary gland cells, it interacts with the enzyme galactosyl transferase to assist in the transfer of galactose from UDP galactose to glucose in the synthesis of lactose, but is not itself the catalytic enzyme. Two genetic variations of the lactalbumin proteins occur in dairy milk, called A and B. These two variants differ in a single amino acid substitution at residue 10. In the A variant, residue 10 is a Gln, and in the B variant it is an Arg. Only the A variant of α-lactalbumin is found in the milk of European breeds of cattle, but both forms are found in the milk of Indian cattle. This protein binds calcium ions stoichiometrically, one per protein molecule (Figure 5.27). It is much more heat stable than blg and does not denature after extensive cooking at almost 80 °C. It has four stabilizing disulfide bonds, but no free, unpaired cysteines.

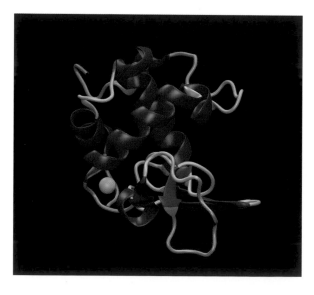

Figure 5.27. The conformational structure of α-lactalbumin. The folding of this protein is very similar to that of lysozyme (Figure 5.50), with which it is homologous. In this conformational cartoon, alpha helices are colored red, beta sheets are colored dark blue, reverse turns are aqua colored, and coil regions are colored orange. The green sphere illustrates the position of a bound calcium ion. (A.C. Pike, K. Brew, and K.R. Acharya, *Structure* 4:691–703, 1996.)

Figure 5.28. The conformational structure of bovine lactoferrin. The bound ferric iron atom is shown in red. (N. Singh, S. Sharma, and T.P. Singh, unpublished data.)

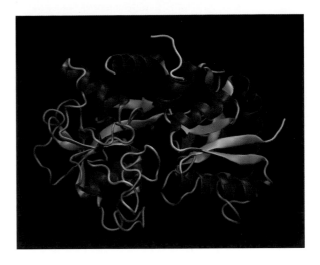

Serum albumin, a blood serum protein, is also present in milk, but its function, if any, is equally unknown. It has recently been shown that whey from bovine milk also contains nearly 300 other soluble proteins in low concentrations (Le et al. 2011), including many antibodies, proteases, and protease inhibitors. The various immuno-globulins apparently serve an antimicrobial effect. One of these, IgM, acts as an agglutinin against *Streptococci* and, when it precipitates onto fat globules at low temperatures, leads to the cold agglutination of these globules. In the dairy literature it is thus often referred to as a **cryoglobulin**. Whey also contains **lactoferrin** (Figure 5.28), an antimicrobial defense glycoprotein homologous to ovotransferrin (see Figure 5.48) that strongly binds ferric iron (Fe^{3+}). The concentration of lactoferrin in bovine milk is low (1% or less of total protein), but it is much higher in human milk (~15%). Lactoferrin is sold commercially as a nutraceutical or dietary supplement as an antioxidant and for its putative ability to boost immune function.

Caseins: Random Coil-Like Chains

The caseins are a group of proteins found in milk that are not similar to one another in sequence (that is, they are not homologous), but which are classified together because of their functional behavior. The caseins are operationally defined as those proteins that precipitate out of milk when the pH is lowered below 4.6. All of the caseins share the important feature that they are not globular (or at least not entirely globular), without well-defined tertiary structures. There are several different types of casein molecules. The four major types are called α_{s1}-, α_{s2}-, β-, and κ-casein. The dairy literature also often refers to γ-caseins, which are various C-terminal fragments of β-casein that result from cleavage at several positions in the sequence by the enzyme plasmin. In milk the casein proteins are organized into roughly spherical particles on the colloidal size scale that are suspended in the milk plasma as a

Table 5.5 Properties of bovine milk proteins

Property	Caseins				Whey proteins		
	α_{s1}	α_{s2}	β	κ	α-Lactalbumin	β-Lactogloblin	Serum albumin
molecular weight	23,614	25,230	23,983	19,023[a]	14,176[a]	18,363	66,267
amino acids	199	207	209	169	123	162	582
prolines	17	10	35	20	2	8	34
cystines	0	2	0	2	8	5	35
disulfides	0	?	0	?	4	2	17
phosphates	8	11	5	1	0	0	0
% charged residues	34	36	23	21	28	30	34
net charge/residue	−0.10	−0.07	−0.06	−0.02	−0.02	−0.04	−0.02
distribution	uneven	uneven	very uneven	very uneven	even	even	
A_{280}	10.1	14.0	4.5	10.5	20.9	9.5	6.6

Source: Data from P. Walstra and R. Jenness, *Dairy Chemistry and Physics* (New York: Wiley-Interscience, 1984).
[a]Does not include carbohydrate residues.

relatively stable dispersion. For historical reasons these particles are called casein "micelles," but they are not true micelles of the type discussed in Chapters 2 and 3. The role of the casein micelles is to keep the high concentration of proteins in the milk from making the fluid too viscous, and to provide calcium and phosphate as nutrients without the danger of these ions precipitating out as an insoluble salt. Some of the general properties of the caseins are summarized in Table 5.5 and compared with those of the globular whey proteins. Information about concentrations of proteins in milk appears in Table 5.6.

As noted, none of the casein proteins is thought to have compact, globular tertiary conformations. Both κ and β-casein contain an unusually large number of proline residues scattered throughout their chain sequences. The high proportion of irregularly but fairly evenly distributed proline residues in these two caseins prevents them from folding up, and so they may be somewhat like more extended random coils. In addition, in β-casein the charged and polar residues are concentrated at the N-terminal end of the chain while the rest of the chain, with a large number of nonpolar residues, is predominately hydrophobic, giving β-casein something of the character of a detergent or surfactant molecule. Like surfactant molecules, pure solutions of β-casein will form true micelles, exhibiting a critical micelle concentration.

κ-Casein also has a very polar, hydrophilic C-terminal end. In part this is because this end of the molecule has a number of Thr and Ser residues with polar hydroxyl groups (see Figure 6.11), as well as other polar residues. More significantly, the single polar phosphate group of κ-casein is esterified to Ser 149 in this region. Finally, κ-casein is an example of a glycopeptide or glycoprotein, one in which an oligosac-

Table 5.6 Concentrations of proteins in bovine milk

Protein	Concentration in milk		Percentage of total protein (w/w)
	g/kg	mmol/m^3	
total protein	33.0	~1490	100.0
total casein	26.0	1170	79.5
whey proteins	6.3	~320	19.3
fat globule membrane proteins	0.4	—	1.2
α_{s1}-casein	10.0	440	30.6
α_{s2}-casein	2.6	110	8.0
β-casein	9.3	400	28.4
γ-casein	0.8	40	2.4
κ-casein	3.3	180	10.1
α-lactalbumin	1.2	90	3.7
β-lactoglobulin	3.2	180	9.8
blood serum albumin	0.4	6	1.2
immunoglobulins	0.7	~4	2.1
miscellaneous	0.8	~40	2.4

Source: Data from P. Walstra and R. Jenness, *Dairy Chemistry and Physics* (New York: Wiley-Interscience, 1984).

charide is covalently linked to one of the side chains of a protein. In the case of κ-casein, a branched tetrasaccharide, shown in Figure 5.29, is bound to the side chain of Thr 133. In this figure, NANA is *N*-acetyl neuraminic acid (see Figure 4.34), Gal is galactose, and GalNAc is *N*-acetylgalactosamine. The functional role of this oligosaccharide is not completely known, but one function must be to make the C-terminal end of the κ-casein more hydrophilic, facilitating its role in the steric stabilization of the micelles.

κ-Casein is an example of an *O*-linked glycoprotein, in which the oligosaccharide is bound through the anomeric group of the reducing end to the oxygen of a serine or threonine. κ-Casein has several other threonine and serine residues in its C-terminal end, which is thought to be more exposed to the serum solution, and in human milk these residues are all sites for *O*-linked glycosylation with several variant oligosaccharides, producing a large number of possible combinations of glycosylation, which can be separated by gel electrophoresis. While it was earlier thought that bovine κ-casein was glycosylated only at Thr 133, more recent gel electrophoresis studies, combined with the use of the protease pepsin (see Chapter 6) to cleave the bovine κ-casein into smaller fragment polypeptides, has identified several variants in bovine milk as well, with from 0 to 3 oligosaccharides on their different C-terminal threonine residues (Holland, Deeth, and Alewood 2005). Trisaccharide variants of

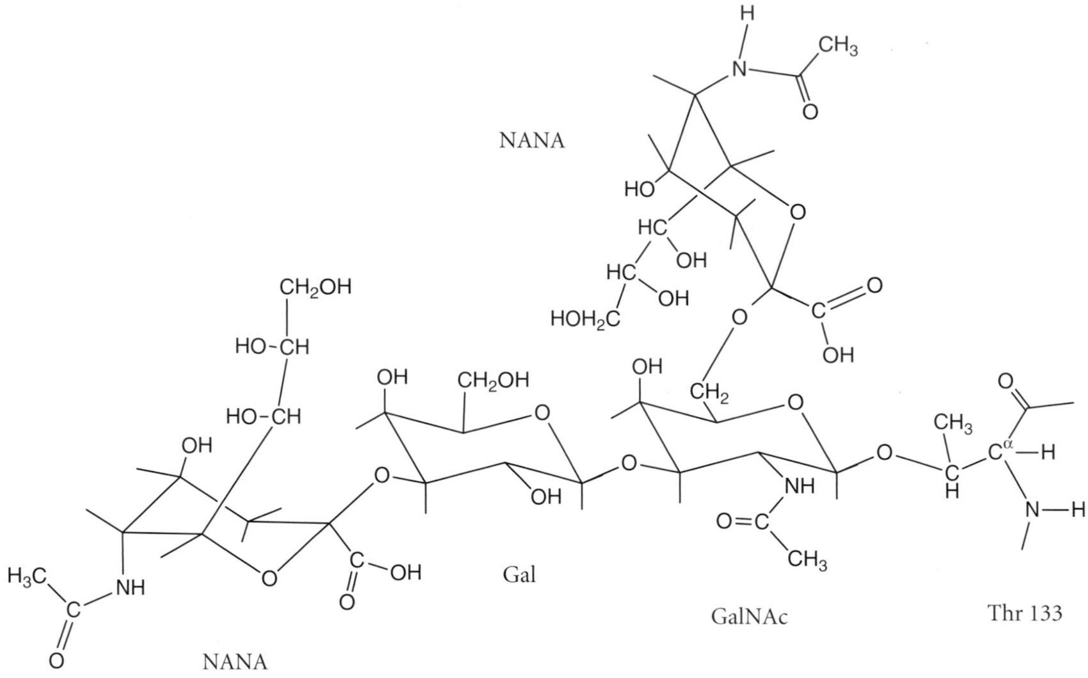

Figure 5.29. The structure of the oligosaccharide bound to residue Thr 133 of κ-casein in bovine milk.

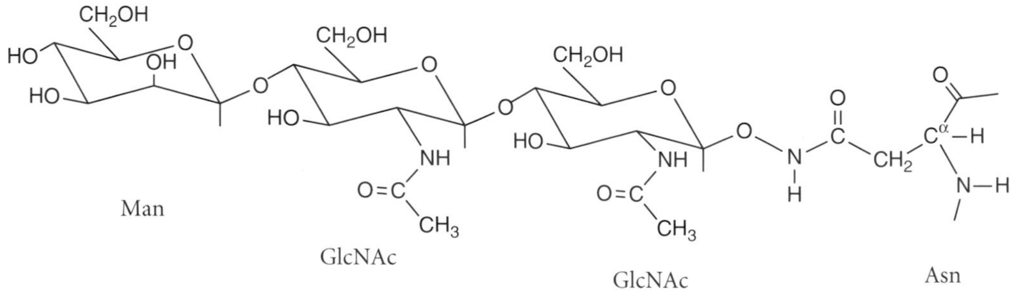

Figure 5.30. The structure of the most typical trisaccharide core for *N*-linked glycoproteins.

the oligosaccharide shown in Figure 5.29 also occur in which one or the other NANA group is missing.

There are also glycoproteins of a different type in other foods that have *N*-linked oligosaccharides, where a sugar is linked to the protein through the nitrogen atom of an asparagine (Asn) residue (Figure 5.30). This linkage type resembles a peptide bond. Many naturally occurring *N*-linked glycoproteins have oligosaccharides with a common trisaccharide core, Manβ(1→4)GlcNAcβ(1→4)GlcNAc-Asn, shown in Figure 5.30, with various branches off of these core sugars. Ovalbumin from eggs (see Figure 5.45) is a food example of such a glycoprotein with an *N*-linked oligosaccharide.

Casein monomer solutions are relatively stable; for example, boiling causes no change. Casein is sensitive to Ca^{+2}; small amounts can lead to precipitation by screening the ester phosphate groups. Adding $CaCl_2$ raises Ca^{+2} activity and colloidal calcium phosphate and decreases pH, all detrimental to stability. Adding NaCl increases stability because it decreases calcium phosphate content in the micelles (Na^+ competes with the Ca^{+2} somewhat and causes the micelle to fall apart, the opposite of what one might expect from DLVO theory). Lowering the pH leads to a dissolution of the colloidal calcium phosphate and thus some dissolution of micelles. As the pH is lowered below 5.5–5.2, the zeta potential goes to zero and all colloidal phosphate is lost. Lowering the pH to the isoelectric point of 4.6 will cause casein to precipitate out, because it loses its charge, and its high hydrophobicity then leads to aggregation. Indeed, this precipitation is used to define the caseins; as we have already observed, the caseins are considered to be those proteins which precipitate out of milk when the pH is lowered below 4.6.

For many years, the accepted picture of the structure of the casein micelle was based on the so-called subunit or submicelle model, in which the larger "micelles" consist of a number of smaller subunits that resemble true micelles because their behavior is assumed to be dominated by the β-casein components. In this model, in raw milk at its normal casein concentration, pH, and ionic strength but low calcium concentration, small aggregates are thought to form that are roughly 10 to 20 nm in size which contain between 10 and 100 casein molecules. These submicelle aggregates resemble true micelles, with a hydrophobic core and a hydrophilic surface. All of the different caseins are present in these submicelles, possibly in variable ratios; the overall molar ratios of the caseins in the micelles is $\alpha_{S1}:\beta:\kappa:\alpha_{S2}$ of 8:8:3:2. Some of these submicelles may contain no κ-casein; the amount of the κ-casein chains in the submicelles appears to be random. The proteins in these subunits are held together by hydrophobic forces and by colloidal calcium phosphate clusters bridging between phosphate groups on different casein molecules. These submicelles aggregate by hydrophobic association into the larger aggregates usually called "micelles", with calcium phosphate also acting as the "glue" cementing the submicelles together. The presence of the κ-casein serves to halt the aggregation through steric stabilization before total coagulation occurs. The hydrophilic C-terminal ends of the κ-caseins stick out into the serum like flexible hairs, behaving as random-coil polymer chains. Submicellar surfaces with these molecules on them would then be sterically stabilized. In this picture, aggregation would continue at surfaces uncoated by κ-casein chains until, by random arrangement, all of the outer surface of the aggregate casein micelle particle was covered by the κ-casein hairs. The majority of the micelle's κ-casein would then be on the outside. These κ-casein chains, being one molecule thick, are too small to be seen by an electron microscope; the effective thickness of the "hairy" layer would be around 5 nm. Only about half of the volume of such an open structure would be made up of casein submicelles; the rest would consist of serum in the interstitial spaces. This model is illustrated schematically in Figure 5.31.

Casein micelles are highly polydisperse, with diameters generally between 10 and 300 nm, and with the distribution in size being caused by the random arrangements

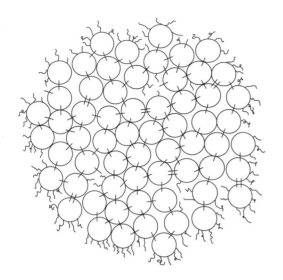

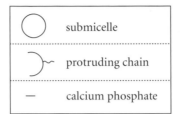

○   submicelle

)~ protruding chain

—   calcium phosphate

Figure 5.31. Schematic illustration of the submicelle model for the structure of the casein micelle. (Figure adapted and redrawn from P. Walstra and R. Jenness, *Dairy Chemistry and Physics*. New York: Wiley-Interscience, 1984.)

that would result from the association of submicelles partially covered with the κ-casein protective chains. Some workers have found a small number of very large particles ~800 nm in diameter. The size distribution is sharply peaked at low particle size due to a large number of small submicelles, ~1 to 2% of the total casein at room temperature. The number average diameter is 25 nm (65 nm if particles less than 20 nm are excluded).

In this subunit model there is a dynamic equilibrium in milk between the micelles and their constituents: between free casein molecules and submicelles, between dissolved and colloidal calcium and phosphate, and between free submicelles and micelles. Casein micelles are dispersed into smaller units by any treatment that dissolves a considerable portion of the calcium phosphate, including adding a calcium binder such as sodium citrate or dialyzing against a calcium-free solution. Raising the pH causes calcium phosphate to precipitate, leading to partial disintegration of the micelle as the caseins go back into solution. Casein micelles also disintegrate when the submicelles dissociate due to increased solubility of the casein molecules, for example, from the addition of large amounts of urea or guanidinium chloride or small amounts of sodium dodecyl sulfate. Lowering the temperature produces some changes, because hydrophobic interactions become weaker, causing part of the casein, particularly β-casein, to dissociate.

In the subunit model of casein micelle structure, treatment of the milk with the chymosin enzyme clips off the hydrophilic tails of the κ-casein molecules by hydrolyzing the peptide bond between residues Phe 105 and Met 106 (see Chapter 6). The C-terminal polypeptide then floats away into the serum as a fragment called a **macropeptide** or glycomacropeptide. The N-terminal portion of the κ-casein remains embedded in the micelle, but without the steric stabilization imparted by their hydrophilic tails sticking out into solution, the micelles clump together, adhering through hydrophobic interactions and further calcium phosphate bridging between submicelles. This clumping of micelles creates a matrix of highly associated casein proteins, with the larger fat globules entrapped between the casein strands formed by this association.

This older submicelle model for the nature of casein aggregates is consistent with nearly all of the known behavior of caseins in milk and provides a good explanation for the formation of curds when treated with chymosin (Chapter 6). The publication of high-magnification electron micrographs of casein micelles that had the appearance of a "fuzzy" clump of grapes, much like the schematic illustration in Figure 5.31, seemed to definitively confirm this standard model.

In recent years, however, a number of doubts about this model have arisen. In the scientific community it is now thought that the seemingly confirmatory structure observed in the electron micrographs was actually an artifact of the manner in which the sample was prepared. Newer electron microscopy studies have failed to produce similar images, and it is no longer universally believed that casein "micelles" have a substructure of smaller discrete particles like the postulated submicelles. No new consensus model has emerged to replace the previously accepted description.

However, one new model that has gained wide acceptance postulates that the interior of the casein micelle consists of a complex lattice of the α_{S1}, α_{S2}, and β caseins intertwined or entangled with one another, perhaps somewhat like a gel, interacting by hydrophobic association and again being cemented together by the interaction of their phosphate groups with colloidal calcium phosphate clusters (Holt et al. 2003; Holt 2004; Horne 2006; McMahon and Oommen 2008). As in the subunit model, this aggregate ceases to grow when the outer surface is covered with κ-casein C-terminal chains sticking out into solution and imparting steric stabilization. This new model is also consistent with the known properties of casein micelles, but its validity has not yet been conclusively established, and the structure of the casein "micelle" is still an active area of research.

SWEET-TASTING PROTEINS

The majority of proteins are tasteless, but several proteins have been discovered that are intensely sweet. These included **thaumatin**, **monellin**, curculin, **brazzein**, pentadin, and mabinlin. Hen egg white lysozyme (HEL or HEWL) has also been reported to be sweet (see Figure 5.50). Three-dimensional structures from crystallography or NMR are available for several of these, including the two best known, thaumatin and monellin. Most of these proteins contain no carbohydrates bound to them, and their sweet taste arises solely from the three-dimensional arrangement of their amino acids, because, as with most protein functionalities, their sweet taste is lost if the proteins are denatured. Such proteins could potentially be quite useful as sweeteners in diet drinks and products for diabetics. Several of these proteins have been isolated from the fruits of West African plants, but neither the plants nor the proteins appear to be closely related. Each of the proteins whose structure has been determined by crystallography or NMR has a different folded pattern and no significant sequence homology. Monoclonal antibodies have been found that cross-react with several of the proteins, suggesting a similarity in some part of their structure that is presumably responsible for the sweet taste, but as of this writing, the exact details of the sweet taste epitope are still uncertain.

Thaumatin is a 207–amino acid globular protein found in the fruit of the West African katemfe plant, *Thaumatococcus daniellii*. This protein is from 2,000 to 3,000 times sweeter than sucrose on a weight basis and almost 100,000 times sweeter on a mole basis. Thaumatin is an all β-sheet protein that contains eight disulfide bonds and a prominent cleft lined with positively charged lysine and arginine residues (Figure 5.32). Because of its extensive disulfide cross-linking, thaumatin is stable at elevated temperatures and over a range of pHs, which makes it attractive for commercial applications. However, it aggregates above 70 °C and loses its sweetness. It can also be freeze-dried without loss of activity. It is quite soluble in both water and alcohol.

Apparently, some of the eleven lysine residues in this protein play a role in its sweet taste, because the sweetness can be modified by acetylation of the ε-amino groups of these residues. There are at least five different related forms of thaumatin. The best studied of these is thaumatin I (Figure 5.32). In this form, five of the lysine residues, along with three arginine residues, mostly clustered on the protein surface around the cleft and on the "wings," or lobes, bracketing the cleft, have been identified using site-directed mutagenesis studies, coupled with taste panels, as important for the sweet taste of the protein (Ohta et al. 2008). One residue in particular, Arg 82, was found to be especially important; when it was replaced with alanine, the resulting mutant thaumatin was 25 time less sweet than the wild-type protein.

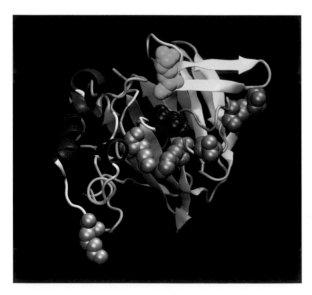

Figure 5.32. The conformation of thaumatin. The essential Arg 82 residue is shown in red, and Lys 67, which also has a significant effect on sweetness, is shown in green. The other important Lys residues are shown in pink, and the other important Arg residues are shown in orange. (S.H. Kim, A. de Vos, and C. Ogata, *Trends Biochem. Sci.* 13:13, 1988.)

Thaumatin is marketed by the Talin Food Company of the United Kingdom, under the trade name of Talin. It is used as a flavor enhancer as well as a sweetener. Its sweetness is synergistic when combined with other sweeteners. This protein is regarded as safe for human consumption and has been used to sweeten chewing gum and milk products. On the negative side, thaumatin has a licorice-like aftertaste that fades very slowly.

Monellin (named after the Monell Institute in Philadelphia, a research institute focusing on sensory science) is a smaller protein complex (MW ~11,500 Da), also from a West African fruit, the so-called serendipity berries of the *Dioscoreophyllum cumminsii* plant. Monellin consists of two separate, noncovalently bound polypeptide chains labeled A and B. The A chain is 44 amino acids in length, and the B chain is 50 residues in length. The conformation consists of an extended β-sheet made up of four antiparallel strands, with a single α-helix perpendicular to the direction of the sheet strands. This protein is also 2,000 to 3,000 times sweeter than sucrose on a per weight basis. Monellin probably has little commercial potential because it has low thermal and pH stability and undesirable features in its taste profile (it takes several seconds for the sweet taste to register, which then persists for up to an hour). Genetically engineered variants of monellin have been produced in which the two chains are fused into a single polypeptide chain. The conformation of one of these proteins, which is essentially the same as for natural monellin, is shown in Figure 5.33.

The folded conformation of the protein brazzein has also been determined, by two-dimensional NMR in solution, rather than by X-ray crystallography (Caldwell et al. 1998). This protein also comes from the fruit of a West African plant, *Pentadiplandra brazzeanna*, and has a clean, sweet taste with no lingering aftertaste. It is from 500 to 2,000 times sweeter than sucrose on a weight basis. Brazzein is the smallest

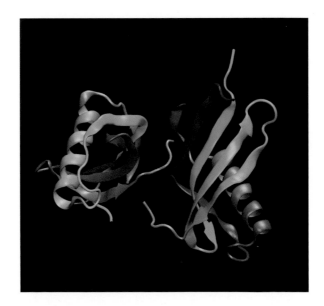

Figure 5.33. The conformation of single-chain monellin (there are two molecules per asymmetric unit, seen at different angles). (J.R. Somoza et al., *J. Mol. Biol.* 234:390–404, 1993.)

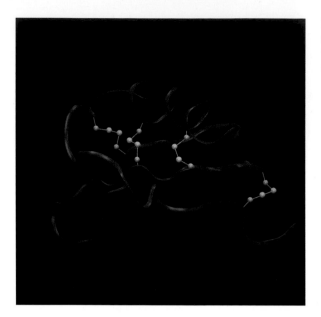

Figure 5.34. A ribbon diagram of the backbone conformation of brazzein, also illustrating the side chains for the eight Cys residues involved in the four stabilizing disulfide bonds.

of the sweet proteins, with a single chain containing only 54 amino acid residues. This protein is one of the most thermostable of the class due to a network of four disulfide bonds; sweetness is retained even after being heated to 98 °C (still lower than cooking temperatures, however) and being held at that temperature for two hours. This small globular protein contains a single helical segment and three strands of a twisted antiparallel beta sheet (Figure 5.34). Mutagenic studies suggest that certain specific charged residues are also important in the sweetness of this protein as well (Lee et al. 2010; Yoon et al. 2011). Interestingly, the protein can be made significantly sweeter by mutants that introduce new arginine resides (Yoon et al. 2011).

Yet another protein that affects sweetness has been identified that differs considerably from the others, and which has been the subject of growing interest in the food industry. This protein is **miraculin**, found in the small, ellipsoidal red miracle berry or miracle fruit from another West African plant, formerly called *Synsepalum dulcificum,* but now called *Richadella dulcifica*. Miraculin was first identified in 1968 (Kurihara and Beidler 1968). This globular glycoprotein has 191 amino acid residues, and is *N*-glycosylated at Asn 42 and Asn 186 with incompletely characterized oligosaccharides, which are known to contain mannose, galactose, fucose, xylose, and *N*-acetylglucosamine. The active protein is a dimer of two globular modules linked by a disulfide bond.

The taste properties of this protein are quite unusual. It has no strong taste of its own and is not sweet by itself, but it has the remarkable ability to change the perception of acid-triggered sourness to a strong sweet taste. It can thus turn sour-tasting foods to sweet ones. Apparently, it accomplishes this seemingly miraculous feat (the origin of the name) by binding in some way to the protein(s) of the sweet-taste T1R1-T1R3 receptors and perhaps changing their conformation or specificity in such a way that they are triggered by acid. Experiments have found that the glycosylation is not necessary for the sweet taste stimulation effect, but the dimer structure is required; the individual monomers produced by reducing the disulfide bond do not function to modify acid sourness to sweetness.

Miraculin has long been used in West Africa to sweeten traditional local foods. Eating a single berry before consuming sour foods such as lemons and other citrus fruits will render them sweet tasting. The effect is persistent and can last for up to an hour, and routinely persists for half an hour. Miraculin has already been approved for human consumption in Japan and is being marketed in tablet form. The ability to induce a sweet taste is dependent on the conformation of the protein, which unfortunately denatures and loses its tertiary structure upon heating up to or above 100 °C during cooking. The native conformation of this protein has not yet been reported from crystallographic or NMR studies, but it has been cloned and expressed in lettuce (Sun et al. 2006), and transgenetic tomatoes (H.-J. Sun et al. 2007). The potential for the use of miraculin as a low-calorie sweetener is enormous, and active research is ongoing.

Another taste-modifying protein called curculin, or neoculin, has been isolated from berries of *Curculigo latifolia*, which is native to Malaysia. Like miraculin, it has the capacity to cause acid/sour-tasting foods to be perceived as sweet, but unlike miraculin it also has a sweet taste of its own. Also like miraculin, this protein consists of two disulfide-linked subunits, but unlike miraculin, the subunits are different, with one being acidic and one basic (Shimizu-Ibuka et al. 2006). The subunits are not separately taste-active. Curculin is homologous with several mannose-binding lectins from monocot species such as garlics and daffodils. The subunits have about 74% sequence identity to one another and are predominantly beta sheet.

BOTULINUM TOXIN

Botulism is caused by consuming food contaminated with the neurotoxin produced by the anaerobic bacterium *Clostridium botulinum*. As this organism grows, it produces a metalloprotein endopeptidase, homologous to the tetanus toxin, which is excreted into the extracellular environment. This protease cleaves a protein involved in neurotransmission in mammals, rapidly blocking the release of the neurotransmitter acetylcholine at the synaptic junction, producing paralysis and death. Untreated, the death rate can be above 60%, but with aggressive treatment, sometimes involving months on a respirator, survival rates can be much higher. A trivalent horse serum antitoxin, much like snake **antivenins** or antivenoms, is available against serotypes A, B, and E (see below). This protein is one of the most toxic poisons known and is the most toxic found in nature (see Table 11.1). Fortunately, the botulinum protein is heat labile and denatures irreversibly at high temperature, so that even foods that have been extensively contaminated by *Clostridium botulinum* as the result of incomplete sterilization before canning can, in principle, be rendered safe by boiling for 10 to 15 minutes (however, *any* food that is suspected of being contaminated with *botulinum* toxin should be discarded). While botulinum toxin is easily irreversibly denatured by heating, the bacterial spores are much more heat stable and require boiling at 100 °C or above for long periods in order to render them inactive.

Classic botulism is an intoxication, rather than an infection, in that the symptoms are caused by ingesting the toxic protein rather than the organism. A rare variation occurs in infants up to 20 weeks of age who have been fed honey contaminated with *Clostridium botulinum* spores. These spores grow in the gut, producing the toxin and the associated disease symptoms. Only infants are affected because of their lack of a competing intestinal flora and higher gastric pH. This form of botulism is generally not fatal. A number of different serotypes of *C. botulinum* toxin are known, labeled A through G. Types C and D rarely cause disease in humans.

The botulinum toxin protease contains an essential zinc atom in its catalytic site. The enzyme is synthesized as a single polypeptide chain, but is posttranslationally cleaved by another protease, producing two chains, one with a molecular weight of about 100 kDa that is about twice as long as the other. The two polypeptide fragments remain covalently linked by a disulfide bond between the heavy and light chains. The zinc-binding site and the active site for the protease activity are located on the light chain of the molecule. Figure 5.35 shows a schematic illustration of the principal structural features of the botulinun toxin. As is the case with insulin, the cleaved botulinum toxin chain is unable to refold once denatured, as in cooking. The post-translationally modified protein is kinetically trapped in a conformation that was the lowest free energy state for the original intact protein but not for the modified version; that is, the missing bond is necessary for proper folding. The three-dimensional conformation of the light chain of serotype B has been determined by crystallographic methods and is shown in Figure 5.36.

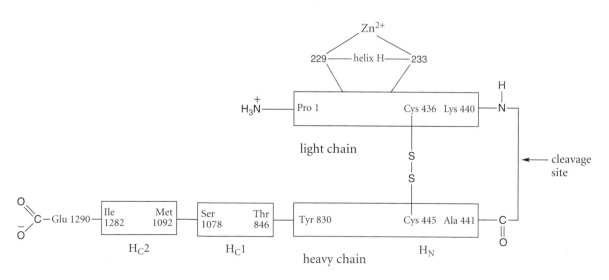

Figure 5.35. A schematic illustration of the botulinum toxin.

Figure 5.36. The conformational structure of the catalytic light chain of the serotype B botulinum toxin has been determined by X-ray crystallographic diffraction studies. The zinc atom is shown in gray. (M.A. Hanson and R.C. Stevens, *Nat. Struct. Biol.* 7:687, 2000; and M.A. Hanson et al., *J. Am. Chem. Soc.* 122:11268, 2000.)

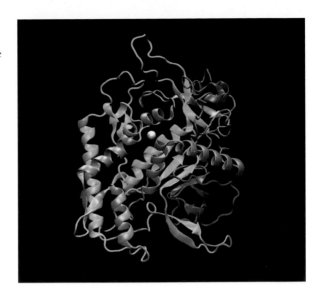

In the human body, the C-terminal portion of the toxin's heavy chain "recognizes" and binds to a protein receptor on the surface of the axon terminals, where it is incorporated into the cell by endocytosis. The light chain containing the catalytic site is then incorporated directly into the cytoplasm, where it cleaves the so-called SNAP-25 protein (the 25-kD synaptosome-associated protein) involved in the release of the neurotransmitter acetylcholine necessary for nerve impulse transmission, leaving the muscle unable to contract for months.

LEGUME PROTEINS

Soy Proteins

The seeds of the legumes (plants of the Fabaceae family), which include the beans, peas, lentils, and peanuts, contain large amounts of protein and thus are important in balanced vegetarian diets as a source of amino acids. Soybeans are approximately 35% protein, while peas and lentils are only about 25% protein. The United States is the largest producer of soybeans in the world, and soybeans are the second largest crop grown in the United States. Soy proteins are used in a large variety of products, from tofu and various texturized meat substitutes to miso, soy sauce, and soy milk. The proteins of legumes in general tend to be deficient in the sulfur-containing amino acids. Soy protein is low in methionine, and also tryptophan, which are essential amino acids, but it is a good source of lysine, which makes it complementary to maize protein (see below).

The storage proteins of soybeans are contained in small grains or granules about 2 to 20 μ in diameter. The proteins are generally classified into so-called Osborne fractions, based on their solubility behavior; these fractions are called **albumins** (water soluble), **globulins** (salt water soluble), **prolamins** (soluble in ethanol), and

glutelins (insoluble). About 10% of soy protein is made up of albumins, which are poorly characterized. The globulins of soybeans make up about 90% of the total protein content. The globulin fractions of all of the legumes can be further separated by centrifugation into two major fractions, called 7 S and 11 S (where S is the Svedberg, a measure of their sedimentation velocity), and two minor fractions. The major protein constituent of the 7 S fraction of legumes in general is called **vicilin**, and in soybeans in particular is called **β-conglycinin**. The main 11 S globulin in legumes in general is called **legumin**, and in soybeans in particular is termed **glycinin**. In soybeans the 7 S fraction makes up about a third of the total soy protein, and the 11 S fraction makes up about 40% of the total. The two minor globulin fractions are a 2 S component and a 15 S fraction. The 2 S fraction makes up from 6 to 20% of the total protein and contains the soybean trypsin inhibitors, including the **Kunitz** and **Bowman-Birk inhibitors**, cytochrome c, and several other proteins. The component molecules of the 15 S fraction of soy protein, which make up about 10% of the total, are unknown.

Table 5.7 shows the principal proteins found in soybeans. Note that the terminology is somewhat inconsistent and potentially confusing.

The conformation of vicilin from the 7S fraction of jack beans has been determined by X-ray crystallography (Figure 5.37). Because all of the legume vicilins are highly homologous, it is quite probable that the conformation of the soybean protein (conglycinin) will be similar. The structure of the soybean glycinin homotrimer has also been solved by X-ray crystallographic studies (Adachi et al. 2001). The glycinin is very large, with a molecular weight of 380,000 Da, and a hexameric structure (Figure 5.38). The glycinin complex consists of five subunits, each made up of two polypeptide chains, an acidic chain with a mass of 32,000 Da and a smaller basic chain with a mass of 20,000 Da. The two chains are linked together covalently by disulfide bonds; glycinin subunits have two or three disulfide bonds, while β-conglycinin has none. These subunits consist of two jellyroll β-barrel glob-

Table 5.7 Principal soy proteins, divided according to their sedimentation behavior

Globulins (soluble in salt water) 90% of total soy protein						Prolamins (soluble in ethanol)
2 S (~6–20% of total soy protein)			7 S (~30% of total soy protein)		11 S (40% of total soy protein)	15 S (10% of total soy protein)	
Kunitz inhibitor	Bowman-Birk inhibitor	cytochrome c	β-conglycinin (a vicilin)	agglutinin	glycinin (a legumin)	unknown proteins	hydrophobic seed protein

The majority of these proteins are globulins, although one prolamin is also known.

Figure 5.37. A Richardson-type cartoon representation of the backbone conformation of vicilin from jack beans, which is homologous to conglycinin. (T.P. Ko, J.D. Ng, and A. McPherson, *Plant Physiol.* 101:729, 1993.)

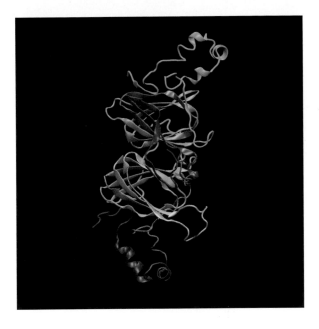

Figure 5.38. A stereo view of the crystal structure of the proglycinin trimer from soybeans, with each chain shown in a different color. (M. Adachi et al., *J. Mol. Biol.* 305:291–305, 2001.)

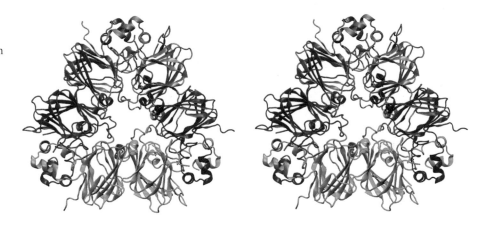

ular domains and two extended alpha-helical domains. The subunits are synthesized as a single chain called proglycinin, but are cleaved posttranslationally by a protease and then are collected into protein storage vacuoles, which become the protein grains visible by microscopy, where they assemble into the hexameric structures. As can be seen by comparison with the vicilin structure in Figure 5.37, this is essentially the same folding pattern of the conglycinin from the 7S fraction. Glycinin can form transparent gels upon heating that possibly involve intermolecular disulfide bonds.

The functions of the trypsin inhibitors are unknown, but they apparently serve in a defensive role. The Kunitz and Bowman-Birk inhibitors both show inhibitory activity toward both trypsin and chymotrypsin. These inhibitors are themselves glob-

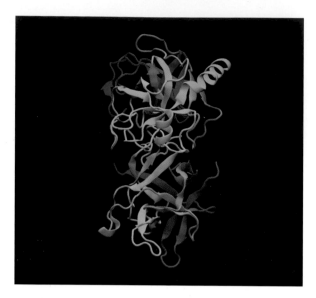

Figure 5.39. The crystal structure of the Kunitz soybean trypsin inhibitor (bottom) complexed with trypsin (top). (H.K. Song and S.W. Suh, *J. Mol. Biol.* 275:347, 1998.)

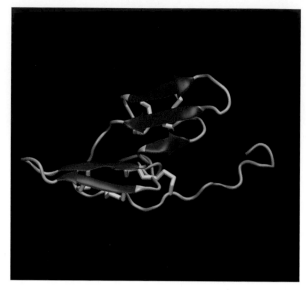

Figure 5.40. The solution structure of the Bowman-Birk type trypsin inhibitor from soybeans as determined by 2D NMR. The seven disulfide bonds are shown in yellow. (M.H. Werner and D.E. Wemmer, *Biochemistry* 31:999, 1992.)

ular proteins that bind tightly to the protease active site and cover it up, but are not themselves rapidly cleaved and released. The structures of both have been solved and are shown in Figures 5.39 and 5.40. The **Kunitz inhibitor** consists of 181 amino acid residues, in a closed β-barrel variant conformation called a β-trefoil (Figure 5.39). It is stabilized by two disulfide bonds. The **Bowman-Birk inhibitor** is much smaller, consisting of only 71 amino acid residues, but this small protein contains seven disulfide bonds, giving it much greater thermal stability. As can be seen in Figure 5.40, this protein contains two similar subdomains, one of which (residues 7–26) binds to and inhibits trypsin, and the other of which (residues 27–53) binds to and inhibits chymotrypsin. Each subdomain has a β-hairpin region, with five residues in the antiparallel β–sheets on either side of the turn, and with a *cis* proline residue in the turn.

These serine protease inhibitors are a nutritional problem when eating soy protein. The Kunitz inhibitor is partially denatured and inactivated by the gastric juices of the human stomach, but the Bowman-Birk inhibitor is not and interferes with the digestion of both the soy protein consumed with it and any other protein as well. In animal tests these inhibitors were found to cause pancreatic blisters and are thought to lead to enlargement of the pancreas and other pancreatic disorders, because the inactivation of trypsin and chymotrypsin by these inhibitors stimulates the pancreas to overproduce more of these enzymes. These inhibitor proteins can be heat denatured and made inactive by cooking. Thus, soy protein should be thoroughly cooked before eating, at temperatures close to 100 °C for 15 or 20 minutes, which will inactivate the bulk of the protease inhibitors. The Kunitz inhibitor denatures at a relatively low temperature, but the rate for the denaturation is slow. Because the Bowman-Birk inhibitor, with its large number of disulfide bonds, is more thermostable than the Kunitz inhibitor, heat-treated soy protein may retain some Bowman-Birk activity. Heating for an hour or more may be required to completely destroy this inhibitory activity, unless a reducing agent such as cysteine is added. Similar serine protease inhibitors are found in the seeds of other legumes, which also should be thoroughly cooked before eating.

The 7S protein fraction from soybeans also contains globular glycoproteins called **lectins**, or hemagglutinins, that recognize and bind with high affinity to specific carbohydrates. This specific binding has made these proteins useful in laboratory work, where they are used to aggregate particular proteins or cells with glycoproteins on their surface. Since hemagglutinins complex with surface glycoproteins in red blood cells, they are widely used to aggregate these erythrocytes, which is where they get their name. The actual function of these proteins in the plant seed is unknown. The legumes are nitrogen-fixing plants due to a symbiotic relationship with *Rhizobia* species in their roots, and the lectins may serve to protect plant seeds from these organisms.

Some lectins are extremely toxic. Perhaps the most notorious example of this toxicity is the poison **ricin** B chain, a lectin from castor bean seeds, which was used by the Bulgarian Secret Police to carry out the assassination of the dissident Georgi Markov in London during the Cold War. Soybean lectins are not as toxic as ricin but nonetheless can be antinutritional when consumed, because they will bind to the surface glycoproteins of mucosal cells in the intestine and interfere with adsorption of nutrients. The same cooking regimen that inactivates the soybean protease inhibitors will also inactivate the soy lectins. The conformational structure of soybean lectin has been determined by X-ray crystallography and is shown in Figure 5.41. The lectins from various legume seeds are similar in structure. All require Ca^{+2} and Mn^{+2} ions for activity. The soybean agglutinin is a glycoprotein with 253 amino acid residues that associates into a noncovalently bound tetramer. Each monomer has one carbohydrate binding site, which involves the calcium ion directly coordinating to the carbohydrate ligand. The core of the structure of the globular monomer is made up of two extended β-sheets, one with six strands and the other with seven individual strands, as can be seen in Figure 5.41.

Soybeans also contain a curious small hydrophobic protein, which can be extracted from ground soy meal using an organic solvent such as ethanol, making it a prolamin by Osborne's classification scheme. This protein, whose structure is shown in Figure 5.42, has only 80 amino acid residues, with four disulfide bridges, but a high proportion of nonpolar amino acids. Three of the disulfide bonds appear to be necessary for the conformational integrity of this protein. Its conformation is a four-helix bundle with a single twisted strand in a β-sheet conformation. The function of this protein is unknown. It has a poor nutritional profile, because it contains no

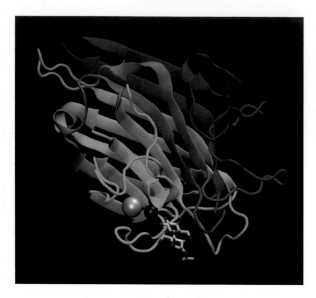

Figure 5.41. Crystal structure of a lectin, the soybean agglutinin complexed with a galactose-*N*-acetyl glucosamine disaccharide. The Ca^{+2} ion is shown in black, and the Mn^{+2} in gray. (L.R. Olsen et al., *Biochemistry* 36:15073, 1997.)

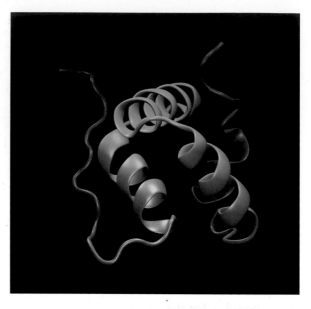

Figure 5.42. Crystal structure of the small hydrophobic seed protein from soybeans. (F. Baud et al., *J. Mol. Biol.* 231:877, 1993.)

methionine, phenylalanine, tryptophan, histidine, or lysine residues, but it makes up a small portion of the total protein in soybeans.

Soy protein does not contain gluten (the important protein found in wheat flour, responsible for the special properties of wheat dough; see above) and thus is not useful for baking without additives, but it is used in a wide range of other food products. Soluble soy proteins extracted from soy meal by soaking in mildly basic water (pH 8 to 9), and then precipitated by subsequent acidification, are flavored and texturized to produce meat substitutes. **Soy milk** is prepared by grinding the water-swollen bean in a large excess of water, then heating the resulting suspension to almost boiling for about 20 minutes. The heating not only pasteurizes the suspension, but also inactivates by heat denaturation the endogenous lipoxygenases that hydrolyze triglycerides to produce free fatty acids (Chapter 6; soy meal also contains significant amounts of lipids), and also denatures the protease inhibitors. The suspension is then enriched with vitamins and calcium. **Tofu** (bean curd) is a protein gel precipitated from soy milk by the addition of calcium sulfate or other salts such as $MgSO_4$, $CaCl_2$, or $MgCl_2$, followed by squeezing to express excess fluid and washing with water. Because $CaSO_4$ is not very soluble, coagulation using this salt is slow and produces a cohesive curd, while use of the more soluble salts gives faster reaction rates and a less cohesive, more fragmented curd. When pressed, the $CaSO_4$ curd gives a softer texture, while the curds produced with the other salts are firmer in texture. About 55% of the nonwater content of tofu is made up of protein, and about 28% is lipid triglycerides, made up mainly of linoleic and oleic acids.

Peanut Allergies

Several million people in the United States suffer from various types of food allergies. Common allergies include those to wheat and wheat products, dairy products, chocolate, eggs, strawberries, certain seafoods, and various types of nuts. One of the most common of these allergies is to peanuts, which affects more than 0.6% of the population in the United States and the United Kingdom. Peanut allergies are also possibly the most dangerous food allergy, causing more than 100 deaths in the United States per year. These allergies are produced by an immune system response to specific protein allergens in the peanuts. Peanuts are legumes, and as noted above, the various legumes have similar types of storage proteins in their seeds. Thus, peanuts contain a 2S protein fraction similar to that found in soybeans. Several of these 2S proteins are responsible for the common peanut allergies. Seven of these have so far been identified, and are called Ara h 1 through Ara h 7 (Figure 5.43). Peanut allergies are IgE-mediated allergies. The first time that a susceptible person is exposed to these peanut proteins, the immune system produces IgE (immunoglobulin E) antibodies that cover the surfaces of the individual's mast cells. When that individual is again exposed to the allergen, these IgE antibodies complex with the protein and trigger the release of histamine (see Chapter 11), which in extreme cases can lead to anaphylaxis and death.

The Ara h 2 and Ara h 6 allergens contain protease-resistant cores that are also stable toward temperature denaturation under oxidizing conditions. Thermal

Figure 5.43. The backbone trace of the peanut allergen Ara h 6, with the core alpha helices colored purple and with the stabilizing disulfide bonds illustrated as well. (K. Lehmann et al., *Biochem. J.* 395:463–472, 2006.)

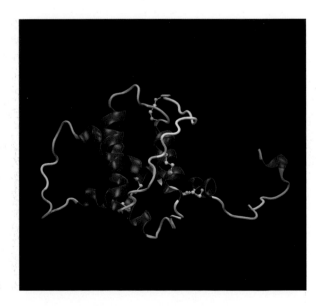

stability and protease resistance are required for a molecule to be a food antigen, because it might otherwise be inactivated by digestion. These characteristics make peanut allergens difficult to deal with in industrial processing, because they are not inactivated by cooking and make equipment and processing lines used for multiple products problematic because these denatured proteins are hard to clean away. Because very small quantities of the allergens can induce dangerous allergic reactions in sensitive individuals, it is necessary to rigorously segregate production lines intended for hypoallergenic products from those that contain any amount of peanuts, and products that are not produced under these careful conditions should be and usually are labeled to make it clear that they could possibly contain peanut allergens.

ZEIN PROTEINS

The major storage proteins in the kernels of corn (maize) are called zeins. Four main types of zeins, designated as α-, β-, γ-, and δ-zeins, occur in small protein accretion bodies in the endosperm cells of maize. These proteins are highly hydrophobic, insoluble in water, but soluble in alcohol, and thus are classified as prolamines using the Osborne classification system. They are present in the seeds as a means of nitrogen storage for the developing plant embryos. Zein proteins have a high proportion of proline, glutamine, and nonpolar amino acids such as leucine, and none of the essential amino acids lysine or tryptophan, making them a poor nutritional source for humans. In the kernels, the zeins are generally found as disulfide-linked dimers. Due to their hydrophobicity, the structures of the zeins have not yet been definitively determined, but they apparently have the shapes of elongated prolate ellipsoids, with axial ratios estimated to range between 25:1 to the most recent estimate of 6:1. Circular dichroism experiments indicate that they have a high alpha-helical content, around 50 to 60%. One recent study (Momany et al. 2006) proposed, on the basis of computational modeling, that α-zein consists of a supercoil of three intertwined helices, with this triple helix having a hydrophobic core into which the corn's zeaxanthin fits. Zeaxanthin is the long, hydrophobic carotenoid pigment molecule that gives corn its yellow color (see Figure 9.16). Although speculative, this model explains the helical content of zein, its geometry, and its tenacious binding to zeaxanthin.

As natural products from one of the most common of foods, zeins have GRAS status. They can be used as the starting material to manufacture confectioner's glaze, or even plastics. Zeins were used to manufacture fibers for clothing before the widespread introduction of synthetic fibers. These prolamine zeins make up more than 50 to 70% of the endosperm protein in maize, but there are other proteins as well, including a glutelin fraction that makes up around 34%, and albumin and globulin fractions that constitute about 3% each. The proteins of the glutelin, albumin, and globulin fractions have balanced amino acid distributions, but the deficiency of lysine and tryptophan in the dominant zein fraction can lead to serious nutritional disorders for people living exclusively on maize diets, contributing to the nutritional disease pellagra (see the discussion of niacin availability and nixtamilization in Chapter 12). Genetic variants of the traditional maize strains, some of which are referred to as quality protein maize (QPM), have been developed that have much higher proportions of lysine and tryptophan.

Maize was first domesticated in North America, in the Valley of Mexico, and traditional Native American cultures practiced a form of agriculture based on maize that combined it with two other complimentary crops, the so-called three sisters: maize, beans, and squash. In the field, the maize was planted first, followed by the nitrogen-fixing legumes, which used the corn stalks as beanpoles, while simultaneously fertilizing the corn plants. The squash was planted between the corn/bean stalks, and by covering the ground, inhibited the growth of weeds. These crops were also complimentary in the diet, because the lysine and tryptophan content of the beans supplemented the protein content of the maize, which also contributed the bulk of the calories through its starch, while the squash provided vitamins, phytochemicals, and fiber.

EGG WHITE PROTEINS: ALBUMEN

Egg white, often called **albumen**, is a 10% to 15% aqueous solution of various proteins, with the approximate compositions of the most important of these proteins listed in Table 5.8. The albumen proteins of egg whites are albumins (note the difference in spelling; see the section on Soy Proteins) using the Osborne classifications, because they are soluble in water. Several of these proteins, including ovalbumin, conalbumin, and lysozyme, are readily denatured by heating to temperatures above 60° to 80°, which is central to their functionality in foods. Egg whites constitute the cytoplasm of the original egg cell. Mature eggs actually have an onion-like structure, with successive layers in both the yolk and the whites. In the intact egg, the alternating

Table 5.8 Principal proteins of egg white (albumen)

Protein	% of total protein	Molecular weight (kilodaltons)	Isoelectric point (pH)	Denaturation temperature (°C)
ovalbumin	58.0	45	4.6	85
conalbumin (ovotransferrin)	13.0	80	6.6	63
ovomucoid	11.0	28	3.9	70
lysozyme	3.5	14.6	10.7	78
ovoglobulins G_2 and G_3	8.0	30–45	5.5–5.8	93
ovomucin	1.5	210	4.5–5.0	—
avidin	0.05	62.4	10.5	85
miscellaneous others	5.0	—	—	—

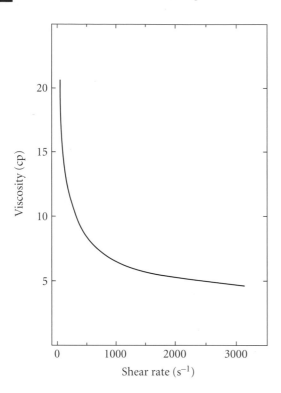

Figure 5.44. Like xanthan gum, egg albumen exhibits shear thinning. (Graph redrawn from data given in W.J. Stadelmann and O.J. Cotterill, eds., *Egg Science*. Westport, CT: AVI, 1977.)

albumen layers are divided into so-called thick white and thin white layers, with two of each, referring to the viscosity of the component solutions rather than the dimensions of the layers, and chalaziferous material, which is actually part of the inner thick layer. The chalaza, or chalaza chords, are two thick white fibers formed from the twisting of the inner thick layer of the egg white as the egg rotates while moving through the oviduct. They serve to suspend the yolk of the egg in the interior, away from the shell, regardless of the orientation of the egg. Egg albumen is slightly basic (pH 7.6 to 7.9), but the pH gradually rises on storage, to around 9.4 after three weeks, as CO_2 dissolved in the aqueous solvent diffuses out through the shell. Egg whites are used for a number of purposes in foods, particularly in making foams such as soufflés and meringues. They are also exploited for their ability to bind to other molecules, as in the fining of wines and in clarifying consommés. Collectively egg albumen is a thick, viscous liquid that exhibits shear thinning, as can be seen in Figure 5.44, making it additionally useful for preparing suspensions.

Most of the proteins of albumen are glycoproteins, and several of them seem to function to protect the egg contents from bacteria. Lysozyme is a glycosidase, which hydrolyzes the cell wall polysaccharides of Gram-positive bacteria, and ovomucoid is a trypsin inhibitor that probably prevents bacterial enzymes from attacking the egg proteins. Conalbumin can also inhibit the growth of bacteria by binding iron and rendering it unavailable. Ovomucin is a high molecular weight glycoprotein from the thick white of albumen. This very large protein consists of two domains, one heavily glycosylated with a molecular weight of approximately 400–610 kDa, and a smaller, less glycosylated domain with a molecular weight of 254 kDa, extensively cross-linked by disulfide bonds, with a total molecular weight as high as 860 kDa. This complex is very viscous in solution, and makes a large contribution to the overall thickness (viscosity) of albumen (Offengenden et al. 2011).

Several fractions of egg albumen are surface active. Egg albumen is, therefore, a good foaming agent used in cooking to produce foams by whipping. By adsorbing on the interface between the continuous liquid phase and the gas in the air bubbles of foams, these proteins raise the viscosity in the thin liquid layers between the bubbles, retarding the drainage of liquid away from the interface under the action of gravity, thus stabilizing the foam. (Incidentally, such foams are considered colloidal, even though the air bubbles are much larger than the upper size limits usually chosen to define colloids, because the intervening liquid layers between the bubbles are very thin, with thicknesses in the colloidal size range.) The ovoglobulins G2 and G3 are excellent foaming agents and, while together making up only 8% of the total protein content of albumen, contribute significantly to its overall usefulness in stabilizing foams. Unfortunately, these two proteins are poorly characterized, and it is not known in structural terms why they are so effective.

The most abundant of the albumen proteins is ovalbumin, which makes up more than half of the protein found in egg white. Ovalbumin is a globular glycoprotein with a molecular weight of 45kDa, made up of 385 amino acids. The protein contains four cysteine residues but only one disulfide bond, so it has two unpaired, reactive sulfhydryl groups. Ovalbumin is also phosphorylated with two phosphate groups, at serines 87 and 350. In electrophoresis experiments it can be separated into three forms which have both, one, or neither of these phosphate groups. There are, of course, two possible forms of the protein with a single phosphate group. About 85% of the proteins have two phosphates per molecule, while about 12% have only one phosphate, and about 3% have no phosphate.

The function of ovalbumin is unknown. It belongs to a family of trypsin inhibitor proteins called serpins, but it has no inhibitory activity toward trypsin or any other known protease. It has been suggested that ovalbumin is simply an amino acid food supply for the embryo as it grows, or that it serves to transport metal ions, since it can bind metals.

The tertiary structure of ovalbumin has been solved at 1.95 Å resolution by X-ray diffraction and is shown, as a dimer, in Figure 5.45. The folded conformation of the monomer is ellipsoidal in shape, roughly 70 Å × 45 Å × 50 Å. It contains extensive secondary structure, with three β-sheets and nine α-helices, as well as three shorter α-helical segments one turn in length.

Ovalbumin is only a weak foaming agent and is relatively resistant to surface denaturation. Compared with some of the other albumen proteins, it is also somewhat more resistant to heat denaturation, exhibiting only about 5% denaturation when heated to 62 °C at the slightly basic pHs prevailing in egg white, and only fully denaturing around 80 °C (see Figure 5.46). In addition to a disulfide bond, ovalbumin contains free Cys residues. During storage for lengthy periods, ovalbumin

Figure 5.45. Cartoon representation of hen ovalbumin. (From P.E. Stein et al., *J. Mol. Biol.* 221:941, 1991.)

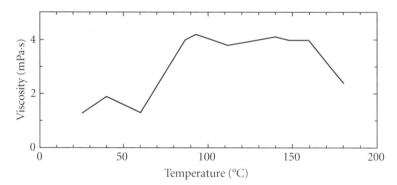

Figure 5.46. The viscosity of a 10 mg/ml solution of ovalbumin, under a shear rate of 383.0 s^{-1}, as a function of temperature. The protein denatures around 80 °C, and the entanglement and aggregation of the denatured polymers leads to significant increase in the viscosity of the solution. (S. Photchanachal, A. Mehta, and N. Kitabatake, *Biosci. Biotechnol. Biochem.* 66:1635–1640, 2002.)

is gradually transformed into an even more heat-stable form called S-ovalbumin, which is thought to result from an internal rearrangement involving a disulfide-free thiol exchange. Because ovalbumin makes up a large fraction of the total protein of egg whites, its properties contribute substantially to those of egg whites overall, and this stabilization therefore means that fresh egg whites cook more readily than those a few days old.

The oligosaccharide portion of the ovalbumin glycoprotein is linked to the polypeptide at Asn 298. There is considerable variability in this oligosaccharide, which has been characterized in at least six different forms, but all share a common trisaccharide core,

$$\text{Man}\beta(1 \rightarrow 4)\text{GlcNAc}\beta(1 \rightarrow 4)\text{GlcNAc}-\text{Asn}^{298}$$

often with another mannose linked to the O3 group of the middle GlcNAc (Figure 5.47). Frequently the core mannose residue is bound to additional mannose groups.

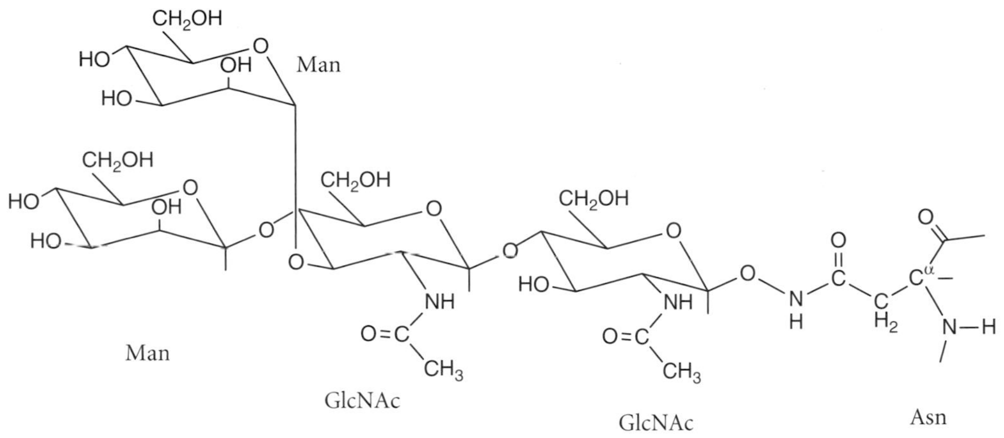

Figure 5.47. The oligosaccharide fraction of the ovalbumin glycoprotein. The carbohydrate portion is linked to the side chain of the asparagine 298 residue through a β(1→4) glycosidic linkage from the terminal *N*-acetylglucosamine residue. The first three-residue core of this oligosaccharide is a common structural motif of so-called *N*-linked oligosaccharides.

Ovalbumin acts as an allergen for some individuals, like the Ara h proteins of peanuts, and can produce a potentially dangerous allergic reaction. Egg allergies are one of the most common types of food allergies, although they occur mainly in the young; about half of those who suffer from egg allergies outgrow them by the time they reach adulthood. Allergies to egg white proteins are more common than those to egg yolk proteins, with ovalbumin, ovomucoid, ovotransferrin (conalbumin), and lysozyme being the usual antigens. Of these, ovalbumin is the most common allergen. Some vaccines, particularly those against influenza (flu), are incubated in hen eggs, and those who are allergic to egg proteins will likely experience an allergic reaction from the flu vaccine.

Conalbumin, or **ovotransferrin**, which makes up around 12 to 13% of the protein of albumen, is also an interesting molecule. The conformational structure of this protein has also been solved by X-ray crystal diffraction (Figure 5.48), and it is homologous to the closely related blood plasma glycoprotein **transferrin**. As its name suggests, the function of transferrin is to reversibly bind and transport ferric iron, Fe^{+3}. Ovotransferrin thus stoichiometrically binds one mole of Fe^{3+}, Mn^{3+}, or Cu^{2+} per mole of protein. The binding of the ions produces colored complexes with absorption maxima in the visible range. Conalbumin/ovotransferrin bound to Fe^{3+} has a reddish color and is sometimes responsible for the red discoloration of some processed foods containing egg whites. Complexes of conalbumin with Mn^{3+} and Cu^{2+} are yellowish in color. The apparent function of ovotransferrin in egg whites is to inhibit the growth of *Salmonella enteritis* and other bacteria by sequestering the iron of the egg (which is needed later for the development of blood in the embryo). Ovotransferrin has the lowest denaturation temperature, 63 °C, of the albumen proteins (Table 5.8), and therefore is the first to coagulate when eggs are cooked, and thus determines the

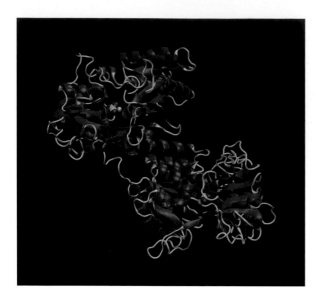

Figure 5.48. The structure of diferric hen ovotransferrin (conalbumin) as determined using X-ray crystallography. The protein's alpha helices are shown in red, and the beta sheet strands are indicated in dark blue. The reverse turns are colored aqua, and coil sections without regular secondary structure are colored orange (note that this does not mean that these segments do not have a well-defined conformation). The protein has two homologous lobes, each with one Fe^{3+} binding site (one iron can be seen as a gray sphere in the upper left) and a noncovalently bound carbonate ion (CO_3^{2-}, visible as van der Waals spheres colored red for oxygen and green for carbon). (H. Kurokawa, B. Mikami, and M. Hirose, *J. Mol. Biol.* 254:196–207, 1995.)

temperature at which the eggs begin to "set." When bound to iron or copper ions, ovotransferrin is more stable toward heat denaturation than the uncomplexed form due to the tight binding, raising the T_m to 84 °C. The binding apparently stabilizes those loops of the protein involved with coordinating the ions. Thus, when the protein is in films on the surfaces of air bubbles in whipped foams, it is less susceptible to complete denaturation if complexed with a metal ion, and so can slow the drainage of the water in the liquid film between the bubbles, stabilizing the foam.

The principal culinary use of egg whites is to create foams by beating with a whisk, as in soufflés. As the albumen is beaten, air is worked into the viscous liquid, creating bubbles. Several of the albumen proteins will undergo partial denaturation on the air/liquid interfaces of these bubbles, helping to stabilize the foam by forming a protective matrix on the bubble surface. This coagulation develops, in part, as the result of intermolecular disulfide bonds forming under oxidizing conditions between the free sulfhydryl (S—H) groups on different proteins that are exposed upon denaturation. Further cross-linking results from disulfide exchange between free sulfhydryl groups and the sulfur atoms of existing, native disulfide linkages, and between two such native disulfide linkages. A foam will degrade if the water in the liquid layers between the bubbles drains out quickly. If the surface-denatured proteins coagulate too thoroughly and tightly as the result of extensive disulfide formation, the viscosity of the remaining liquid medium will be adversely affected and will subsequently drain rapidly from the foam. This happens if an egg white foam is overbeaten, and the albumen proteins, and particularly the ovotransferrin, completely denature and become extensively cross-linked in tightly coagulated masses. When this happens the foam will become grainy, and significant amounts of thin liquid will bleed out. In traditional French cooking this undesirable result was avoided by whisking egg whites in copper bowls, which produced firm, glossy foams even when inexpertly overbeaten. In 1984, Harold McGee and coworkers (McGee, Long, and Briggs 1984) postulated that this effect resulted from copper ions from the bowl binding to the ovotransferrin proteins, stabilizing them toward heat denaturation and coagulation. McGee himself has now revised this picture of the process, however. In the newest edition of his classic book, *On Food and Cooking: The Science and Lore of the Kitchen*, 2nd ed. (New York: Scribner, 2004), he describes his observation that whipping egg whites in a silver bowl produced a good foam, rather than the expected grainy failure. Because ovotransferrin does not bind silver ions, he expected that the protein would more completely denature, as in overbeating. In his new model, the silver and copper in the bowls serve to bind to the free sulfhydryl groups of the denatured proteins, preventing them from tightly coagulating. The plausibility of this explanation is demonstrated by the fact that adding a small amount of acids such as lemon juice or cream of tartar (the potassium salt of tartaric acid; Chapter 10) also produces good stable foams even when beaten in a glass bowl. This is because the excess protons of the acids help keep the sulfhydryl groups of free cysteine residues protonated and unavailable for the formation of cross-linking disulfide bonds.

A number of proteins are present in small amounts in egg whites, with a variety of functions. One of the most interesting of these is avidin, which makes up only about 0.05% of the total protein in albumen. This molecule, which occurs as a non-covalent tetrameric complex (Figure 5.49), tightly binds the enzymatic cofactor and vitamin biotin (Vitamin B_7, see Chapter 12). The binding is so tight that biotin complexed in this way is not bioavailable. Uncomplexed avidin denatures at 85 °C and subsequently is unable to bind to biotin. Unfortunately, however, avidin complexed with biotin is more stable and does not denature, releasing the biotin, until the temperature reaches 132 °C (Donovan and Ross 1973).

Egg whites also contain large amounts of lysozyme, an enzyme that functions to protect the embryo (in fertilized eggs) from bacteria. Lysozyme is a glycosidase, an enzyme that hydrolyzes the glycosidic linkages in the polysaccharides found in the

Figure 5.49. A backbone cartoon representation of two molecules of avidin, shown in different colors, with two bound biotin molecules shown as liquorice traces, using standard atomic coloring. Avidin actually occurs as a tetramer consisting of two dimers of this type. (L. Pugliese, A. Coda, M. Malcovati and M. Bolognesi, *J. Mol. Biol.* 231:698–710, 1993.)

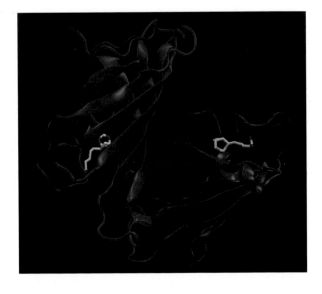

cell walls of Gram-positive bacteria. These polysaccharides consist of an alternating copolymer of β(1→4)-linked *N*-acetylmuramic acid and *N*-acetyl-D-glucosamine. Lysozyme occupies a special place in the history of protein biochemistry. It was originally discovered by Alexander Fleming, who was also the discoverer of penicillin. Lysozyme was the first enzyme to have its three-dimensional conformational structure determined at a useful atomic resolution by X-ray crystallography, by D.C. Phillips at the Royal Institution in London (Figure 5.50). As such, it has been one of the most extensively studied of all of the known proteins. Phillips used the structural information from the diffraction experiments to propose a mechanism for how lysozyme functions, explaining why the enzyme could produce such a large increase in the hydrolysis reaction rate (See Chapter 6). In this classic model, when the substrate is bound into the active site cleft of the enzyme (Figure 5.50), the sugar residue preceding the glycosidic bond to be cleaved is distorted into a higher-energy half-chair conformation resembling the reaction transition state and promoting the hydrolysis. In cooking, lysozyme denatures at 78 °C, and can serve to help stabilize egg foams.

Like the sweet-tasting protein thaumatin, hen egg white lysozyme (referred to either as HEL or HEWL) is a basic molecule with positively charged surface residues, and like thaumatin, it has also been shown to be sweet-tasting (Masuda, Ueno, and Kitabatake 2001). The surface positive charges of lysozyme are localized in one region of the surface of the native folded conformation, which is required for the sweet taste. Because lysozyme is a natural food protein with GRAS status, it may have potential applications in foods as a sweetener in the future.

Egg whites are used to clarify consommés and to "fine" wines. Both of these processes involve removing suspended colloidal-size particles or macromolecules from an aqueous suspension. Consommés are clear broths made from meats which have had their suspended cellular material removed by the addition of egg whites as they are being cooked. The high temperatures denature the albumen proteins, which

Figure 5.50. The crystallographic structure of hen egg white lysozyme as determined by A.C.T. North, D.C. Phillips, and coworkers (R. Diamond et al., *J. Mol. Biol.* 82:371–391, 1974) using X-ray diffraction. The protein has a number of alpha helices (shown in red) but only one small antiparallel beta sheet consisting of three short strands (shown in yellow; the blue sections are short sections of 3_{10} helices). It has two main globular domains separated by a deep cleft containing the active site, and into which the polysaccharide substrate binds.

then bind, presumably in part by hydrophobic association, to various macromolecules and larger nanometer-scale particles, as well as other albumen proteins, creating a matrix with a density less than that of water, so it rises to the surface as a mat, further removing more particles as it rises. In the fining of wines, the albumen proteins bind to the tannins in solution. Tannins are large polyphenolic polymers (see Chapter 9) defined by their tendency to bind to proteins. In this case, however, the resulting complex is slightly denser than water, and slowly settles to the bottom of the aging barrels. The supernatant wine is subsequently drawn off, leaving the albumen-tannin complexes in the barrels. Not all red wines are fined, and not all fining uses egg albumen. Those wines that do employ egg albumen are unacceptable to vegans because of their use of animal proteins.

EGG YOLK PROTEINS: LIPOPROTEINS

All of the lipids of eggs are located in the yolks, along with the lipid-soluble carotenoid pigments such as lutein (Chapter 9), which give the yolk its characteristic yellow color. There is a considerable amount of protein in the yolk as well. Egg yolks are somewhat like an oil-in-water emulsion, with a fat composition of about 65%, but also with a protein composition of around 30%. Among the proteins found in egg yolks are the water-soluble **livetins**, which are globular glycoproteins that make up about 10% of the total solid weight of the yolk; the phosphoprotein **phosvitin**, which constitute about 4% of the dry weight; and the **lipovitellins**, which are lipoproteins that constitute the largest fraction of yolk proteins. There are several different types of livetins, called α, β, and γ, and these proteins correspond to blood serum proteins of the chicken. The γ-livetins are IgY immunoglobulins. The lipovitellins are lipid and metal storage proteins. The conformational structure of lipovitellin from a species of lamprey has been solved by X-ray crystallography (Figure 5.51) and has been found to have a large hydrophobic cavity in one of its domains. Nearly all of the lipids of egg yolks are noncovalently bound in the interior of this large binding cavity of lipovitellin. These lipovitellin lipid–protein complexes may be analogous to the so-called low-density lipoprotein (LDL) complexes in the bloodstream in humans and other higher organisms (see Chapter 7).

In cooking, the yolk proteins that denature at temperatures less than 75°C are γ-livetin and α-livetin, while the other yolk proteins require heating above 75° for denaturation. Thus, for achieving "runny" egg yolks in frying or poaching eggs, temperatures in the range 60 to 65°C are believed to be optimal (Le Denmat et al. 1999; Vega and Mercadé-Prieto 2011).

MEAT PROTEINS

Historically, one of the richest sources of protein in the human diet has been meat, generically defined as the flesh and some internal organs of animals. Our

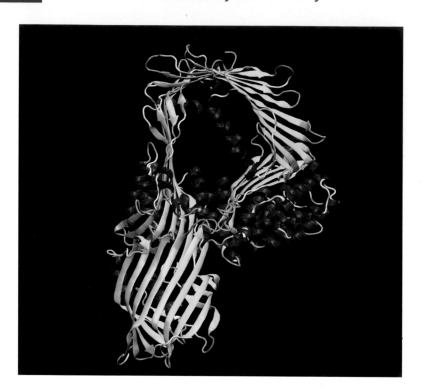

Figure 5.51. The conformational structure of lipovitellin from the yolks of silver lamprey (*Ichthyomyzon unicuspis*) eggs. Close homologs of this protein are found in the egg yolks of all egg-laying species, including chickens, but the structure of the hen protein is not yet available. The large domain of this protein seen in the upper part of the figure consists of many strands of antiparallel beta sheets surrounding a large cavity lined with hydrophobic functional groups. Nearly all of the lipids of the egg yolk are bound to these proteins within this cavity. (J.R. Thompson and L.J. Banaszak, *Biochemistry* 41:9398–9409, 2002.)

hunter-gatherer ancestors regularly consumed small quantities of lean meats acquired through hunting or scavenging the kills of other predators, practices followed by their prehuman ancestors such as *Homo erectus* before them, and by several of our modern primate relatives such as chimpanzees and baboons. While meat can mean various things, including internal organs and smooth muscle tissues such as intestines, it most commonly refers to skeletal muscle tissue. Some cultures and traditions make distinctions that exclude the flesh of fish and shellfish from being considered as meat, while vegans would include any animal flesh in this category. In the modern US context, the most common meats found in the market are from various domesticated species such as cows, pigs, chickens, turkeys, sheep, goats, ducks, and geese. In many countries including the United States, other domesticated species including yaks, camels, reindeer, horses, donkeys, or water buffaloes are eaten. Semidomesticated wild species such as bison, deer, elk, wild boar, ostrich, crocodiles, and alligators are often raised on special farms for their meat. In addition, almost all forms of wild game, including even primates, whales, and dolphins, are eaten either in the United States or in other countries.

Skeletal muscles are complex fibrous tissues which, in addition to connective tissues such as collagen and adipose tissue (depot fats), are primarily made up of muscle cells, which in addition to the usual cellular apparatus, contain large amounts of proteins (approximately 50% of such muscle tissue is protein). The muscles are attached to the bones by tendons made up of collagen. Muscle fibers are actually large single cells as much as 20–30 cm in length, with diameters in the tens of microns, that are often visible even to the naked eye. Each such cell is multinucleate, possibly containing more than 200 individual nuclei and many mitochondria that produce ATP. More than half of the body mass of most animals is made up of such muscle cells. The outer membrane of the muscle cells is called the sarcolemma. It contains many invaginations called T-tubes that constitute a network reaching throughout the cell. Within the muscle cells are myofibrils made up of hundreds of protein filaments bundled together. Each of these myofibrils is surrounded by its own membrane, called the sarcoplasmic reticulum, which is internal to the cell's outer bilayer membrane. The sarcoplasmic reticulum stores and pumps calcium, the release of which triggers muscle contraction. Muscle contraction requires ATP, while relaxation is a passive process that does not involve ATP breakdown. Energy is also required to pump calcium ions back into the sarcoplasmic reticulum after contraction.

The basic repeating unit of the skeletal muscle cell myofibril is the sarcomere, consisting of so-called thick and thin filaments which are made up of myosin and actin proteins, respectively. The sarcomeres are visible under a phase contrast microscope, where they appear to consist of light and dark bands at regular intervals. The dark bands are called the Z lines, and individual sarcomeres are bounded by two successive Z lines. During contraction, the thick and thin filaments slide past one another, shortening the sarcomeres and thus the fibril, a process which is driven by the consumption of energy produced in the filaments by the cleavage of the phosphate–phosphate bond in ATP to produce ADP. The filaments are made up of many proteins bound noncovalently to one another, some of which undergo conformational transitions triggered by increasing calcium concentrations, which initiates the series of changes that lead to the contraction of the fibers.

The major protein component of the thick filaments is myosin. The term "myosin" now refers to a large family of actin-binding proteins, but in the context of muscle proteins, it refers to ATPase enzymes responsible for the contraction of muscles. The principal myosin of muscles is called myosin II, a complex of two long heavy chains and four light chains. This large protein complex consists of three domains, often referred to as the "head," "neck," and "tail" domains (see Figure 5.52a). The ATPase active site and the actin binding site are located in the head group of the myosin, which is composed primarily of the heavy chain, while the tail is made up of the intertwined light chains (see Figure 5.52a). The neck is the bent portion of the T-shaped complex connecting the head group to the tail. Two of the light chains bind to the heavy chains in the neck portion that joins the head portion of the complex to the tail. The thick filaments consist of about 300 myosin molecules with their tail groups bound together and with the head groups on the sides of the filaments, where they can bind to actin filaments and split off phosphate from their bound ATP to release energy and produce ADP and inorganic phosphate. Another very large protein complex called titin, or connectin, connects the ends of the thick filaments to the Z line.

The three principal proteins of muscle thin filaments are actin, tropomyosin, and troponin. Actin, which is also found in nonmuscle cells and is one of the most abundant proteins in eukaryotic cells, is a multifunctional, 375-residue globular molecule that contains the myosin-binding site of the thin filaments. Its structure and confor-

(a)   (b)

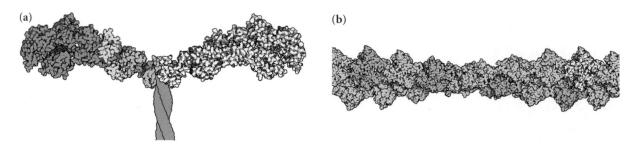

Figure 5.52. (a) An illustration of the head and neck regions of the myosin II complex, along with a schematic indication of the top portion of the intertwined tail chains. (b) An actin filament, composed of individual actin molecules, one of which is shown in atomic coloring to indicate the size of a single actin monomer. All other actin monomers are colored either entirely green or entirely blue to illustrate how they aggregate into two intertwined helices. Each actin monomer binds an ATP molecule as well. Tropomyosin binds in both of the grooves along the sides of the filament. (Images from the RCSB PDB June and July 2001 Molecule of the Month features by David Goodsell [doi: 10.2210/rcsb_pdb/mom_2001_6 and 10.2210/rcsb_pdb/mom_2001_7]; used with permission.)

mation are highly conserved across a wide range of species. Individual actin monomers associate into oligomeric complexes that constitute long, noncovalent filaments consisting of a great many individual actin units, as seen in Figure 5.52b. The actin monomer is often referred to as G-actin, while actin molecules organized into filaments are called F-actin. Two such noncovalent filaments intertwine to form a double helix, as can be seen in the figure, with two grooves along the sides of the filaments. The actin monomer has two domains separated by a deep cleft that contains an ATP/ADP binding site as well as a site that binds divalent cations (calcium or magnesium); thus, each G-actin monomer can bind one ATP molecule. The tropomyosin component of the thin filament is made up of two long alpha helical chains twisted around each other into a coiled-coil (Figures 5.53a and 5.53b), which then nestles into each of the two grooves in the actin helix.

The third major protein of the thin filament complex is the calcium-binding protein troponin, with one troponin for every tropomyosin molecule. Troponin controls the interaction between myosin and the actin filaments in response to the changes in calcium concentration. The troponin complex consists of three subunits, referred to as troponin C, troponin I, and troponin T. The troponin C binds calcium ions, causing it to undergo a conformational change, which in turn changes the conformation of troponin I, which releases the inhibition of the interaction of myosin with the actin filaments, leading to muscle contraction. Troponin C has a dumbbell shape and consists of two globular domains (see Figure 5.53c) connected together by a long alpha helical segment. The globular domains of the dumbbell consist mostly of alpha helices, with nine helical segments in all in troponin C. The globular domains each have two calcium binding sites. The two calcium binding sites in the C-terminal globular domain have a high ion binding affinity and are always occupied with either calcium or magnesium ions. The two calcium-binding sites in the N-terminal globular domain bind calcium at high concentrations but are unoccupied at lower ion concentrations (10^{-5}–10^{-7} M). When troponin C binds calcium into these N-domain

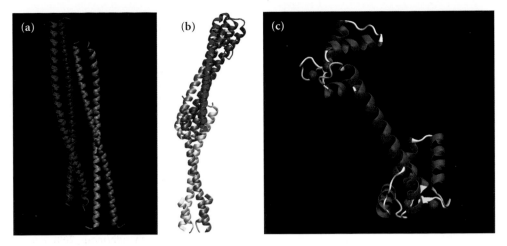

Figure 5.53. (a) Two tropomyosin molecules, each consisting of two long alpha helices intertwined as a coiled-coil. (J.H. Brown et al., *Proc. Natl. Acad. Sci. USA* 98(15): 8496–8501, 2001.) (b) The overlap of two tropomyosin pairs illustrating their juncture zone. (J. Frye, V.A. Klenchin, and I. Rayment, *Biochemistry* 49:4908–4920, 2010.) (c) A schematic illustration of the conformational structure of troponin C, with the bound calcium ions shown as red spheres. (J. Soman, T. Tao, and G.N. Phillips, Jr., *Proteins: Struct. Funct. Genet.* 37:510–511, 1999.)

sites, it undergoes a significant conformational change, exposing a previously covered hydrophobic patch to which the troponin I binds, causing the tropomyosin to also undergo a conformational change, uncovering the myosin binding site of the actin filament. Lowering the calcium concentration reverses the ion binding in the N-terminal domain, restoring the original conformation.

Thus, muscle contraction is initiated by an increasing calcium concentration, causing two additional calcium ions to bind to the unoccupied binding sites in the N-terminal domain of the troponin C, causing conformational changes that somewhat displace the tropomyosin helices from the groove of the actin filaments, exposing the myosin binding sites. This initiates the binding of the catalytic headgroups of the myosin molecules to the actin. Cleavage of the ATP molecule bound to the myosin head causes the myosin to bend as shown in the left portion of Figure 5.54. The myosin then flexes as shown in the right portion of Figure 5.54, releasing the product inorganic phosphate, as the myosin "walks" along the actin fibril in a ratchet-like motion, uncovering the actin monomer and releasing the product ADP, allowing its replacement by another ATP, which has a higher affinity for the myosin molecule.

When an animal dies, its muscle tissues will stiffen as rigor mortis, or the stiffness of death, sets in. Most people who have watched murder mysteries or police process dramas on TV are aware that the progression of this stiffness as a dead body cools can be used to estimate the time since death. This stiffness results from muscles becoming locked into a contracted state as cellular respiration ceases and their mitochondria run out of energy and are unable to generate further ATP to pump calcium against the concentration gradient in the membranes of the sarcoplasmic reticulum.

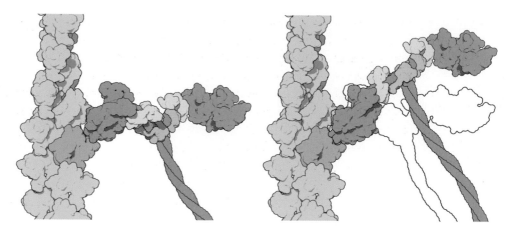

Figure 5.54. A schematic illustration of the interaction of a myosin head group with an actin fibril. Left: the initial interaction that cleaves the ATP and flexes the head group. Right: the myosin head group moves along the actin fibril via a conformational shift. (Images from the RCSB PDB June 2001 Molecule of the Month feature by David Goodsell [doi: 10.2210/rcsb_pdb/mom_2001_6]; used with permission.)

As a result, the binding of the myosin to the actin filaments cannot be released, and the muscles remain contracted. Further uncontrolled calcium diffusion throughout the muscle tissues results in widespread myosin–actin binding and extensive contraction-induced rigidity that cannot be relaxed. Rigor mortis also affects the muscle tissues of animals slaughtered for meat, making the tissues difficult to process as they lose their flexibility. Meat that is cooked immediately after slaughter will be tender, but if cooked after rigor sets in, it will be extremely tough. Ultimately, however, the flexibility of the muscle tissues is recovered in a process referred to as the resolution of rigor. This change is due to a proteolytic hydrolysis that cleaves the connectin at the Z junction, releasing the myosin–actin complex, restoring the flexibility of the meat, but which results in the slaughtered, post-rigor meat being slightly different from the living or immediately postslaughter flesh.

MAD COW DISEASE

By far the worst disaster faced by British farmers since at least the end of World War II was the near total destruction of the national beef and dairy cattle herds necessitated by the discovery of **mad cow disease** in the 1980s and its transmission to humans. Mad cow disease is a new and extremely unconventional food-borne illness, unlike any other with which food safety experts have had to contend. The reason this disease is discussed here is that the infectious agent involved in mad cow disease appears to be a denatured protein. Fortunately, the outbreak of this novel disease was brought under control, but because of its importance, the topic deserves attention. The threat of mad cow disease has not disappeared; wild cervid (deer) populations in the United States are infected with a similar disease today, which could potentially enter the human food supply, and cases of mad cow disease amongst cattle were found in Alberta in 2003 and Japan in 2001. On December 23, 2003, it was announced that a dairy cow slaughtered in Washington state on December 9, 2003, had been found to have mad cow disease, the first such case reported in the United States. Within a few days of this announcement, more than two dozen countries had imposed bans on the importation of American beef, including Japan and Korea, the largest overseas customers. Although it was determined in little more than a week that the cow in question had been imported from Canada and was born before the North American ban on the use of animal offal in cattle feed, the economic impacts of this event were quite serious.

The story of mad cow disease is long and complex. Perhaps somewhat surprisingly, one strand of the story begins in the remote highlands of Papua New Guinea, at a time when few or no cattle were being raised there at all. Among one group of Papuans, the South Foré of the Okapa Subdistrict of the Eastern Highlands, western doctors in the 1950s noted an unusual number of cases of an invariably fatal degenerative neurological disease called **kuru**, unknown elsewhere, which seemed to affect predominantly women and children. Autopsies of victims of kuru found that their brains had become riddled with cavities or vacuoles that made them resemble sponges. In a number of ways this disease was similar to a rare hereditary disease found in many populations called Creutzfeldt-Jakob Disease (CJD), which is characterized by a fatal dementia and which also leaves its victims with brain tissues that resembled sponges. Because of their effects on brain tissues, these diseases are classified as spongiform encephalopathies. CJD differed from kuru in several important respects, however, because CJD usually only affects the elderly, is rare (generally less than one case per million people per year), and tends to occur in both males and females. Ultimately, it was discovered that the cause of kuru was related to a form of ritual cannibalism practiced by the Foré people. In their funerary practices, the bones of dead family members were cleaned of flesh in preparation for burial by the female members of the community, who would also eat the flesh and particularly brain tissues of the departed, perhaps as a way of incorporating the spirit of the deceased into the living. Only the female members of the community and young children took part in this practice, and whatever the infectious agent of kuru was, it was being transmitted by this ritualized cannibalism. The epidemic of kuru among the South Foré reached its peak in the late 1950s and 1960s, until the connection to cannibalism was discovered. Government suppression of cannibalism has almost completely eliminated kuru. The actual cause of kuru, and its connection to CJD, remained unexplained, however.

Other spongiform encephalopathies are known among animals, with the most important being the common disease of domestic sheep called scrapie. Like kuru, scrapie can be transmitted from an infected individual to others, so both diseases are classified as transmissible spongiform encephalopathies (TSEs), which differ from CJD, which is hereditary and cannot be transmitted to others (except for one's genetic descendants through inheritance). While scrapie can be quite devastating to sheep and goat flocks, it apparently cannot be transmitted to humans in any conventional way, so it was and still is mainly considered to be an economic and agricultural problem rather than a public health concern. The disease, while uncommon, is still found in American sheep herds, with several hundred cases reported each year. The method of transmission is unknown.

In 1986, a new form of spongiform encephalopathy, called bovine spongiform encephalopathy, or **BSE**, was identified in dairy cattle in the United Kingdom. BSE was associated with a fatal dementia that soon came to be called "mad cow disease," because in the latter stages of the disease the affected cattle behaved abnormally. For epidemiological reasons it was believed that the origin of this disease was protein dietary supplements fed to British cattle which were made using ground-up offal from slaughtered sheep and cattle. This offal included brain and spinal nerve tissue. Some of these sheep were apparently infected with scrapie, and there seems to have been a "species jump" from sheep to cattle, which in a sense were being converted into unwitting carnivores by this feeding practice. Mad cow disease was invariably fatal and spread rapidly through Great Britain in the late 1980s and early 1990s until tens of thousands of animals were infected. The use of sheep and cattle offal in cattle feed was proscribed in Britain in 1988, but the practice continued in places until strict enforcement was instituted around 1991. Because the disease has a long incubation period, perhaps as much as 5 years in some instances, the number of cases continued to rise even after this practice was discontinued. Concerns arose as the number of infected cattle grew that their meat might not be safe for human consumption, but in the absence of direct evidence that a danger existed, agricultural and medical authorities in the British government continued to assure the public that it was safe to consume British beef. In March of 1996, however, a panel of scientists told Parliament that ten human patients had developed a neurodegenerative disease resembling CJD, but which appeared to primarily affect younger people who had no other known risk factors for CJD. The best explanation for the origins of this new disease appeared to be the consumption of meat from cattle suffering from mad cow disease. This new human disease was initially called new variant CJD, and now usually just variant CJD, or vCJD, but is widely referred to in the press simply as mad cow disease. A mass panic ensued that made British beef almost unsellable. Unfortunately, the repeated government denials and assurances in the face of growing public concern, some given only a short time before the announcement of the human cases, led to a major loss of confidence in British regulatory agencies.

More than 160,000 British cattle died from mad cow disease. Beginning in July of 1988, the British government began a program to destroy all potentially infected cattle in the United Kingdom. Eventually over 4.5 million cattle were destroyed (*New York Times*, Dec. 30, 2003) and their carcasses either buried or burned. Following the announcement in 1996 that BSE could infect humans and produce vCJD, the Continental nations of the European Union banned the importation of British beef, as did the United States and Japan. The export ban on British beef was not lifted for three and a half years, until 1999. The cost of this epidemic and the subsequent slaughter is difficult to quantify, but surely was in the billions of pounds sterling.

Reuters reported on May 23, 2003, that as of that date, 139 cases of vCJD in humans had been recorded, mostly among the residents of Great Britain or people who had spent extended periods there, although some cases on the European continent had also been reported, as had incidences of mad cow disease in herds in France and Germany, among other countries. All of the human victims of vCJD died, and no effective treatment has been found. A single case has been reported in the United States, in a woman who lived for years in Great Britain, where she almost certainly contracted the disease, before immigrating to the United States.

In spite of intense efforts, no bacterial, viral, or fungal agent has been found that is responsible for the transmission of mad cow disease, scrapie, or kuru. It thus is impossible to test a food for "infection" by culturing samples. The disease also has a long incubation time, on the order of months or longer, which makes the usual techniques of investigating outbreaks of food-borne illness almost useless. After the failure of efforts to find a viral or bacterial agent responsible for transmission, or even the necessity of nucleic acid material, T. Alper was led to propose the radical hypothesis that the infectious agent was actually a protein. J.S. Griffith further proposed the "protein-only" hypothesis, that the infectious protein might be a normal protein that has been modified in some way. Stanley Prusiner of the University of California at San Francisco isolated such a protein in sheep scrapie, which he called a **prion** (from proteinaceous infectious particle), abbreviated as PrP.

Experiments have demonstrated that cellular material taken from infected animals can transmit spongiform encephalopathies to normal animals. This was true even when the injected material contained only proteins and no detectable amounts of nucleic acids. Such material cannot be rendered safe by heating, and ultraviolet radiation has little effect on the infectiousness. It was also found that some species of animals could cross infect other species in this manner, but that other animals, usually more distantly related, were immune. In those cases where the spongiform encephalopathy of one species of animal cannot infect another, there is said to be a **species barrier**. Such a barrier apparently exists between sheep with their scrapie and humans, because there are no known cases of humans being infected from sheep suffering scrapie, but there apparently is no species barrier between sheep and domestic cattle. Prusiner was awarded the Nobel Prize for his work on prions in 1997, and while uncertainties still exist in the protein-only hypothesis, this model is currently the most widely accepted explanation for TSEs.

Much effort has gone into characterizing the prion proteins. It has been found that they are glycoproteins found on the extracellular surfaces of neurons. The function of prions, as of this writing, has not been determined. So-called knockout mice, in which the prion gene is removed, appear to develop normally and to be fairly healthy, except for minor neurological problems. Prion proteins have been found in all vertebrates studied, however, and it is almost impossible that a nonfunctional protein would be so widely preserved evolutionarily. It has recently been found that prions are involved in neuron function in a cell-signaling pathway, which is reasonable because it is on the outside surface of the neuron cells, but the ultimate purpose of the signaling remains unknown.

The genes for prions in a number of species have been identified and sequenced, and the conformational structure of the murine (mouse) prion has been partially characterized. It has been determined that it consists of a long, flexible, and unstructured, or disordered, perhaps random-coil-like chain from residues 23 to 120, with a small compact globular domain at the C-terminal end from around residue 121 to 231. In healthy, functioning cells this globular domain has a regular tertiary native structure that has been determined in aqueous solution using NMR techniques (Riek et al. 1997) to consist of three alpha helices and a short antiparallel beta sheet region (shown on the left in Figure 5.55). This native conformation for the healthy cell protein is usually abbreviated as PrP^C. This protein is susceptible to proteolytic digestion, and can be solubilized. It is also posttranslationally glycosylated, with normal cells containing at least three forms: unglycosylated, singly glycosylated, and diglycosylated. Because covalently linked oligosaccharides on glycoproteins are widely involved in molecular recognition, it has been proposed that the sugar residues may play an important role either in prion function or in TSEs.

In the protein-only hypothesis of TSE transmission, the "infectious" agent is a modified or abnormal form of the normal prion protein, widely designated as PrP^{Sc}. This proteinaceous material is insoluble, and contains at least one other molecular component, called an aggregating compound, which is covalently bound to fatty acids. The PrP^{Sc} form of the prion protein is apparently an aggregating denatured

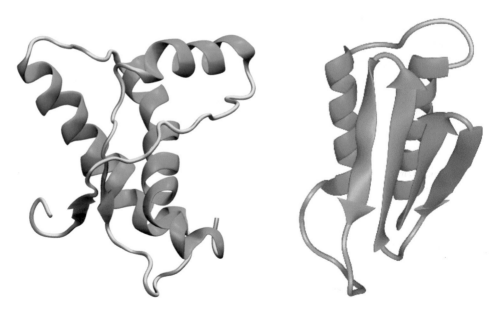

Figure 5.55. On the left is a cartoon representation of the globular domain of the native PrP protein as determined from NMR (R. Rick et al., *Nature* 382:180, 1996.) As can be seen, this protein is predominantly alpha helical (shown in green), with only a very small portion of beta sheet (shown in blue). On the right is the conformation of the globular domain of the denatured PrP^{Sc}. This form of the protein consists mostly of a four-strand antiparallel beta sheet with two alpha-helical segments. (Adapted from *J. Mol. Biol.* 293, F.E. Cohen, "Protein misfolding and prion diseases," 313–320 [1999], with permission from Elsevier.)

version of the normal PrP protein, and as such is not susceptible to inactivation by heating. It is also resistant to digestion by proteases. It is believed that the conversion of PrP^C to PrP^{Sc} is promoted by direct contact with another denatured PrP^{Sc} protein, which serves as a template for the further denaturation of native proteins in an auto-catalytic process, somewhat like the growth of a crystal out of a supersaturated liquor once a small seed crystal has nucleated or been added. The conformation of this denatured form of the prion protein has also been characterized (shown on the right in Figure 5.55), and found to be different from the mostly alpha helical native con-formation. In the insoluble, infectious conformation the globular domain of the prion protein has approximately 47% beta sheet and 37% reverse turns, with only two alpha helices making up only 17% of the globular domain.

One way to explain this denaturing transformation would be to assume that the native conformation is not the lowest free energy form, and that it is only kinetically stable, as in insulin or botulinum toxin. In this model, then, interaction with an already denatured PrP^{Sc} protein provides a template that somehow lowers the transi-tion state energy barrier and thus promotes the denaturation of the native protein (see Figure 5.56). Normal hereditary CJD would result from a mutant form of the prion protein which raises the energy of the "normal" PrP form sufficiently, relative to the barrier height, such that over the course of many years some proteins can spontaneously surmount the lowered barrier, initiating the spongiform encephalopa-thy, which would thus likely only develop in old age. Alternatively, in a thermody-namic control model, the native state might indeed be the lowest energy form for the isolated protein, but with the denatured PrP^{Sc} form able to aggregate irreversibly into the insoluble multimeric cluster, shifting the equilibrium and allowing more PrP to convert (see Figure 5.56), thus driving the conversion to completion. In this model

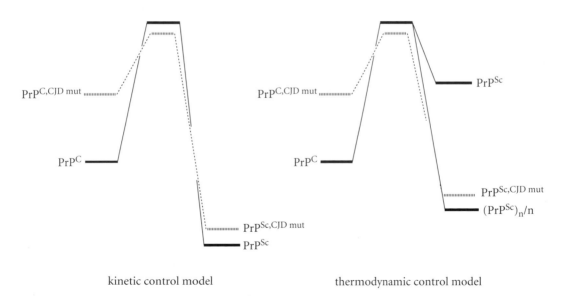

kinetic control model   thermodynamic control model

Figure 5.56. Two possible models for the energetics of prion denaturation and aggregation. (F.E. Cohen, *J. Mol. Biol.* 293:313–320, 1999.)

then, the energy per molecule in the cluster is lower than that of the energy of the individual prion molecule in the native state, even though the energy of a single denatured prion might be higher than that of the native conformation.

It is possible that sometime around the end of the nineteenth century or the beginning of the twentieth century, a member of the Foré people who suffered from conventional hereditary CJD died and his or her brain tissues were consumed as part of funerary rituals. The denatured prions in this individual's brain transmitted the disease to the majority of those who took part in this ritual, resulting in the appearance of kuru. It has been determined, however, that some women of the Foré possessed a genetic resistance to prion disease that protected them from developing kuru. This genetic resistance obviously would have rapidly become much more common in the surviving population, had ritual cannibalism not been suppressed, as those women and children without it perished from kuru. The mad cow epidemic in Europe presumably resulted either from scrapie-infected sheep proteins making a species jump to cattle when fed to the cows in protein supplements, or again from a cow with a spontaneous mutation producing a bovine version of CJD being fed to other cows as a supplement.

Worldwide, it would appear that this type of food-borne illness is rapidly declining as the ban on using vertebrate offal in feed intended for food animals becomes nearly universal. Presumably millions of people were exposed to bovine PrP^{Sc} before and during the height of the mad cow epidemic in Britain, but fewer than 200 people there contracted the disease, suggesting that the transmission probability to humans is low and possibly dependent on some uncommon genetic susceptibility in the victims. Such a resistance may result from the majority of humans possessing a prion sequence that differs from that of cows in some critical point. It has been suggested that the majority of the British population, as well as the majority of the human population worldwide, is protected by the same sort of genetic resistance that protected some of the Foré women, and that this genetic signature is so widespread in the world population that it probably is indicative of frequent cannibalism among our human ancestors in the Pleistocene. For those susceptible to the disease, increasing the level of exposure increases the risk of infection, so vigilance in the future will be needed to keep this dangerous agent out of the human food supply. While the US cases of mad cow disease have been traced to Canada and contaminated feed, deer and elk in the western United States suffer from a similar TSE called chronic wasting disease, and many of these animals share pastures with cattle on open ranges, raising the possibility of a species jump. To date, there still is no effective cure or treatment for vCJD, which is invariably fatal.

REACTIONS OF PROTEINS

Acid Hydrolysis

As with the glycosidic linkages of polysaccharides, the peptide bonds of polypeptides can be hydrolyzed under acid conditions, particularly at elevated temperatures, cleaving the bond by the addition of water. Humans produce a series of proteolytic enzymes to deconstruct proteins consumed in foods by hydrolyzing their peptide bonds to ultimately yield free amino acids. These proteases include pepsin, produced in the stomach, and trypsin, chymotrypsin, and carboxypeptidases, produced by the pancreas (see Chapter 6). The low pH of the stomach, resulting from digestive acids, also promotes peptide bond hydrolysis. **Hydrolyzed vegetable protein** (**HVP**) is a mixture of free amino acids and short polypeptides produced industrially by treating soy, maize, or wheat protein with acid at high temperature. Usually the soy product is separately referred to as **hydrolyzed soy protein**. This process is the principal method for generating glutamic acid (which of course must be separated from the other amino acids) in the industrial production of monosodium glutamate (MSG). Because of its glutamate content, HVP, and especially hydrolyzed soy protein, is becoming widely used as a flavor enhancer because its use avoids the requirement of putting MSG on the ingredient label.

Racemization and Cross-linking

When heated under basic conditions amino acids in proteins can undergo isomerization to produce racemic mixtures of D and L amino acids. D amino acids are less easily hydrolyzed by proteases during digestion and are adsorbed less efficiently across the intestinal wall, so this process leads to a loss of nutritional value. In addition, the formation of a carbanion at the alpha carbon can lead to β elimination reaction with a loss of the side chain beyond the β carbon, giving dehydroalanine. These processes can also occur at neutral pH at temperatures above 200 °C. The mechanism for the base-initiated reaction is shown in Figure 5.57.

Dehydroalanine (DHA) is highly reactive and can react with nucleophilic groups of other amino acid side chains to form cross-links within, or more commonly between, proteins (Figure 5.58). These nucleophilic groups include amines and thiols, such as the ε-amino group of lysine, the δ-amino group of ornithine (from arginine), and the thiol group of cysteine.

Heating of proteins in the absence of water can lead to changes in a number of amino acids that result in a loss of protein quality; for example, such heating of fish proteins may damage tryptophan, arginine, methionine, and lysine. Heating of some proteins leads to a cross-linking reaction between the ε-amino groups of lysines and the amide groups of asparagine and glutamine residues, with the formation of an amide bond similar to the peptide bonds of the backbone (Figure 5.59).

Reactions with Nitrites and the Formation of *N*-Nitrosamines

Bacon, ham, and hot dogs have traditionally been cured using salts of nitrate (NO_3^-) and nitrite (NO_2^-), along with table salt, NaCl. Usually sodium nitrate, $NaNO_3$, or potassium nitrate (KNO_3, also called saltpeter, or niter, one of the principal ingredi-

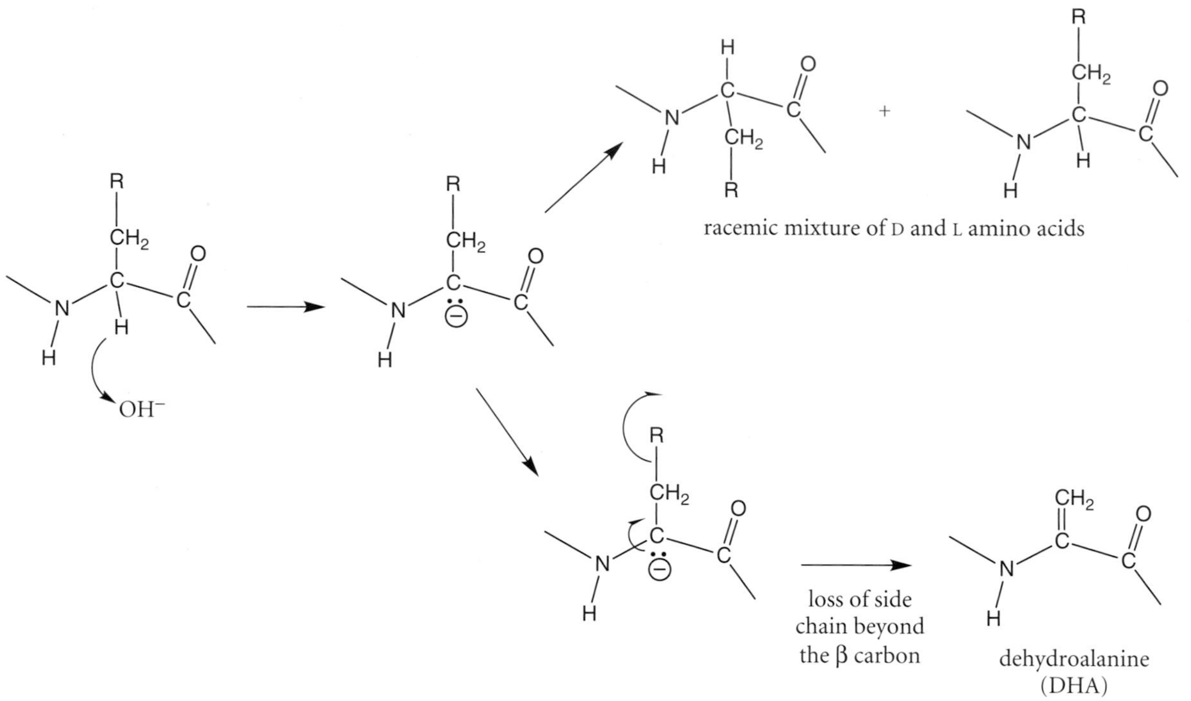

Figure 5.57. The mechanism of the formation of DHA and of D/L racemization in amino acids.

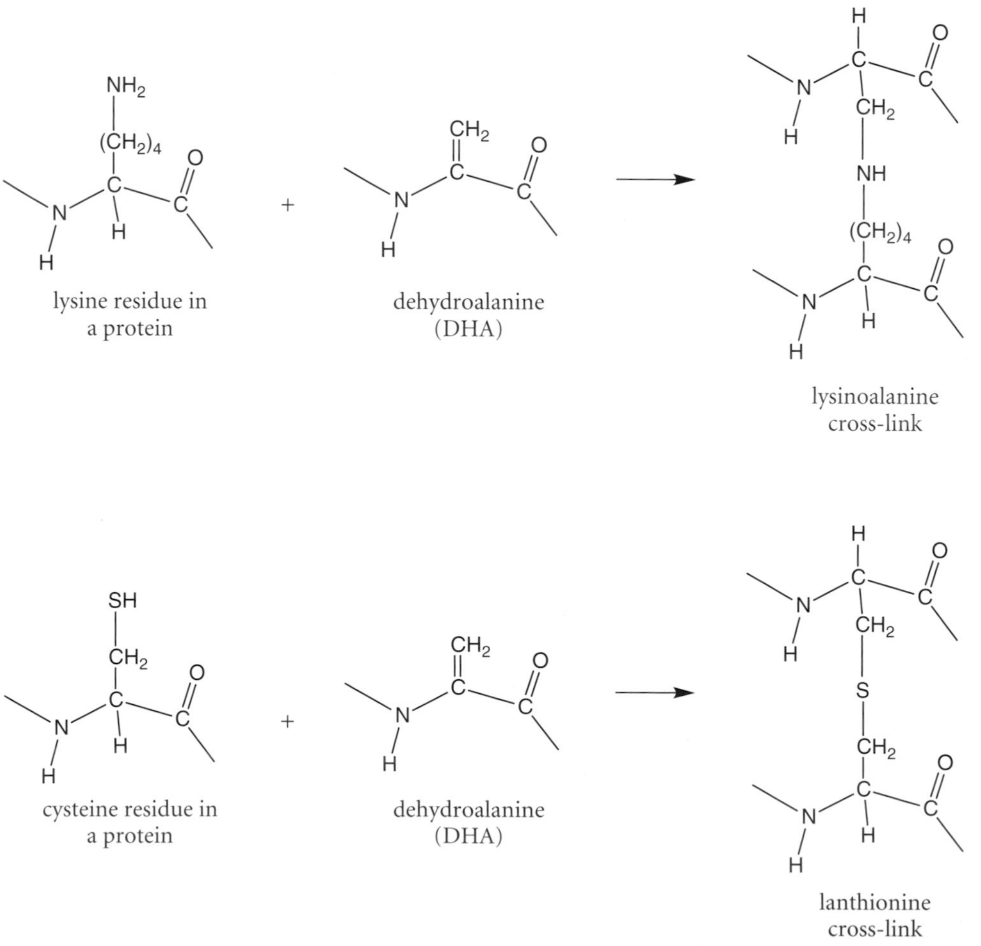

Figure 5.58. The mechanism of formation of lanthionine and lysinoalanine.

Figure 5.59. Amide-amino group side chain cross-linking in amino acids.

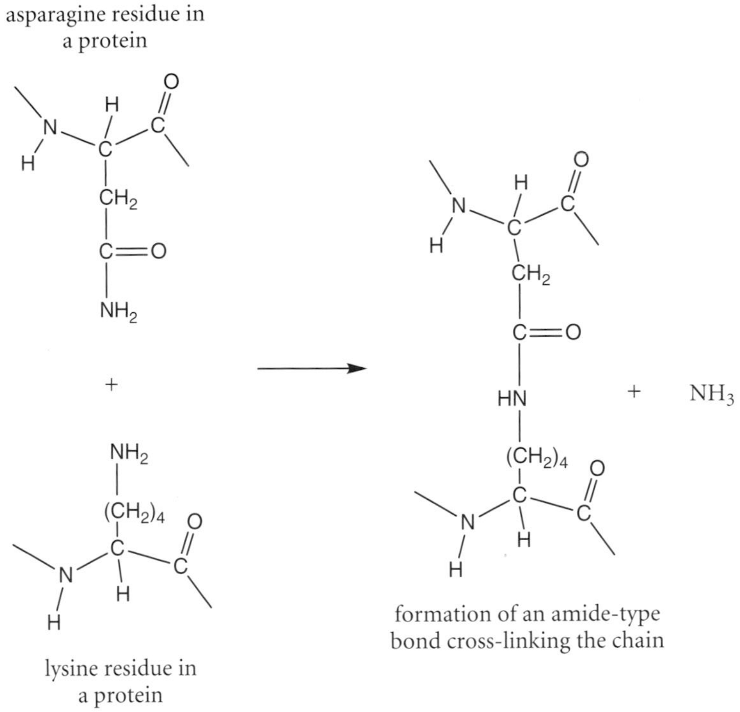

asparagine residue in a protein

formation of an amide-type bond cross-linking the chain

lysine residue in a protein

ents of gunpowder) is employed. Sodium nitrate also has antimicrobial properties and is therefore widely used as a food preservative. The main purpose of adding nitrates and nitrites in curing meats is to prevent the growth of the anaerobic bacterium *Clostridium botulinum*, but it also is used to fix the familiar pink color associated with ham and bacon, as will be discussed in Chapter 9. The principal concern that arises from this practice is the fact that nitrites can react with the amines in proteins to form compounds that are carcinogenic.

During curing, some of the added nitrate can be reduced to nitrite, which may also be originally present in the salt used for curing. When nitrites are heated under acidic conditions, as in cooking, they are converted to nitrous acid (HONO, or more properly, HNO_2) that can react with secondary amines, and to a lesser extent with tertiary amines, to form carcinogenic **N-nitrosamine** compounds (Figure 5.60). The reaction with primary amines simply produces nitrogen gas and hydroxide by deamination. The formation of nitrosamines occurs primarily with the amino acid residues tryptophan, proline, and histidine, but also with arginine, tyrosine, and cysteine. The products in the cases of Trp and Cys are illustrated in Figure 5.61. The secondary amine proline reacts with nitrous acid to produce N-nitrosopyrrolidine. These nitrosation reactions can be inhibited by the addition of reducing agents such as **ascorbic acid** or **erythorbate**.

Sodium nitrate is found naturally in many foods, particularly in green vegetables such as spinach, celery, and onions. The levels found in some of these vegetables can be several times the levels used in curing meats. Nitrate levels in celery juice are high

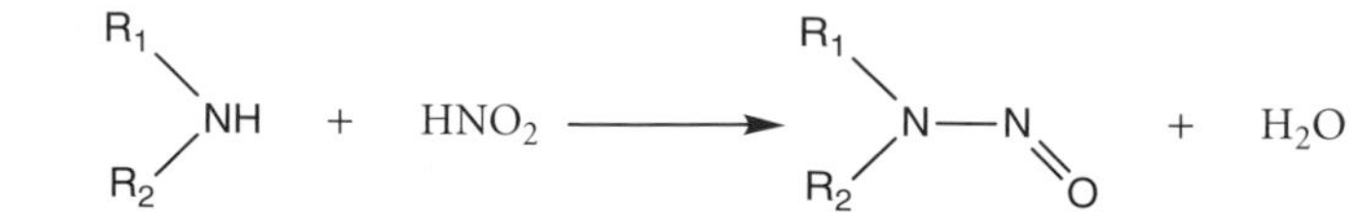

Figure 5.60. The formation of N-nitrosoamines.

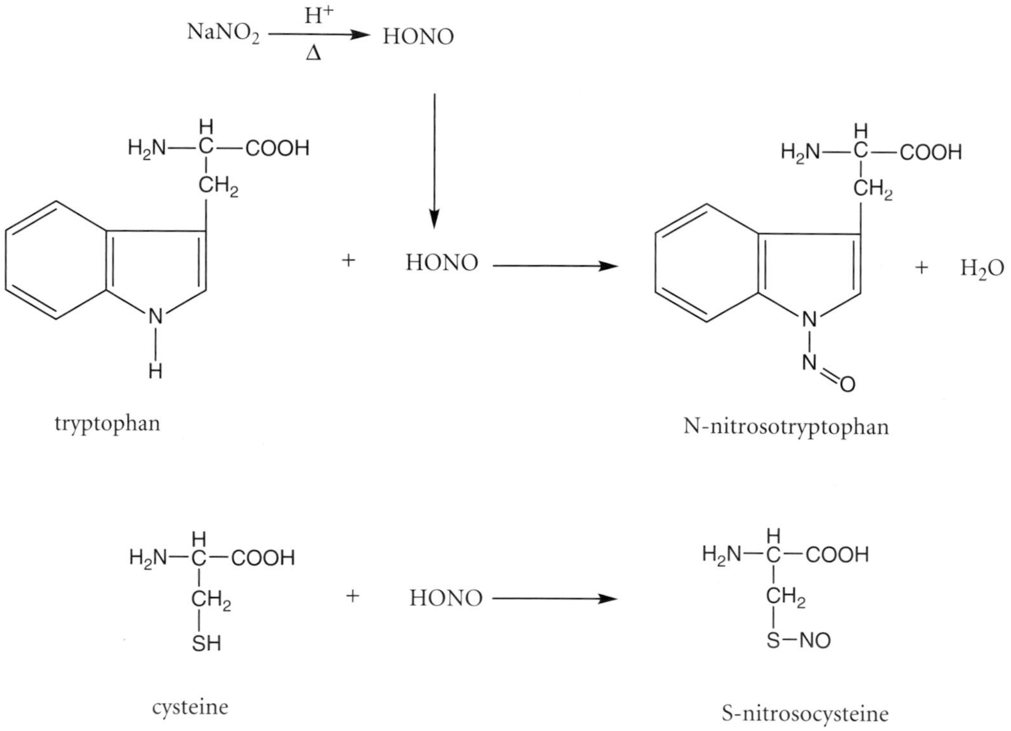

a secondary amine

Figure 5.61. The formation of nitrosoamines from tryptophan and cysteine.

enough that it can be used in combination with lactic acid and onions to cure meat, eliminating the necessity of direct addition of nitrates. Some hot dogs are advertised or labeled as "natural" or "organic," with no added nitrate, implying that this form of addition of nitrate is significantly different from direct addition of nitrate.

In modern industrial processing of meats, good manufacturing practice, the use of ascorbate and erythorbate, and the use of lower amounts of nitrates has significantly decreased the levels of nitrites and nitrosamines in cured meats to those that might be expected from natural sources.

Methionine Oxidation

Methionine has a relatively hydrophobic side chain, but the thioether is fairly reactive as a nucleophile and can react with alkylating agents such as methyl iodide to add a

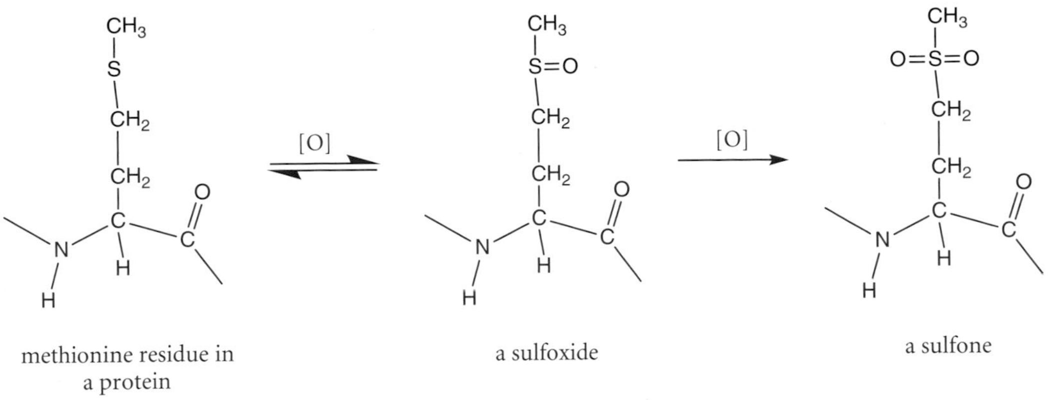

Figure 5.62. The formation of sulfoxides and sulfones from methionine.

methyl group to the sulfur atom. The sulfur atom is also easily oxidized, either by oxygen in the air during heat processing of foods or more readily by stronger oxidants such as hydrogen peroxide, to produce first the sulfoxide and then the sulfone (Figure 5.62). Production of the sulfoxide is easily reversible, but production of the sulfone is not and results in the nutritional loss of this essential amino acid.

The Kjeldahl Method

Food technologists often need to make a simple estimation of the total protein content of a food. It is actually quite difficult to determine exactly the total amount of protein present in a biological sample. A reasonable approximation can be made by determining the total nitrogen content and multiplying by the factor 6.25, on the assumption that on average proteins are made up of about 16% nitrogen. This process is called a **Kjeldahl determination**, after the Danish chemist Johan Kjeldahl, who first developed this method. In practice, Kjeldahl determinations are made by heating the food sample in sulfuric acid, which converts all organic nitrogen to ammonium sulfate. Sodium hydroxide is added when the conversion is complete to neutralize the ammonium sulfate, producing ammonia as a vapor, which is distilled and quantitatively collected in mild acid. The amount produced is then determined by titration.

$$\text{Protein} + H_2SO_4 \xrightarrow{\Delta} (NH_4)_2SO_4 + CO_2 + H_2O + SO_x$$

$$(NH_4)_2SO_4 + 2NaOH \rightarrow 2NH_4OH + Na_2SO_4 + H^+$$

Ammonium hydroxide does not actually exist as a molecular species and is a shorthand notation for an ammonium ion solvated in water, from which ammonia can be distilled. Correction factors are needed for samples known to contain proteins

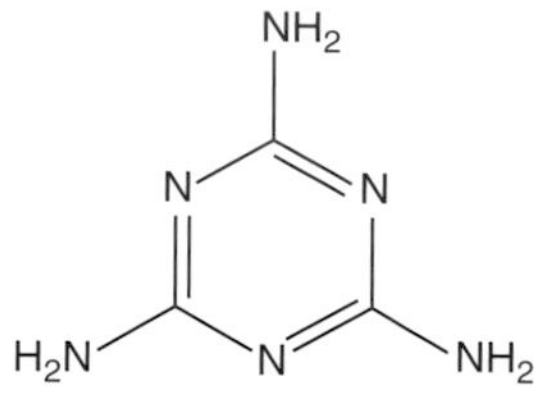

NH$_2$

N   N

H$_2$N   N   NH$_2$

1,3,5-triazine-2,4,6-triamine
(melamine)

Figure 5.63. The structure of melamine.

with sequences particularly rich or deficient in nitrogen-containing amino acids. Kjeldahl determinations take at least an hour, and usually much longer, to complete, and require dangerous reagents. An alternate method, **Dumas analysis**, based on the combustion of the sample to produce nitrogen oxide, is quicker to perform and uses less dangerous reagents, but is still a total nitrogen test and not a direct measure of protein content.

Keep in mind that as a test for total nitrogen rather than for protein itself, the Kjeldahl method can be fooled by the presence of other nitrogen-containing compounds such as urea. An example of this limitation made international news in 2007 when wheat gluten purchased from China by a Canadian manufacturer for use in pet foods was found to be contaminated with melamine ($C_3H_6N_6$, Figure 5.63). This inexpensive, high-nitrogen compound (66% by weight), used industrially as a flame retardant and in the manufacture of plastics and fertilizers, was not previously known to be particularly toxic, but was not permitted in human or animal food in the United States. Many pets were killed or sickened, perhaps as the result of the conversion of melamine to ammonia after being eaten, and massive nationwide recalls of pet foods were necessary in both the United States and Canada. Because it costs much less than protein, it was reportedly a common practice in China prior to this event to adulterate animal feed with melamine to make the feed appear more protein rich in Kjeldahl determinations. China subsequently banned the use of melamine in food products.

Unfortunately, this action did not end the practice of food adulteration with melamine in China. In the fall of 2008, following the Beijing Olympics, tens of thousands of infants were hospitalized with serious illnesses, including kidney failure, as the result of having consumed infant formula made from powdered milk contaminated with melamine. Unfortunately, six of the affected children died. While the majority of the contaminated infant formula was sold in China and Taiwan, some was also exported to various developing nations. The problem was subsequently found to have spread even more widely than infant formula, as melamine-contaminated milk and yogurt products from several manufacturers were found in a number of markets throughout China. Chocolates manufactured in China and marketed in a number of Asian countries were also found to be contaminated with melamine, and even Chinese eggs were determined to contain traces of melamine as the result of adulterated chicken feed. Some contaminated products were even found in specialty Asian markets in North America and had to be removed from shelves. Dozens of people were arrested and charged with adulterating milk or other products or knowingly selling milk that had been adulterated with melamine to various manufactures of baby formula; two of these individuals were subsequently executed in 2009, and many more were sentenced to prison. To overcome the problems of quantitating the protein content of food samples based on nitrogen content, alternate methods have been proposed that are based on the binding of dyes such as Comassie Brilliant Blue G-250, which changes color from red to blue upon binding to proteins (Bradford 1976).

Maillard Reactions

As was discussed in Chapter 4, proteins heated under neutral or basic pHs in the presence of a reducing sugar undergo a complex series of reactions between the free amino groups of lysine residues and the aldehyde or ketone groups of the sugars which eventually degrade the lysine–sugar adduct to melanoidin pigments. This irreversible reaction results in the destruction of the essential amino acid lysine and thus decreases the nutritional quality of the protein. This reaction can be inhibited by avoiding reducing sugars or by lowering the pH of the system, which inhibits the formation of the initial carbonyl-amine complex. This reaction can also occur in living systems and may be an important part of the changes that take place with aging.

Reactions with Aldehydes

Both primary and secondary amines will react readily with aldehydes and ketones to give α-hydroxyamines. This reaction is in fact the basis for the Maillard reactions discussed above and in the carbohydrates section. As we saw before, this reaction can be reversed and is inhibited by acidic conditions. Amines in protein can undergo such reactions, particularly with small volatile aldehydes, and several other examples occur in which such reactions are important in food processing. One of these is in preserving foods by smoking. Wood smoke contains formaldehyde in significant concentrations, which is used in sterilizing and preserving biological specimens, and which can kill bacteria. This is why smoking can preserve foods. The mechanism of action appears to be such a reaction with protein amines, as shown in Figure 5.64.

Malonaldehyde, an oxidation product of some lipids, has two aldehyde groups and can serve to cross-link protein chains, as shown in Figure 5.65. This reaction might occur in fish protein during storage, as well as in other meats, even when frozen, and may cause the muscle protein to toughen.

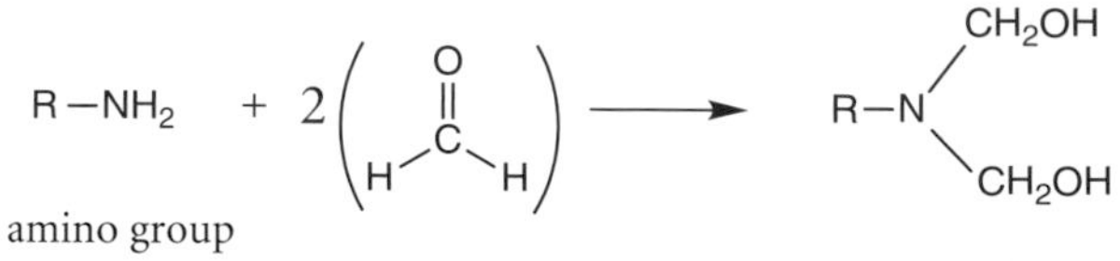

$$R-NH_2 \;+\; 2\left(\underset{H}{\overset{O}{\underset{\displaystyle |}{\overset{\displaystyle \|}{C}}}}\overset{}{}H\right) \longrightarrow R-N\overset{CH_2OH}{\underset{CH_2OH}{<}}$$

amino group

Figure 5.64. The reaction of aldehydes with amino groups, as occurs in smoking.

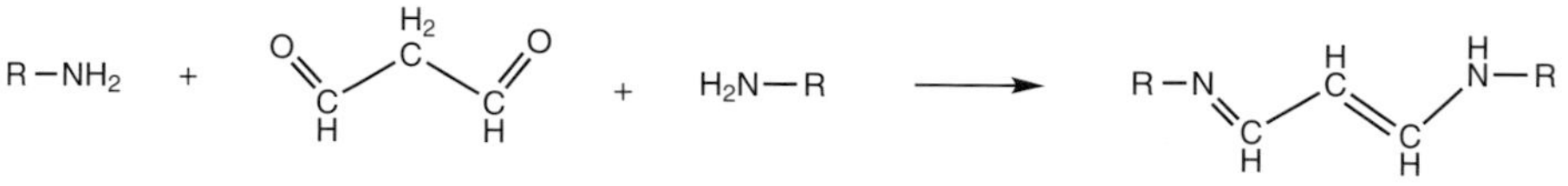

$$R-NH_2 \;+\; \text{(malonaldehyde)} \;+\; H_2N-R \longrightarrow R-N\text{...}N-R$$

Figure 5.65. The reaction of amines with malonaldehyde.

$$H_2O \rightarrow H_2O^+ + e^-$$
$$H_2O^+ + H_2O \rightarrow H_3O^+ + {}^{\bullet}OH$$
$$P + {}^{\bullet}OH \rightarrow P^{\bullet} + H_2O$$
$$P^{\bullet} + P \rightarrow P-P$$

Figure 5.66. The mechanism for protein polymerization induced by ionizing radiation.

Ionizing Radiation

In the irradiation of food for sterilization at ambient temperatures, the ionizing radiation can produce free radicals by the ionization of water, which can in turn produce protein free radicals, leading to polymerization (Figure 5.66).

Acylation

Proteins can be acylated by treatment with acetic anhydride or succinylated by treatment with succinic anhydride. The groups susceptible to these acylation reactions include tyrosine, lysine, and cysteine, as illustrated in Figure 5.67. This reaction is often used to increase the solubility of proteins or to shift the pH of maximum solvation (Figure 5.68). Acylation and succinylation of lysine are irreversible and lead to a nutritional loss of this amino acid.

Homocysteine

Homocysteine is an amino acid found naturally in the bloodstream, although it is not one of the 20 commonly occurring amino acids found in proteins and coded for genetically. In the blood it is produced as an intermediate product of the enzymatic conversion of methionine to cysteine (Figure 5.69). Homocysteine differs from cysteine only in having an additional CH_2 group in the side chain. Elevated levels of serum homocysteine are now thought to be harmful, perhaps playing a

succinic anhydride
maleic anhydride
acetic anhydride
tyrosine
lysine
cysteine
succinic anhydride
tripeptide segment of a protein
pH 8-9
pH 9
CH_3COOH
lysine residue in a protein
acetic anhydride

Figure 5.67. Protein acylation.

Figure 5.68. Solubilities for native (---) and succinylated casein (— 50% and --- 76%) as a function of pH, illustrating the pH shift of maximal solubility upon modification. (Reproduced from H.-D. Belitz and W. Grosch, *Food Chemistry*. Berlin: Springer-Verlag, 1987.)

Solubility (%)
100
60
20
2
3
4
5
pH

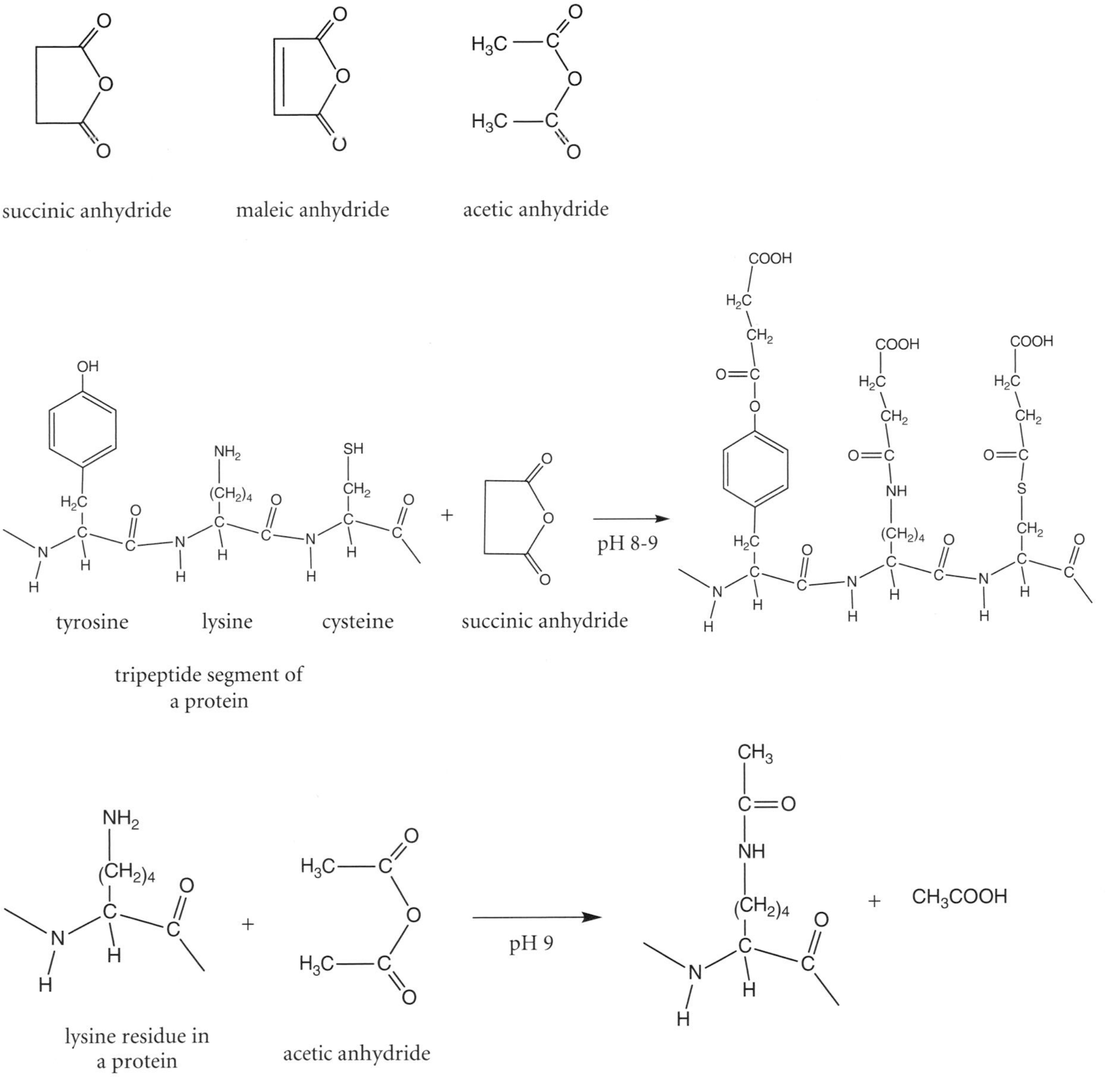

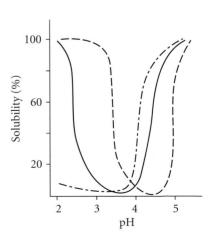

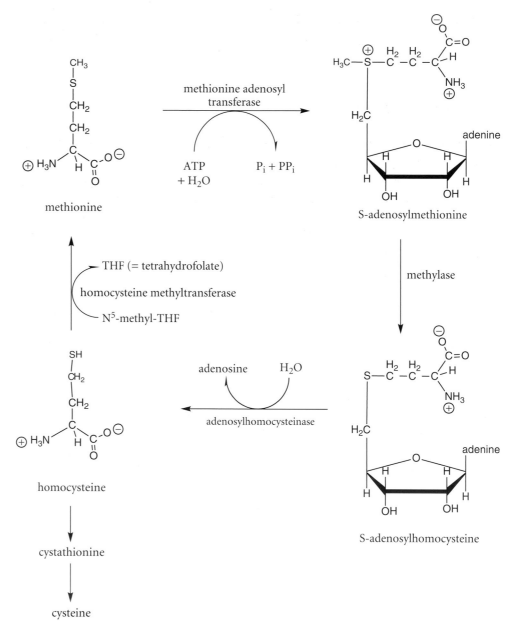

Figure 5.69. The mechanism for the formation of homocysteine. (Diagram redrawn from D. Voet and J.G. Voet, *Biochemistry*, 2nd ed. New York: Wiley & Sons, 1995.)

role in a variety of diseases, including heart attack, stroke, Alzheimer's disease, birth defects and miscarriages, and presbyopia. Serum homocysteine levels have been reduced by the widespread fortification of flour with folate. Serum levels can also be reduced by taking supplements of B vitamins (folate, vitamin B_6, and vitamin B_{12}). It has been suggested, however, that the fortification of flour with folate could hide the anemia symptoms of vitamin B_{12} deficiency, which can also produce neurological damage if undiagnosed (see Chapter 12). The B vitamins are cofactors for the homocysteine methyltransferase enzyme that converts homocysteine back to methionine.

Aspartame Degradation

Because of its clean, sweet taste, similar to that of sucrose, aspartame (Figure 5.7) has become popular as a lower calorie sweetener. Use of aspartame is limited by its chemical instability, which makes it unsuitable for many applications. As a dipeptide, aspartame is susceptible to acid hydrolysis under conditions that favor peptide bond hydrolysis, such as low pH and elevated temperatures. Under such conditions, the molecule is cleaved to produce equimolar amounts of both amino acids and methanol (which is toxic),

$$\text{L-Asp-L-Phe-OMe} \rightarrow \text{L-Asp} + \text{L-Phe} + \text{MeOH}$$

Aspartame is reasonably stable under the mildly acidic conditions of carbonated soft drinks. Such drinks are sometimes stored under conditions of high temperatures, as during the summer months in the southwestern deserts. Long storage of bottled soft drinks sweetened with aspartame or storage under such high-temperature conditions leads to a gradual loss of sweetness from hydrolysis, with a half-life of less than a year at room temperature, and of course, less at elevated temperatures.

Under neutral or basic conditions, aspartame can undergo an intramolecular condensation at elevated temperatures to produce a diketopiperazine, 5-benzyl-3,6-dioxo-2-piperazine acetic acid, and again methanol (Figure 5.70). This reaction is inhibited at lower pHs because the amino terminus is then protonated, and its lone pairs are less available for reaction with the carbonyl group.

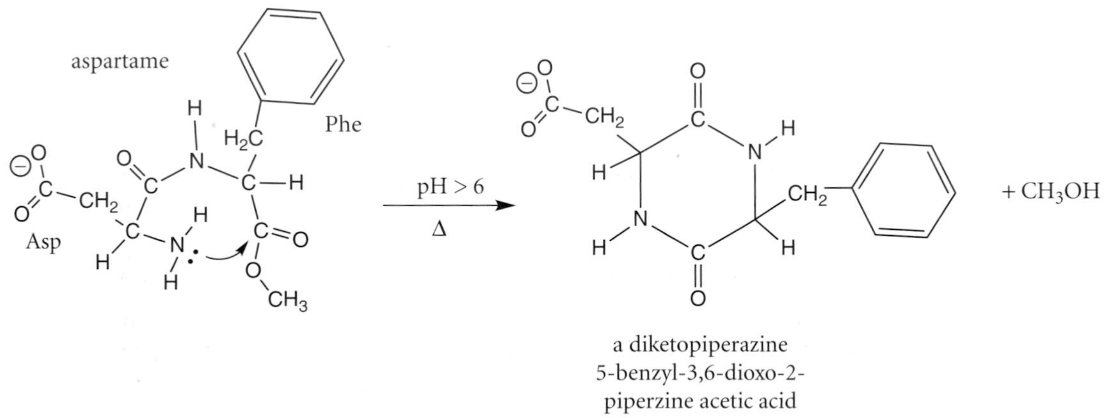

Figure 5.70. The thermal degradation of aspartame.

Because of its susceptibility to degradation when heated, aspartame is not suitable for foods that must be cooked or stored for long times even at room temperature if the pH is acidic. As a peptide, aspartame is also susceptible to microbial degradation.

Heterocyclic Amine Formation

One reaction of particular concern that some amino acids undergo during cooking at high temperatures is the formation of **heterocyclic amines**, often abbreviated as HA in the popular literature. These compounds are cyclic organic molecules containing at least one nitrogen atom in one of their rings. Several of the most important of these molecules are 2-amino-1-methyl-6-phenylimidazo[4,5-*b*]pyradine (**PhIP**); 2-amino-3-methyl-imidazo[4,5-*f*]quinoline (**IQ**); 2-amino-3,4-dimethylimidazo[4,5-*f*]quinoline (**MeIQ**); 2-amino-3,8-dimethylimidazo[4,5-*f*]quinoxaline (**MeIQx**); and 2-amino-3,4,8-trimethylimidazo[4,5-*f*]quinoxaline (**DiMeIQx**), shown in Figure 5.71.

These molecules are produced when muscle meats are cooked at high temperatures, particularly in barbecuing, but also in frying and broiling (Murray et al. 1988). Although the mechanisms of their formation are incompletely known, they are produced by the reaction of certain amino acids with **creatine**, a high-nitrogen compound found in muscle tissues and involved in supplying energy to the muscles

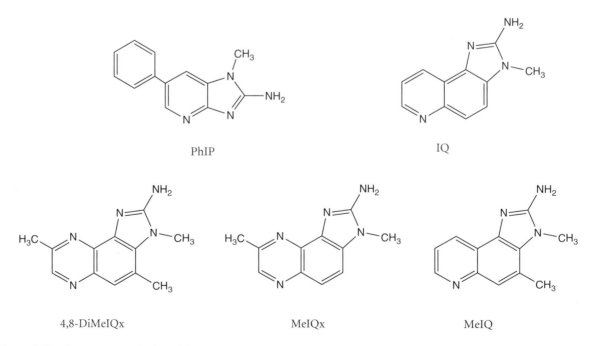

Figure 5.71. The structures of selected heterocyclic amines.

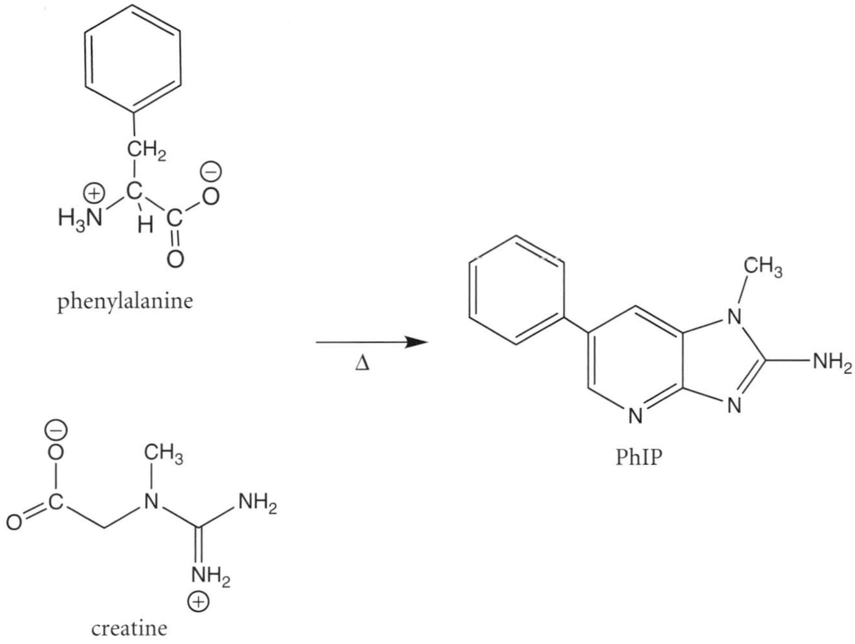

Figure 5.72. The formation of PhIP from phenylalanine and creatine.

(Figure 5.72). Creatine is synthesized endogenously, but dietary creatine from muscle foods also contributes to the energy supply of muscle tissues in humans. Creatine is not found in fruits and vegetables.

Heterocyclic amines in the diet are a serious concern because most of them are known to be powerful mutagens and are potentially carcinogenic, causing various types of cancer in lab animals. Epidemiological data has implicated meats cooked at a high temperature as a cause of stomach cancer, as well as pancreatic, breast, and colorectal cancers. Obviously one way of avoiding these compounds would be to follow a strictly vegetarian diet; heterocyclic amines are not produced in high levels in plant foods even when vegetables are barbecued, although some are produced in high-heat–processed gluten-containing foods. Organ meats such as liver also contain little creatine and thus do not produce significant amounts of heterocyclic amines when cooked. Muscle meats that are baked or roasted in ovens at lower temperatures contain significantly less heterocyclic amines; however, gravies made from the drippings of meats cooked in this manner do have larger amounts of heterocyclic amines. The National Institutes of Health's National Cancer Institute recommends that muscle meats be partially cooked in a microwave oven for 2 minutes before being cooked by other higher-temperature methods, because this approach has been shown to reduce the formation of heterocyclic amines by 90%. Direct contact of the meat with flame should also be avoided, and particularly the charring that occurs in "well-done" barbecuing over high flames. Gravies made from meat juices should be avoided if one wants to minimize consumption of heterocyclic amines.

GMO FOODS: "FRANKENFOODS"

Foods containing genetically modified organisms (GMOs), derisively called "Frankenfoods" by some, are becoming increasingly common in North America and some other markets. While the term GMO can be somewhat vague, in its most general usage, it refers to organisms whose genomes have been artificially manipulated, generally using some form of recombinant technology, in a manner that exceeds the variations that can be achieved through more traditional plant and animal breeding techniques. Because genes directly code for proteins, the most direct effect of such genetic engineering will be that the engineered organism will possess proteins that differ in some way from those of a more "conventional" organism. However, because protein enzymes carry out the synthesis of the other molecules in an organism, as well as their metabolism and deconstruction, and make the hormones that regulate growth and death, all other aspects of an organism's molecular makeup can also in principle be affected, at least indirectly, by genetic engineering.

Genetic engineering could be as simple as making a single point mutation change in a particular protein to alter its function. This is also the type of change that can be achieved by conventional breeding methods, but genetic engineering is much faster and more efficient, because it does not rely on the desired trait arising spontaneously by random mutation, or on randomly inducing such a change by subjecting the organism to mutagenic doses of radiation, for example. A slightly bigger type of change through genetic engineering might involve removing a particular protein from an organism's genome altogether, which could be particularly useful in preparing nonallergenic organisms as foods (providing that the organism is viable without the deleted protein). For example, such an approach might be used to remove the allergenic Ara h proteins from the peanut genome, producing peanuts that would be perfectly safe for even hypersensitive individuals to eat. Already this approach has reportedly been used by at least two companies to create hypoallergenic cats that lack the protein allergen that makes many pet owners allergic to cats, but these claims remain controversial at the time of this writing and have not been independently verified. Many other possible applications could be foreseen, however, such as removing bitter-tasting compounds from certain vegetables or the antinutritional proteins from soybeans.

Probably the most controversial aspect of genetic engineering involves taking genes from one species and introducing them into the genome of a completely unrelated species. A particularly striking example of this practice was the creation of a new strain of zebra danio fish, a popular aquarium "pet," by introducing the gene for a red fluorescent protein from a type of coral into their genomes, to produce a new type of fluorescent pink zebrafish for the pet trade, with the trademark name GloFish. Green fluorescent protein and yellow fluorescent protein versions are also available. Two similar types of transgenic crop organisms are much more relevant for food science. One is so-called Bt-corn (maize), which contains the gene to produce a protein toxin, called Bt toxin, from a type of soil bacteria, *Bacillus thuringiensis* (Bt), that kills insects from the order Lepidoptera (butterflies

Figure 5.73. The conformational structure of Bt toxin as determined by X-ray crystallography. Note the two domains, one dominated by alpha helices and the other consisting of beta sheets. (N. Galitsky et al., *Acta Crystallogr., Sect. D* 57:101–1109, 2001.)

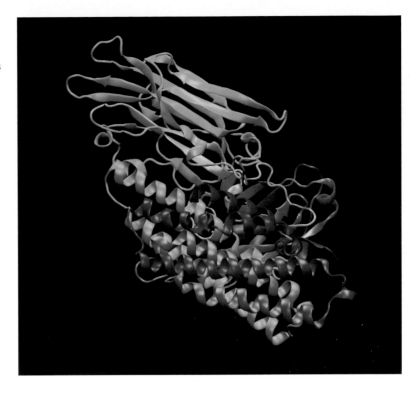

and moths; see Figure 5.73). The transgenic corn produces this bacterial toxin, which then protects the corn from one of its most destructive pests, the caterpillars of the European corn borer moth (*Ostrinia nubilalis*). The Bt toxin is harmless to the corn itself, to other types of insects, other animals, and to humans, at least as far as is known. That is to say, no scientific evidence has been reported in the reviewed technical literature demonstrating any harm from Bt toxin in the diet. The other widely planted GMO crop is Roundup Ready soybeans. These are soybeans that have been transgenetically modified by the insertion of a gene from bacteria of the genus *Agrobacterium* to give them resistance to the herbicide Roundup produced by Monsanto. The active ingredient of Roundup is glyphosate (Figure 5.74), an inhibitor of one of the enzymes involved in the synthesis of aromatic amino acids in growing plants. A number of other Roundup-resistant food crops have now also been developed and widely planted, including corn, cotton, sugar beets, and alfalfa. The advantage of this modification is that it allows the farmer to use herbicides for weed control without killing the crop plant. These crops have potential environmental benefits such as reduced soil erosion through no-till or minimum-till practices and reduced fuel use, as well as reduced labor costs. GMO crops, including corn, soybeans, and cotton, have also been developed by Bayer CropScience to be resistant to another broad-spectrum herbicide, glufosinate (Figure 5.74). These crops are marketed as LibertyLink crops. As with Bt corn, no definitive evidence exists that soybeans from GMO-engineered herbicide-resistant plants are in any

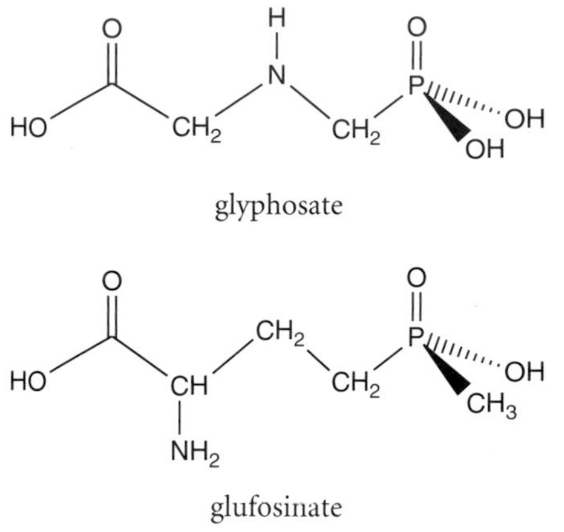

Figure 5.74. The structure of glyphosate, the active ingredient of Roundup, and of glufosinate, another widely used herbicide.

way harmful in the diet, but as with Bt corn, the current lack of such evidence does not mean that no harm exists.

The larger concern about GMO crops such as Bt maize and Roundup ready soy is not that these crops are harmful in the human diet, but that they may represent an environmental threat. For example, there have been concerns that Bt corn could harm nontarget Lepidoptera species such as monarch butterflies (*Danaus plexippus*). Evidence was presented to this effect, while other studies suggested that monarch butterfly populations might actually have increased due to its use, because this resulted in a significant decrease in broad-spectrum pesticide usage to control corn borers. Because ecosystems are so complex and typically experience natural fluctuations that are not always well understood, the full environmental impact of the large-scale shift to Bt corn in the United States is not yet known. For example, monarch butterfly populations are also very sensitive to environmental factors unrelated to Bt crops, such as severe winters and illegal forest clearance in the mountain pine forests of Central Mexico where the butterflies overwinter. Almost nothing is known about how Bt corn might impact Eurasian ecosystems and nontarget Lepidoptera on those continents.

More generally, widespread distribution of Bt corn raises the possibility that target pest species might quickly develop resistance to the Bt toxin, which worries farmers producing "organic" foods, because Bt toxin is permitted as an "organic" pesticide on "organic" produce. Additional concerns exist about the use of Roundup and Roundup Ready crops in North America. On one hand, after 20 years of the widespread use of glyphosate as an herbicide, a number of virulent glyphosate-resistant weeds have evolved in the United States and are eliminating any advantage to using Roundup Ready seeds. On the other hand, the planting of Bt corn is now so widespread that corn borer populations have fallen even on non-Bt crops because they are generally surrounded by Bt-protected fields. As of this writing, it remains to be seen whether the corn borer will develop resistance to the Bt toxin.

The escape of the genes from both types of modified crops into non-GMO crops or even wild plants is also a significant concern, perhaps particularly for a wind pollinated plant such as corn, although the heavy pollen grains of maize do not travel far from the parent plant. Unfortunately, it seems virtually impossible to determine in advance whether these vague fears are justified, and it is entirely possible that unforeseen dangers might exist that could prove to be even larger problems than those potential threats already identified. As with all such environmental threats, including the introduction of plant diseases and pests (like the European corn borer) to new continents as the result of international trade, the harmful effects only become apparent when it may be too late to do anything to correct the problem.

In general, neither starch produced by Bt corn nor the lecithin produced from Roundup Ready soybeans differs chemically in any way from that obtained from non-GMO crops, which is referred to as "substantial equivalence" by the FDA for regulatory purposes. Cornstarch, corn meal, corn syrups, and high-fructose corn syrup (HFCS), as well as the emulsifier lecithin from soy, are among the most commonly used ingredients in manufactured foods. In North America, the overwhelming majority of these commodities now come from genetically modified crops. In addition, GMO corn is used as feed for hogs, cattle, chickens, and even some types of farmed fish, which means that many people would consider the meat, eggs, and milk from these animals to be GMO foods. As a result, it is becoming increasingly difficult to find food products in the middle aisles of most US supermarkets (that is, in those aisles containing various manufactured or highly processed products) that do not contain at least one of these ingredients and would thus be considered as a GMO food. In Europe, consumer resistance to any GMO-containing food product is much greater than in North America, contributing to the difficulty in marketing US-manufactured food products in the European Union. This situation has also led to such incidents as the government of Zambia refusing food aid from the United States during a period of drought in 2002, because the shipments would have contained maize that was not certified GMO-free, raising the remote possibility that these genes might somehow have escaped into local crops, making them unmarketable in Europe when the drought abated.

The potential of genetic engineering raises the prospect of remarkable breakthroughs in agricultural productivity (Borlaug 2000), such as nitrogen-fixing wheat, rice, or corn; crops that can grow on saltwater, or are resistant to insect pests, drought or frost; or crops that have dramatically increased yields. Trees for wood or fruit production might be developed that have the growth rates of bamboo, and perennial varieties of staple crops such as wheat and corn would reduce soil erosion by reducing the need for tilling and plowing. Applications have already sought to increase the vitamin A content of rice and to introduce vaccine antigens into bananas. These applications would potentially bring the benefits of genetic engineering to the consumer, while current GMO crops tend to benefit primarily the farmer and the companies that produce the modified strains (and associated products such as glyphosate). The potential benefits of such GMO crops, however, will have to be weighed against any possible dangers to the environment or, less probably, to the consumer, that these crops might pose.

SUGGESTED READING

Aguzzi, A., and C. Weissmann. 1997. "Prion research: The next frontiers." *Nature* 389:795–798.

Borlaug, N.E. 2000. "Ending world hunger: The Promise of biotechnology and the threat of antiscience zealotry." *Plant Physiol.* 124:487–490.

Cantor, C.R., and P.R. Schimmel. 1980. *Biophysical Chemistry*. San Francisco: Freeman.

Caughey, B., and D.A. Kocisko. 2003. "Prion diseases: A nucleic-acid accomplice?" *Nature* 425:673–674.

Chinachoti, P., and Y. Vodovotz. 2001. *Bread Staling*. Boca Raton, FL: CRC Press.

Cohen, F.E., and S.B. Prusiner. 1998 "Pathologic conformations of prion proteins." *Annu. Rev. Biochem.* 67:793–819.

Deleault, N.R., R.W. Lucassen, and S. Supattapone. 2003. "RNA molecules stimulate prion protein conversion." *Nature* 425:717–720.

DeMarco, M.L., J. Silveira, B. Caughey, and V. Daggett. 2006. "Structural properties of prion protein protofibrils and fibrils: An experimental assessment of atomic models." *Biochemistry* 45:15573–15582.

Horne, D.S, 2006. "Casein micelle structure: Models and muddles." *Curr. Opin. Colloid Interface Sci.* 11:148–153.

Mead, S., et al. 2003. "Balancing selection at the prion protein gene consistent with prehistoric kurulike epidemics." *Science* 300:640–643.

Regenstein, J.M., and C.E. Regenstein. 1984. *Food Protein Chemistry. An Introduction for Food Scientists*. Orlando, FL: Academic Press.

Riek, R., S. Hornemann, G. Wider, R. Glockshuber, and K. Wuthrich. 1997. "NMR characterization of the full-length recombinant murine prion protein, mPrP(23–231)." *FEBS Lett.* 413:282–288.

Rudd, P.M., M.R. Wormald, D.R. Wing, S.B. Prusiner, and R.A. Dwek. 2001. "Prion glycoprotein: Structure, dynamics, and roles for the sugars." *Biochemistry* 40:3759–3766.

Voet, D., and J.G. Voet. 1995. *Biochemistry*, 2nd ed. New York: Wiley & Sons.

6.
Enzymes in Foods

ENZYMES IN FOODS

Enzymes are protein catalysts. In living systems they orchestrate and control the vast array of reactions that constitute life, such as metabolism, growth, mobility, and reproduction, and thus are present in all foods derived from animal and plant materials. Often they can be quite important in determining the postharvest or postslaughter properties of these materials as foods. Although they are much more efficient than nonbiological catalysts, like all other catalysts, enzymes work by speeding up the rate at which chemical equilibrium is reached, and not by changing the position of the equilibrium in any way. As such, they are not consumed or changed during the course of the reaction and after each catalytic step are available to catalyze the process again (although they may become inactivated by incidental processes or inhibited by product or other chemical species). For this reason, many enzymes in living systems are present in very small amounts, although some, such as the digestive glycosidases and proteases, are produced in large quantities.

Enzymes are becoming commercially available in higher purities and at lower costs than in the past, making the routine use of enzymes in food processing much more practical. The specificity and gentle reaction conditions of enzymes allow specific functional changes to be effected without resorting to such crude methods as heating or acid treatment, which may have many other undesirable effects on the component molecules. Naturally occurring enzymes are also present in many foods, and controlling their activity to prevent undesirable changes in fresh products is often important.

In general, the site in the enzyme where catalysis takes place is thought to have a close structural complementarity, in the distribution of its functional groups, to that of the substrate. The reactants then fit snugly into a specific binding position at this catalytic site, in the so-called lock-and-key model of enzyme–substrate interaction, as is illustrated schematically in Figure 6.1. The protein does not bind strongly to the reaction products, so that after the reaction takes place, the products are released, allowing the enzyme to catalyze the reaction again.

Enzymes can accelerate reaction rates by many orders of magnitude. They do this by promoting attainment of the transition state for the reaction. By specifically binding reactants, enzymes increase the local concentration of these reactants and, more important, ensure that they approach with the correct orientation for reaction. The actual binding is designed to be strongest not for the reactants, however, but for

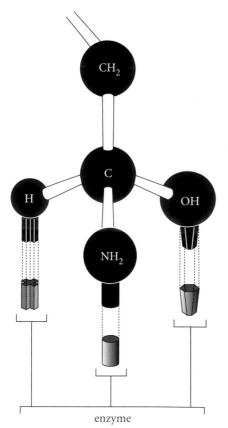

Figure 6.1. A schematic representation of the lock-and-key model for enzyme–substrate interactions.

the transition state complex, which is how the system is promoted to the transition state by the enzyme. Exploiting this principle, it is sometimes possible to make artificial enzymes called catalytic antibodies, or abzymes (antibody enzymes), by employing the response of animal immune systems to foreign molecules. If one could synthesize a nonreactive molecule that had many of the topological and functional features of what is believed to be the transition state for a particular reaction, this compound could be injected into the bloodstream of an animal such as a rabbit or mouse. The animal's immune system might recognize this molecule as an invading outside antigen and raise antibodies against it with binding sites that were the functional complements of this molecule and, hopefully, also of the actual transition state of interest. When isolated, these antibodies often are then found to indeed catalyze the desired reaction.

Enzymes are found naturally in many foods, and their actions are important in food quality in a variety of ways. For example, in Chapter 1 we mentioned the thiopropanal sulfoxide molecule produced by the enzyme alliinase when onions are cut, which is the compound responsible for the tears associated with chopping onions. Many natural enzymes catalyze degradative reactions involved in processes such as ripening or digestion. Often the consequences of these reactions are undesirable, and the goal of the food technologist is to control or prevent these reactions, but in other cases such natural reactions may be considered to contribute to the final quality of the product, as in the aging of cheeses. Commercially available enzymes are now also used for a variety of purposes in the manufacture of processed foods (Table 6.1).

The advantages of the use of enzymes in industrial processing are clear from their ability to accelerate reactions, but drawbacks to their use that do not

Table 6.1 Some examples of the uses of enzymes in the food industry

Enzyme	Food	Use
amylases	syrups	hydrolysis of starch in the production of high-fructose corn syrups
	baked goods	increase sugar content for yeast fermentation
cellulases	coffee	hydrolysis of cell-wall cellulose during drying of beans
	fruits	peeling of apricots, tomatoes
invertase	artificial honey	conversion of sucrose to glucose and fructose
lactase	ice cream	prevents crystallization of lactose, which makes ice cream have a sandy or grainy texture
	lactose-free milk	removal of lactose
pectic enzymes	fruit juices	improve yield of press juices, prevent cloudiness
proteases	meats and fish	tenderization
	cheese	casein coagulation; aging
	"milk"	production of soy milk
lipases	cheese	aging, ripening, development of general flavor

necessarily arise with other catalysts are also apparent. Unlike many other catalysts, enzymes are often rather fragile; many denature and thus become inactive at elevated temperatures, at surfaces, or in the presence of certain other denaturant molecules. They are also subject to acid-catalyzed hydrolysis and possible chemical modification under some conditions. The pH or temperature optimum of enzymes might be inconvenient for industrial processing, or their turnover rate might be unsatisfactorily slow for optimal industrial throughput. In some cases, these problems can be overcome by redesigning the enzyme through the techniques of modern genetic engineering and site-directed mutagenesis; in other cases, it may be necessary to modify the overall process or accept suboptimal performance. As difficult-to-isolate natural products produced in living tissues in only small quantities, enzymes tend to be expensive commercially.

To avoid the inconvenience of being limited to batch processing when using an expensive ingredient like an enzyme, **immobilized enzymes** are often used in industrial applications. If an enzyme can be covalently linked to a polymer matrix without interfering with its ability to function catalytically, then a flow bed or elution column can be packed with this enzyme-linked matrix, and the reaction carried out in a continuous fashion by passing the reactants over the matrix. It may also be possible to immobilize an enzyme by trapping it noncovalently in a polymer matrix, unless the substrates are also polymers that are so large that they cannot penetrate the pores and spaces in the lattice of the polymer matrix to reach the enzyme.

ENZYME KINETICS

As catalysts, enzymes change the rates at which reactions progress, but do not affect the final equilibrium—just the speed at which equilibrium is reached. Usually they are present in low concentrations in relationship to the substrates (reactants) in the reaction that they catalyze. The classic treatment of enzyme kinetics is the Michaelis-Menton equation, described in every biochemistry textbook. Enzyme catalyzed reactions exhibit the feature that, when the concentration of the substrate S, $[S]$, is very much higher than the concentration of the enzyme E, the reaction rate becomes approximately independent of the reactant concentration. This results from a reaction mechanism in which an enzyme–substrate complex ES is initially formed that subsequently converts to product, releasing the enzyme for further reaction.

$$E + S \underset{k_1}{\overset{k_2}{\rightleftharpoons}} ES \tag{6.1}$$

$$ES \xrightarrow{k_3} E + P \tag{6.2}$$

The initial rate of formation of product, $\dfrac{d[P]}{dt}$, or the reaction velocity v, is also the rate of disappearance of the substrate, or the reaction rate, which can also be expressed as the rate of disappearance of the intermediate complex ES, $k_3[ES]$,

$$-\frac{d[S]}{dt} = \frac{d[P]}{dt} = v = k_3[ES] \tag{6.3}$$

If $[E]$ is the total enzyme concentration added to the system (free enzyme plus that bound up in the transition state complex ES), then the concentration of free enzyme is $[E] - [ES]$, so the rate of formation of ES in the second order forward process is

$$\frac{d[ES]}{dt} = k_1([E]-[ES])[S] \tag{6.4}$$

The rate of breakdown of ES is

$$-\frac{d[ES]}{dt} = k_2[ES] + k_3[ES] \tag{6.5}$$

When a steady state is reached, $\dfrac{d[ES]}{dt} = 0$, so

$$k_1([E]-[ES])[S] = k_2[ES] + k_3[ES] \tag{6.6}$$

$$\frac{([E]-[ES])[S]}{[ES]} = \frac{k_2+k_3}{k_1} = K_m \tag{6.7}$$

Solving for $[ES]$,

$$[E][S] - [ES][S] = K_m[ES] \tag{6.8}$$

$$[ES] = \frac{[E][S]}{K_m + [S]} \tag{6.9}$$

so, substituting into Equation 6.3,

$$-\frac{d[S]}{dt} = \frac{d[P]}{dt} = v = k_3[ES] = \frac{k_3[E][S]}{K_m + [S]} \tag{6.10}$$

The maximum possible velocity occurs when the substrate is present in such a high concentration that essentially all of the enzyme is complexed as much of the time as possible, $[S] \gg [E]$ and $[S] \gg K_m$, so

$$v_{max} = k_3[E], \quad \text{pseudo 1st order in } [E], \text{ 0th order in } [S] \tag{6.11}$$

So, one can substitute v_{max} into the equation for the initial velocity above, giving the so-called Michaelis-Menton Equation

$$v_0 = \frac{v_{max}[S]}{K_m + [S]} \qquad (6.12)$$

If substrate is present in very low concentration, $[S] << K_m$, then Equation 6.10 gives

$$v = \frac{k_3}{K_m}[E][S] \qquad (6.13)$$

One can invert the Michaelis-Menton equation for v_0 to give

$$\frac{1}{v_0} = \frac{K_m + [S]}{v_{max}[S]} \qquad (6.14)$$

$$\frac{1}{v_0} = \frac{K_m}{v_{max}}\frac{1}{[S]} + \frac{1}{v_{max}} \qquad (6.15)$$

If one plots $1/v_0$ versus $1/[S]$, a straight line will result, with a slope of K_m/v_{max}, a y-intercept of $1/v_{max}$, and an x-intercept of $-1/K_m$. Such a plot, illustrated in Figure 6.2, is called a **Lineweaver-Burk plot**.

ENZYME THERMODYNAMICS: THE PRODUCTION OF HIGH-FRUCTOSE CORN SYRUPS

There is great interest in the food industry regarding the conversion of corn starch, which is available in abundance, into fructose for use as a sweetener (Chapter 4). The basic steps involved are the hydrolysis of the glycosidic linkages in the amylose and amylopectin polymer chains to produce free glucose, and then the isomerization of the aldohexose D-glucose into the ketose D-fructose. The endpoint is a viscous **high-fructose corn syrup** (HFCS). The hydrolysis step can be accomplished by acid treat-

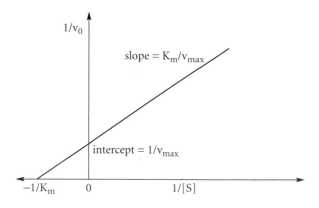

Figure 6.2. A Lineweaver-Burk plot illustrating Michaelis-Menton enzyme kinetics.

Figure 6.3. The general reaction scheme for the enzymatic production of fructose from corn starch.

Figure 6.4. A cartoon representation of the backbone structure of xylose isomerase from *Streptomyces rubiginosus*, which catalyzes the conversion of xylose to xylulose, and of glucose to fructose. As the sugar binds in the active site, the pyranose ring opens up to the linear form; a product xylulose molecule in its open chain form is also indicated bound in the active site. There are two essential magnesium or manganese ions in the active site, shown as small orange spheres, that are bridged by the acid group of a glutamic acid residue, which effect the hydride shift (Kovalevsky et al. 2008).

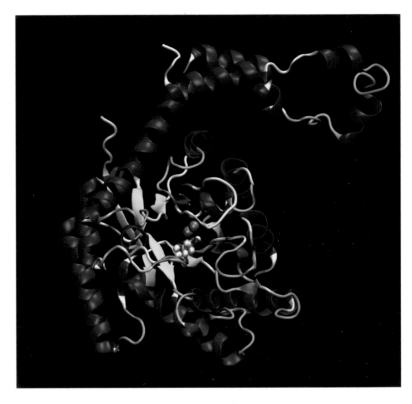

ment at elevated temperatures, but this also leads to the production of undesirable reversion products with a bitter flavor, as we saw in Chapter 4. Currently the best means available for producing high-fructose corn syrups employs enzymes to accomplish both tasks (Figure 6.3). Amylases are used to cleave the glycosidic linkages, and **xylose/glucose isomerase** (Figure 6.4) is used to convert the aldose into the ketose (this enzyme evolved to isomerize xylose to xylulose, but also isomerizes glucose to fructose).

A great deal of effort has gone into the use of modern biotechnological methods to develop improved glucose isomerases that have greater thermal stability or optimal activity over a broader range of pH. These efforts can only go so far in increasing the yield of fructose, however, because of the thermodynamics of the scheme shown in Figure 6.3. The principal problem is that both glucose and fructose have similar free energies (in fact, glucose is slightly lower in energy; see Figure 6.5). Since enzymes are catalysts, which do not affect the underlying chemical equilibrium, the enzymatic

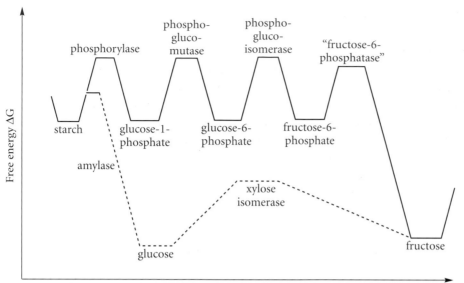

Figure 6.5. A schematic representation of the relative energies of starch, glucose, fructose, and various phosphorylated glucose intermediates. (Redrawn from A. Moradian and S.A. Benner, *J. Am. Chem. Soc.* 114:6980–6987, 1992.)

reactions will not in general go to completion, instead typically converting less than half of the glucose to fructose even under ideal conditions—unless some way can be found to continuously remove the fructose as it is produced, to drive the reaction to the right. Incomplete conversion is undesirable because the two sugars are not equally sweet. Fructose is sweeter than sucrose, while taste panels rate glucose as being only about 70% as sweet as sucrose, which is why high-fructose corn syrups are sought (Chapter 4). For this reason more complete isomerization would be desirable. In the industrial production of HFCS, high final fructose levels (90%, HFCS 90) can be achieved by using liquid chromatography to separate out the two sugars, and then subjecting the glucose portion to further isomerization.

It has been proposed that a more efficient conversion could be accomplished without the need for chromatographic separation if the isomerization reaction could be carried out in a fashion similar to the way in which reactions are organized in living cells (Moradian and Benner 1992). In vivo, reactions are coupled and segregated in space and time as required, and often carried out as a series of steps leading to the final product as a way of imposing greater control. In the conversion of starch to high-fructose corn syrups, the step with the largest energy change is the hydrolysis step, which as a result of the large energy decrease leads to a near complete conversion to glucose (in principle; in actuality, difficulties resulting from insolubility, branching, and the packed structure of starch granules prevent complete hydrolysis in an actual industrial application). Moradian and Benner have suggested storing the energy of the glycosidic bond in the product glucose by using the enzyme **phosphorylase** to cleave the glycosidic bond and simultaneously phosphorylate the anomeric carbon, to give glucose-1-phosphate (Figure 6.5). Because of the high energy of the glucose–phosphate bond, nearly the same as the glycosidic bond, this reaction will produce an equilibrium between this product and glucose, with roughly equal amounts. Another naturally occurring enzyme, phosphoglucomutase, could then be used to move the phosphate group from the anomeric carbon to the exocyclic C6 carbon, again with little change in free energy, so that an equilibrium will exist between all three molecules, again with roughly equal amounts. Yet another naturally occurring enzyme, phosphoglucoisomerase, can convert the product glucose-6-phosphate into fructose-6-phosphate, again with little change in free energy. Because each of these phosphorylated molecules has almost the same energy, the reaction will not be driven to any one of these species, but an equilibrium will exist with roughly equal populations of each. Fructose-6-phosphate is close to the desired end product, fructose, and if an enzyme could be found that removed the phosphate group from this molecule, the large change in energy would essentially drive the reaction to completion, causing the entire coupled sequence to shift to near complete conversion, subject to the difficulties of starch granule digestion (Figure 6.6). Unfortunately, no "fructose-6-phosphatase" enzyme has been found, but it might be produced by protein engineering or by using abzymes, and might be a more profitable target for further research in this area than trying to make a more efficient glucose isomerase.

With this introduction to the kinetics and thermodynamics of enzymes, let us turn to a brief survey of some of the enzymes important in foods and food processing. These enzymes can be classified into a number of broad categories based on their functions. The most important of these are the proteases, glycosidases, lipases, and polyphenol oxidases.

PROTEASES

Proteases, or proteinases, are various enzymes that cleave peptide bonds and function to digest polypeptide chains. Proteolytic enzymes can be classified as **exopeptidases**, which remove amino acid residues from the ends of polypeptide chains, and **endopeptidases**, which can cleave a chain at various bonds interior to the chain. Humans produce a number of proteases to digest proteins in food. The acid-resistant endopeptidase pepsin is produced by chief cells in the stomach. Trypsin and chymotrypsin, produced by the pancreas, digest peptides and proteins in the duodenum of the small intestine. These enzymes are available commercially and are used primarily for research purposes. Trypsin is sometimes used commercially to predigest protein in baby foods as a digestive aid. The pancreas also produces the carboxypeptidases, which are exopeptidases. Many other species also produce versions of these digestive enzymes.

Endopeptidases

Generally the active sites of endopeptidases are such that these enzymes have characteristic specificities and will only cleave bonds following or preceding certain

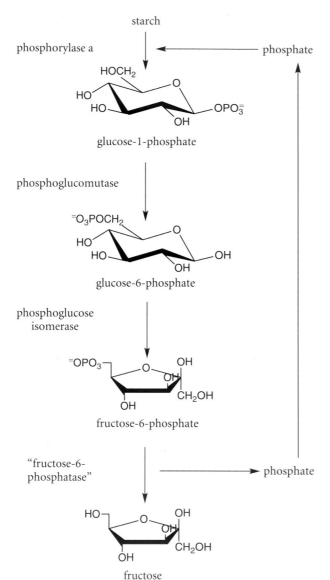

starch

phosphorylase a

glucose-1-phosphate

phosphoglucomutase

glucose-6-phosphate

phosphoglucose isomerase

fructose-6-phosphate

"fructose-6-phosphatase"   → phosphate

fructose

phosphate

Figure 6.6. A proposed alternate scheme for the conversion of starch to fructose. Because the large energy release is in the final step, the reaction would be driven essentially to completion without the need for separating glucose from fructose. (Redrawn from A. Moradian and S.A. Benner, *J. Am. Chem. Soc.* 114:6980–6987, 1992.)

amino acids (Table 6.2; Figure 6.7). For example, the enzyme **trypsin** will hydrolyze only peptide bonds that follow the charged basic amino acids arginine and lysine. If one of these residues is followed by a proline, however, the bond cannot be cleaved. Because of this specificity, hydrolysis of a protein by trypsin will break up the chain into a series of smaller oligopeptides called tryptic peptides. Trypsin generally can only attack exposed peptide bonds as in denatured proteins, and often such bonds in native, folded proteins are protected from attack.

The enzyme **chymotrypsin** is a related endopeptidase that only cleaves peptide bonds that follow large hydrophobic residues such as phenylalanine, tyrosine, or tryptophan. This specificity for bonds following residues with large hydrophobic side

Table 6.2 Endopeptidase specificities

Enzyme	Specificity
chymotrypsin	R_{n-1} = Phe, Trp, Tyr, other bulky hydrophobic residues; $R_n \neq$ Pro
trypsin	R_{n-1} = Arg, Lys; $R_n \neq$ Pro
pepsin	R_n = Leu, Phe, Trp, Tyr; $R_{n-1} \neq$ Pro; a preference for Phe, Trp, Tyr, Ile
thermolysin	R_n = Ile, Met, Phe, Trp, Tyr, Val; $R_{n-1} \neq$ Pro
chymosin	R_{n-1} = Phe; $R_n \neq$ Pro

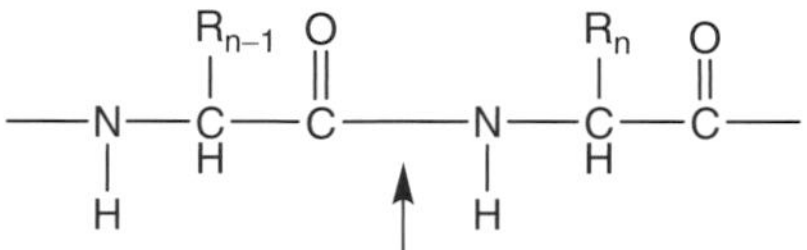

Figure 6.7. Endopeptidases in general will cleave only certain peptide bonds in a substrate protein, with specific requirements on the identities of the residues preceding the bond to be cleaved or on the residue following the scissile bond (see Table 6.2).

chains is because of the presence of a nonpolar pocket into which these side chain groups fit (see Figure 6.8), properly positioning the next bond relative to the catalytic residues. Both of these enzymes are examples of a class of proteases called **serine proteases**, or serine endopeptidases, because they contain an essential serine residue in their active sites. Enzymes in this class also contain histidine and aspartic acid residues in the active site in a conserved relative arrangement called a catalytic triad (see Figures 6.8 and 6.9). The nucleophilic serine initiates the cleavage by coordinating to the carbon atom of the scissile peptide bond (that is, the one being cleaved). The lone pair of the nearby histidine δ^1 nitrogen atom simultaneously abstracts the proton of the serine hydroxyl group and, stabilized by the aspartic acid, donates it to the amide nitrogen of the peptide bond, thereby effecting the cleavage. The lone pair of the histidine δ^1 nitrogen atom then abstracts a proton from a nearby water molecule and transfers it to the serine oxygen at the same time as the resulting hydroxyl group attaches to the carbonyl carbon atom, completing the hydrolysis and restoring the catalytic triad to its original state (Figure 6.9).

Pepsin is the principal protein-digesting enzyme found in the very acidic medium of the stomach and is therefore stable in an acid environment. Trypsin and chymotrypsin are secreted into the duodenum by the pancreas, where the pH is much higher than in the stomach (a low pH would protonate Asp 102 and prevent the initial step of the mechanism shown in Figure 6.9). Proteins that are stable toward degradation by acid and pepsin for periods of up to an hour have the potential for being allergens, because they will arrive in the intestines essentially intact (of course, most such proteins will not be allergenic; also, some digestion fragments may also be allergenic;

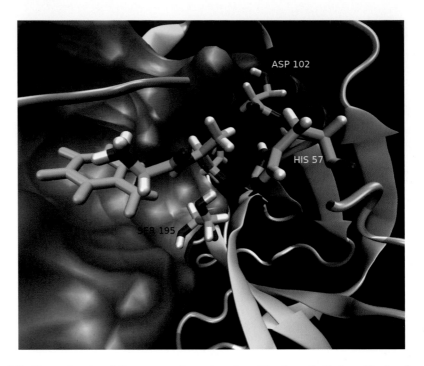

Figure 6.8. The active site of the enzyme chymotrypsin, with a hypothetical positioning for a bound Phe-Ala-Ala tripeptide substrate. The side chains of the catalytic triad residues are shown in atomic detail as a liquorice representation, as are the residues of the substrate chain, and the van der Waals surface of the atoms making up the binding pocket for aromatic residues in the R_{n-1} site is shown in orange. The remainder of the enzyme is shown as a backbone cartoon, with yellow arrows for antiparallel beta sheets and a stretch of alpha helix indicated in purple. Nitrogen atoms are in blue, oxygen in red, carbon in green, and hydrogen atoms in white. Notice how the proton of the Ser 195 points toward the lone pair of the nearby His 57, which is in turn stabilized by the nearby acid group of Asp 102. (Substrate coordinates adapted and built from information given in M.R. Groves et al., *Protein Eng.* 11:833–840, 1998.)

these are called **incomplete allergens**). Pepsin was the second enzyme to be crystallized, after urease, and its crystal structure is well known. It is not a serine protease, but rather belongs to a class of enzymes called aspartic acid proteases. Pepsin has a pair of aspartic acids in its active site, with one serving as a general base that abstracts a proton from a water molecule, generating a hydroxyl group that then attacks the carbonyl atom of the scissile bond, producing a negatively charged oxyanion that is stabilized by the other acid group. The original acid group, now protonated, then donates its proton to the nitrogen atom of the leaving group, leading to bond scission. Because of their activity in degrading proteins, it is, of course, important for the action of proteases to be tightly controlled. For this reason, all three of these enzymes are initially produced as inactive proenzymes or **zymogens**, with additional peptide sequences covering up their active sites. These extra polypeptide sequences are removed by other proteases, activating the enzyme, when they are needed.

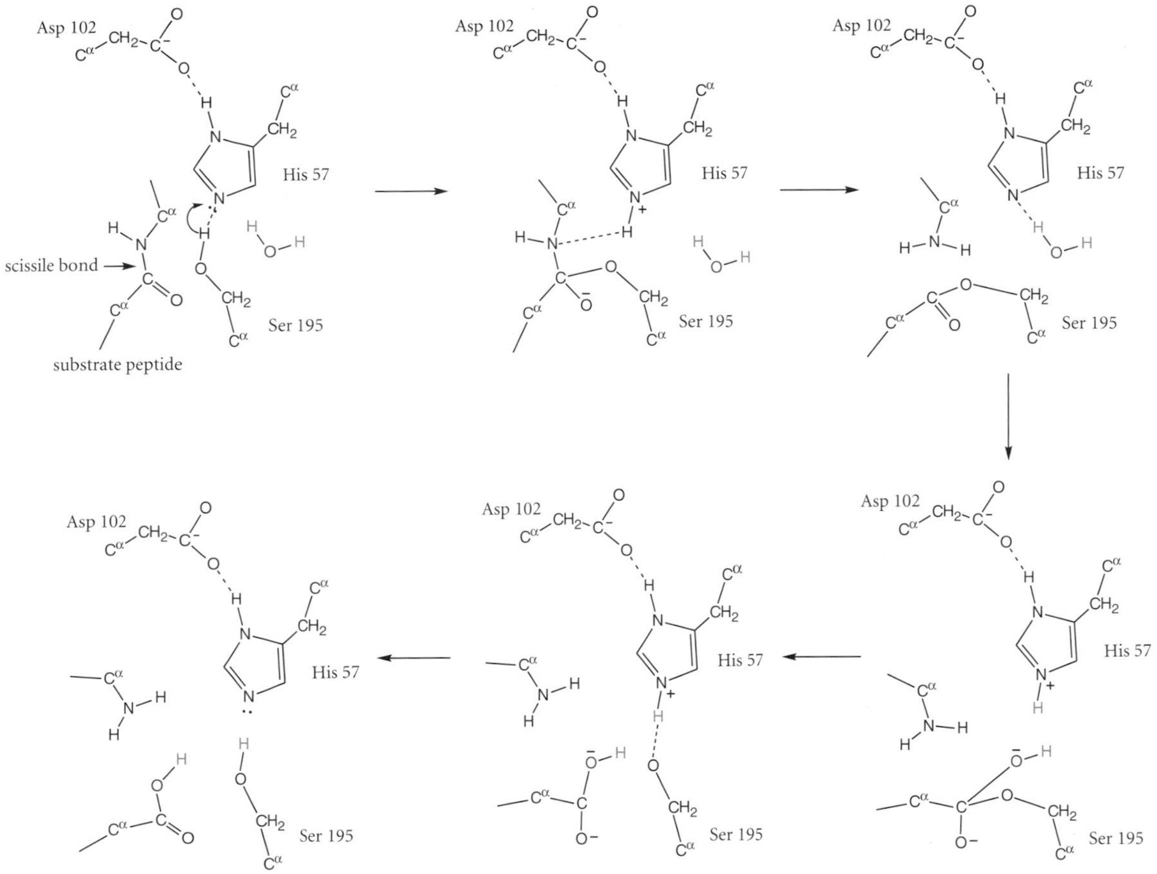

Figure 6.9. The catalytic mechanism of chymotrypsin.

Perhaps the most important of the endopeptidase proteases for food processing is the acid protease **chymosin**, or **rennin**, which is the major enzymatic component of **rennet**, produced from the fourth stomach of young milk-fed calves and used in cheese making. In young mammals, the agglutination of the casein micelles in milk that is produced by the chymosin enzyme slows the passage of the milk proteins through the intestines of the infant, thus allowing more thorough digestion. Chymosin is not a serine protease, and is similar in its catalytic mechanism to pepsin. It also belongs to the class of aspartic proteases, with two catalytic aspartic acid residues in the active site (see Figure 6.10). Like pepsin, chymosin is active at the low pHs found in the stomach, unlike the serine proteases, and the two carboxylic acid groups in its active site serve as the proton donor and as the nucleophile in a double displacement mechanism.

Chymosin cleaves several types of peptide bonds at low pH, but at pH 6.7 it cleaves only the peptide bond after phenylalanine residues. This specificity is again because there is a large nonpolar pocket into which the phenylalanine side chain fits by hydrophobic association (Figure 6.10). In the case of κ-casein, which is partially

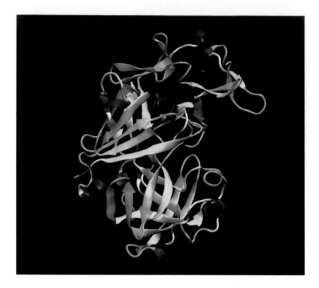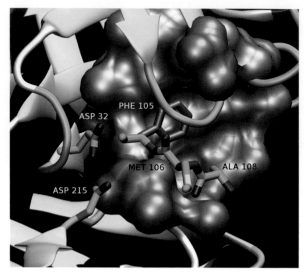

Figure 6.10. On the left is shown a cartoon representation of the three dimensional backbone conformation of bovine chymosin, showing the prominent active site cleft, as determined by X-ray diffraction. On the right is a view of the active site from this structure, with a Phe-Met-Ala tripeptide built into the active site, using as a guide the atomic positions of an inhibitor cocrystallized with the protein in an X-ray study. The Phe 105 residue in the R_{n-1} position is shown in pink, and most of the pocket into which this residue fits is shown as a van der Waals surface. Those atoms making up this pocket that are nonpolar are colored gray, and those atoms that are polar are shown in gold. The reverse turn loop consisting of residues 75–77, shown as a blue loop in the foreground, is part of this hydrophobic pocket, but is not shown as van der Waals surfaces because they would obscure the view of the Phe ring. As can be seen, those atoms that are directly adjacent to the Phe, which constitute a recognition site for this type of side chain, are hydrophobic. The catalytic Asp residues are also shown. (Based on the crystallographic coordinates determined by Groves et al., *Protein Eng.* 11:833–84, 1998).

buried in the casein micelle, with its C-terminal end sticking out into solution, chymosin cleaves the exposed bond between Phe 105 and Met 106 to produce the so-called **para-κ-casein** (residues 1–105) and a soluble **macropeptide** (residues 106–169). The para-κ-casein remains embedded in the submicelle, while the macropeptide goes into solution. The loss of its steric stabilization leads to the clumping of the casein micelles into curds, as the essential step in cheese making (see Chapter 5). The rest of the water-soluble macropeptide fragment (Figure 6.11), contains no Phe, and is not susceptible to further attack by chymosin at physiological pHs. Several Phe residues occur in para-κ-casein, but further cleavage of this polypeptide is inhibited because the chain is buried inside the submicelles (or within the micelle matrix, depending on which of the casein micelle models is correct), and may be partially globular in character (due to the cross-linking of the disulfide bond between Cys 11 and Cys 88), and thus protected from hydrolysis. Chymosin is easily deactivated by heat denaturation. Other acid proteases produced by molds growing on the surface or throughout certain soft cheeses, such as Roquefort, Stilton, Camembert, and Brie, play important roles in the ripening of these cheeses, but these proteins have not yet had their conformational structures determined. Because supplies of calf rennet are limited, much industrial cheese making now employs rennet produced from bacteria that have been

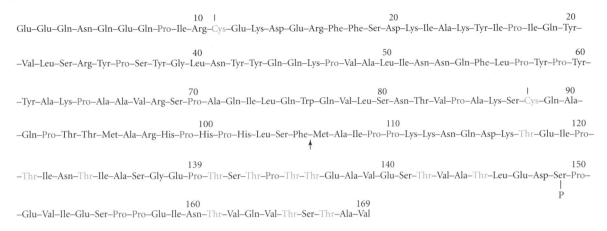

Figure 6.11. The sequence for κ-casein data (taken from J.C. Mercier, G. Brignon, and B. Ribadeau-Dumas, *Eur. J. Biochem.* 35:222–235, 1973; and H.M. Ferrell, *J. Dairy Sci.* 87:1641–1674, 2004). The chymosin cleavage site is indicated by an arrow; the glycosylated Thr residues are indicated in green, and the single phosphate group at Ser 149 is indicated by a "P". Prolines are indicated in purple, phenylalanine residues in red, and the two cysteine residues in yellow, with their cross-linking implied by vertical red lines. Threonine residues, which are possible sites for *O*-linked glycosylation, are colored green.

genetically engineered to produce calf chymosin. Such cheeses might thus be considered by some to be GMO foods.

Papain, extracted commercially from papaya latex, is a **sulfhydryl protease**, with the reactive thioalcohol S—H group of a cysteine residue as its catalytic residue, which cleaves bonds following Phe, Asn, Glu, and Tyr. It is a 212–amino acid, single-chain protein with a two-domain structure, as shown in Figure 6.12. The active site of the enzyme is located in the cleft between the two subdomains. It is similar to the sulfhydryl proteases **bromelain** from pineapple and **ficin** from tropical figs, which are both sequence homologs of papain. The three-dimensional conformations of bromelain and ficin have not been determined as of this writing, but they are presumably similar to that of papain. The structure of a small inhibitor molecule of bromelain, bromelain inhibitor VI, has been determined and is also shown in Figure 6.12. This protein is one of several that the pineapple plant makes to control the activity of bromelain. Papain and bromelain have found commercial use as meat tenderizers. Enzymes that make only a few specific cuts in polypeptide chains are more useful for this type of application than the more general proteases such as carboxypeptidase C (next section), which would essentially liquefy the meat. They can be injected in solution into a carcass, or applied on the surface of smaller cuts of meat. Injecting can cause overdigestion of some parts of the carcass due to their uneven distribution, because the enzymes do not readily perfuse through the meat tissues, while surface treatment would primarily produce surface digestion for the same reason.

Bromelain is found in both the fruits and stems of pineapple, but is produced commercially from the stems, which would otherwise be a waste product. Because of its presence in pineapple fruit, marinating meat with fresh pineapple and its juice

 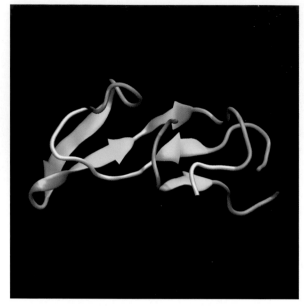

Figure 6.12. On the left is shown a conformational cartoon representation of the structure of papain, showing the two-domain structure, with one domain dominated by beta sheets and the other domain predominantly alpha helical. On the right is illustrated the solution conformational structure of bromelain inhibitor VI, a small predominantly beta sheet regulator of the activity of bromelain (Kamphuis et al. 1984, Hatano et al. 1996).

can have a tenderizing effect, but this enzyme is heat denatured by cooking or pasteurization, so processed pineapple juice or canned pineapples do not have this effect. Bromelain is also the reason that fresh pineapple cannot be used to make Jell-O desserts, because the enzyme hydrolyzes the collagen strands and prevents the gelatin from setting. Papain is also used commercially to prevent haze formation in beer, because it hydrolyzes large protein fragments remaining after the malting of barley, resulting in smaller soluble fragments.

One can do an easy experiment in a home kitchen to see the effects of bromelain hydrolysis. Modern marshmallows are generally made from sugar and/or glucose (dextrose), corn syrup, and maltodextrins, trapped in a matrix of gelatin (collagen). All of the carbohydrate ingredients are water soluble, so that hydrolysis of the polypeptide matrix leads to the complete dissolution of the marshmallow. Fresh, raw pineapple juice is rich in bromelain, so by cutting up a fresh pineapple and expressing the juice at room temperature, a solution of active bromelain (among many other things) can be obtained. Taking four small bowls, place a generous sample of such raw juice into two of the bowls. In one of the bowls containing pineapple juice, place a small-size mini-marshmallow (approximately 1 cm diameter). In the second bowl containing pineapple juice, place a small sliver of fresh meat (beef, chicken, or pork, for example). Fill the third bowl with white vinegar, and place a small marshmallow in it as well. Fill the final bowl with tap water and again place a small marshmallow in it as a control. Let all four bowls sit at room temperature overnight. At the end of this period, the marshmallow in the fresh pineapple juice will be found to have completely disappeared as the collagen was hydrolyzed into soluble peptide fragments by the bromelain, and the released carbohydrates dissolved. The marshmallow in the vinegar will also be seen to have disappeared, in this case as the result of acid hydrolysis by the acetic acid of the vinegar. The small piece of meat will still be present, but its aspect will have changed. The surface will probably have a "fuzzy" appearance, and the meat fibers can be easily separated by touching them with a fork or spoon, if they have not already separated spontaneously. The bromelain cannot fully penetrate the muscle tissue, so its activity will mostly involve the surface and the spaces between the muscle fibers, and the collagen in these regions. The control system, the marshmallow in tap water, will be seen to be unchanged over this time period. A second control could also be performed using the juice from canned pineapples, which will have denatured bromelain due to the thermal processing before canning. Again, no effect on an immersed marshmallow should be seen in this control system.

Exopeptidases

Exopeptidases cleave off single amino acids from the ends of polypeptide chains (Figure 6.13). The best known examples of this class are the carboxypeptidases, which remove amino acids from the C-terminal end of polypeptide chains. Carboxypeptidase A is a metalloenzyme that contains an essential zinc atom in its active site, and which cleaves peptide bonds of C-terminal aliphatic residues and particularly prefers the aromatic residues phenylalanine and tryptophan. Because of the importance of the zinc atom, it can be inhibited by EDTA (Chapter 10), a chelating agent that complexes with zinc atoms. Carboxypeptidase B cleaves off the basic residues lysine and arginine. The action of carboxypeptidase C can lead to complete digestion of a protein because this enzyme has no specificity requirements and is not deterred by proline either as the terminal residue or preceding it, as the other two types are (Table 6.3).

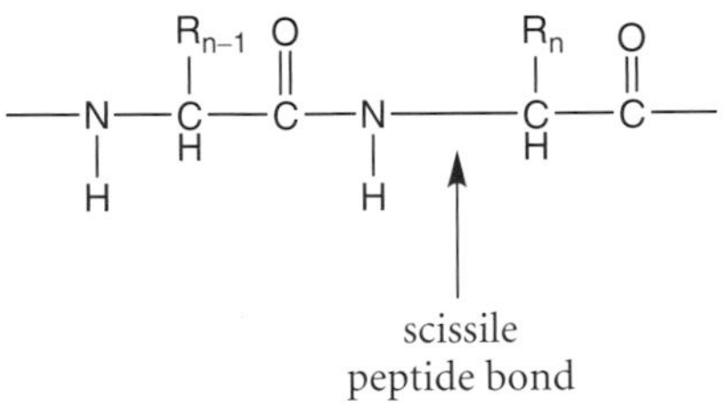

Figure 6.13. Like endopeptidases, exopeptidases in general will cleave only certain peptide bonds in a substrate protein, with specific requirements on the identities of the residues preceding the bond to be cleaved or on the residue following the scissile bond, as specified in Table 6.3.

scissile
peptide bond

Table 6.3 Exopepdiase specificities

Enzyme	Specificity
carboxypeptidase A	$R_n \neq$ Arg, Lys, Pro; $R_{n-1} \neq$ Pro; prefers Phe, Trp, and Ile
carboxypeptidase B	$R_n =$ Arg, Lys; $R_{n-1} \neq$ Pro
carboxypeptidase C	all amino acids, including proline

GLYCOSIDASES AND OTHER ENZYMES WITH CARBOHYDRATE SUBSTRATES

Several glycosidases are important in foods, either for their use as industrial catalysts, or because controlling their activity is important for food quality.

Amylases

As already discussed, amylases are widely used in the food industry to produce corn syrups, and glucose/xylose isomerase (which is not a glycosidase) is used to convert the liberated glucose monomers into fructose in the production of HFCS.

Pectic Enzymes

Pectic enzymes are found in plants and microorganisms and are enzymes capable of degrading pectin. They are divided into pectinases, pectin esterases, and pectin lyases. These enzymes are responsible for the softening of fruits as they ripen and also cause generally undesirable softening during postharvest storage. **Pectin esterase**, or **pectin methyl esterase**, converts methoxyl groups to acids in pectin (Figure 6.14) and is used to produce low methoxyl (LM) pectins. It is thus not a glycosidase.

The rate of this reaction can be followed by monitoring the rate of methanol production. The removal of the methyl groups by pectin methyl esterases allows the pectin chains to more readily form calcium-bridged juncture zones with other pectin molecules, leading to a partial gelling and an increase in stiffness or firmness in the vegetable tissues in the presence of calcium.

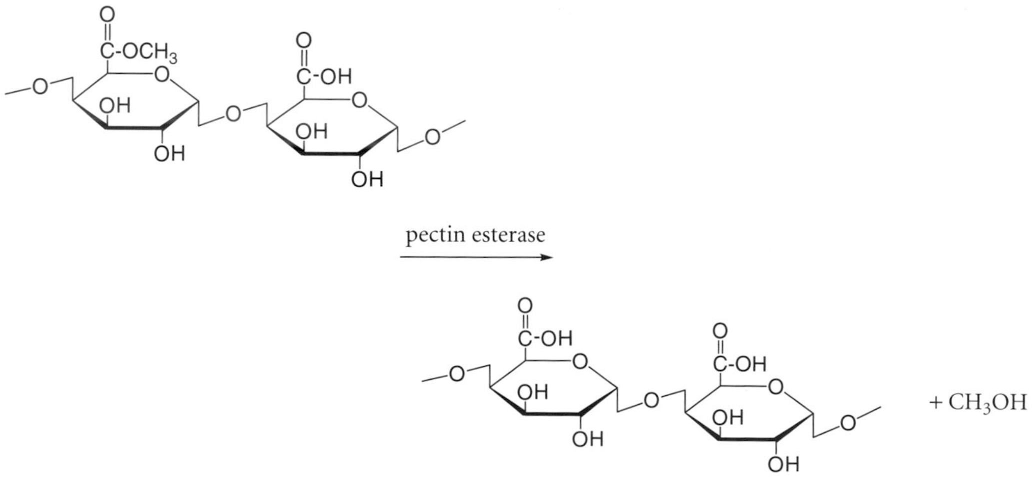

Figure 6.14. The reaction catalyzed by pectin methyl esterase.

Figure 6.15. The reaction catalyzed by pectinase.

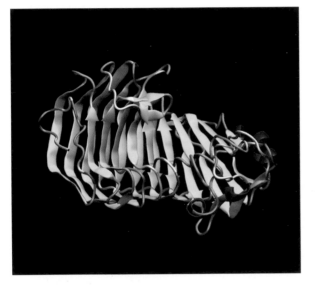
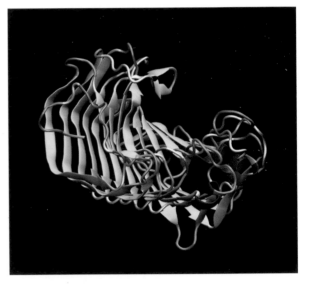

Figure 6.16. Two views of the pectinase endopolygalacturonase II from *Aspergillus niger*, showing both the active site groove and the unusual β-helix structure composed of four parallel β-sheets. (Y. van Santen et al., *J. Biol. Chem.* 274:30474–30480, 1999.)

Pectinase, or **polygalacturonase**, hydrolyzes the glycosidic linkages in pectins (Figure 6.15). Unlike the action of the pectin methyl esterases, the hydrolysis of the glycosidic linkages in the pectin of plant tissues leads to a significant softening or decrease in firmness of the plant material and is often undesirable.

The crystal structures of two pectinases from *Aspergillus niger* have been determined by X-ray crystallography; one of these is shown in Figure 6.16. These proteins are made up largely of parallel β-sheets twisted into a helix called a β-helix. They make random cleavages of glycosidic bonds anywhere in the pectin chain. The active

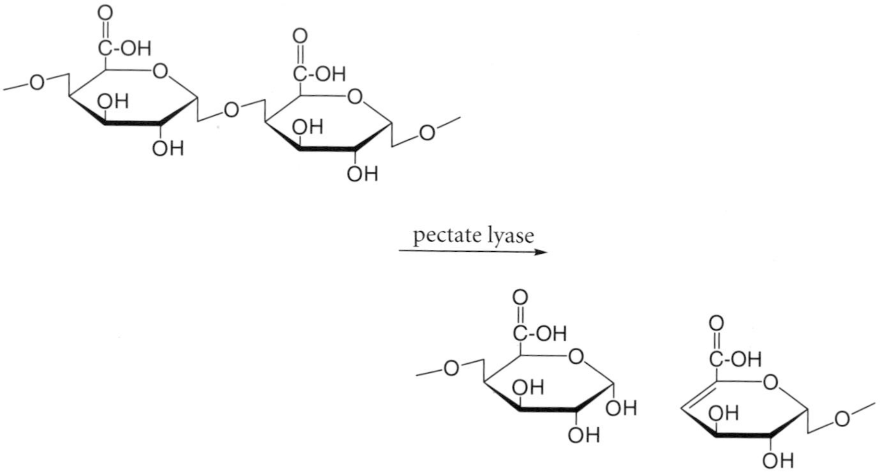

Figure 6.17. The cleavage of the pectin glycosidic bond by pectin lyase.

site groove contains a number of essential aspartic acid residues that serve to bind the polysaccharide chain into the site as well as function as the catalytic acid and base residues.

A third class of pectic enzymes consists of the **pectin lyases**, which also cleave the glycosidic linkage between two successive galacturonic acid residues, but without the addition of a water molecule, so that the cleavage reaction is not a hydrolysis. Instead, these enzymes cleave the bond by a *trans* elimination of hydrogen from the C4 and C5 carbon atoms of the nonreducing, aglycon end of the bond, resulting in a dehydration at the C4 position and a C4 = C5 double bond in the terminal residue of the nonreducing end of one of the product chains (Figure 6.17). Pectate lyases are generally bacterial enzymes and are not present in fruit and vegetable tissues.

Pectic enzymes in fruits such as tomatoes are found in the cytoplasm in the interior of the cells, so that the pectin of the intercellular matrix is separated and protected from these enzymes and is not degraded by them. When the cellular structure is broken, however, these enzymes are released to attack the pectin, rapidly degrading the structural integrity of the intercellular matrix. Pectic enzymes are used commercially in the production of fruit juices and in the production of low methoxy pectins. They are also used in the clarification of fruit juices to hydrolyze the large pectin polymers into smaller chains and eventually galacturonic acid, making them water soluble, reducing the viscosity and thus their ability to suspend particles.

In the production of tomato juice and tomato puree, however, cloudiness is important for an acceptable appearance, and thus it is desirable for the high levels of pectin esterase found naturally in tomatoes to be controlled; otherwise, they will quickly destroy the pectins when the tomato is broken up. To prevent this, a **"hot-break" method** is used in which the tomatoes are heated to high temperatures (~95 °C) before being broken up to denature these enzymes. Similarly, stewed tomatoes will become less firm ("mushy") due to the action of these enzymes upon slicing and peeling unless first heated. In addition to thermal denaturation of the pectic enzymes, $CaCl_2$ is usually added to cause the gellation of the pectin (Chapter 4), with the egg-box juncture zones restoring some of the firmness lost in cooking. The "cold break" method, where the tomatoes are only heated to around 65 °C, produces a less viscous paste due to pectinase activity but retains more flavor since destruction of heat-labile flavor components is reduced.

Pectic enzymes in fruits and vegetables can cause excessive softening in these products. In addition, many bacteria and fungi produce pectic enzymes, and bacterial degradation of pectin in plant tissues causes a kind of spoilage called "soft rot."

Lactases

Lactases from yeasts and fungi have several commercial applications. Because many people are lactose intolerant, lactases are used commercially to hydrolyze the lactose of milk and milk products. They are also used in the manufacture of some ice creams and condensed milk, not only to render them acceptable for intolerant individuals, but also to prevent sandiness from developing as the result of lactose crystallization. Lactose is less soluble than glucose (Table 4.7), although twice as soluble as galactose. An equimolar mixture of glucose and galactose is sweeter than an equivalent weight of lactose (see Table 4.4; recall that one mole of lactose produces two moles of mono-saccharides—one of glucose and one of galactose). Thus, milk, yogurt, or ice cream treated with lactase is sweeter than untreated products. For lactose-intolerant individuals, commercial preparations of lactase in capsule form are available that can be taken immediately before eating lactose-containing foods, reducing symptoms.

LIPASES

Lipases belong to the class of enzymes called esterases that catalyze the hydrolysis of ester linkages; lipases hydrolyze the ester linkages in lipids (see Chapter 7), producing an alcohol and a free fatty acid (Figure 6.18). Humans produce a pancreatic lipase as

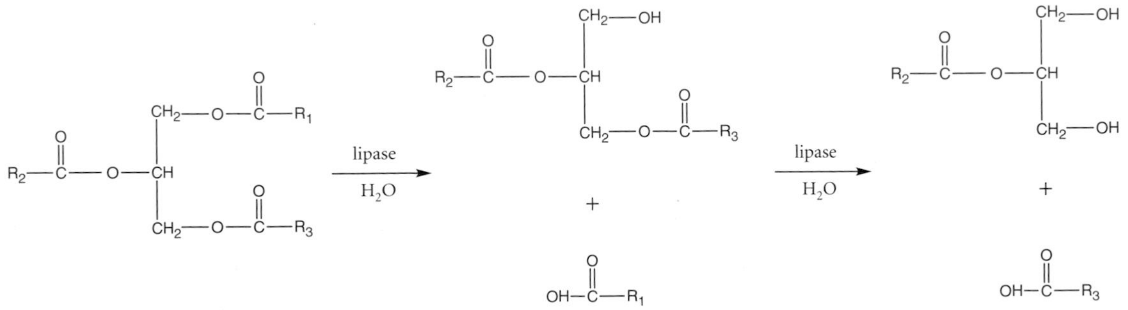

Figure 6.18. Lipases hydrolyze the ester bonds of the 1 and 3 fatty acid chains of triacylglycerols (triglycerides), but usually not the 2 chain linkage.

one of the main fat-digesting enzymes. Lipases are also produced by some bacteria and molds, certain plants and animals, and are found in milk. Lipases in milk are responsible for the development of rancidity because free fatty acids in milk produce an unpleasant off flavor and odor even at low concentrations. In the aging and ripening of some cheeses, however, the action of these enzymes on short-chain fatty acids is desirable and contributes to the characteristic flavor of the cheese. Lipases are used particularly for flavor development in cheeses such as feta, Romano, Parmesan, pecorino, provolone, or asiago. The characteristic flavor of Roquefort and other blue cheeses from sheep's milk is due to the action of lipases produced by the fungus *Penicillium roqueforti* in the blue-colored veins of the cheese. Commercial lipases used in cheese making are derived from pregastric tissues of animals, including goat kids, lambs, and calves, although microbial lipases are also commercially available. Lipases from animals are more specific for the short and medium-length fatty acids of dairy milk (see Table 7.6) than are microbial lipases. These short-chain fatty acids contribute strongly to the flavor and aroma of cheeses, so the animal lipases are preferred, except when considerations such as religious dietary laws make microbial enzymes required.

The lipases are water soluble, but their substrate lipids are not, so these enzymes attack lipids on the surfaces of milk fat globules. The active sites of lipases contain a catalytic serine residue in a three–amino acid triad containing serine, histidine, and aspartic acid residues (Figure 6.19), similar to the serine proteases, with a catalytic

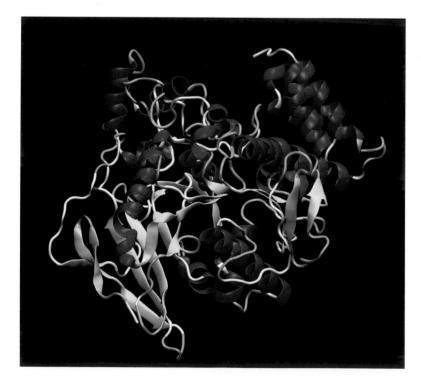

Figure 6.19. The conformational structure of bovine pancreatic cholesterol lipase (Chen et al. 1998).

mechanism similar to that of chymotrypsin (Figure 6.9). Many lipases require the simultaneous binding of bile salt to activate the enzyme.

Foaming in Beer

Lipids called lysophosphatidyl cholines, which are found in grains as inclusion compounds among the starch amylose, reduce foam stability when these grains are used to brew beer. The concentration of lysophosphatidyl cholines in the mash can be determined by controlling the temperature because of the differing thermal stabilities of two enzymes in the system. The enzyme α-amylase releases the lysophosphatidyl cholines from their close association with the amylose. Upon release, another enzyme, phospholipase B degrades this lipid. Because phospholipase B is less thermostable than α-amylase, processing at temperatures above 65 °C favors the buildup of lysophosphatidyl cholines because the enzyme that breaks them down is being denatured. Thus, high-temperature processing will lead to less head on the resulting beer.

ENZYMATIC BROWNING: POLYPHENOL OXIDASES

As we have previously observed, browning reactions are among the most important processes in foods. In Chapter 4 we considered nonoxidative browning reactions—caramelization and Maillard browning. Another major class of browning reactions involves enzymatic oxidation of phenolic compounds found in plant tissues (these phenolic compounds are the subject of considerable interest due to their possible health benefits).

Plants contain a class of copper-containing oxidoreductase enzymes called **polyphenol oxidases (PPOs)**, which can oxidize phenolic compounds to o-quinones (Figure 6.20). These enzymes are responsible for the rapid browning of certain fruit and vegetable tissues, such as apples, potatoes, and bananas, when they are cut and exposed to oxygen.

Dihydroxyphenols are colorless. Quinones are lightly colored, but are highly reactive, and quickly react with amino groups of proteins to form dark brown-pigmented

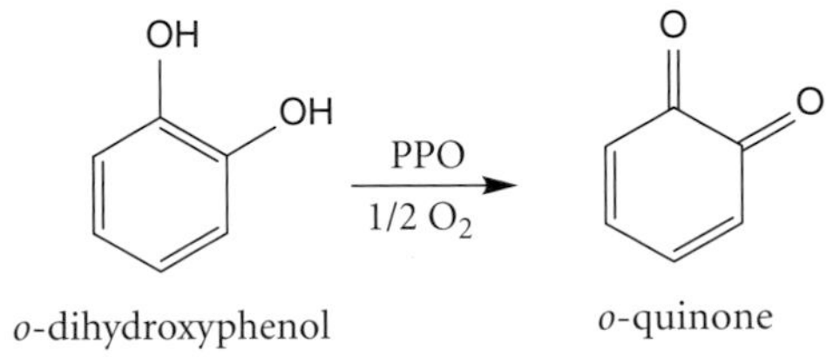

Figure 6.20. The general reaction catalyzed by polyphenol oxidases.

o-dihydroxyphenol o-quinone

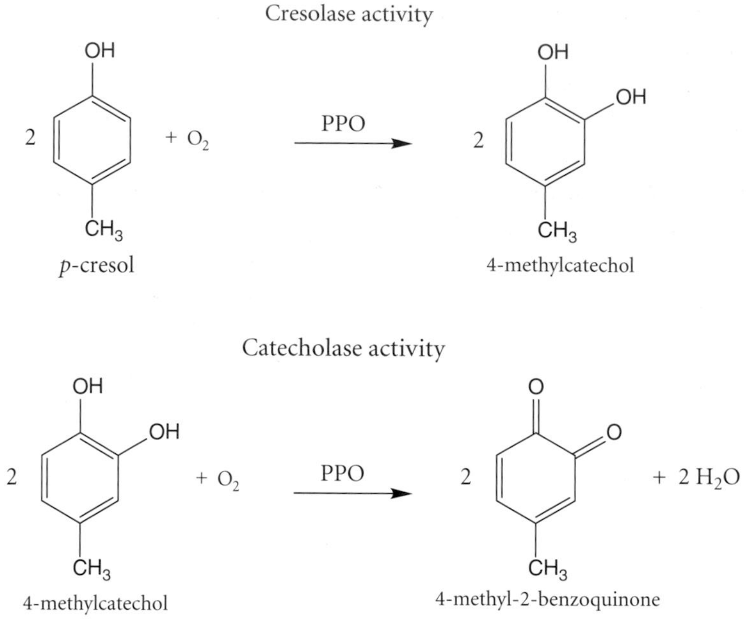

Cresolase activity

Figure 6.21. The reactions catalyzed by polyphenol oxidases.

Catecholase activity

melanin complexes. These enzymes are found at relatively high concentrations in certain fruits and vegetables, such as apples, bananas, and potatoes, and are present in nearly all plants in lesser amounts, as well as in fungi such as mushrooms. In intact plant tissue the PPOs are mostly segregated into the plastids, but when the tissue is damaged these enzymes are able to mix with polyphenolic substrates and are responsible for the browning that is observed when plants are cut or bruised and exposed to oxygen.

PPOs exhibit two types of activity (Figure 6.21). One type is referred to as cresolase activity (also sometimes called tyrosinase activity) and involves an *ortho*-hydroxylation of monophenols; that is, it adds a second alcohol group to a phenol functional group at the carbon position adjacent to the first hydroxyl group. The second type of activity is called catecholase activity, which is the oxidation of dihydroxyphenols to *ortho*-quinones. Because the side chain of the amino acid tyrosine is a phenol like cresol, PPOs can oxidize this side chain to an *ortho*-quinone by the successive application of both activities. The principal substrates of PPOs, however, are the various polyphenolic compounds found in plants, particularly flavonoids such as anthocyanins, catechins, and flavonols (Chapter 9). The reactive *o*-quinones produced by PPO activity can subsequently polymerize, or react with other types of molecules, to produce colored compounds. PPOs are also involved in the synthesis of the so-called condensed tannins (Chapter 9).

A number of different PPOs are found in fruits and vegetables. They may be involved in wound response in plants. PPOs exist in a stable latent state until activated by tissue damage. Once activated, these enzymes undergo self-inactivation after a limited number of catalytic cycles, as would be expected in a wound response. These enzymes differ in their substrate specificity, thermal stability, and temperature and pH optima. The amount of PPO activity in plant tissues varies between species and cultivars, and with maturity or ripeness and postharvest age. PPOs can be denatured by heating above 70 to 90 °C, and this method is often used to prevent enzymatic browning in vegetables and fruits. The exclusion of oxygen, where possible, is also effective. Ascorbic acid prevents browning by changing *o*-quinones back to diphenols. Sodium bisulfite, $NaHSO_3$, can prevent enzymatic browning, but its use on fresh fruits and vegetables is banned in the United States because bisulfite can induce a serious, even potentially fatal, allergic reaction in susceptible individuals. This type of browning can also be prevented by 4-hexylresorcinol, but the mechanism by which it does so is different from that of ascorbic acid, because it functions as an inhibitor of the PPO enzymes.

Polyphenol oxidases contain two copper ions in their active sites, incorporated into a Cu_2O molecule. The structure of one of these enzymes, from grapes, has been solved by X-ray crystallography (Figure 6.22). This 339-residue enzyme is a mainly alpha-helical protein of the so-called orthogonal bundle type, with only a few strands of beta sheet and significant stretches of coil. Each copper atom is bound to three histidine residues, with one of the histidines making a thioether linkage with a cysteine residue. How the enzymes are activated is not completely clear, but it has been proposed that the formation of this Cys–His thioether linkage is the activation step in the catalytic mechanism (Figure 6.23).

Figure 6.22. The backbone cartoon of the conformational structure of the polyphenol oxidase (PPO) enzyme from the chloroplasts of Grenache (*Vitis vinifera*) grapes. The alpha helices are shown in purple, the beta sheets are shown in yellow, and short segments of 3_{10} helix are shown in dark blue The protein contains a Cu_2O molecule complexed in the active site, shown as van der Waals spheres here. (V.M. Virador, J.P. Reyes-Grajeda, A. Blanco-Labra, E. Mendiola-Olaya, G.M. Smith, A. Moreno, and J.R. Whitaker, *J. Agric. Food Chem.* 58:1189–1201, 2010.)

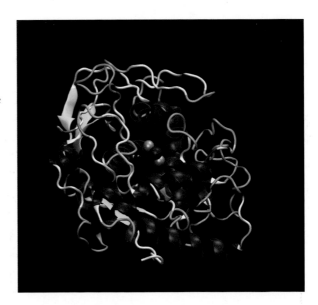

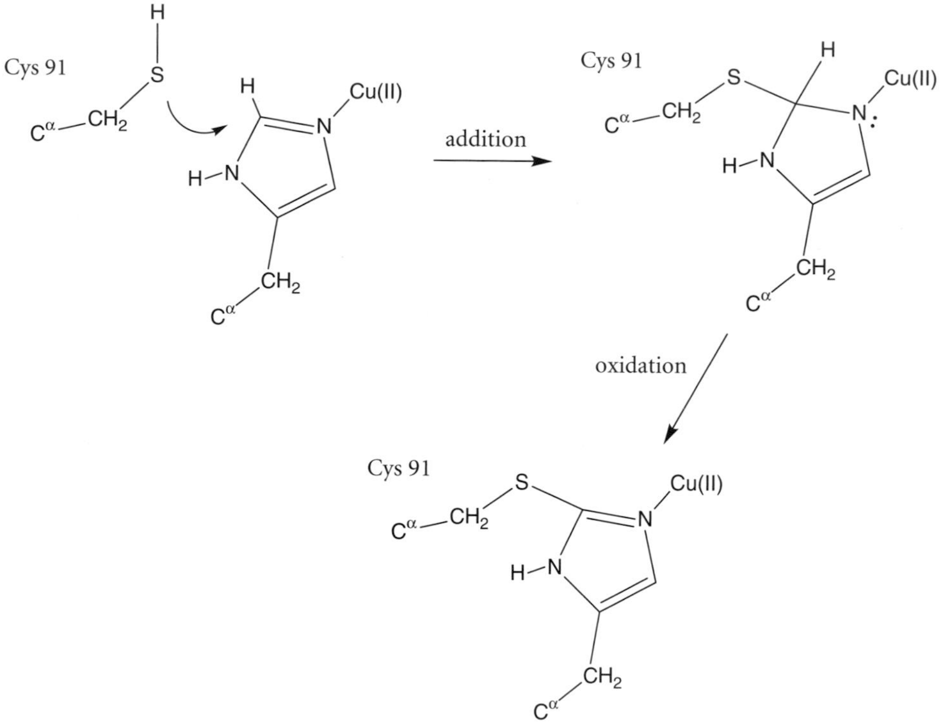

Figure 6.23. The hypothesized activation process for the latent PPO enzyme. (V.M. Virador, J.P. Reyes-Grajeda, A. Blanco-Labra, E. Mendiola-Olaya, G.M. Smith, A. Moreno, and J.R. Whitaker, *J. Agric. Food Chem.* 58:1189–1201, 2010.)

SUGGESTED READING

Voet, D., and J.G. Voet. 2011. *Biochemistry*, 4th ed. Hoboken, NJ: John Wiley & Sons.

Wong, D.W.S. 1995. *Food Enzymes: Structure and Mechanism*. New York: Chapman & Hall.

7.
Lipids

FATS IN FOODS

Lipids, from the Greek word *lipos*, meaning fat, are biological molecules that are insoluble or only sparingly soluble in water but are soluble in hydrophobic organic solvents such as chloroform. The term encompasses several different types of molecules important in foods, including various triacylglycerol esters of fatty acids generically referred to as **fats** and **oils**. These two terms in the food science context simply refer to the solid and liquid phases of these compounds, respectively.

Lipids are the third main category of food molecules, along with carbohydrates and proteins. As with these other two classes of molecules, fats are extremely important to the properties of foods. In living organisms, fats serve the function of energy storage, but they also have many functional roles in foods. Because of their high energy content, and the purported relationship between lipids in the diet and cardiovascular diseases, they have been the focus of dietary controversies and concerns for decades, but nutritionists generally agree that around a third of one's daily caloric intake should come from lipids, because many essential nutrients are only soluble in fats, and some of the fats are themselves essential nutrients. In this chapter, we will review the structures and properties of lipids, as well as their principal reactions, with particular attention on lipid oxidation, which is one of the most important reactions affecting quality in foods.

Probably the most fundamental of lipids are the fatty acids, which are simple hydrocarbon chains with a single carboxylic acid functionality on one end. These fatty acids are the primary constituents of soaps and detergents, but are usually not found free in foods in large amounts. In foods, they are generally found esterified to glycerol, a simple three-carbon polyol that is itself used as a food ingredient. Glycerol can have one, two, or three fatty acids esterified to it, with triaclyglycerols being the most abundant. Phospholipids are esters of glycerol where one of its hydroxyl groups is esterified to a phosphate group. These molecules are the primary constituents of cell membranes and thus are common in foods. They are also widely used in food processing as emulsifiers. Cholesterol, an essential component of animal cell membranes, is also classified in foods as a lipid, and is abundant in some foods of animal origin.

FATTY ACIDS

Fatty acids are carboxylic acids with long hydrophobic hydrocarbon chains attached, R—COOH, where the R group is usually 10 or more carbons long. Most natural fatty acids have an even number of carbon atoms, because they are synthesized through the addition of C_2 units to the chain. Many of the fatty acids are unsaturated; that is, they contain one or more double bonds in their chains. Those that contain a single double bond are called monounsaturated, while those that contain more than one double bond are called **polyunsaturated fatty acids** (PUFAs). In most food fatty acids the first double bond occurs between the C9 and the C10 carbon atoms, referred to as a Δ9-double bond, and subsequent double bonds generally occur at every third carbon toward the terminal CH_3 group. Fatty acids are numbered from the acid group, as the highest priority carbon, because it is bonded to oxygen atoms. A shorthand designation is frequently used to refer to fatty acids, which consists of two numbers separated by a colon; the first number gives the total number of carbon atoms, and the second number refers to the number of double bonds. This designation can be augmented by specifying the positions of the double bonds in parentheses, where the numbers refer to the first atoms of the double bond. For example, the eighteen carbon linoleic acid (see Table 7.1), which has two double bonds, is designated as 18:2 (9,12), with the first double bond being between carbon atoms 9 and 10, and the second double bond being between carbon atoms 12 and 13.

The double bonds of fatty acids are usually in the *cis* configuration; if they are not, an additional "tr" is added in the parentheses of the shorthand notation to indicate the nonstandard configuration, as in linolelaidic acid, 18:2 (tr9,tr12); see Figure 7.3. The predominant *cis* configuration puts an approximately 30° to 40° bend in the hydrocarbon chains at each double bond that interferes with their efficient packing in a crystalline state (see Figure 7.1). For this reason, the presence of double bonds lowers the melting point, as can be seen from Table 7.1, which can be very significant in food product formulation.

Table 7.1 Structures of several of the most common fatty acids found in foods

Designation	Structure	Common Name	Systematic name	mp (°C)
14:0		Myristic acid	Tetradecanoic acid	52
16:0		Palmitic acid	Hexadecanoic acid	63
18:0		Stearic acid	Octadecanoic acid	70
18:1 (9)		Oleic acid	9-Octadecenoic acid	13
18:2 (9,12)		Linoleic acid	9,12-Octadecadienoic acid	−9
18:3 (9,12,15)		α-Linolenic acid	9,12,15-Octadecatrienoic acid	−7

There are three methods of naming these molecules. All of the fatty acids in foods have common names, as well as systematic IUPAC names. In addition, each can be specified using a numerical shorthand, shown in the first column.

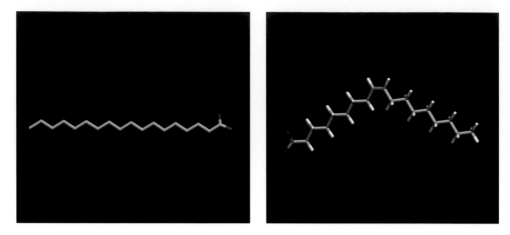

Figure 7.1. Liquorice bond molecular representations of: left, stearic acid; right, oleic acid. The hydrogen atoms are omitted for clarity. Note how the *cis* double bond in oleic acid changes the direction of the fatty acid chain.

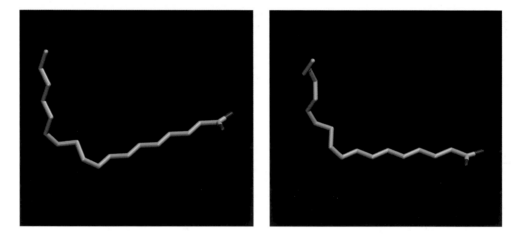

Figure 7.2. Liquorice bond molecular representations of, left, linoleic acid; right, α-linolenic acid. The hydrogen atoms are omitted for clarity.

Figures 7.1 and 7.2 illustrate the effects of *cis* double bonds on the overall shape of the fatty acids that contain them. These figures illustrate how the *cis* configuration can change the direction of the hydrocarbon chain, while the *trans* configuration somewhat restores the overall direction of the chain to that found in a fully saturated fatty acid (Figure 7.3). Because each *cis* bond can bend the direction of the chain by 30° to 40°, two successive *cis* bonds such as occur in linoleic acid can substantially bend the chain by as much as 60° to 80°. Because the *trans* bond angle is 120° rather than 109.5°, the *trans* bond does introduce a slight kink in the direction of the chain, but it is much smaller than the change in direction resulting from a *cis* bond. It can readily be appreciated how the more bent configuration of the *cis* fatty acids could interfere with packing in a crystalline lattice and thus lower the melting point if present in significant amounts.

Figure 7.3. *Cis* double bonds in lipid chains introduce a substantial angle in the direction of the polymer chain, while a *trans* double bond more closely resembles the original direction of the chain. Fatty acid chains can rotate about single bonds, but rotations about the double bonds, regardless of whether they are *cis* or *trans*, require too much energy to take place easily.

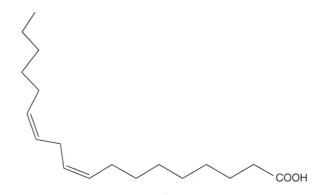

18:2 (9,12); linoleic acid; both double bonds are shown in the normal *cis* configuration

18:2 (tr9,tr12); linolelaidic acid; both double bonds are shown in the *trans* configuration

In examining Figure 7.2, one might wonder how any system of fatty acids with chains like linoleic acid mixed in could ever crystallize, because the deviation in direction is so great. Actually, this figure overemphasizes the distortion, since it should be remembered that the chains can rotate about all of the single bonds, such as the C10–C11 and C11–C12 bonds, which for some rotational values would have the effect of partially restoring the chain direction.

A major role of fats is energy storage. As can easily be seen, the structure of a fatty acid differs little from that of the molecules in gasoline, for example, which are primarily simple hydrocarbon chains. Carbohydrates, the principal energy storage molecules of plants, with many carbon–oxygen and oxygen–hydrogen bonds, are already partially oxidized. As a result, they are soluble in water, but contain much less energy than fats. In general, fats contain more than twice as much energy on a per weight basis than either carbohydrates or proteins and are thus an ideal energy storage molecule for mobile organisms: a cheetah cannot afford to drag along unnecessary weight as it sprints after a gazelle (and neither can the gazelle!), while a redwood has no such concerns. On average, dietary fat yields around 8.8 kcal/mol (37 kJ/g) upon complete oxidation to water and carbon dioxide, while dietary carbohydrate yields only around 3.8 kcal/mol (16 kJ/g).

The one case where most plants do need to have a high-energy content in a small package is in their seeds, where they are imparting an energy legacy to their offsprings. Thus, plant oils are most often pressed from seeds, such as peanuts, corn, soybean, flax seeds, sunflower seeds, or olives.

Table 7.2 lists the names and classifications of the principal food fatty acids.

Table 7.2 Names of the principal fatty acids found in foods

Designation	Structure	Common name	Systematic name	Mp (°C)
4:0	$CH_3(CH_2)_2COOH$	butyric acid	butanoic acid	−8
6:0	$CH_3(CH_2)_4COOH$	caproic acid	hexanoic acid	−4
8:0	$CH_3(CH_2)_6COOH$	caprylic acid	octanoic acid	16
10:0	$CH_3(CH_2)_8COOH$	capric acid	decanoic acid	31
12:0	$CH_3(CH_2)_{10}COOH$	lauric acid	dodecanoic acid	44
14:0	$CH_3(CH_2)_{12}COOH$	myristic acid	tetradecanoic acid	54
16:0	$CH_3(CH_2)_{14}COOH$	palmitic acid	hexadecanoic acid	63
18:0	$CH_3(CH_2)_{16}COOH$	stearic acid	octadecanoic acid	70
20:0	$CH_3(CH_2)_{18}COOH$	arachidic acid	eicosanoic acid	75
22:0	$CH_3(CH_2)_{20}COOH$	behenic acid	docosanoic acid	80
5:0	$CH_3(CH_2)_3COOH$	valeric acid	pentanoic acid	−35
7:0	$CH_3(CH_2)_5COOH$	enanthoic acid	heptanoic acid	−8
9:0	$CH_3(CH_2)_7COOH$	pelargonic acid	nonanoic acid	12
17:0	$CH_3(CH_2)_{15}COOH$	margaric acid	heptadecanoic acid	61

Fatty Acids with *cis* double bonds

Designation	Structure	Common name	Mp (°C)
	ω-9 family		
18:1 (9)	$CH_3(CH_2)_7$—CH=CH—CH_2—$(CH_2)_6COOH$	oleic acid	13
22:1 (13)	$CH_3(CH_2)_7$—CH=CH—CH_2—$(CH_2)_{10}COOH$	erucic acid	35
24:1 (15)	$CH_3(CH_2)_7$—CH=CH—CH_2—$(CH_2)_{12}COOH$	nervonic acid	43
	ω-6 family		
18:2 (9,12)	$CH_3(CH_2)_4$—(CH=CH—$CH_2)_2$—$(CH_2)_6COOH$	linoleic acid	−9
18:3 (6,9,12)	$CH_3(CH_2)_4$—(CH=CH—$CH_2)_3$—$(CH_2)_3COOH$	γ-linolenic acid	
20:4 (5,8,11,14)	$CH_3(CH_2)_4$—(CH=CH—$CH_2)_4$—$(CH_2)_2COOH$	arachidonic acid	−50
	ω-3 family		
18:3 (9,12,15)	CH_3CH_2—(CH=CH—$CH_2)_3$—$(CH_2)_6COOH$	α-linolenic acid	−7
16:3 (7,10,13)	CH_3CH_2—(CH=CH—$CH_2)_3$—$(CH_2)_4COOH$		
	Δ^9 family		
18:1 (9)	$CH_3(CH_2)_7$—CH=CH—CH_2—$(CH_2)_6COOH$	oleic acid	13
16:1 (9)	$CH_3(CH_2)_5$—CH=CH—CH_2—$(CH_2)_6COOH$	palmitoleic acid	1

Fatty acids with *trans* double bonds

Designation	Structure	Common name	Mp (°C)
18:1 (tr9)	$CH_3(CH_2)_7$—CH$\overset{tr}{=}$CH—CH_2—$(CH_2)_7COOH$	elaidic acid	46
18:2 (tr9,tr12)	$CH_3(CH_2)_7$—CH$\overset{tr}{=}$CH—CH_2—CH$\overset{tr}{=}$CH—$(CH_2)_7COOH$	linolelaidic acid	28

OMEGA-3 FATTY ACIDS

In Table 7.2, the unsaturated fatty acids are designated by specifying the location of the first double bond, counting either from the acid end or from the opposite end. This second method, counting from the end of the chain, designates the unsaturated fatty acids as ω-3, ω-6, ω-9, etc., where the number indicates the atom number of the first carbon in the first double bond, counting from that end. The reason for adopting this unconventional system in describing food lipids is that considerable evidence is accumulating to indicate that long-chain ω-3 fatty acids (n-3 PUFAs) can significantly reduce the risk of coronary heart disease. While rare in most foods, these types of fatty acids are found in significant amounts in fish and fish oils (Table 7.3). The two most important of these fatty acids in marine foods are **eicosapentaenoic acid** (**EPA**), 20:5 (5,8,11,14,17), and **docosahexaenoic acid** (**DHA**), 22:6 (4,7,10,13,16,19); Figure 7.4 shows the structures of EPA and DHA (DHA should not to be confused with dehydroalanine, also abbreviated DHA; see Figure 5.57). As can be seen from Table 7.2, the 18-carbon α-linolenic acid, which is more common in oils from the seeds of some land plants, is also an ω-3 fatty acid, although the evidence for health benefits from consuming this shorter-chain fatty acid are less certain.

The relatively uncommon long-chain ω-3 fatty acids are found in significant amounts in fish and other animal seafoods because they are synthesized by the phytoplankton that form the base of the marine food chain. Marine kelp and microalgae produce large amounts of DHA. Land plants also produce some ω-3 fatty acids, mostly the 18-carbon α-linolenic acid, but generally in smaller amounts than marine algae, and these lipids are thus rare in non-fish-eating animals. Very little ω-3 fatty acids are present in most of the grains used as animal feeds, particularly corn, (see Table 7.6), although soybeans and flax seeds have significant amounts. As a result, beef from grass-fed cows will have more ω-3 fatty acids than beef from grain-fed cows (but still much less than fish). Grass-fed beef generally has roughly twice as much ω-3 fatty acids, relative to ω-6 fatty acids, as grain-fed beef.

The ω-3 fatty acid α-linolenic acid is an essential nutrient, as are ω-6 fatty acids, because humans cannot directly synthesize either. DHA is an important component of phospholipids in the brain, and because it is necessary for brain development, it is added to the infant formula given to premature babies who do not necessarily receive enough in the womb before birth. In recent years many health claims have been made for DHA and EPA in the diet, many of which are somewhat speculative. There does seem to be good evidence that DHA lowers serum cholesterol and many even help prevent Alzheimer's disease. Both the EPA and DHA ω-3 fatty acids are thought to be particularly effective in reducing heart disease. The mechanisms of action of ω-3 fatty acids in the body are still incompletely known, but progress is being made in characterizing their roles (Calder 2012). DHA produced from algae is available commercially as a neutraceutical, although most commercial supplements are produced from fish oils, and occasionally even from seal oil.

Table 7.3 Amounts of ω-3 fatty acids present in various types of seafoods

Seafood	Amount in a cooked 4 oz serving (g)
Fish with 2 or more grams of ω-3 fatty acids in a cooked 4 oz serving	
Pacific herring	2.4
Atlantic herring	2.3
Atlantic salmon	2.1
Sablefish	2.0
Fish with 1–2 g of ω-3 fatty acids	
canned pink salmon	1.9
whitefish	1.9
Pacific oysters	1.6
pink salmon	1.5
Atlantic mackerel	1.4
sockeye or red salmon	1.4
Coho salmon	1.2
bluefish	1.1
trout	1.1
Fish with less than 1 g of ω-3 fatty acids	
freshwater bass	0.9
blue mussels	0.9
swordfish	0.9
rainbow trout	0.8
white canned tuna	0.8
canned sardines	0.7
flounder	0.6
halibut	0.5
rockfish	0.5
shrimp	0.4
snapper	0.4
light canned tuna	0.3
yellowfin tuna	0.3
Atlantic cod	0.2

CONJUGATED LINOLEIC ACID (CLA)

In general, synthetic *trans* fatty acids, those with at least one *trans* double bond produced artificially by partial hydrogenation of polyunsaturated fatty acids from vegetable oils, are now thought to be harmful, because they may increase the risk for heart disease in humans. Small amounts of *trans* fatty acids do occur naturally in

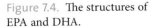

Figure 7.4. The structures of EPA and DHA.

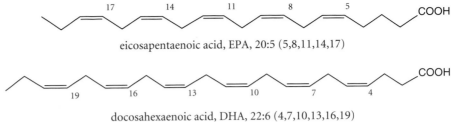

eicosapentaenoic acid, EPA, 20:5 (5,8,11,14,17)

docosahexaenoic acid, DHA, 22:6 (4,7,10,13,16,19)

Figure 7.5. CLA, conjugated linoleic acid, 18:2 (9,tr11); the C9–C10 double bond is shown in the normal *cis* configuration, but the C11–C12 double bond is *trans*. Compare this structure with that of linoleic acid (Figure 7.2).

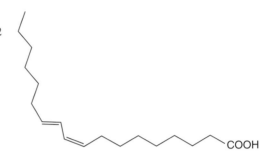

some foods, however, such as the **conjugated linoleic acid (CLA)**, 18:2(9,tr11), also called rumenic acid, present in dairy products (Figure 7.5). Rumenic acid constitutes approximately 85 to 90% of the CLA in milk, while a second isomer, 18:2(tr10,12) makes up the remaining 10 to 15%. Conjugated *trans* double bonds such as these, separated by only one intervening single bond, are not thought to be as harmful in the diet. CLA is formed as an intermediate in the biohydrogenation that takes place during fermentative digestion in ruminates such as cows, and can be present in small amounts in any ruminant fats, including milk and tallow (see below). The effects of CLA on heart disease are not clear, but accumulating evidence suggests that small amounts of CLA are beneficial in preventing certain types of cancer, at least in lab rats. The recently enacted prohibition against *trans* fats in manufactured foods in New York City does not include naturally occurring *trans* fatty acids such as CLA, which in any event are present only in small amounts.

GLYCEROL AND TRIACYLGLYCEROLS (TRIGLYCERIDES)

The fats and oils from animals and plants do not generally contain large amounts of free fatty acids, but rather are primarily made up of fatty acid triesters of **glycerol** (shown in Figure 7.6) called **triacylglycerols**, or **triglycerides**. Glycerol (also called glycerin, glycerine, or 1,2,3-propanetriol) is a viscous, colorless liquid polyalcohol, or polyol, with a sweet taste, which is about 60% as sweet as sucrose, or a little less sweet than glucose. It is soluble in water and very hygroscopic and is used as a moisturizer, or emollient, in many personal care products such as skin care lotions and creams, hair care products, soaps, and shaving creams. Glycerol is also approved for use as a food additive, and is used as a humectant or filler, and as a solvent for water insoluble

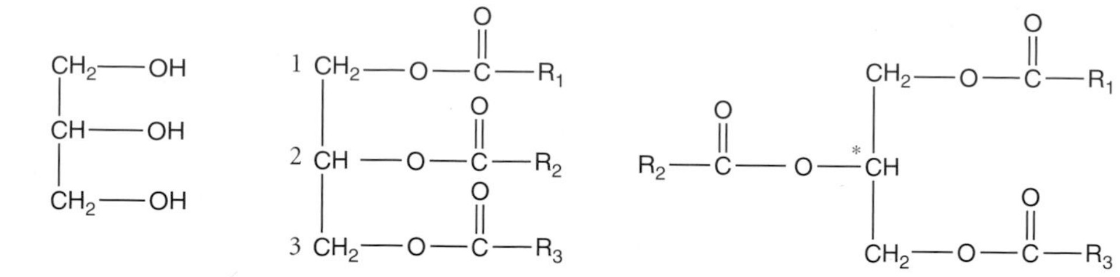

Figure 7.6. The structures of glycerol and a generalized triacylglycerol, with fatty acid chains esterified to each hydroxyl group of the glycerol. If the three fatty acids are different, the C2 carbon, indicated by an asterisk on the right, is chiral.

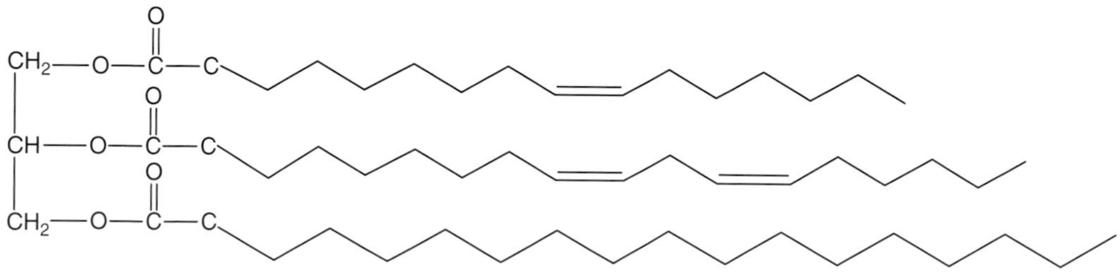

a mixed triacylglycerol - 1-palmitoleoyl-2-linoleoyl-3-stearoyl-glycerol

Figure 7.7. An example of a mixed triacylglycerol molecule.

colors and flavors such as vanilla (particularly useful in predominantly Muslim nations where ethanol is not allowed in foods). In the liver, glycerol can be converted to glucose or used to synthesize the triglycerides of depot fats.

Triacylglycerols are most abundant in animals because they serve as energy storage molecules. Because they are less oxidized than carbohydrates (with their large numbers of OH groups), they give up more energy when oxidized metabolically and can be stored anhydrously, while glycogen requires twice its weight in waters of hydration in a cell. Triacylglycerols in which all three fatty acid chains are the same are called **simple triacylglycerols**. For example, triolein, or trioleoyglycerol, has three oleic acid chains.

If the chains are different, the triacylglycerols are called **mixed triacylglycerols** (Figure 7.7). In such a situation, it is necessary to use so-called stereospecific numbering to indicate the exact structure (see below).

DIGLYCERIDES AND MONOGLYCERIDES

Milk contains enzymes called **milk lipases** that catalyze lipolysis of the milk triglycerides, removing fatty acid chains primarily from the outer end of the triglycerides,

Figure 7.8. The structures of prototypical di- and monoglycerides.

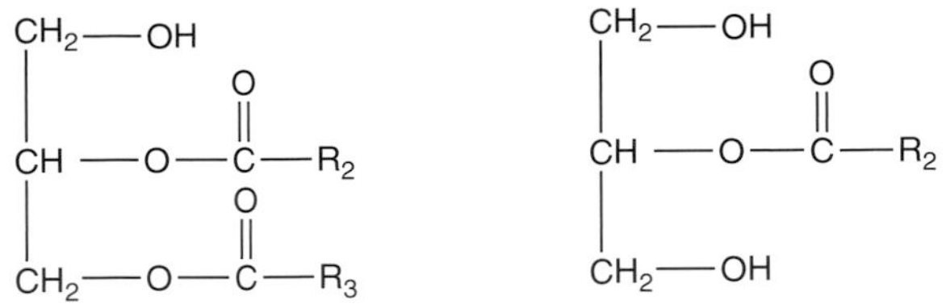

diacylglycerol
diglyceride

monoacylglycerol
monoglyceride

Figure 7.9. The stereospecific numbering system for triacylglycerols.

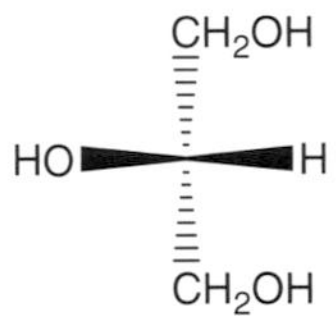

sn-1

sn-2

sn-3

glycerol

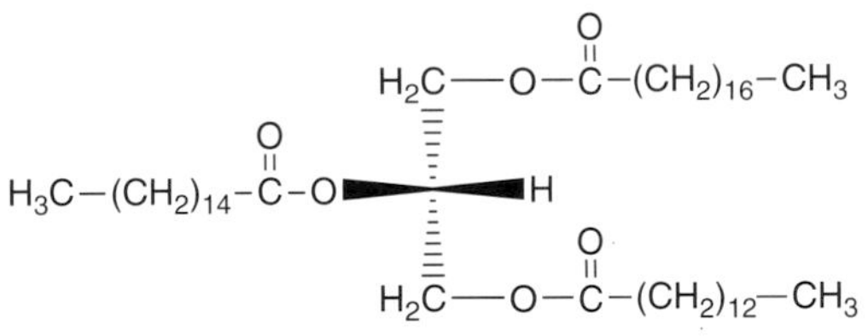

sn-1   stearic

sn-2   palmitic

sn-3   myristic

1-stearoyl-2-palmitoyl-3-myristoyl-*sn*-glycerol

producing 1,2 and 2,3 diacylglycerols or **diglycerides**. Further action of these enzymes gives **monoglycerides**, usually with the remaining fatty acid chain at the 2 position (Figure 7.8). Monoglycerides are much more polar than triglycerides, due to their free hydroxyl groups, and are surface active; that is, they tend to accumulate at oil/water and air/water surfaces. For this reason they can be used as emulsifiers in food preparations. In a pure fat phase they form reverse micelles containing water in their interior.

STEREOSPECIFIC NUMBERING

The central carbon atom of a triglyceride is in general chiral, and a system called the stereospecific numbering, or *sn* system, has been developed to label the carbons of the glycerol skeleton. In this system, the Fisher projection formula of the triglyceride is oriented such that the alcohol or ester group of the middle carbon atom is on the left, as shown in Figure 7.9, which by the Fisher convention means that it is projecting up out of the plane of the paper. When oriented in this fashion, the carbon on the top is numbered "1", and labeled "*sn*-1", and the carbon on the bottom is numbered "3" and labeled "*sn*-3". If the *sn*-1 and *sn*-3 substituents are the same, then the *sn*-2 carbon atom is no longer chiral and there is no need to distinguish between *sn*-1 and *sn*-3.

PHOSPHOLIPIDS

Glycerophospholipids (Figure 7.10) are the main lipid constituent of biological membranes. They are derived from *sn*-glycerol-3-phosphate esterified at its C1 and C2 positions with fatty acids and with another group "X" at its phosphoryl group. The simplest of these classes of molecules is when "X" is a proton, H, giving **phosphatidic acids**. When "X" is choline, $-CH_2CH_2N(CH_3)_3^+$, the phospholipid is called a **phosphatidylcholine**, which has the trivial name **lecithin**. Because of their long nonpolar

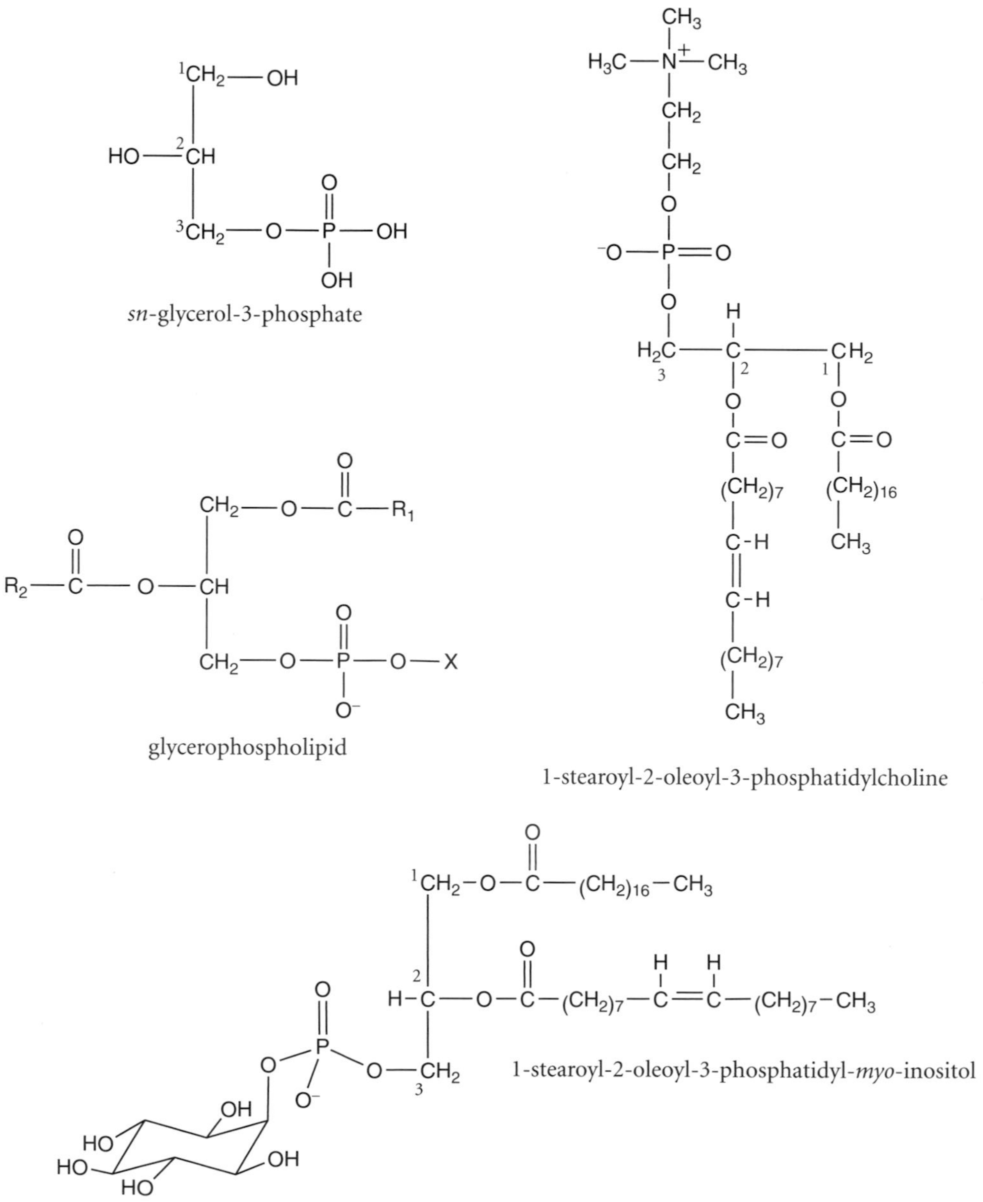

sn-glycerol-3-phosphate

glycerophospholipid

1-stearoyl-2-oleoyl-3-phosphatidylcholine

1-stearoyl-2-oleoyl-3-phosphatidyl-*myo*-inositol

Figure 7.10. The structures of some prototypical phospholipids.

Figure 7.11. A schematic illustration of the functional components of a typical phospholipid molecule; the choline, phosphate, and glycerol groups are polar in character and will interact favorably with water, while the acyl tails are hydrophobic.

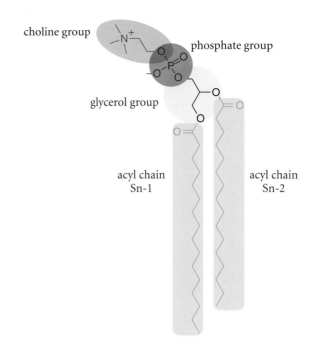

tails and their highly polar, charged head groups, phospholipids are amphiphilic in character (see schematic illustration, Figure 7.11).

Phospholipids are more polar than monoglycerides, and in an aqueous environment they organize so as to associate their tails together with their head groups pointing out into the solvent, leading to the formation of bilayers, micelles, and vesicles (Figure 7.12). Like monoglycerides, in an oil environment, such amphiphilic molecules can form reverse micelles, with an aqueous interior and the polar head groups on the inside, and with the hydrophobic tails in contact with the continuous oil phase.

Lecithins are widely used in foods as surfactants and emulsifiers. Lecithin from eggs is used as an emulsifier in the preparation of mayonnaise and baked products containing eggs. In the commercial context, lecithin as a separate ingredient usually refers to a mixture of lipids refined from soybean oil, which includes not only phosphatidylcholines but other phospholipids as well. Lecithin makes up about 35% of the phospholipids in soybeans. These soybean phospholipid mixtures have high hydrophile–lipophile balance (HLB) numbers that are the weighted average over the HLB numbers of the individual molecules in the mixture (see Chapter 3). The HLB value for commercial soy lecithin has been determined to be ~9.2 to 9.5 (Kunieda and Ohyama 1990). High HLB numbers make lecithin mixtures effective at solubilizing hydrophobic species in water (see Table 3.5). In several molecular gastronomy recipes, soy lecithin is used to create savory foams as a substitute for liquid sauces.

The majority of soybeans grown in the United States are now genetically modified to be "Roundup ready"; that is, they are not harmed by the commercial herbicide Roundup (see Chapter 5). The lecithin produced by these soybean plants is not known to be different in any way from lecithin from nonengineered plants, but those

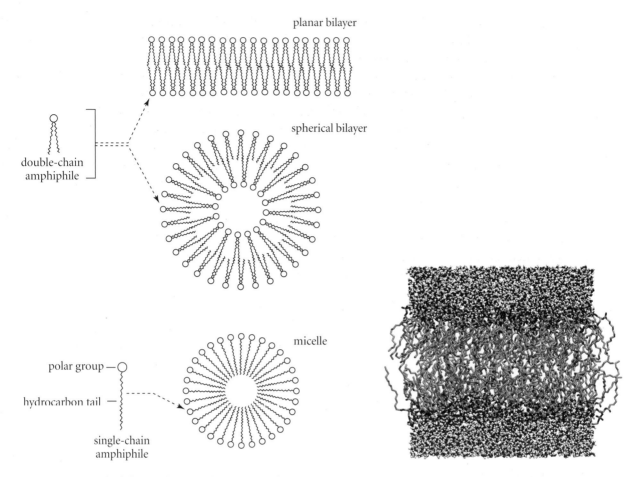

planar bilayer

spherical bilayer

double-chain
amphiphile

micelle

polar group —

hydrocarbon tail —

single-chain
amphiphile

Figure 7.12. On the left are schematic illustrations of a phospholipid bilayer, a bilayer closed into a spherical vesicle, and a micelle formed from a single-chain amphiphile. On the right is an illustration of a computer model of a section of a dipalmitoyl phosphatidylcholine (DPPC) phospholipid bilayer, with water layers on the top and bottom (compare to the cartoon in the upper left).

who object in general to genetic engineering would also object to such lecithin, and consider any manufactured product made with lecithin from these plants to be "genetically engineered."

WAXES

Waxes are the result of esterifying a long-chain alcohol (a so-called higher alcohol) with a fatty acid, as in the cerotyl cerotate that must be removed from the hulls of sunflower seeds in extracting sunflower oil, $CH_3—(CH_2)_{24}—CO—O—CH_2—(CH_2)_{24}—CH_3$. Because of their regular and hydrophobic nature, waxes can aggregate into glossy, water-impervious layers. Most waxes have melting points higher than common lipids. They also are usually more resistant to hydrolysis and require higher temperatures or more extreme pHs than lipids. Waxes are produced by plants as a covering for leaves and fruits to reduce the loss of water and to protect the underlying tissues from attack by microorganisms. **Carnauba wax**, from the cuticle of the leaves of the carnauba palm, or Brazilian wax palm (*Copernicia cereferia*) is an example of a commercial plant wax with food applications. Widely used in furniture and shoe polishes, car waxes, and cosmetics, Carnauba wax is also used in foods as a surface covering and glaze on a variety of candies and chewing gums as well as in a number of other products. Waxes are also produced by animals, most notably the beeswax secreted by bees for the construction of the honeycombs in their colonies. **Shellac** is a wax obtained from the secretions of the *Laccifer lacca* insect, which is used as a confectioner's glaze.

Waxes have a variety of uses in food processing, including as a glaze for candies and as a coating or polish for fruits to improve their appearance and consumer acceptability. When used for this purpose they are considered as ingredients by the FDA, and the label must state "coated with food-grade animal-based wax, to maintain freshness," or "coated with food-grade vegetable-, petroleum-, beeswax-, and/or shellac-based wax or resin, to maintain freshness." Beeswax is GRAS (generally recognized as safe), and few direct health concerns are associated with consumption of waxes. The practice of coating fruits with waxes, particularly those that will be eaten with their peels, such as apples, raises the possibility of sealing in pesticides that may have been sprayed on these fruits in the orchard. Because of the impermeable wax barrier, these pesticides might not be removed by simple hand washing, even if the pesticides themselves are water soluble.

CHOLESTEROL AND PHYTOSTEROLS

Among the unsaponifiable fractions of fats in foods (those components that cannot be converted into free fatty acids by treatment with alkali) are various molecules classified as steroids. **Steroids** are derivatives of the molecule cyclopentanoperhydrophenanthrene, shown in Figure 7.13. A number of steroid derivatives are important in biology. The estrogen and testosterone sex hormones are examples of steroids. Steroid alcohols that have a hydroxyl group substituted onto the carbon at the 3-position of ring A are referred to as sterols. The estrogens are examples of sterols; the term estrogen actually refers to several related sterols, including estriol, shown in Figure 7.13, estradiol, and estrone. Estradiol is identical to estriol except that it lacks the hydroxyl group on the 16-position of ring D, and estrone is identical to estradiol except that the hydroxyl group on the 17-position of ring D is replaced by a ketone. The male sex hormone testosterone (actually possessed by both genders, but in different amounts, as are the estrogen female sex hormones) is a similar steroidal compound, but is not classified as a sterol because it has a ketone functionality at the 3-position on ring A instead of a hydroxyl group. Sterols are found in the phospholipid membranes of both plants and animals, and play important roles in membrane fluidity.

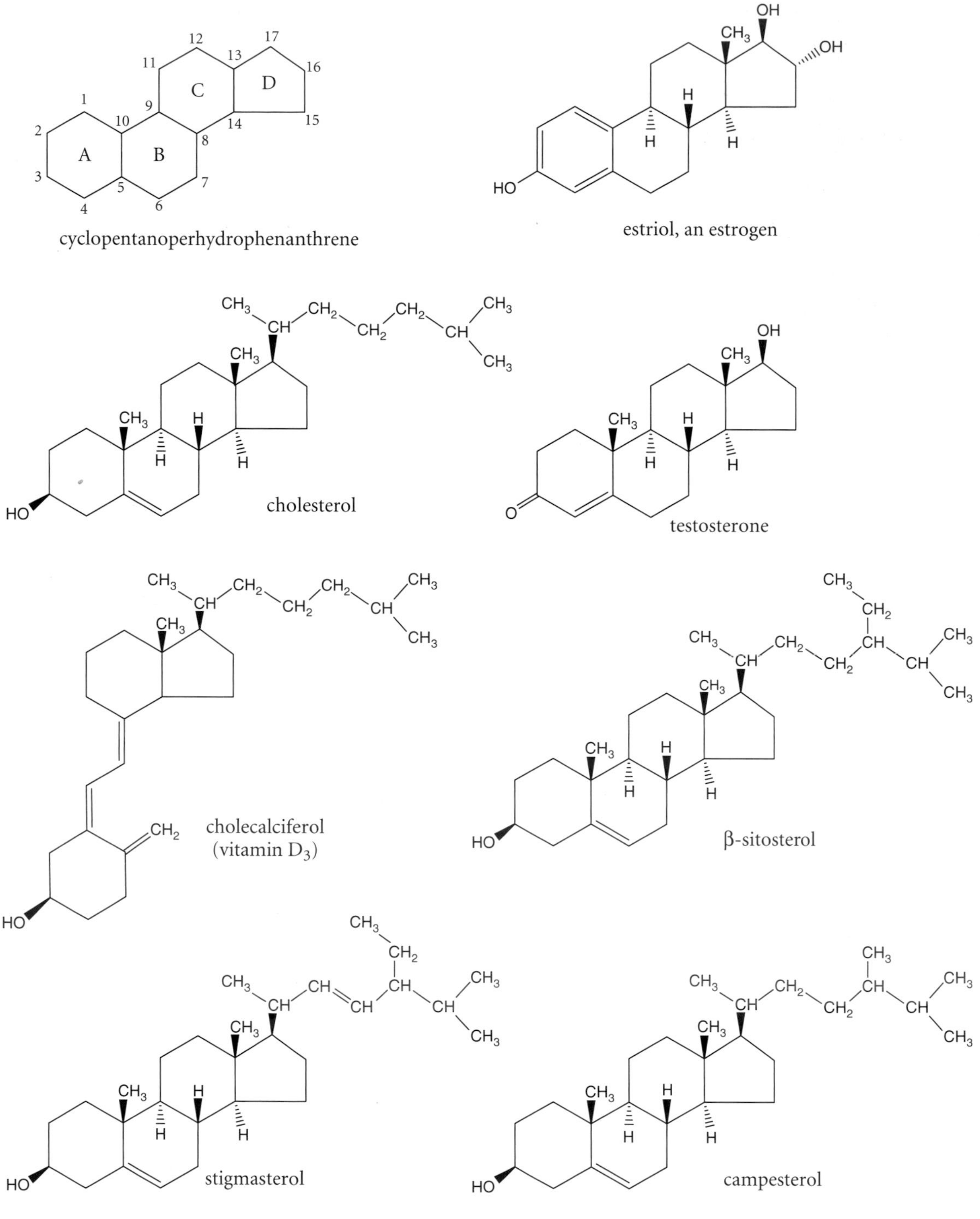

Figure 7.13. The structures of selected steroid molecules.

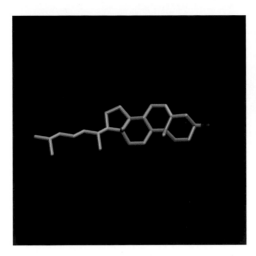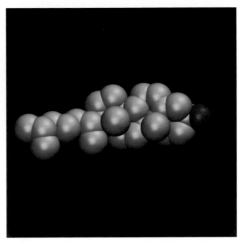

Figure 7.14. Two molecular representations of the cholesterol molecule. On the left is a "liquorice" bond representation, with the oxygen atom colored red, the carbon atoms colored blue, and the protons omitted. On the right is a van der Waals representation with the same color scheme, and again with the protons omitted. Note that the molecule is not flat, since the fused rings are not aromatic.

One steroid of particular interest is **cholesterol**, the most abundant steroid in animals. It is a hydrophobic, waxy sterol molecule, insoluble in water, with only its C3 hydroxyl group being polar and hydrophilic, giving the molecule a weak amphiphilic character. Cholesterol is produced endogenously in the liver and is used to synthesize bile, which emulsifies the free long-chain fatty acids produced by lipase digestion in the intestine. Note that the fused rings in cholesterol are not aromatic, so the rings are puckered, and the molecule is thus not planar (Figure 7.14). The single double bond in the B ring does reduce the pucker of the rings somewhat. The molecule possesses a number of chiral centers, whose substituents project above and below the approximate average plane of the rings as indicated in Figure 7.13. Cholesterol is incorporated easily into the phospholipid bilayers of cell membranes, where its rigidity due to its fused rings can significantly affect the fluidity of the membrane. In phospholipid membranes, cholesterol generally resides under the phosphate head groups, restricting the conformational fluctuations of the lipids. Cholesterol is necessary in animal cell membranes, but it is not an essential nutrient because it can be synthesized endogenously. It is present in all animal foods, but is largely absent from plant foods (it actually is present in plants, but in such small amounts as to be effectively zero). In its place, plants produce other sterols, called phytosterols, which are similar to cholesterol. Several of the most important of these are also shown in Figure 7.13, including β-sitosterol, stigmasterol, and campesterol. Campesterol, for example, differs from cholesterol only in a single methyl group on the hydrocarbon chain.

Cholesterol is a precursor in the synthesis of a number of other essential molecules, including estrogen, testosterone, and vitamin D. Controlling serum cholesterol through the diet has become an important issue in nutrition, because serum cholesterol levels are strongly related to coronary heart disease (Keys 1980a,b; Brown

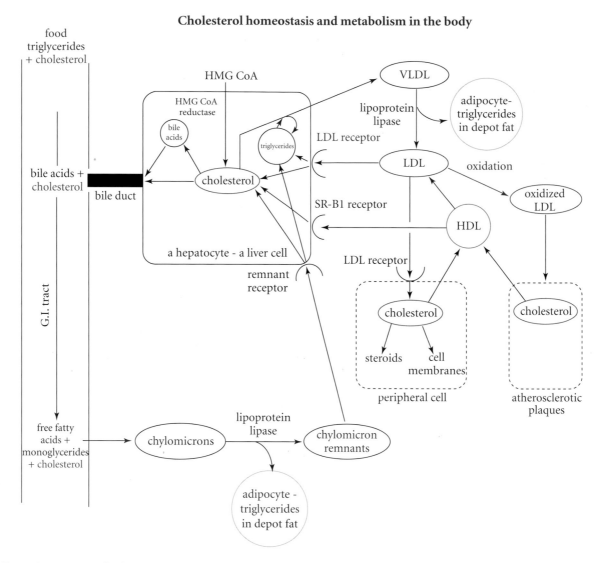

Figure 7.15. A general schematic representation of cholesterol homeostasis in the human body. (Courtesy of D.D. Miller, printed with permission.)

and Goldstein 2006). The levels of serum cholesterol are a complex function of both dietary intake and internal regulation. It should be noted, however, that when attempting to reduce the amount of cholesterol in the diet, trimming excess fat off of meats does not eliminate the cholesterol, because it is present in all cell membranes of the meat.

Figure 7.15 illustrates the currently accepted scheme for how cholesterol is both synthesized endogenously and absorbed from food. Cholesterol is essential for healthy membranes, but as already mentioned, is not an essential nutrient because enough is produced in the liver for normal biological requirements. Cholesterol is synthesized via the mevalonate metabolic pathway, a rate-limiting step of which is reduction of 3-hydroxy-3-methylglutaryl-coenzyme A (HMG CoA) to mevalonic acid by the enzyme HMG CoA reductase. The statin family of drugs lower serum cholesterol by inhibiting the HMG CoA reductase enzyme, thus reducing cholesterol production. Among other things, cholesterol is used to synthesize bile salts and is itself a component of bile. Bile is required for the digestion of fatty foods in the intestine. Cholesterol is insoluble in water, so circulating cholesterol in blood serum is contained in large, nanoscale structures called **lipoproteins** that consist of surfactant phospholipids, triglycerides, proteins, and cholesterol. These colloidal-size molecular aggregates resemble micelles, with the phospholipids contributing the main structure, and with the proteins embedded in the surface of these particles and the triglycerides in the interior. As in a membrane bilayer, the cholesterol is insinuated between the phospholipids, just under the head groups. These lipoprotein particles are classified into five different types based upon the way they behave in an ultracentrifuge, which also means they are classified according to their densities. The lowest density lipoproteins are called chylomicrons, and the other four types of particles, in order of increasing density, are very low-density lipoproteins (**VLDL**), intermediate-density lipoproteins (**IDL**), low-density lipoproteins (**LDL**), and high-density lipoproteins (**HDL**).

Chylomicrons are large micellar lipoproteins (75 to 1200 nm in diameter) that form in the small intestine and are transported across the intestine walls and into the lymphatic system, and eventually into the bloodstream, where they transport their lipids to the liver and to adipose tissues where depot fat is stored. The enzyme **lipoprotein lipase** then releases the triglycerides into fat cells (adipocytes); after depletion of their triglycerides the chylomicron remnants are then taken up by the liver via a receptor on the surface of liver cells. In the liver, both the food cholesterol absorbed from the chylomicron remnants and some of that synthesized endogenously are packaged into very low-density lipoproteins along with more triglycerides. These VLDLs are then secreted into the blood serum, where they are subsequently converted into IDL particles by the removal of triglycerides into adipocytes by lipoprotein lipase, and these IDLs are themselves subsequently converted into low-density lipoproteins (LDL) by further removal of triglycerides.

The LDL lipoprotein particles are the main cholesterol-carrying lipoprotein in the blood and are believed to be the most atherogenic of the lipoproteins. The risk for coronary heart disease is strongly correlated with the levels of serum LDL. For this reason they are colloquially referred to as "**bad cholesterol**." They are removed from the blood by specific **LDL receptors** located on the surface of the liver, as well as by receptor-independent processes in other organs. The fatty acids of the triglycerides in the LDL particles are apparently susceptible to oxidation, and the resulting oxidation products contribute to the formation of atherosclerotic plaques in coronary arteries. This process is thought to be the principal mechanism by which circulating cholesterol affects heart disease. Furthermore, various dietary antioxidants are believed to help prevent heart disease by inhibiting lipid oxidation in the low-density lipoproteins. The fatty plaque deposits in arteries are composed primarily of cholesterol. These plaque deposits can build up until they block the flow of blood through

Table 7.4 Cholesterol content of selected foods

Food	Amount (mg/100 g)
egg yolks	1000
butter	240
lean pork	70
lean beef	60
halibut	50

Source: Data taken from H.-D. Belitz and W. Grosch, *Food Chemistry*, 2nd ed. (Berlin: Springer, 1999).

the arteries. When the blocked arteries are those that deliver blood to the muscle tissues of the heart itself, a heart attack results. Saturated fatty acids and *trans* fatty acids are thought to affect the serum LDL concentrations by somehow inhibiting the reabsorption of LDL by the liver, perhaps by blocking the LDL receptors.

The high-density lipoproteins, which may be formed in the intestines or liver, promote reverse cholesterol transport, taking it from peripheral tissues back to the liver, where they also are recognized by a specific receptor site, the SR-B1 receptor. These particles are thought to be beneficial with respect to atherosclerosis, and are referred to colloquially as "**good cholesterol**." It may also be possible that HDL particles are able to remove cholesterol from plaque deposits and return it to the liver, but this has not yet been firmly established.

As can be seen from Figure 7.15, one source of cholesterol in the blood is dietary cholesterol, and at least in some individuals, controlling the amount of cholesterol consumed in food can result in a lowering of serum cholesterol. Different foods contain differing amounts of cholesterol; plant foods contain essentially none, while some animal foods contain large amounts. Table 7.4 lists the cholesterol content of several common foods. As can be seen from the table, eggs and butter are particularly rich in cholesterol.

REDUCED-CALORIE SYNTHETIC FAT SUBSTITUTES

Synthetic triacylglycerols in which one or two of the fatty acid chains are significantly shortened possess many of the physical properties of naturally occurring triglycerides, while giving fewer calories when metabolized due to the smaller number of carbon atoms available for oxidation. This idea is the basis for the reduced-calorie fat substitute **Salatrim** (from *Short* and *Long* *A*cyltriglyceride), shown in Figure 7.16. When metabolized, this molecule yields around 4.7 kcal/g (20 kJ/g) in energy, compared with regular triglycerides, which give ~9 kcal/g (38 kJ/g). Another example of this type of approach is Caprenin, with caprylic, capric, and behenic acid chains (Table 7.2), and 5 kcal/g.

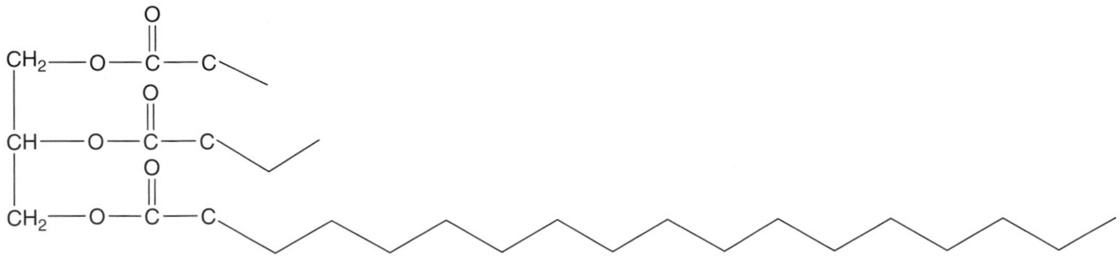

1-propionyl-2-butyryl-3-stearoyl-glycerol
Salatrim

Figure 7.16. The structure of the reduced calorie fat substitute Salatrim.

Figure 7.17. Olestra is a sucrose molecule in which every free hydroxyl group is esterified with a fatty acid chain.

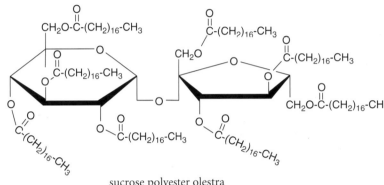

sucrose polyester olestra

SYNTHETIC FAT REPLACERS

Synthetic compounds that have similar physical properties to fats but which cannot be hydrolyzed enzymatically provide no dietary calories and thus can be used to replace fats in low-calorie foods. The best known and most tested of these compounds is **sucrose polyester**, which has the trade name **Olestra** (Figure 7.17). Olestra has been approved for food use by the FDA and is available in a number of commercial products. Sucrose polyester can be easily prepared in pyridine at 0 °C using trifluoroacetic acid and fatty acids from soybean or cottonseed oils (Figure 7.18).

Simplesse is another type of fat substitute produced and marketed by CP Kelco. It consists of denatured and cross-linked whey proteins, primarily β-lactoglobulin (Chapter 5), in colloidal-size particles approximately the dimensions of fat globules in milk.

FAT COMPOSITIONS

Fats from different food sources have different compositions in terms of the fatty acids that make up their triglycerides. In some cases, such as cocoa butter, these differences can also extend to the distribution of the fatty acids within the

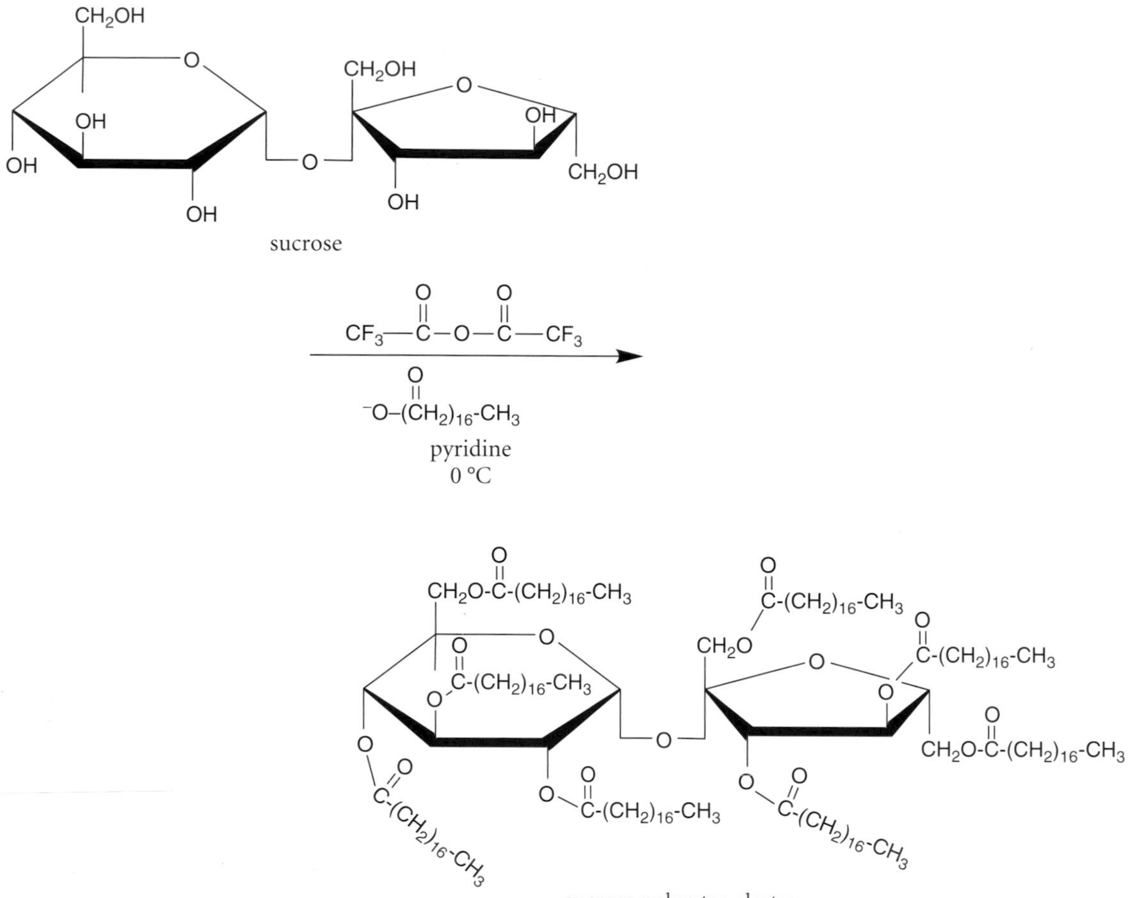

Figure 7.18. The formation of sucrose polyester from sucrose.

triacylglycerols. These differences can affect commercially important characteristics such as their melting temperature, susceptibility to oxidation, or stability during frying. In addition, it is widely believed that different fatty acids have different health effects when consumed regularly in the diet. For these reasons, briefly surveying the principal differences in composition of fats from different sources is worthwhile.

Animal Fats

Animal depot fats have long been an important source of fat for commercial purposes and remain so in spite of health concerns about the use of lipids from animal sources. Animal fats are high in saturated fats (Table 7.5), primarily palmitic and stearic acids, which is why they are solid at room temperature, but which have been linked to possible increases in heart disease in recent years. Nevertheless, even in animal fats, the most abundant fatty acid is oleic acid, which is thought to be the most healthy fatty acid. In addition, fats from animal sources also contain cholesterol (discussed below),

Table 7.5 Average fatty acid composition of fats from selected animal sources

Animal fat type	Fatty acid content (% by weight)											
	12:0 lauric	14:0 myristic	14:1(9) myristoleic	16:0 palmitic	16:1(9) palmitoleic	18:0 stearic	18:1(9) oleic	18:2(9,12) linoleic	18:3(9,12,15) α-linolenic	20:0 arachidic	20:1 & 20:2	others
beef tallow	0	3	0.5	26	3.5	19.5	40	4.5	0	0	0	3
lard	0	2	0.5	24	4	14	43	9	1	0.5	2	0
goose fat	0	0.5	0	21	2.5	6.5	58	9.5	2	0	0	0
sheep tallow	0.5	2	0.5	21	3	28	37	4	0	0.5	0.5	3

Source: Data taken from H.-D. Belitz and W. Grosch, Food Chemistry, 2nd ed. (Berlin: Springer, 1999).

which is not found in oils from plant sources. In industrial production, fats are recovered from adipose tissues by simple heating either with hot water or steam, a process called **rendering**.

Commercial fat from cattle is called **beef tallow**. Beef tallow is around 40% oleic acid, 26% palmitic acid, 20% stearic acid, and about 8% linoleic and palmitoleic acids combined. It melts around 45 to 50 °C. Beef tallow is slightly yellow in color due to carotenoids dissolved in the lipid that are derived from the plants eaten by the cattle. Prime beef fat, which has a low acid content (free fatty acids less that 0.65%), is produced by heating fat trimmings from beef in water (wet rendering) to around 50 to 55 °C. At around 30 to 35 °C, this extracted fat separates into two fractions: a solid fraction called **oleostearine** and a liquid fraction called **oleomargarine**. When cooled to room temperature, oleomargarine has a soft consistency, like melted butter, and is used in baking and the production of margarines. Oleostearine, which has a higher melting temperature range than oleomargarine, is used in the manufacture of shortenings.

Fat from hogs is commonly referred to as **lard**; its fatty acid composition is somewhat similar to that of beef tallow except that it has a higher composition of unsaturated fatty acids. Lard has fewer triglycerides of the fully saturated SSS type, and more mixed triglycerides such as SUU and USU, as well as more unsaturated triglycerides, UUU (here S means a saturated fatty acid and U means an unsaturated fatty acid). As a result of the higher unsaturated fatty acid composition, lard melts at a lower temperature than beef tallow, and over a broader range of temperatures, due to its more mixed composition.

Sheep tallow is not used in food processing because it has an unpleasant odor that is difficult to remove by fractionation. A small specialty market exists for goose fat, primarily in Europe; this fat can be almost 60% oleic acid in composition, with substantially less stearic acid than other animal fats. There is a market for fish oils made from members of the herring family, which includes herrings, sardines, anchovies, and Atlantic menhaden. As already mentioned, these oils contain ω-3 fatty acids, and have a much larger proportion of long-chain fatty acids, including large fractions of the 20:1, 20:5, 22:1, and 22:6 fatty acids.

Dairy Fats

If you're afraid of butter, use cream.
—Julia Child

Lipids make up slightly less than 4% by weight of bovine whole milk. The vast majority of these lipids consist of triacylglycerols (triglycerides) found in the fat globules. Table 7.6 shows the distribution of different fatty acids in the milk from a range of species. As can be seen, palmitic acid (16:0) and oleic acid (18:1) are common in almost all of these species. Bovine milk has a relatively small amount of linoleic acid (18:2), but in some species such as horses and cottontail rabbits, this fatty acid is

Table 7.6 Distribution of fatty acid types in the milk of a number of species

Species	Fatty acid content (% by weight)												
	4:0 butyric	6:0 caproic	8:0 caprylic	10:0 capric	12:0 lauric	14:0 myristic	16:0 palmitic	16:1 palmitoleic	18:0 stearic	18:1 oleic	18:2 linoleic	18:3 α-linolenic	C_{20}–C_{22}
cow	3.3	1.6	1.3	3.0	3.1	9.5	26.3	2.3	14.6	29.8	2.4	0.8	T
water buffalo	3.6	1.6	1.1	1.9	2.0	8.7	30.4	3.4	10.1	28.7	2.5	2.5	T
sheep	4.0	2.8	2.7	9.0	5.4	11.8	25.4	3.4	9.0	20.0	2.1	1.4	—
goat	2.6	2.9	2.7	8.4	3.3	10.3	24.6	2.2	12.5	28.5	2.2	—	—
blackbuck	6.7	6.0	2.7	6.5	3.5	11.5	39.3	5.7	5.5	19.2	3.3	—	—
elephant	7.4	—	0.3	29.4	18.3	5.3	12.6	3.0	0.5	17.3	3.0	0.7	—
human	—	T	T	1.3	3.1	5.1	20.2	5.7	5.9	46.4	13.0	1.4	T
horse	—	T	1.8	5.1	6.2	5.7	23.8	7.8	2.3	20.9	14.9	12.6	—
cottontail	—	—	9.6	14.3	3.8	2.0	18.7	1.0	3.0	12.7	24.7	9.8	0.4
harp seal	—	—	—	—	—	5.3	13.6	17.4	4.9	21.5	1.2	0.9	31.2
elephant seal	—	—	—	—	—	2.6	14.2	5.7	3.6	41.6	1.9	—	29.3
polar bear	—	T	—	T	0.5	3.9	18.5	16.8	13.9	30.1	1.2	0.4	11.3
grizzly bear	—	T	—	—	0.1	2.7	16.4	3.2	20.4	30.2	5.6	2.3	9.5

Source: Data taken from P.F. Fox, ed., *Advanced Dairy Chemistry*, Volume 2: *Lipids*, (London: Chapman & Hall, 1982).
T = trace amount.

quite abundant. In human milk, this acid is also fairly abundant, replacing stearic acid (18:0) as the third most abundant component. Almost half of all of the fatty acid in human milk is made up of oleic acid. Approximately one third of the fats of bovine milk are monounsaturated fatty acids, primarily oleic acid and a little palmitoleic acid (16:1). It is important to note that two-thirds of the fats of human milk are unsaturated, compared with bovine milk, in which almost two-thirds (63%) of the fats are fully saturated, which is one of the reasons that dairy fats are thought by some to be less healthy in the human diet. Because it is generally believed that milk has evolved to be the perfect food for infants of each species, this difference may suggest that a high ratio of unsaturated to saturated fats is more healthy for humans, at least as infants. The fat distributions of the major dairy species (cows, buffaloes, goats) are relatively similar, and distinguishable, for example, from those of many nondairy species, particularly seals and cetaceans and other marine species. In particular, the milks of dairy species contain nonnegligible amounts of very short-chain fatty acids, while these are rare in human milk. Such very short-chain fatty acids are not generally thought to cause health problems and are important to the development of the flavors and odors of some cheeses. When liberated by the action of lipases, the volatile butyric acid (4:0), with a low odor detection threshold, is particularly important in the aroma of Parmesan cheese, cultured butter, and in the development of rancidity. Some species, primarily those that live in cold environments (e.g., marine mammals and arctic species including polar and grizzly bears) have significant amounts of longer-chain fatty acids (C_{20}–C_{22}) in their milk.

Fairly unique among common food fats, milk from ruminants contains small amounts of fatty acids with odd numbers of carbon atoms. These particularly include pentadecanoic acid (15:0) and heptadecanoic acid (17:0), but these two together make up only a little more than 1% of the total fat content of dairy milk. They result from bacterial action of the rumen flora, which is the reason these fatty acids are absent from the milk of nonruminant species. They are sufficiently unique that their presence could be used as a diagnostic tool to distinguish dairy milk from that of nondairy species.

Oils From Plants

Plant oils are isolated from the seeds or fruits of a wide range of plants, and many contain a higher proportion of unsaturated fatty acids than lipids from animal sources (Table 7.7) and, thus, are usually liquid at room temperatures. Many triglycerides from plant oils have an SUS-type (saturated-unsaturated-saturated) structure; that is, the fatty acids in the 1 and 3 positions are more likely to be saturated, while the fatty acids in the 2 position are more likely to be unsaturated. Diets rich in polyunsaturated fatty acids are believed to produce better blood cholesterol profiles than saturated fats, and lipids from plant sources also do not contain cholesterol, which is thought to contribute to coronary diseases, so that vegetable oils are generally thought to be healthier for human diets than animal fats. Some contain high proportions of

Table 7.7 Plant oil compositions, with beef tallow and dairy fat for comparison

Oil	Fatty acid composition (% by weight)											
	10:0 capric	12:0 lauric	14:0 myristic	16:0 palmitic	18:0 stearic	20:0 arachidic	22:0 behenic	18:1 (9) oleic	18:2 (9,12) linoleic	18:3 (9,12,15) α-linolenic	20:1 and 20:2	22:1(13) erucic
sunflower oil	—	—	—	6.5	5	0.5	0	23	63	<0.5	1	—
soybean oil	—	—	—	10	5	0.5	0	21	53	8	3.5	—
peanut oil	—	—	—	10	3	1.5	3	41	35.5	0	1	—
canola oil	—	—	—	4	1.5	0.5	0	63	20	9	1	~0.5
olive oil	—	—	0	11.5	2.5	0.5	—	75.5	7.5	1.0	—	—
palm oil	—	—	1	44	5	0.5	—	39	10	0.2	—	—
palm kernel oil	4	47	16	8	2.5	—	—	14	2.5	—	—	—
coconut oil	6	47	18	9	2.5	—	—	7	2.5	—	—	—
corn oil	—	—	0	10.5	2.5	0.5	—	32.5	52	1	—	—
flax seed oil	—	—	—	6.5	3.5	0	0	18	14	58	0	0
grape seed oil	—	—	—	7.4	3.9	—	—	15.6	72.2	0.2	—	—
cotton-seed oil	—	—	1	22	2	1	—	29	45	—	—	—
beef tallow	—	—	3	26	19.5	0	—	40	4.5	0	0	—
bovine milk fat	3	3.1	9.5	26.3	14.6	—	—	29.8	2.4	0.8	—	—

Sources: Data taken from H.-D. Belitz and W. Grosch, *Food Chemistry*, 2nd ed. (Berlin: Springer, 1999); B.S. Kamel et al., *JAOCS* 62:881–883 (1985); T.P. Hilditch, *Brit. J. Nutrition* 3:347–354 (1949).

monounsaturated fatty acids, which are believed to be the healthiest form of fat for human diets. For those purposes that require a firmer, solid fat, it is necessary to increase the proportion of saturated fats in vegetable oils, which is often accomplished by metal-catalyzed hydrogenation (discussed below). The high level of unsaturation in vegetable oils makes them more susceptible to oxidation, leading to **rancidity**, the production of unpleasant odors and flavors. For this reason, they are less suitable for frying, particularly in industrial and restaurant applications, and are more susceptible to degradation on storage, limiting product shelf-life.

Because high temperatures are detrimental to the stability of polyunsaturated fatty acids, the steam rendering used to extract fats from animal carcasses is less effective for plant fats. Consequently, plant oils are most commonly extracted from seeds and other plant tissues using the organic solvent hexane. Hexane is highly volatile and is subsequently vaporized away, leaving no trace. At high levels, hexane can have neurotoxic effects, but it is not a concern in food oil production because it leaves no residue in the product oils. Plant oils can also be produced by cold pressing plant materials, but with much lower efficiency. This method is generally used for olive oils and those to be labeled as organic. Oils produced by cold pressing without further refinement will contain other components besides triglycerides, which is the reason for the characteristic flavor, color, and aroma of olive oil, for example. In the case of soybean oil and other oils from legumes, these can include antinutritional components such as the trypsin inhibitors. Plant oils can also be extracted using supercritical carbon dioxide, but at greater expense.

The most widely produced, and cheapest, vegetable oils in the United States are **corn oil** and **soybean oil**. Corn oil is pressed from the corn germ and is produced as a by-product of the production of corn starch and corn meal. Corn oil is over 50% linoleic acid, as are the oils from cottonseed oil, wheat germ oil, and pumpkin seed oil. These four oils are also all unusually rich in palmitic acid, containing between 10 and 20% of this short chain fatty acid. Corn oil also is almost 33% oleic acid, so that these three fatty acids (18:2 (9,12); 18:1 (9); and 16:0) make up over 95% of the fatty acids in corn oil. Soybean oil is similar to corn oil in its composition of these three fatty acids, and its principal difference from corn oil is that it contains substantial amounts of α-linolenic acid (around 8%) and 20-carbon fatty acids. These oils are often partially hydrogenated to decrease the level of unsaturation and raise the melting point to produce fats suitable for the crisp or flaky textures needed in baked products or for the production of margarines. Monsanto has recently developed a new strain of soybeans whose oil is approximately 60% oleic acid, making it more like olive oil and canola oil.

Soybean oil is unstable when exposed to light, because it contains furan fatty acids (Figure 7.19). Fatty acids that contain a furan molecule embedded in their hydrophobic tails are susceptible to oxidation when exposed to light, producing the volatile aroma compound 3-methyl-2,4-nonandione, which gives the oil a "beany" aroma. Soy milk is marketed in opaque cartons and containers to inhibit this reaction.

A number of other seed oils have composition profiles (high in oleic and linoleic acids and low in palmitic acid) similar to soybean oil; these include sunflower oil,

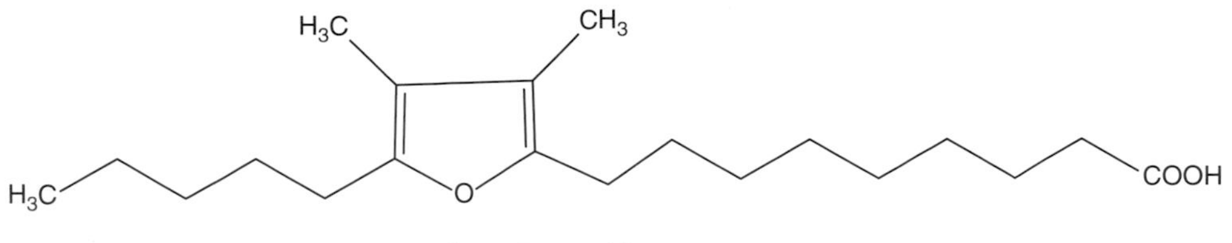

H₃C   CH₃

H₃C   O   COOH

a furan fatty acid

Figure 7.19. The furan ring embedded in the hydrocarbon tails of some soy lipids results in instability when exposed to light.

peanut oil, sesame seed oil, safflower oil, and linseed oil. **Linseed oil**, or **flax seed oil**, is the oil cold pressed from the seeds of the flax plant, *Linum usitatissimum*, whose fibers have been used since ancient times to produce linen cloth. Flax seed oil differs from the other seed oils in that it contains high proportions, around 50%, of the ω-3 fatty acid α-linolenic acid. For this reason, it is very susceptible to oxidation. Because it has a strong odor and flavor, linseed oil is not generally used in foods; however, cold-pressed flax seed oil is used as a nutritional supplement due to the supposed health benefits of consuming ω-3 fatty acids, and strains of flax have been developed that have high α-linolenic acid contents. Linseed oil is more widely used in nonfood industries, as a finishing agent for treating wood, and traditionally as a binder or carrier base for the oil paints used by artists. It functions well in these applications because it hardens on exposure to air as the result of oxidative polymerization, a reaction that limits its utility as a food.

Cottonseed oil is a relatively inexpensive oil that is produced from the seeds of cotton, which is not grown solely for food purposes. It has a high proportion of unsaturated fatty acids, although less of the monounsaturated oleic acid than some of the other vegetable oils, and of the 26% of its fatty acids that are fully saturated, the largest proportion consists of the relatively short-chain palmitic acid, which is believed to elevate LDL levels in the blood. Because of its price advantage, this oil is popular for product formulations, although many purposes may require that its saturation be increased by hydrogenation (see next section).

The two oils with the highest monounsaturated fatty acid content, and thus the healthiest nutritional profile, are **olive oil** and **canola oil**. Olive oil is pressed from the pulp of the olive fruit (*Olea europaea sativa*); virgin olive oil is the result of cold pressing and has not been heated. The monounsaturated oleic acid (18:1 (9)) makes up more than 75% of olive oil; palmitic acid makes up another 11 or 12%, and around 8% is linoleic acid. Because olives, which are primarily produced in the Mediterranean region, are harvested by hand, and the pressing is also labor intensive, olive oil is expensive and thus not as widely used in food formulation as it perhaps should be.

Canola oil is obtained from strains of rapeseeds that have been bred to be free of erucic acid (22:1(13); see Table 7.2), which is bitter-tasting and antinutritional for humans and may cause damage to internal organs, including the heart and kidneys. Strains of rapeseed containing less than 1% erucic acid were originally developed in Canada, and their oil was marketed as canola, from "Canadian oil," to distinguish it from ordinary rapeseed oil with a high erucic acid content. Canola oil is nearly two-thirds oleic acid, around 20% linoleic acid (double that of olive oil), and about 9% α-linolenic acid. Canola oil has perhaps one of the healthiest fatty acid compositions of any of the widely available commercial fats or oils, since it has such a high content of unsaturated fats, with saturated fatty acids making up less than 5% of the total, unlike olive oil, which contains almost twice this content of palmitic acid alone.

Palm oil is obtained from the fruits and seeds of the oil palm (*Elaeis guineensis*), native to West Africa, which is now widely grown in Southeast Asia, particularly in Indonesia and Malaysia. No other crop produces as many calories per hectare of cultivated land as the oil palm, and huge tracts of lowland forest in Southeast Asia have been cleared and converted to oil palm plantations. Significant amounts of palm oil are now also produced in Columbia, and several African nations have long produced palm oil for domestic consumption. Two different types of oil are derived from the fruits of this palm tree: palm oil, pressed from the pulp of the fruit, and palm kernel oil, pressed from the seed. Unlike olive oil, palm oil is only around 40% oleic acid, and is nearly 45% palmitic acid, which took its name from the fact that this short-chain fatty acid is so abundant in palm oil, from which it was initially identified. This is not true of the palm kernel oil, however, which has little palmitic acid (8%) but which contains an even higher proportion (47%) of the still shorter chain saturated fat lauric acid, much like **coconut oil**, which is made up almost entirely of short-chain fatty acids. Palm oil is the most widely used cooking oil in many developing nations, and is an essential commodity for many poorer families in the developing world. Due to its high saturated fatty acid composition, palm oil is semisolid at room temperature, which makes it much more useful for formulating products such as crackers and cookies than other vegetable oils. Raw palm oil is orange in color due to a high β-carotene composition (see Chapter 9); it can be rendered colorless by boiling to destroy this carotenoid. The semisolid palm oil can be made even more solid by using simple thermal fractionation to increase the proportion of saturated fatty acids, thus raising the melting point without hydrogenation (see below).

The use of palm oil has become controversial for several reasons. Some of these issues are nutritional. Because shorter-chain saturated fatty acids are now thought to contribute to atherosclerosis and heart disease, many nutritionists advise against the use of palm oil, with its high content of palmitic acid, in foods. Nevertheless, as *trans* fatty acids are removed from products in the United States (see below), they are often being replaced by palm oil because of its low price and higher melting temperature, with unknown health implications. Another issue relating to the growing use of palm oil is the way in which it is produced. The swampy, humid tropical lowlands of Southeast Asia are ideal for the commercial production of palm oil, and they have been extensively converted to palm oil plantations in recent years. Unfortunately, these lowland forests are also among the richest and most biodiverse ecosystems on the planet, exceeded only by the Amazon rain forests, and perhaps not even by them. Even more unfortunately, these forests are now nearly gone, and the few remaining tracts are rapidly disappearing for palm oil plantations. Palm oil production is so lucrative that even some of the few national parks protecting this swampy habitat are

Table 7.8 Comparison of cocoa butter with beef tallow and Borneo tallow

Fatty acid	Cocoa butter	Beef tallow	Borneo tallow
16:0 (palmitic)	25	36	20
18:0 (stearic)	37	25	42
20:0 (arachidic)	1		1
18:1 (9) (oleic)	34	37	36
18:2 (9,12) (linoleic)	3	2	1
Triglyceride composition			
SSS	2	29	4
SUS	81	33	80
SSU	1	16	1
SUU	15	18	14
USU	—	2	—
UUU	1	2	1

Source: Data taken from H.-D. Belitz and W. Grosch, *Food Chemistry*, 2nd ed. (Berlin: Springer, 1999).

being invaded and cleared for illegal planting of oil palms. A number of endangered species, ranging from orangutans and Sumatran rhinoceroses to countless plants, birds, and insects, are being threatened with outright extinction. Finally, palm oil can be used to make so-called biodiesel fuel for internal combustion engines in trucks and cars. Such fuel was initially promoted as being "green," because it is renewable, but in addition to the potential great harm to endangered species, the clearing of these swampy forests, which often contain extensive peat swamps, may actually be contributing substantially to global warming. Because automotive fuel sells for a higher price than cooking oil, the resulting diversification of the palm oil market led to such steep price rises for cooking oil that many poorer consumers in developing countries are facing considerable hardship in purchasing this essential product.

Cocoa butter is of course obtained from cocoa beans. As can be seen from Table 7.8, this highly prized fat, which is actually tasteless, consists of almost equal parts stearic, oleic, and palmitic acids, but with 81% of the triglycerides in cocoa butter being of the SUS type, with oleic acid in the middle. These SUS (saturated-unsaturated-saturated) triglycerides are 1,3-dipalmito-2-olein (POP), 1-palmito-3-stearo-2-olein (POS), and 1,3-distearo-2-olein (SOS) in the nearly constant ratio of 22:46:31. Another 15% of the total triglycerides in cocoa butter is of the SUU type. As can be seen from the table, this composition profile differs considerably from that seen in beef tallow, and most other fats and oils as well. Because cocoa butter is so uniform, it has a narrow melting temperature range, approximately between 28 and 36 °C (close to the human body temperature of 37 °C), which is fairly high for a plant source. This narrow melting range at a high temperature is due to the high proportion of saturated fatty acids and the regular packing of the nearly uniform triglycerides, allowing for a high crystallinity. Because this narrow melting temperature range is just below body temperature, chocolate made from cocoa butter melts in the mouth, absorbing the latent heat of fusion from the tongue and thus producing a cooling sensation as the cocoa butter melts.

Borneo tallow is also shown in Table 7.8 for comparison. This oil, from the fruits of a *diterocarp* tree of the *Shorea* genus, has a similar triglyceride compositional profile to that of cocoa butter, and therefore a similar melting behavior. For this reason it can be used as a substitute for cocoa butter, particularly in cosmetics applications.

REACTIONS OF LIPIDS

Fractionation

Mixtures of fats can be partially separated out by exploiting the differences in melting temperature of the different triglycerides. A melted mixture of several fats can be slowly cooled to a temperature where one or more of the fats will solidify, but where others are still liquid. These lower-melting triglycerides can then be separated by decanting or filtration to remove the liquid phase. Because fats will not crystallize out as a pure phase under such conditions, the remaining solid phase may need to be reheated to allow lower-melting fats that became intermixed with the higher-melting crystals to melt and separate out. Industrial fractionation is often used on palm oil to separate out palm olein and palmatin, which together make up nearly 85% of the fatty acids in palm oil. This separation is facilitated by the widely differing melting points of triglycerides containing different amounts of these two fatty acids (for example, OOO: −12 to +5 °C; PPP: 57 to 66 °C; POP: 30 to 35 °C. Note the similarity of POP to cocoa butter in structure and melting behavior).

Interesterification

In the industrial processing of lipids, it is possible to change the positions of the fatty acid chains in a triglyceride using sodium methoxide as a catalyst to produce triglycerides with different physical properties, such as melting temperature and crystallization behavior. This exchange of acyl radicals between triacylglycerol molecules is called **interesterification**. The resulting triglycerides produced by this process are, in general, random combinations of the possible positions for the various chains involved. For example, if one started out with a 50:50 mixture of triolein, OOO, and tristearin, SSS, six different mixed triglycerides would result from interesterification: the original SSS and OOO, along with SOS, OSS, SOO, and OSO. The relative proportions of each triglyceride would be OOO (12.5%); SSS (12.5%); SOS (12.5%); OSO (12.5%); OSS (25%); and SOO (25%). The proportions of OSS and SOO are double those of the others because two equivalent ways of making these combinations are

Figure 7.20. The hydrolysis of triglyceride ester linkages in saponification.

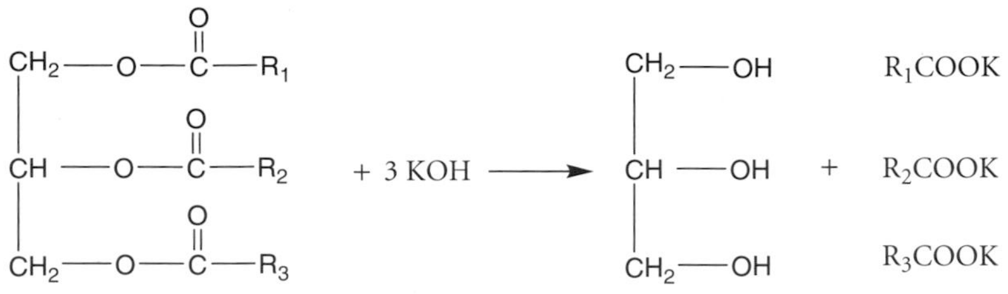

possible, due to their symmetry. Interesterification followed by fractionation can be used to produce *trans*-free fats from vegetable oils following partial hydrogenation, which produces high levels of *trans* fatty acids (see below).

Saponification

The alkaline hydrolysis of the ester bonds in acylglycerides is called **saponification** and produces alkali salts of the free fatty acids (Figure 7.20). These fatty acid salts are the principal components of soap. They work to solubilize and clean hydrophobic molecules away in water by forming micelles around the "dirt" with the carboxylic acid group pointing out into the aqueous solution. This hydrolysis is also the basis of the compositional analysis of fats and oils.

Hydrogenation of Vegetable oils

Vegetable lipids are high in unsaturated fatty acids, which is the reason that they are oils (that is, liquid at room temperature). Because they are derived from plant sources, they contain no cholesterol and are thought to be generally healthier, but liquid oils are not as suitable as solid fats for some applications, such as the manufacture of margarines and shortenings. In addition, the unsaturated bonds in these oils increase the susceptibility of these lipids to undesirable oxidation reactions. For these reasons, vegetable oils are often hydrogenated partially or fully in industrial applications by treatment with gaseous hydrogen in the presence of a solid nickel catalyst.

The diatomic H_2 molecules adsorb dissociatively onto the surface of the nickel metal, as do the double bonds of the fatty acids. Hydrogen atom radicals can migrate across the metal surface to add to the bound carbons, saturating the atom as it desorbs. If the hydrogenation process is incomplete, the lipid molecule can desorb again with the reformation of the double bond. Both *cis* and *trans* isomers are produced, however, and the unsaturated bond can form either to the right or left of the bound carbon atom. Because either of the carbon atoms of the original unsaturated bond can be the one that fails to capture a hydrogen atom while adsorbed, this can lead to six possible isomers, including three *trans* forms (Figure 7.21).

Nutritionists now generally believe that *trans* fatty acids contribute to cardiovascular disease when consumed in significant amounts, raising serum low-density

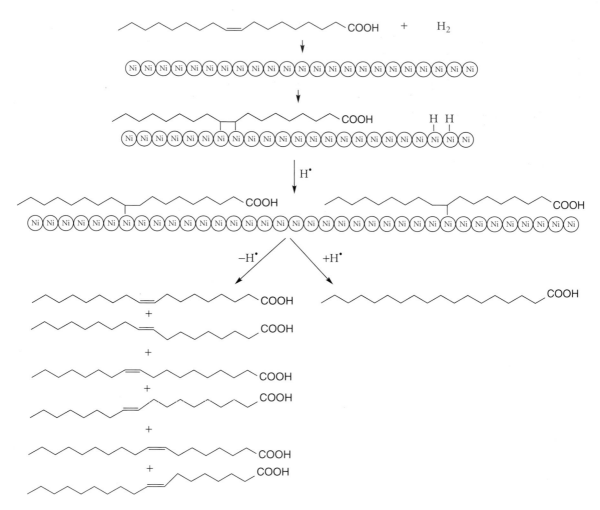

Figure 7.21. The mechanism of hydrogenation of fatty acids using nickel as a catalyst, using oleic acid as an example, showing how *trans* fats can be produced by the process.

lipoprotein (so-called bad cholesterol) levels while lowering high-density lipoprotein (so-called good cholesterol) levels. Their production by this mechanism is the chief drawback to the use of partially hydrogenated vegetable oils. Batches of such oils can be completely hydrogenated, producing saturated fats with no *trans* fatty acids, which can then be combined with unmodified samples of the vegetable oils in the appropriate proportions to give mixtures with the desired melting temperatures. Because this operation is done as a batch process, it can be more expensive, but offers a straightforward way to precisely control the proportion of saturated fat in a partially hydrogenated vegetable oil without introducing *trans* fatty acids.

Since January of 2006, the FDA has required that food product labels in the United States include the amount of *trans* fatty acids (if the amount per serving exceeds 0.5 gm). This change in regulations is leading to a decrease in *trans* fatty acids in manufactured foods in the United States. It should be noted, however, that "0 grams of *trans* fat" on a product label does not have the same meaning as the mathematical definition of zero. In fact, a food could be legally labeled this way and have as much as 0.49 gm of *trans* fatty acids per serving.

Product developers are of course faced with the dilemma of choosing what *trans* fats should be replaced with when a fat that is solid at room temperature is needed. As we have already seen, the higher melting temperatures needed to keep baked goods crisp and not "oily" are promoted by regular triglyceride packing, which in turn is promoted by increasing the proportion of saturated fatty acids. Until the mid-twentieth century, this objective was achieved using animal fats such as lard as shortenings. Following preliminary studies linking consumption of animal fats to heart disease, these fats began to be replaced by vegetable oils, which had to have their proportions of saturated fats increased to achieve the same temperature behavior. This was logical because, even if industrial modification simply achieved an identical distribution of saturated fatty acids as is found in animal fats, it would still have no cholesterol, which is thought to be beneficial for those individuals who have high blood cholesterol. The resulting product would also be acceptable to vegans because no animal products would be used in its manufacture. Palm oil became a common substitute for animal fat because it was cheap, and already has a high proportion of saturated fatty acids, with a 44% content of the short-chain fatty acid palmitic acid. In the 1970s many commercial crackers, cookies, and other baked products were reformulated to use palm oil.

With the publication in 1980 of the Seven Countries Study of Ancel Keys (Keys 1980a,b), however, saturated fatty acids came to be identified as the principal concern, rather than the actual source of the fat. In particular, short-chain saturated fatty acids are now believed to play a negative role in circulating cholesterol regulation and thus to contribute to atherosclerosis and coronary artery disease. Accordingly, palm oil was removed from many products in the 1980s and 1990s. Now that *trans* fats have been identified as being even more harmful than saturated short-chain fatty acids, palm oil is once again being used as one of the most common replacements for hydrogenated vegetable oils in many manufactured food products. As might be expected, all such choices are highly controversial.

Bromination and Brominated Vegetable Oil (BVO)

Brominated vegetable oils (BVOs) are modified vegetable oils used as emulsifiers in certain beverages. Alkenes such as the unsaturated fatty acid chains of vegetable oils will readily add Br_2 across their double bonds in an inert solvent such as carbon tetrachloride (Figure 7.22); in fact, this reaction is used as a test for the presence of carbon–carbon double bonds, because the bromine solution is red, but the halogenated hydrocarbon is colorless. The resulting brominated fatty acid is less saturated; by controlling the extent of the reaction, the average degree of substitution can be set at a level that gives a desired molecular weight and density.

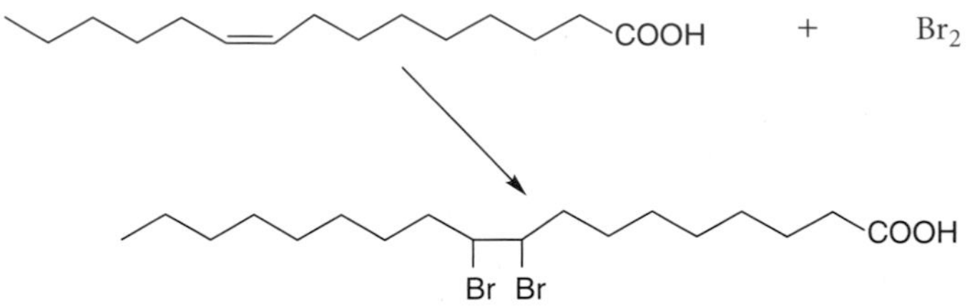

Figure 7.22. The addition of bromine to an unsaturated bond in a lipid.

BVO is used in some citrus-flavored soft drinks as an emulsifier to uniformly disperse citrus oils in the aqueous medium without having them cream out and pool at the top surface. The addition of the heavy bromine atoms increases the density of the vegetable oils, allowing the emulsified oil droplets to be uniformly dispersed in the aqueous phase. The colloidal size of the dispersed droplets gives these drinks a slightly cloudy appearance from the light scattered by the particles.

Concerns have been expressed about the safety of BVO. There have been reports in the literature of rare cases of bromism, a neurological disorder, resulting from excessive consumption of sodas containing BVO. High levels of bromine can have a number of undesirable physiological and neurological effects. The use of BVO as a food additive is limited or not allowed in a number of countries, although it is permitted for soft drinks by the US FDA.

Lipid Oxidation

One of the most important reactions of lipids in foods is the oxidation of unsaturated bonds to form hydroperoxides and their decomposition products. This reaction is the basis for the development of rancidity in fats and oils. Because they are higher in energy, the double bonds of unsaturated fatty acids are inherently unstable and thus susceptible to destruction by oxidation. This instability is clearly greatest for polyunsaturated fatty acids. Lipid oxidation can take place by several mechanisms. These include (1) **autoxidation**, (2) **photosensitizer-induced singlet oxygen oxidation**, and (3) **enzymatic lipid oxidation**. Autoxidation is the nonenzymatic reaction of unsaturated fatty acids with triplet (ground state) oxygen $^3O_2$ resulting in the breakdown of the lipid. Oxidation can also occur if the more highly reactive singlet oxygen $^1O_2$ is generated by the interaction of light with photosensitizers. Finally, lipid oxidation can be catalyzed by enzymes called lipoxygenases found in many plant, and some animal, tissues. Each of these types of lipid oxidation will be discussed in more detail below.

Autoxidation

Lipid autoxidation to form hydroperoxides and their further degradation can be thought of as consisting of four stages:

$$RH \xrightarrow{\text{initiator}} R^{\bullet} + H^{\bullet}$$

$$RH + HO^{\bullet} \longrightarrow R^{\bullet} + H_2O$$

$$RH + Fe^{+3} \longrightarrow R^{\bullet} + H^+ + Fe^{+2}$$

Figure 7.23. The initiation step of lipid oxidation.

$$Fe^{+2} + H_2O_2 \longrightarrow Fe^{+3} + HO^{\bullet} + OH^-$$

$$RH + HO^{\bullet} \longrightarrow R^{\bullet} + H_2O$$

Figure 7.24. The Fenton reaction. Ferrous iron can initiate oxidation through the production of a hydroxyl radical.

1. **Initiation**
2. **Propagation**
3. **Decomposition**
4. **Termination**

These stages describe the sequence by which a lipid free radical is generated, its reaction with oxygen in a form of chain reaction that produces another alkyl radical while leading to the decomposition of the first fatty acid, and the eventual termination of the reaction sequence by combination with other radicals. We will consider each of these steps in more detail as they apply to a specific lipid, linoleic acid, which, as a highly unsaturated fatty acid, is particularly susceptible to oxidation.

1. Initiation

Autoxidation is initiated through the formation of a lipid free radical as the result of one of several processes, such as activation by absorption of visible or ultraviolet radiation, heat, or the interaction with hydroxyl radicals or metal catalysts (Figure 7.23).

Ferric iron Fe^{+3} can initiate autoxidation by direct attack on a fatty acid to produce an alkyl free radical. Ferrous iron Fe^{+2} can also indirectly initiate autoxidation by producing a hydroxyl radical via a pathway called the **Fenton reaction**, illustrated in Figure 7.24.

Once an alkyl free radical is generated from a lipid molecule by one of the processes described above, it can readily react with ground state (triplet) oxygen to produce alkylperoxyl radicals. The methylene group between two double bonds is particularly susceptible to loss of a hydrogen atom, because the abstraction of a proton from such a position requires less energy than any other in the fatty acid chain due to the possibility of resonance stabilization of the resulting alkyl free radical. Table 7.9 lists the energies required to abstract a hydrogen atom from several types of sites in an alkyl chain such as in a fatty acid.

Thus, in linoleic acid, the C11 methylene is the most susceptible to proton loss and is the initial site of abstraction in autoxidation. This alkylperoxyl radical is resonance stabilized by at least two other major forms, as is illustrated in Figure 7.25. The reaction can proceed from each of these three resonance forms, leading to a multiplicity of final products, but for simplicity we will focus on the reactions of the resonance form with the alkoxy group on the C13 position.

Table 7.9 Proton abstraction energies for different types of aliphatic protons

	$D_{R\text{-}H}$ (kJ/mole)
H | CH_2—	422
H | CH_3—CH—	410
H | —CH—$CH{=}CH$—	322
H | —$CH{=}CH$—CH—$CH{=}CH$—	272

2. Propagation

In the propagation stage of autoxidation, the alkyl free radicals generated in the initiation stage react with ground state (triplet) oxygen to produce alkylperoxyl radicals, which can abstract a hydrogen atom from another fatty acid alkyl chain to generate a hydroperoxide (ROOH) and another alkyl radical (Figure 7.26).

For the case of the linoleic acid with the alkoxy group on the C13 position, this general mechanism is illustrated in Figure 7.27.

3. Decomposition

The decomposition of the unstable hydroperoxide molecules proceeds in two stages. The first step is the breaking of the O—O bond in the hydroperoxide to give an alkoxy lipid radical RO˙ and a hydroxyl radical HO˙ (Figure 7.28).

In the second step the alkoxy radical cleaves via a β-scission of one of the C—C bonds on either side of the alkoxy group. In the illustrated linoleic acid example in Figure 7.29, a scission to the "right" of the alkoxy group at position "b" leads to the formation of hexanal and a fatty acid radical, which then either can react with a hydroxyl radical to produce an alcohol or can abstract a proton from another lipid to give a fatty acid terminated by a double bond, with the production of another fatty acid radical. Similarly, bond scission at position "a" can lead to the formation of pentane and pentanol, as well as a fatty acid terminated by a double bond to an aldehyde group.

4. Termination

The radical-induced chain reaction can be terminated in several ways, illustrated by the steps shown in Figure 7.30.

The total autoxidation scheme for the case of linoleic acid is illustrated in Figure 7.31, focusing on the bond scission at position "b" pathway to produce hexanal. The production of hexanal, which can be detected and quantified chromatographically, can be used as an operational measure of the extent to which lipid autoxidation has progressed.

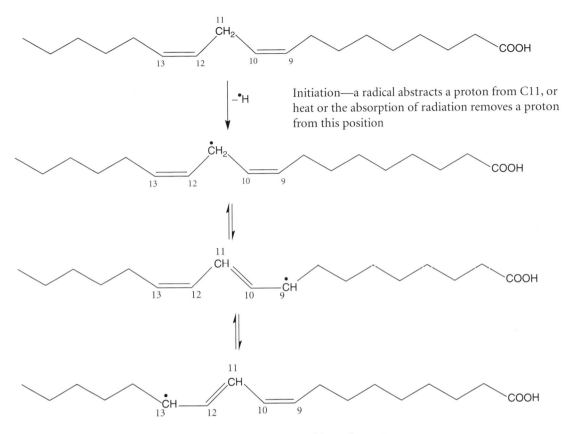

Initiation—a radical abstracts a proton from C11, or heat or the absorption of radiation removes a proton from this position
–•H
COOH
11
CH
13
12
10
9

Figure 7.25. The proton on C11 in linoleic acid is the most susceptible to abstraction.

Figure 7.26. The general scheme for the propagation phase of lipid oxidation.

$$R^{\bullet} + O_2 \longrightarrow ROO^{\bullet}$$

$$ROO^{\bullet} + RH \longrightarrow ROOH + R^{\bullet}$$

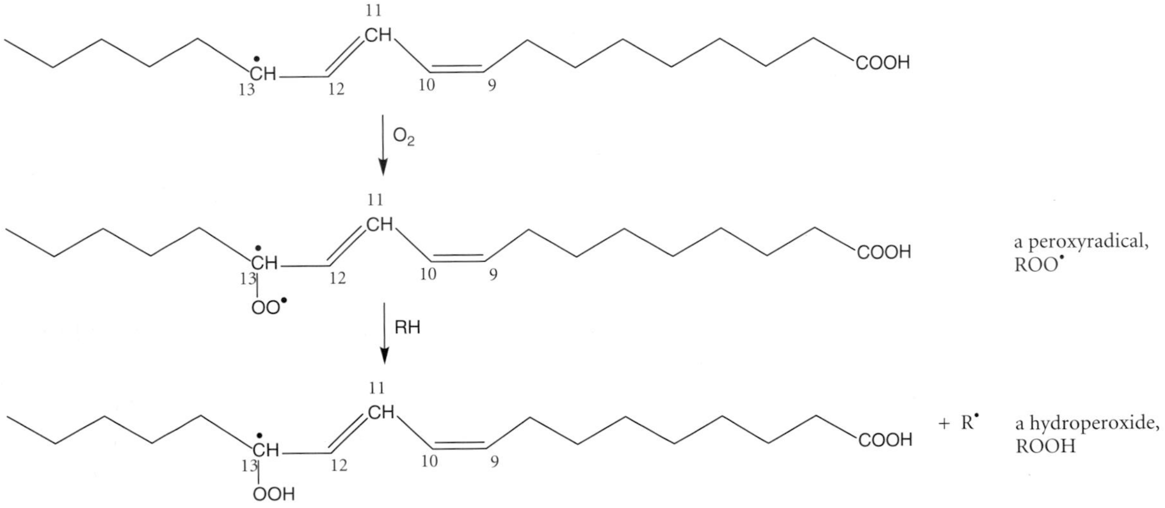

O₂
RH
COOH
OO•
OOH
+ R•
a peroxyradical, ROO•
a hydroperoxide, ROOH

Figure 7.27. The propagation phase in the oxidation of linoleic acid.

$$ROOH \longrightarrow RO^{\bullet} + {}^{\bullet}OH$$

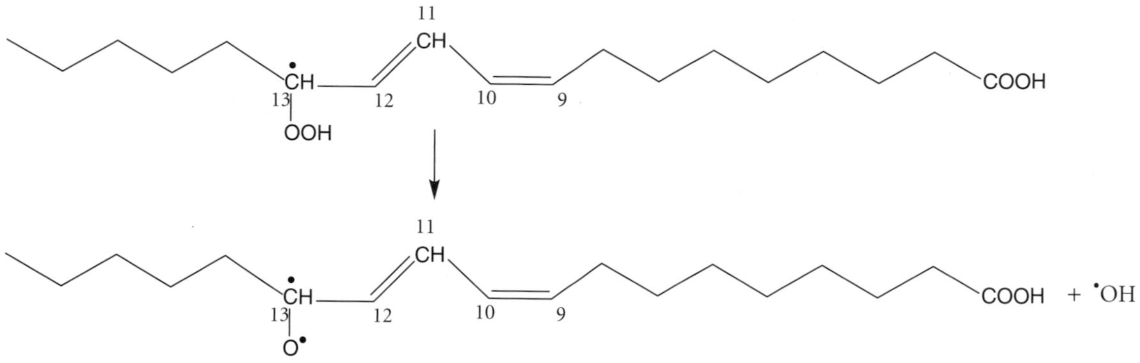

Figure 7.28. The decomposition phase of oxidation.

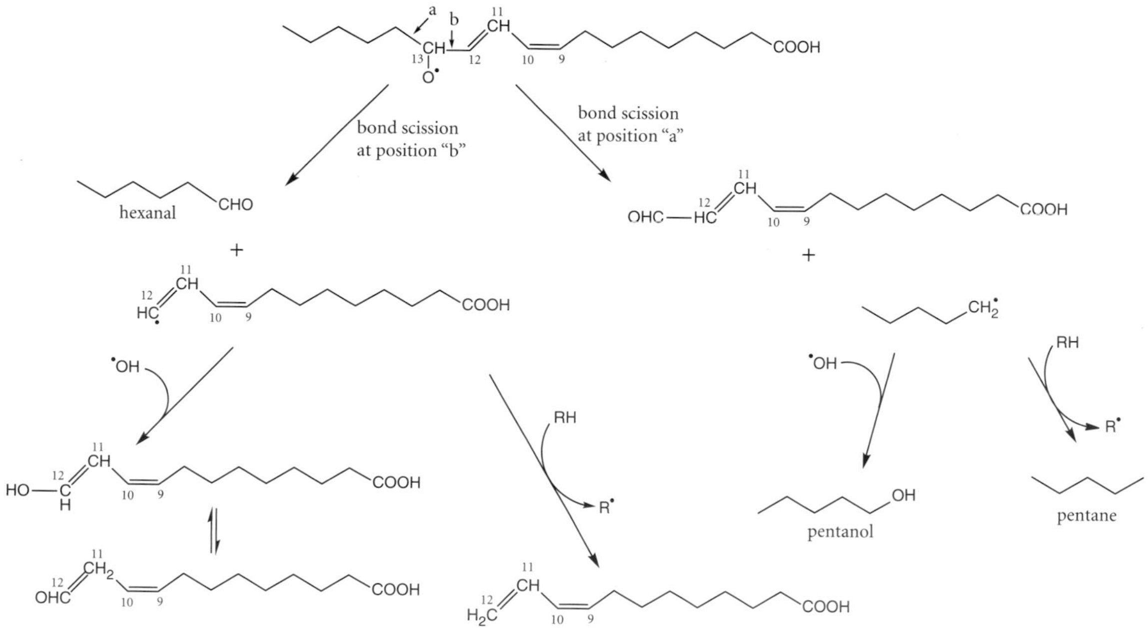

Figure 7.29. The decomposition of linoleic acid illustrating the various possible products.

$$R^{\bullet} + R^{\bullet} \longrightarrow RR$$

$$R^{\bullet} + ROO^{\bullet} \longrightarrow ROOR$$

$$ROO^{\bullet} + ROO^{\bullet} \longrightarrow ROOR + O_2$$

$$RO^{\bullet} + RO^{\bullet} \longrightarrow ROOR$$

$$R^{\bullet} + RO^{\bullet} \longrightarrow ROR$$

Figure 7.30. The termination process in lipid oxidation.

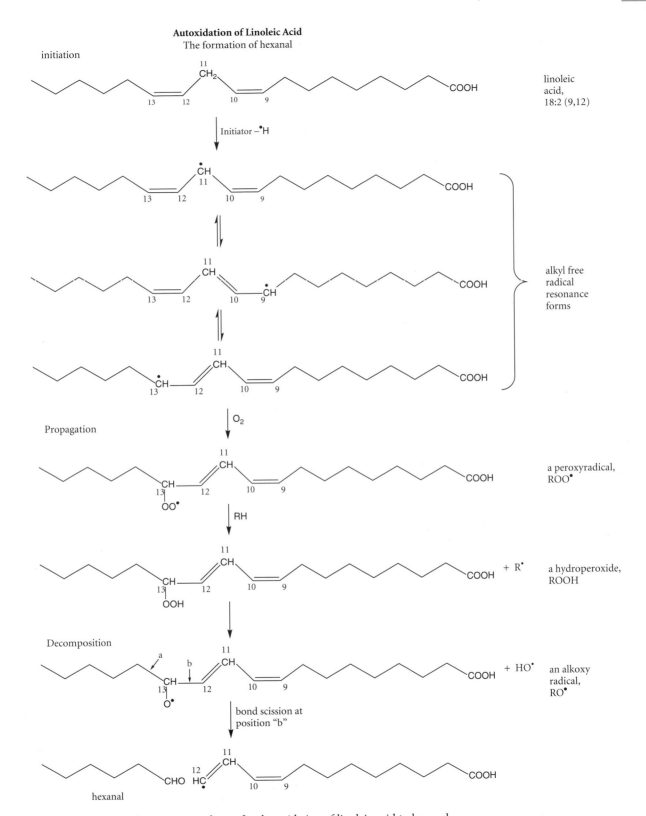

Figure 7.31. The complete reaction scheme for the oxidation of linoleic acid to hexanal.

Photosensitizer-Induced Singlet Oxygen Oxidation

In the molecular orbital description of the bonding in the oxygen molecule O_2, the lowest-energy ground state is an open shell triplet state (shown on the left in Figure 7.32). The molecule has two degenerate (that is, same energy) excited singlet states (shown on the right in Figure 7.32). The energy difference between these two states is 22.5 kcal/mol (94.2 kJ/mol). The excitation of triplet oxygen to its singlet state is a forbidden transition and thus does not generally occur spontaneously in atmospheric oxygen. When singlet oxygen gas is generated, however, it is long-lived in the gas phase, but in solution, as in a food sample, it is quickly quenched by energy transfer to surrounding molecules, which unfortunately in the case of lipids results in the oxidation of unsaturated fatty acids. For this reason, the generation of singlet oxygen in foods is quite undesirable.

Singlet oxygen can be generated by interaction of $^3O_2$ with various photosensitizer molecules that absorb light energy as they are promoted to their own excited states and then transfer this energy to triplet oxygen, promoting it to the excited singlet state,

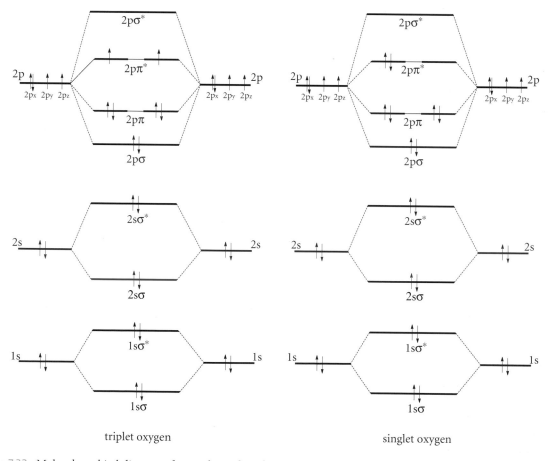

Figure 7.32. Molecular orbital diagrams for singlet and triplet oxygen.

Figure 7.33. The generation of singlet oxygen by photosensitizers.

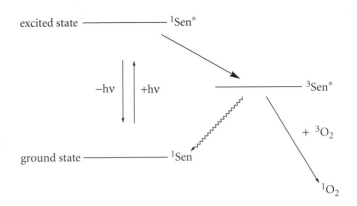

$$^1Sen + h\nu \rightarrow {}^1Sen^\star \rightarrow {}^3Sen^\star$$

$$^3Sen^\star + {}^3O_2 \rightarrow {}^1Sen + {}^1O_2$$

as illustrated in Figure 7.33.

Various dye molecules are able to promote the generation of singlet oxygen, including the porphyrins of common food molecules, such as chlorophyll, myoglobin, and hemoglobin (discussed in Chapter 9), which all contain metal ions, as well as the vitamin riboflavin. Singlet oxygen can also be generated by the direct interaction of triplet oxygen with ferrous iron, Fe^{+2},

$$Fe^{+2} + O_2 \rightarrow Fe^{+3} + {}^1O_2$$

Once generated, singlet oxygen can rapidly react with unsaturated fatty acids to oxidize the double bonds (Figure 7.34).

In controlling photoxidation of lipids in foods, various $^1O_2$ quenchers can be used to scavenge or eliminate the $^1O_2$ before it can react with the unsaturated lipids in the food. Carotenoid pigments such as zeaxanthin, astaxanthin, and lutein (discussed in Chapter 9) are quite effective in reacting with singlet oxygen. Other effective $^1O_2$ quenchers are butylated hydroxyanisole (BHA) and tocopherol (vitamin E, Chapters 10 and 12).

Enzymatic Lipid Oxidation

Fatty acids can also be oxidized as the result of the action of the enzyme **lipoxygenase**, which removes hydrogen atoms from methylene carbon atoms between nonconjugated double bonds, such as the C11 carbon atom of linoleic acid, to produce the C11 free radical. The oxidation chain reaction then proceeds as outlined above for autoxidation. Lipoxygenases are found in many plant and animal tissue cells, and their presence can therefore be a problem for foods that contain whole pieces of meats or vegetables, particularly if they have not been inactivated by heat denaturation from cooking.

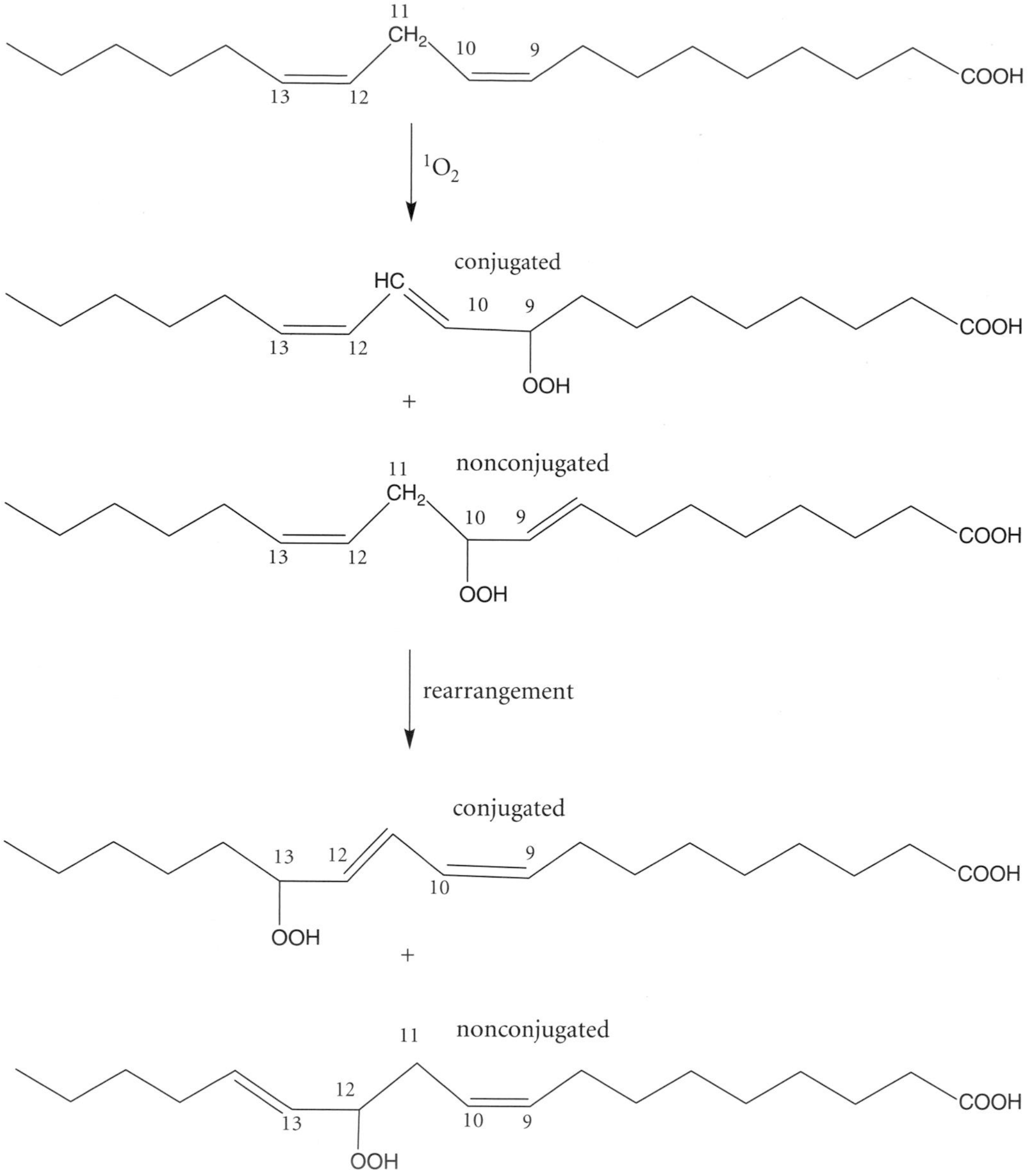

Figure 7.34. The oxidation of lipids (in this case linoleic acid) by singlet oxygen.

Tests for Lipid Oxidation

Because the volatile aldehyde hexanal is one of the most characteristic products of the autoxidation of linoleic acid (see Figure 7.31), the analysis of the concentration of hexanal in the "headspace" above a food product can be used as a measure of the progress of lipid oxidation. Malondialdehyde (Figure 7.35) is a breakdown product of the autoxidation of α-linolenic acid. This aldehyde is a so-called thiobarbituric acid reactive substance (**TBARS**), in that it reacts with thiobarbituric acid to produce

Figure 7.35. The structures of malondialdehyde and thiobarbituric acid.

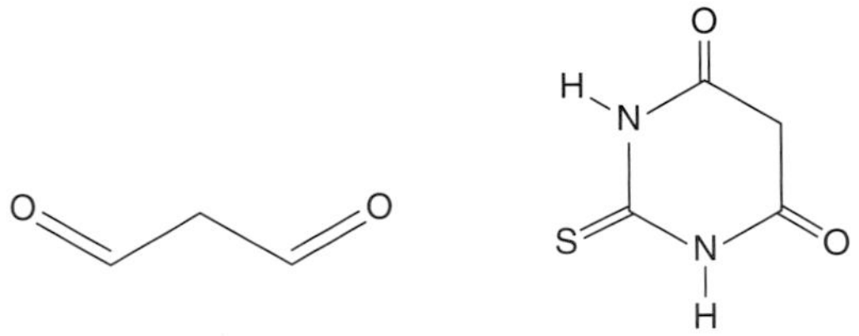

malondialdehyde

thiobarbituric acid

a fluorescent product, which can be quantitated spectrophotometrically to give a measure of the extent of α-linolenic acid oxidation (Raharjo et al. 1993). A third test is the so-called peroxide value (**PV**) test, which measures the amount of iodine released when hydrogen iodide is added to a sample and reacts with the lipid hydroperoxides resulting from oxidation.

The Control of Lipid Oxidation

Most foods contain lipids, and the oxidation of those lipids is one of the most important limitations on the quality and shelf life of food products. Accordingly, the control of lipid oxidation is of paramount importance in food manufacture. A number of approaches can be taken to deal with this problem. One of the simplest is to use saturated fatty acids, because reducing the number of susceptible double bonds in the food will naturally reduce the amount of oxidation possible. Unfortunately, this solution conflicts with the nutritional goal of increasing the proportion of unsaturated fats in the diet, and particularly of reducing the proportion of short-chain saturated fat.

Another method to control lipid oxidation is to reduce the production of radicals through the Fenton reaction (Figure 7.24) by the control of iron, copper, and other pro-oxidant metals. Eliminating transition metals can also prevent them from catalyzing the decomposition of hydroperoxides, generating singlet oxygen, or reacting with unsaturated fatty acids. Metal ions can be sequestered using chelating agents such as citric acid and particularly ethylenediamine tetraacetic acid (EDTA; see Chapter 10). These molecules bind to metal ions and prevent them from interacting with lipids. Because oxidation proceeds faster at higher temperatures, like all conventional reactions with an activation barrier, it can also be controlled by lowering the temperature at which food products are stored. Storage below −18 °C is necessary to prevent enzymatically catalyzed oxidation by lipoxygenases. Lipoxygenases can also be inactivated by heating, although other oxidation reactions may be promoted by the high temperatures necessary for lipoxygenase denaturation. As we saw in Chapter 2 (Figure 2.40), the rate of lipid oxidation varies strongly with water activity, being high at $a_W > 0.55$, then falling to low values at a_W around 0.3, but rising to high rates again for $a_W < 0.1$, so that lipid oxidation is lowest at low-intermediate moisture contents.

The oxidation of lipids can also be controlled by antioxidants (Figure 7.36). There are a number of such molecules of various types, some of which will be discussed in

$$—ROO^{\bullet} + AH \longrightarrow A^{\bullet} + ROOH$$

$$—RO^{\bullet} + AH \longrightarrow A^{\bullet} + ROH$$

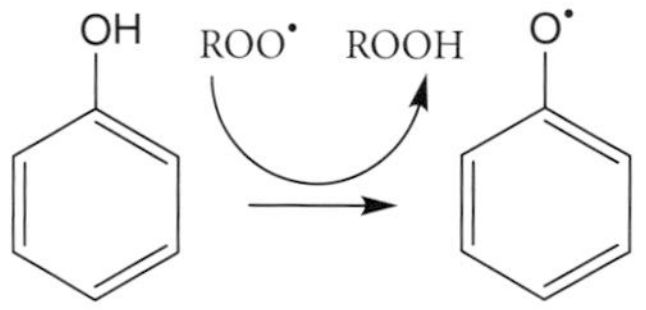

Figure 7.36. The mechanism of action of antioxidants (AH), including phenols.

Chapters 8, 9, and 10. Some are natural products, and some are synthetic food additives. Among the natural products are α-tocopherol (vitamin E) and L-ascorbic acid (vitamin C). One important class of antioxidants includes phenolic compounds, particularly the so-called polyphenolic compounds including the flavonoids that are abundant in various fruits and vegetables. The most common synthetic antioxidants in manufactured foods are butylated hydroxyanisole (BHA) and butylated hydroxytoluene (BHT; see Chapter 10). Other synthetic antioxidants include D-ascorbic acid, propyl gallate, ascorbyl palmitate, and tertiary butylhydroquinone.

FRYING

Oils are used for frying foods and have an advantage over using water-based solutions as a cooking medium because they have a much higher boiling point than water, allowing foods to be heated to temperatures much higher than 100 °C. Typical frying temperatures are in the range from 160 to 195 °C. At higher temperatures the oil may begin to break down. Food fried in oil will absorb the oil, sometimes in substantial amounts (from 10 to 40% by weight), with longer frying times leading to greater oil absorption. For this reason fried foods may be less healthy than the same item boiled or eaten raw. To minimize the absorption of oil, higher frying temperatures and thus shorter cooking times are desirable, but higher temperatures increase the rate at with the oil is degraded. Also, frying at high temperatures increases the production of potentially carcinogenic heterocyclic amines in meats (see Chapter 5).

As we have seen, unsaturated fatty acids are susceptible to destruction by oxidation. Frying releases water from the cooked food as steam, which produces hydrolysis of acylglycerols to give free fatty acids, and also removes volatile antioxidant molecules. Of the commonly occurring fatty acids, the unsaturated linoleic and linolenic acids are the most susceptible to oxidation. The suitability of a particular oil for frying will be related to its content of these molecules, because the more of these lipids that an oil contains, the more rapidly it will break down at the elevated temperatures of frying. For this reason the vegetable oils, such as soybean and sunflower oil, are poor choices for frying unless partially hydrogenated, and lard was a popular choice until recent decades when health concerns about saturated fats led to a search for alternatives. A number of palm oils are also good choices because of their low concentrations

Table 7.10 Smoke and flash points of selected oils (C°)

Oil	Smoke point	Flash point
butter	135	—
coconut oil	195	290
lard	200	240
canola	225	275
olive oil	210	—
corn oil	230	335
soybean oil	235	330

Source: N. Myhrvold, *Modernist Cuisine* (Bellevue, WA: Cooking Lab).

of unsaturated and polyunsaturated fatty acids (although the abundance of the short-chain palmitic acid in palm oil again raises health concerns). Because of the accumulation of oxidation products with time, frying oil in commercial and industrial operations must be regularly replaced, or is constantly replaced in continuous industrial frying as oil is lost through absorption into the food.

Smoke Points

The **smoke point** of an oil is the temperature at which visible smoke can be seen rising from the oil, usually as it is being used in frying. Because oils vary considerably in composition as a result of both origin and processing, their smoke points are not precise numbers but general ranges (Table 7.10). In addition, the history of an oil can change the smoke point; for example, an oil that is repeatedly heated to high temperatures or used for frying will have a lower smoke point than it did when new and unused. In general, unrefined oils will have much lower smoke points than refined oils due to impurity-mediated catalysis. The related **flash point** is the approximate temperature at which the oil will burst into open flame, an event to be avoided in any kitchen!

FAT CRYSTALLIZATION

The thermal properties of fats are important in their functional roles in foods. Fats have a low thermal conductivity, which is why they are good as thermal insulators; extensive fat deposits are essential for ocean-dwelling mammals such as whales, seals, and polar bears that must maintain a high body temperature in nearly freezing waters. Fats have a high heat capacity, as one might expect for a nearly freely jointed polymer, because absorbed heat can promote numerous conformational transitions rather than going to increase the molecular kinetic energy, and thus the temperature.

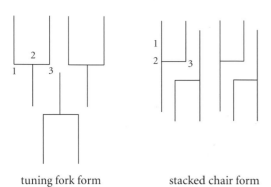

Figure 7.37. Two ways that triglycerides could be stacked together in crystals. The tuning fork form on the left has not been observed in actual lipid crystals.

tuning fork form   stacked chair form

Different triglycerides have differing melting temperatures, depending upon their fatty acid chains. In general, unsaturation lowers the melting temperature because the kinks introduced into the chains by the *cis* double bonds interfere with the ordered packing necessary for crystallization. Most fat-based foods consist of a mixture of fats with different structures and properties, so that these fats do not have sharp melting temperatures, but rather exhibit a gradual softening over a temperature range, as with butter.

Crystal Phases

The crystallization and melting behavior of triglycerides is extremely complex, not only because of the great heterogeneity in composition of naturally occurring triglycerides but also because even pure samples of a completely regular saturated triacylglycerol (i.e., a triglyceride of the type SSS where all three saturated fatty acids are the same) have polymorphic solid phases (that is, the triglyceride can be organized into a crystal lattice structure in more than one way). Tuning fork arrangements, as illustrated schematically in Figure 7.37, might be favored where all of the fatty acids of the triglyceride are the same, or where the two outer fatty acids are the same but the middle one is different. For triglycerides where the chains are not all the same, chair type packing is favored, and in fact is often found in symmetrical molecules as well. The majority of crystal structures that have been determined experimentally are for homotriacylglycerols, in which the individual triglycerides adopt a chair-like conformation, which then stack as is shown in Figure 7.37, although with a slant as seen in Figure 7.39.

The regular SSS triglycerides can crystallize into at least three crystal types, designated α, β′, and β, which are now thought to have the stacked chair type arrangement for individual antiparallel pairs of triglycerides, but which differ in the packing of these chains (Figure 7.38). In the crystal structures studied so far, the lipids in the stacked chair arrangement are tilted, as shown in Figure 7.39 for crystalline tricaprin in the β phase. The highest energy phase is α, which thus has the lowest melting

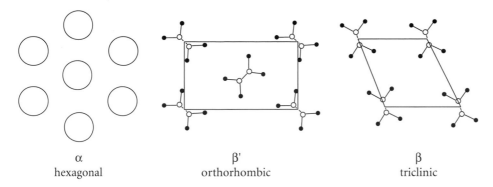

Figure 7.38. Generalized representations of the packing of triglycerides in the three principal natural crystal forms. In each case the view is looking down the length of the triglyceride molecules, so that the acyl chains are pointing up and down. For the α phase the chains are represented as circles to indicate that they can freely rotate about their long axis.

temperature. In this phase the orientation of the chains is not fixed, and they can still rotate around their long axis, so that the α phase is similar to a liquid crystal. This is indicated in Figure 7.38 by representing the fatty acid chains as circles. The β' phase is intermediate in energy and has orthorhombic symmetry for the packing of the chains, with two chains per unit cell and a stacking angle of ~70°. The orientations of the chains alternate in successive layers so that the planes of the chains are approximately perpendicular to those on either side. The lowest energy phase is the β, which thus has the highest melting temperature. This phase has triclinic symmetry with only one chain per unit cell and a stacking angle of 59°.

When cooling a sample of liquid fat consisting of a regular SSS triglyceride, it will first crystallize into one of the three phases, with the cooling regimen, such as the rate of cooling, determining which phase is produced. Slow cooling is likely to allow the system to crystallize into the lowest energy β form when its freezing point is reached. Faster cooling can trap the system in one of the other phases. If the α phase is heated, it can reorganize into the β' phase. With further heating the fat can reorganize again into the β phase. Crystal packing will be less regular in the case of a mixed triacylglycerol, and much less regular if any of the chains are unsaturated. In addition, in any real food lipid sample, there will be many different types of triglycerides present, further interfering with regular crystalline packing. For this reason, lipids in foods display a broad melting range and often are softer than the crystals of a pure sample of an SSS triglyceride.

The triacylglycerols of butter and margarine are generally crystallized in the intermediate energy, and thus less stable, β' phase. For this reason, they have a lower melting point than they would have if their triglycerides were in the more stable β phase. In the β' phase the packing is looser than in the β phase, resulting in a softer consistency for butter. In chocolate, the fats generally are in the most stable β phase, which gives them a harder texture and a higher melting range.

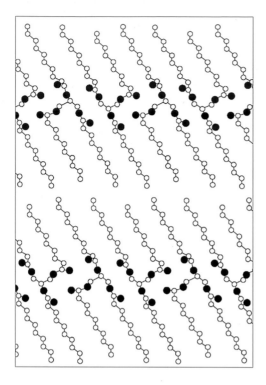

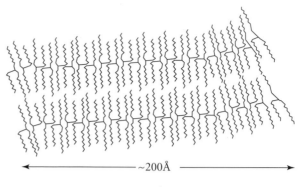

Figure 7.39. At left is a picture of the triclinic crystal structure of tricaprin projected into the bc plane. Notice the stacked chair arrangement of the molecules. On the right is a schematic picture of the proposed structure of the liquid crystal state of triglycerides, where small pseudolamellar domains up to 200 Å in extent remain even in the liquid. (Both figures adapted and redrawn from D.M. Small, *The Physical Chemistry of Lipids: From Alkanes to Phospholipids*. New York: Plenum, 1986.)

Chocolate

Chocolate is probably the most popular confection worldwide, with many larger cities having dozens of stores that sell essentially nothing else. The English word "chocolate," possibly derived from a Nahuatl (Aztec) word, refers generally to several foods made from the seeds of the tropical cocoa tree, *Theobroma cacao*, native to Mexico and Central America. The peoples of Central America have used cocoa for several thousand years, probably dating back to the Olmecs. It was highly prized by the Mayan and Aztec civilizations, with important ceremonial and religious uses, particularly as a bitter-tasting drink but also as a general food ingredient. After the Spanish Conquest, chocolate was introduced to Europe, where it was first consumed as a beverage sweetened with sugar, and only much later was developed into solid candies. Much of the world's modern production of chocolate comes from West Africa, with the largest single producer being Cote d'Ivoire. Together, Cote d'Ivoire and Ghana account for more than half of world production. Significant production also comes from Latin America and Indonesia.

Chocolate is made from the fruits of the cocoa tree, referred to as "beans." Up to forty of these seeds are borne in large pods, surrounded by a sticky pulp. When the pods are ripe, they are cut open, and the seeds and pulp are removed and fermented together for about a week, which develops the flavor and color of the beans. After fermentation, the seeds are generally washed, to separate them from the pulp, and are then roasted, which dries the beans and promotes various reactions, further developing the flavor. The roasted, dried beans are broken open, and the outer shells are separated from the inner "nibs" and discarded, while the nibs are heated and ground up to produce a relatively homogenous **cocoa liquor**. This paste can be liquefied by heating and consists of two fractions: lipids, which make up more than half of the total, and **cocoa solids** suspended in the lipids. Cocoa powder can be produced by separating these solids from the liquid fat by filtration, with the remaining purified fat being called **cocoa butter**. As we have seen, cocoa butter is special among common food fats, with a fairly uniform SUS composition that allows for regular crystalline packing as it solidifies. As with most regular triacylglycerols, it can adopt several crystalline polymorphs of differing energies, and with different properties. Thus, the production of quality chocolate involves producing the best crystalline polymorph, as well as maintaining that structure through proper handling and temperature control after production.

Cocoa powder is often treated with alkali (base) in a process called "Dutching," since it was first developed in Holland in the nineteenth century. Dutching raises the normally acidic pH of the product and lightens the brown color while reducing the bitter taste. However, the alkali treatment also destroys a significant fraction of the polyphenolic flavanol content of the cocoa, reducing its antioxidant capacity (Miller et al. 2008).

Chocolate candy is produced by blending cocoa liquor with other ingredients, including additional cocoa butter, sugar to soften the bitter taste of the cocoa, and depending on the desired product, vanilla, milk, or milk powder. Dark chocolates contain no milk, while milk chocolates of course contain milk or milk solids. Because the liquid mixture constitutes an emulsion, an emulsifier such as lecithin is often added to further promote uniformity and a smooth texture in the final product. Nevertheless, the texture of this liquid suspension is still grainy due to the large and nonuniform size of the cocoa solid particles and sugar crystals. Thus, the suspension is further refined by a grinding process called **conching** that reduces the size of the suspended particles such that they cannot be detected by the tongue, resulting in a smooth liquid suspension.

The three typical triglyceride crystal polymorphs discussed in the previous section, α, β, and β', are idealized states that generally would be possible for homogenous samples of fully saturated triglycerides. As we have seen, food fats are highly heterogeneous, and usually contain significant amounts of unsaturated fatty acid chains. Thus, the crystallization behavior of such mixtures of triacylglycerols would be much more complex and significantly less regular. As we have noted cocoa butter is much more regular than other typical food lipids; nonetheless it does not consist of a uniform sample of all saturated fatty acids. The actual composition of cocoa butter is approximately 35% POS, 23% SOS, and 15% POP, where P, O, and S refer to palmitoyl, oleoyl, and stearoyl chains, respectively. This still produces a saturated-unsaturated-saturated profile for around three quarters of the triglycerides in the cocoa butter. The remaining triglycerides have a variety of other sequences. Thus, the crystallization of cocoa butter would be expected to be more complex than just producing the expected three phases α, β, and β′. In fact, at least six important solid phases of cocoa butter have been identified. These are variously referred to using either a version of the Greek labeling system (Vaeck 1960), a system of Roman numerals (Willie and Lutton 1966), or a combination of the two (Schenk and Peschar 2004). The six phases characterized by Willie and Lutton have the following characteristic melting temperatures: I (17.3 °C), II (23.3 °C), III (25.5 °C), IV (27.5 °C), V (33.8 °C), and VI (36.3 °C). They identify the I and II states with α crystalline packing forms, and the III state as a mixture of α and β′ forms. The IV state they identify solely as a β′ form, and the V and VI states with the most stable β packing arrangement. The most desirable of these crystalline forms for chocolate candies is the V form of the β crystalline packing (the VI state is too hard). The V crystalline polymorph has a firm texture that breaks with a crisp snap and has a glossy surface texture. Note that it melts at 34 °C, just below the normal human body temperature of 37 °C, which produces the cooling sensation on the tongue mentioned earlier. The other crystalline polymorphs either melt too easily, have a soft texture, or crumble instead of breaking cleanly with a snap.

Producing the best solid chocolate candies thus requires controlling the crystallization process to maximize the production of the type V β crystals in the finished product. This is accomplished by cooling the heated liquid fat suspension to 27 °C, just above the melting point of the types I–III forms, to prevent crystals of these undesirable polymorphs from forming. This will allow crystals of the forms IV–VI to nucleate and grow. Although this temperature allows crystals of the undesirable type IV to form, the lower temperature will also produce a larger number of nucleation sites, and thus smaller crystals of the types V and VI (See Chapter 3), favoring a smoother and more uniform product. The product can then be "tempered" by heating it up to a temperature in the range 30 to 32 °C, which will melt the undesirable crystals of type IV without affecting the type V and VI crystals. Formation of more type V crystals can be encouraged by seeding the liquid at the initial cooling stages with previously tempered solid chocolate already in the type V form, which produces a larger number of smaller type V crystals, and thus a smoother texture. The emulsifier molecules are primarily located on the surfaces of the sugar granules with their hydrophobic tails pointing away from the surface, promoting fat crystallization on the surfaces of the coated granules, and thus also producing a smoother texture (Svanberg et al. 2011).

The β crystalline form of the V and VI phases of chocolate is the most stable packing form and thus has the highest melting temperature. If improperly stored, fine chocolates can undergo various degradative transformations. Chocolate that has been subjected to temperature abuse may exhibit a grayish-white surface film, consisting of poorly crystallized triglycerides or sugar crystals, that is called **fat bloom** in the industry (Chapman et al. 1971). Chocolate can also absorb moisture if the ambient relative humidity is too high, and if it is stored with other foods, fat soluble aroma molecules can also absorb into the cocoa butter, changing the flavor or aroma profile of the product.

THREE-COMPONENT PHASE DIAGRAMS

When formulating foods that contain both oil and water, which do not ordinarily mix in stable phases, it is necessary to add a third surfactant component in order to produce a stable uniform phase, for example by the formation of micelles such as illustrated in Figure 7.40. These three components cannot be mixed in arbitrary amounts without at least one of the components partially separating out in a pure phase. To characterize the possible stable mixtures, three-component phase diagrams such as are shown in Figure 7.41 are used to indicate the range of stable compositions and the type of microstructures produced.

Figure 7.40. A schematic illustration of a micelle with hydrophobic molecules (represented by spheres) in the oily interior.

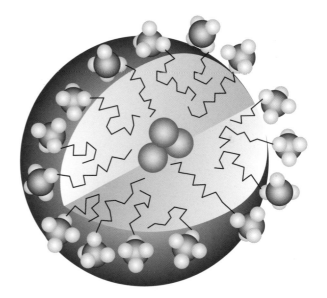

Figure 7.41. A generalized three-component phase diagram. The dotted lines denote lines of constant composition of the component at the opposite vertex.

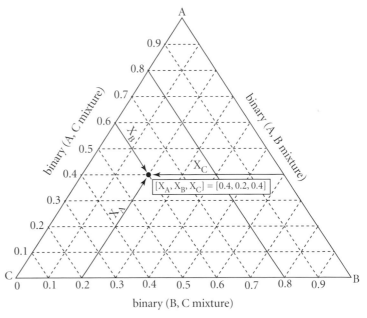

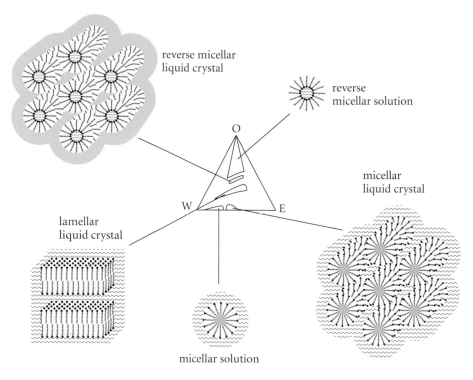

reverse micellar
liquid crystal

reverse
micellar solution

micellar
liquid crystal

lamellar
liquid crystal

micellar solution

Figure 7.42. In a three component system, not all compositions can be mixed into a single homogenous system; most compositions will cause one or more of the components to separate out as a separate phase. Those compositions that do lead to uniform systems are contained within the regions outlined in the triangular phase diagram. The actual structure of the system will depend on the relative compositions. For example, a small amount of oil but a large amount of water and emulsifier will lead to a dispersion of discrete micelles in a continuous phase of water, with the oil sequestered in the micelle interiors, as indicated schematically on bottom center. As the proportion of oil increases, these micelles elongate like cigars, ultimately into long tubes, with the oil in their centers, and as the proportions of oil and emulsifier increase further, these cylindrical tubes pack together with the water in the spaces between them, as indicated in the lower right. Reverse micelles, with water in the interior, can from under conditions of high oil composition and limited water and emulsifier (top right), which can also form inverted cylinder packed together with the oil between the cylinders as in the upper left. Lamellar liquid crystals, consisting of stacked bilayers, with oil in the center, and separated by water layers, can form when all three components are present in roughly equal amounts. (Adapted from S. Friberg, *J. Am. Oil Chem. Soc.* 48:578–581, 1971, with kind permission from Springer Science and Business Media.)

Each of the vertices of a triangular three-component phase diagram such as the one shown corresponds to one of the components as a pure phase. Any point in the triangle corresponds to a particular composition of the system, where the composition of each component is read from the lines parallel to the opposite face of the triangle. The dashed lines running down the diagram correspond to mole fractions of B and C in the same ratio. For example, in Figure 7.41, the composition of the indicated point corresponds to a mole fraction for component A of 0.4, or 40%, a mole fraction for component B of 0.2, or 20%, and a mole fraction for component C of 0.4 or 40%.

In the three-component phase diagram shown in Figure 7.42, O, W, and E stand for oil, water, and emulsifier, and at each vertex of the triangle you have only that phase. Only those compositions inside the regions enclosed by the solid lines will produce uniform phases. For example, at a low concentration of oil and emulsifier and a high concentration of water, corresponding to the lower left region on the diagram, micelles will form with the oil enclosed in the non-polar interior of the micelle. At a high concentration of oil but low to intermediate concentrations of water and emulsifier, corresponding to the uppermost region near the top vertex, reverse micelles are formed, with the oil on the outside, and the water and polar head groups of the emulsifier on the inside. In the middle regions of the diagram, where the proportions of all three phases are nearly the same, liquid crystals are formed, either as laminar lipid bilayers stacked on top of on another with intervening water layers, or long, log-shaped tubular micellar structures stacked together. At higher water and lower oil compositions, these tubes have the oil on the inside of the micellar log and water on the outside, occupying the spaces between the long cylindrical structures. At lower water and higher oil compositions, the structure is reversed.

SUGGESTED READING

Akoh, C.C., and D.B. Min, eds. 2002. *Food Lipids: Chemistry, Nutrition, and Biotechnology*, 2nd ed. New York: Dekker.

Marangoni, A.G., and S.S. Narine, eds. 2002. *Physical Properties of Lipids*. New York: Dekker.

Mulder, H., and P. Walstra. 1974. *The Milk Fat Globule*. Farnham Royal, Bucks, England: Commonwealth Agricultural Bureaux.

Schenk, H., and R. Peschar. 2004. "Understanding the structure of chocolate." *Rad. Phys. Chem.* 71:829–835.

Small, D.M. 1986. *The Physical Chemistry of Lipids: From Alkanes to Phospholipids*. New York: Plenum Press.

The sweetest
honey is
loathsome in its
own deliciousness,
and in the taste
confounds the
appetite.
WILLIAM SHAKESPEARE,
ROMEO AND JULIET

8.
Flavors

TASTE AND ODOR PERCEPTION

Our senses are the means by which we receive information about the outside world around us, detecting sound and electromagnetic waves, and through touch, the presence and texture of nearby objects. The senses of taste and smell detect chemicals, and thus give our brains a wealth of information about our molecular surroundings. Our sense of smell is able to detect and often distinguish thousands of different molecules, frequently at very low concentrations. Primates such as us tend to rely more on sight than smell, and compared with many other animals, such as ursids (bears) and canids (dogs), our sense of smell is relatively less sensitive. Nevertheless, we can detect some odorants at quite low threshold concentrations, sometimes in the parts per billion or even parts per trillion, and our sense of smell serves important functions, including identifying those things that might be good to eat. The sense of taste of course also helps us to identify foods that are desirable to eat and to reject those that might be potentially harmful. In "lower" animals, odors and flavors can elicit strong voluntary and involuntary responses, and their effects on humans are only slightly less compelling. It is thought that odor molecules have a powerful ability to stimulate memories and feelings because olfactory signals are received in the part of the brain close to the limbic system, which is involved with emotions. In the modern context, the taste and aroma of foods constitute the bulk of their culinary appeal (appearance and texture also contribute—and even sounds do as well, such as the snap of fresh vegetables or the sizzle of a fried steak). For those who do not suffer from food shortage, the enjoyment of a meal extends beyond simply quelling hunger pangs and is crucially dependent on the taste and aroma of the food.

The sensations of taste and smell are the result of molecules of various chemicals binding to chemoreceptors in our mouths and noses and triggering neural signals that are interpreted in the brain. Thus, the properties of taste and smell are not strictly the properties of the molecules themselves, but of how they interact with the receptor complexes and even the neural/psychological systems of a living organism. As a result, there is no chemical probe that can determine the flavor of a particular type of molecule; this must be accomplished using controlled taste panel experiments with human subjects. For psychologically complex, self-aware organisms such as ourselves, this psychological component in particular is poorly understood and far beyond the scope of this book. Even the mechanistic details of the interactions of flavor and odor molecules with the chemoreceptors are poorly characterized and only now beginning to be explored in detail.

Most humans, at least as children, have a natural preference for sweet flavors, which is probably an evolutionary inheritance from our distant primate ancestors, since sweetness is usually an indicator of the presence of calorie-rich sugars. Similarly, many people dislike bitter tastes, especially as children, which may be an inherited defense mechanism, because many bitter-tasting molecules are poisonous plant alkaloids, and an instinctive avoidance of these dangerous plants would have clear survival value. Often, a preference for bitter flavors like caffeine or some vegetables is an acquired taste, which illustrates the complex nature of taste preferences. Because of individual genetic differences in the structures of the chemoreceptors, as well as psychological differences between individuals, and many other factors, tastes must be evaluated as statistical averages over the responses reported by a number of individuals.

The chemoreceptors for taste in humans are located in structures called **taste buds** located on the surface of the tongue. Taste buds are specialized collections of epithelial cells connected to nerve endings. Most of the taste buds are located around the edges of the tongue. The tongues of young children are covered with active taste buds, while those of adults are generally located along the edges. Since the number of active taste buds decreases with age, sensitivity to taste can also decrease with age, particularly after age 45, although odor perception is generally more significantly impacted by age than is taste perception. It was previously thought that taste buds in different parts of the tongue responded to different types of flavors, but this model, which is still found in older textbooks, is no longer thought to be strictly true, and all parts of the tongue can perceive all four of the basic tastes. Taste buds for certain flavors do seem to be concentrated in particular parts of the tongue, which was the source of the earlier observations, but the differences are small (Collings 1974).

Classically, tastes were divided into four broad categories: sweet, salty, sour, and bitter, sometimes referred to as the SSSB model. Salty flavor is of course produced by sodium chloride and other small inorganic salts, with the flavor depending on both the cation and anion. Sour flavors are generally produced by H_3O^+ and, thus, by acid foods; however, the hydronium ion is not the sole determinant of sourness, since different acids producing the same pH will have different levels of sourness, so that the intact acid is the determinant of the taste. Bitter tastes are often related to extensive hydrophobic functional groups or surface area in a tastant molecule. The taste receptors for sweetness are located in taste buds somewhat more concentrated on the tip of the tongue, while saltiness is detected by taste buds slightly more concentrated on the outer edges of the tongue just behind the region on the tip that detects sweetness. Sourness is detected by taste buds more concentrated on the rear edges of the tongue. Bitterness is detected by receptors primarily on the middle of the base of the tongue.

This SSSB model of tastes has been revised in recent decades. In the early twentieth century Japanese scientists identified a fifth, salty-sweet type of savory flavor, called **umami**, associated with monosodium glutamate (MSG). Specific taste receptors for umami have now been identified. This taste is also sometimes referred to as "meaty." Thus, the four-taste model becomes a five-taste SSSBU model. In addition, two other chemical qualities of foods, astringency and pungency (spiciness or "hotness") also contribute substantially to taste perception, leading to seven basic food tastes. The last two of these are not detected by the taste buds but are actually the result of chemical irritation of pain receptors, and such effects are referred to as "**chemesthesis**." Thus, astringency and hotness are not detected solely on the tongue but throughout the mouth.

Odors are detected by olfactory receptors in the nasal cavity. Volatile odorant molecules bind to odor-binding proteins in these olfactory receptors, triggering an electrical nerve signal to the brain. The odor-binding proteins belong to the G-protein-coupled receptor class of transmembrane receptor molecules spanning the cell bilayers. Volatile odorants enter the nasal cavity by two different routes: directly through the nostrils, called **orthonasal**, or from the oral cavity (which is connected to the nasal cavity at the back of the mouth), which is referred to as **retronasal** stimulation. It is important to note that the senses of taste and smell are thus connected, and information from these two types of chemisensors is integrated in the brain to produce the enormous range of flavors that we perceive. It is also worth noting in this context that sensory scientists do not use the common definition of the word "flavor," which is generally considered to be synonymous with taste in standard dictionaries, but define it more broadly to mean the collective sensation produced by food in the mouth as the result of stimulation of both taste and odor receptors and the stimulation of pain, tactile, and temperature receptors (Ternishi 1971). It is this collective stimulus that is interpreted in the brain as the "flavor" of a food.

On the molecular level, the actual chemical receptors for taste and odor are complexes of membrane-bound proteins, and the flavor molecules bind transiently with these proteins as ligands and trigger a neuron firing to signal the brain that a flavorant has been detected. As with all protein–ligand binding, there must presumably be some form of close molecular complementarity between the binding site of the receptor complex and the functional groups of the flavor and odor molecules. We have already discussed the so-called AH/B-γ model for sweet taste in Chapter 4, which assumes a specific geometry for the functional groups of the receptor binding site and therefore requires a complementary arrangement of functional groups in the sweet molecule. Molecules that have a specific taste are called **sapid**, and the specific collection of functional groups that produce a particular taste are referred to as the **saporous units**.

Membrane-bound protein complexes are particularly difficult to isolate, crystallize, and characterize, which may be one reason why it has proven so difficult to identify them. In the last few years good progress has been made in identifying the proteins responsible for detecting bitter (Adler et al. 2000), umami and sweet tastes (Ozeck et al. 2004). In 2001, several groups reported the discovery of the protein T1R3, part of a heterodimeric complex (T1R3/T1R1), which has been identified as the sweet receptor molecule (Bachmanov et al. 2001; Kitiagwa et al. 2001; Li et al. 2001, 2002; Temussi 2006). T1R3 belongs to a class of proteins called metabotropic, or Class C, G protein–coupled receptors (GPCR). The three-dimensional conformation of T1R3 has not been determined as of this writing, but the structure of a homologous member of its class, metabotropic glutamate receptor, mGluR1, is known. mGluR1 has two binding sites for the side chain of glutamate (glutamic acid), and these are also thought to exist in the T1R3 protein, which would explain why among the amino acids glutamic acid is unique in its flavor properties and how MSG (the monosodium salt of glutamic acid) can serve as a flavor enhancer. Computer modeling studies have attempted to deduce the conformation of the T1R3 protein based on homology with mGluR1, and to then use this information to deduce the interactions of the protein with sweet molecules (Figure 8.1), but a more definitive understanding of the detailed molecular basis of sweetness will have to await the determination of the structure of sweet molecules complexed with T1R3, as well as the development of a model for how the formation of this complex triggers the neural signal interpreted by the brain as sweetness. It seems likely that this type of information will be available in the near future. Hopefully other flavor receptor proteins will also be identified soon. These developments mark exciting breakthroughs in flavor science, and surely the coming years will produce remarkable advances in the knowledge of flavors.

While particular molecules may make a disproportionate contribution to a particular taste or aroma, it is important to remember that the sensory impact of most foods is the result of contributions from many different molecules with different concentrations, sensory thresholds, and stability lifetimes, which is difficult to reproduce artificially, and which is why synthetic vanillin differs from natural vanilla, or why grape-flavored gum or an orange popsicle do not taste like an actual grape or orange. The overall sensory experience of a fresh strawberry may involve not only the sweetness of several sugars, but also organic acids, many volatiles, and even some molecules that may add bitter notes. Commercial synthetic banana flavoring may

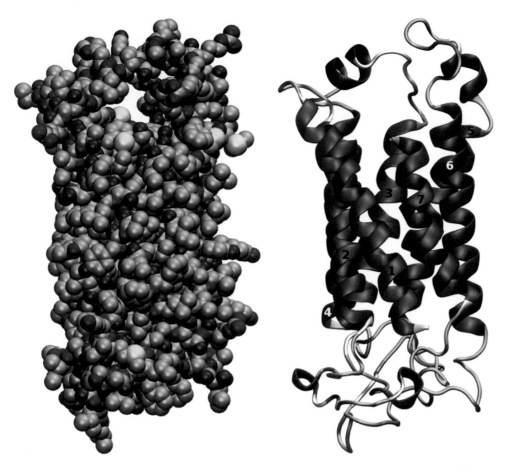

Figure 8.1. The proposed conformational structure of the odor receptor protein as predicted from computational modeling. The seven helices that span the membrane bilayer are clearly visible in the cartoon on the right. The top of the molecule, in this orientation, is outside of the cell membrane and contains the binding site for the odor molecules; the bottom is inside the cell cytoplasm. The odorant-binding site is clearly visible at the top in the van der Waals representation on the left. Coordinates courtesy of W. Goddard and coworkers.

consist of more than twenty individual components (Coultate 2009), and the flavor of an actual banana would result from contributions from very many more.

Different aroma molecules will have different perception thresholds, with some being detectable in very low concentrations while others may have very high perception threshold levels. Thus, the contribution that a particular type of molecule may make to the aroma profile of a food will depend not only on the concentration of the molecule in the sample but also its vapor pressure and the perception threshold, so that a simple chemical analysis of a sample will not completely characterize its aroma without an analysis of its threshold values. Similarly, the concentration of a tastant in a food will not necessarily directly correlate with the taste impact if its molecules are trapped in an internal matrix such as a gel or plant cell wall matrix, or if they differentially partition between fat and aqueous phases in the food. Interactions with other flavor molecules will also affect the way in which a flavorant is perceived. For example, umami not only has its own specific taste but is also a flavor enhancer, modifying the taste elicited by other tastants; the protein miraculin introduced in Chapter 5 is another such example.

Characterizing the full taste and aroma of various types of foods has required heroic efforts in analytic chemistry to separate and identify all of the potential flavorants present in foods, their concentrations and lifetimes (shelf-lives), and in sensory science, to determine their sensory impacts and properties, individually and in combination. These subjects are vast and far beyond the scope of this textbook; here we will survey only a few of the more important flavorants, concentrating on their chemistry rather than on their sensory evaluation. In the process, we will also consider various other miscellaneous molecules that are either chemically related to important tastants or are found in particular herbs or spices being discussed.

The number of molecules that elicit a taste or aroma response is very large, and they are chemically quite diverse; only a small sample of such compounds can be discussed here. Like the animals exhibited in a zoo, these molecules could be arranged in various ways, such as all those that share chemical similarities, or all those found in the same food, or all those that impart similar tastes (analogous to grouping all the bears in a zoo together, or all the animals from India, or all of the predators). The following stroll through the "zoo" of flavor molecules will in part employ all three types of groupings, with apologies to those with more systematic minds. We will begin with a discussion of sweeteners.

SWEETENERS

Many molecules have a sweet taste, including, of course, sucrose and many other carbohydrates, a number of the L-amino acids, certain peptides and proteins such as thaumatin, various artificial sweeteners such as saccharine and cyclamate, and even small molecules such as ethylene glycol. The "gold standard" for sweetness is sucrose, which has the "cleanest" and most widely accepted sweet taste. All other sweeteners are therefore rated on a sweetness scale relative to sucrose. The complete taste profile of a sweetener includes more than simply the concentration needed to produce a detectable sweet taste. Among the factors that are considered in evaluating the acceptability of a sweetener are the threshold value, which is the minimum amount that will be perceived as sweet; how quickly the taste is registered; how quickly it fades away; and whether or not any sort of aftertaste is detected. In selecting a sweetener for a specific purpose, other factors must be considered as well, including solubility, thermal and pH stability, possible interactions with other ingredients in the product, cost, and any potential toxicity or other health concerns. In human taste test panels evaluating the sweetness of various molecules, subjects are often asked to rate whether any sweet taste is detected in samples that are prepared by serial dilution, and the lowest threshold concentration at which any sweetness is detected is then compared with the threshold concentration for sucrose to give a shorthand ranking of the sweetness of the molecule relative to sucrose. In the sections that follow on sweeteners, as well as in the previous sections on sugars (e.g., Table 4.4) and sweet proteins, these numbers are used to give a quick indication of the potency of various sweeteners, but it must be remembered that sweetness is a complex property that is insufficiently captured by this crude comparison.

Nutritive Sweeteners

The nutritive sweeteners are those that can be metabolized to yield dietary calories (in general, however, they have no other nutritional value, and are usually considered to be "empty calories"). They include sucrose, fructose, invert sugar, and various sugar alcohols, including primarily sorbitol, mannitol, and xylitol. The sugar alcohols such as sorbitol and xylitol have only about two-thirds of the calories on a weight basis as sugars. The intensely sweet proteins monellin and thaumatin could also be considered nutritive because they can be hydrolyzed to produce amino acids, as is also true of aspartame, but they are present in such low concentrations as to impart negligible calories. For this reason, we will consider them to be nonnutritive. In many foods, particularly soft drinks and dessert and snack foods, a substantial part of the total calories may come from sucrose or high-fructose corn syrup (HFCS). A typical US soft drink sweetened with HFCS or sucrose contains around 100 to 110 calories in an 8-oz serving. The same size serving of orange juice contains about the same number of calories, ~110 calories, while the most popular US brand of Concord grape juice contains 170 calories in an 8-oz serving, some 70% more than a soft drink! For comparison, due to its fat content, an 8-oz serving of whole milk contains 150 calories, while the same amount of 2% milk has 130 calories, and skim milk has 110 calories, equivalent to a glass of orange juice.

In addition to its high energy content, sucrose is cariogenic; that is, it causes dental caries (cavities). Increasing concern about the excessive consumption of calories in Western societies, particularly those viewed as empty calories, as well as the need by diabetics to avoid sucrose and glucose, has led to an extensive search for noncaloric sweeteners. We will briefly review a few of the better known of these noncaloric substitutes. The sweet-tasting proteins have already been discussed in Chapter 5 and will not be covered again here.

Nonnutritive Sweeteners

Saccharin

> Anyone who thinks saccharin is dangerous is an idiot.
>
> —Attributed to Teddy Roosevelt

Saccharin (Figure 8.2) is synthesized by the Remsen-Fahlberg process from toluene or occasionally from the methyl ester of anthranilic acid via the Maumee process

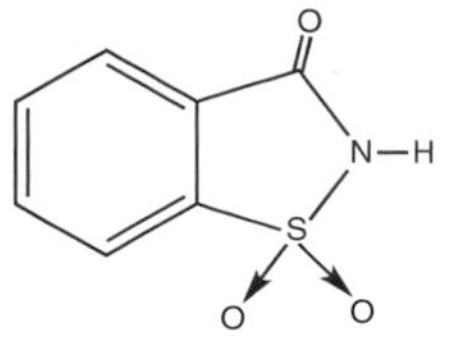

Figure 8.2. The structure of saccharin.

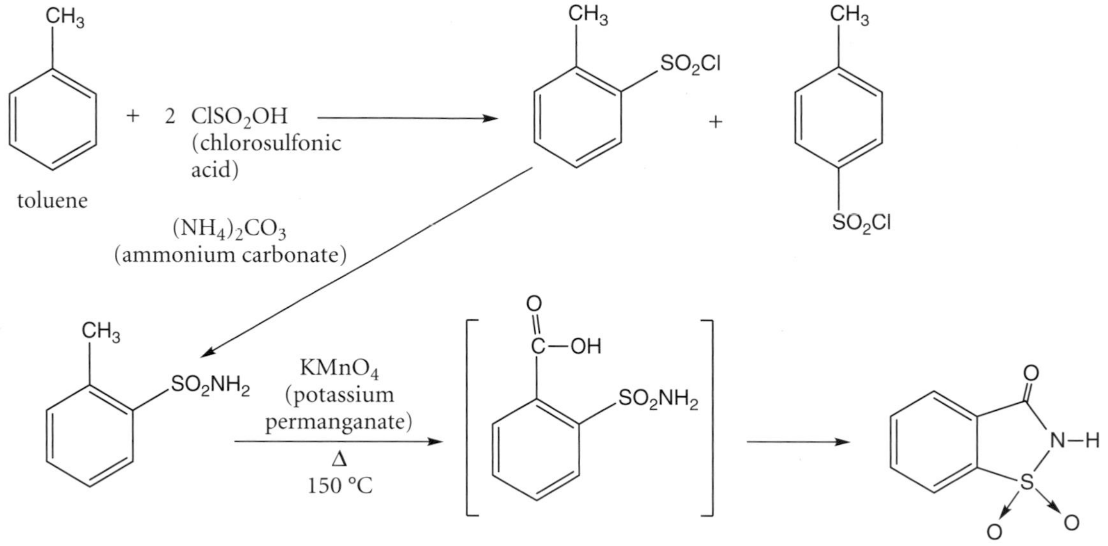

Figure 8.3. The Remsen-Fahlberg Synthesis of saccharin.

(Figure 8.3). The amide proton is weakly acidic and is easily abstracted, and saccharin is thus usually crystallized as a sodium or calcium salt. Saccharin is more than 300 times as sweet as sucrose and has become synonymous in everyday usage with intense, and false, sweetness. Saccharin leaves an aftertaste that can be bitter at high concentrations. It is stable at pHs from 2 to 7, and at temperatures up to 150 °C, which makes it superior for many uses to aspartame, but the aftertaste is more noticeable in foods that have been heated.

Saccharin has been used in foods for more than 100 years. In spite of the confident assertion quoted above from that noted food scientist, President Theodore Roosevelt, the safety status of saccharin remains unclear. Laboratory tests in Canada in the 1970s found that saccharine could cause bladder cancer in rats. As a result, since US regulations prohibit food additives that have been found to cause cancer in laboratory animals, the FDA proposed banning saccharin for food use in 1977. By that time however, saccharin was in widespread use in diet drinks and had become popular among dieters, as well as with the soft drink and diet food industries. As a result of public and industry outcry, Congress intervened and specifically exempted saccharin from this regulation, allowing it to be sold as long as it is labeled with the warning "Use of this product may be hazardous to your health. This product contains saccharin, which has been determined to cause cancer in laboratory animals." Saccharin is readily excreted in urine in humans, and does not accumulate in the body, and it is not clear whether these tests on laboratory rats are relevant for human health. Other more recent studies have also suggested that saccharin can cause various types of cancer in laboratory rats and mice, but little data has been found that would indicate that the widespread use of saccharin for decades has resulted in any human cancers. In the late 1990s the food industry began actively agitating for the removal of saccharin from the list of cancer-causing agents. The US law requiring warning labeling for saccharin-containing products was subsequently repealed in 2000. As of this writing, saccharin still may not be used for foods in Canada, and can only be sold there as a tabletop sweetener, must only be sold in pharmacies, and must carry a warning label. Currently the acceptable daily intake (ADI) for saccharin in the United States is set at 2.5 mg/kg of body weight.

Cyclamate

Cyclamate (Figure 8.4), the sodium salt of cyclohexane sulfamic acid, is synthesized by sulfonation of cyclohexylamine. Cyclamate is thirty times more potent than sucrose, but like saccharin can also have a bitter aftertaste. Although not as sweet as some other artificial sweeteners, cyclamate has good thermal stability that makes it useful for processed foods. When mixed with other nonnutritive sweeteners, cyclamate exhibits a synergistic effect, giving a sweeter taste than the sum of the separate concentrations of the two individual molecules. Cyclamate is soluble in aqueous solution.

Cyclamate was first approved for food use in 1950, but subsequently concerns about its safety arose. For example, some intestinal bacteria are able to convert this molecule back into the carcinogenic cyclohexylamine from which it is synthesized. A controversial feeding study in the 1960s of rats that consumed large doses of cyclamate indicated that this sweetener could cause bladder cancer, prompting the US FDA to ban the use of cyclamate as a food ingredient in 1970. Based on more recent studies, more than 50 nations allow the food use of cyclamate, including the European Union. An FDA review in the mid-80s concluded that cyclamates do not cause cancer in humans, although it did not prove that other possible health dangers might not occur, and cyclamates remain banned in the United States. Cyclamate is permitted in Canada as a tabletop sweetener but is not allowed as an additive in foods.

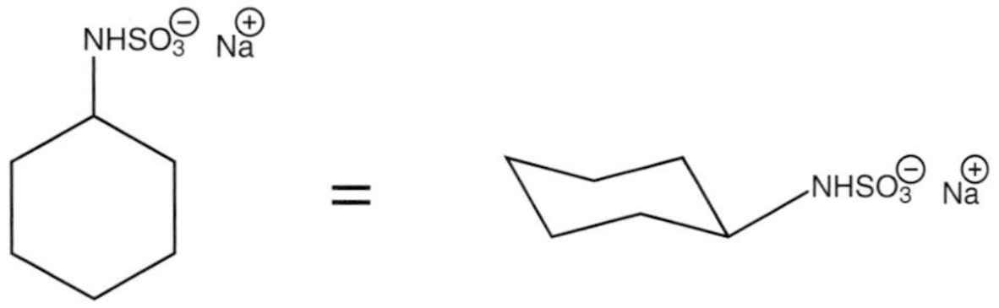

Figure 8.4. The structure of cyclamate. Note that the ring is not aromatic, so that it is not planar, as is shown on the right.

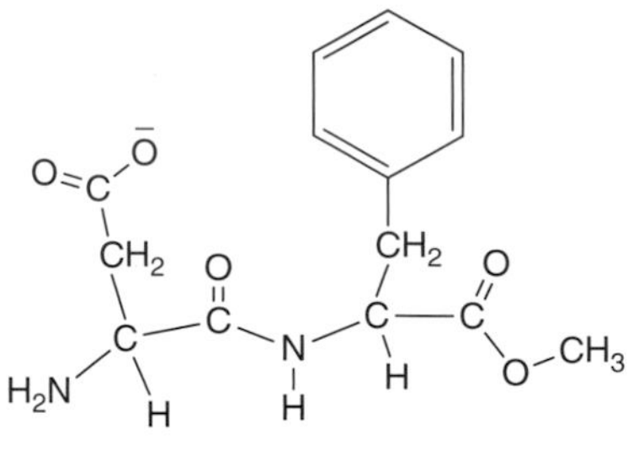

Figure 8.5. The structure of the dipeptide sweetener aspartame.

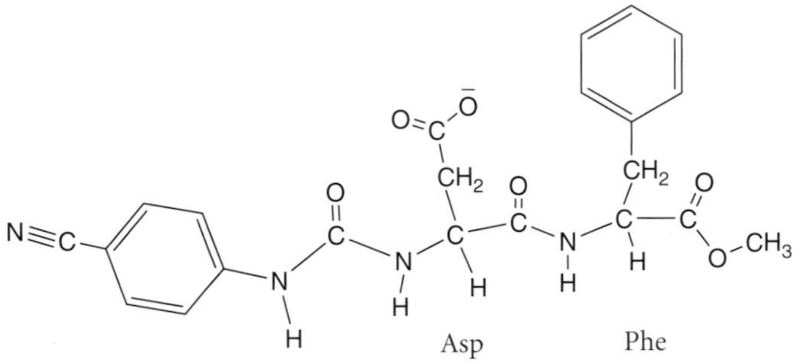

Figure 8.6. The structure of the dipeptide sweetener superaspartame. This sweetener has not been approved for use in the United States.

Aspartame

Aspartame (Figure 8.5), which is about 200 times more potent than sucrose, was approved for food use in the United States in 1981 and is the active ingredient of NutraSweet. It is widely used in soft drinks, but as previously noted in the proteins discussion, aspartame is not stable at high temperature for extended periods, particularly under acid conditions, which makes it unsuitable for cooking or extensive thermal processing. The safety of aspartame remains unclear. It was approved by the FDA for human consumption in spite of early data that linked it to brain cancer in laboratory rats. As a methyl ester of phenylalanine, the hydrolysis of aspartame by esterases produces methanol, which is toxic, but only in very small amounts, because so little aspartame is used due to its intense sweetness. Individuals who suffer from **phenylketonuria**, an uncommon genetic disorder resulting from a deficiency of the enzyme phenylalanine hydroxylase, which converts phenylalanine to tyrosine, are unable to metabolize phenylalanine properly, and thus should avoid products sweetened with aspartame.

Superaspartame, which has not been approved for food use, is around 100 times sweeter than aspartame (~20,000 times sweeter than sucrose!). It contains a *p*-cyanophenyl-carbamoyl residue on its *N*-terminus (Figure 8.6).

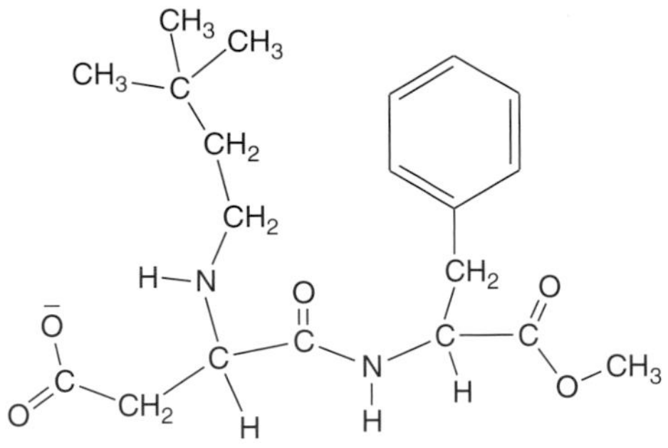

Figure 8.7. The structure of the dipeptide sweetener neotame.

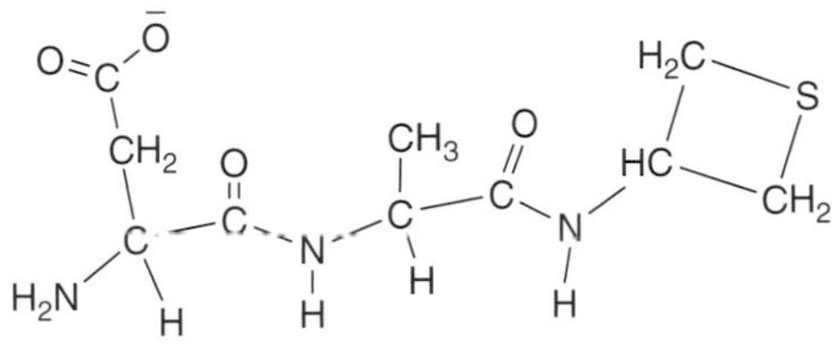

neotame

Figure 8.8. The structure of the dipeptide sweetener alitame.

L-Asp   D-Ala

alitame

Neotame

In 2002 the US FDA approved a new sweetener called neotame, which is similar to aspartame but has several superior attributes (Figure 8.7). It is much sweeter than aspartame, being approximately 8,000 to 13,000 times sweeter that sucrose. It is also more thermostable than aspartame, and because the aspartic acid residue is derivatized by the addition of a dimethylbutyl chain to its amino group, it is not hydrolyzed to produce phenylalanine, so that it is tolerated by people suffering from phenylketonuria.

Alitame

Alitame, L-α-aspartyl-N-(2,2,4,4-tetramethyl-3-thietanyl)-D-alaninamide, is a dipeptide derivative of L-aspartic acid and D-alanine (Figure 8.8). It is around 2000 times sweeter than sucrose. While storage under acid conditions can lead to the hydrolysis of the peptide bond in alitame, it is more thermostable than aspartame and can be used in baked goods. It has been approved for use in several countries, including Australia, New Zealand, and Mexico, but is not currently permitted in the United States or Canada.

The Safety of Peptide Sweeteners

Popular controversy remains concerning the safety of all of the dipeptide sweeteners (aspartame, alitame, and neotame), and there is considerable consumer resistance to

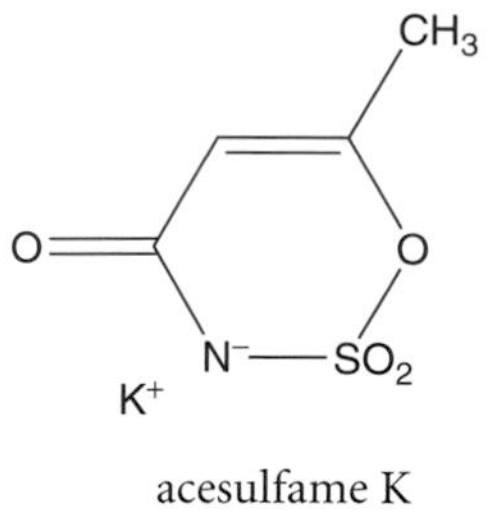

Figure 8.9. The structure of acesulfame potassium, or acesulfame K (Ace K).

acesulfame K

their use in many quarters. Some have alleged improper procedures and inadequate testing (and even conspiracies in some overheated imaginations) relating to the approval of these sweeteners by the FDA, although these compounds have also been approved in dozens of other countries, and aspartame has been in widespread use for almost three decades. As with so many other similar issues, definitive resolution of the questions concerning the safety of these sweeteners may have to await data from further studies, but all available evidence supports the safety of aspartame when consumed at normal levels.

Acesulfame K

Acesulfame K (Figure 8.9) has a clean sweet taste, at low concentrations, with no aftertaste, and is about 200 times sweeter than sucrose, but has a bitter metallic taste at high levels. Acesulfame K is thermostable and is stable over a range of pHs, making it a suitable choice for cooked products and even baked foods.

Acesulfame K was approved for use in foods by the FDA in 1988 and is widely employed in soft drinks, baked goods, dessert mixes, chewing gums, and many other products (its use in soft drinks was not approved until 1998). It is also approved for use in the European Union and many other countries. Currently, no generally accepted studies demonstrate adverse health effects from consuming this sweetener. The Center for Science in the Public Interest and other consumer groups claim that the safety testing for this molecule was inadequate and have urged that more extensive testing be conducted.

Sucralose

Sucralose is a trichloro-substituted derivative of sucrose (Figure 8.10). It consists of a disaccharide of galactose linked α-(1→4) to the C5 carbon of fructose, where both of the exocyclic primary alcohol groups of fructose (C1 and C6) and the C4 hydroxyl group of the galactose are replaced with chlorine atoms. It is about 600 times sweeter than sucrose, but has a similar sweetness profile to that of sucrose. It has much better thermal stability than aspartame, which allows it to be used in cooked foods such as baked goods.

Sucralose is synthesized in a complex series of steps using sucrose as a starting material (Figure 8.11). Thus, it is advertised as being made from sugar, which is liter-

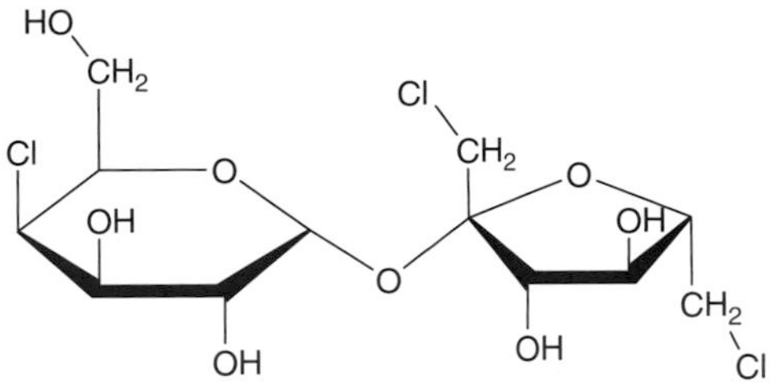

4,1',6'-trichloro-4,1',6'-trideoxy *galacto*sucrose
sucralose

Figure 8.10. The structure of sucralose.

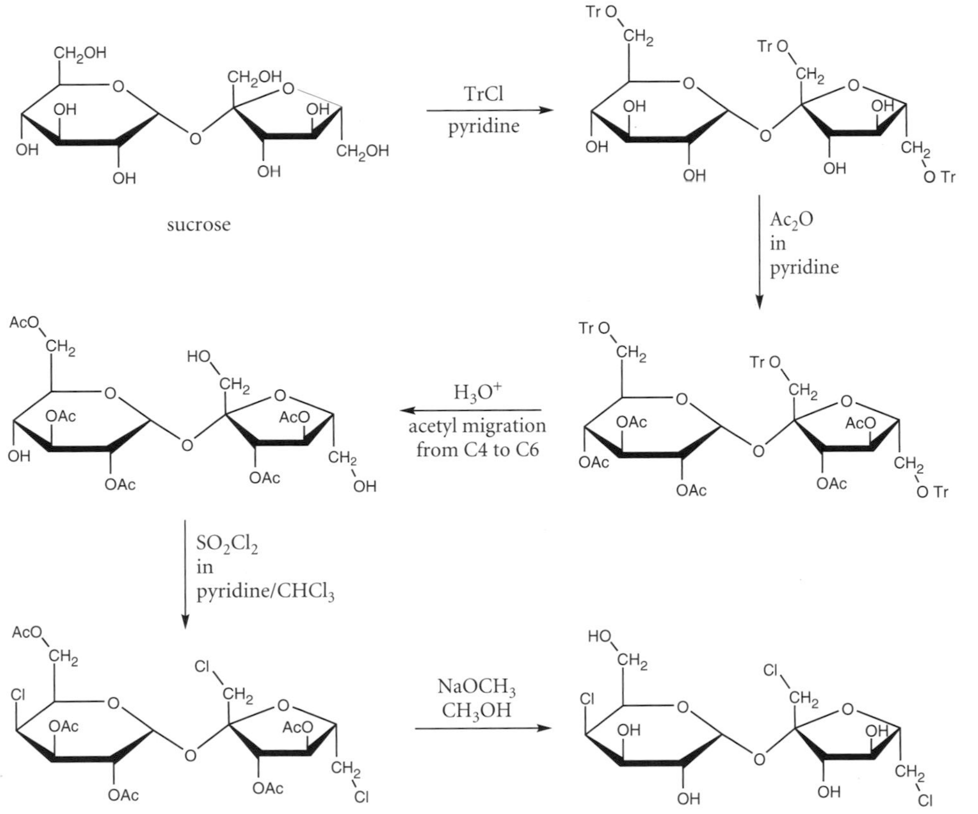

4,1',6'-trichloro-4,1',6'-trideoxy *galacto*sucrose
sucralose

Figure 8.11. The synthesis of sucralose.

ally true, if perhaps somewhat misleading. Remember that the pyranose sugar ring of sucrose is D-glucose, not D-galactose. The synthesis requires beginning by protecting the primary alcohols by reaction with trityl chloride, $(C_6H_5)_3CCl$ (TrCl), followed by acetylation of the secondary alcohols. Detritylation in acid causes a migration of the glucose C4 acetyl group to the C6 carbon. Reaction of the unprotected hydroxyl groups with SO_2Cl_2 is accompanied by an epimerization at the C4 position of the glucose, producing the chlorinated galactose derivative.

Sucralose was approved by the FDA for food use in the United States in 1998 and is gaining increasing acceptance. It is available as a tabletop sweetener as well as being used in a number of products. In consumer use it is marketed as Splenda. Its largest use is in soft drinks, as might be expected. No safety concerns for sucralose are currently known. It is only partially absorbed, the majority passing through the intestines, while the passively absorbed fraction is excreted by the kidneys without being broken down. Sucralose is thus a noncaloric sweetener. Unlike sucrose, sucralose also does not promote the development of dental caries.

Choice of Sweeteners

The choice of sweetener for a particular product depends on many factors and varies with the nature of the product and the target consumer groups. Sensory attributes such as any possible aftertastes, and how they might interact with other flavor notes in the product, must be considered. In many cases now, food manufacturers choose to use a combination of several sweeteners to produce a more sucrose-like sweetness profile and to mask aftertastes. Processing stability must certainly be taken into account; recall for example that the poor heat stability of aspartame precludes it from being used in products that must be cooked at high temperatures. The removal of sucrose as a sweetener from a product and its replacement by high-potency sweeteners, present in only very small amounts, can change the physical properties of a food or drink, such as water activity, surface tension, viscosity, and density. The regulatory status for the various artificial sweeteners varies from country to country, which may affect decisions about products intended for international marketing. Table 8.1 summarizes and compares the properties of a number of common sweeteners.

NONSWEET (SAVORY) FLAVORS AND ODORS

Although most acids can elicit a sour flavor, not all salts taste "salty." A diverse range of molecules taste bitter, as we have already seen when discussing small hydrophobic peptides. Many molecules produce other flavors besides the five of the SSSBU taste model, with the odor stimulation combining with the taste signal to give rise to the wide range of flavors we experience in foods. These taste and odor molecules are of many different types and are far too numerous to attempt to catalogue here. In the following sections, a few of the more important and better known molecules will be briefly introduced.

Volatile Sulfides and Mercaptans: Onions and Garlic

A **mercaptan** is a molecule containing a sulfur alcohol, or thiol (—S—H), group, while a **sulfide** is a sulfur-containing compound in which the sulfur is in the bivalent,

Table 8.1 Comparison of the properties of various sweetener

Sweetener	Relative sweetness	Aftertaste/off tastes	Processing stability	Regulatory status as of 2011	Benefits/problems
Nutritive					
sucrose	100 ("the gold standard")	none	hydrolyzes under acid conditions; caramelizes at high temperature	GRAS	cariogenic
fructose	100–175	none	stable	GRAS	
glucose	40–80	none	stable	GRAS	
HFCS	variable depending on composition	none	stable	GRAS	
invert sugar	>sucrose	none	heat and acid stable	GRAS	
sorbitol	63		good stability under processing conditions	GRAS	noncariogenic; laxative effect in high amounts
xylitol	90	rapid sweetness impact; cooling sensation	good stability under processing conditions	not GRAS for food use, but approved for use as a sweetener in the US	noncariogenic; laxative effect in high amounts
maltitol	68–90	none; very similar to sucrose	good stability under processing conditions	approved for use as a sweetener, GRAS status under consideration	noncariogenic; laxative effect in high amounts
Nonnutritive					
saccharin	300× sucrose	metallic or bitter aftertaste	not suitable for baking; degrades above 150 °C; stable at pH 2–7	permitted in the US; regulated as a tabletop sweetener in Canada	possible carcinogen; evidence ambiguous
cyclamate	30× sucrose	bitter aftertaste	good thermal stability	not permitted in the US; allowed in the EU; permitted as a tabletop sweetener only in Canada	originally suspected of being carcinogenic, apparently incorrectly
aspartame	200× sucrose	little aftertaste for most individuals; for some a bitter aftertaste	not heat stable or acid stable		cannot be used for cooking; is a problem for phenylketonurics
superaspartame	20,000× sucrose			not approved for food use	
neotame	7,000–13,000× sucrose		more heat stable than aspartame, but cannot be used for cooking	approved for use in the US	not a problem for phenylketonurics
alitame	2,000× sucrose	no aftertaste	more heat stable than aspartame, but cannot be used for cooking	not approved for use in the US	not a problem for phenylketonurics
acesulfame K	200× sucrose	fairly clean taste, bitter or metallic taste at high levels	thermostable and pH stable; useful for cooking and baking	permitted in the US and EU	
sucralose	600× sucrose		thermostable; can be used for cooking and baking	approved for use in the US	good shelf life
thaumatin	2,000–3,000× sucrose	licorice aftertaste	loses sweetness above 70 °C	approved for use as a flavor enhancer in the US, but not as a sweetener	
stevia (rebaudio-side A)	300× sucrose	licorice or bitter aftertaste		GRAS	

Note: GRAS = generally recognized as safe.

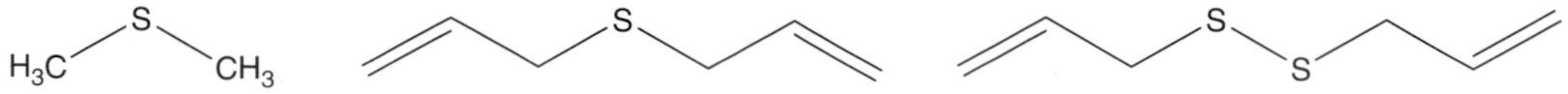

Figure 8.12. The structures of several sulfides contributing to the aromas and flavors of foods.

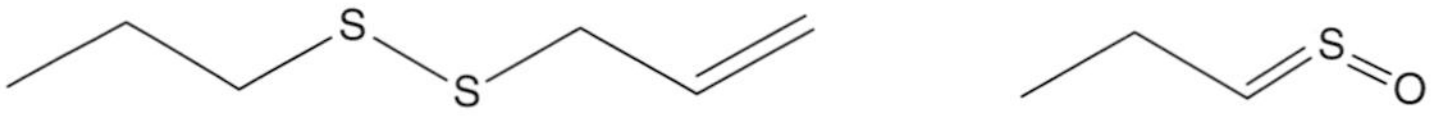

Figure 8.13. The structures of allyl propyl disulfide, from onions, and thiopropionaldehyde-S-oxide (thiopropanal sulfoxide), the lachrymatory factor produced by cutting onions.

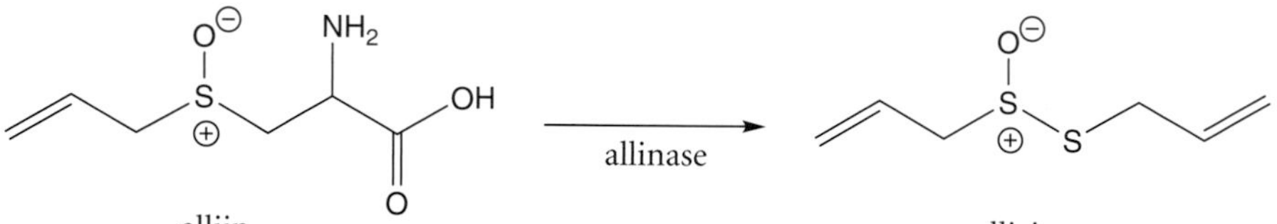

Figure 8.14. The enzymatic production of allicin from alliin, catalyzed by allinase.

−2 oxidation state. Many of the volatile mercaptans and other sulfur-containing molecules have strong sulfur odors that can be readily detected by humans, generally in small amounts. Often these are quite unpleasant, particularly at high concentrations, but some are also important components in food aromas or flavors. Two examples are diallyl sulfide and diallyl disulfide (Figure 8.12), which are important components of oil of garlic and contribute to the aroma of garlic (the allyl group, C_3H_5, was first identified from garlic, from which it takes its name, since *allium* is the Latin word for garlic). The simpler dimethyl sulfide has the aroma of cabbage or sauerkraut.

The related molecule allyl propyl disulfide (Figure 8.13) is responsible for the odor of freshly cut onions. Yet another volatile sulfur-containing molecule, thiopropionaldehyde-S-oxide, or thiopropanal sulfoxide, is the tear-producing (lachrymatory) irritant in onions. It is not related to the pungency of the onion's flavor, however.

When garlic, which is in the onion family, is cut or crushed, the sulfoxide alliin, derived from cysteine, is converted by the enzyme alliinase into allicin (Figure 8.14), the most important of the flavor molecules in fresh garlic. Allicin is unstable and degrades quickly in dried garlic powder or when garlic is cooked.

The volatile compounds allicin and thiopropanal sulfoxide are not present in the living garlics and onions. They are produced by enzymatic degradation of sulfur-containing amino acids that takes place when cutting or crushing disrupts the cellular structure of the bulbs that previously segregated the enzymes from their substrates. The lability and volatility of these molecules is the reason that freshly cut onion and garlic are so different from dried onion or powdered garlic. A wide variety of related sulfur-containing molecules that result from the reactions and rearrangements of

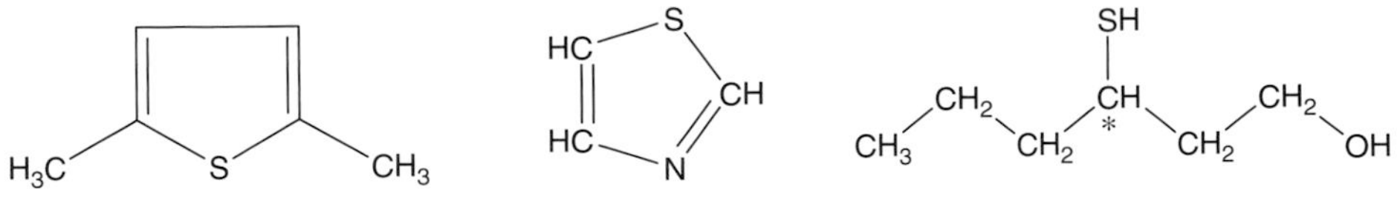

2,5-dimethylthiophene   1,3-thiazole   3-mercaptohexanol

Figure 8.15. The structures of 2,5-dimethylthiophene, 1,3-thiazole, and 3-mercaptohexanol.

Figure 8.16. The structure of furfuryl mercaptan.

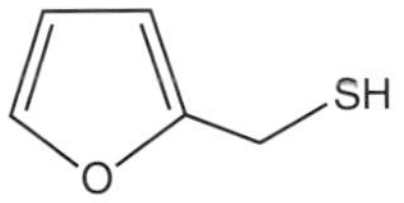

these sulfides and sulfoxides, and similar mercaptans, are responsible for the aromas of fried, caramelized and cooked onions and garlic. The most important of these in the aroma of fried onions are dimethylthiophenes such as 2,5-dimethylthiophene (Figure 8.15).

Other foods also contain sulfides. The heterocyclic aromatic ring compound thiazole (Figure 8.15), which contains both a nitrogen and a sulfur atom and is produced by frying meat proteins at high temperatures, has a nut-like odor and is an important component of the odor of fried meat. The thiazole ring moiety is also a part of the structure of the vitamin thiamine (vitamin B1). 3-Mercaptohexanol (Figure 8.15), formed during alcoholic fermentation, is a component of the bouquets of several varieties of wines, including Cabernet Sauvignons, Cabernet francs, and Merlots. This molecule contains a chiral carbon, indicated by the asterisk, with two possible isomers. The R stereoisomer has the aroma of boxwoods, while the S stereoisomer has an aroma characterized as "passion fruit."

Furfuryl Mercaptan

Furfuryl mercaptan, or 2-furylmethanethiol, C_5H_6OS, is a volatile mercaptan largely responsible for the fresh roasted smell of coffee (Figure 8.16). It has been shown that, when ferrous iron Fe^{2+} is present in the water used to brew coffee, hydrogen peroxide, also present in the coffee as a result of the normal brewing process, can lead to the production of hydroxyl radicals through the Fenton reaction, which in turn leads to the loss of furfuryl mercaptan via the formation of various difurfuryl molecules, many of which are also volatile.

Aldehydes, Esters, and Ketones

Many simple ketones and acetates have characteristic odors and flavors and are components of a number of "fruity" fragrances and the tastes of fresh fruits (Figure 8.17). For example, the ester **isoamyl acetate** (also called isopentylacetate in the food industry) is the most important flavor and odor compound in ripe bananas, which

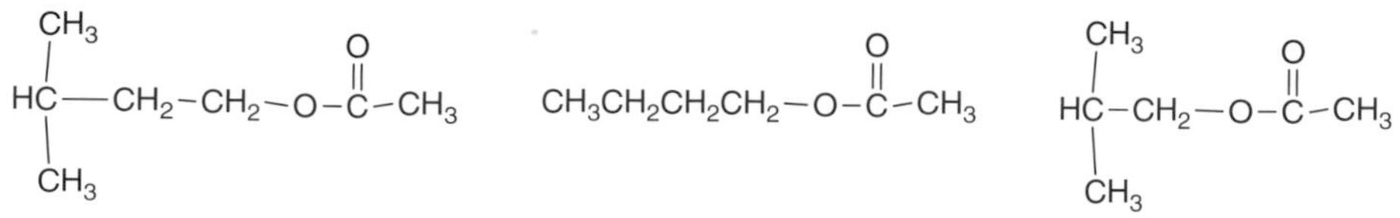

Figure 8.17. The structures of several acetates that contribute to "fruity" fragrances and flavors.

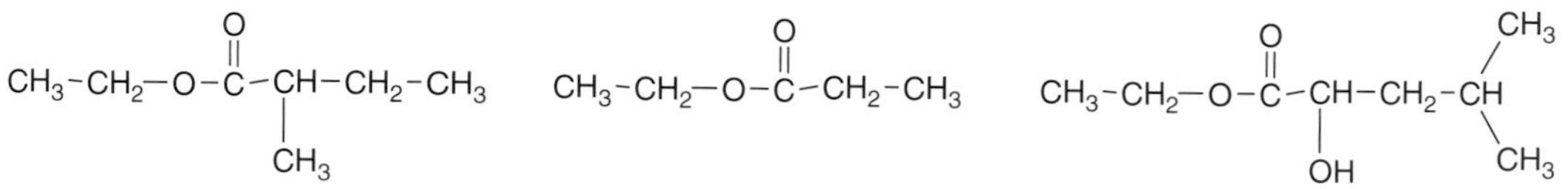

Figure 8.18. The structures of ethyl 2-methylbutanoate and ethyl *n*-propionate.

also contain significant amounts of **butyl acetate** and **isobutyl acetate**. These molecules also contribute to the odor of ripe apples and pears, and often the aroma of isoamyl acetate is described as that of pears (it is sometimes called "banana oil" or "pear oil"). Isoamyl acetate is an ester of acetic acid and isoamyl alcohol. Isobutyl acetate is similarly the ester of acetic acid with isobutyl alcohol. Amyl acetate, the ester of acetic acid and *n*-pentanol, is yet another component of the "fruity" aromas of apples and bananas.

Ethyl 2-methylbutanoate, shown in Figure 8.18, is a small volatile ester that is important in the aroma of apples, while ethyl *n*-propionate also has a fruity aroma and is found in a number of fruits. These two molecules are esters but are not acetates. Both molecules are insoluble in water but are soluble in alcohol. The ester ethyl DL-leucate (DL-ethyl 2-hydroxy-4-methylpentanoate; Figure 8.18) is present in a variety of fruits and berries, particularly blackberries, and is thought to contribute blackberry aroma notes to wines.

All of these compounds have seven or fewer carbon atoms, and esters of about this size tend to possess fruity odors. They are synthesized from the breakdown products resulting from the oxidation of fatty acids during the ripening process. In general, these molecules are only slightly soluble in water and are more soluble in nonpolar organic solvents, and because in pure form they are liquids at room temperature, they can be used as organic solvents themselves. These molecules are toxic in large amounts but are present in fruits in such low doses that they pose no safety concerns.

As we have seen in the previous chapter, volatile straight-chain unsaturated aldehydes are produced by lipid oxidation. The odors of these aldehydes are generally unpleasant and often are detectable in low concentrations. Their production is the principal reason why the oxidation of unsaturated fatty acids is such a problem in the preservation of foods containing lipids.

Various ketones are also important as odor molecules (Figure 8.19). Ionones and damascones are breakdown products from carotenoids (Chapter 9). α-Ionone

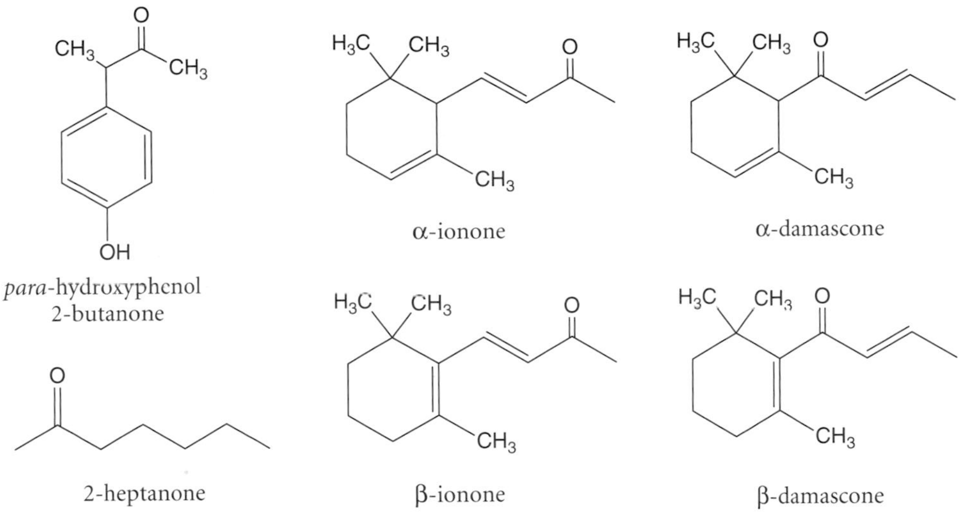

Figure 8.19. The structures of various ketones important in food aromas.

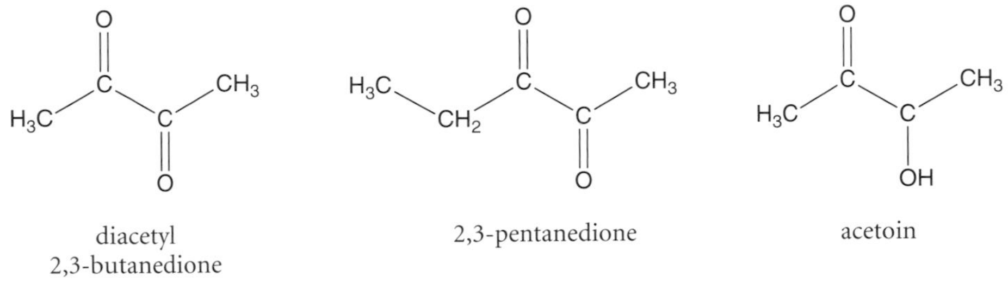

Figure 8.20. The structure of diacetyl, familiar as a dominant aroma note in artificial butter.

is found in raspberries, vanilla, and black teas and contributes to their aromas. β-Ionone is an important component of the fragrance of violets and roses, while *para*-hydroxyphenol-2-butanone is another principal odor component of raspberries, and 2-heptanone is the odor of oil of cloves. Raspberries also contain *para*-hydroxyphenol-2-butanone, and synthetic *para*-hydroxyphenol-2-butanone is the compound used in artificial raspberry flavor. β-Damascone contributes to the bouquets of black teas and roses. Because of the great expense of oil of violet, the ionones used in the food and fragrance industries are usually synthetic.

Diacetyl

Diacetyl, or 2,3-butanedione, is a small volatile ketone that is a liquid at room temperature (Figure 8.20). This molecule is a byproduct of fermentation and a flavor component at low concentrations in some wines and beers, where it is said to impart "buttery" notes and is usually considered an off-flavor note. Diacetyl occurs naturally in low levels as a fermentation by-product in European-style cultured butter and is an important component of the aroma of melted butter, along with the related molecules acetoin and 2,3-pentanedione (Figure 8.20). It has a butterscotch or butter flavor at higher concentrations and has been widely used to flavor artificial butter, particularly the artificial butter used on commercial popcorn (in addition to diacetyl, commercial imitation butter flavor contains butyric acid, ethyl propionate, and a small amount of vanillin). Almost anyone who has been to a movie theater in the United States has smelled the odor of diacetyl. Concerns have arisen, however, that when inhaled in large amounts this molecule might be hazardous and may cause an incurable form of bronchiolitis. The US Occupational Safety and Health Administration (OSHA) and Environmental Protection Agency (EPA) are investigating the safety of breathing large quantities of diacetyl, and workers in factories producing microwavable popcorn have won court cases alleging that they were made sick by diacetyl vapors. As a result of these concerns, several popcorn producers have reformulated their products to remove diacetyl from artificial butter flavors. The very similar 2,3-pentanedione also has a buttery aroma and is being substituted for diacetyl in some products. It is possible that further investigations of the safety of diacetyl may result in it being regulated or even banned as an additive in the future.

Essential Oils

Essential oils are hydrophobic liquids composed of small volatile organic molecules obtained from plants by pressing leaves, petals, fruits or fruit peels, or extracted using organic solvents. Many of the small molecules that make up these oils have distinctive odors and flavors and are widely used in the perfume and fragrance industry, but are also used to flavor foods.

Essential oils are often extracted from the plant tissues that contain them by steam distillation. When the distillate is condensed to liquid water, the oils form a surface layer that can be collected by skimming. Although this method is cheap and easy, the thermal processing can lead to undesirable changes in some cases, such as hydrolysis, polymerization, or oxidation. For this reason, some oils are obtained by cold pressing the plant tissues under high pressures; this method is used to extract orange and lemon oils. Because they are hydrophobic, essential oils are usually marketed to the consumer as solutions in alcohol or propylene glycol.

Many of the active compounds in essential oils are **terpenes**, small polymeric compounds based on the **isoprene** unit (Figure 8.21), which is also a repeat unit in many of the color molecules in plants (Chapter 9).
Often these molecules have the compositional formula $C_{10}H_{16}$. Among the terpenes found in foods are carvone from oil of spearmint, cinnamaldehyde, vanillin, geranial and neral from lemon oil, and limonene from various citrus oils.

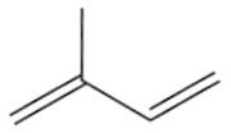

isoprene

Figure 8.21. The structure of isoprene, the basic molecular building block in terpenes.

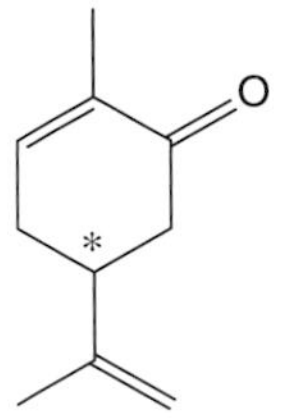

Figure 8.22. The structure of carvone; the chiral carbon is indicated by an asterisk.

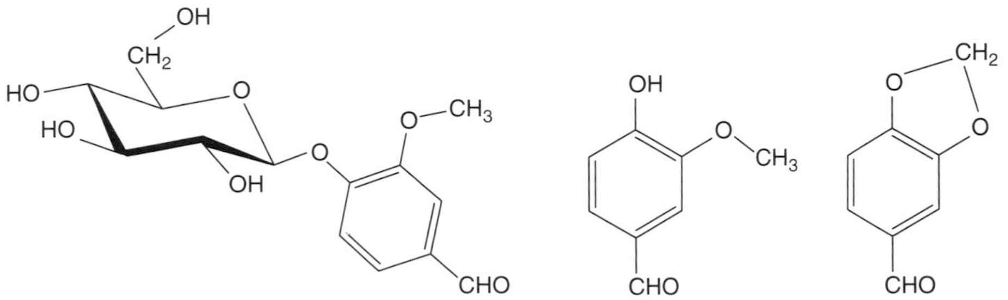

carvone

vanillin glycoside

vanillin

piperonal

Figure 8.23. The structures of vanillin and piperonal.

Carvone: Spearmint and Caraway

Carvone (Figure 8.22) is the main flavor and odor note of oil of spearmint obtained by distillation from the leaves of the spearmint plant (*Mentha spicata*). It is widely used to flavor spearmint chewing gum. Carvone contains an asymmetric carbon atom, and the R-(—)-carvone stereoisomer exhibits the spearmint flavor, while the S-(+)-carvone stereoisomer is the principal flavor note of caraway and dill seed.

Vanillin

Vanillin (Figure 8.23), or 3-methoxy-4-hydroxylbenzaldehyde, is the main component of oil of vanilla. Vanilla oil is extracted from the fermented seed pods of the so-called vanilla bean plant, which is not actually a leguminous "bean," but rather an orchid, *Vanilla fragrans*, which grows as a climbing vine on trees. It was originally native only to Central America, where it was highly prized by both the Aztecs and Mayas. It is now also cultivated in Madagascar, Indonesia, Zanzibar, the Comoros and other Indian Ocean Islands, and Tahiti. Vanillin is synthesized in the plant as a glycoside of glucose, but is released enzymatically by glycosidic hydrolysis during the fermentation process.

Pure vanillin is only sparingly soluble in water, with a solubility limit of 10 g/liter at room temperature, but it is more soluble at elevated temperatures (Figure 8.24). It is, however, soluble in alcohol. Pure vanillin exists as an off-white crystal at room temperature (mp 81–83 °C). It is also somewhat volatile at elevated temperatures, imparting a pleasing aroma. Vanillin is active as a flavor and odorant at very low

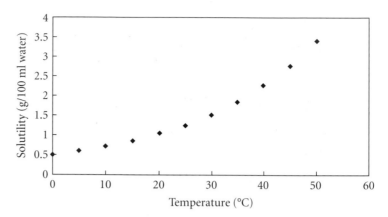

Figure 8.24. The solubility of vanillin in water as a function of temperature. (L.C. Cartwright, *Agric. Food Chem.* 1:312–314, 1953.)

concentrations but saturates quickly so that increasing the concentration beyond a certain level produces little increase in flavor or odor, which is important to keep in mind because natural vanilla extract is the second most expensive common food ingredient, after saffron. Natural vanilla extract contains many other flavor molecules in small amounts. One of these is piperonal (Figure 8.23), closely related to vanillin in structure, which contributes to the aroma of natural vanilla.

In the United States, only products containing pure extract of the vanilla bean can be labeled as "natural vanilla." Products flavored with a mixture of natural vanilla and synthetic vanillin must be labeled as vanilla-flavored, while products that contain only synthetic vanillin must be labeled as artificially flavored.

Because of the expense of natural vanilla, vanillin is synthesized commercially in large amounts. It can be prepared by the oxidation of eugenol (see below), or from petrochemicals, or by the hydrolysis-oxidation of lignosulfonates from the lignin in wood, produced as a byproduct of paper manufacture. Small amounts of vanillin and vanillin derivatives are present in wood as natural breakdown products of lignin, and vanillin that leaches out of the wood of the European oaks used to make wine barrels is a component of the flavor of aged wine.

Citrus Oils

Terpenes are also the main flavor and odor components of citrus oils such as oil of orange and oil of lemon. **Limonene**, shown on the left in Figure 8.25a, is a principal component of citrus oils, including oranges, limes, and oil of lemon, from which it takes its name (from an obsolete French word for lemon, not from the English "lime"). This cyclic terpene contains an asymmetric carbon atom, like carvone, and one stereoisomer, (+)-limonene, or D-limonene, is the important odor molecule in these fruits. It also contributes to the distinctive aroma of limes. Limonene is used as a fragrance in many cosmetic and household products, as well as a flavorant in foods. It is produced commercially by steam extraction from citrus peels left over

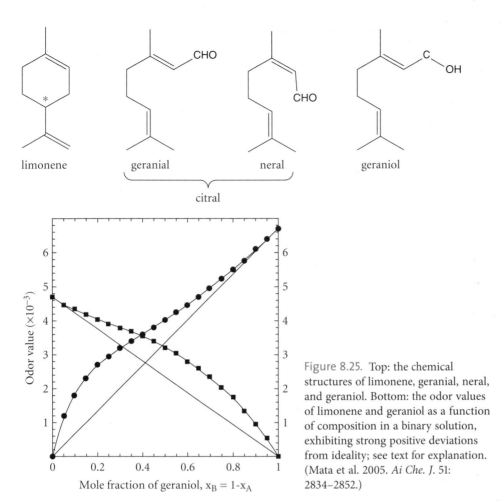

limonene geranial neral geraniol

citral

Figure 8.25. Top: the chemical structures of limonene, geranial, neral, and geraniol. Bottom: the odor values of limonene and geraniol as a function of composition in a binary solution, exhibiting strong positive deviations from ideality; see text for explanation. (Mata et al. 2005. *Ai Che. J.* 51: 2834–2852.)

from juice production. **Citral** is the name given to the mixture of the two structural isomers **geranial** and **neral**, also shown in Figure 8.25a, which contribute strongly to the characteristic odor of lemon oil. Geraniol, the alcohol that results from the reduction of geranial, is the principal odor molecule in citronella and is also found in the leaves of geraniums, from which it takes its name. It is also present in lemons and other citrus fruits and is used as a component in many fruit flavors. Limonene is relatively stable to heat but can undergo oxidation to produce carvone or the alcohol of carvone, carveol, leading to the development of an "off flavor" in orange juice during storage.

Limonene is hydrophobic and almost insoluble in water, while geraniol has one polar hydroxyl functional group but is nonetheless also fairly insoluble in water. Both molecules are soluble in nonpolar solvents, and as liquids at ambient temperatures, these species are themselves used as solvents and are mutually soluble in one another in all proportions. However, because of their difference in polarity, they do not form an ideal solution when mixed, as can be seen from Figure 8.25b. This figure is essentially a phase diagram for the two species, augmented with sensory information (Mata et al. 2005). For both species, the odor value, OV, is plotted as a function of mole fraction of geraniol (which in a binary solution is equal to 1 minus the mole fraction of limonene). The odor value, related to the activity of each component (Equations 1.92 and 1.93), is a quantitative measure of the fragrance intensity of the compound and is defined as the concentration of the molecule in the headspace of a closed reaction vessel, divided by the threshold concentration for that component, which is the minimum concentration at which it can be detected by humans. The sensory threshold for geraniol, at $2.48 \times 10^{-5}\,g/m^3$, is two orders of magnitude lower than that for limonene, at $2.45 \times 10^{-3}\,g/m^3$, on a weight basis (note that geraniol has a higher molecular weight). The straight lines in Figure 8.25b indicate the behavior that would be expected if the solution was ideal; the positive deviations from ideality indicate that the activity coefficients for these solutes are greater than 1, meaning that there is more of each molecule in the vapor than there would be for the pure liquids at that temperature. This is because of the self-association promoted by the hydrogen bonding capacity of the geraniol, which is lacking in the limonene. Thus the two species tend to push each other out of solution as they avoid each other. It is not known if this ability to hydrogen bond is the reason that geraniol can be detected at much lower concentrations.

Cinnamon, Cloves, Nutmeg, and Bay

Cinnamaldehyde (Figure 8.26), also called β-phenylacrolein, is the main flavor and odor molecule of oil of cinnamon, from the bark of the cinnamon tree *Cinnamomum zeylanicum,* native to Southern Asia, and related species from Southeast Asia. The oxidized form of this molecule, cinnamic acid, is another important component of

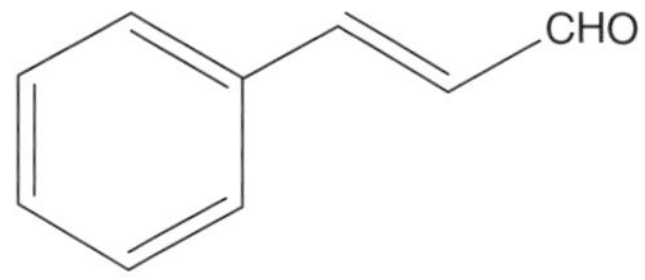

cinnamaldehyde

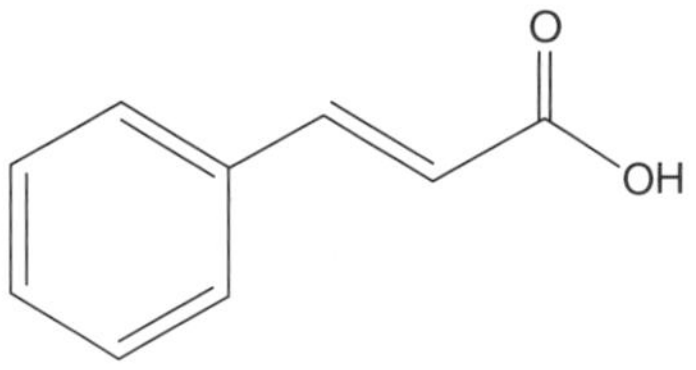

cinnamic acid

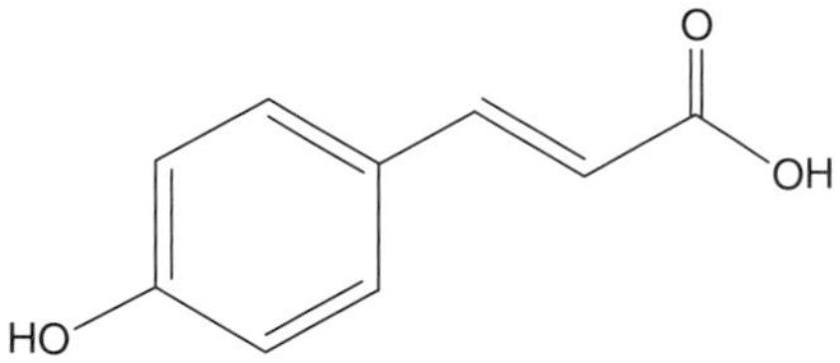

p-coumaric acid

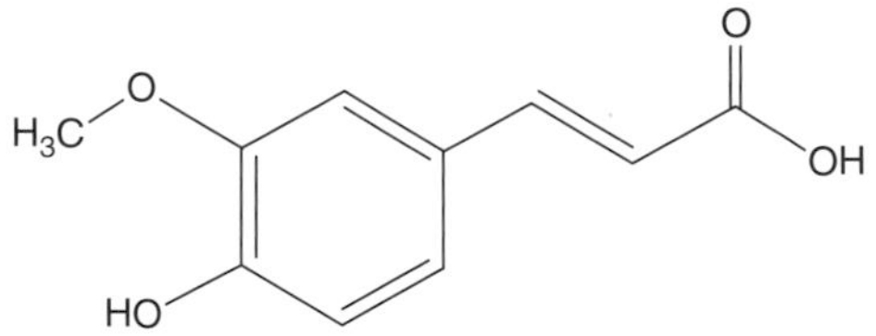

ferulic acid

Figure 8.26. The structures of cinnamaldehyde and selected hydroxycinnamates.

oil of cinnamon, and also serves as a template for a class of cinnamate derivatives called **hydroxycinnamates**. These include *p*-coumaric acid, found in garlics, tomatoes, and peanuts; and ferulic acid, found in coffee, peanuts, artichokes, apples, and oranges, as well as in some grains such as wheat and rice. Caffeic acid, discussed below, also belongs to this class. Cinnamaldehyde has strong antimicrobial, particularly antifungal, properties, which have been exploited since ancient times. Recently, it was demonstrated that cinnamaldehyde can be incorporated into active packaging to inhibit the spoilage of bread (Rodríguez, Nerín, and Batlle 2008). The related **eugenol** molecule (Figure 8.27) is a primary component of oil of bay, extracted from bay leaves, and also of oil of cloves, from the dried flower buds of the clove tree *Eugenia aromatica*, now called *Syzygium aromaticum*. Eugenol has a spicy aroma and is also found in tomato paste, cherries, and plums. Isoeugenol, which differs from eugenol in structure only in the location of the double bond in the hydrocarbon chain, changes the odor of the molecule from that of cloves to nutmeg. Both cinnamaldehyde and eugenol are carminative, which means that they produce gases in the digestive system, including hydrogen, hydrogen sulfide, and methane.

Thyme and Thymol

The herb thyme, *Thymus pulegioides*, has been used in Mediterranean cuisines for thousands of years. It was known to the ancient Egyptians and was used ritually by the Greeks and Romans. It is used in many types of dishes, especially to flavor meats, and is one of the components of "*herbes de Provence*." It retains its flavor well when dried. Oil of thyme, an essential oil extracted from thyme leaves, contains large quantities of the monoterpene thymol, shown in Figure 8.28. Thymol has antiseptic

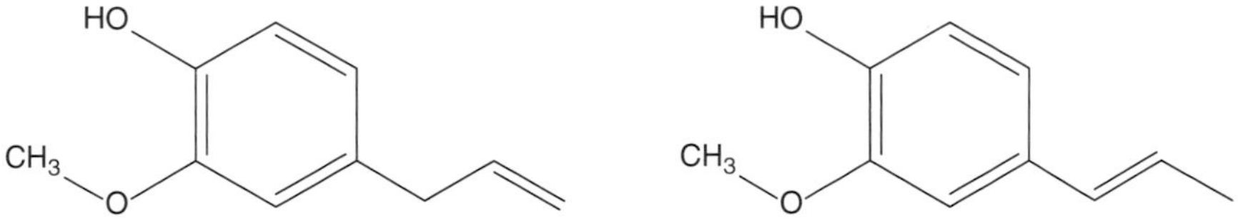

Figure 8.27. The structures of eugenol and isoeugenol.

Figure 8.28. The structure of thymol.

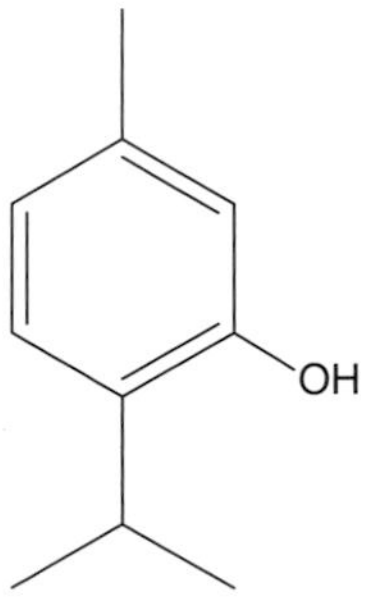

properties and was formerly used to sterilize bandages and treat wounds, and to control fungi in bookbinding. It is a principal active ingredient in many commercial mouthwashes.

Menthol

Menthol (Figure 8.29) is a terpene found in peppermint oil from the Japanese pep-permint plant (*Mentha arvensis*) that has a clean, sweet, and cooling taste. It is also found in common mint (*Mentha piperita*) and can be extracted from turpentine from pine trees. It is a predominantly hydrophobic molecule, soluble in alcohol but only sparingly soluble in water, containing one polar hydroxyl group and three chiral carbon atoms. The stereochemistry at these chiral centers can vary for menthol from different sources, and the flavor and aroma of this molecule depend upon the con-figuration. The form used commercially is L-(—)-menthol, with the configuration (1R, 2S, 5R); D-menthol has the opposite configuration at each of the three chiral centers. L-menthol can trigger cold temperature sensors in the tongue even at ele-vated temperatures and produce the cooling sensation associated with mint (note that the mechanism of this cooling sensation is neurological and differs from that produced by cocoa butter, which results from an actual lowering of the temperature due to the adsorption of heat from the tongue to melt the crystalline cocoa butter). Menthol, which is toxic at high doses (Table 11.1), is used in low concentrations in candies, chewing gum, cough drops, and toothpastes.

Menthol can be prepared synthetically from *m*-cresol obtained from petrochemi-cals. The synthesis is a straightforward Friedel-Crafts electrophilic substitution alkyl-ation followed by hydrogenation, but the product is a racemic mixture of the stereoisomers (Figure 8.30).

The more difficult step involves separating the stereoisomers. Very high purities are required, because the threshold for detection of the unpleasant thymol is quite low. Recently, an asymmetric synthesis using rhodium- or ruthenium-diphosphine complexes of the asymmetric catalyst BINAP (*R*- or *S*-2,2′-diphenylphosphino-1,1′-binaphthyl, Figure 8.31), have been developed in Japan to produce enantiomerically

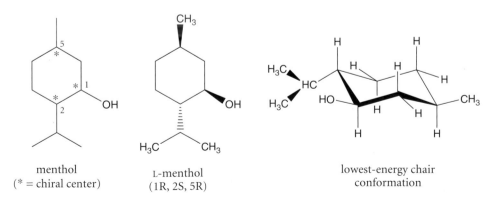

Figure 8.29. The structure of menthol, indicating its chirality and three-dimensional shape.

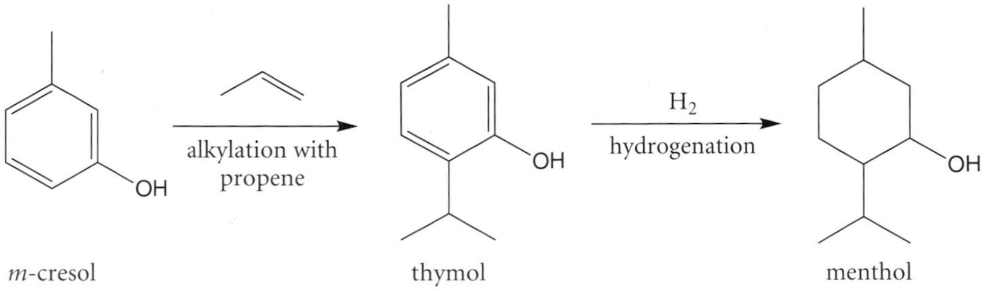

Figure 8.30. The synthesis of thymol and menthol from cresol.

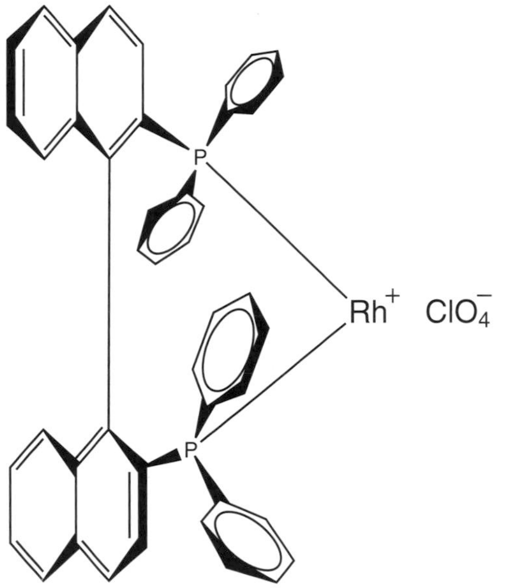

Figure 8.31. The rhodium-BINAP catalyst for synthesizing (–)-menthol. (S. Akutagawa, *Topics Catal.* 4:271–274, 1997.)

pure L-menthol from β-pinene (Figure 8.33), found in turpentine. At least two other commercial synthesis methods have been developed, including a direct chiral catalytic hydrogenation of citral, and a multistep synthesis starting from myrcene (McCoy 2010).

Phellandrene

The phellandrenes are cyclic monoterpenes found in eucalyptus oil from *Eucalyptus dives* and also in some herbs such as fennel and dill. Two structural isomers, α- and β-phellandrenes (Figure 8.32), resemble thymol and limonene. Unlike thymol, the rings in the phellandrenes are not aromatic. Like thymol, however, the ring in α-phellandrene is planar, while one atom of β-phellandrene is displaced from the plane of the other atoms in the ring. These molecules contain no polar functional groups and are insoluble in water.

α-phellandrene β-phellandrene

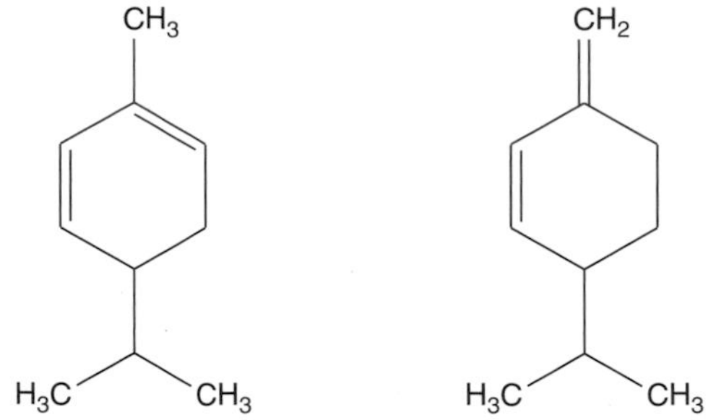

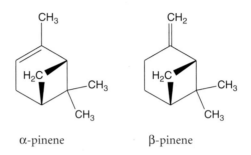

Figure 8.32. The structures of the phellandrenes.

α-pinene β-pinene

Figure 8.33. The structures of the pinenes.

Pinene

Another important pair of terpenes contributing to the aromas of many different herbs are the two related bicyclic isomers of $C_{10}H_{16}$ called pinenes. These two isomers, shown in Figure 8.33, are referred to as α- and β-pinene. Both are liquids at ambient temperatures. Pinene takes its name from the resin of pines, where it is abundant, and gives pine resin and other conifer saps, as well as turpentine, their characteristic aromas. As can be seen from their structures, both molecules are completely hydrophobic and insoluble in water. The two isomers are important components of the aromas of a number of herbs and spices, such as nutmeg, black pepper, bay, and juniper berries, and are more minor contributors to many others such as rosemary and dill. Two chiral carbon atoms in the pinene molecules occur at the branch points where the two rings join.

Sage

Common **sage** (*Salvia officinalis*) is another Mediterranean herb widely used to flavor various dishes, particularly poultry stuffings, sausages, cheese, and soups. The essential oil of sage contains a number of important active constituents, including several that are bicyclic terpene derivatives. These molecules include **eucalyptol**, **borneol**,

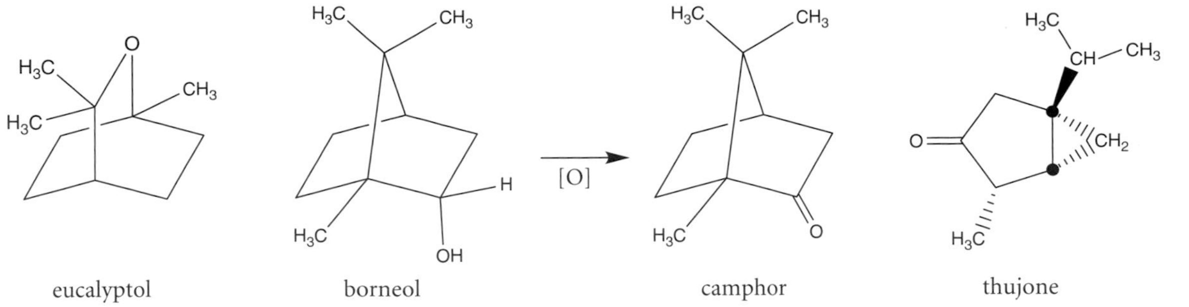

Figure 8.34. The structures of some bicyclic terpenes in foods.

and **thujone**. As its name implies, eucalyptol (1,8-cineol) is the principal ingredient in the eucalyptus oil cineol from many types of eucalyptus trees, but it is also found in rosemary, basil, and bay, as well as in sage. It contains two fused six-atom hetero-cyclic ether rings and, as can be seen from Figure 8.34, is nonpolar and insoluble in water, but is soluble in alcohol and other organic solvents. Eucalyptol is a liquid at room temperature (mp ~ 1.5 °C), but is volatile and highly flammable, with a flash point of only 48 °C, and is partially responsible for the intensity of the fires in euca-lyptus forests in California and in its native Australia. Eucalyptol has a camphor aroma and spicy taste but is mildly toxic if consumed in significant quantities. Only a few species such as koalas have evolved the resistance to its toxic effects necessary to subsist on eucalyptus leaves. Eucalyptol is used as an additive in cigarettes and as an active antimicrobial ingredient in mouthwashes, and as a flavorant in various medications.

Borneol contains two fused five-carbon rings and also has a camphor aroma. It has a single hydroxyl group, and camphor can be produced from borneol by oxidizing this hydroxyl group to a ketone (Figure 8.34). The carbon to which this hydroxyl is attached is chiral, and two isomers exist; the other isomer is called isoborneol. Thujone also has two stereoisomers, both of which have a menthol-like aroma. In addition to being found in sage, thujone is also present in absinthe, the supposedly hallucinogenic alcoholic drink popular with artists and other "bohemians" in late nineteenth-century France. Like eucalyptol and borneol, thujone is toxic when con-sumed in significant amounts.

Dill Ether

The herb dill (*Anethum graveolens*) from Central Asia, in the carrot family, has a particularly recognizable and unique taste. As we have just seen, two of the molecules that contribute to the flavor and odor of dill are phellandrene and pinene. However, the most important component of the characteristic flavor of dill is the so-called dill ether molecule (Figure 8.35) found in the dill oil pressed from the fresh dill herb.

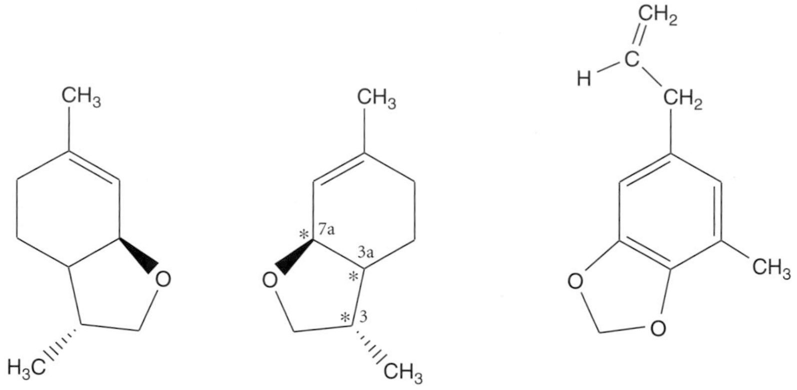

dill ether,
(3S,3aS,7aR)-3,6-dimethyl-2,3,3a,4,5,7a-
hexahydrobenzofuran   myristicin

Figure 8.35. The structure of dill ether (middle) and its mirror image (left). On the right is the structure of myristicin, also found in dill oil. (M. Wüst and A. Mosandl, *Eur. Food. Res. Technol.* 209:3–11, 1999.)

This bicyclic compound contains three chiral carbon atoms, and the flavor of the molecule depends strongly on the stereochemistry at each of these centers. Only one of the possible stereoisomers, shown in Figure 8.35, is found in dill oil and has the characteristic dill aroma and flavor. As with phellandrene and pinene, dill ether is completely hydrophobic and insoluble in water.

A number of other volatile compounds also contribute to the aroma of dill, including myristicin, another bicyclic ether compound (Figure 8.35, right) classified as a phenylpropene (see below), eugenol, vanillin, limonene, and 2-methylbutyric acid methyl ester, but dill ether and α-phellandrene are the two most important notes. Dry dill can have an order of magnitude less of these two components than is found in the fresh herb.

Parsley and Apiol

The common herb parsley (*Petroselinum crispum*) contains a molecule similar to estragole, anethole, and chavicol (see below) in its essential oil. This molecule, **apiol**, illustrated in Figure 8.36), also present in celery, is an important component of oil of parsley, and has a pungent taste. Apiol is a derivative of myristicin (Figure 8.35), which is also present in parsley. It has been known for centuries, and can have significant effects on human physiology in high concentrations. It is a known abortifacient, and as a result, parsley perhaps should not be eaten by pregnant women. High doses can damage kidneys and the liver, but the small amounts present in parsley used as a seasoning or garnish probably will not cause long-term harm to those who are not pregnant.

Figure 8.36. The structure of apiol from parsley.

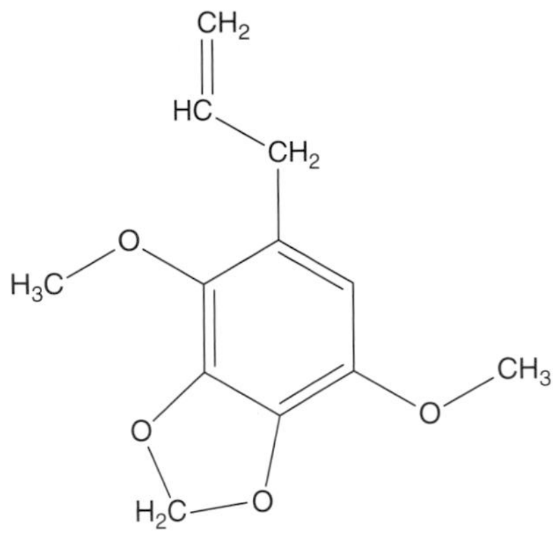

apiol

Figure 8.37. The structure of phenol.

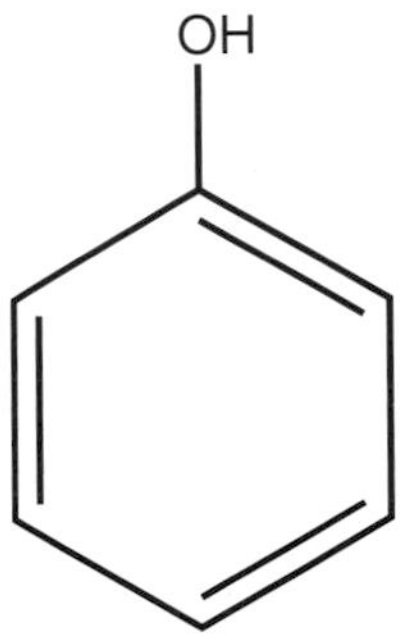

Substituted Aromatics, Phenolics, and Polyphenolics

Many plant molecules are substituted benzene derivatives, particularly phenol and polyphenolics, molecules that contain two or more phenol moieties (Figure 8.37). This category includes a diverse range of molecules such as tannins, which have astringent, bitter tastes, lignins, and flavonoids. Some of these molecules are thought to have significant antioxidant effects when consumed as part of the diet, as well as a variety of other putative health benefits.

Tumeric

Because many of the polyphenolics are colored, they will be discussed in greater detail in Chapter 9. One that is important as a flavor molecule is **curcumin**, the main flavor molecule of the spice **turmeric**, which is the principal component of curry. Tumeric is ground from the dried rhizomes of the turmeric plant (*Curcuma longa*) of Southern Asia. The familiar taste of turmeric is often described as "earthy" and slightly bitter. As with many other herbs and spices, the flavor of turmeric is primarily due to a variety of volatile terpenes, but curcumin contributes to the taste of this spice, as well as giving it its characteristic color.

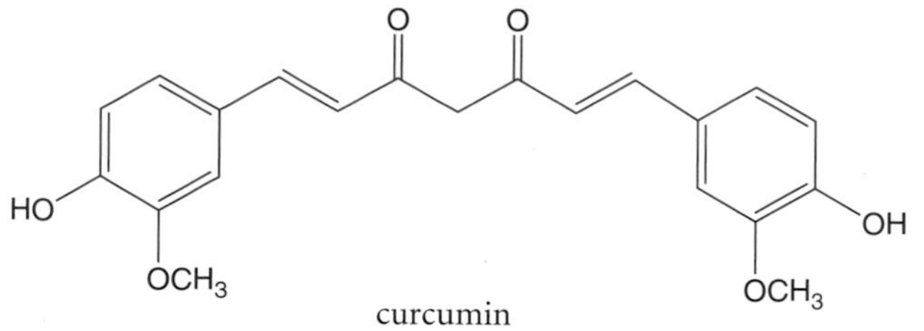

curcumin
keto form

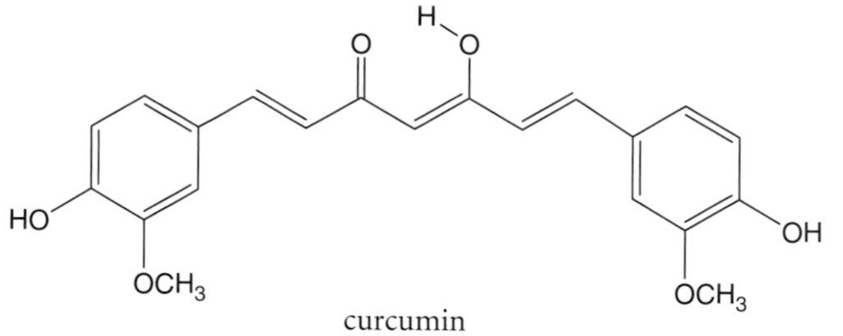

curcumin
enol form

Figure 8.38. The structures of the two forms of curcumin.

Curcumin exists in two forms (Figure 8.38), a keto form and an enol form, with the enol form being lower in energy. Curcumin, also called Natural Yellow, has a strong yellow color and is approved for use as a food coloring additive. Although dietary curcumin is poorly absorbed (simultaneous consumption of piperine, from black pepper reportedly increases the adsorption significantly), many health benefits have been ascribed to the consumption of the curcumin in curry. It is thought to have strong antioxidant and anti-inflammatory activity, to possibly have anti-tumor properties, and it has been found to inhibit the formation of the amyloid β oligomers whose accumulation in the brain cause Alzheimer's disease (Yang et al. 2005). Curcumin has antimicrobial activity that makes curry useful for preserving foods in warm tropical climates.

Rosemary

> There's rosemary, that's for remembrance; pray, love, remember.
>
> —William Shakespeare, *Hamlet, Prince of Denmark*

The common herb rosemary, *Rosmarinus officinalis*, native to the Mediterranean, is well known as an aromatic spice widely used in Provençal and other Mediterranean cuisines. It is one of the main components of "*herbes de Provence*." It contains a number of phenolic, polyphenolic, and terpenoid compounds thought to have a variety of beneficial health effects. These molecules include **caffeic acid**, **rosmarinic acid**, **carnosic acid**, and **carnosol**. Because of the presence of these compounds,

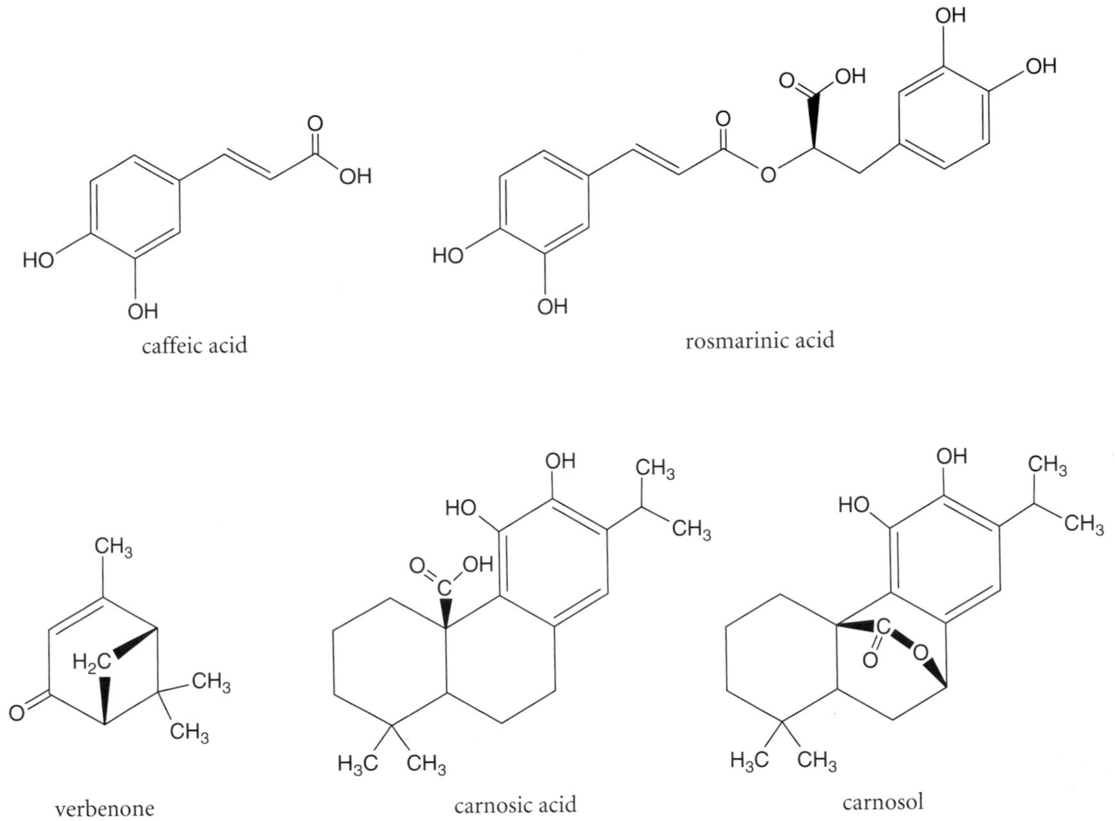

Figure 8.39. Antioxidants found in rosemary.

consumption of rosemary is thought to be particularly healthy, with possible benefits ranging from fighting cardiovascular disease and cancer to helping preventing dementia and memory loss.

Much of the characteristic flavor of rosemary is due to various related volatile monoterpenes. These include the pinenes (Figure 8.33), verbenone (Figure 8.39), a very similar molecule resulting from the oxidation of α-pinene, and eucalyptol (Figure 8.34), also called 1,8-cineole. All three of these molecules are bicyclics and are found in other herbs and spices besides rosemary, including sage, bay, and basil. Verbenone is found in abundance in the oil pressed from the flowering plant Spanish verbena, from which it gets its name. Borneol (Figure 8.34) and the linear monoterpene myrcene also contribute to the flavor of rosemary.

The phenolic compound caffeic acid, a hydroxycinnamic acid, is a powerful antioxidant (Figure 8.39) found in a number of fruits, vegetables, and herbs besides rosemary, including coffee, from which it gets its name. It is soluble in alcohol and hot water but is not very soluble in water at room temperature. It is also believed to be anticarcinogenic at low concentrations, although caffeic acid at high concentrations is itself a suspected potential carcinogen based on older experiments on laboratory animals. Rosmarinic acid, which of course takes its name from rosemary, is a polyphenolic antioxidant formed from caffeic acid that is also soluble in alcohol but only sparingly soluble in water. It may contribute astringent and bitter notes to the taste of rosemary. As can be seen from its structure (Figure 8.39), rosmarinic acid resembles curcumin, and like curcumin, it absorbs light in the visible range; it is colored red/orange. And like curcumin-containing turmeric, rosemary can also inhibit the growth of microorganisms, and thus helps increase the shelf life of foods as well.

Carnosic acid and carnosol have been shown to significantly increase the synthesis of nerve growth factor (NGF) in human cells (Kosaka and Yokoi 2003). NGF is a 118-residue protein that is essential for nerve cell growth and maintenance. It has thus been suggested that these compounds in rosemary can function against Alzheimer's disease and other forms of dementia, and thus could be the basis for the medieval European belief that rosemary was good for improving memory function.

Carnosic acid is also a strong antioxidant, as is carnosol, and thus may also be effective in helping to prevent cardiovascular disease. In addition, both of these molecules are inhibitors of lipases, the enzymes that digest fats, so that they might also be able to play a role in controlling weight.

The reason carnosic acid is so effective as an antioxidant is because of its high oxidation potential. It can initiate a series of oxidation reactions producing other intermediate molecules that are themselves antioxidants because they can be further oxidized. This series of oxidation reactions is referred to as the **carnosic acid cascade** (Wenkert et al. 1965). The general sequence of these reactions is shown in Figure 8.40.

The antioxidants in rosemary are sufficiently powerful that it can be used as a natural substitute for antioxidant additives such as butylated hydroxytoluene (BHT), Figure 10.4, in controlling rancidity and lipid oxidation in foods, and its antimicrobial properties make it useful in inhibiting the spoilage of meats as well. Because of its sensory properties, however, such applications are limited to foods where the flavor of rosemary is desired or acceptable.

Phenylpropenes: Tarragon, Anise, Estragole, and Anethole

Tarragon, *Artemisia dracunculus*, native to Europe, Northern Asia, and Western North America, is another common aromatic herb widely used in Western cooking, particularly in certain French dishes. It is the main flavor, along with chervil, in Béarnaise sauce. The aromatic compound **estragole** (Figure 8.41) is an important component of tarragon oil extracted from the leaves of this plant. Estragole, also called *p*-allylanisole, is related as an isomer to **anethole**, one of the main flavor components of the sweet aromatic herb anise (*Pimpinella anisum*). Anethole has a licorice-like taste and is the source of the sweet flavor of anise and fennel (*Foeniculum vulgare*). **Chavicol** is another molecule that is similar to estragole; it is an important component of betel oil, from the betel "nuts" of the betel palm (*Areca catechu*). All three of these molecules belong to the class of molecules called phenylpropenes, and may be potentially carcinogenic in large amounts, although the quantities in tarragon and anise used in cooking are probably not particularly harmful (indeed, it is often

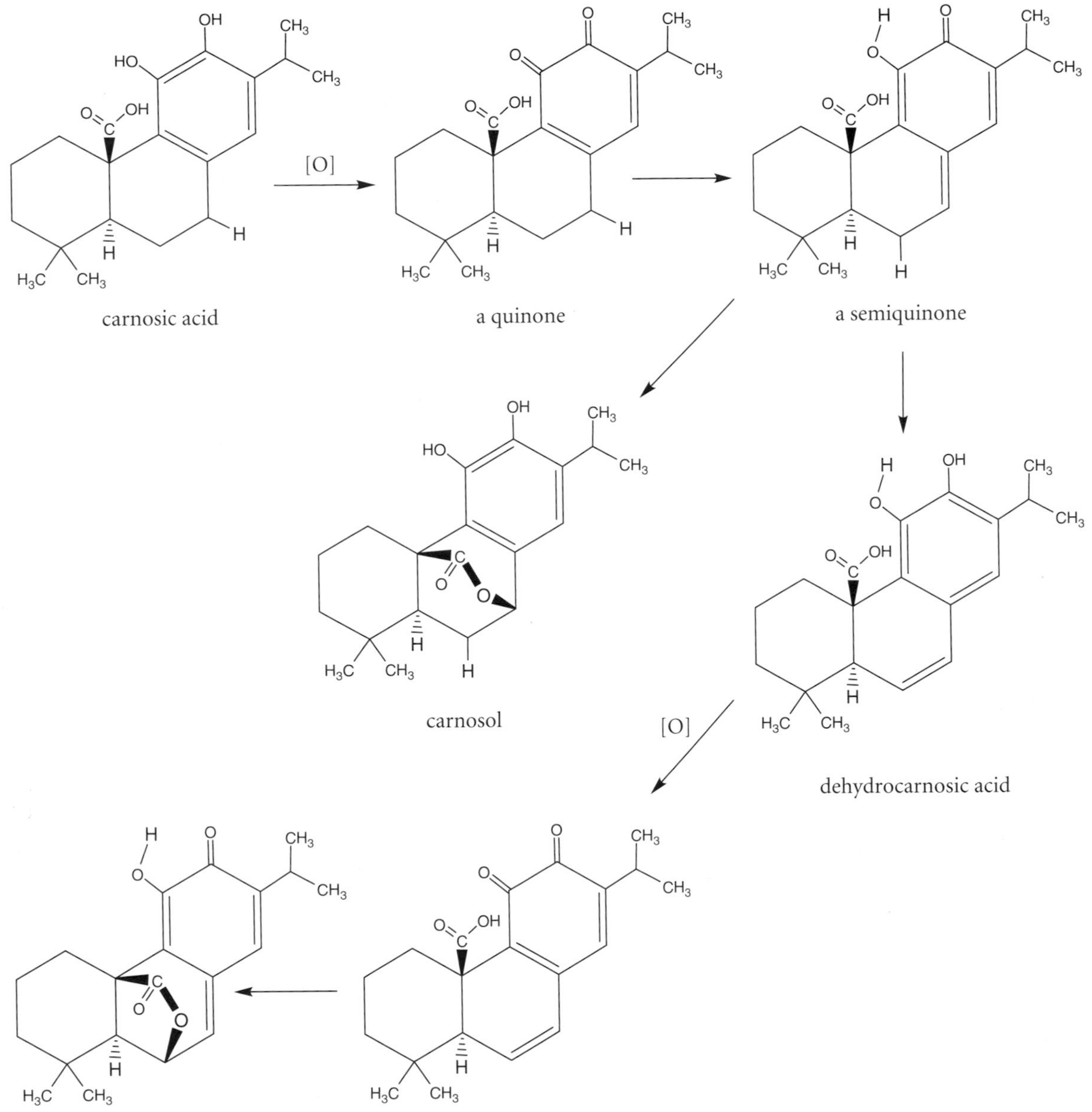

Figure 8.40. The carnosic acid cascade. (E. Wenkert, A. Fuchs, and J.D. McChesney, *J. Org. Chem.* 30:2931–2934, 1965.)

asserted that tarragon actually prevents cancer). The chewing of betel nuts, popular in Southern Asia, can cause oral cancer, however, and is similar to the effects of chewing tobacco.

Brett Character in Wines and Beers

Yeasts of the genera *Brettanomyces* and *Dekkera* in the Saccharomycetaceae family are an occasional wine spoilage organism responsible for a wine defect in flavor and

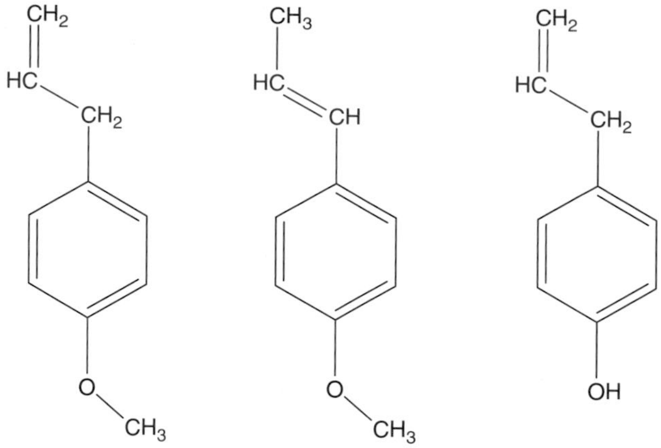

Figure 8.41. The structures of estragole, anethole, and chavicol.

odor usually referred to as a "Brett character" or "Bretty." The Brett aroma is often compared to a wet animal, Band-Aid, horse sweat, or barnyard aroma, which conveys some idea of the way that most people feel about this defect. At very low levels Brett compounds are sometimes considered to add beneficial complexity, but usually Brett notes are considered undesirable. Like many flavors and aromas, Brett character is actually due to a large number of compounds, which can be separated and identified using gas chromatography and mass spectrometry. In a special form of gas chromatography called CHARM analysis (Acree et al. 1984), individual peaks in a gas chromatography separation can be directed to the nose of a trained individual for subjective characterization. Using this approach, various compounds were identified as contributing to Bretty character (Figure 8.42), with the two most important ones being 4-ethylphenol and 4-ethylguaiacol, with the 4-ethylphenol contributing the most undesirable notes (Band-Aid, horses, barnyard). Brett notes are also usually considered undesirable in most beers as well, but in a few beers such as some Belgian ales and specialty and microbrewery American beers, they are actually part of the intended character and are produced through the deliberate inoculation with *Brettanomyces*.

Ethylene, Carrots, and Isocoumarin

The ripening of fruits is controlled by a number of plant hormones, such as gibberellins and auxins, but one of the most important hormones controlling ripening and respiration is the simple hydrocarbon gas **ethylene**, C_2H_4. This hormone affects not only the ripening of fruits but also the induction of flowering, the abscission of plant parts, and the yellowing of tissues through the loss of chlorophyll. Ethylene is widely used by food processors to induce ripening in fruits, particularly in so-called climacteric fruits. Climacteric fruits, which include apples, avocados, bananas, papayas, tomatoes, peaches, pears, figs, plums, blueberries, mangos, and apricots, are fruits

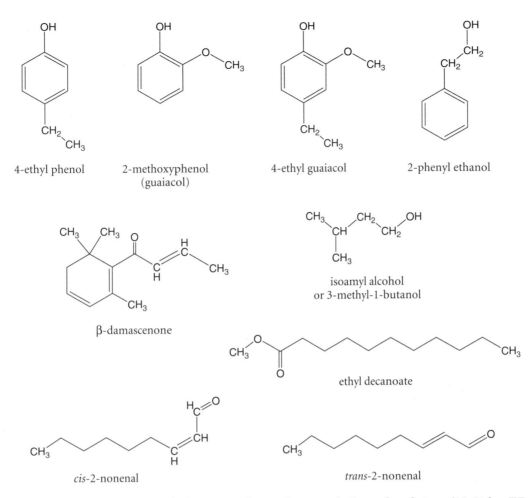

Figure 8.42. The structures of various volatile compounds contributing to the Brett odor of wines. (J.L. Licker, T.E. Acree, and T. Henick-Kling, in *Chemistry of Wine Flavor*, ACS Symposium Series 714, A.J. Waterhouse and S.E. Ebeler, eds. Washington, D.C.: American Chemical Society, 1998.)

that experience a sharp increase in rates of growth, development, and respiration at the end of their development, as they rapidly ripen in a burst of physical and physiological changes characterized by a significant increase in respiration rates. Nonclimacteric fruits, such as cherries, grapes, lemons, oranges, strawberries, cucumbers, and pineapples, also contain ethylene in their tissues, but in much lower concentrations, and they undergo a slow, steady growth and development, characterized by relatively constant respiration rates, throughout their ripening.

In climacteric fruits, the production of ethylene increases significantly in a short time as they fully ripen. The most dramatic increases in internal ethylene concentrations in fruits are in apples, which can reach levels of over 2000 μl/l. Avocados also reach high levels of ethylene at ripeness. Because of their high ethylene production, which escapes from the fruits as a gas, these fruits can affect other plant foods stored with them under confined conditions. For example, storing sweet potatoes with apples

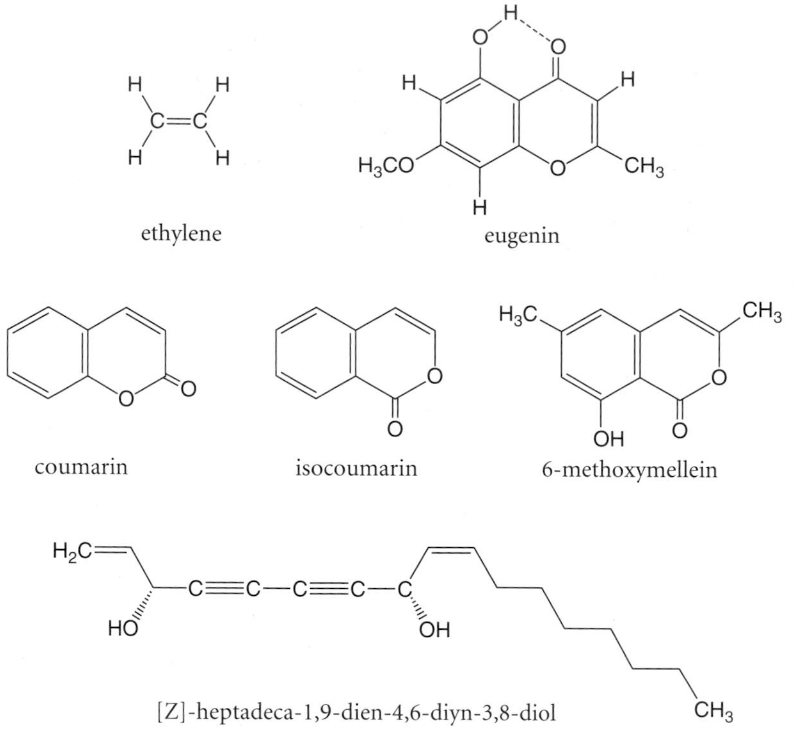

Figure 8.43. The structures of ethylene, coumarin, isocoumarin, and eugenin.

results in the production of bitter flavors in the potatoes, and can cause yellowing in green tissues in cabbage, broccoli, and brussels sprouts. When apples are stored with carrots in the same compartment of a refrigerator, the high concentrations of ethylene produced by the apples induces the enzymatic synthesis in the carrots of eugenin, isocoumarin, and derivatives of isocoumarin like 6-methoxymellein (Figure 8.43), bitter-tasting and somewhat toxic phenolic compounds that may contribute bitter flavors in the carrots, making them unpalatable (Chalutz, DeVay, and Maxie 1969; Kader 1985; Seljasen et al. 2001). It has recently been suggested, based on concentrations and recognition threshold levels, that the majority of the resulting bitter flavor in carrots is due to bis-acetylenic compounds such as [Z]-heptadeca-1,9-dien-4,6-diyn-3,8-ol (Figure 8.43) and related molecules (Hofman 2009). A structural isomer of isocoumarin, coumarin, is found in a number of grasses and is responsible for the familiar odor of freshly cut grass and hay. Coumarin is also found in the bark of the cassia tree, one of the several species that are a source of cinnamon.

Pyrazines and Methoxypyrazines

Pyrazines are generally volatile molecules that often contribute to the aromas of burned or roasted foods, as in baked bread, toast, roasted meats, and roasted coffee

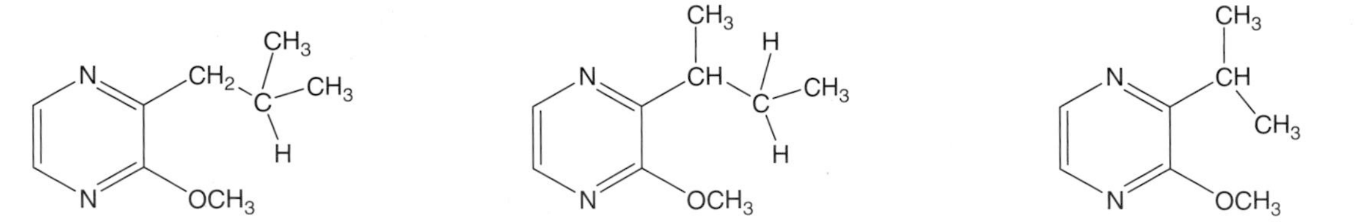

isobutylmethoxypyrazine 2-methoxy-3-(2-methylpropyl) pyrazine

sec-butylmethoxypyrazine 2-methoxy-3-(1-methylpropyl) pyrazine

isopropylmethoxypyrazine 2-methoxy-3-(1-methylethyl) pyrazine

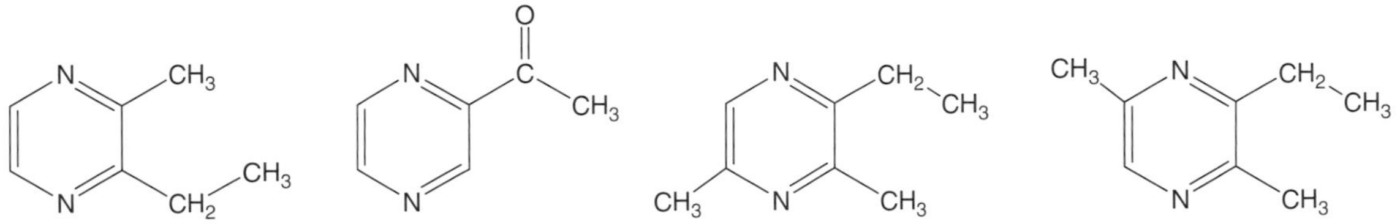

2-methyl-3-ethylpyrazine

acetylpyrazine

2-ethyl-3,5-dimethylpyrazine

2-ethyl-3,6-dimethylpyrazine

Figure 8.44. On the top row are the structures of the three most common methoxypyrazines found in grapes and wines. (M.S. Allen and M.J. Lacey, in *Chemistry of Wine Flavor*, ACS Symposium Series 714, A.J. Waterhouse and S.E. Ebeler, eds. Washington, D.C.: American Chemical Society, 1998). On the bottom row are several pyrazine derivatives that contribute burned notes to the aromas of roasted foods.

beans. Often they are generated from Maillard browning through the Strecker reaction (see Chapter 4) and as pyrolysis products of amines. Common examples of pyrazine derivatives are 2-methyl-3-ethyl-pyrazine, acetylpyrazine, 2-ethyl-3,5-dimethyl-pyrazine, and 2-ethyl-3,6-dimethylpyrazine (Figure 8.44). All four of these molecules have a burned or roasted aroma or an earthy smell.

Methoxypyrazines are a class of pyrazine derivatives synthesized enzymatically in a wide range of food plants including unripe fruits. Molecules of this class (Figure 8.44), with a methyl group bound to the aromatic ring through an ether linkage, are also among the many types of molecules that contribute to the aromas, or "bouquets," of wines. These molecules impart an earthy or vegetative/herbaceous aroma, like that of bell peppers, to Sauvignon Blanc, Semillon, and Cabernet Sauvignon wines. Many methoxypyrazines occur in the grapes used to make these wines, but three of them in particular make significant contributions to the aromas of these wines; they are isobutylmethoxypyrazine, *sec*-butylmethoxypyrazine, and isopropylmethoxypyrazine (Figure 8.44). The aroma of these compounds can be detected at low concentrations but can be too strong or even unpleasant at higher concentrations, and while they are an important signature in Sauvignon blanc wines, they are often considered to be undesirable in other types of wines.

Pyrazines also contribute to the aromas of various peppers, from sweet bell peppers to chilies and paprika. The *sec*-butylmethoxypyrazine molecule is important in the aroma of carrots. Bacteria-produced isopropylmethoxypyrazine is sometimes responsible for off-flavor notes in fish, dairy products, and eggs.

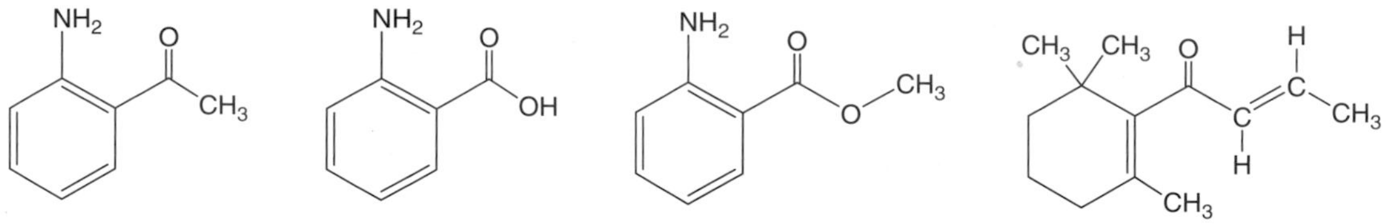

Figure 8.45. Some compounds responsible for "foxy" aromas in grapes.

o-aminoacetophenone   anthranilic acid   methyl anthranilate   β-damascenone

"Foxy" Grape Aromas

The methyl ester of anthranilic acid, methyl anthranilate (Figure 8.45), is a relatively hydrophobic, aromatic molecule found in North American grapes such as Concord grapes (*Vitis labruscana*), as well as in a number of other fruits, that plays a major role in the flavor and aroma of grapes. It is also widely used as a food additive and has US FDA GRAS (generally recognized as safe) status. It is used as an artificial grape flavor additive in manufactured foods such as candies, soft drinks, and chewing gums. Methyl anthranilate is also sprayed on fields as a bird repellant to deter wild geese and ducks.

Methyl anthranilate has been identified as one of the aroma components that contributes to the so-called foxy bouquets of American grapes and the wines made from them. "Foxy" has been described as floral or fruity in character. However, many grapes that exhibit "foxy" notes do not contain methyl anthranilate, and many other compounds also apparently contribute to this bouquet, including the damascenones. β-damascenone is the product of the oxidative degradation of the carotenoid β-carotene (Figure 9.16), and is present in most fruits. Other molecules that contribute to the "foxy" bouquet include numerous aromatic molecules similar to methyl anthranilate, such as ethyl anthranilate and several related compounds where the amine group is replaced by either a hydroxyl group or a carboxymethyl group. A number of short-chain ester polymers such as ethyl *trans*-2-butenoate, ethyl *trans*-2-hexenoate, and ethyl *trans*-2-octenoate may also contribute, as well as sulfur compounds such as ethyl-3-mercaptopropanoate. However, one of the most active components contributing to "foxiness" is O-amino acetophenone, which differs from methyl anthranilate only in that it is a methyl aldehyde instead of a methyl ester. This molecule, which was also found to contribute a grape aroma to the Japanese weasel (*Mustela itatsi*), is present in relatively high concentrations in Concord grapes (Acree et al. 1990).

Hot (Spicy) Tastes

Certain spicy foods are generally referred to as "hot" because the sensation they elicit is similar in some respects to the response to high temperature, but in fact is an irritation caused by chemicals from plants that are probably defensive, in that they

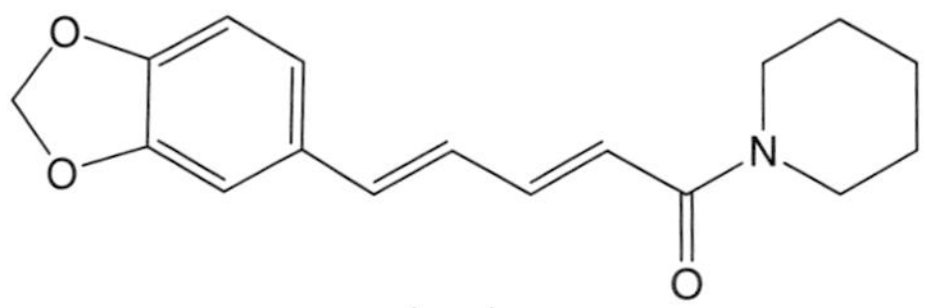

piperine

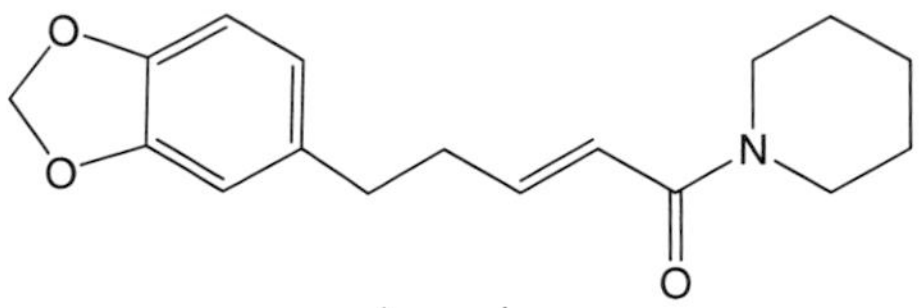

piperanine

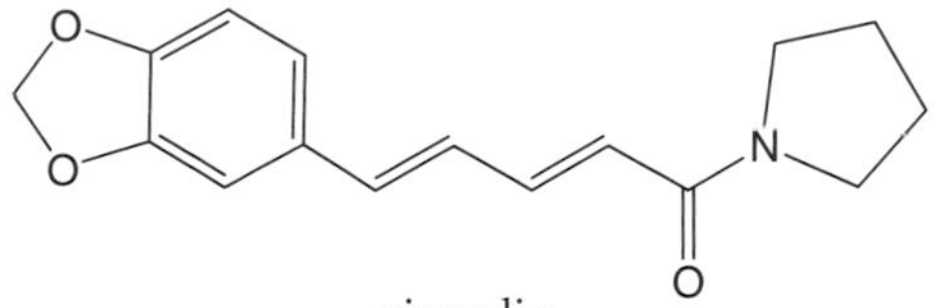

piperylin

Figure 8.46. The structure of the alkaloid piperine and related molecules that contribute to the spiciness of black pepper.

discourage some species from eating them. Many of these spicy condiments also inhibit bacterial growth, like curcumin.

Piperine

The principal active ingredient that gives black and white pepper, from the *Piper nigrum* plant, its spicy flavor is **piperine**, shown in Figure 8.46. It is an example of an **alkaloid**, a complex group of heterocyclic organic bases containing nitrogen atoms, many of which are poisonous, often manufactured by plants as protection against herbivores. (Caffeine, discussed later, is also such an alkaloid.) The diene double bonds in piperine are both in the *trans* configuration, but when exposed to light easily isomerize to the *cis,trans* configuration of the tasteless molecule **isochavicin**. For this reason, pepper should be protected from exposure to strong light. Piperine is hydrophobic and insoluble in water. Two other related molecules also contribute to the "heat" or spiciness of black pepper. One of these is piperylin, which is identical to piperine except that the heterocyclic nitrogen-containing ring has only four carbon atoms, while piperanine has the identical structure as piperine except there is only one *trans* double bond in the carbon chain linking the ring structures.

Capsaicin

Capsaicin is the molecule responsible for the pungent, spicy flavor of both red and green chili peppers. It is also the active ingredient found in paprika. Capsaicin is an irritant on mucous membranes and is the active ingredient in defensive pepper sprays used by police to subdue violent detainees and by hikers as a defense against bears. Like allicin from garlic and menthol from peppermint, capsaicin is an irritant of the trigeminal nerve, the main nerve of the mouth, nose, and eyes. Because it is so nonpolar, capsaicin is not soluble in water. Several molecules with related structures, collectively called capsaicinoids, contribute to the pungent flavors of chilies. One such

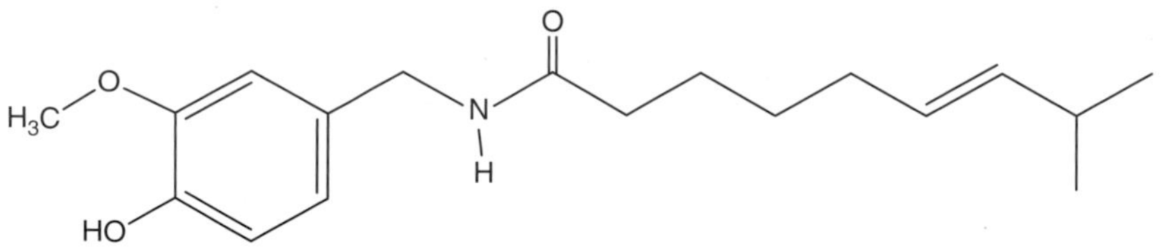

capsaicin

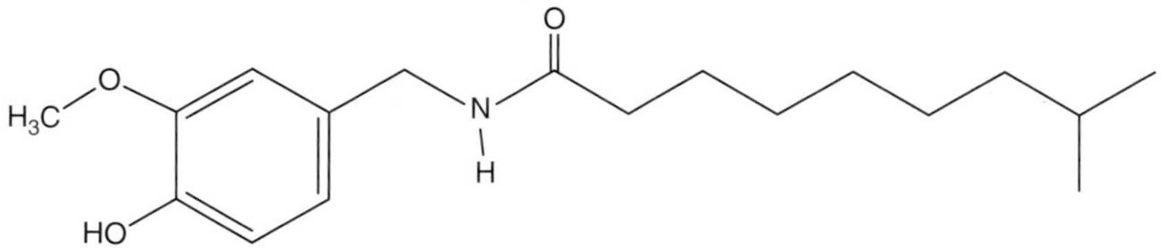

dihydrocapsaicin

Figure 8.47. The structures of capsaicin and dihydrocapsaicin.

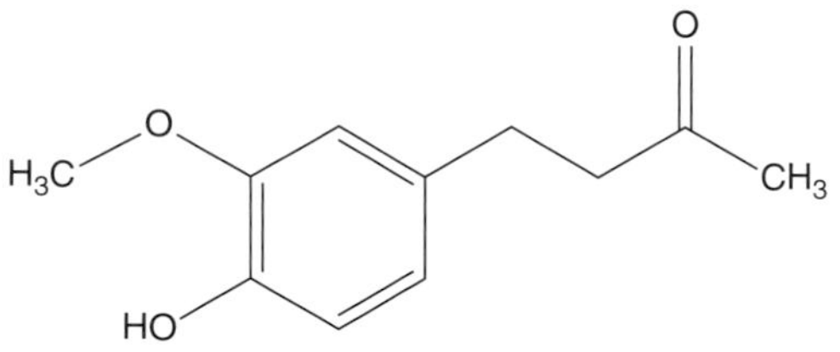

zingerone

Figure 8.48. The structure of zingerone.

molecule found in chilies is the related dihydrocapsaicin molecule, in which the double bond of the nonaromatic part of the molecule is saturated (Figure 8.47).

The spicy taste of capsaicin is due to its triggering the vanilloid receptor 1 in the dorsal root ganglion neurons. Both capsaicin and piperine are perceived as "hot" because they activate the TRPV1 (transient receptor potential vanilloid subtype 1) ion channel on the pain sensing trigeminal nerve cells that are responsible for detecting heat. Because capsaicin is hydrophobic, it is not efficiently cleaned from the taste buds by drinking water. Thus, if a bite of food is too "hot," fat-containing foods such as milk or yogurt are more effective at cooling the fires than a drink of water. Some evidence exists that capsaicin may have anticancer activity by inhibiting the proliferation of various types of cancer cells by inducing apoptosis, and it also serves as an antimicrobial in foods.

Ginger

Zingerone is the active, spicy-sweet ingredient in cooked or processed ginger from the *Zingiber officinale* plant. As can be seen from Figure 8.48, this molecule is similar

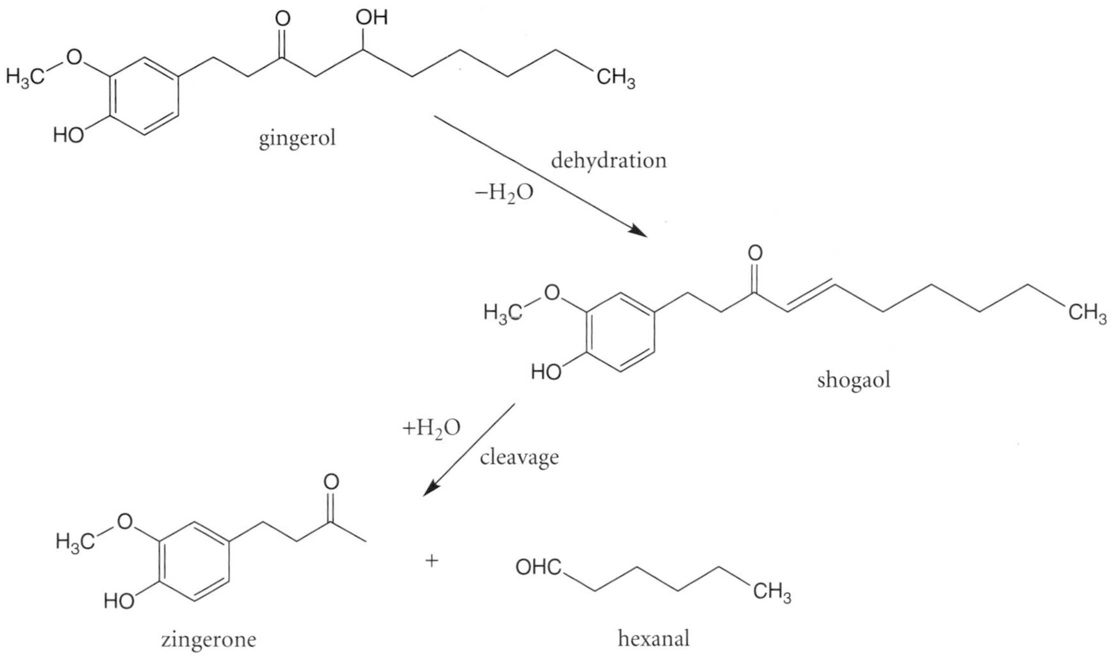

Figure 8.49. The conversion of gingerol to shogaol and zingerone.

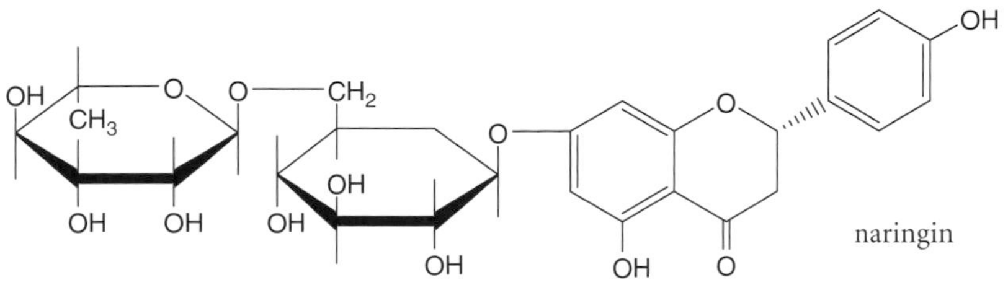

Figure 8.50. The structure of naringin.

to capsaicin (Figure 8.47) in several of its structural features, and it is also able to elicit a hot or stinging sensation. Pure zingerone is solid at room temperature and is only sparingly soluble in water.

In fresh ginger the predominant pungent flavor compound is **gingerol**, which during processing and storage dehydrates to produce shogaol, which is more pungent than gingerol (Figure 8.49). Shogaol in turn can undergo hydrolytic cleavage to give the more stable zingerone and the volatile hexanal.

Bitter Tastes

The bitter taste of grapefruit is due to the glycoside **naringin**, shown in Figure 8.50. Naringin is produced in the fruit as a glycoside of the carbohydrate rutinose, which is a disaccharide of L-rhamnose and D-glucose, 6-O-α-L-rhamnopyranosyl-D-glucopyranose. The aglycon unit of naringin is the flavanone **naringenin** (see Chapter

9). Naringenin is an intermediate in the biosynthetic pathway for the anthocyanin pigments. Naringin is even more bitter than quinine (below), but requires the glycosyl group for its bitter flavor. Hydrolyzing the glycosidic linkage by boiling in acid produces the free sugar and the naringenin aglycon, which is not bitter. Although naringin is not a heterocyclic alkaloid like caffeine and quinine, it does contain fused rings that may form part of the bitter flavor recognition functionality. Certain fungi and microorganisms produce an enzyme called naringinase that hydrolyzes the glycosidic linkages in naringin and are finding commercial application in reducing the bitterness of grapefruit juices.

Alkaloids

Alkaloids are alkaline (basic) amine compounds found in plants. Many are bitter tasting and serve a defensive function, discouraging herbivores from eating the plant. They are bases due to the lone pairs of the nitrogen atoms. Quinine, capsaicin, and piperine are classified as alkaloids, as are the drugs caffeine, nicotine, cocaine, morphine, and mescaline.

Quinine

Quinine (Figure 8.51) is an example of a bitter-tasting alkaloid found in the bark of various species of *Cinchona* tree from South America. This molecule is famous as a preventative of malaria, and as such, it was one of the most important drugs of the last 150 years. Unfortunately, a number of quinine-resistant strains of malaria have arisen that can only be combated through the use of more powerful drugs. Quinine has an extremely bitter taste and is used to flavor soft drinks and to make tonic water. FDA regulations limit the concentration in tonic water to subclinical doses. Quinine fluoresces under UV light and is used as a fluorescence standard.

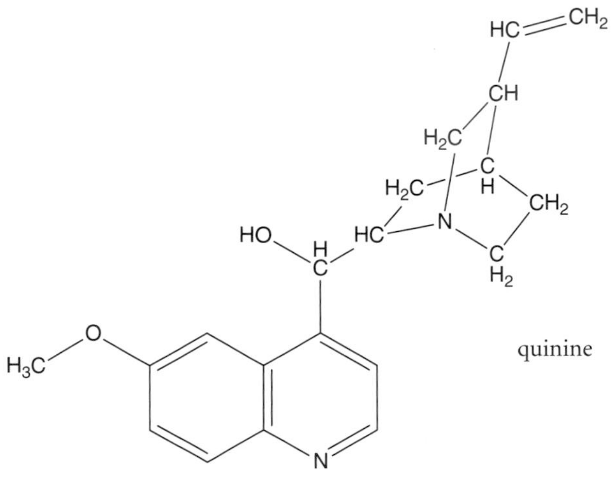

Figure 8.51. The structure of quinine.

Caffeine

Caffeine is a bitter-tasting alkaloid stimulant found in coffee and tea and cola beverages. It is a methylated derivative of the purine nucleotide base xanthine, synthesized from 7-methylxanthine by the bifunctional enzyme **caffeine synthase**, which successively adds methyl groups to 7-methylxanthine (Figure 8.52). The intermediate compound **theobromine** (3,7-dimethylxanthine) is the principal stimulant in chocolate. There is approximately ten times as much theobromine as caffeine in chocolate and cocoa. A typical cup of American-style drip coffee might contain around 100 mg of caffeine, while a cup of tea might contain half that amount, although the amount of caffeine in both types of drinks varies greatly with type and brewing method; dark black teas, for example, contain considerably more caffeine than lapsang souchong. An average 12-oz soft drink might contain approximately 10 to 50 mg of caffeine. Caffeine can interfere with normal sleep and can cause insomnia. It can also lead to hypertension and gastrointestinal distress. Partly for these reasons, some consumers avoid drinks that contain caffeine, and several religions teach that caffeine should not be consumed.

As a pure substance caffeine is bitter tasting and moderately toxic, being more than four times as toxic as morphine (see Table 11.1). It contributes to the taste of tea and kola nut-flavored drinks (in such soft drinks it also replaces cocaine, which was once added to these drinks as a stimulant). While caffeine contributes somewhat to the bitterness of coffee, it is not an important factor in that beverage, as might be assumed from the observation that decaffeinated coffees are still bitter. Various analytical studies using liquid chromatography coupled with mass spectroscopy, as well as one and two-dimensional nuclear magnetic resonance, have identified chlorogenic acid hydrolysis products such as caffeic acid (Figure 8.39), and a series of hydrophobic

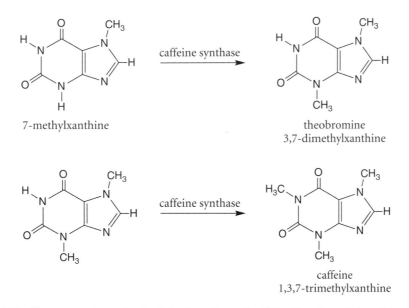

Figure 8.52. The enzymatic synthesis of theobromine and caffeine from 7-methylxanthine.

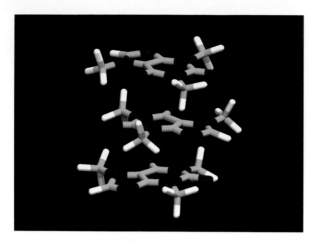 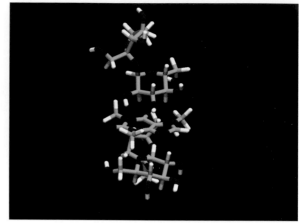

Figure 8.53. Left: three caffeine molecules stacked against one another in aqueous solution, as calculated in a molecular dynamics computer simulation; right: two sucrose molecules stacking against the faces of a caffeine molecule, as in coffee sweetened with sugar. L. Tavagnacco, A. Cesàro, and J.W. Brady, unpublished data.

lactone quinides and phenylindanes, as being primarily responsible for the bitter taste of coffee (Hofmann 2009).

Caffeine is only sparingly soluble in water, with a solubility limit of 22 mg/ml at 25 °C, although this increases to 180 mg/ml at 80 °C (Cesàro et al. 1976). Due to its somewhat hydrophobic character, caffeine molecules tend to form stacked dimer pairs and even higher-order aggregates in aqueous solutions (Figure 8.53), and will preferentially partition into a non-polar phase in contact with an aqueous solution. Thus, caffeine can be removed from coffee (decaffeination) by soaking the beans in organic solvents or superheated steam under high pressure. The organic solvent most commonly used for decaffeination of coffee is dichloromethane, or methylene dichloride, CH_2Cl_2. It is volatile, and thus completely vaporizes from the coffee after treatment, leaving no detectable trace, which is fortunate, because it is a suspected carcinogen. In another process, all water-soluble compounds are removed by soaking in water, which is then mixed with an immiscible organic solvent such as CH_2Cl_2 to remove the caffeine. After this solvent is vaporized, the remaining water and solutes are soaked back into the coffee beans. Simple water soaking can be used without the organic solvent treatment, but many desirable water-soluble flavor compounds may be lost along with the caffeine with this procedure. Recently, supercritical carbon dioxide has been used to remove caffeine from coffee beans as well as teas. Because the CO_2 vaporizes when the beans are returned to normal pressures, no residual CO_2 is found in the beans. The drawback to this method is that the high pressures necessary make it an expensive batch process. As with the other methods, molecules besides caffeine are removed. In teas, the catechins (Chapter 9), which make important contributions to the taste of teas, can also be lost. It has also been proposed that

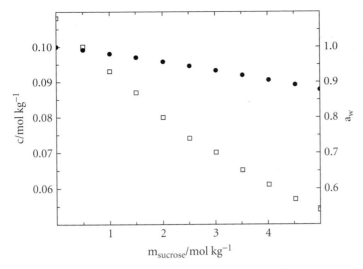

Figure 8.54. The solubility of caffeine in water as a function of added sucrose concentration. As the sugar concentration increases, the caffeine becomes less soluble, and the water activity decreases, due to the strong interactions of the sugar with water. (T.H. Lilley, H. Linsdell, and A. Maestre, *J. Chem. Soc. Faraday Trans.* 88:2865–2870, 1992.)

coffee and tea could be decaffeinated by selectively removing the gene for caffeine synthase from the coffee and tea plants (*Coffea ssp.* and *Camellia sinensis,* respectively) by genetic engineering, rendering them incapable of synthesizing caffeine (Kato et al. 2000).

In beverage solutions such as coffee and tea, to which table sugar might be added, the added sugar decreases the solubility of the caffeine, even though there is a weak association of the sucrose rings with the hydrophobic planar faces of the caffeine (see Figure 8.53). This hydrophobic association in principle should make the caffeine more soluble, but the effect is out-balanced by the significant increase in chemical potential, and surface tension, of the aqueous solution as the sugar is added, due to its strong interactions with the water molecules through hydrogen bonding (see Figure 8.54). This effect thus increases the tendency of caffeine to self-associate, as is also shown in Figure 8.53.

Humulone

The bitter quality of beer is largely contributed by the molecule humulone from hop resin from the female hop plant (*Humulus lupulus*). During brewing, humulone is isomerized to isohumulone, which is even more bitter (Figure 8.55).

On exposure to light, isohumulone reacts with hydrogen sulfide from yeast fermentation to produce volatile, aromatic sulfur compounds, including 3-methylbutane-1-thiol, a component of the defensive odor of skunks, which give the beer a "skunky" aroma, referred to as "lightstruck" flavor (Gunst and Verzele 1978).

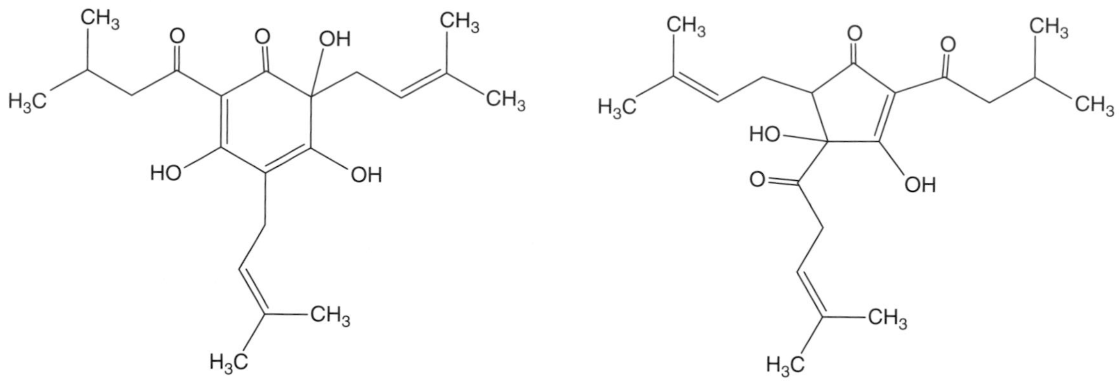

Figure 8.55. The structures of humulone and isohumulone, found in beer.

Protein Decomposition Products

Several volatile nitrogen-containing compounds with unpleasant odors can result from the breakdown of animal proteins. Two of the most important of these amines are **putrescine** and **cadaverine** (Figure 8.56), which, as their names suggest, are significant odor components of the smell of rotting flesh. They both also contribute to the odor of urine and bad breath. Cadaverine is produced from lysine, and putrescine is produced from ornithine. The ornithine is made from arginine by the enzyme argininease as part of the urea cycle. Putrescine is then produced by the action of the enzyme ornithine decarboxylase, which removes the acid group and which also acts on lysine to give cadaverine. The heat denaturation of these enzymes is the reason that little putrescine and cadaverine are produced in cooked meats. Both molecules can be present in very low concentrations in fresh meat as the result of autolysis and are produced in much larger amounts by bacterial decomposition.

Skatole, shown in Figure 8.57, is a breakdown product of the tryptophan side chain, as can readily be appreciated from the figure. This foul smelling molecule is partially responsible for the unpleasant odor of feces, as the name implies, but it is also present in small amounts in some foods, as an off flavor odor produced by bacterial metabolism. The sulfur-containing compound dimethyl sulfide (Figure 8.12) also contributes to the odor of feces, as do methyl mercaptan and hydrogen sulfide, which result from bacterial enzymatic decomposition of the amino acids cysteine and methionine.

Other amines important in the odor of decomposition of meat proteins are methylamine, dimethylamine, and **trimethylamine**, C_3H_9N (Figure 8.58). These gaseous derivatives of ammonia are major components of the smell of rotting fish (along with the oxidative degradation products of unsaturated fish lipids), because they are given off as the result of the degradation of fish proteins by bacterial enzymes.

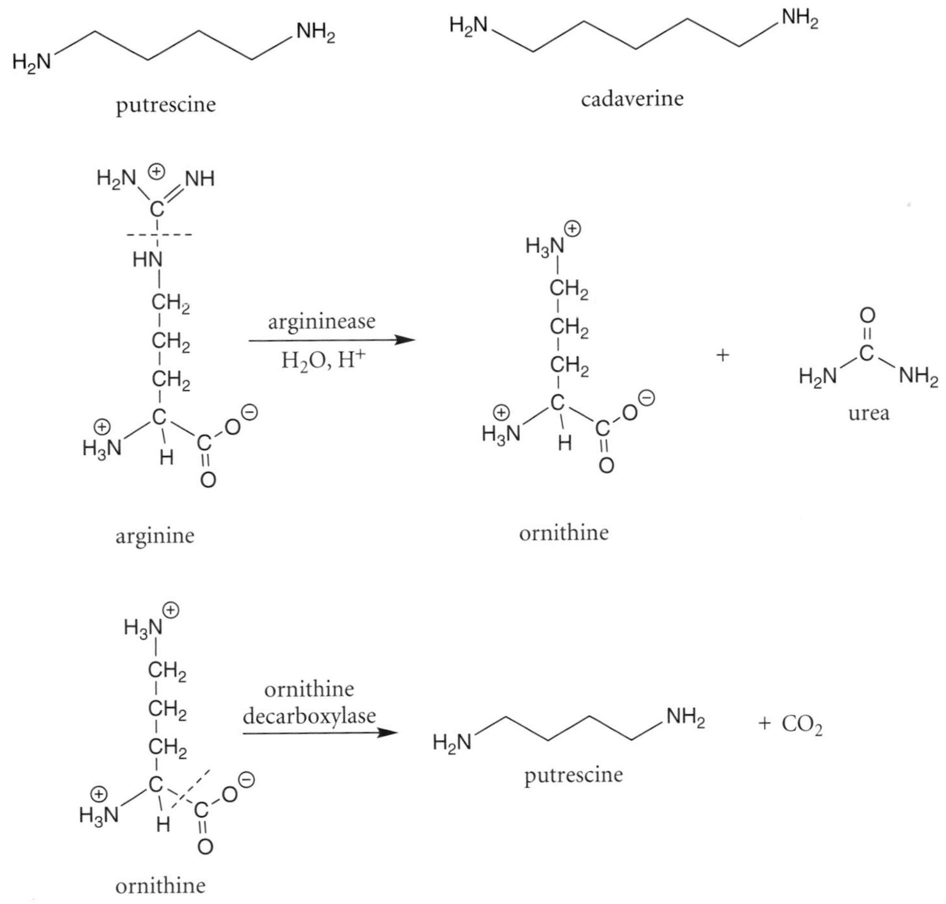

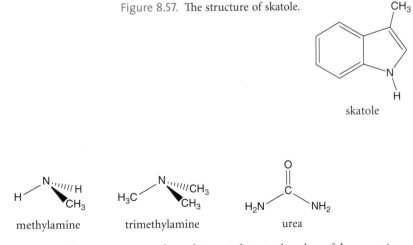

Figure 8.56. The structures of several of the protein decomposition products contributing to the odors of decaying protein foods.

Figure 8.57. The structure of skatole.

skatole

methylamine   trimethylamine   urea

Figure 8.58. Three protein decomposition products that contribute to the odors of decomposing proteins.

Urea present in the muscle tissue of sharks and rays also contributes to the odor of the decay of these meats, along with the ammonia (NH_3) produced by the conversion of urea to ammonia by bacterial ureases.

SUGGESTED READING

Attokaran, M. 2011. *Natural Food Flavors and Colorants.* Hoboken: Wiley-Blackwell.

Coultate, T. 2009. *Food: The Chemistry of Its Components,* 5th ed. Cambridge: RSC Publishing.

Gunst, F., and M. Verzele. 1978. "On the sunstruck flavor of beer." *J. Inst. Brew.* 84:291–292.

Lawless, H.T., and H. Heymann. 1998. *Sensory Evaluation of Food: Principles and Practices.* New York: Chapman and Hall.

Mata, V.G., P.B. Gomes, and A.E. Rodrigues. 2005. "Engineering perfumes." *AIChE J.* 51:2834–2852.

Reineccius, G. 2006. *Flavor Chemistry and Technology,* 2nd ed. Boca Raton, FL: CRC Taylor & Francis.

Taylor, A.J., and D.D. Roberts, eds. 2004. *Flavor Perception.* Ames, Iowa: Blackwell Publishers.

'Tis the voice of the Lobster; I
heard him declare, "You have
baked me too brown, I must
sugar my hair."

LEWIS CARROLL, *ALICE'S ADVENTURES
IN WONDERLAND*

9.
Food Colors

THE IMPORTANCE OF FOOD COLORS

Color is an important property of both manufactured foods and natural unprocessed foods and can play a large part in determining consumer acceptability. Not only are people perhaps instinctively pleased by the bright reds, oranges, purples, and yellows of fresh fruits and the bright green of fresh vegetables (Figure 9.1), but they may also from experience associate a loss of the bright red color of fresh meat with spoilage, or at least a lack of freshness. Manufactured foods may also be brightly colored, especially if they are intended to appeal to children; this may not only extend to candies and breakfast cereals, but even to such child-oriented products as green-colored ketchup. Understanding how to preserve the natural colors of foods during processing can be extremely important for the acceptability of a product, and knowing how to mix color additives to produce a particular pigment can also be useful, particularly in designing confectionary products and those intended to appeal to children. It also is fortuitous that many of the natural pigments in raw food are quite beneficial as antioxidants, and in many fruits a deep red or purple color is indicative of a high concentration of these compounds.

Figure 9.1. Left, apples and pears in a department store supermarket in Osaka; the apples in the middle are priced at 1000¥ each—almost $13US; right, cherries, strawberries and other fruits at an open air market in Paris.

VISIBLE LIGHT AND COLOR

Perceived colors result from electromagnetic radiation in the wavelength range from around 380–400 nm up to around 780–800 nm (Figure 9.2). Wavelengths greater than 800 nm are in the infrared range, and wavelengths less than 400 nm are in the ultraviolet range. These limits are not precise and may vary from individual to individual; some writers think that Sir Isaac Newton, who first used prisms to analyze the color spectrum of visible light (Figure 9.3), could see farther into the ultraviolet range than most people, and for this reason divided the spectrum into seven colors, including indigo. Others believe that he was influenced by the analogy with the seven-note chromatic musical scale used in Western music (Ball 2002). It is now more conventional to divide the visible continuum into six principal colors (see Figure 9.4), but all such divisions are arbitrary because the colors intergrade continuously. Normal "white" light contains all wavelengths of light in the visible range, while light of a particular color will correspond to radiation of a single wavelength, or more generally, to a narrow range of wavelengths, or a selective combination of particular wavelengths.

The retina of the human eye is covered with cells called rods and cones, which are specialized photoreceptor cells that are sensitive to light. The cone cells are responsible for color perception, while the rods are more generally sensitive to light

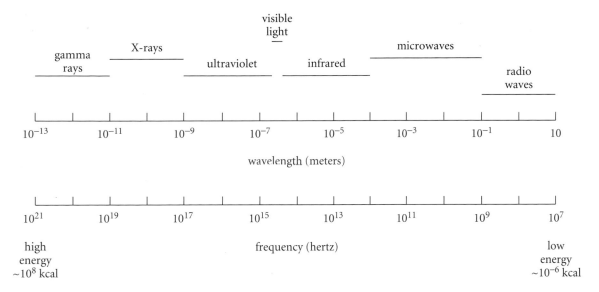

Figure 9.2. The electromagnetic spectrum.

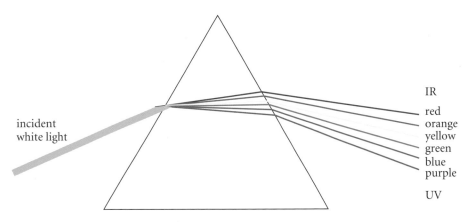

Figure 9.3. A schematic illustration of the separation of visible white light into component colors using a prism.

of all wavelengths. There are three types of cone cells that respond optimally to light of different wavelengths; these are the S, M, and L types, for short, medium, and long, referring to the wavelengths to which they respond. The S-type cones have an absorption maximum centered around 440 nm, the M-types maximally absorb photons with wavelengths around 540 nm, and the L-types respond most strongly to wavelengths centered around 570 nm, with considerable overlap in the absorption ranges of the three types, so that the entire visible range is covered. The ratios of the varying responses of the different cone types are then integrated into the perception of a particular color. Ultimately, color perception, like taste and odor perception, takes place in the brain and is subject to psychological factors that further complicate the picture. For example, the context of surrounding colors can affect the perception of a particular hue.

Perceived color can result from emission, including incandescence, which is the emission of light of all wavelengths by a very hot object, from filtered transmission (light passing through a medium), and by reflectance, including selective reflection resulting from the absorption of particular wavelengths or ranges of wavelengths by the reflecting object. Traditionally, colors have been considered to have complements, or opposites, and are often arranged in an artist's **color wheel**, with the **complementary colors** placed opposite to one another on the wheel. Thus, as can be seen from the color wheel (Figure 9.4), green is the complementary color to red, orange is complementary to blue, and yellow is the complementary color to violet. Because white light contains all of these colors, the selective adsorption of one color results in the perception of the opposite color in the reflected light. As an example, when visible white light strikes most plant leaves, the abundant chlorophyll molecules of the chloroplasts absorb violet and red light, so that the reflected light appears green, because it is complementary to these colors. In this system, red, blue, and yellow are considered to be "**primary**." By mixing of these primary pigments, as with paint, the secondary colors can be produced; for example, green is produced by mixing blue and yellow. This arrangement of the colors on a wheel, first done by Newton, is quite arbitrary, and there is, of course, no continuous intergradation of violet into red in a cyclic fashion, with the implied jump in wavelength from 400 to 800 nm. There is, however, a continuous intergrading of each color into the next at all of the other boundaries, rather than the abrupt change at the boundaries implied by Figure 9.4a.

This discussion is extremely simplistic, and ignores such factors as the variation of hue with brightness. While the color of light reflected by pigments works by

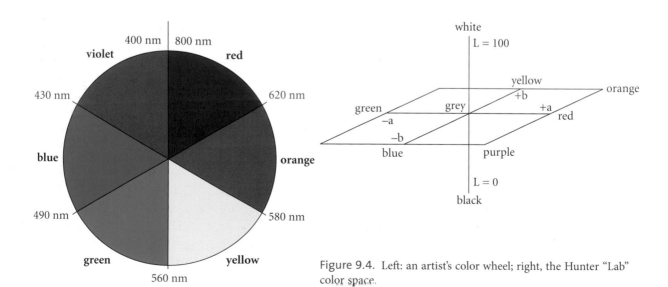

Figure 9.4. Left: an artist's color wheel; right, the Hunter "Lab" color space.

subtraction, the direct mixing of light works differently (mixing the three primary pigments produces brown, not white, for example). In mixing light directly, the three primary colors are red, green and blue (the now-familiar **rgb** format of many HDTV screens and computer files, whose images are constituted on computer screens by mixing light of different colors).

In the early decades of the twentieth century, an international standard system of color description was established by the Commission International de l'Eclairage, called the CIE System of theoretical primaries, which replaces the primitive color wheel of Figure 9.4 with a more complex chromaticity diagram based on three chromaticity coordinates x, y, and z that quantitate the relative amounts of the three theoretical primaries and specifies the mathematical rule for their combination. For practical problems of specifying color, various other systems are now used; one of the most common is the Hunter "Lab" color space (Figure 9.4), which describes any particular color as a point in a three-dimensional space defined by the coordinates L, a, and b (hence the name "Lab"; see Figure 9.4). The coordinate L describes the lightness of the color and ranges from a value of 0, which designates black, through grey, to $L = 100$, which corresponds to white. The coordinate a runs from $-a$ to $+a$, ranging from green to red. The b coordinate ranges from yellow ($+b$) to blue ($-b$), with orange at ($+a,+b$) and purple at ($+a,-b$). Colors can be compared against standard samples, much like matching house paints against paint chips, but color matching is still difficult. In matching reflected colors, it is even necessary to control the light source, since light from incandescent, fluorescent, and LED bulbs have different spectra, all of which differ from natural sunlight.

The visible spectrum is only a small portion of the total electromagnetic spectrum (Figure 9.2). Remembering that $v = c/\lambda$, where c is the speed of light, and that $\Delta E = hv$ (Equation 1.5), the visible spectrum range of wavelengths, between 400 and 800 nm, corresponds to energies in the range from 36 to 72 kcal/mol. Energies in this range are commonly due to molecular electronic transitions, either of the n $\rightarrow$ π^* type, between a nonbonding molecular orbital and an antibonding orbital, or of the $\pi \rightarrow \pi^*$ type, between a bonding orbital and an antibonding orbital. Thus, among organic molecules, those likely to adsorb electromagnetic radiation in the visible region are those possessing π electrons, or containing heteroatoms such as oxygen or nitrogen with nonbonding valence shell electron pairs. Groups of atoms that absorb visible light are called **chromophores**. The double bond between two carbon atoms, —C=C—, as in ethylene, would be such a chromophore, with a $\pi \rightarrow \pi^*$ transition in the range 170 nm. This wavelength is outside the visible range, in the ultraviolet, and hydrocarbon molecules with only one double bond, such as ethylene, are colorless. Conjugation of such double bonds, as in —C=C—C=C—, which extends the π system and allows electron delocalization, shifts the absorption of the chromophore toward longer wavelengths by approximately 30 nm for each additional conjugated double bond. Thus, a system of approximately seven or more conjugated double bonds will absorb in the visible range, and so will be colored. In general, the more conjugated double bonds a system possesses, the more its absorbance will be shifted toward the red (longer wavelengths).

NATURAL COLORS IN FOODS

Perhaps the two most important natural colors of foods are the green of chlorophyll in many plant foods and the bright red of myoglobin in the muscle tissue of much fresh meat. Both of these colors are taken as indicators of freshness, because they can be affected by aging and spoilage as well as by processing, so that controlling these changes is quite important. Both of these colors arise from similar large heterocyclic organic molecules called tetrapyrroles complexed with metal ions.

TETRAPYRROLE PIGMENTS

An important class of natural pigments includes several molecules containing large tetrapyrrole rings called **porphins**, which are built from four **pyrrole** molecules linked together (Figure 9.5). Porphins are heterocyclic aromatic molecules in which all of the atoms lie in the same plane. Because of the extensive conjugation of the double bond system, the porphrin rings absorb light in the visible range and are colored. Substituted porphin rings are called **porphyrins**. The most important of these pigment molecules in foods are the heme group of hemoglobin/myoglobin and chlorophyll. In these molecules, transition metal ions are bound in the center of porphyrin groups, coordinated to the nitrogen atoms with the loss of two protons.

Heme Compounds and Meat

The molecule responsible for the red color of muscle tissue is the protein myoglobin (Figures 5.24 and 9.7), a predominantly alpha helical globular protein containing 153 amino acids whose fold incorporates a noncovalently bound complex called a **heme group**, which is a porphyrin ring bound to a ferrous iron atom (Figure 9.6). The function of myoglobin is to store oxygen for use by the muscle mitochondria during contraction.

Figure 9.7 shows the general folding pattern of the myoglobin molecule, as determined by X-ray crystallography. The position of the heme group in the molecule is

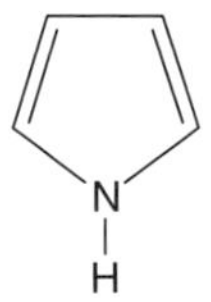

Figure 9.5. The structures of pyrrole and porphin.

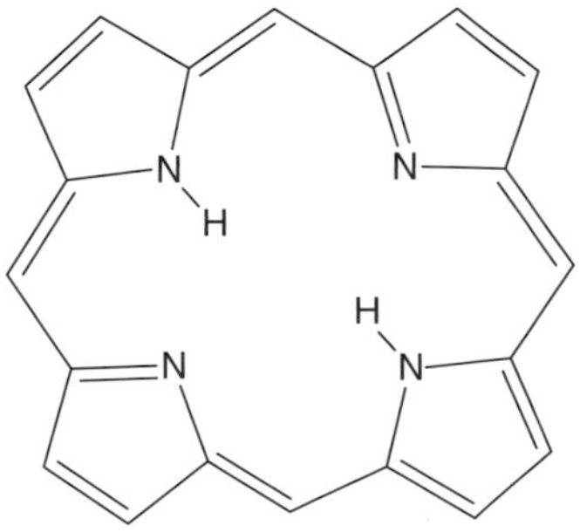

pyrrole

porphin

Figure 9.6. The heme group of hemoglobin and myoglobin.

CH₂

CH₃

H₃C

A   N   N   B   CH₂

Fe⁺²

N   N

C   D   CH₃

H₃C

COO-   -OOC

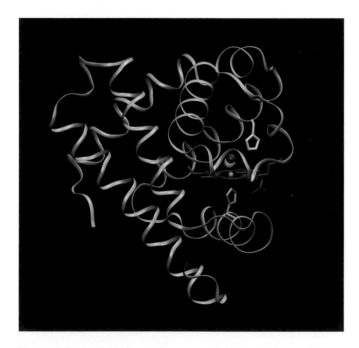
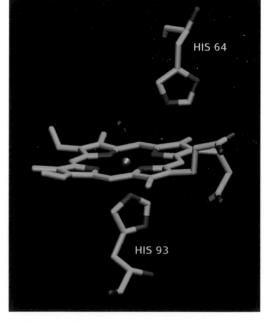

Figure 9.7. On the left, a ribbon trace of the backbone conformation of the myoglobin molecule bound to a carbon monoxide molecule (i.e., the carboxymyoglobin complex), with the prosthetic heme group shown in red. The iron atom is indicated as an orange sphere. Two histidine side chains are shown in atomic detail. His 93, which coordinates to the heme iron atom is also shown in orange. On the right, a close-up of the heme group bound in the protein, with two of the histidine residues indicated. A carbon monoxide molecule coordinated to the central heme iron atom is also indicated. (F. Yang and G.N. Phillips, *J. Mol. Biol.* 256:762–774, 1996).

indicated in red. Figure 9.7 also shows how the iron atom is coordinated to lone pairs from four nitrogen atoms of the porphyrin ring and to a fifth nitrogen lone pair from a His residue of the protein chain. The sixth position can be occupied by various ligands such as O_2, CO_2, or a water molecule. The binding of these ligands is reversible, and their competition for this site depends on the local partial pressures of these various species.

Hemoglobin is a heterotetrameric complex of four noncovalently associated subunits (two α subunits and 2 β subunits), each of which is similar to an individual **myoglobin** (Mb) molecule. The function of hemoglobin is to transport oxygen in the blood from the lungs to the cells, where CO_2 is picked up and transported back to the lungs to be exchanged again for oxygen. Myoglobin is a monomer found in large amounts in muscle tissue, making up 2.5% of the dry weight of muscle, and at low O_2 partial pressures, as occurs in muscles, hemoglobin gives up O_2 to myoglobin. The iron in the center of the heme group is in the ferrous (Fe^{2+}) state, and the iron-O_2 complex, called **oxymyoglobin** (MbO_2), absorbs in the green region of the visible spectrum, in two strong peaks at 542 and 580 nm (see Figure 9.8), and thus is colored bright red. If the oxygen molecule is replaced by a water molecule, the complex adsorbs in a single broad band centered in the green region ($\lambda_{max} = 555$ nm) of the spectrum, and the myoglobin appears purple. If the iron atom is oxidized to the ferric state, the molecule is called **metmyoglobin** (MetMb). Metmyoglobin absorbs

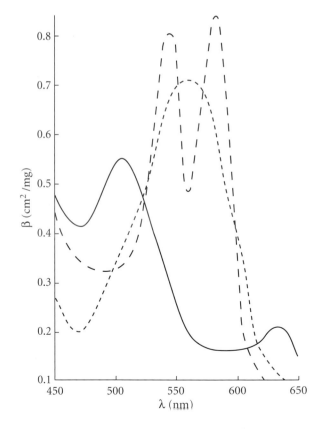

Figure 9.8. The absorption spectrum (extinction coefficient vs. wavelength) of myoglobin (short-dashed line), oxymyoglobin (long-dashed line), and metmyoglobin (solid line). (Adapted with permission from F.M. Clydesdale and F.J. Francis, in *Principles of Food Science. Part I. Food Chemistry*, O.R. Fennema, ed. New York: Marcel Dekker, 1976.)

primarily in the blue region, at 505 nm, with a smaller peak at 635 nm, and MetMb is colored brown. The oxidized metmyoglobin cannot bind O_2. Freshly cut meat is colored bright red on the surface due to the abundance of oxygen, and thus the presence of oxymyoglobin. The interior tissues are purple when just cut, due to a relative scarcity of oxygen, but will quickly turn red on exposure to air. The red color of the meat surface slowly turns progressively brown as more of its myoglobin is oxidized to metmyoglobin,

$$\underset{\substack{\text{red} \\ \text{oxygenated myoglobin (oxymyoglobin)}}}{MbO_2\,(Fe^{2+})} \rightleftarrows \underset{\substack{\text{purple} \\ \text{deoxymyoglobin}}}{Mb\,(Fe^{2+})} \rightleftarrows \underset{\substack{\text{brown} \\ \text{metmyoglobin (oxidized)}}}{MetMb\,(Fe^{3+})} \qquad (9.1)$$

In living tissue and fresh red meat, reducing agents in the tissue reduce the metmyoglobin back to the ferrous myoglobin form. As these are used up, the brown color comes to dominate. Packaging that excludes oxygen will cause the oxygen pressure to drop and favor oxidation and the development of brown color, so permeable packaging films that allow the exchange of at least 5 liters of O_2 per square meter per day are used to slow this process. Low pH and trace metals also favor oxidation. The growth of bacteria on the surface of meat may reduce the partial pressure of oxygen, inducing the brown color and indicating spoilage. In general, there is no degradation in the quality of a cut of meat from a culinary standpoint as the surface myoglobin is converted to metmyoglobin; however, there is considerable consumer resistance to brown colored meat. In part, this is somewhat reasonable, because the development of the brown color indicates the passage of some period of time, and because the growth of bacteria also results in the development of brown color.

Because of this consumer preference, retail sellers strive to keep fresh meat in the red-colored oxygenated state. Reducing agents such as **sodium ascorbate** (vitamin C; Figure 9.9) and **sodium erythorbate** prevent oxidation and maintain the fresh red color of meat, and delay lipid oxidation as well, but because they do not inhibit bacterial growth, they may mask spoilage. Modified-atmosphere storage can also be used to inhibit discoloration, but may also permit the growth of anaerobic bacteria (Manu-Tawiah et al. 1991).

When meat is cooked the myoglobin proteins denature, and the iron atoms are oxidized as a gray/brown pigment called **hemichrome** is produced. This is the familiar process that can be easily observed when frying hamburger or steak, for example. Curing meat with nitrites produces a pink color because nitric oxide, NO, binds irreversibly to the myoglobin to give **nitrosylmyoglobin**. When cooked, this complex

Figure 9.9. The structure of the reducing agent ascorbic acid (Vitamin C, see Chapter 12).

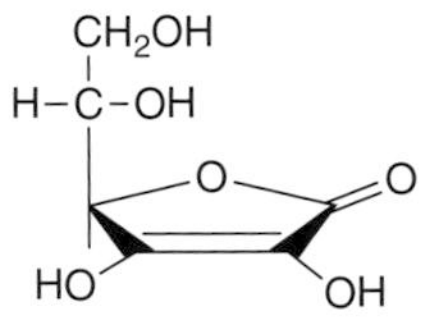

ascorbic acid
(vitamin C)

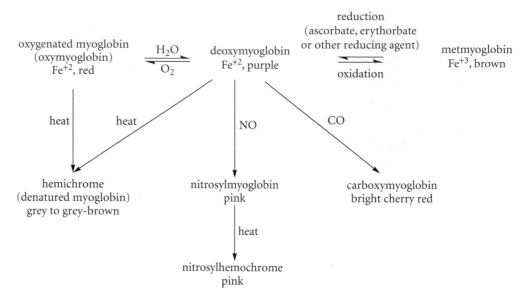

Figure 9.10. A schematic illustration of the various processes that can take place in fresh red meats and their effects on the color of the myoglobin of the meat.

denatures to give **nitrosylhemochrome**, which is also pink/red and stable. Carbon monoxide (CO) poisons myoglobin by binding irreversibly to it to give the cherry-colored **carboxymyoglobin**. This is why breathing carbon monoxide can be quickly fatal and requires that cooking and other forms of combustion must always be carried out under conditions of surplus oxygen to ensure complete combustion to CO_2. The development of the cherry red color of carboxymyoglobin in cooked meats can take place, for example, when meat is cooked under insufficient oxygen. Figure 9.10 summarizes the different processes that can affect the state of myoglobin in meats and the colors that the products exhibit (Figure 9.11).

Chlorophyll

Chlorophyll is the light-gathering molecule found in the chloroplasts of green plants that captures visible light and uses it to synthesize glucose from CO_2 and water, releasing oxygen (O_2) as a by-product (Chapter 4). It is noncovalently bound to the proteins of the photosystem II complex in plants or the photosynthetic reaction center of purple bacteria. The photosystems are membrane-bound complexes of enzymes and cofactors in the thylakoid membranes of the chloroplasts of green plants. Chlorophyll is responsible for the green color of leaves and green fruits. It is a tetrapyrrole pigment similar to the heme porphyrin group, but is not actually a porphin because one of the pyrrole groups is reduced and contains only a single double bond (Figure 9.12). Such a ring structure is called a **chlorin ring**. The chlorin ring has extensive electron delocalization through the conjugated double bonds, but

Figure 9.11. Left: a fresh rib eye steak, showing the characteristic bright red color of oxygenated myoglobin (oxymyoglobin); right: the same cut of meat after storage at 4 °C in a sealed impermeable plastic storage bag for one week, illustrating the characteristic brown color of metmyoglobin which gradually developed as the iron atoms of the myoglobin were oxidized to Fe^{3+} in the absence of liganded oxygen.

is not completely aromatic through the entire ring due to the presence of the single bond in the fourth pyrrole ring. In chlorophyll, the chlorin ring is complexed with a magnesium atom. The chlorin ring is a dicarboxylic acid in which both acid groups have been esterified, one with a methanol and one with the much longer phytol chain (see Figure 9.12). Because of the presence of the hydrophobic phytol chain, chlorophylls are lipophilic and insoluble in water, but can be extracted from their protein complexes and dissolved in organic solvents such as methanol or acetone. The structure of the cyanobacterial photosynthetic reaction center has been determined by X-ray diffraction; it was the first membrane-bound protein complex to have its conformation determined by crystallography, for which H. Michel, J. Deisenhofer, and R. Huber were awarded the Nobel Prize in 1988.

Chlorophyll absorbs light in the red and blue-violet regions of the visible spectrum, so that the light it reflects is green or green-yellow, from the middle portion of the visible spectrum. The green color of chlorophyll is so intense that it can mask the hues of other pigment molecules present in leaves, such as xanthophylls or quercetin (see below). In the autumn, when the production of chlorophyll stops and that already present deteriorates, the colors of these other pigment molecules are revealed, and the leaves change to yellow, red, orange, and even purple. There are two types of chlorophyll, a and b, which differ in their substituents at the 3 position on the second or "B" pyrrole ring (see Figure 9.12), with chlorophyll a having a methyl group at this position and chlorophyll b having an aldehyde group. These two molecules differ slightly in their absorption spectra. Chlorophyll a, the more common form, has absorption maxima at 428 and 660 nm in diethyl ether solution and is blue-green, while chlorophyll b has maxima at 452 and 642 nm and is more yellow-green. The different absorption maxima of these two forms allows a wider range of the visible spectrum to be utilized for photosynthesis.

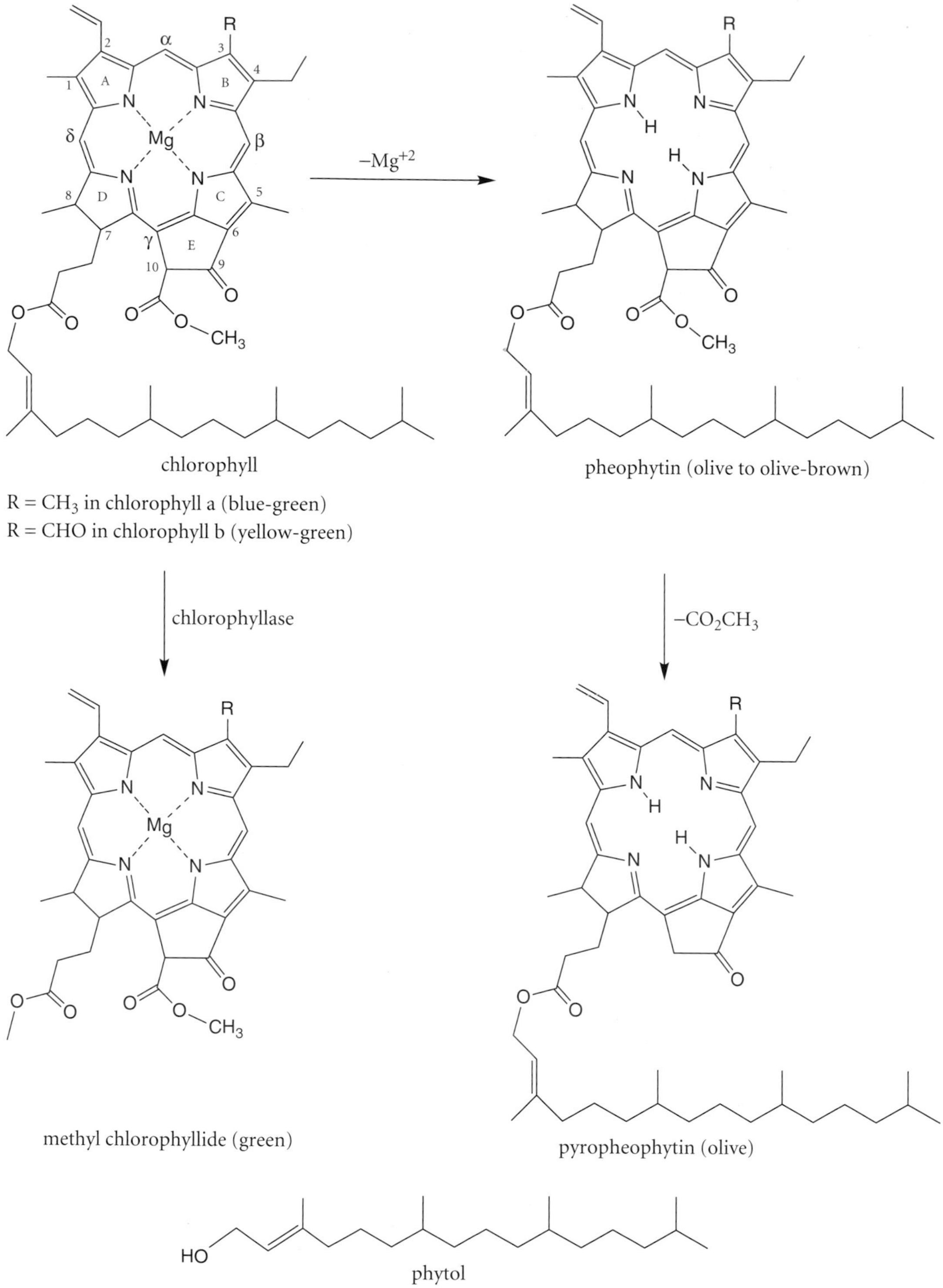

−Mg+2
chlorophyll
R = CH3 in chlorophyll a (blue-green)
R = CHO in chlorophyll b (yellow-green)
pheophytin (olive to olive-brown)
chlorophyllase
−CO2CH3
methyl chlorophyllide (green)
pyropheophytin (olive)
HO
phytol

Figure 9.12. The structures of chlorophyll and related molecules; on the bottom is the structure of phytol.

Chlorophylls are sensitive to acid degradation, but are protected in the chloroplast by the complexing proteins. Chlorophylls are also degraded enzymatically by **chlorophyllase**, which catalyzes the cleavage of the phytol chain from the porphyrin ring to produce **chlorophyllides**, which are still green. This enzyme is inactivated by heating above 100 °C. Due to the removal of the hydrophobic phytol chain, the chlorophyllides are water soluble.

When chlorophyll in plant tissues is heated in the process of cooking at acid pHs, the central magnesium ion of the complex is lost and replaced by hydrogen to produce **pheophytin**. This molecule is less green, and more an olive-brown in color. Chlorophyll's absorption maximum in the red shifts more to the red in pheophytin, and its absorption maximum in the blue shifts more to the blue, producing the observed change in hue. This reaction is irreversible in aqueous solution. Chlorophyll b is more heat stable than chlorophyll a. Chlorophyll conversion to pheophytin is sensitive to pH, because at pH 9 the reaction does not proceed. When heated at high temperatures, the CO_2CH_3 group on C10 is replaced by a proton to give methanol, CO_2, and pyropheophytin, which is also olive colored, and which has the same maxima for adsorption in the red and blue as pheophytin.

The two protons that replace the magnesium atom in the tetrapyrrole ring in the pheophytin molecule are easily displaced by zinc and copper ions to give green complexes that are stable under acid conditions. Copper complexes of this type are used as food colorant additives in the European Union, but are not permitted in the United States.

The loss of green color due to the formation of pheophytin and pyropheophytin is a problem in the thermal processing of vegetables. Blanching, which is necessary to inactivate enzymes even in frozen vegetables, can result in almost complete conversion of chlorophyll to pheophytin. This process occurs even without the addition of acid due to the low pH resulting from cellular acids.

Several approaches are used to try to control the loss of green color from chlorophyll destruction in the processing of green vegetables such as spinach. Because the loss of magnesium occurs at low pH, one approach is to maintain or raise the pH by the addition of compounds such as calcium oxide and sodium dihydrogen phosphate to the blanch water. Magnesium carbonate or sodium carbonate combined with sodium phosphate have also been used. These approaches cause the softening of the plant tissues and can produce an alkaline flavor. Sterilization by higher temperatures for a shorter time (HTST treatment) is thought to result in less initial color loss, but for long-term storage this advantage is offset by a decrease in product pH. Processing at lower than normal temperatures has also been tried, to allow chlorophyllase to convert more of the chlorophyll to chlorophyllides, in the belief that these compounds are more thermostable, but the improvement in color retention from this approach is not great. In Europe, commercial colorants derived from zinc or copper complexes with chlorophyllin (the chlorophyllide resulting from the removal of the phytol chain from chlorophyll by chlorophyllase) are used to improve the green color of processed vegetables, but this is not allowed in the United States. In the United States, a process of adding zinc chloride to the blanching solution in order to produce zinc pheophytin in situ has been developed and patented.

CAROTENOIDS

Carotenoids are widespread plant pigments (Figure 9.13), generally consisting of symmetrical polymers of **isoprene** units, often containing two β-ionone rings (without the ketone groups; see Figure 9.14). They are fat-soluble, intensely yellow, red, or orange colors manufactured only by plants, but which are found in animal fat as a result of being consumed as part of the diet. Many of those carotenoids that contain at least one β-ionone ring exhibit so-called provitamin A activity, because under enzymatic cleavage they can yield a vitamin A molecule.

The colors of these molecules result from their systems of conjugated double bonds; the more of these conjugated double bonds that are present, the further the

Figure 9.13. A summertime open-air market in Paris. The asparagus and green beans are green due to chlorophyll. The cherries are colored red primarily because of peonidin and cyanidin, which also contributes to the red of the strawberries, along with pelargonidin. The tomatoes, however, are red due to the presence of the carotenoid pigment lycopene, while the oranges are colored by several carotenoids, including lutein, β-carotene, and β-apo-8′-carotenal. The white asparagus lack chlorophyll because they were grown in the dark.

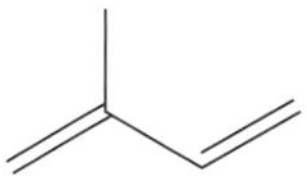

Figure 9.14. Isoprene and β-ionone.

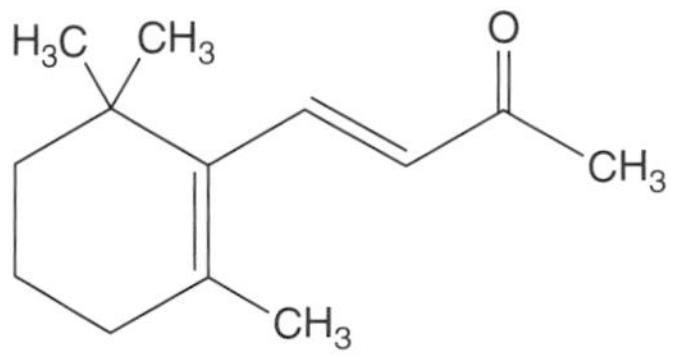

isoprene

β-ionone

absorption bands will be shifted to longer wavelengths, so the more red the molecule will appear. Generally, around seven conjugated double bonds are required before a detectable yellow color is present. Food carotenoids are usually all *trans*. All-*trans* compounds have the deepest color; *cis* double bonds lighten the color. The absorption maxima of carotenoids will also be affected by their environment, such as solvent, cosolutes, or by binding with other molecules, as with astaxanthin in crustaceans (see Figures 9.15 and 9.17)

The carotenoids are generally classified into two main categories, **carotenes** and **xanthophylls**. The carotenes are pure polyene hydrocarbon chains, while the xanthophylls contain oxygen atoms as hydroxyl, carbonyl, or epoxy groups. Both types of carotenoids are insoluble in water, but are soluble in fats and oils and nonpolar organic solvents.

The archetypal food carotenoid **β-carotene** (see Figure 9.16) is found in carrots, where, as its name would imply, it is the predominate carotenoid, and many other foods as well. β-Carotene contains two 6-atom ionone rings, symmetrically placed at each end. As already noted, many carotenoids contain ionone rings or other six-carbon atom rings, but these rings are not aromatic because they contain one or at most two double bonds. β-Carotene contains eleven conjugated double bonds and is yellow to orange in color. The absorption shift produced by this many conjugated double bonds gives β-carotene an absorption maximum at 451 nm, in the blue-violet. Absorption in the blue-violet allows the reflection of its complementary colors, yellow-orange (Figure 9.4), which is why β-carotene has this color. The astaxanthin π

Figure 9.15. "Downeast" lobster feast, Fisherman's Festival, Stonington, Maine. The red of the boiled lobster comes from the carotenoid astaxanthin, while the yellow color of the corn is due to another carotenoid, zeaxanthin.

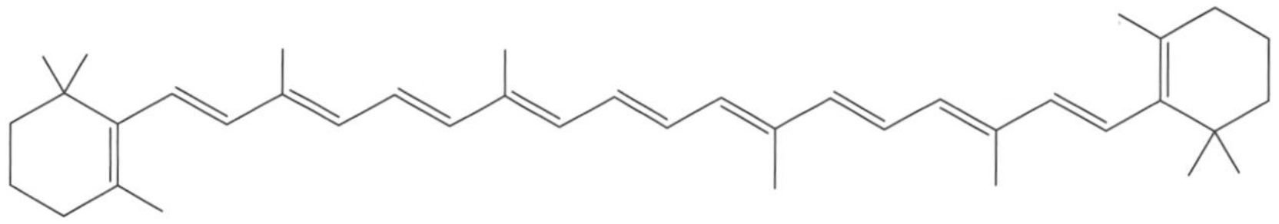

β-carotene, $C_{40}H_{56}$ (light yellow to orange)

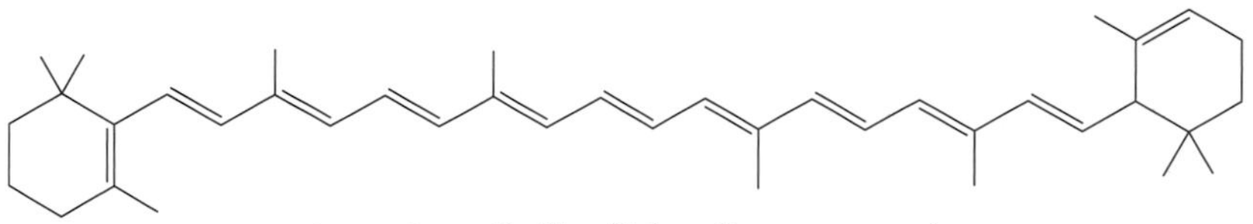

α-carotene, $C_{40}H_{56}$ (light yellow to orange)

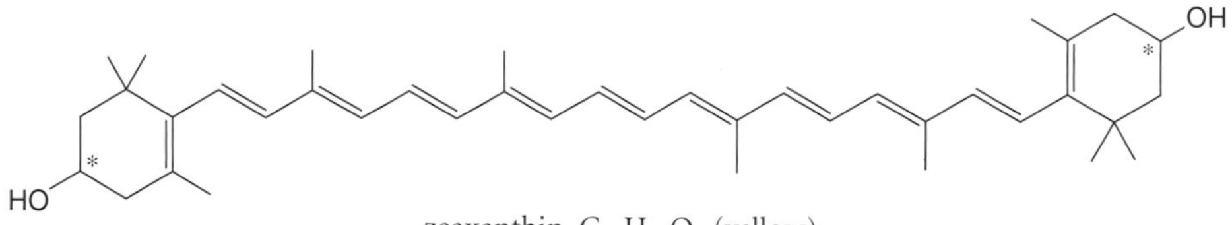

zeaxanthin, $C_{40}H_{56}O_2$ (yellow)

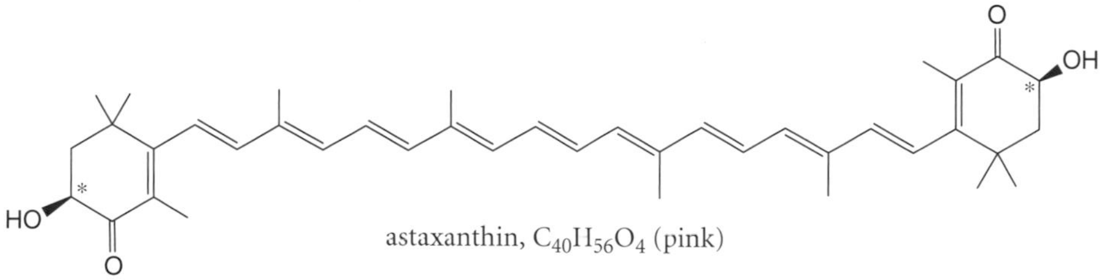

astaxanthin, $C_{40}H_{56}O_4$ (pink)

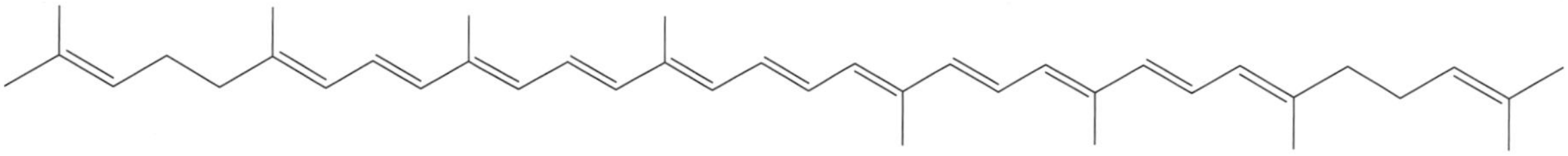

lycopene, $C_{40}H_{58}$ (red)

Figure 9.16. The structures of several carotenoid pigments.

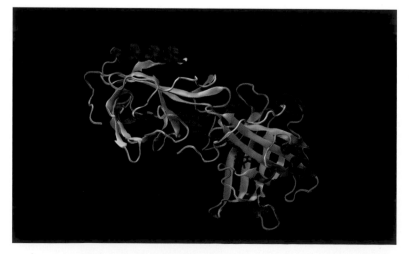

Figure 9.17. The crystallographically determined structure of a dimeric unit of the protein β-crustacyanin complexed with two molecules of astaxanthin, shown in red—one for each protein of the dimer. The alpha helices are indicated in purple, the beta sheets in yellow, and the reverse turns are shown in blue. The color of this complex is dark blue to black.

bond system is extended by two more double bonds and absorbs in the green, which is the complement of red, so that this molecule is pink/red. β-Carotene has an asymmetrical structural isomer, α-carotene, in which the position of the double bond in one of the ionone rings is shifted over one bond, so that it contains only ten conjugated double bonds.

Lycopene, found in abundance in tomatoes, watermelons, red peppers, and some other fruits, is red and is responsible for the deep red color of these fruits when ripe. This carotenoid consists of eight isoprene units and 13 double bonds, but has no β-ionone rings. Note, however, that the terminal double bonds in lycopene are not part of the conjugated system, which thus extends only over 11 double bonds. All of the double bonds in lycopene have the *trans* configuration. Although a 40-carbon carotenoid like β-carotene, without any rings, it has a smaller end-on cross section and thus can more readily insinuate itself into the pores in the polymer matrices of plastics, explaining why lycopene in tomato paste more readily colors plastic containers than other carotenoids. Because it does not possess a β-ionone ring, lycopene does not have any provitamin A activity. It is, however, extremely effective as an antioxidant and a singlet oxygen quencher, and thus is considered to be a highly valuable component in the diet. As one might expect from its exclusively hydrocarbon composition, lycopene is completely insoluble in water, in spite of its abundance in fruits with high water contents. In the tissues of these fruits, the lycopene is strongly bound up with the cellular fiber, which reduces its bioavailability. Processing and cooking operations that disrupt the cellular structures make the lycopene more bioavailable, and cooking with oils, as in spaghetti sauces and other such sauces, significantly increases the bioavailability of lycopene.

Astaxanthin and **zeaxanthin** are examples of xanthophylls, partially oxidized carotenoids containing hydroxyl or carbonyl groups. As the result of the presence of these functional groups containing oxygen atoms, the xanthophylls are more polar than other carotenoids, but are still insufficiently polar to be soluble in water; like other carotenoids, they are soluble in fats. As is true for all carotenoids, these molecules are produced only by plants, even though some can be found in animal foods as the result of their presence in the foods consumed by the animals. Many, but not all, of these molecules are yellow in color. Zeaxanthin is yellow, and is responsible for the yellow color of corn (*Zea mays*), from which it takes its name. (The blue color of the increasingly popular blue corn does not come from carotenoids, but rather from anthocyanins, discussed in the next section). Lutein, shown in Figure 9.19, is also yellow, while other xanthophylls are orange, pink, or violet.

Astaxanthin, 3,3′-dihydroxy-β,β′-carotene-4,4′-dione, shown in Figure 9.16, is produced in large amounts by marine algae and obtained in the human diet primarily from marine animal foods. Two chiral carbon atoms occur in the molecule, indicated in Figure 9.16 by asterisks, at the 3 and 3′ positions in the rings. There are thus three possible stereoisomers of this xanthophyll, with the configurations (3R,3′R), (3R,3′S), and (3S,3′S). The S,S-configuration is shown in Figure 9.16. Astaxanthin is orange to pinkish red and is the source of the red color of cooked shrimp and lobster. In the live animals, however, the color is usually masked because the molecule is bound to the protein α-crustacyanin, which consists of eight heterodimeric pairs of 40-kDa globular β-crustacyanin units, with each of these sixteen subunits bound to a single astaxanthin molecule (Figure 9.17). This binding results in a complex with a black color, but which denatures when cooked, releasing the astaxanthin, shifting its absorption maximum and allowing it to regain its red color (Figure 9.18). It has one of the highest antioxidant capacities of any of the carotenoids.

Canthaxanthin (Figure 9.23), which lacks the hydroxyl groups of astaxanthin, is found in Chanterelle mushrooms, which get their color from this carotenoid, as well as in some crustaceans and fish. Astaxanthin and the related canthaxanthin are also responsible for the pink color of wild salmon and flamingos as a result of the presence of these molecules in the shrimp and similar small crustaceans in the diets of these animals (Figure 9.18). Flamingos kept in zoos and fed nonmarine bird food lose their pink color and are white. Similarly, farmed salmon that are fed various forms of artificial protein meal are not pink, and are thus less acceptable to consumers. To overcome this problem, the feed for farmed fish is usually supplemented with carotenoid pigments, which must now be labeled as an artificial coloring. Sometimes this added astaxanthin is synthetic, in which case all three stereoisomers are present, in the proportions 1:2:1, with twice as much of the mixed configuration. Canthaxanthin is also sometimes added to poultry feed to produce eggs with more intensely colored yolks. Because neither astaxanthin nor canthaxanthin contain an unmodified ionone ring, neither has any provitamin A activity.

Lutein, shown in Figure 9.19, is a xanthophyll closely related to zeaxanthin, differing from it only in the location of one of the ring double bonds (note that as a result, lutein is not symmetrical, unlike zeaxanthin). It contains three chiral carbon atoms, indicated in the figure by asterisks, with the naturally occurring form having the R configuration at both the 3 and 3′ positions. Lutein, which is also yellow, is the

Figure 9.18. (a) Cooked shrimp; (b) fresh salmon.

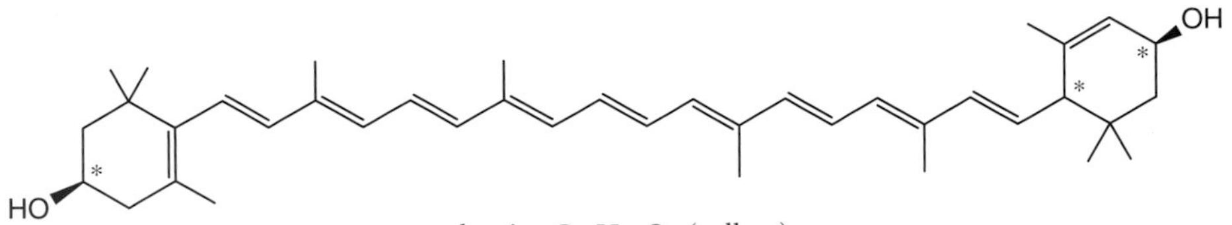

lutein, $C_{40}H_{56}O_2$ (yellow)

Figure 9.19. The yellow-colored xanthophyll lutein is found in egg yolks and in oranges and orange juice. Chiral carbon atoms are indicated by asterisks.

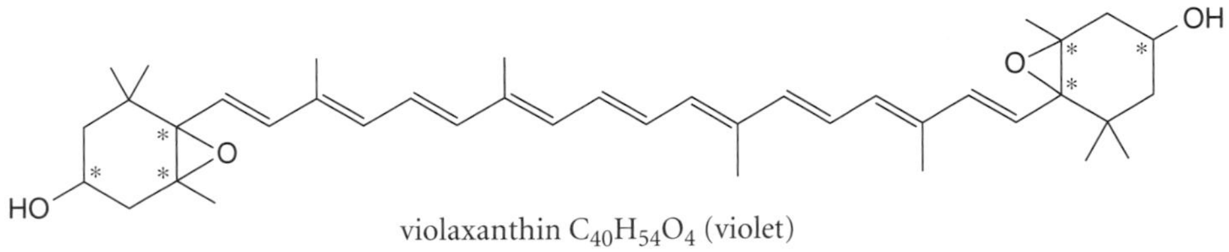

violaxanthin $C_{40}H_{54}O_4$ (violet)

Figure 9.20. Violaxanthin is an example of an epoxy xanthophyll. Six chiral carbon atoms are indicated by asterisks.

principal pigment of egg yolks, where it is found with zeaxanthin, and is also found in orange juice and jalapeño peppers. Present in the leaves of trees, lutein is masked by the green of chlorophyll until autumn, at which time chlorophyll synthesis ceases, the existing supply breaks down, and the underlying more stable xanthophyll pigment colors are revealed. This pigment is also present in the tissues of other leafy vegetables such as spinach. Like canthaxanthin, it is sometimes fed to chickens to produce more intensely colored yolks in their eggs.

Violaxanthin, shown in Figure 9.20, is another variant on the carotenoid pattern that differs in that it has epoxy oxygens on each of its rings. This orange-colored pigment is found in many plant leaves and in orange juice and jalapeño peppers. The presence of the epoxy groups increases the number of chiral carbon atoms in this xanthophyll to six (with the configuration 1S,4R,6R in both rings), while making this molecule more susceptible to acid-induced degradation during processing.

Capsanthin (Figure 9.21) is the xanthophyll pigment responsible for the red color of paprika and red peppers. It is sensitive to destruction by oxidation, and this is partly responsible for the gradual change in the color of stored paprika from red to brown. (Maillard browning also contributes to this browning process.)

While they are natural pigments, three of the carotenoids, β-carotene, β-apo-8′-carotenal (shown in Figure 9.22, found naturally in the skins of citrus fruits such as oranges and tangerines), and canthaxanthin (Figure 9.23), are also produced synthetically and used commercially as food colors. Because they are considered to be "nature identical" molecules, they are exempt from certification. They are used to color products such as margarine, ice cream, cheeses, sauces, confectionery, and bakery products. They are even used to ensure consistent color in butter.

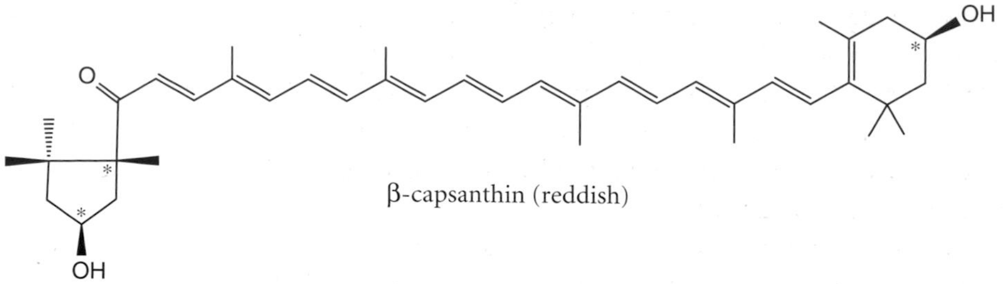

Figure 9.21. The structure of β-capsanthin from peppers. Three chiral carbon atoms are indicated by asterisks.

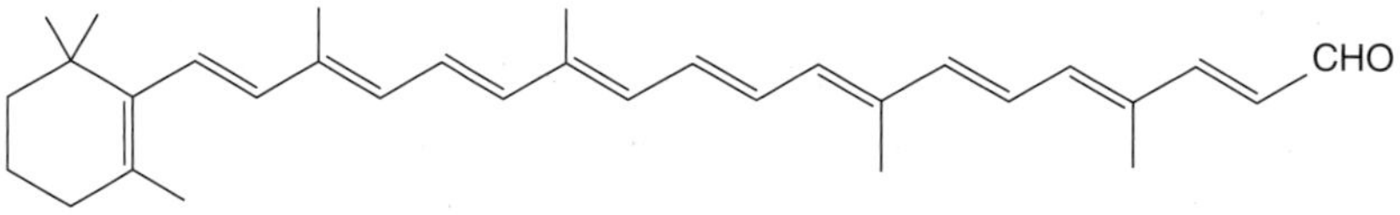

β-capsanthin (reddish)

Figure 9.22. The structure of β-apo-8′-carotenal from peppers.

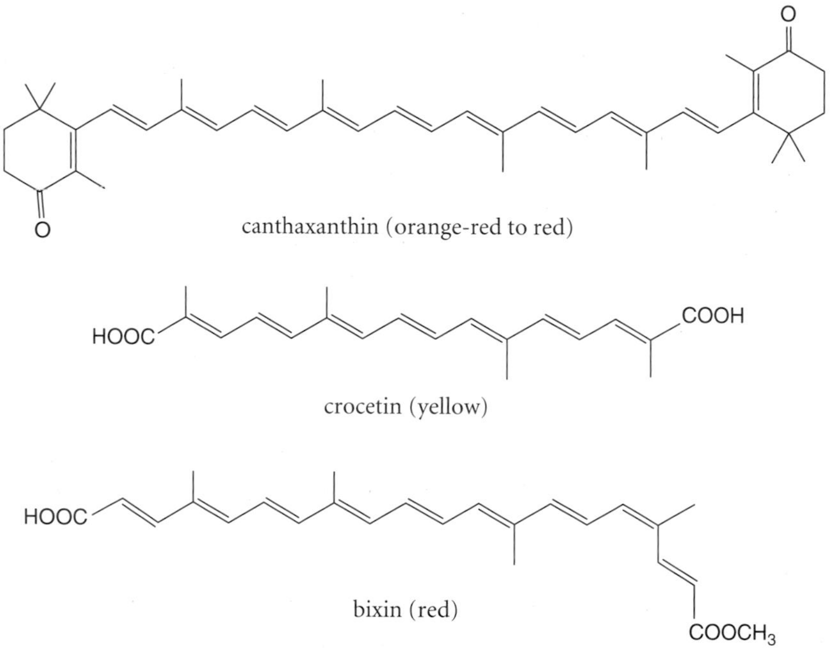

β-apo-8′-carotenal (light orange to reddish orange)

Figure 9.23. The carotenoids canthaxanthin, crocetin (from saffron), and bixin (from annatto), all from natural sources, are used as colorants and do not require certification as additives in the United States. Canthaxanthin is also produced synthetically, but is considered by the FDA to be a "nature identical," still exempt from certification.

canthaxanthin (orange-red to red)

crocetin (yellow)

bixin (red)

Several other carotenoids from natural sources are used as colorants (Figure 9.23), and do not require certification as additives (Chapter 10). **Crocetin**, which is responsible for the yellow color of saffron, is extracted from the stigmas of the crocus flower, *Crocus sativus*, and is used to color saffron rice (saffron is the most expensive common food ingredient by weight, since enormous numbers of these stigmas must be gathered, generally by hand). In the plant it is found as a diglycoside, with both carboxylic acid groups esterified with molecules of the disaccharide gentiobiose (the β-(1→6)-linked disaccharide of glucose; see Figure 4.48). In this form, the pigment is called crocin and is water-soluble. **Bixin** is the main carotenoid pigment of **annatto**, which is extracted from the seeds of a tropical shrub, *Bixa orellana*, found in the Caribbean. Unlike the other commonly occurring carotenoids in foods, bixin contains a *cis* double bond. Bixin is red-colored (as is annatto).

Carotenoids are reasonably stable during common unit operations of food processing, but because of their highly unsaturated structures, are very susceptible to oxidation. Because these molecules adsorb visible and UV light, they are broken down by prolonged exposure to light and can be destroyed by acids. They can also be destroyed by excessive high heat treatment, but most are fairly stable up to the boiling point of water. In intact cells, the reducing environment of the cytoplasm protects carotenoids from oxidation. Their stability in dried foods is poor, but in general they are stable in frozen foods and during heat sterilization. Because they are easily oxidized, carotenoids can serve as antioxidants, and diets rich in these molecules are thought to be beneficial in helping to prevent cardiovascular disease. They quench singlet oxygen and can inhibit lipid peroxidation. Lipoxygenases can catalyze the oxidative decomposition of carotenoids, and blanching, which denatures and inactivates this enzyme, can lead to increased levels of carotenoids. Because boiling does not destroy carotenoids but does break down the cellular structures in which they are found, the bioavailability of carotenoids from boiled or steamed vegetables is generally greater than from raw ones. In general, many of the carotenoids of vegetables are not bioavailable when they are eaten raw by themselves, because these molecules are insoluble in water. Thus, eating raw tomatoes or carrots with a little olive oil dressing in the salad to solubilize these nutrients might significantly increase their bioavailability.

Provitamin A Activity of Carotenoids

β-Carotene exhibits provitamin A activity because it consists of two retinol, or vitamin A, molecules linked together. As the most common of the carotenoids in plant foods, is β-carotene an important source of this vitamin. In order to have vitamin A activity, a carotenoid must have at least one β-ionone ring. Thus, while lycopene is an extremely good antioxidant, it has no vitamin A activity because it lacks an ionone ring. The β-carotene molecule can be split enzymatically into two retinol molecules by carotenoid oxygenase in the intestines (Figure 9.24). In humans, however, only about half of the adsorbed β-carotene is converted to vitamin A. The vitamin A value of fruits and vegetables (see Table 9.1) is further reduced by the fact that only about one-third of the amount that is consumed is adsorbed across the intestinal lumen. Carrots, spinach, and sweet potatoes have the highest concentrations of β-carotene. Consumption of too much of these foods, as sometimes occurs in young children, can lead to high concentrations of carotenoids in the blood and to a yellowing of the skin in a mild condition called carotenaemia.

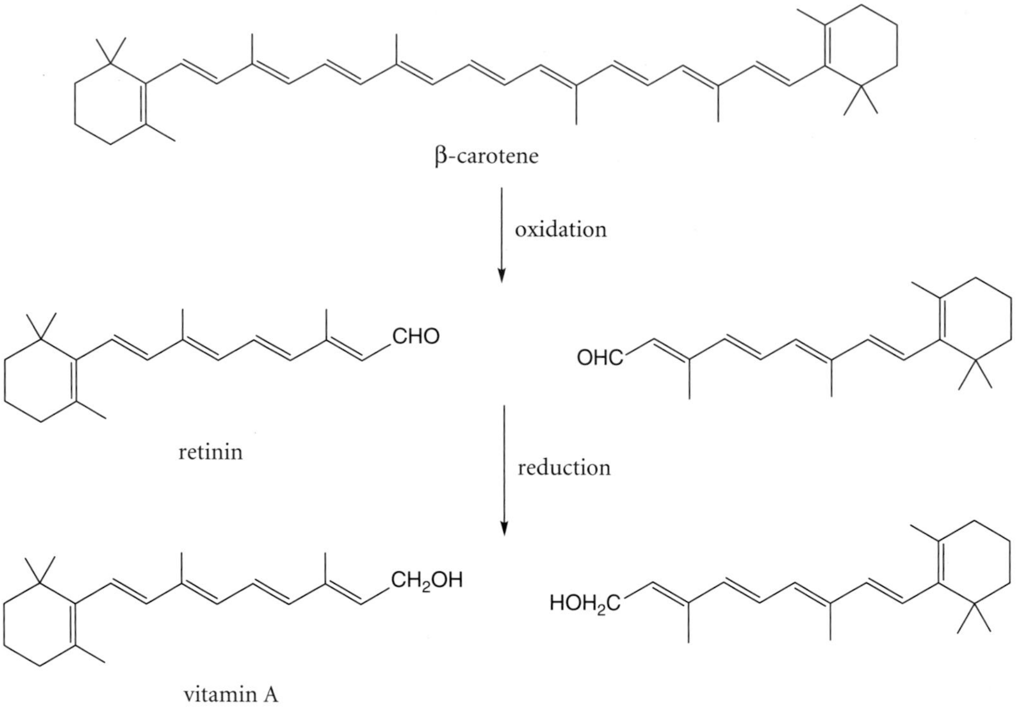

Figure 9.24. The provitamin A capacity of β-carotene.

Table 9.1 Provitamin A values for selected fruits and vegetables

Source	IU/100 g
mature carrots	20,000
spinach	13,000
sweet potato	6000
apricots	2000
tomatoes	1200
peaches	800
cabbage	500
bananas	400
orange juice	200

Source: Data from B. Borenstein and R.H. Bunnell, "Carotenoids: Properties, Occurrence, and Utilization," in *Advances in Food Research*, vol. 15, C.O. Chichester et al., eds. (New York: Academic Press, 1967).

Figure 9.25. The crystallographically determined conformation of the enzyme caretenoid oxygenase from cyanobacteria of the genus *Synechocystis*, with an uncleaved β-carotene substrate molecule bound in the actuve site tunnel. Alpha helices are shown as purple cylinders, 3_{10} helices as blue cylinders, and beta sheets as yellow arrows. (Kloer et al., *Science* 308:267–269, 2005.)

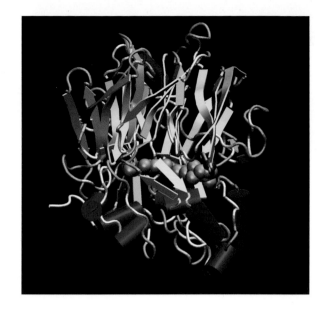

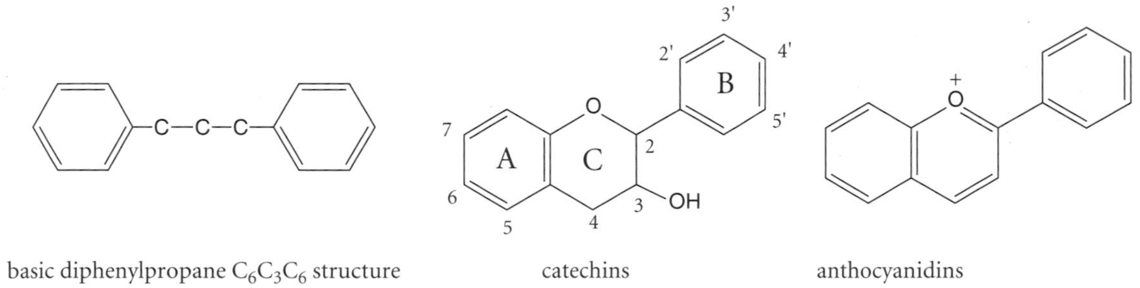

basic diphenylpropane $C_6C_3C_6$ structure   catechins   anthocyanidins

Figure 9.26. Flavonoids are built off of a basic diphenylpropane unit (left), usually with the linking propane chain and one of the rings linked into a fused bicyclic structure through bonding to an oxygen atom to give a prototypical three-ring skeleton, as in the catechin illustrated in the center, with the conventional atomic numbering indicated; on the right is shown the backbone structure of the anthocyanidins.

The structure of the human caretenoid oxygenase enzyme that cleaves β-carotene to produce Vitamin A has not yet been reported, but the conformation of the enzyme from a cyanobacterium has been determined (Kloer 2005; Figure 9.25).

FLAVONOIDS

Flavonoids are a class of polyphenolic compounds that share a common diphenyl-propane structure of two aromatic rings linked together by a three-carbon chain, as in the structure on the left in Figure 9.26. Often, this linker chain, along with an oxygen atom, forms a third ring as in the structures on the right. These molecules are often polyphenolic because the aromatic rings contain hydroxyl substituents, usually at the 4′ and 7 positions of the B and A rings, respectively.

The flavonoids are categorized into groups such as catechins, anthocyanidins, flavones, or flavonols, depending on the oxidation state of the three-carbon linker between the two aromatic rings (next section). They are important in plant foods as antioxidants, and diets rich in these molecules are thought to be particularly healthy. In some of the flavonoids in which the three-carbon linker sequence forms a heterocyclic ring, double bonds in this C ring allow an extended conjugated sequence linking the two aromatic rings, with electronic transitions in the visible range of the spectrum giving rise to colored molecules. One of the most important classes of such polyphenolic pigments are the anthocyanins, found in many fruits and in red wines.

Anthocyanins

Anthocyanins are glycoside pigments found in plant cells that are responsible for the colors of many flowers, fruits, and vegetables. These molecules generally have structures of the type shown in Figure 9.27 for cyanin, which gives the class its name. In these glycosides the glycosyl group is often glucose, α-linked to the 3-position hydroxyl on the C ring of the catechin aglycon.

When the glycosidic bond is hydrolyzed to free the sugar molecule, the remaining aglycon is called an **anthocyanidin** (Figure 9.28).

Anthocyanins generally have a lower absorption maximum than the corresponding anthocyanidins, as can be seen from Table 9.2, in which the R group refers to the structure in Figure 9.28, where the different molecules will have different groups at R1, R2, and R3 (see Figure 9.29).

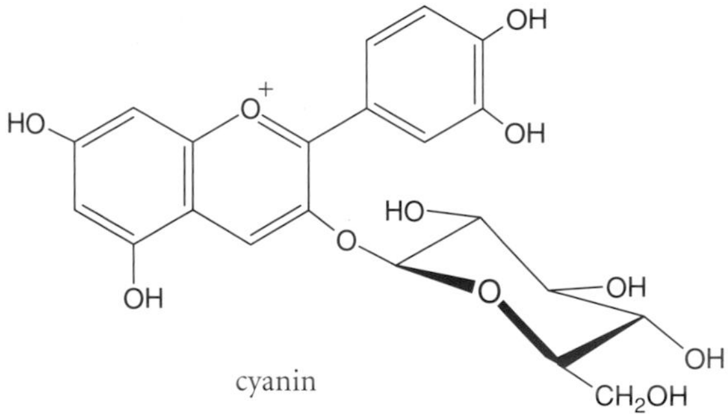

Figure 9.27. The structure of cyanin.

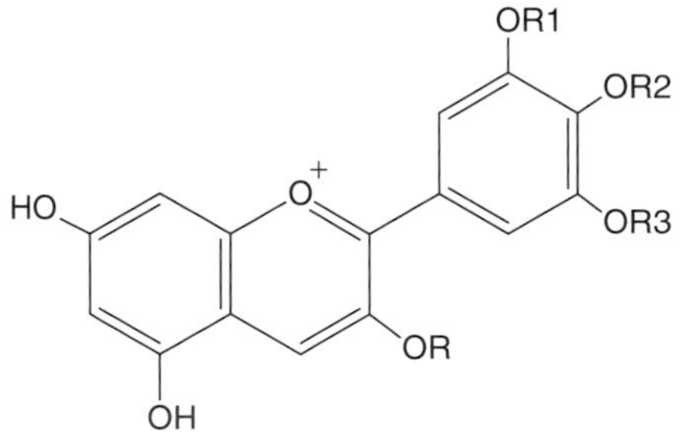

cyanin

Figure 9.28. The template structure for the anthocyanidins; see Figure 9.29.

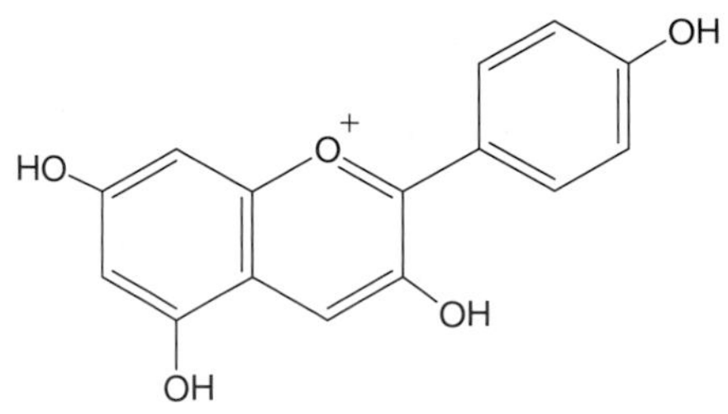

pelargonidin (scarlet)

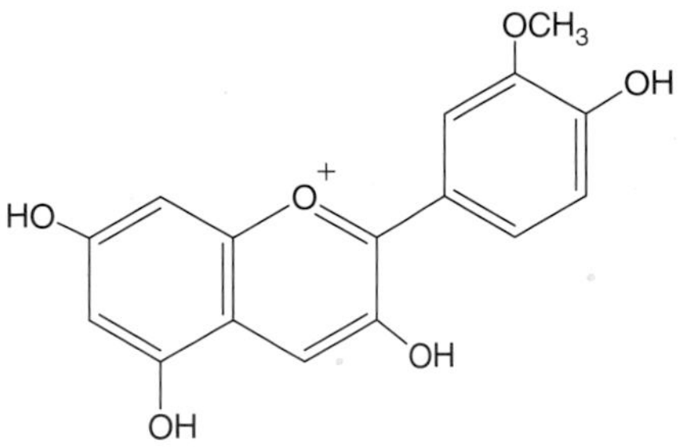

peonidin (purple-red)

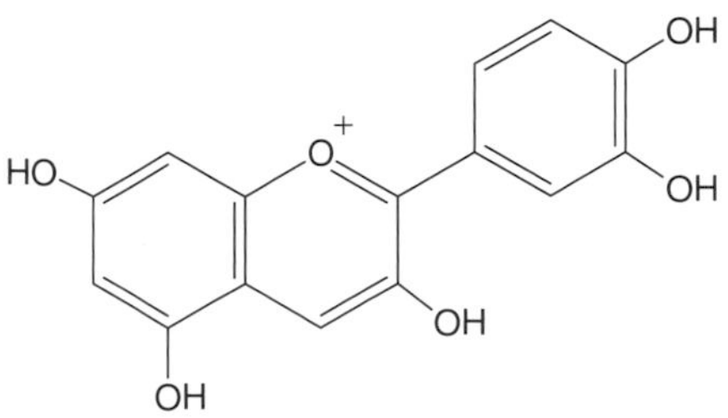

cyanidin (crimson)

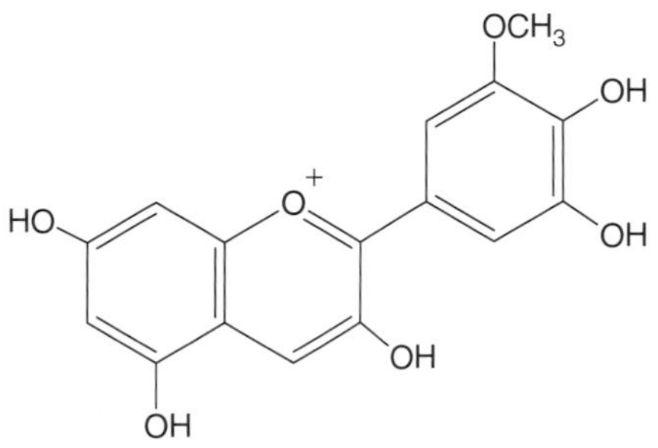

petunidin (red-purple)

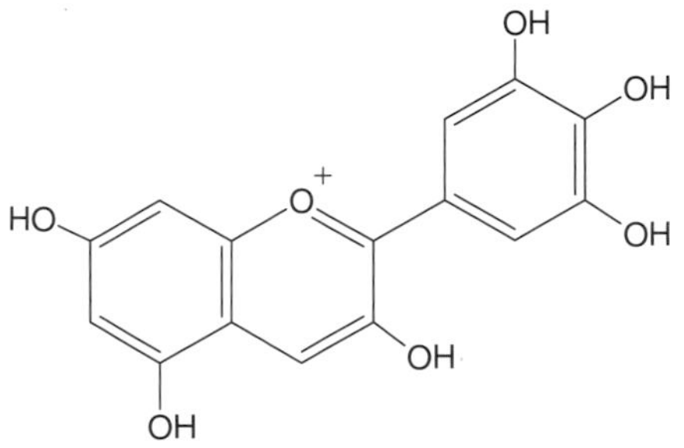

delphinidin (blue)

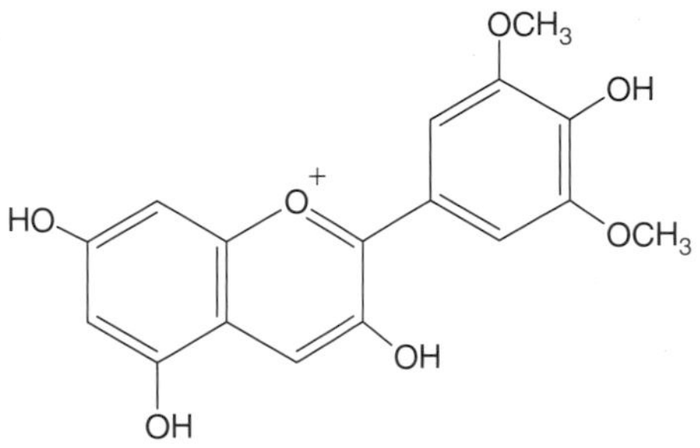

malvidin (red)

Figure 9.29. The structures of the common anthocyanidins.

Table 9.2 Absorption maxima of the common anthocyanins and their corresponding anthocyanidins

	λ_{max} (nm) in methanol with 0.01% HCl	
Molecule	**R = H**	**R = glucose**
pelargonidin	520	506
cyanidin	535	525
peonidin	532	523
delphinidin	544	535
petunidin	543	535
malvidin	542	535

Source: H.-D. Belitz, W. Grosch, and P. Schieberle, *Food Chemistry*, 4th rev. and extended ed. (Berlin: Springer, 2009).

At least sixteen anthocyanidins are known from plant sources, but only six of these occur in foods in significant amounts. These pigments are mostly named after the plants from which they were first isolated; they are pelargonidin, cyanidin, delphinidin, peonidin, malvidin, and petunidin. The structures of these molecules are shown in Figure 9.29.

At low pHs anthocyanins exist as **flavylium cations**, as shown in Figure 9.29, that are generally colored red, with the color shifting toward the blue as the number of substituents on the B ring increases (from pelargonidin to malvidin).

One of the most common of the anthocyanidins is cyanidin, found in fruits such as apples, strawberries, raspberries, blueberries, plums, cherries, and many others as cyanin, the anthocyanin glycoside of D-glucose. The charge state and color of this molecule, as for all of the anthocyanins, are a function of pH. The charged (protonated) cationic or oxonium salt form, a flavylium cation, is colored red, but in aqueous solution this form is in equilibrium with a neutral hemiketal pseudobase, ROH, which is colorless, due to the disruption of the conjugated double bond system (Figure 9.30). As the pH of a solution is raised, more of the pseudobase is formed and the color becomes weaker. Under basic conditions, the proton on the 4′ hydroxyl group is abstracted to give a ketone, and with the expulsion of the hydroxyl group on position 2, leads to the establishment of a new conjugated π system colored blue,

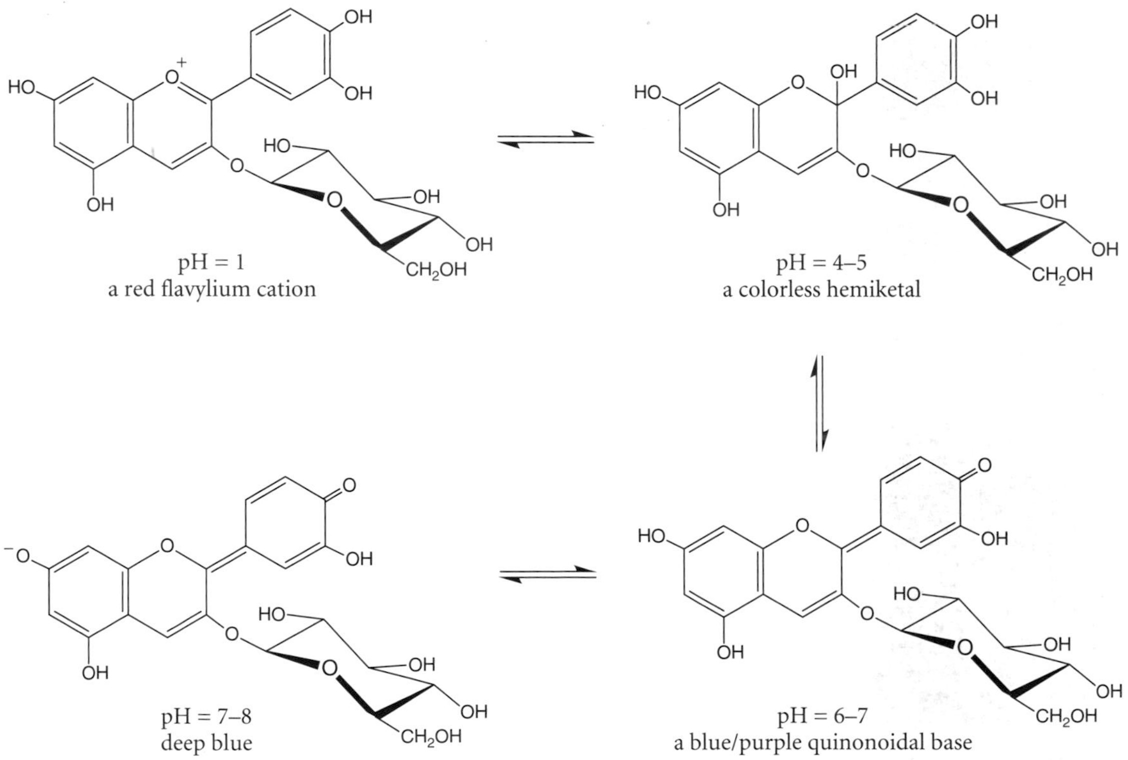

Figure 9.30. The protonation states of cyanin as a function of pH

which deepens as the pH is raised further and the proton of the hydroxyl group on carbon 7 is lost. Thus, in acid solution anthocyanins are colored red, while under basic conditions they are blue (Table 9.3).

As a result of this pH shift, the cyanin molecule illustrated in Figure 9.30 is the primary pigment found in both cornflowers (*Centaurea cyanus*) and poppies (*Papaver rhoeas*), even though these flowers are different in color (Figure 9.31). In cornflowers the sap is alkaline, producing a blue color, while in poppies the sap is acid, giving the familiar red color of these flowers. This molecule is also responsible for the red color of rhubarb because oxalic acid present in this vegetable keeps the cyanin protonated.

Cyanin is also the principal source for the purple/red color of red cabbage. Red cabbage grown on basic soils will be less red, and more green/yellow, due to the disappearance of the flavylium and hemiketal forms. When red cabbage is cooked, it will lose its red color, and may even change color to blue. This is particularly true if it is cooked with baking soda, which neutralizes the acids present in the sap of the raw plant. Pickling red cabbage in vinegar (acetic acid) turns it from purple to strongly red as the pH is lowered by the acid. The juice of cooked red cabbage can be used as a home-made pH indicator in "kitchen chemistry" experiments because of this color sensitivity of anthocyanins to pH. For example, if this juice is added to

Figure 9.31. Poppies and cornflowers, both colored by cyanin, seen here growing in a wheat field in the Roussillon region of Southern France.

egg whites as they are cooked, it will turn them green/blue in color. This is because the albumen proteins that are the primary components of egg whites produce a basic solution.

Along with carotenoids, anthocyanins are responsible for the reds, oranges and purples of maple and oak leaves in autumn, as well as the colors in the petals of many flowers. Anthocyanidins in grape skins are the source of the color in red wines, where they react with tannins as the wines age, reducing their bitter flavor (discussed in more detail later). Anthocyanidins are found in the skins of many other fruits, as shown in Table 9.4. For example, they are responsible for the red color of apples, the blue color of blueberries, and the purple of plums.

Blueberries are particularly rich in anthocyanins (see Table 9.4) and are believed to be one of the healthiest of the fruits due to their high concentration of these antioxidant compounds. These anthocyanins are responsible for another well-known color change in cooking. When blueberries are incorporated into products such as pancakes or muffins that are leavened with baking soda, a greenish color often develops in the matrix adjacent to the berries, or even in the berries themselves (Figure 9.32). This color change is also due to a pH-induced shift from the flavylium to hemiketal and quinonoidal forms in the presence of the alkaline baking soda.

Table 9.3 Absorption maxima of selected anthocyanidins in 0.01% HCl in methanol solution

Anthocyanidin	Visible absorption maximum (nm)	Color
pelargonidin	520	scarlet
cyanidin	535	crimson
delphinidin	544	blue-mauve

Table 9.4 Anthocyanidins commonly found in various fruits

Fruit	Anthocyanidin
apple	cyanidin
black currant	cyanidin and delphinidin
blueberry	cyanidin, delphinidin, malvidin, petunidin, and peonidin
plum	cyanidin and peonidin
cherry	cyanidin and peonidin
grape	malvidin, peonidin, delphinidin, cyanidin, petunidin, and pelargonidin
orange	cyanidin and delphinidin
raspberry	cyanidin
strawberry	pelargonidin and a little cyanidin

Figure 9.32. A commercial blueberry muffin leavened with baking soda, showing the characteristic green color that indicates that the anthocyanins in the vicinity of the berries are in the quinonidal state.

Anthocyanins are powerful antioxidants (see next section) and thus are thought to help prevent heart disease and stroke (Wang et al. 1997). Evidence from experiments on laboratory animals suggest that they can also inhibit the development of various cancers. These molecules thus might be an important component of the supposed health benefits of eating brightly colored fresh fruits, although more exhaustive clinical studies will be needed to convincingly establish their actual effects in the diet, including their bioavailability.

POLYPHENOLIC FLAVONOID ANTIOXIDANTS

The anthocyanins of the previous section are specific examples of a much larger category of important plant molecules called phenolics, which are derivatives of the basic organic template phenol (Figure 8.37). We have also already encountered these molecules when considering flavors, because the curcumin of curry, the caffeic acid of coffee, and the rosmarinic acid of rosemary are all examples of this broad category. Many other examples of this category of molecules are important in foods; Figure 9.33 illustrates the classification scheme used to organize these molecules.

Many of these phenolic compounds are polyphenolics, compounds that contain two or more phenol moieties linked together. The anthocyanins are examples of this category, as are curcumin and rosmarinic acid, but caffeic acid is not. Not all polyphenolics are colored, but many are. These molecules will be considered here whether they are colored or not, due to their chemical similarities.

The flavonoids, to which the anthocyanins belong, are a large and diverse group with a number of classificatory subdivisions (Figure 9.33). Many of the polyphenolic flavonoids have been the subjects of a great deal of interest recently due to their antioxidant capacities. Dietary antioxidants are reducing molecules that react with,

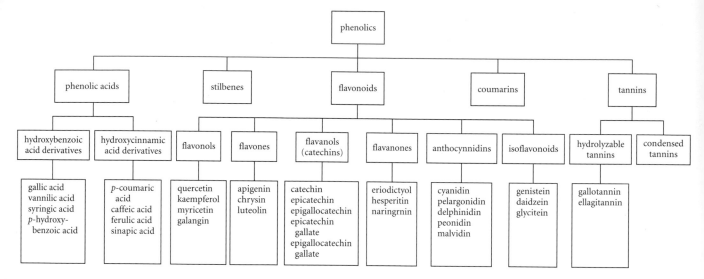

Figure 9.33. The classification scheme for some of the phenolic compounds from plants that are important in foods. (R.-H. Liu, *J. Nutr.* 134:3479S–3485S, 2004.)

and are oxidized by, free radicals, thereby preventing them from oxidizing other more important molecules, such as lipids. These antioxidants have been found to help prevent coronary heart disease and cancer. They function by scavenging free radicals and reducing oxidative stress, inhibiting the oxidation of lipids that contributes to heart disease and the DNA damage that can lead to cancer. Several methods for evaluating the antioxidant potential of compounds in vitro are possible; one of the most widely accepted is the oxygen-radical absorbance capacity (**ORAC**) assay, which monitors spectrophotometrically the fluorescence of phycoerythrin proteins as they are degraded by radicals by determining how various added antioxidants extend the time that the system fluoresces by protecting the proteins (Cao et al. 1993; Huang et al. 2005). Relative antioxidant effectiveness is often expressed as "vitamin C equivalent antioxidant capacity," or **VCEAC** (Kim and Lee 2004).

Flavanols

Nutritionists have focused recent attention in particular on the antioxidant capacity of the **flavanols** (flavan-3-ols) or **catechins**. The structures of several of the catechins are shown in Figures 9.34 and 9.35, including EGC ((–)-epigallocatechin), EC (epicatechin), EGCG ((-)-epigallocatechin gallate), and ECG ((–)-epicatechin gallate). All of these substances are found in various plant materials, with the greatest concentrations being found in pine bark and grape seeds, as well as in many fruits and vegetables. Infusions of plant materials such as teas also contain these compounds in large amounts, which is the reason for the recent flurry of interest in the health benefits of tea. Green teas in particular are a rich source of these compounds, with the

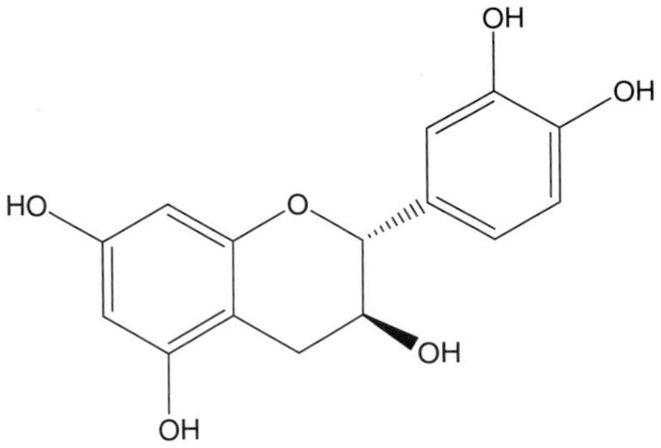

catechin

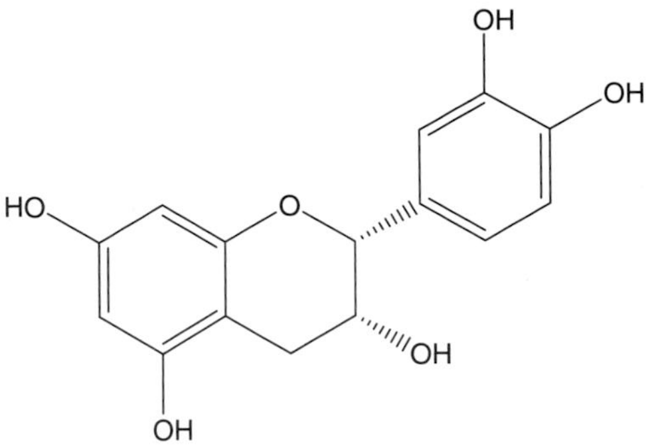

epicatechin
EC

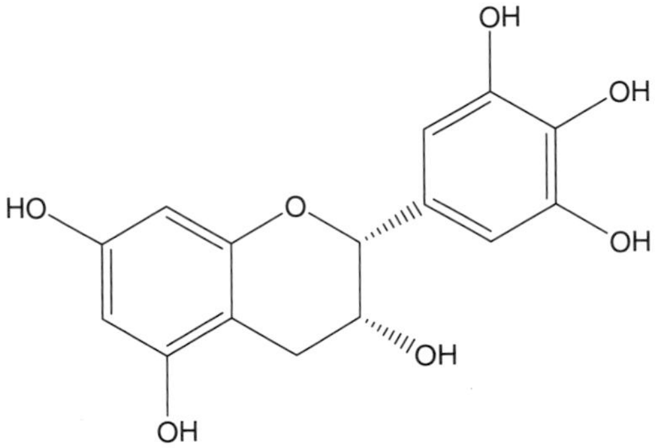

(-)-epigallocatechin
EGC

Figure 9.34. The structures of several of the most important catechins.

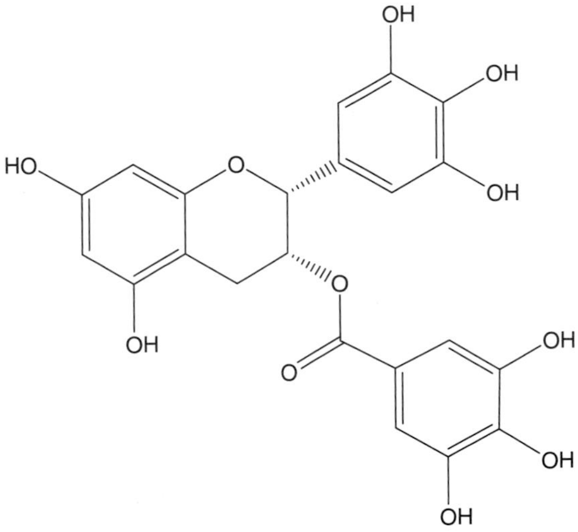

(-)-epigallocatechin gallate
EGCG

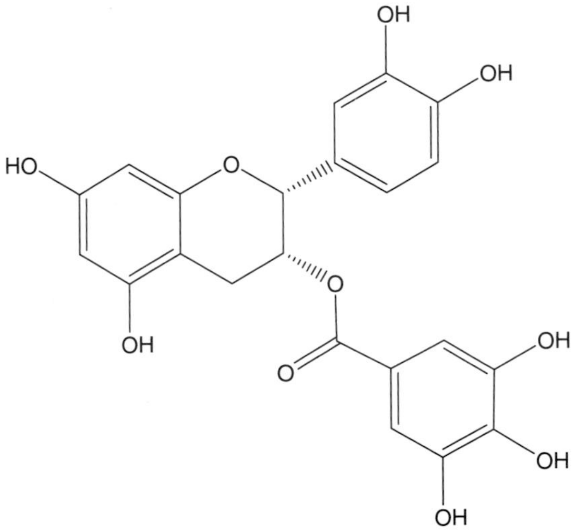

(-)-epicatechin gallate
ECG

Figure 9.35. The structures of additional catechins.

most common antioxidant of this series, EGCG, present in green teas at about five times its concentration in typical black teas. More of these compounds are in green teas than in black teas because they are susceptible to destruction by oxidation during the fermentation process that produces black teas. Unlike the anthocyanidins, the catechins do not occur in plants as glycosides linked to sugars. Also, there is no electronic delocalization between the A and B aromatic rings in the catechins as there is in the anthocyanidins, because the heterocyclic C ring is completely saturated (contains no double bond). For this reason, these molecules are colorless.

Along with their role in preventing heart disease due to their antioxidant potentials, some of these compounds are thought to play a role in preventing certain cancers as well. Part of this anticancer effect is probably due to a reduction in DNA damage caused by excess free radicals. In addition, a recent study of the catechin EGCG has determined that it prevents cancer cell growth by binding to the enzyme dihydrofolate reductase (DHFR) and inhibiting its action (Navarro-Perán et al. 2005). This enzyme is involved in the synthesis of DNA in proliferating cells. The structure of EGCG is similar to that of the anticancer drug methotrexate, which binds strongly to DHFR and leads to the death of cancer cells. EGCG also binds to DHFR, but less strongly, suggesting that this catechin might have less severe side effects for noncancerous cells.

The catechin molecules possess two chiral carbon atoms in their heterocyclic C rings. These two chiral carbons allow four possible stereoisomers if synthesized in vitro, but natural catechins have specific stereochemistries.

Flavanones, Flavones, Flavonols, and Isoflavonols

Other categories of flavanoid antioxidants include the **flavanones** and **flavones** (Figure 9.36), which are similar to the flavanols but differ from these molecules in having a ketone functionality at the 4-carbon position of the C ring. Flavanones contain a single asymmetric carbon in the 2-position of the C ring, while the flavones have an additional double bond in the C ring, with no chiral carbon atoms in the principal template skeleton. When a flavone has a hydroxyl substituent on the

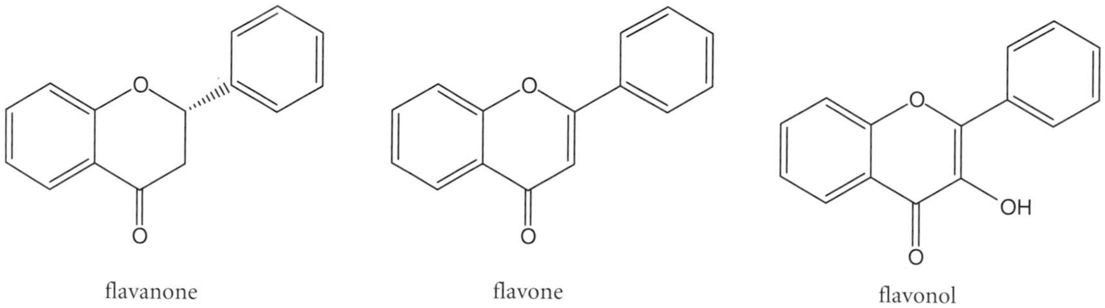

flavanone   flavone   flavonol

Figure 9.36. The structural templates for flavanones, flavones, and flavonols.

3-carbon position of the C ring, the molecule is called a **flavonol** (a 3-hydroxyflavone). The unfortunate similarity of the names of the various flavonoids can cause confusion among nonspecialists and requires particular attention to spelling when discussing these molecules.

The most important flavanone is naringenin, the aglycon unit from the bitter-tasting naringin found in grapefruits (see Chapter 8). The related flavanone hesperitin is like naringenin except that both the 3′ and 4′ substituents of the B ring are hydroxyl groups. Hesperitin is also found in grapefruits and other citrus fruits as the aglycon of a glycoside called hesperidin. Both molecules are bitter-tasting and colorless, and a number of health benefits are suspected for diets containing them, where they are believed to have antioxidant and anticancer properties.

Flavones and flavonols have a double bond in their C ring between carbons 2 and 3, giving an extended conjugated double bond system, so these molecules absorb in the visible range and are colored. Flavonols (Figure 9.37) are found widely in foods from plant sources, such as fruits and vegetables (Miean and Mohamed 2001). The flavonol **quercetin**, which was first identified from the bark of oak trees (the genus

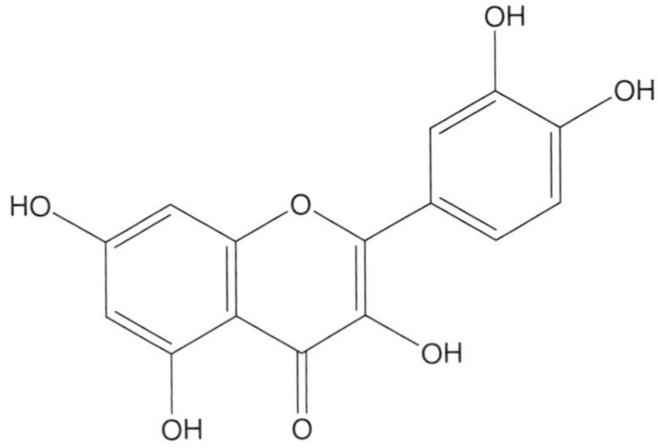
quercetin

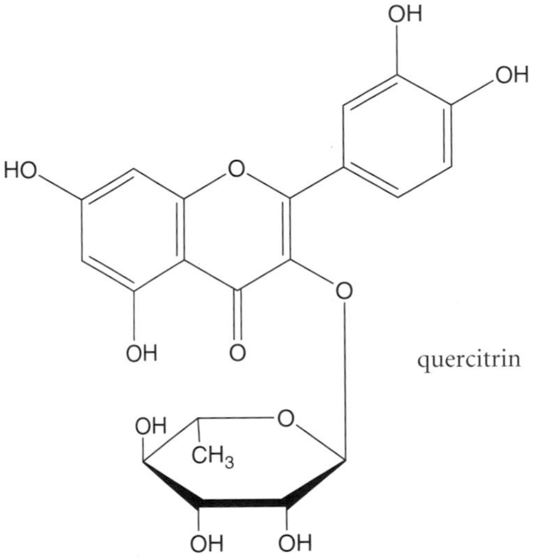
quercitrin

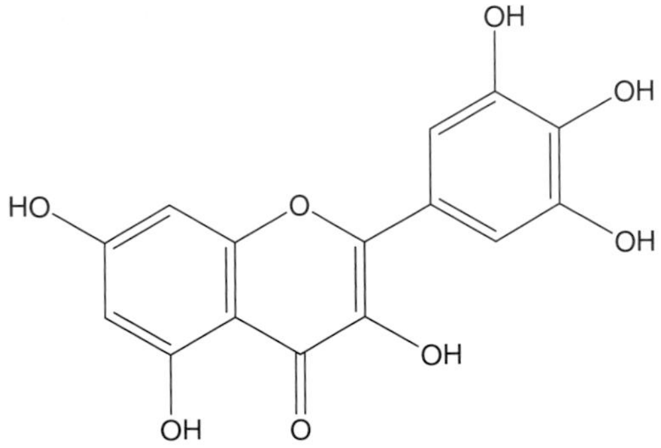
myricetin

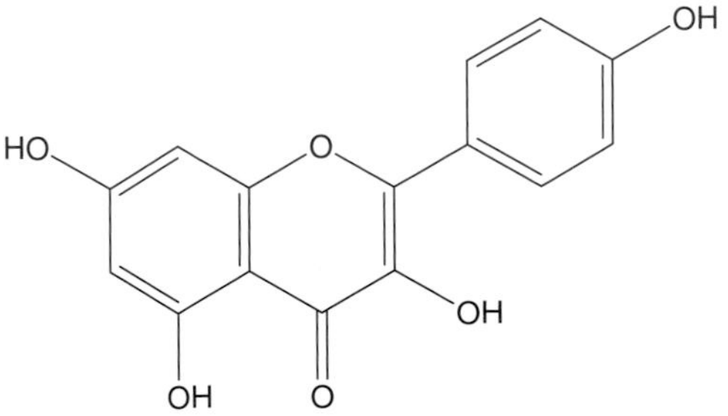
kaempferol

Figure 9.37. The structures of some flavonols found in foods.

name for oaks is *Quercus*), is also found in a number of fruits and vegetables such as apples, onions, grapes, lettuce, tomatoes, teas, and red wines (Crozier et al. 1997). It is also considered to be effective as a dietary antioxidant, comparable to EGCG and ECG. Quercetin is the aglycon unit of the glycoside **quercitrin**, which is actually present in most plant materials in greater concentrations than quercetin. The glycosyl moiety of quercitrin is an α-linked L-rhamnose, the 6-deoxy derivative of L-mannose. The antioxidant capacity of quercetin is much greater than that of quercitrin. Both molecules are colored yellow. Quercetin is insoluble in water, although it is soluble in methanol and DMSO. The polarity of the rhamnose in quercitrin, however, makes the glycoside water-soluble. Quercitrin, which is bitter-flavored, is used as a dye. In addition to its antioxidant activity, quercetin is thought to also have anticancer, anti-inflammatory, and even antiviral activities. Two other flavonols found in significant amounts in fruits and vegetables are myricetin and kaempferol. Myricetin has a higher antioxidant activity than quercetin, while kaempferol has only half the anti-oxidant activity of quercetin.

Yet another class of polyphenolics thought to be effective as dietary antioxidants are the **isoflavones**, flavonoid structural isomers in which the B aromatic ring is attached to the 3 position of the C ring rather than to the 2 position as in the flavo-noids (see Figure 9.38). Because isoflavones can exhibit estrogen-like activity, they are called phytoestrogens. Isoflavones are found in a number of grains, vegetables, and legumes, with the greatest concentrations being found in soybeans and soy-containing foods such as tofu and miso. They are thought to be at least partially responsible for the health benefits of soy-based foods. These molecules are believed to prevent certain types of cancer and are active as antioxidants. Several such mol-ecules are illustrated in Figure 9.38. Genistin is the β-linked D-glucopyranose

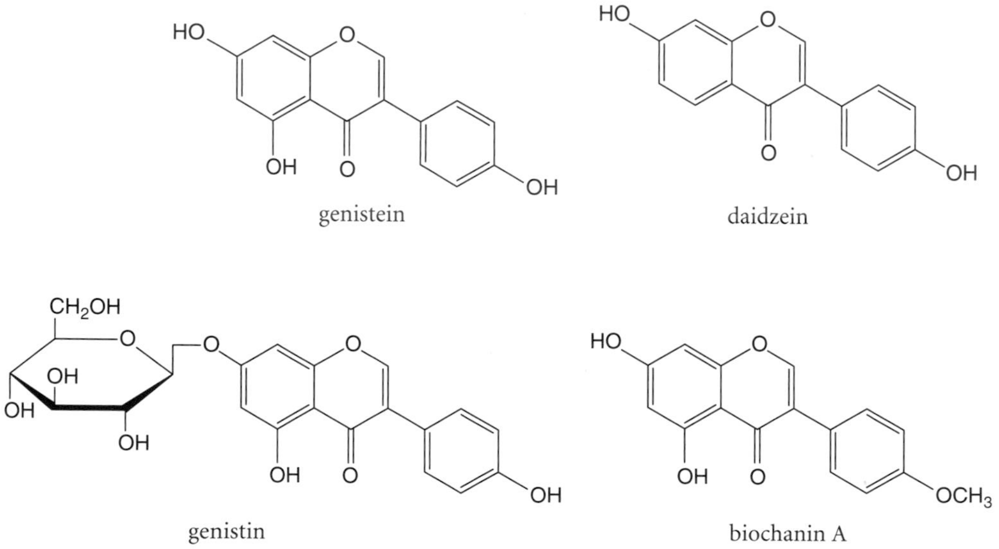

genistein

daidzein

genistin

biochanin A

Figure 9.38. The structures of the isoflavones.

glycoside of genistein bound to the carbon-7 hydroxyl of the A ring. Genistein is soluble in DMSO and ethanol but has limited solubility in water, while the glucose moiety of the glycoside genistin makes it more water soluble. As with the flavones, the additional double bond in the C ring of the isoflavones causes them to adsorb in the visible spectrum, and genistein is colored yellow. Isoflavones are relatively thermostable and generally are not degraded by the normal cooking temperatures needed to inactivate the antinutritional factors in soy protein.

Resveratrol

One of the most interesting of the polyphenolic molecules is **resveratrol**, a nonflavonoid derivative of **stilbene**, *trans*-1,2-diphenylethylene, both illustrated in Figure 9.39. Resveratrol and several other molecules, such as pterostilbene and piceatannol, with the same basic molecular framework as stilbene, are referred to generally as stilbenes. Resveratrol is found in several foods, including peanuts, mulberries, blueberries, raspberries, and various pine nuts, but the most common source for resveratrol in Western diets is from the skins of the grapes of *Vitis vinifera*, the grapes used to produce wine. High concentrations of resveratrol are also found in the seeds, vines and roots of this species. Resveratrol is present in relatively high concentrations in purple grape juices and in red wines, which typically contain 1.5 to 3.0 mg/liter.

A flurry of research activity has suggested that consuming resveratrol has a variety of health benefits. Studies have found that resveratrol can offset the bad effects of eating a high-fat diet, and it has been postulated that this molecule in red wine is the

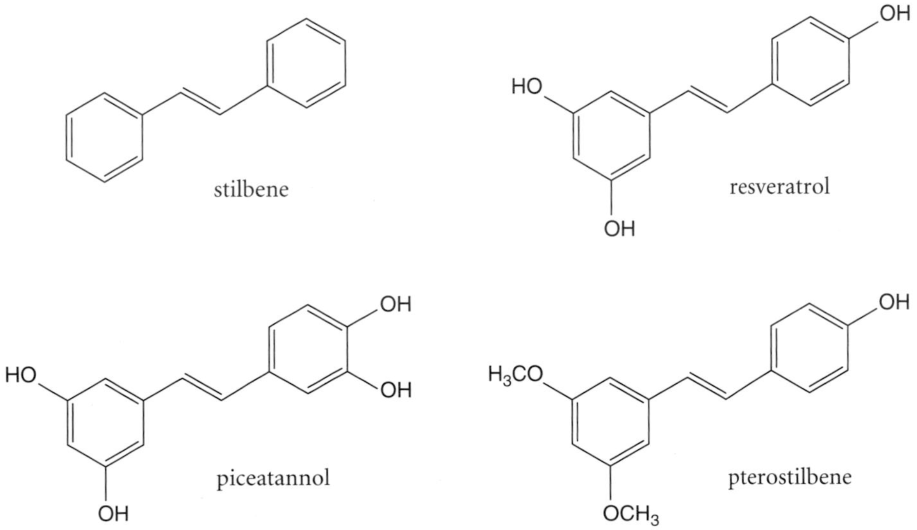

Figure 9.39. Structures of important food stilbenes.

explanation for the so-called French Paradox, the fact that in spite of a national diet high in animal fats, the French have a significantly lower rate of heart disease than Americans. A 2006 study (Baur et al. 2006) in which mice were fed high-fat diets found that large daily doses of resveratrol prevented the mice from developing diabetes and significantly extended their life spans, such that they equaled those of a control group of mice on a standard low-fat diet. Even if the results of these lab animal studies are directly applicable to humans, it is unlikely that this effect alone explains the French Paradox, because the resveratrol doses consumed by the mice in this study would be the equivalent of drinking 750 to 1500 *bottles* of red wine *per day*, but the findings are nevertheless interesting. More recently, another study found that much lower doses of resveratrol, which could readily be achieved in the human diet, could mimic many of the effects of calorie-restricted diets, which have been shown in mice to increase life spans significantly. Resveratrol apparently works in this capacity by inhibiting the expression of genes related to heart and skeletal muscle aging, possibly by altering chromatin structure and transcription (Barger et al. 2008). Resveratrol is already being marketed as a nutritional supplement, although its utilization is hampered by a low bioavailability, limited water solubility, and instability toward oxidation.

Stilbenes such as piceatannol and pterostilbene are also found in fruits such as blueberries and are thought to contribute to the general antioxidant capacity of these berries.

Several of the plant polyphenols such as quercetin and resveratrol have been demonstrated to bind to the protein F(1)-ATPase and inhibit both the synthesis of ATP and its hydrolysis (Figure 9.40). It has been postulated that the putative anticancer effects of resveratrol may result from this inhibition of ATP synthesis in the mitochondria of cancer cells, leading to their death by apoptosis, or programmed cell death (Gledhill et al. 2007).

Figure 9.40. The crystal structure of F(1)-ATPase, shown as a van der Waals representation, complexed with a resveratrol molecule.

Tannins

Tannins are a broad class of acidic plant polyphenols, generally characterized by their ability to bind to and precipitate various proteins, which are responsible for the astringent taste of wines and certain fruits, particularly before they ripen. This class of molecules, generally referred to collectively as tannic acid, is found in large amounts in the bark of some trees, such as oaks, tanoaks, and chestnuts, as well as other plants, and was used traditionally in the tanning of animal hides to produce leather. The term tannin is applied to a diverse set of molecules, and the exact definition is somewhat vague. In general, however, it refers to higher-molecular-weight polyphenolics that form strong complexes with proteins. All tannins are water soluble. In addition to proteins, they will bind to both starch and cellulose. Many of the tannins are complex molecules consisting of a core D-glucose molecule with many or all of its hydroxyl groups esterified to gallic acid or oligomers and derivatives of this phenolic compound, such as ellagic acid (Figure 9.41). Tannins of this type are referred to as **hydrolyzable tannins**, because under mildly acidic conditions in aqueous solution their ester linkages are hydrolyzed to give the freed sugar core and phenolic acids.

The apparent steric clashes in the structure of ellagitannin are an artifact of a two-dimensional representation, because in the hexahydroxydiphenic acid form of ellagic acid the two aromatic rings do not lie in the same plane, as indicated in Figure 9.42.

Many of the tannins in wines are so-called **condensed tannins**, which consist of up to 50 flavan-3-ol units, often catechin, epicatechin, gallocatechin, gallocatechin gallate, and anthocyanidins, linked together into a polymer by carbon–carbon single bonds (Figure 9.43). Condensed tannins are not hydrolyzed in weakly acidic aqueous solutions like the hydrolysable tannins are, but they can be oxidized when heated in more strongly acidic alcohol solutions to give free anthocyanidins. As with the ellagic acid, the steric clashes that appear to be present in the condensed tannin structure are avoided by rotations about the C—C bonds that link the individual flavonoid units so that successive monomers do not lie in the same plane and have a helical conformation.

Because they complex with macromolecules, tannins in foods can interfere with their digestion and reduce the overall digestibility of protein consumed with tannins. This problem is more severe for animal feeds than for human diets, given the relatively small amounts of tannins consumed by people and the excess amounts of protein eaten by most people, particularly in the United States. A number of studies, however, have suggested or actually demonstrated significant health benefits from consuming tannins. For example, a recent clinical study on prostate cancer patients found that consuming pomegranate juice containing ellagitannins, which are hydrolyzed upon digestion to give ellagic acid, led to a substantial slowing of the progress of the disease (Pantuck et al. 2006; Seeram et al. 2007). Note that such a study does not prove that any observed effect is due to the ellagitannins.

Tea contains a number of astringent phenolic compounds often referred to as tannins, but these are generally lower-molecular-weight catechins or other phenolic

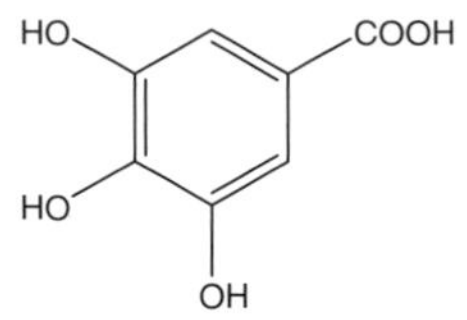

gallic acid

ellagic acid

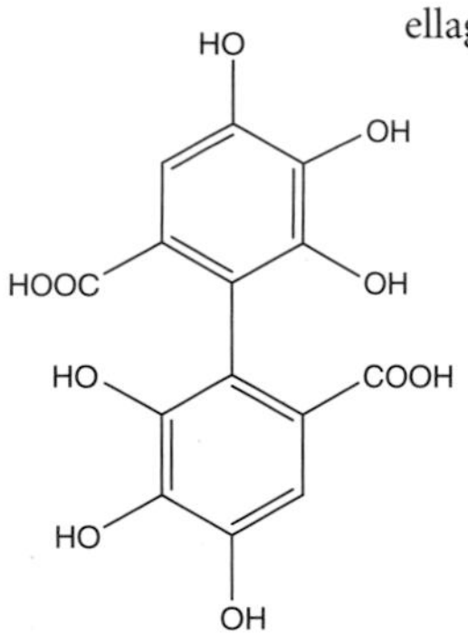

hexahydroxydipenic acid form

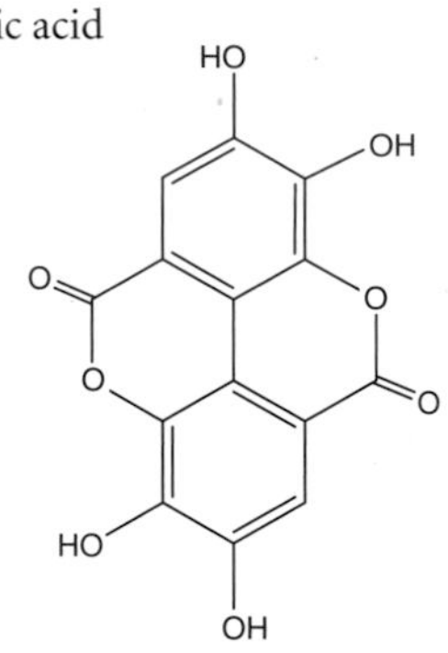

dilactone form

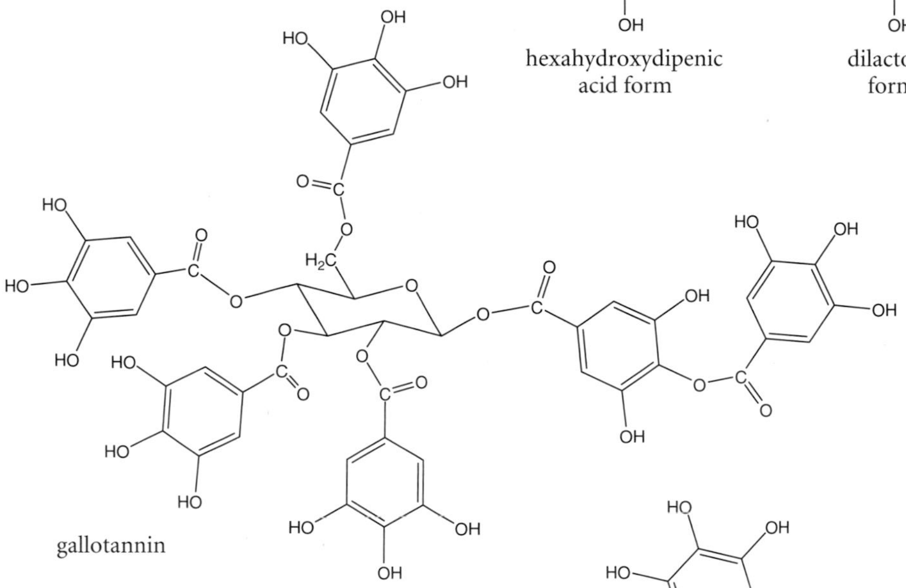

gallotannin

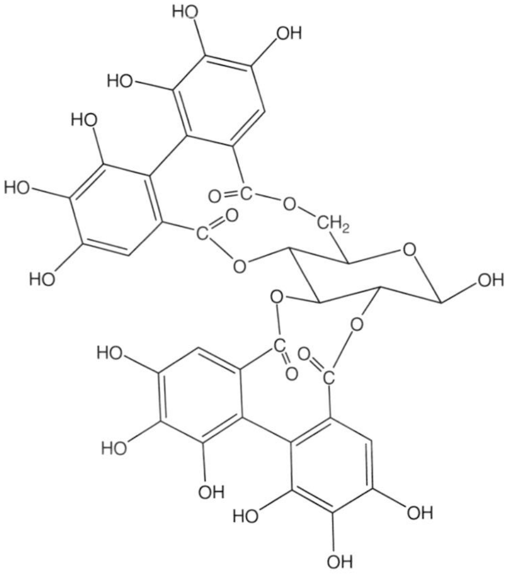

ellagitannin

Figure 9.41. The structures of the hydrolysable tannins gallotannin and ellagitannin and the gallic acid and ellagic acid molecules from which they are built.

Figure 9.42. The structure of ellagic acid.

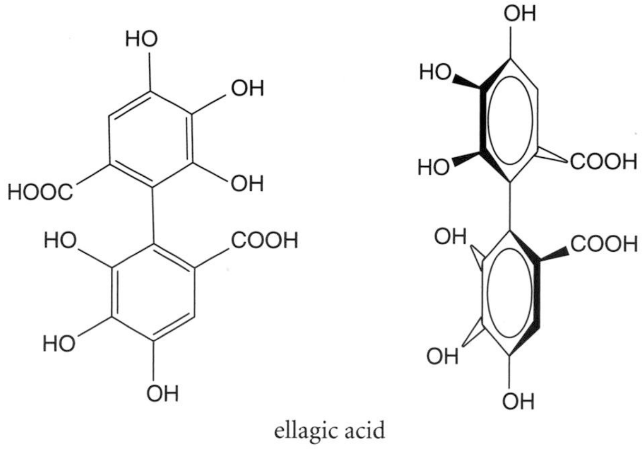

ellagic acid

Figure 9.43. The general structure of a condensed tannin.

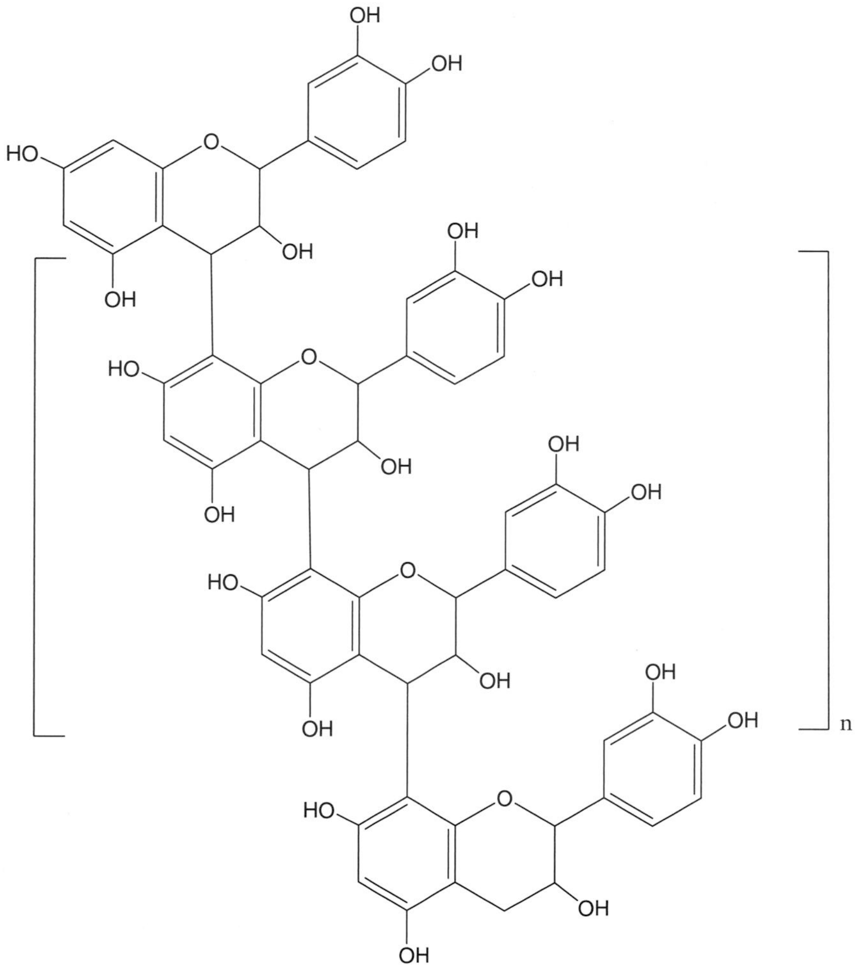

compounds referred to as theaflavins, rather than tannic acid or similar molecules. The archetype of this class, **theaflavin**, is shown on the left in Figure 9.44. Another small phenolic found in tea is the gallic acid ester theogallin, 3-galloyl quinic acid, also shown in Figure 9.44.

The **thearubigins** are another class of polyphenolic molecules found in black teas whose structures have not yet been completely characterized but are believed to be produced from theaflavins. These red-brown pigment molecules give black teas their dark color. The color of thearubigin is pH dependent, and its protonation at lower pH is responsible for the fading of the brown color of tea that can be observed with the addition of lemon juice, which is very acidic. One can reverse this effect by adding baking soda (which is basic) to tea after lemon has been added, restoring the dark brown color; further addition of vinegar (acetic acid) will again cause the color to fade as the pH once again falls (of course, if you try this kitchen experiment at home, do not drink the resulting "tea"—it will not be harmful, but probably will not taste good!).

Both green and black teas are produced from the leaves of the same plant, *Camellia sinensis*. They are processed differently, however, which affects their contents of various compounds such as phenolic antioxidants and caffeine. Black teas are produced by a process called fermentation, which does not involve bacterial fermentation, but rather is actually an oxidation process under controlled environmental conditions. Much of this oxidation is PPO- and peroxidase-catalyzed enzymatic oxidation of catechins, although some results from direct exposure to air (Subramanian et al. 1999). When the oxidation has proceeded to the desired level, the process is halted by heating, followed by drying. In general, black teas contain more caffeine than green teas. In addition, black teas contain higher concentrations of theaflavins than green teas, while green teas have far more catechins (Table 9.5). The most significant theaflavin in black teas is theaflavin-3,3′-digallate, while a major catechin in green teas is EGCG. Both classes of molecules are antioxidants, but the measured total antioxidant capacity per serving of green teas is nearly twice as high as in black teas, in terms of free radical scavenging activity, as might be expected from the

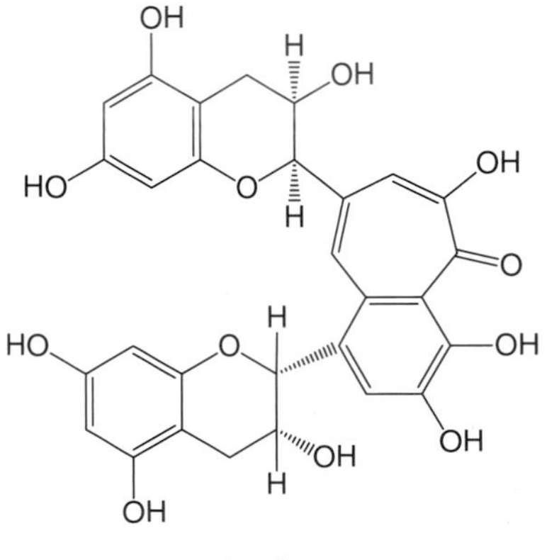

Figure 9.44. The structures of theaflavin and theogallin.

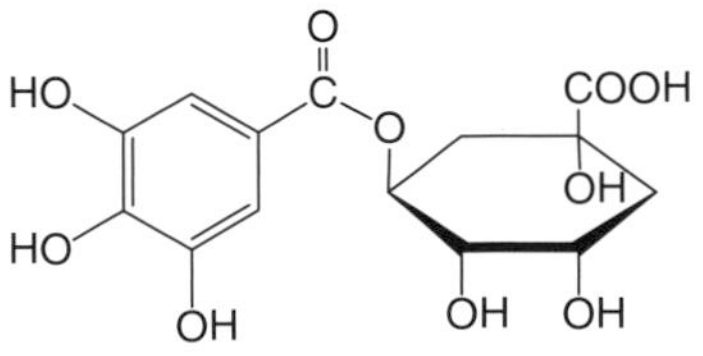

theaflavin   theogallin

Table 9.5 Flavonoid composition of teas

Tea product	Total catechins	Total theaflavins	Total thearubigins	Total flavonols
	mg/100 g			
dry black tea leaves	3605.6	603.3	5918.9	371.3
decaffeinated black tea	242.3	123.1	4412.6	398.5
brewed black tea	34.3	6.1	73.4	3.9
dry green tea leaves	12516.0	6.8	132.0	515.7
decaffeinated green tea	3942.2	26.7	972.5	444.9
brewed green tea	132.1	0.1	1.1	112.1

Source: S. Bhagwal, G. Beecher, D. Haytowitz, J. Holden, J. Dwyer, J. Peterson, S. Gebhardt, A.L. Eldridge, S. Agarwal, and D. Balentine, "Flavonoid composition of tea: Comparison of black and green teas," *12th World Congress of Food Science and Technology, Chicago, IL, July 13–16, 2003.*

oxidation that takes place in the production of black teas (Lee, Lee, and Lee 2002). As a result, green teas are generally considered to be healthier than black teas, although both contain significant levels of antioxidants.

Polyphenolic Chemistry in Wines

The *vinifera* grapes used to make most wines contain a number of anthocyanin pigments in their skins, and these pigments color the juice from the grapes used in the manufacture of red wines. These anthocyanins include delphinidin, cyanidin, petunidin, peonidin, and malvidin, with malvidin-3-glucoside being the most important. Wines contain many other chemical species as well, including many other polyphenolic molecules such as catechins and tannins. A number of chemical reactions take place in wine that affect its appearance, bouquet, and most importantly, its taste. Many of these reactions are acid catalyzed because the pH of wine is usually around 3–4. We will briefly consider a few of the most important of these reactions here (Cheynier et al. 2006; Brouillard et al. 2003; Rovner 2006).

The color of the grape juice used to make red wines is originally purple/violet due to the malvidin and other anthocyanins but undergoes a progressive shift in color during the maturation period as the taste is simultaneously softened. In a typical red Bordeaux wine, for example, which is usually a blend of the juices from Cabernet Sauvignon and Merlot grapes, the color of the wine when young is a ruby red, which gradually shifts to garnet red and finally brick red when fully mature. A wine past its optimal maturity begins to turn brown in color. In young wines the free anthocyanin pigments responsible for the red color exist in an equilibrium between several different states, including a flavylium cationic state, the colorless neutral hemiketal form, and the blue-colored neutral quinonoidal base form. Because the pH of wine is low, the predominant color is red (Figure 9.45).

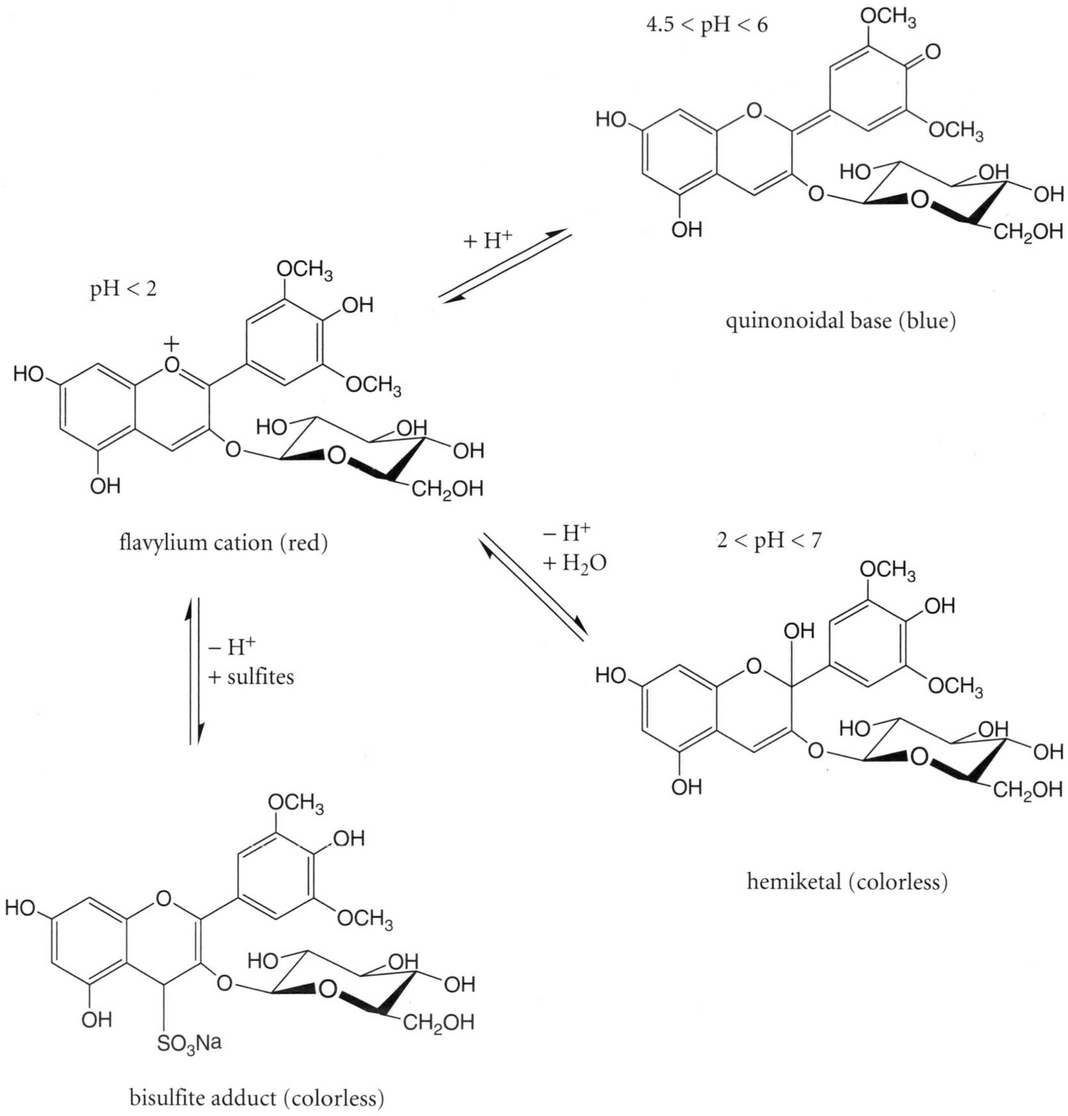

Figure 9.45. The equilibrium between the various states of the anthocyanins in wines (Cheynier et al. 2006).

Yet another form also participates in this equilibrium due to sulfites in the solution. A small amount of sulfites are present as products of the fermentation process, but additional sulfites result from the addition of sulfur dioxide, SO_2, as a preservative. Sulfur dioxide is allowed as a wine additive to function as an antimicrobial and antioxidant. Because some people are allergic to sulfites, wine that has been treated with sulfur dioxide must be labeled as "contains sulfites." Under alkaline conditions the sulfur dioxide forms bisulfite ions, HSO_3^-,

$$SO_2 + NaOH \rightarrow NaHSO_3 \tag{9.2}$$

which combine with the anthocyanin pigments to form a colorless bisulfite adduct. This process, also illustrated in Figure 9.45, is called sulfite bleaching.

The tannins in wines are primarily responsible for their astringent taste. Astringency may be due to the hydrophobic association between the tannins and salivary proteins in the mouth that lead the tannins to precipitate out (Zanchi et al. 2008). Thus, if the tannins become less hydrophobic, their astringency is reduced. Over time, under acid conditions, the anthocyanin pigments initially present in wine react with the tannins to form covalent flavanol–anthocyanin complexes that are red in color and which are more hydrophilic than the original tannins, thus softening the taste of the wine (Figure 9.46). During the aging process, in the presence of oxygen from the air, the free anthocyanins progressively undergo such condensation reactions with the tannins to form more and more of the polymeric pigments, sometimes called proanthocyanidins.

Wine also contains small amounts of pyruvic acid, CH_3COCO_2H, present as a yeast metabolite resulting from the fermentation process. This pyruvate can react with anthocyanins to give pyranoanthocyanins, another class of pigments that are also called **vitisins**, illustrated in Figure 9.47. The flavylium form of the vitisins is more stable than the colorless hemiketal form, unlike the case for the free anthocyanins, thus producing more red color as the wine ages. After about a year of aging, almost all of the free anthocyanins are gone as the result of these various reactions. These processes are responsible for the development of the brick red color of mature red wines.

Figure 9.46. The structure of a proanthocyanidin of malvidin; only two flavonoid units are shown in the tannin portion of this molecule for brevity; tannins actually contain many more of these flavonoid units.

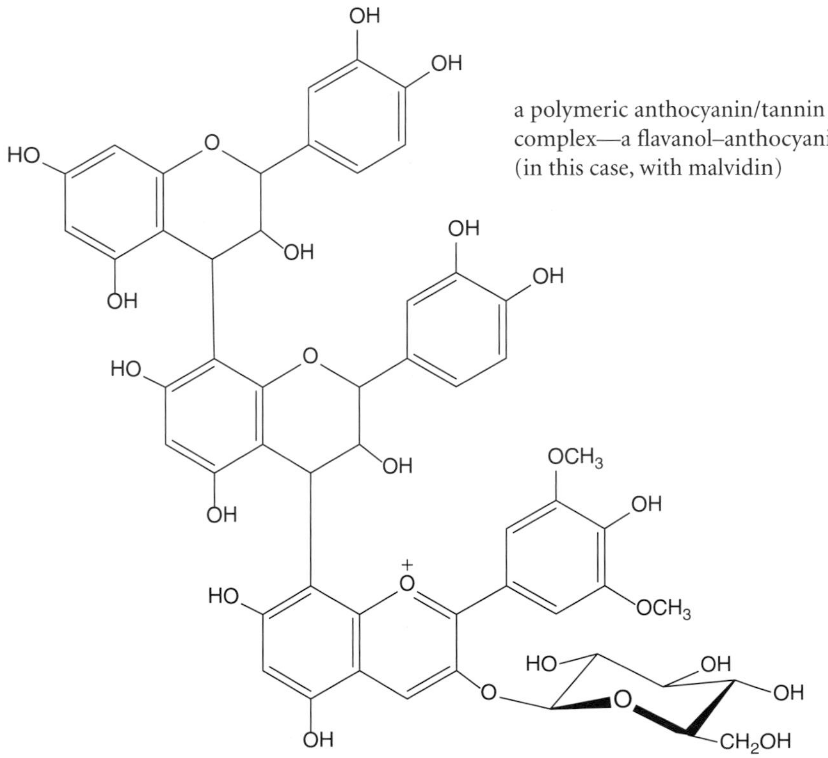

a polymeric anthocyanin/tannin complex—a flavanol–anthocyanin (in this case, with malvidin)

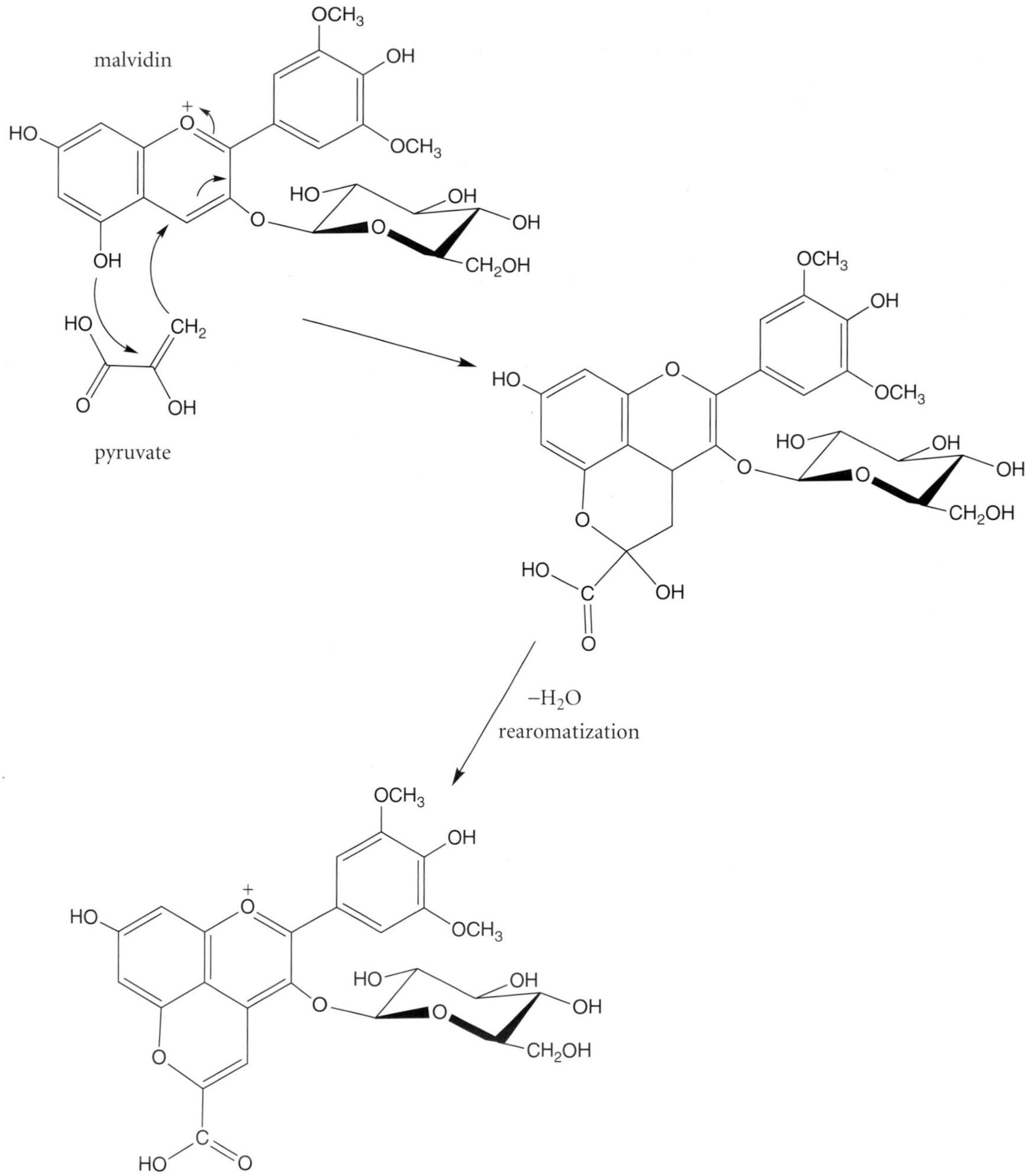

Figure 9.47. The formation of vitisins, contributing to the brick red color of mature wine.

In wines that have aged too long, the process of polymerization of the tannins and anthocyanins leads to large molecules that precipitate out of the wine and diminish both the color and taste.

Betalains

The betalains are a class of pigment molecules that resemble the anthocyanins, but are not themselves anthocyanins or even flavonoids. Betalains are glycosides with heterocyclic, nitrogen-containing indole rings in their aglycon group. They are colored dark red or red-purple, and are found in a number of red-colored plants, but not in plants that contain anthocyanins. In particular, betalains are responsible for the red of beet roots and stems. Beet juice and beet powder, which are allowed for use as food additives as natural food colorants, contain several betalains, including **betanin**, the most important, isobetanin, probetanin, and neobetanin. The archetype of the class, betanin, shown Figure 9.48, is a glycoside of glucose (as are the other three listed); the aglycon of betanin is called betanidin. Like cyanidin, the color of betanin is pH dependent; at weakly acidic pHs it is bluish-red, but at weakly basic pHs it is violet.

Betalains are easily degraded by various treatments. When heated under acidic or basic conditions, the glycosidic bond can be hydrolyzed to release the sugar and aglycon components. Betalains are susceptible to oxidation in the presence of oxygen and can be degraded by light. Because of their instability, beet juice pigments are most useful in either frozen or dry products, such as the pasta illustrated in Figure 9.50. Because of their susceptibility to oxidation, betalains can serve as dietary

Figure 9.48. The structure of betanin.

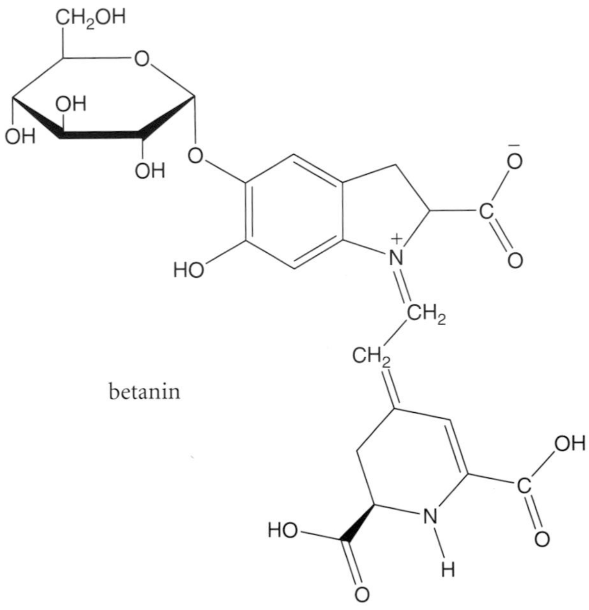

betanin

antioxidants, and beets have long been considered in traditional folklore to have beneficial health effects. Betalains are found in other plants besides beets, most notably in cactus fruits and flowers.

Cochineal

Carmine, or cochineal extract, is a natural colorant allowed for food use, that has a crimson or bright red color. Cochineal dye is extracted from the bodies of the female cochineal insect, *Dactylopius coccus*, a type of scale insect native to Central and South America that feeds on various species of cactus of the genus *Opuntia*. It was prized by the Aztecs and Mayas, who used it to dye cloth, and it was one of the most important of Mexico's exports during the colonial period. The dye molecule responsible for the red color of carmine is carminic acid, $C_{22}H_{20}O_{13}$, illustrated in Figure 9.49. Note that while this molecule contains a glucose moiety, it is not an O-linked glycoside like betanin, because the anthraquinone chromophore component of the complex is directly bonded to the anomeric carbon atom of the glucose portion, with no intervening oxygen atom.

As might be expected from the large number of polar hydroxyl groups on both portions of the molecule, as well as the carboxylic acid group of the anthraquinone portion, carminic acid is soluble in water. Carmine, or cochineal dye, is quite stable to normal processing operations. It does not oxidize under normal conditions (because it is already extensively oxidized), is heat stable, and is not degraded by light. It is used to color a wide variety of products, from sausages and surimi, to desserts, beverages, cheeses, and confectionary products, as well as cosmetics. Because carmine is extracted from insects, products containing cochineal may be unacceptable to vegans and some Jews and Muslims. There is also a certain degree of general consumer resistance in the United States to food ingredients made from insects.

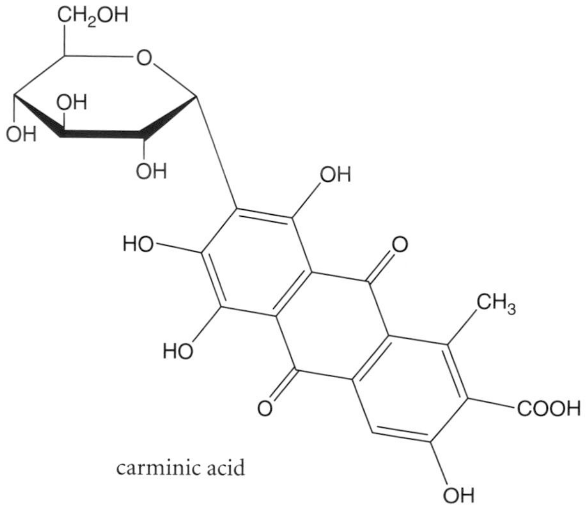

Figure 9.49. The structure of carminic acid.

carminic acid

Some individuals are also allergic to cochineal, although carminic acid may not be the actual allergen, because trace amounts of other molecules are present in the extracts from the insects. Carminic acid can now be produced synthetically (P. Allevi et al. 1991).

COLORANTS

In the United States, two types of color additives are available, those obtained from plant, animal, or mineral sources that are also found naturally in the diet (even if these molecules are synthetically produced) and which are exempt from federal certification, and color additives from synthetic sources whose use in foods must be certified by the US Food and Drug Administration (FDA). Only nine of these are now permitted in the United States; all are water-soluble pigments synthesized from fossil hydrocarbons such as coal. Synthetic molecules of permitted naturally occurring colorants are called "**nature-identical colors**." The synthetic dyes and pigments used in foods in this country are given approved names such as FD&C Blue No. 1 that basically combine the color name and a number. In general, these synthetic colorants are more stable than natural or nature-identical colors under processing and storage conditions, are less variable in their colors, and are less likely to produce undesirable flavors as the result of decomposition. Almost any desired color can be produced by blending the nine allowed synthetic colorants, so that there really is no need for a wider range of color molecules.

The list of colorants that do not require certification includes vegetable and fruit juices, which are natural food products themselves (Table 9.6). For example, the colorants turmeric, annatto, and paprika are also used as spices, while β-apo-8′-carotenal is found in oranges and tangerines (Figure 9.50).

For coloring foods based on the hydrophobic fats, an alternative to the hydrophilic synthetic dyes is needed. For this purpose the dyes are combined with an small insoluble particle to produce a pigment called a lake. Thus, a **lake** is a water-insoluble pigment made from an organic dye absorbed onto the surface of an inert carrier particle such as alumina (aluminum oxide, Al_2O_3). This is accomplished by precipitating the water-soluble FD&C dyes with aluminum, calcium, or magnesium salts onto an aluminum hydroxide substrate. The term is derived from the French term "*laque*" meaning lac, the word for the reddish insect secretion (from the insect *Laccifer lacca*) used to make shellac, and also used as a red pigment.

Titanium Dioxide

Titanium dioxide, TiO_2, is the most widely used white pigment in almost all industrial applications, including the food industry. Nanoparticles of TiO_2 are an extremely bright white with a high refractive index. Titanium dioxide is chemically stable, relatively inert biochemically, and insoluble in water and organic solvents. It is widely used

Table 9.6 Color additives not requiring certification

Colorant	Restrictions
annatto (red)	—
β-apo-8′-carotenal (orange)	33 mg/kg
β-carotene (yellow to orange)	—
beet powder (red)	—
canthaxanthin (red)	66 mg/kg
caramel (brown)	33 mg/kg
carrot oil	—
cochineal extract (carmine)	—
ferrous gluconate (black)	ripe olives only
grape color extract	nonbeverage use only
grape skin extract (enocianina)	beverages
paprika (red)	—
riboflavin (yellow)	—
saffron (yellow)	—
titanium dioxide (white)	1%
turmeric (yellow)	—

Source: From J.M. deMan, *Principles of Food Chemistry* (Gaithersburg, MD: Aspen Publishers, 1999).

in paints, inks, ceramics, and cosmetics, as well as in foods. It also has many other industrial uses, for example as a catalyst or semiconductor. Titanium dioxide is a completely inorganic mineral that is mined from ores. It has four crystal forms, including rutile and anatase, and in mined ores is usually bound to impurities that are removed in processing before use in foods. The anatase crystal form is less dense, and thus "softer," and is the preferred form for most food uses. Because of its color, it is often used in milk and other dairy products, but is permitted in a wide range of additional products. Titanium dioxide is not known to be toxic, carcinogenic, or mutagenic, and it appears to be safe to consume it in the amounts used in foods, although feeding experiments on mice have suggested that nanoparticles of TiO_2 can cause DNA breakage and chromosomal damage. It has recently been suggested, based on laboratory experiments on rats, that the finely powdered TiO_2 to which workers in production facilities are exposed could cause lung cancers when inhaled in large amounts.

Synthetic Colors

The list of approved synthetic colors allowed in the United States has decreased over the years (Table 9.7). All synthetic fat-soluble dyes have been removed from the list,

Figure 9.50. Italian dried farfalle pasta, colored with natural pigments. The red comes from tomatoes (lycopene) and beet powder (betalains), the olive green from spinach (pheophytin), the yellow from carrots (β-carotene) and tumeric (curcumin), and the blue/black from sepia, the ink of cuttlefish.

which now includes only nine water-soluble dyes. Two are only approved for limited uses: Citrus Red No. 2 is allowed only for external use on oranges, and FD&C Orange B is only allowed for the coloring of sausage and frankfurter casings, with no more than 150 ppm by weight allowed in the final product. Three of the remaining seven are azo compounds, and two are triphenylmethane compounds, while the remaining two dyes are chemically unrelated to the others.

Azo Dyes

Aromatic **azo compounds**, containing R—N=N—R groups, where the R groups are aromatic rings, form the basis of the azo dyestuff industry. Only a handful are approved for consumption as food additives. These dyes are thermally stable and relatively stable toward oxidation, but can be reduced to a colorless compound. The three allowed for food use, FD&C Yellow No. 5 (tartrazine), FD&C Yellow No. 6 (sunset yellow), and FD&C Red No. 40 (allura red), are shown in Figure 9.51. The N=N double bond in these molecules extends the conjugated double bond system and shifts the adsorption spectrum into the visible range through extensive

Table 9.7 Seven synthetic dyes allowed for general food use in the United States

FD&C name	Type	Stability at pH				Stability in light
		3	5	7	8	
FD&C red No. 3 erythrosine	xanthine	insoluble	insoluble	stable	stable	considerable fading
FD&C red No. 40 allura red	azo	stable	stable	stable	stable	fades very slightly
FD&C yellow No. 5 tartrazine	azo	stable	stable	stable	stable	fades very slightly
FD&C yellow No. 6 sunset yellow	azo	stable	stable	stable	stable	appreciable fading
FD&C green No. 3 fast green	triphenylmethane	fades slightly	fades slightly	fades slightly	fades slightly	appreciable fading
FD&C blue No. 1 brilliant blue	triphenylmethane	fades slightly	fades very slightly	fades very slightly	fades very slightly	appreciable fading
FD&C blue No. 2 indigotine	indigoid	appreciable fading	appreciable fading	considerable fading	fades	fades

Source: J.H. von Elbe and S.J. Schwartz in Food Chemistry, 3rd ed., O.R. Fennema, ed. (New York: Marcel Dekker, 1996).

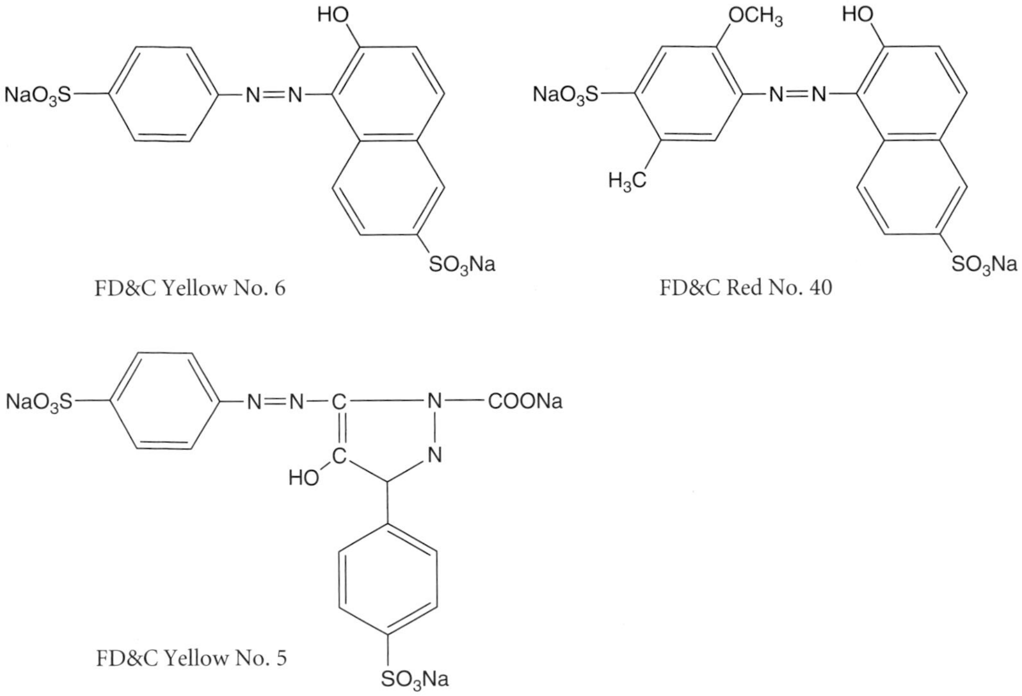

Figure 9.51. The structures of the azo dyes approved for food use in the United States.

delocalization. Red No. 40 is the most commonly used artificial red colorant in US food products.

Triphenylmethane Dyes

Two triphenylmethane dyes are allowed for use in foods in the United States (Figure 9.52), FD&C Blue No. 1, brilliant blue, and FD&C Green No. 3, called fast green. Fast green is the only approved green-colored food dye, and of all the synthetic colors, it may be the hardest to replace with a natural substitute, because, as we have seen, the most obvious choice, pheophytin, is olive-colored rather than bright green. Greens can also be produced by combining yellow (such as tartrazine) and blue (such as brilliant blue).

Reduction of triphenylmethane dyes, or reaction with bases, produces colorless derivatives, because both will disrupt the extensive system of conjugated double bonds found in these dyes (Figure 9.53).

Indigotine

Indigotine, FD&C Blue No. 2 (Figure 9.54), is a synthetic dye related to the traditional blue dye indigo, used since ancient times to dye cloth (and even to dye the skin of British warriors who fought against invading Roman legions (Huxtable 2001)).

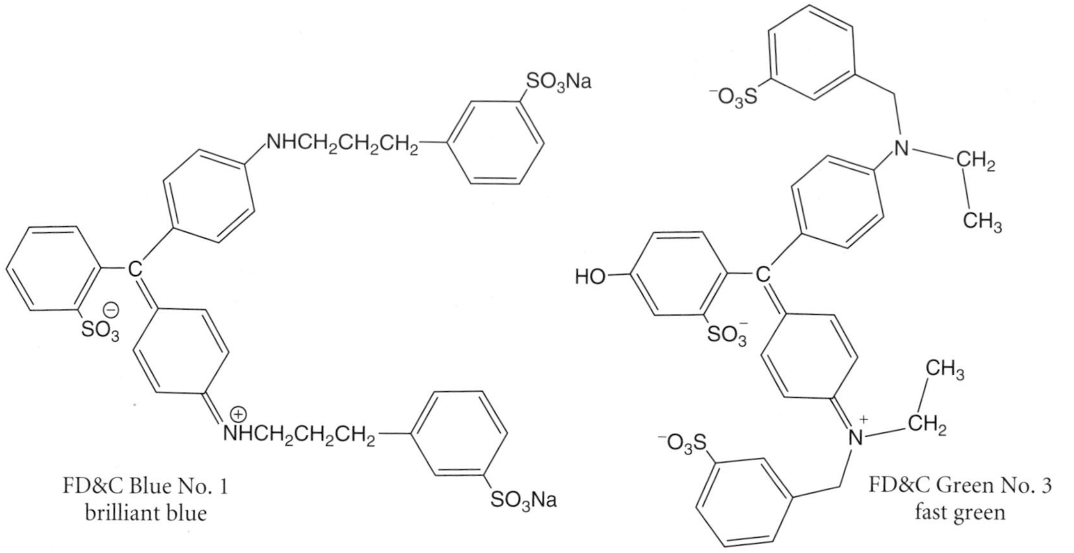

Figure 9.52. The structures of the triphenylmethane dyes brilliant blue and fast green.

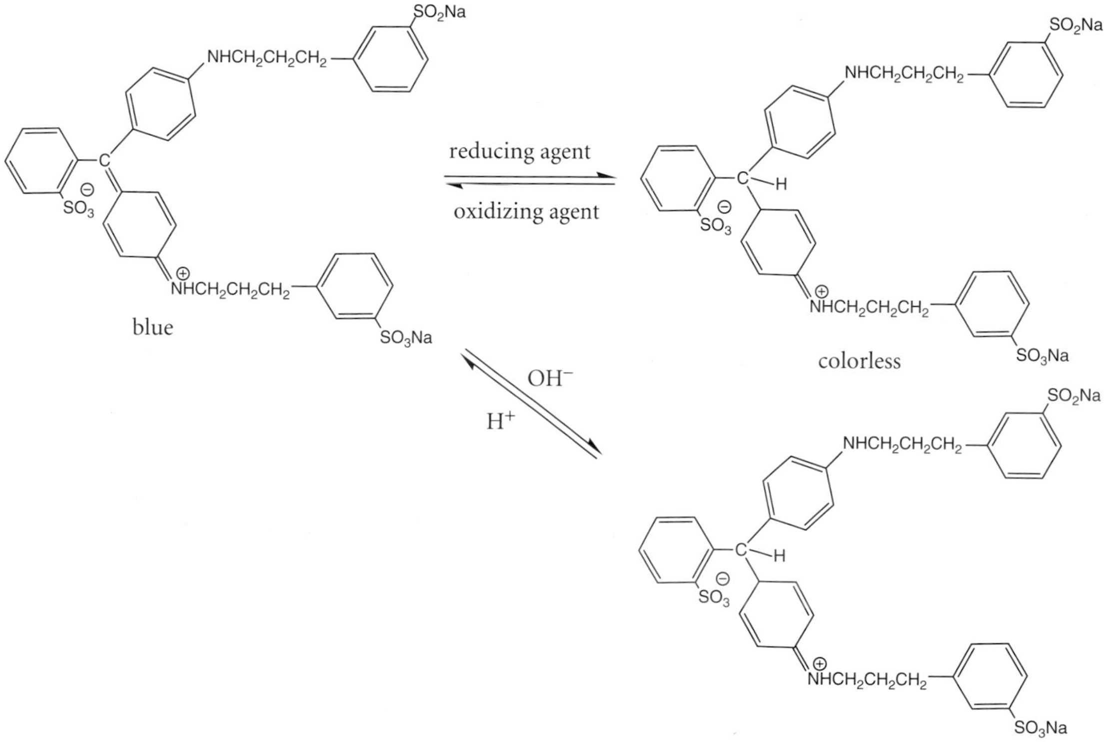

Figure 9.53. Reduction or protonation of triphenylmethane dyes produces colorless derivatives.

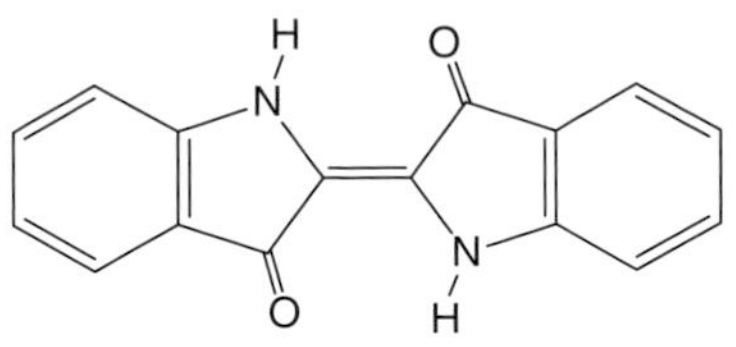

indigo blue
(a nonfood dye)

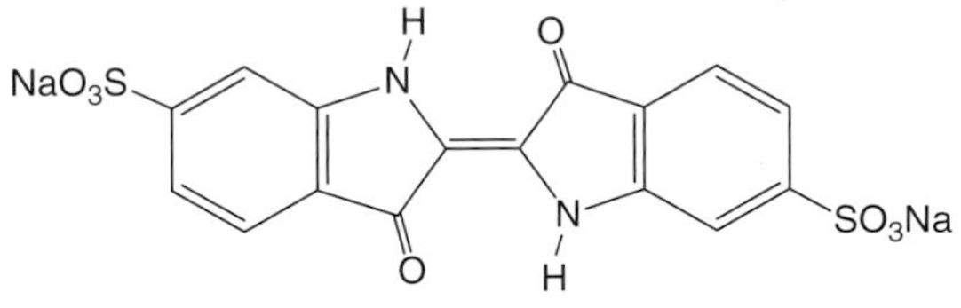

FD&C Blue No. 2
indigotine

Figure 9.54. The structure of the nonfood plant dye indigo, and the related food dye brilliant blue, FD&C Blue No. 2.

Figure 9.55. The structure of erythosine, or FD&C Red No. 3.

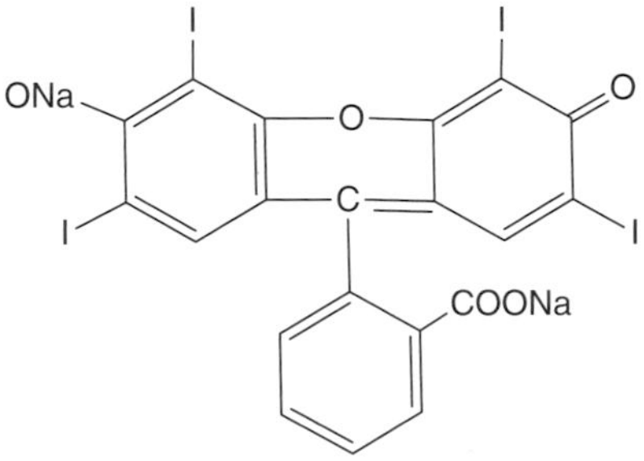

FD&C Red No. 3
a xanthine

Indigo, which was traditionally produced in Europe from dyer's woad (*Isatis tinctoria*), and later from the indigo plant (*Indigofera tinctoria*) from Asia, is a nonfood dye that is completely insoluble in water in its ordinary oxidized form. The double bond between the two fused heterocyclic ring systems creates an extended delocalized conjugated system that absorbs in the orange region, with a λ_{max} of 602 nm, so that it reflects a deep blue color. Indigotine, FD&C Blue No. 2, is made water soluble as the disodium sulfonate derivative of indigo (Figure 9.54). At very high pHs (> 13) this molecule turns yellow in color, but it is fairly heat stable, and soluble in both water and alcohol. Apart from a few berries such as blueberries, few natural foods are blue, and the principal use for this dye is in manufactured products such as candies and gum that are primarily marketed toward children (Gilman 2003).

Xanthines

Only one xanthine is allowed for food use in the United States, FD&C Red No. 3, also called erythosine, which is shown in Figure 9.55. The rings of erythosine have four covalently bound iodine atoms, and as with the other FD&C dyes, it is colored due to an extensive set of conjugated double bonds. High doses of erythosine are suspected of causing cancer in lab rats, and its use has been limited by the FDA, and its status is under review. Xanthine is primarily used to color confections, cherries, and sometimes the shells of pistachio nuts.

Commercial Use of Synthetic Colors

Combinations of these synthetic dyes can be used to produce virtually any shade desired, in much the same way that an artist's palette of basic pigments can be combined to produce any other color. For example, an artificial grape drink could be colored "naturally" with grape skin extract, but a similar dark purple hue can be produced artificially through a combination of FD&C Red No. 40 and FD&C Blue No. 1.

All of these water-soluble synthetic food dyes are produced from fossil hydrocarbon sources such as petroleum or coal (indeed, in the dyeing industry they are referred to as coal tar dyes). In general, they are primarily used to color highly manufactured foods, particularly those marketed toward children, such as artificial "fruit" cereals and soda drinks. Although no data exists to prove that these dyes cause cancer (or else they would be banned), these nine are the surviving members of a much larger list, most of whose members are forbidden for food use because they were discovered to be harmful when consumed. It has been suggested that these coal tar dyes, and particularly the azo compounds, give rise to allergic reactions in some individuals, especially in those who exhibit aspirin intolerance. Caution might thus be warranted, particularly given the marginal need for these colors in the first place, and given the fact that many of the products that employ them are targeted toward children. In this context, a recent study in the United Kingdom (McCann et al. 2007) found that drinks colored with a mixture of food dyes and with sodium benzoate added resulted in an increase in hyperactivity in young children, in both 3 year olds and 8/9 year olds, as evaluated by teachers and parents in double-blind tests. While no lasting physical harm was demonstrated, such hyperactivity could lead to learning problems in school. Unfortunately, because this study included a mixture of four colorants (FD&C Yellow No. 6, FD&C Yellow No. 5, carmoisine [E122], and ponceau 4R [E124], also called brilliant scarlet 4R), the latter two of which are banned in the United States, as well as the preservative benzoate, it is impossible to say which of the compounds caused the hyperactivity or if it was caused by some combination of more than one of them. Although the use of artificial food dyes could be considered frivolous, the same is not true for benzoate, which performs a vital safety function in many food products (see Chapter 10). These results should be quickly subjected to independent confirmation that also tests the effects of each of the components separately.

SUGGESTED READING

Atkins, P. 2003. *Atkins' Molecules*, 2nd ed. Cambridge: Cambridge University Press.

Attokaran, M. 2011. *Natural Food Flavors and Colorants*. Hoboken:Wiley-Blackwell.

Carmen, S. 2007. *Food Colorants: Chemical and Functional Properties*. Boca Raton: CRC Press.

Lee, D. 2007. *Nature's Palette: The Science of Plant Color*. Chicago: University of Chicago Press.

10.
Food Additives

FOOD ADDITIVES AND THE LAW

A variety of different chemical substances are added to foods because of their functional properties. These substances can be synthetic or may even be found naturally in some foods, but when they are added to processed foods they are referred to as food additives and are subject to government regulation. These additives may serve many different roles, but they may not be used to deceive consumers or to conceal spoilage. In general, they also should not be used for purposes that can otherwise be accomplished by good manufacturing practices. Additives in processed foods have some reason for being in the food, because the law requires that they have some useful function and because they cost money and would decrease profits if they did not increase quality and sales.

Among the general public, food additives are often viewed with suspicion or prejudice. The presence of an additive on a product ingredient label, particularly if it has a "chemical"-sounding name, can result in considerable consumer resistance. This attitude is part of the larger suspicion of "chemicals" in foods and arises from the fear of being somehow poisoned by unscrupulous food manufacturers for the sake of increased profits. Conversely, the food scientist and toxicologist Joseph

Hotchkiss has said, only half in jest, that if one wants to be really safe, one should eat only additives, because foods themselves are not tested for safety, while additives are extensively tested. In the United States and most other countries, any scientific demonstration that a chemical is harmful results in it being banned for human consumption. The benefits to the consumer resulting from the presence of additives, such as the inhibition of spoilage or preventing the development of off-flavors and odors from lipid oxidation, can be considerable, and the increase in safety unquestionably saves, at a minimum, thousands of lives annually compared to a regime without food additives.

Many different purposes can be accomplished using food additives. We have already discussed a variety of these in previous chapters. For example, a number of different polysaccharides and gums are used as thickeners and stabilizers. Other molecules, such as monoglycerides and fatty acids, lecithin, tweens, and spans, are added as emulsifiers. Some sugars and other molecules are used as sweeteners, both caloric and noncaloric. Other molecules are used as fat substitutes. Many other molecules are added as flavorants as well, and we have discussed a variety of molecules that are added to foods as colorants, or to affect color, such as zinc chloride to react with pheophytin or reducing agents to convert metmyoglobin back to the red-colored reduced form. Even the CO_2 added to carbonated beverages is considered to be an additive. Additives can have more than one function, such as sugar, which can not only sweeten but also control water mobility, affect texture, and in some cases, as when added to pectin, promote gelation. Table 10.1 lists a few of the more common food additives.

In the United States, food additives are regulated by the Food and Drug Administration (FDA) under the authority of the Food, Drug, and Cosmetic Act of 1958. The Food Additives Amendment to that act defines a food additive as

> any substance the intended use of which results, or may be reasonably expected to result, directly or indirectly, in its becoming a component or otherwise affecting the characteristics of any food (including any substance intended for use in producing, manufacturing, packing, processing, preparing, treating, packaging, transporting, or holding food; and including any source of radiation intended for such use), if such a substance is not generally recognized, among experts qualified by scientific training and experience to evaluate its safety, as having been adequately shown through scientific procedures (or, in the case of a substance used in food prior to January 1, 1958, through either scientific procedures or experience based on common use in food) to be safe under the condition of its intended use; except that such a term does not include pesticides, color additives, and substances for which prior sanction or approval was granted.

Food colorants and agricultural pesticide residues are covered by other laws. Note that, under the amendment quoted above, the irradiation of food is treated as a food additive, even though no chemical is being directly added to the food, because the irradiation process can generate chemicals in situ that are not originally present in

Table 10.1 Some common food additives and their principal uses

Additive	Uses*
acetic acid, sodium acetate, etc.	antimicrobial agent, buffering salts
ascorbic acid (vitamin C)	acidulant, antimicrobial agent, antioxidant, reducing agent to maintain red color in meats, inhibits enzymatic browning in fruits and vegetables, flour dough improver
benzoic acid, sodium benzoate	antimicrobial
calcium chloride	firming agent for canned tomatoes and other vegetables
calcium propionate	antimycotic in baked goods
calcium stearate	anticaking agent
calcium sulfate	desiccant, coagulant
caseinate	binder, extender, clarifying agent, emulsifier, stabilizer
cellulose gum (the sodium salt of carboxymethyl cellulose)	anticaking agent, binding agent, thickener, stabilizer
citric acid, citrate	acidulant, tart flavor, antimicrobial agent, antioxidant, chelating agent, inhibits enzymatic browning by lowering pH
DATEM (Diacetyl Tartaric Ester of Monoglycerides, a mixture of mono- and diglycerides esterified with tartaric acid and/or acetic anhydride)	emulsifier for baking
dextrose (glucose)	sweetener, humectant, texturizing agent
dipotassium phosphate (K_2HPO_4)	buffering agent, sequestrant
EDTA	chelating agent, antioxidant preservative
ethoxylated mono- and diglycerides	emulsifier
ferrous sulfate	iron fortificant
glycerin (glycerol)	humectant, solvent, plasticizer, bodying agent
lactic acid	antimicrobial, flavor improver, inhibits enzymatic browning
locust bean gum, guar gum	thickeners, stabilizers
maltitol	lower calorie nutritive sweetener
maltodextrin	anticaking agent to promote free flowing, bulking agent, stabilizer and thickener, surface-finishing agent
methyl cellulose	thickener, stabilizer, emulsifier, film former, bulking agent, binder
monocalcium phosphate (calcium phosphate, monobasic)	buffer, dough conditioner, firming agent, leavening agent
potassium bromate	flour improver
potassium sorbate	antimicrobial preservative
polydextrose	Bulking agent, humectant, texturizer
polysorbates (e.g., Poly 60, Poly 80)	emulsifier, stabilizers
silicon dioxide, SiO_2	anticaking agent
sodium diacetate	sequestrant, preservative, mold inhibitor
sodium erythorbate	reducing agent, chelating agent
sodium phosphate (mono-, di-, and tribasic	emulsifier, buffer, nutrient, texturizer (dibasic only), water binding
sodium stearoyl lactylate	emulsifier, humectant
sulfur dioxide, SO_2	bleaching agent, preservative, antioxidant in wines
whey	formulation aid, processing aid, flavor enhancer, texturizer
xylitol	nutritive sweetener

*Sources: Food and Nutrition Board, Institute of Medicine, National Academy of Sciences, *Food Chemicals Codex*, 4th ed., (Washington, DC: National Academy Press, 1996); O.R. Fennema, ed., *Food Chemistry*, 3rd ed., (New York: Dekker, 1996).

the food. The law also recognizes two special categories of additives, those that are "generally recognized as safe" (**GRAS**), and those that were approved for use before the passage of this act in 1958; these are referred to as "**prior sanctioned**" additives. Prior sanctioned substances are approved for specific applications and in specific amounts. Examples include BHA and BHT as antioxidants; calcium propionate, sodium benzoate, sodium propionate and sorbic acid as antimicrobials; and sodium nitrate and potassium nitrate for the curing of red meat products and cured poultry products. Many of the most commonly used food additives have already been discussed in previous chapters. This chapter will introduce a few selected food additives that were not already discussed in those sections.

ACIDS

A number of simple acids are used as additives in foods (Figure 10.1). One of their most important uses is as a preservative, because many microorganisms either cannot survive or cannot proliferate in acid media. Weak acids are also extensively used as flavorants, because their tart or sour taste is one of the major categories of tastes in classical flavor theory.

The single most widely used acid as a food additive is **citric acid**, which is a good example of a polyfunctional additive. It has an important inhibitory function in controlling enzymatic browning in fruits and vegetables. Citric acid also functions as a chelating agent, to sequester metals, particularly calcium, and is widely used as

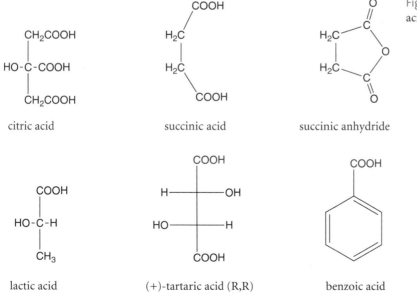

Figure 10.1. The structures of common acids used as food additives.

a flavorant, to add tartness. It is a weak acid, found naturally in small amounts in a number of vegetables and fruits, particularly citrus fruits such as lemons and limes, from which it gets its name. It can be isolated as a crystalline powder in two crystalline forms, an anhydrous crystal, and a monohydrate crystal incorporating one water molecule per citric acid molecule. Because it has three acid groups, it becomes progressively harder to remove each successive proton, with pK$_a$s of 3.15, 4.77, and 6.4. Often it is used as its sodium salt, **sodium citrate**.

Benzoic acid is frequently used as an antimicrobial agent and is one of the most important and widely used food preservatives, particularly in carbonated beverages, juices, jams and jellies, and salad dressings. It is usually added as an alkali salt such as sodium benzoate or potassium benzoate because of the low solubility of the protonated, neutral acid, and in weakly acidic foods it is then converted to benzoic acid. It suppresses not only the growth of bacteria but also molds and yeasts. Benzoic acid works by inhibiting the enzymes in the citrate cycle.

Tartaric acid, found naturally in wines and various fruits, as well as in juices and other foods, is used as an acidulant and as a chelating agent. Succinic acid is used as a plasticizer in bread dough and also as a chelating agent. Succinic anhydride is used to control water in dehydrated foods and baking powders by binding to water molecules.

The tribasic **phosphoric acid**, H_3PO_4, accounts for around a quarter of all the acids used in food processing. It is the only inorganic acid widely used in food manufacture, because it is only a moderately strong acid, unlike the other inorganic acids such as sulfuric or hydrochloric acids. Phosphoric acid is primarily used in the formulation of carbonated soft drinks to add tartness and as a preservative. Phosphoric acid has been linked to the development of kidney stones, and a recent study reported that drinking two cola drinks containing phosphoric acid per day substantially increased the incidence of kidney disease and the formation of kidney stones, while similar drinks flavored with citric acid did not produce the same effect (Saldana et al. 2007).

Acetic acid, CH_3COOH, is the principal acid component of **vinegar** and has long been used in this form to preserve foods as in pickling. It is used as an antimicrobial either as vinegar or as a purified component. Various salts of acetic acid, such as sodium acetate, calcium acetate, and potassium acetate are used in baking, ketchup manufacture, and in pickled vegetables. Of course, vinegar is widely used in prepared foods for its distinctive flavor as well.

Lactic acid is the bacterial fermentation product of glucose and lactose. It is also produced in muscle tissue during conditions of intense activity under oxygen deficit, and is responsible for the burning sensation in muscles that have been heavily exercised. Lactic acid is also responsible for the sour taste of yoghurt and sour milk. Lactic acid possesses one chiral carbon atom and thus occurs in two optical isomers. Only the *levo* isomer is important in animal metabolism. Lactic acid is sometimes used as a food ingredient, as an antimicrobial agent and as a flavorant, and to inhibit enzymatic browning. Commercial lactic acid is usually a mixture of the two isomers.

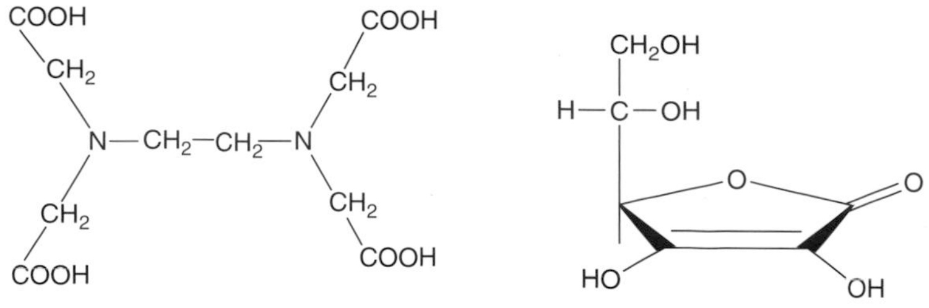

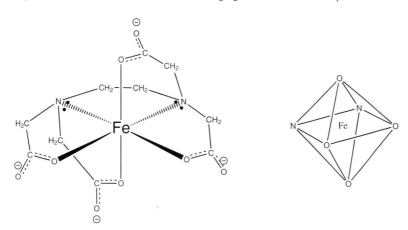

EDTA

ascorbic acid
(vitamin C)

erythorbic acid

Figure 10.2. The structures of the chelating agents most commonly used in foods.

Figure 10.3. Left: A picture of EDTA complexed with an iron atom, showing how the arms of the molecule reach around the ion to occupy the sites of iron ligands in an octahedral complex. Right: An illustration of the octahedral arrangement of these ligands, where only the iron and coordinated atoms are shown.

CHELATING AGENTS

Chelating agents are various molecules that bind tightly with prooxidant metal ions such as iron and copper, or alkaline earth ions such as Ca^{+2}, to sequester them in water-soluble complexes that prevent them from participating in other reactions (Figure 10.2). This can be helpful in inhibiting oxidation because it removes prooxidant metal ions that can serve as the initiator in lipid oxidation, or which can react with O_2 to generate singlet oxygen. The most widely used chelating antioxidants are citric acid, erythorbic acid, and **EDTA**, ethylenediamine tetraacetic acid. These molecules are polydentate and have sufficient conformational flexibility to complex with the target ion and satisfy its coordination requirements.

The most useful of these chelating agents is EDTA, because it can potentially provide six functional ligands separated by enough conformational freedom that they can wrap around a metal ion and occupy the six ligand sites of octahedrally coordinated iron (Figure 10.3). This arrangement is reminiscent of the claws of a crab, which was the origin of the word chelate, derived from the Greek word for "claw." Such a complex is called octahedral, even though there are only six ligands, because the positions of these coordinated atoms define the eight faces of a regular octahedron, as is shown in Figure 10.3. Because of this property, EDTA is widely used in foods containing fats to control iron-promoted oxidation such as occurs through the Fenton

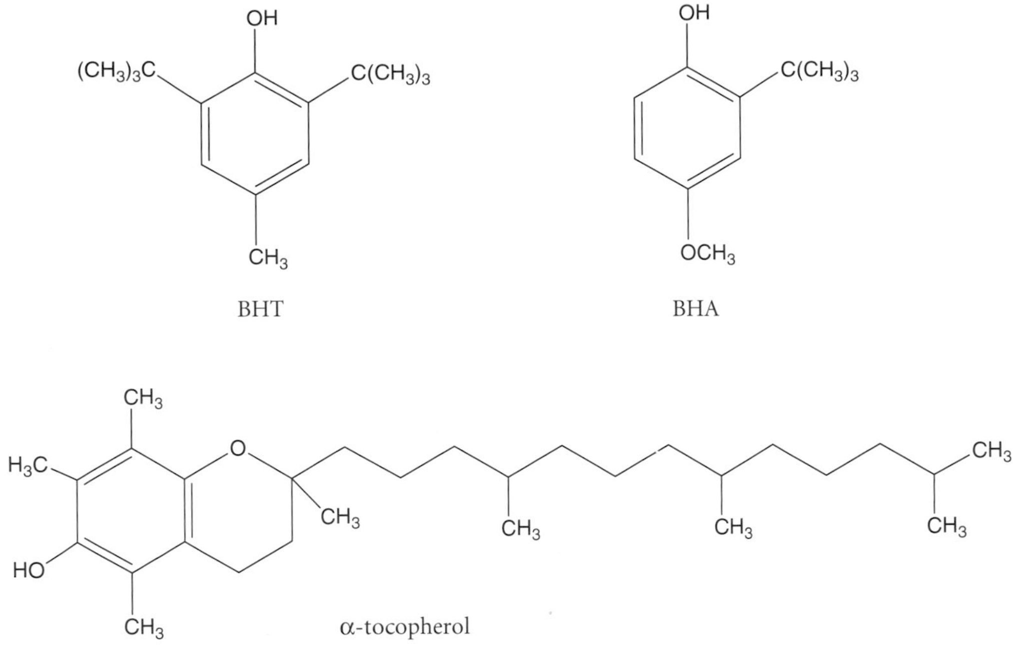

Figure 10.4. The structures of three of the most commonly used antioxidants.

mechanism. EDTA is also used in medical applications as a scavenger to treat heavy metal poisoning. Other available chelators, such as the related EGTA, ethyleneglycol tetracetic acid, which specifically binds calcium over magnesium, have important research applications, but are not approved for food use.

ANTIOXIDANTS

Antioxidants are added to foods to protect sensitive components, primarily unsaturated lipids, from degradation. Chelating agents such as EDTA and citric acid are antioxidants because of their effect of sequestering metal atoms. The synthetic antioxidants **BHT** (butylated hydroxytoluene) and **BHA** (butylated hydroxyanisole) and the naturally occurring vitamins **ascorbic acid** (vitamin C) and **α-tocopherol** (vitamin E) are the most widely used (Figure 10.4). The D isomer of ascorbic acid is **erythorbic acid** (usually used as its sodium salt, sodium erythorbate), which is also widely used as an antioxidant and preservative. These molecules can also serve as oxygen scavengers.

ANTIMICROBIAL AGENTS

A wide range of additives are used as preservatives to help control the growth of microorganisms such as bacteria and molds. Most of the acids used as additives have

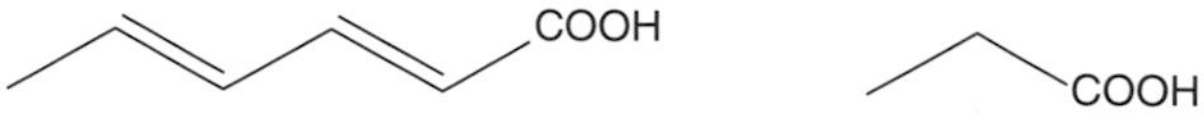

Figure 10.5. The structures of sorbic acid and propanoic acid.

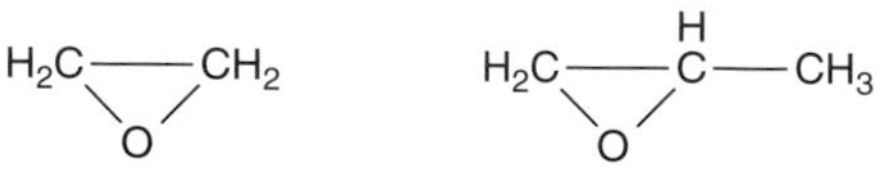

Figure 10.6. The structures of ethylene oxide and propylene oxide.

some measure of preservative function, regardless of what other functionality they possess, because as acids they lower the pH of the food, which inhibits the growth of most bacteria and molds. In addition to the acids discussed above, a number of other molecules are used as antimicrobial preservatives (Figure 10.5).

Sorbic acid is a short-chain (six-carbon) unsaturated fatty acid. It is widely used to control the growth of bacteria and particularly molds. Used most often as an antimycotic, sorbic acid is also quite useful for controlling Clostridium botulinum in fresh meat products. It is usually used in foods as its potassium or calcium salts. The sodium and calcium salts of **propionic acid** (also sometimes called propanoic acid) are used in bakery products such as breads to inhibit mold growth.

Sulfites and sulfur dioxide (SO_2) are used in dried fruits and vegetables, juices, and in wine before fermentation to prevent the growth of other microorganisms. We have already discussed the use of sulfites to prevent pigment formation in the Maillard reaction. Unfortunately, sulfites are also food allergens for some people and have been been banned by the US FDA on foods that will be eaten fresh, as in salads. We have also already discussed the use of nitrate and nitrite to control the growth of Clostridium botulinum in meat products, where they also help maintain the red color of meats.

Epoxides

Ethylene oxide and propylene oxide (Figure 10.6) are also used as antimicrobial agents. The various antimicrobial acids inhibit the growth of bacteria by lowering the pH, but do not necessarily actually kill them. Ethylene and propylene oxides actually kill microorganisms and are considered sterilants. They are powerful and are even capable of destroying spores and viruses. These cyclic ethers are very reactive, but the exact mechanism of their sterilizing effects is not completely known. We have already seen how these molecules can react with free hydroxyl groups of polysaccharides, as in the chemical modification of starches, to alkylate these groups with a hydroxyethyl group. This hydroxyethylation apparently interferes with bacterial metabolism. Ethylene oxide is a gas at room temperature (bp of 13 °C), and treatment of foods with either agent is done by placing the food in a closed chamber under an atmosphere containing 90% carbon dioxide and 10% of the oxide gas, to prevent the

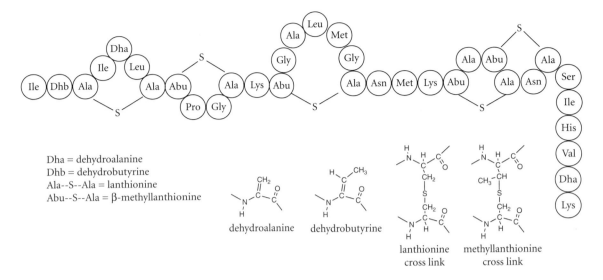

Figure 10.7. The structure of the antibiotic nisin.

flammable oxides from exploding. When propylene oxide is used, the system must be heated, because it has a higher boiling point, 34 °C. Only dry foods, such as spices, can be sterilized using these oxides, because they rapidly react with water to give glycols. This type of treatment is particularly useful for spices because sterilizing them by heating would in most cases lead to a significant loss of product quality as many of the spices contain volatile or thermally labile flavor compounds.

Antibiotics

In general, antibiotics are not used in foods because of the danger of the development of resistant bacteria, but at least two antibiotic-like molecules, nisin and natamycin, have been approved as food additives in the United States.

Nisin is a 34-residue polypeptide antibiotic produced by the bacterium *Lactococcus lactis* as a way of suppressing the growth of its competitors (Figure 10.7). It is a member of the lantibiotic family of antibiotics that contain lanthionine (refer to Figure 5.58 and the section on the reactions of proteins and amino acids). It is post-translationally modified to create several rings closed by thioether bonds. Nisin kills Gram-positive bacteria such as *Streptococcus* and *Staphylococcus*. Nisin is amphiphilic, with several hydrophobic residues at the N-terminus and three polar residues at the C-terminus, and is cationic due to the presence of three lysine residues. It works by forming pores in the membranes of target bacteria, but does not affect human cell membranes. It is approved for use in foods and is used in dairy products such as cheese and evaporated milk to control *Clostridium botulinum*. It is not used for medical purposes and does not lead to the development of resistant strains.

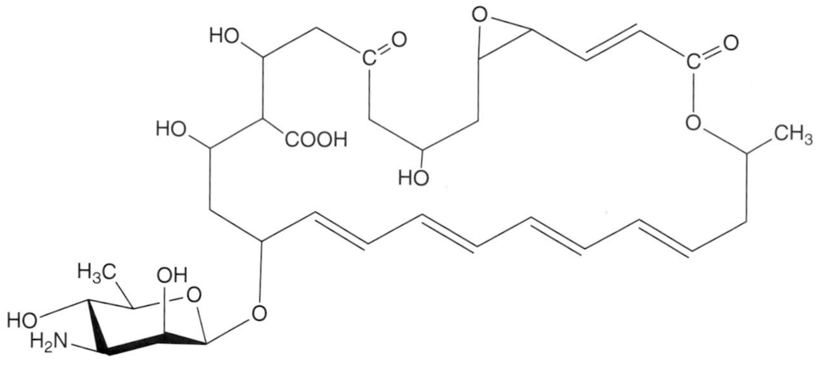

Figure 10.8. The structure of the antibiotic natamycin.

natamycin

Another antibiotic-like molecule that has been approved for use in foods in the United States is natamycin, which is an antimycotic (that is, it kills molds, as well as some yeasts). Natamycin is also used on dairy products to prevent the growth of molds on the surface of cheeses (Figure 10.8).

ANTICAKING AGENTS

A number of food additives are used to keep powdered or granular products able to flow freely. Calcium and magnesium salts of fatty acids, such as calcium stearate, are used in dried or powdered products to keep the ingredients from clumping together, or caking, and to keep them in a form that will flow freely during pouring. Because calcium binds so strongly to the fatty acid, the salt is nearly insoluble in water and binds to particle surfaces, making them water repellant.

Calcium silicate, $CaSiO_3$, is another anticaking agent often used in foods. It works by adsorbing a large amount of water. Because it forms strong hydrate complexes, $CaSiO_3 \cdot xH_2O$, it does not become sticky after adsorbing the water and promotes free flowing. Other anticaking agents are magnesium carbonate, magnesium silicate, aluminum phosphate, kaolin, various powdered starches, and microcrystalline cellulose particles.

CHEMICAL LEAVENING AGENTS

The food additive most commonly used as a chemical leavening agent is sodium bicarbonate, $NaHCO_3$, also called baking soda. Bicarbonate requires acid to release carbon dioxide gas, and in neutral water it simply dissociates. In the presence of acid, however, it undergoes the following reaction,

$$B-H + NaHCO_3 \rightarrow B^-, Na^+ + H_2O + CO_2$$

to produce carbon dioxide gas. A common chemical leavening mixture combines sodium bicarbonate with two acids, tartaric acid (or cream of tartar, its potassium salt) and sodium aluminum sulfate. The use of these two acids results in the release of carbon dioxide in two pulses, because the tartaric acid will react when mixed with water at room temperature, producing an initial burst of small CO_2 bubbles, while the sodium aluminum sulfate reacts at high temperature as the dough is being cooked, swelling the dough as the bubbles heat up, giving an open structure with large cavities before the protein/starch matrix becomes fixed. Traditional baking powder generally also includes cornstarch to adsorb water so as to keep the reagents dry and powdery. Other chemical leavening acids include monocalcium phosphate ($Ca(H_2PO_4)_2$, also called calcium dihydrogen phosphate), sodium acid pyrophosphate (SAPP, $Na_2H_2P_2O_7$), which is soluble in water, and acidic sodium aluminum phosphate, either $NaAl_3H_{14}(PO_4)_8 \cdot 4H_2O$ or $Na_3Al_2H_{15}(PO_4)_8$, which is not soluble in water but is soluble in hydrochloric acid.

CONFECTIONER'S GLAZE

Confectioner's glaze is a sheer glaze used primarily to coat candies and pharmaceuticals to give moisture protection, extend shelf life, and prevent sticking. Generally prepared as an alcohol solution, its primary component is **shellac**, a resinous secretion of the Asian lac beetle *Laccifer lacca*. Shellac is deposited by these insects onto the bark of trees, from which it is collected (lac resin is sometimes referred to as "beetlejuice"). Raw lac resin, which is reddish in color, is crushed, washed, and bleached by dissolving it in aqueous sodium hypochlorite, after which it is collected by evaporation as an off-white granular resin. Bleached shellac is insoluble in water but is soluble in alcohol for application to food products. Lac resin contains a wax that can be removed by filtration during the bleaching process to produce a dewaxed glaze. Although as an insect secretion it is in some ways analogous to honey, the collection process may inadvertently crush some of the insects and incorporate a small amount of insect parts into the product. As a result, some vegans object to the use of traditional confectioner's glaze, and it has led to widespread comments on the Internet that some manufactured candies are made from insects. Various substitute glazes can be produced using alternate lac-free formulations, such as blends of sugar and corn syrup.

POLYDEXTROSE

Polydextrose is a randomly combined polysaccharide of glucose and sorbitol in a roughly 90% to 10% ratio. It is synthesized through acid-catalyzed glycosidic condensation by melting glucose with small amounts of sorbitol and citric acid in the presence of phosphoric acid at elevated temperature. Because all types of glycosidic linkages are present, including branches and $(1\rightarrow6)$-linkages, polydextrose has a low digestibility, yielding only about 25% of the calories of an equivalent weight of sugar. It is water soluble and stable at both low pH and high temperature. Used as a replacement for starch, sugar, or fat in sugar-free, low sugar, or diabetic products, polydextrose is marketed by Danisco under the brand name Litesse.

Because of its low digestibility, polydextrose is marketed as a type of dietary fiber, and it has been shown to function well in this capacity without altering blood chemistry (Jie et al. 2000), although other studies have suggested that it functions differently from more conventional dietary fiber (Yoshioka, Shimomura, and Suzuki 1994). Jie et al. also found no significant laxative effect, although other studies have suggested that consuming large amounts of polydextrose can cause gastrointestinal distress, perhaps due to the osmotic effect caused by the water binding capacity of the undigested polymer in the intestine. Polydextrose is approved for use in foods in the United States, Canada, the European Union, Australia, New Zealand, and a number of other countries. Often used as bulking agent or filler in cakes and other baked foods, candies, and desserts, it is not sweet tasting but can add texture without the sweetness.

SUGGESTED READING

Food and Nutrition Board, Institute of Medicine, National Academy of Sciences. 1996. *Food Chemicals Codex*, 4th ed. Washington: National Academy Press.

Rulis, A.M., and J.A. Levitt. 2009. "FDA's food ingredient approval process: Safety assurance based on scientific assessment." *Regul. Toxicol. Pharm.* 53:20–31.

Alle Ding' sind Gift und nichts ohn' Gift;

allein die Dosis macht, das ein Ding

kein Gift ist.

[All things are poisonous and nothing
is without poison;

the dose alone makes a thing
not poisonous.]

PHILLIP VON HOHENHEIM, KNOWN AS PARACELSUS

11.
Food Toxicology

TOXICITY AND DOSE

The above quotation from the sixteenth-century Swiss-German physician, alchemist, mystic, and astrologer Paracelsus, generally considered to be the father of toxicology, is often taken to be a foundation statement of the field. Although it might be considered somewhat extreme, too much of anything in the carefully orchestrated balance of reactions that constitute a living organism can seriously disrupt its function. The truth of this statement was tragically underlined in 2007 by the death of a young woman in Southern California who participated in a radio station–sponsored contest to see who could consume the most water in a short time. Sadly, her death from water intoxication was not a completely isolated event, because forced overconsumption of water in fraternity hazing has also led to deaths, although the overconsumption of alcohol is a more common cause of hazing fatalities. As we have already noted, however, a lack of water will also lead fairly rapidly to death. Many substances like water are innocuous when consumed in reasonable amounts, while others are extremely dangerous even at very low levels. Thus, as Paracelsus is often paraphrased, the dose makes the poison. The field of food toxicology attempts to study and characterize the toxic effects of foods and their components and contaminants on humans, so that the food supply can be made as safe and nutritious for the consumer as possible.

As the example of water intoxication indicates, the question of what is toxic can be complex. The harm from drinking excessive amounts of water comes from causing ionic imbalances through the disruption of normal osmotic equilibria, somewhat like the effects of the excessive vomiting and diarrhea caused by cholera or the induced vomiting of bulimia. In contrast, the protein neurotoxins of cobra venom (see below), which would fit virtually anyone's idea of a poison if introduced intravenously into the bloodstream, are directly injurious and function to shut down neural transmission, leading to terminal respiratory failure in a short time. If thoroughly cooked and eaten as a food, however, these same proteins might simply be digested and cause no harm at all (providing there are no open sores in the mouth or ulcers in the stomach). Not all proteins are harmless when eaten, because as we have seen, the botulinum toxin is one of the most lethal poisons known (see Table 11.1), but again, thorough cooking renders it harmless. Consuming the denatured prion protein PrP^{Sc}, whether cooked or not, leads to fatal dementia (variant Creutzfeldt-Jakob disease) in some individuals, while producing no effect at all in most others. The chemical element arsenic, which interferes with ATP production, is toxic to all, whether it is eaten or introduced directly into the blood stream, but the lethal dose is more than an order of magnitude greater for oral consumption, because not everything that is eaten is absorbed across the intestinal lumen (that is, is "bioavailable"). Cooking has no effect on the toxicity of arsenic.

The effects of a toxin might be manifest in the short term, which is referred to as acute toxicity, or might require a long time to develop. Many substances might exhibit little acute toxicity and produce no detectable harm if exposure is limited to a few isolated instances, but might produce serious long-term consequences, such as cancer or other life-threatening diseases, from chronic exposure to much smaller amounts. For example, as readers of murder mysteries know, acute arsenic poisoning can lead to fairly rapid death. Low levels of arsenic poisoning will not produce this effect, but long-term low-level exposure to arsenic through food or drinking water may lead to slowly developing health conditions or diseases such as skin cancer. In the case of the cholera mentioned above, the lethal effects are the result of an oligomeric protein toxin secreted by the *Vibrio cholerae* bacterium in infected individuals. Because the toxin is produced inside the individual, however, such an illness would not normally be considered the province of food toxicology, nor would the bacterium itself, because the microbial safety of foods is generally considered to be part of food microbiology.

In light of Paracelsus' assertion that all substances can be poisons, being able to at least characterize what constitutes an acute lethal dose of a given toxin is helpful. This is done using the median lethal dose, the LD_{50}, which is defined as the average dose of a toxic substance needed to kill 50% of a test population, usually of laboratory rats or mice. Being a median quantity, the LD_{50} will not kill some individuals, while a smaller dose might be fatal to others. Obviously the picture becomes more complex if those test animals that are not killed by the LD_{50} doses are nevertheless permanently

Table 11.1 Approximate oral LD$_{50}$ levels in rats for selected substances

Substance	LD$_{50}$ (mg/kg)
ethanol	10,000
NaCl	4000
menthol	3300
vanillin	2800
estragole	1230
morphine sulfate	900
caffeine	200
phenobarbital sodium	150
DDT	113
white arsenic (As$_2$O$_3$)	15
strychnine sulfate	2
nicotine	1
cyanide	0.5
tetrodotoxin	0.1
dioxin	0.001
botulinum toxin	0.00001

harmed by the toxin in some way. The utility of the LD$_{50}$ as a measure of acute toxicity is that it allows a cross-comparison of the danger posed by various substances. For example, the intravenous LD$_{50}$ for western diamondback rattlesnake (*Crotalus atrox*) venom is 4.20 mg/kg, while that for Asian cobra (*Naja naja*) venom is 0.40 mg/kg, and that for the Indian banded krait (*Bungarus caeruleus*) is only 0.09 mg/kg. Note that these LD$_{50}$ doses are expressed as milligrams of the toxin per kilogram of weight of the victim, on the assumption that the results of the animal testing will translate on a per weight basis to other species such as humans. Note also that these LD$_{50}$ doses were specified as being for intravenous injection, because the mode of exposure matters. For foods, of course, oral ingestion (and to a much lesser extent inhalation of vapors) is the form of exposure that is relevant, although in manufacturing a food product it is also necessary to consider the harm that might come to processing workers through other forms of exposure. We have already seen that arsenic is more than an order of magnitude less lethal when eaten than when injected intravenously. Table 11.1 lists measured oral LD$_{50}$ levels for the oral consumption of a number of substances spanning nine orders of magnitude of lethality on a per weight basis. Caffeine, table salt, menthol, vanillin, estragole, and ethanol are included on the list, even though these food substances are consumed regularly by most people without undue concern. As can be seen from the table, caffeine is almost two orders of magnitude more acutely toxic than alcohol. Statistically though, caffeine causes few if any human deaths, while chronic alcohol abuse kills large numbers annually via many mechanisms.

A definition is needed from a toxicological sense of the two related concepts of hazard and risk. A **hazard** is a potential threat or danger, such as being poisoned, struck by lightening, or hit by a car. The **risk** is the statistical probability that some adverse effect or injury will occur from the hazard at a specific dose or level of exposure.

Because it follows from Paracelsus' definition of toxicity that all substances can be safely consumed in some sufficiently small amount, another quantitative measure of toxicity related to the LD_{50} would be the maximum dose level that produces no detectable toxic effect. This measure is called the NOEL limit, the No Observed Effect Level, or the NOAEL, No Observed Adverse Effect Level. Necessarily even more vague and difficult to quantitate than the LD_{50}, the NOEL can be used as a guideline in estimating a safe level of exposure. In the case of foods, the amount of a substance that can be safely consumed is called the acceptable dietary intake, or ADI. Because of the uncertainty in estimating NOELs, a common rule of thumb is to err on the side of safety and employ an arbitrary safety factor that sets the ADI at one-hundredth of the NOEL,

$$ADI = NOEL \times Safety\ Factor \tag{11.1}$$

where the safety factor = 1/100.

Animal testing to determine LD_{50} doses, and even NOELs, is problematic for several reasons. One of these is that not all biological effects scale linearly with mass or size. Another is the assumption that all species respond to toxins in the same way, because there certainly are species-specific toxic effects. For example, the rat poison warfarin is effective in killing rats and mice but is much less toxic to humans, and even has pharmaceutical uses as an anticoagulant in cardiovascular disease. The lipid-like molecule persin (Figure 11.1), found in avocados and foods such as guacamole made from them, is toxic to many animals such as cats ands dogs, but is essentially harmless to humans (apart from being allergenic in some individuals). For this reason, pets should not be allowed to eat avocados. Theobromine and caffeine are also toxic to pets, and particularly cats (which is why your cat should not eat chocolate with her coffee!), as are many of the sulfides in onions and garlics. In yet another example, the Bt toxin discussed before affects primarily the Lepidoptera and is generally harmless to other animal species, including other insects.

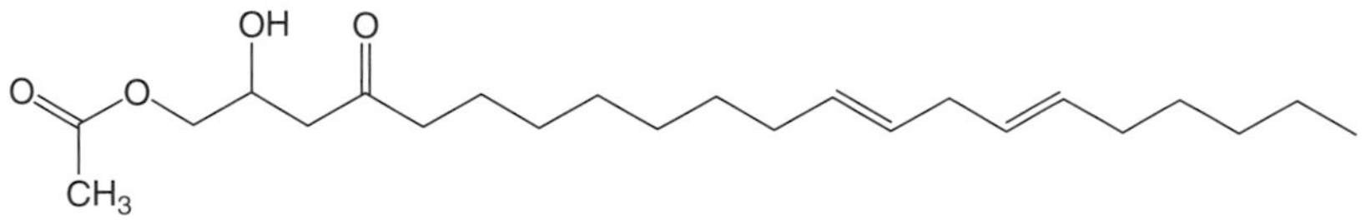

Figure 11.1. The structure of the toxic lipid-like molecule persin from avocados.

persin

Perhaps the most important problem with the determination of LD_{50} doses in the laboratory, however, is that it obviously involves the killing of a great many lab animals, which sometimes experience considerable pain or suffering in the process. Animal rights groups have for many years campaigned for the abandonment of LD_{50} testing, and the United Kingdom has banned oral LD_{50} testing except in cases where no suitable alternative is available, such as in testing foods for botulinum contamination. In the United Kingdom as well as elsewhere, LD_{50} testing is being replaced by the so-called fixed-dose procedure, in which a smaller number of laboratory animals are given smaller doses that result in signs of toxicity without killing the animals, from which attempts are made to estimate what a lethal dose might be based on previous experience and extrapolation. Obviously this test is still unsatisfactory from an ethical point of view for many reasons, because the animals may still suffer even if they do not die, and they are still being objectified, in the sense that they are being used as a means to an end. In addition, the fixed-dose tests are less scientifically useful as well, because they involve estimates and extrapolations. In 2006 the European Union approved alternative methods to animal testing that employ cell cultures to determine the acute toxicity of tested substances. These tests allow a significant reduction in the numbers of laboratory animals used in testing, because a suspected toxin is only tested on animals if it shows no effects on cultured cells. A preliminary finding of acute toxicity for cells is taken as an indication that proceeding to actual tests on animals is not necessary. Most animal testing of cosmetics has been banned in the European Union since the beginning of 2009 and is being replaced by such methods.

In reality, quantitating the level of harm that might result from exposure to a given substance is inherently difficult due to the many ambiguities and many possible forms of harm that might be experienced. A substance does not have to be acutely lethal in any reasonable amount to be considered toxic if it causes some nonfatal harm. For example, a substance that caused mental retardation, blindness, or even just baldness or skin rashes would be considered toxic even if it did not kill its victims. Carcinogenic substances are another class of ambiguous cases, because they do not directly poison, but unquestionably cause harm by inducing the development of serious and potentially fatal disease. Teratogenic agents, which cause malformations of embryos or fetuses, leading to birth defects, would be considered dangerous even if they do not directly harm the mother who consumed these agents. The acute lethality listed for table salt in Table 11.1 is due to osmotic disruption, but a regular diet high in salt can result in high blood pressure, which has been shown to be strongly correlated with heart disease and especially stroke. This type of long-term harm is usually considered to be a nutritional question, but of course such distinctions are arbitrary and, given the vagueness of the definition of toxicity, could be treated as a type of toxicity as well. Similarly, overconsumption of alcohol leading to addiction (alcoholism) is normally considered a medical and psychiatric problem, but attendant physiological conditions such as cirrhosis of the liver could also be considered as toxic effects.

Various types of animal testing other than LD_{50} measurements are widely used to determine whether particular substances or diets produce long-term harmful effects.

Generally these experiments involve feeding studies comparing effects in different groups of test subjects with those in control groups. Often these experiments use rats, mice, or occasionally pigs, whose digestive systems are more similar to that of humans than those of rodents, as well as various other animals for more specific purposes (the proverbial "guinea pig" is actually uncommon as a test subject in most modern animal testing). Tests for carcinogenicity generally involve lab rats or mice, which are often fed doses of the test substances orders of magnitude larger than any human would consume, on a relative weight basis, under the assumption that such doses might substitute for the larger cumulative exposure that a human might experience over a lifetime 20–40 times longer than a lab animal. The validity of such an assumption is not always clear and may well depend in unknown ways on the substance being tested. Differences in metabolism, developmental histories, nutritional requirements, and behavior can also make it difficult to extrapolate from animal tests to humans, particularly if effects on intellectual development are suspected. Since the effects of a substance may interact with those of other components of the diet, it becomes very difficult to identify toxic effects with certainty.

Obviously consumers would be expected to want to avoid food toxins, but this clearly becomes difficult under the Paracelsus definition, because according to it, all substances, including all foods, must be considered as toxins. In manufacturing and preparing foods the obvious objective is then to keep all potentially harmful substances to amounts significantly below the level that might produce any detectable harm; that is, below the ADI limit.

Another factor that must be considered in judging the relative risk of some hazard is how the risk is perceived. For example, probably the most dangerous activity that the average person undertakes on a regular basis is driving. Yet most people do not perceive risk that way and readily choose to drive not only because of the freedom of movement that it gives them, but also because of the perceived personal control they believe they have over their own safety. Many people, in fact, will express fear of flying in airplanes or of riding on trains even though these modes of transportation are demonstrably vastly safer on a statistical basis than driving. The element of choice then is important in determining what level of risk a consumer deems acceptable. A home gardener battling aphids on her tomato plants might choose to spray them with pyrethrin or other pyrethroid insecticide, but might object strenuously to even lower levels of the same pesticide use on tomatoes produced by some large agricultural enterprise, if she feels she has no choice in assuming the attendant risks, and if all of the benefits accrue to the producers rather than to her.

Apprehension about or fear of genuine threats is not illogical, but often subjective and emotional factors affect the way in which hazards are perceived by the public and magnify the fears of possible dangers beyond the level that would be reasonable given the actual risk. For example, in the United States, out of a population of 320,000,000, fewer than two people per year die of spider bites, and only around fifteen people die of snakebites each year (and of those, one study suggested that many of the cases may be due to inebriated individuals trying to kill or even pick up the snakes). In contrast, usually more than 100 individuals die of bee stings each year.

Bees, however, are not regarded with the same degree of fear and dread as snakes and spiders. Bears kill on average less than one person per year in the United States, and the danger from attack by alligators and pumas is similarly extremely low. Most people naturally want to avoid the pain of bee stings, although they are not particularly afraid of bees, yet they might express sheer terror of rattlesnakes, bears, alligators, and pumas (particularly if they are life-long city dwellers), even though the actual risk from these potentially dangerous animals (hazards) is minute. The leading causes of death in our national parks are traffic accidents, drowning, falls, hypothermia, lightning, and heart attacks, rather than attacks by animals (including even bees). Yet the dramatic and primal nature of being killed by a wild animal magnifies the perceived danger of such an event far beyond the fear of much more serious, but mundane, hazards.

While perhaps no fear resonates more deeply in our collective subconsciousness than that of the fangs of a dangerous predator, the fear of being harmed by what we eat is also deep seated, ancient, and powerful, and in many individuals approaches an obsession that transcends ordinary logic. Often such an attitude is coupled with an intuitive personal concept of toxicology that divides substances into foods and toxins. For the latter, in the minds of such people, there are no acceptable levels or degrees of danger. As in the example of the actual risk from wild animals versus the risk from mundane accidents in national parks, the emotional content of fears about being poisoned by food often leads people to ignore real dangers, such as contaminating bacteria, which the CDC estimates kills 3,000 people every year in the United States alone (Scallan et al. 2011).

Given Paracelsus' assertion that all substances are potential toxins, it is clearly impossible to give a comprehensive treatment of food toxicology here, or even a satisfactory survey of the possible toxins. In this chapter we will limit ourselves to consideration of only a few examples of food toxicants that are either serious problems or are particularly interesting cases, or that are of widespread interest among the general public, even if the fears are exaggerated. These cases will include the real danger of methyl mercury in seafoods, pesticides in vegetables and fruits, arsenic in drinking water, and dioxin, as well as a variety of natural food toxins. First, however, let us consider an example of a supposed case of food "poisoning" that illustrates some of the difficulties inherent in this field. This example involves a number of cases reported in the United States, in the late 1960s and 1970s, of various reactions to eating food with monosodium glutamate (MSG) that came to be known as Chinese restaurant syndrome.

CHINESE RESTAURANT SYNDROME

Chinese restaurant syndrome is the name originally applied to a vague collection of nonspecific symptoms reportedly experienced by certain individuals after eating at Chinese restaurants. The symptoms are said to include numbness in the neck and arms, weakness, headache, nausea, and palpitations. These symptoms were first described by R.H.M. Kwok in 1968 (Kwok 1968), based on his own personal and anecdotal experiences, and led to the widespread belief that eating Chinese food could produce some sort of allergic reaction. The belief in this syndrome led to the identification of MSG, a common flavor additive in some Chinese foods, as the causative agent (Schaumburg et al. 1969).

However, the existence of Chinese restaurant syndrome has been challenged based on the available data (Mosby 2009; Williams and Woessner 2009). MSG has GRAS (generally recognized as safe) status, and the studies available to date have not conclusively demonstrated a significant threat from the normal consumption of glutamate. As a nonessential amino acid, significant amounts of glutamate are produced endogenously by humans. In addition, many other common foods contain large amounts of glutamate, including, for example, pizza, but there has been no similar identification of a "pizza parlor syndrome," and the FDA now recommends that the syndrome be referred to as "monosodium glutamate symptom complex.". The fact that glutamate is a neurotransmitter, with an entire class of protein receptors evolved specifically to recognize and bind to it, suggests at least the possibility that consumption of large amounts of glutamate might lead to some sort of temporary imbalance in their functioning. MSG has long been suspected as a trigger for migraine headaches, and some studies report that large doses of MSG could produce headaches (Baad-Hansen et al. 2009; Yang et al. 1997). Nonetheless, scientific studies have not been able to conclusively verify or disprove the existence of any actual Chinese restaurant malady (Mosby 2009; Williams and Woessner 2009), and some workers now consider "Chinese restaurant syndrome" to be an urban myth. Fortunately, the stakes in this case seem rather less serious than cases involving potential carcinogens or agents suspected of causing other serious and lasting harm. Nonetheless, many millions of dollars have been expended on developing products that contain no added MSG, and probably considerable amounts of money and business were lost over this issue by the nation's Chinese restaurants. And as with the cases of the artificial sweeteners discussed in Chapter 8, the resolution of the question of safety about MSG as an additive remains elusive. Some food manufacturers now substitute hydrolyzed soy protein for MSG (Chapter 5), since this mix of amino acids naturally contains glutamate and can accomplish some of the desired flavor enhancement without the necessity of listing MSG on the label.

TOXINS IN FOODS

Let us now turn to a survey of some examples of much more dangerous toxicants that are found in foods. Some of these are general environmental contaminants that are present in foods as the result of industrial pollution. Others, specifically pesticide residues, are present on foods as the result of direct application to crops but also are present in other foods, most notably seafoods, as the result of agricultural runoff into waterways and ultimately the ocean. Some toxicants are naturally occurring poisons in certain species. Yet other toxicants in foods are the result of particular types of processing. All represent potential threats that must be understood and controlled.

Industrial Pollutants

Industrial activity produces large amounts of waste and by-products, many of which are potentially toxic. Often this material is spewed into the atmosphere through smokestacks or discharged into streams and rivers in wastewater, and inevitably some of this pollution ends up in foods. In North America, Western Europe, and Japan, strong antipollution laws have substantially reduced the quantities of these industrial wastes in recent decades, so that, in general, smaller amounts are found in foods from these regions. In developing economies such as China and India, however, pollution laws and their enforcement are much weaker, and enormous quantities of pollution are discharged directly into the environment in these countries. In those places, food contamination by such pollutants as arsenic and heavy metals, which are now rare elsewhere, is still a major problem. This pollution is more than a local problem, however, for several reasons. One of these is that pollution respects no political boundaries. Toxins discharged into rivers anywhere, except those with interior drainage basins, ultimately end up in the oceans, contaminating seafood marketed and consumed worldwide. Similarly, air pollution is carried downwind to other countries, so that toxins discharged into the air in China, for example, contaminate fields and streams in Korea and Japan and become incorporated into the foods produced there. Another reason that such local pollution can be a worldwide problem is that, given the new global nature of markets, foods grown or manufactured in developing countries are sold and eaten around the world, sometimes escaping the tougher regulation that would have applied if they had been produced in the consuming nations. Even in Europe and North America, with their stricter laws, industrial activities such as the burning of fossil fuels discharge significant amounts of mercury into the global environment, resulting in potentially dangerous contamination in seafoods. Literally thousands of possible toxic compounds are produced from industrial pollution, but we will consider only a handful of the most important here, including dioxins, polychlorinated biphenyls, arsenic, and mercury.

Dioxins and PCBs

Dioxins and polychlorinated biphenyls (PCBs) are among the most widespread industrial pollutants in the global environment and represent a serious toxicological problem. Both of these classes of molecules are very dangerous chemicals with the potential for causing serious harm if consumed in significant amounts. Both classes of molecules are particularly problematic because they are generally chemically stable and persist in the environment for long times, contaminating soil and water and becoming incorporated into growing plants. In the animals that eat contaminated plants, these compounds become concentrated in fatty tissues where they neither break down nor are eliminated, so they become progressively more concentrated in animal foods as they move up the food chain.

The term "dioxin" most properly refers to two isomeric monocyclic ring compounds each containing two oxygen atoms, 1,2-dioxin and 1,4-dioxin (Figure 11.2). While toxic, these two molecules are not an important source of environmental pollution. The related molecule diphenyl-*p*-dioxin, or dibenzo-*p*-dioxin, is the parent compound for a series of multiply chlorinated derivatives called polychlorinated dibenzodioxins (PCDDs), loosely referred to collectively as dioxins, which are a significant class of pollutants that are known carcinogens and mutagens. The most dangerous member of this class, 2,3,7,8-tetrachlorodibenzo-*p*-dioxin, was an important contaminant in the Agent Orange defoliant used as a chemical warfare agent in the Vietnam War and became well-known to the general public when veterans of that war began to suffer a variety of maladies as a result of their exposure to Agent Orange. PCDDs are generated whenever organic material is burned with chlorine, and various types of industrial incineration generate the overwhelming majority of the dioxins that are released as pollutants into the environment. The primary source of exposure for the general public is through foods that are contaminated as the result of this air pollution. Dioxins are soluble in fats and accumulate in fatty tissues, where they are neither broken down nor excreted, and have an elimination half-life that is not precisely known, but might exceed the lifetime of the affected individual. As a result, dioxins become concentrated in animal tissues as they move up the food chain. In laboratory animals, exposure to dioxins causes cancers, birth defects, liver problems, and a number of other maladies. The health consequences in humans are less well known, but they are suspected of having these same effects and are also believed to cause diabetes.

Polychlorinated biphenyls (PCBs) are a class of somewhat similar molecules that also contain two benzene rings with multiple chlorine atoms substituting for

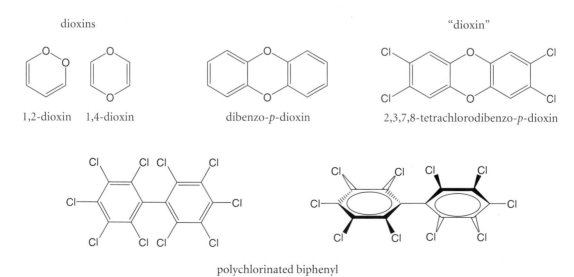

dioxins   "dioxin"

1,2-dioxin   1,4-dioxin   dibenzo-*p*-dioxin   2,3,7,8-tetrachlorodibenzo-*p*-dioxin

polychlorinated biphenyl

Figure 11.2. The structures of the dioxins and polychlorinated biphenyls (PCBs).

hydrogens. Unlike dioxins, however, PCBs contain no oxygen atoms. In the twentieth century PCBs were used for a wide variety of industrial purposes: as a hydraulic fluid and in paints, flame-retardants, adhesives, and sealants. The most important use, however, was as a coolant in electrical transformers and capacitors. Enormous quantities of PCBs were manufactured for these latter purposes in most industrialized nations. Because they are very carcinogenic, PCBs were banned in the United States for open applications (such as paints) in 1973, and Congress outlawed their production in the United States in 1977, although they are still allowed in some closed transformers and capacitors. Most of the industrialized nations of the world have now banned the production of PCBs. Unfortunately, these bans have not eliminated the problem of environmental pollution by PCBs.

PCBs are not very soluble in water, but they are soluble in organic solvents and in fats, and concentrate in fatty tissues in the body when consumed in foods. Due to their large number of heavy chlorine atoms, PCBs are denser than water and thus sink to the bottoms of lakes and rivers. They are stable chemically and do not readily degrade, and thus persist as a significant environmental contaminant in many places decades after they were banned. For example, more than half a million kilograms of PCBs were discharged into the Hudson River in New York during the middle decades of the twentieth century as a byproduct of the manufacture of electrical transformers. Extraordinary sums of money have been expended removing contaminated sediments from the bottom of the river, an effort that continues to this day, but the fish in the Hudson River may still remain too contaminated with PCBs to be safely eaten, particularly by children and pregnant women.

As can be seen from Figure 11.2, PCBs that contain chlorine atoms on the *ortho* positions next to the adjacent ring experience steric clashes that force them to rotate about the carbon-carbon bond joining the rings, forcing the two rings to be non-coplanar. Those PCBs that are not substituted at these *ortho* positions, called non-*ortho* PCBs, have coplanar aromatic rings and behave somewhat like dioxins. These non-*ortho* PCBs tend to be the most toxic members of the class.

Mercury

As a heavy metal, mercury is quite toxic and dangerous. Exposure to liquid mercury metal, and particularly to mercury vapor, can lead to a number of serious health effects, but the most common form of exposure for the general population is through consumption of foods contaminated with mercury. In this context, the contaminant is not pure mercury, but rather methyl mercury, or methyl mercury cation. (Figure 11.3) Burning fossil fuels in automobile engines and industrial activities such as the burning of coal in power plants release trace amounts of elemental mercury as a gas, which cumulatively amounts to many tons annually in the United States alone. This mercury precipitates out of the atmosphere in rainwater, and in aquatic ecosystems such as lakes, rivers, swamps, and the ocean, anaerobic microorganisms convert this mercury into methyl mercury. Additional mercury enters the atmosphere from nonanthropogenic sources such as forest fires and volcanoes, but the principal source in the modern

CH_3-Hg^+ Cl^-

Figure 11.3. Methyl mercury.

world is industrial activity. The ionic methyl mercury cation can complex with nitrate, chloride, or hydroxyl anions to form salts and is soluble in water due to its charge. Methyl mercury has a high affinity for sulfhydryl thioalcohol groups (—S—H) and bonds covalently to the free sulfhydryl groups of cysteine residues in proteins.

Methyl mercury is not easily eliminated from living organisms, so it becomes progressively more concentrated as it moves up aquatic food chains, as plankton consume the bacteria that produce methyl mercury and are themselves then eaten by fish, which are in turn eaten by larger fish. In marine ecosystems in particular, the food chain can be complex, with several levels of progressively larger predators eating smaller fish, leading to high levels of concentration of methyl mercury in the largest predatory fish at the top of the food chain, such as sharks, tuna, sea bass, marlin, king mackerel, and swordfish. Methyl mercury is only slowly eliminated from the body, with a half-life approaching two months in humans. Thus, individuals who regularly consume large amounts of fish, and especially top predators such as large tuna and sharks, may develop high levels of mercury. For this reason, the US FDA recommends that women of childbearing age, and particularly pregnant women and nursing mothers, as well as children, limit their weekly intake of such fish. Pregnant women are particularly advised to avoid such fish, because methyl mercury easily passes across the placenta and the blood–brain barrier, and is known to interfere with neurological development, and thus can lead to a lowering of IQ for exposed fetuses. High levels of methyl mercury poisoning can lead to severe neurological symptoms. Incidents of such severe methyl mercury poisoning occurred in Japan during the 1960s as the result of industrial pollution and were referred to as Niigata Minamata disease, after the localities where outbreaks occurred. More recently, spot studies in sushi restaurants in several US cities, including New York, found surprisingly high levels of mercury in some of the most expensive types of sushi, such as yellowfin tuna. This was because the most expensive cuts often come from the largest fish, which of course grew large by eating many smaller fish and concentrating their mercury. These fish are more expensive because they are more rare, because the numbers of the top predators in any food chain must be much lower than the numbers of their prey.

Methyl mercury in seafoods presents an interesting and troubling dilemma for people trying to eat "healthily." Many nutritionists recommend eating seafoods because they have higher concentrations of omega-3 fatty acids than are generally found in foods from the land. The putative neurological and cardiovascular benefits of these fatty acids may be offset by the deleterious effects of methyl mercury, however. Judging the balance of these effects is made almost impossible by the uncertainty in the data about both omega-3 fatty acids and methyl mercury. One could perhaps achieve the optimum safety/benefit balance by choosing fish from lower tropic levels, such as herring and trout, or smaller individuals of the higher levels. Farm-raised tilapia and salmon usually have very low levels of mercury (although these fish also often have lower levels of ω-3 fatty acids too), as do shrimp. Note from Table 7.3 that yellowfin tuna, which is high in mercury, is also low in ω-3 fatty acids.

Arsenic in Drinking Water

Arsenic is the Group V semimetal element with atomic number 33 (see the periodic table, Figure 1.1). Its chemistry is similar to that of phosphorous, the element immediately above it in Group V. Its most common oxidation states are +3 and +5, and it is often found in the environment as an oxide or sulfide. Arsenic is able to replace phosphorous in the phosphates of biological molecules such as proteins and nucleic acids. One of the most common, and most toxic, forms is arsenic trioxide, or white arsenic, As_2O_3, which is much more toxic than elemental arsenic.

As we have already seen, arsenic is a dangerous poison, and it has been used since ancient times for murder and assassination. It functions as a poison by inhibiting the production of ATP, by interfering with the activity of the enzyme pyruvate dehydrogenase, leading to cell death. Arsenic compounds were formerly used as insecticides and pesticides, and contamination of foods used to occur from its use for these purposes, but this is no longer a significant problem, because these pesticides are now prohibited in most countries. Arsenic has many industrial uses, and industrial arsenic contamination remains a potential problem. Until relatively recently, arsenic in the form of oxides (primarily As_2O_3), mixed with oxides of copper and chromium, was used to pressure treat wood to protect it from insect and fungal damage. Although this practice has now largely been discontinued, older wood that was so treated can still present a danger of arsenic poisoning. Arsenic can leach out of the wood into the soil, and if it is burned, the ash from such wood also is harmful. In spite of this, the primary source of human exposure in most places where it occurs is through natural, low-level contamination of ground water. The contamination results from the leaching of arsenic from certain types of geological formations, which occurs regionally in many places, particularly in Southern Asia and some parts of the Northern United States. The concentrations are orders of magnitude below the LD_{50} dose listed in Table 11.1. The potential danger arises not from the acute toxicity, but because chronic low-level exposure to arsenic has been associated with an increased probability of developing several types of cancers, including skin cancer. Generally, arsenic contamination of foods or food products is not a significant problem, although water used for food processing in some regions may need to have arsenic removed by one of several available methods.

Toxins from Packaging and Containers

Foods sometimes can be contaminated by toxins leaching from the materials used in packaging. It has been suggested that the failure of the famous 1845 Victorian expedition to find the legendary Northwest Passage, led by Sir John Franklin on the HMS *Erebus* and HMS *Terror*, may have been due at least in part to lead poisoning among the crewmen from lead-contaminated solder used to seal the thousands of tin cans of food taken along as provisions. Lead poisoning is the most common and widespread form of heavy metal poisoning. It has even been suggested that such lead poisoning may have contributed to the fall of the Roman Empire, because lead was used extensively to line and seal water pipes in antiquity, and lead cooking utensils were in common use at that time (however, these situations also existed, at least to some extent, during the rise of the Empire). Metal food cans generally no longer use lead sealant, and their interiors are covered by polymer films (which may be a source of contaminants themselves; see below). Fine leaded crystal, which gets its sparkle from the higher index of refraction that results from the addition of lead oxide (PbO) to the silica of ordinary glass, over time can leach lead into acidic foods stored in containers made from it. The quantity that might leach into wine poured into a leaded crystal goblet during a short period is probably not significant, but studies have shown that significant amounts of lead can leach into wine stored for extended periods in crystal decanters. Thus, long-term storage (several days or more) of wines in crystal should probably be avoided. Improperly glazed pottery can also sometimes be a source of lead in home storage of foods.

Not all lead in foods comes from packaging and containers; old lead water pipes and natural groundwater contamination can result in trace amounts of lead in foods. In general, however, lead contamination of processed foods in the developed world is a thing of the past, and most lead poisoning is due to environmental exposure, such as lead-based paints.

Other potential toxins can contaminate foods by leaching out of modern packaging, which has aroused concern among consumer groups as well as regulators. A particular example is bisphenol A, popularly referred to as BPA, shown in Figure 11.4. BPA is used industrially to synthesize several types of plastics, most particularly polycarbonates, polyesters, polysulfones, and epoxy resins. These materials are generally labeled as recycling type 7 plastics. Polyvinyl chloride, recycling type 3, often contains BPA added as an antioxidant to protect the phthalates that are included as plasticizers to this class of plastics (phthalates themselves raise additional toxicological concerns). Polyethylene terephthalate, high-density polyethylene, low-density polyethylene, polypropylene, and polystyrene (recycling types 1, 2, 4, 5, and 6, respectively) do not contain BPA. Polycarbonate is widely used to make water bottles and containers, and thin layers of epoxy resins are used to coat the insides of beverage and food cans.

People eating canned foods or drinking canned sodas or water from polycarbonate bottles are possibly exposed to BPA at potentially significant levels. The leaching of BPA from packaging is greatest for acidic foods and at elevated temperatures. Bisphenol A is an endocrine disruptor, and a number of negative health effects have been alleged to result from exposure to BPA, including neurological disorders and various cancers. It is not yet clear that the levels of exposure that occur from food and water containers are harmful, but possible regulations governing exposure levels are being discussed in many countries.

Pesticides

The term "pesticides" includes a wide variety of agents designed to kill plant, animal, or even fungal species that damage agricultural crops. As such, these chemicals are

Figure 11.4. The structure of bisphenol A (BPA).

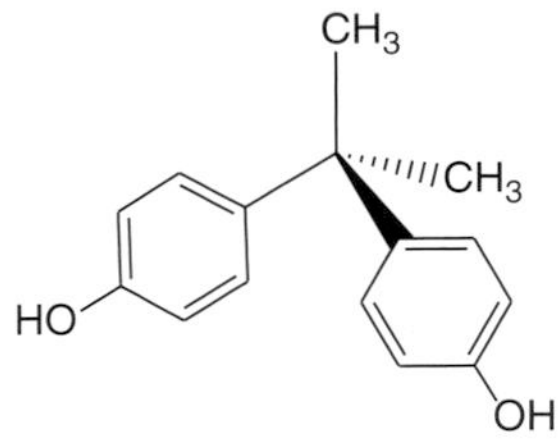

intended for direct application onto food plants, or at least the fields in which they are grown, unlike the general industrial pollutants just discussed.

Organochlorine Compounds

Organochlorine pesticides (Figure 11.5) are neurotoxins that affect not only insects but other animals as well, including humans. The most famous of this class is DDT (dichlorodiphenyl trichloroethane), first developed in the nineteenth century, which was used to make dramatic progress against insect pests and insect-borne diseases such as malaria. DDT is only moderately toxic to humans, with an LD_{50} of 113 mg/ kg. It has been suggested, however, that DDT exposure can also produce developmental effects in infants and possibly increase the likelihood of developing breast cancer later in life, but the evidence for these effects is ambiguous. Unfortunately, DDT and other organochlorines are not easily broken down in the environment and concentrate in organisms as they move up the food chain, having especially devastating effects on birds, making the shells of their eggs too thin to survive. After international campaigns stemming from the publication of Rachel Carson's landmark book *Silent Spring* describing the harmful effects of pesticides, DDT was banned for agricultural use throughout the world, although it is still allowed for controlling disease-carrying insect pests. DDT was banned in the United States in 1972, and as it has gradually been cleared from the North American environment, populations of raptors such as eagles and ospreys that had been severely endangered by DDT have begun to recover. The case of DDT use is particularly complex and difficult. Millions of human lives were saved by DDT spraying campaigns in the tropics, and the incidence of malaria in Africa initially increased dramatically after the ban on DDT use. More targeted indoor spraying has reduced environmental damage from DDT use while also significantly reducing malaria rates but does not eliminate human exposure, including through food stored and consumed indoors.

The organochlorine Aldrin was widely used on such crops as potatoes and corn from the 1950s through the 1970s to control grasshoppers and soil pests such as root worms, termites, and beetles, but is no longer permitted in the United States due to the risk it poses as a carcinogen. Aldrin breaks down to give the closely related dieldrin, but dieldrin is extremely long lived and is only slowly being cleared from the North American environment. It is also a suspected carcinogen. Because these compounds are no longer used, they are rarely found in significant amounts as contaminants in foods in the United States.

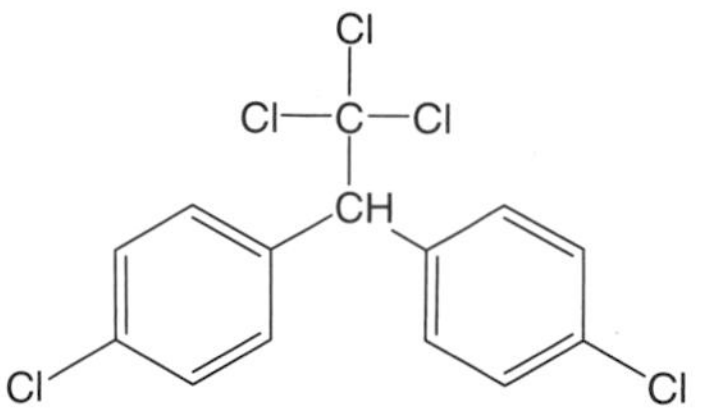

Figure 11.5. Organochlorine pesticides.

DDT
dichloro-diphenyl-trichloroethane

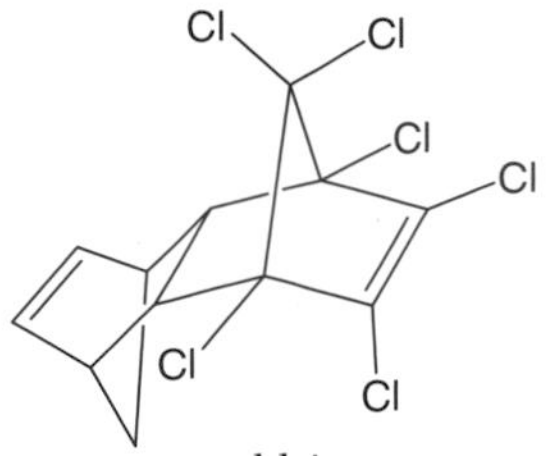

aldrin

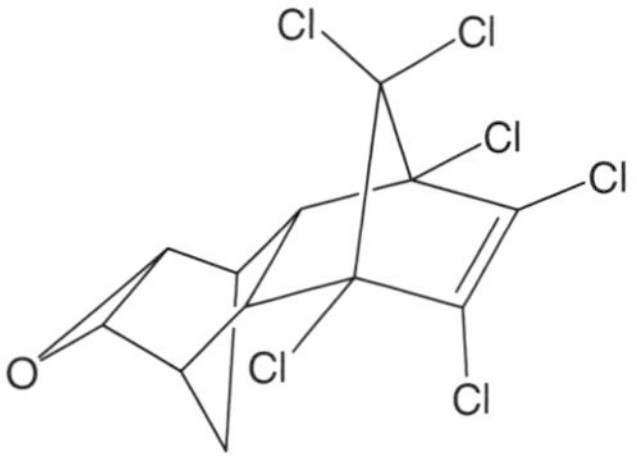

dieldrin

Organophosphates

Organophosphates are esters of phosphoric acid. This class of molecules includes many important biological molecules, including the nucleic acids DNA and RNA. Many neurotoxins, such as the chemical weapons agent Sarin, are also organophosphates that irreversibly inhibit the enzyme acetylcholinesterase, which catalyzes the cleavage of acetylcholine into choline and acetic acid, an essential step in neurotransmission. The organophosphate insecticides, which include the well-known parathion, malathion, phosmet, and methamidophos, function in this same way (Figure 11.6). As neurotoxins, these molecules are potentially very dangerous not only to humans, but to almost all animals. Agricultural workers using these chemicals are at particular risk and have to take precautions to avoid exposure. Parathion is considered to be the most dangerous of these and has been banned in a number of countries, including Germany, but not yet in the United States. Malathion is considered to be safer, and is even used to treat humans for head lice. Organophosphates are unstable and rapidly degrade when exposed to heat, sunlight, or oxygen. Thus they do not accumulate in the environment and usually leave little trace on the food plants with which they are treated. For this reason, they are generally considered to be preferable to organochlorine pesticides such as DDT. This breakdown of organophosphates is not always benign; in the case of malathion, it degrades to malaoxon, which is considerably more toxic. Trace amounts of organophosphate pesticides can sometimes be found on fruits and vegetables and represent a potential danger to consumers. An example of exposure to methamidophos occurred in Japan in 2007–2008 in dumplings imported from China without adequate inspection, in an apparent act of industrial sabotage rather than a contamination that occurred in the field.

Figure 11.6. Organophosphate pesticides.

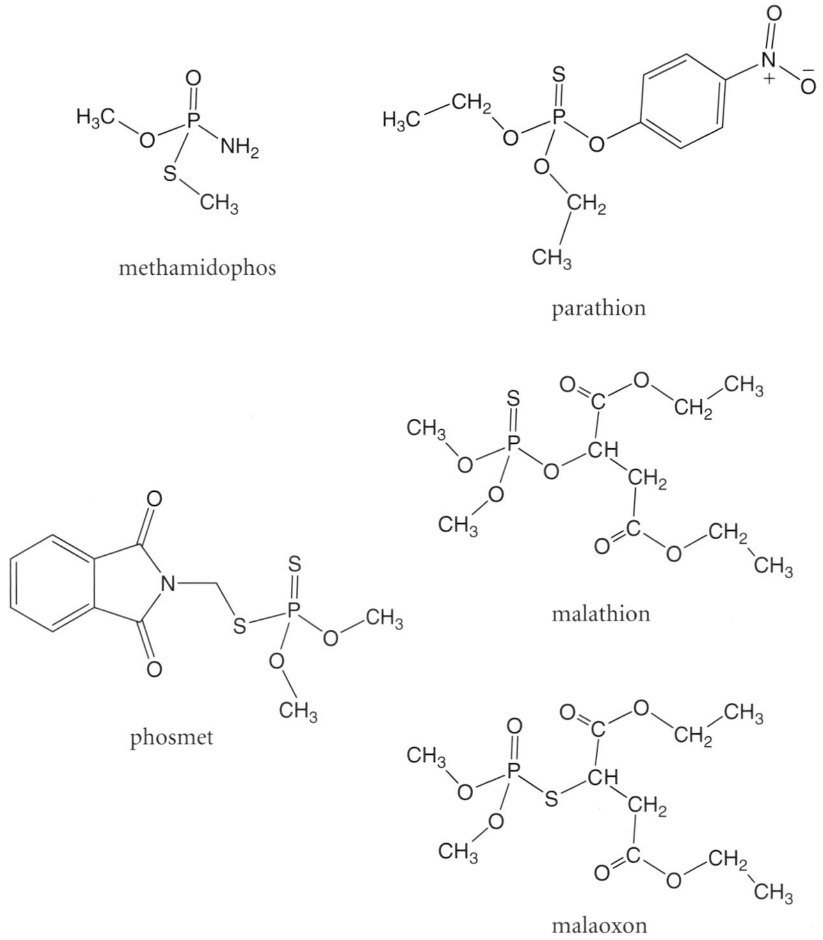

methamidophos

parathion

malathion

phosmet

malaoxon

Pyrethroids

Pyrethrins are naturally occurring plant insecticides found in the seeds of the chrysanthemum plant pyrethrum (*Chrysanthemum cinerariaefolium*) from the Balkan Peninsula of Europe. Two principal forms of natural pyrethrins occur, called pyrethrin I and II, as illustrated in Figure 11.7. They function as neurotoxins for insects and mites but generally are only mildly toxic to mammals and birds, although they are very harmful to fish. Pyrethrins are biodegradable and do not accumulate in the environment, because they readily break down when exposed to sunlight. For this reason pyrethrin insecticides are considered to be among the safest pesticides and are replacing organochlorines and organophosphates in agriculture. In lower concentrations pyrethrins are used as insect repellants, and pyrethrins are considered to be safe enough to be used on household pests.

The pyrethrins serve as templates for an entire class of similar synthetic compounds called pyrethroids that share the same characteristics as the pyrethrins themselves. Pyrethroids generally have a strained cyclopropane ring connected by an ester linkage to other ring moieties. Permethrin is a synthetic, broad-spectrum,

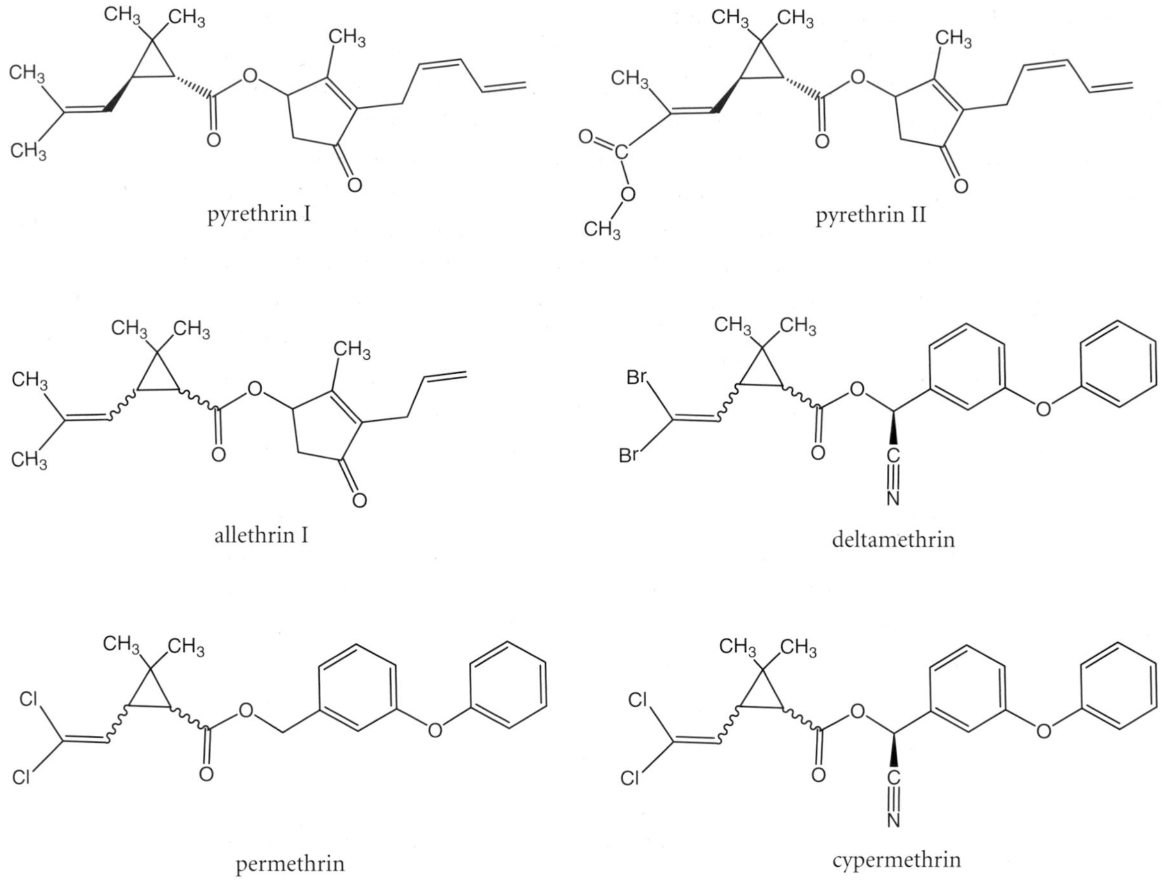

Figure 11.7. Pyrethroid insecticides.

chlorinated pyrethroid used on many food crops, including corn and wheat. It is also used to kill mites and ticks in pet treatments. Like other pyrethroids, it can kill non-target beneficial insects such as bees and is very toxic to fish and even small mammals. Cats are sometimes accidentally sickened or killed by inappropriately large doses of tick treatment intended for large dogs. Cypermethrin is similar to permethrin, containing a cyanide functional group, and deltamethrin is identical to cypermethrin except that it is brominated instead of chlorinated. Because these pyrethroid insecticides are unstable and readily degrade, the main threat that they might pose in foods comes from concentrated surface contamination in fruits and vegetables rather than accumulation in plant or animal tissues.

Naturally Occurring Toxins

It does not follow that just because a substance occurs naturally in foods it is safe or healthy to eat. We have already seen that several of the molecules involved in flavors and aromas, particularly in plants, can be toxic at high levels, even when those same compounds might have very beneficial effects at low levels in the diet. Some plants, like the hemlock used in the execution of Socrates, are very toxic and have been recognized as poisons since ancient times. Many plants produce toxins in their fruits or leaves specifically to deter animals from eating them. Bitter-tasting alkaloids are prime examples of such defensive compounds. The glycoalkaloid **solanine** in potatoes that have turned green from exposure to light is such an alkaloid toxin (although probably does not serve a defensive function). Consumption of high levels of solanine can lead to neurological disruption. Many legumes contain antinutritional factors such as protease inhibitors and lectins, including such dangerous toxins as the ricin of jack beans, that make them unhealthy to eat raw. Fava beans in particular are responsible for many cases of chronic poisoning by two endogenous pyrimidine glycoside toxins called vicine and convicine. Overconsumption of fava beans leads to a serious, even potentially fatal, hemolytic anemia called **favism**. The cyanogenic glycosides discussed in Chapter 4 are another dramatic example of poisons occurring naturally in common foods. Other toxins, like the botulinum toxin from *Clostridium botulinum* contamination, can be present naturally in foods as the result of microbial growth. It is, of course, as important to control these toxins in the food supply as it is to prevent contamination by man-made toxins.

Toxins in Fish and Shellfish

There are several examples of marine toxins in seafoods. In all of these cases, the toxin is not actually produced by the animal itself but rather by other organisms that either contaminate the seafood or were eaten by the animal and thus incorporated into its body. While these toxins are not usually a major public health problem in the United States, poisonings do regularly occur, and we will briefly consider some of the more interesting examples of such toxins.

Scombroid Fish Poisoning

Fish of the family Scombridae, which includes a number of popular commercial food species such as tuna, bonito, sardines, mahi-mahi, and mackerel, are responsible for a type of foodborne illness called Scombroid fish poisoning. This condition results from high contents of histamine (Figure 11.8), the prime mediator of allergic reactions, in the fish muscle tissue as the result of bacterial breakdown by decarboxylation of the amino acid histidine, present in large amounts in Scombridae fish. Because the toxin is histamine, the symptoms of Scombroid fish poisoning, which develop within about two hours and include headache and gastrointestinal pain, vomiting, and diarrhea, can resemble an allergic reaction, so that it is often misdiagnosed. Because the cause of this concentration of histamine is bacterial growth, Scombroid fish poisoning can be prevented through proper storage and handling during processing, including in particular proper refrigeration or freezing, especially for fish caught in tropical latitudes. Contamination occurs not just in commercial operations, however, but can

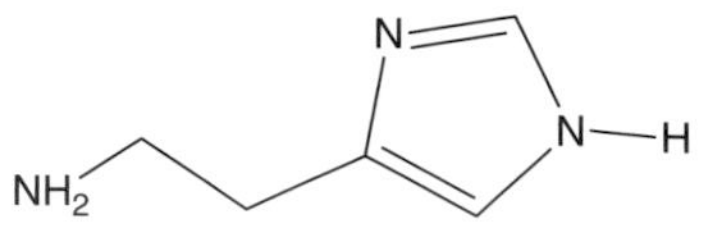

Figure 11.8. Histamine.

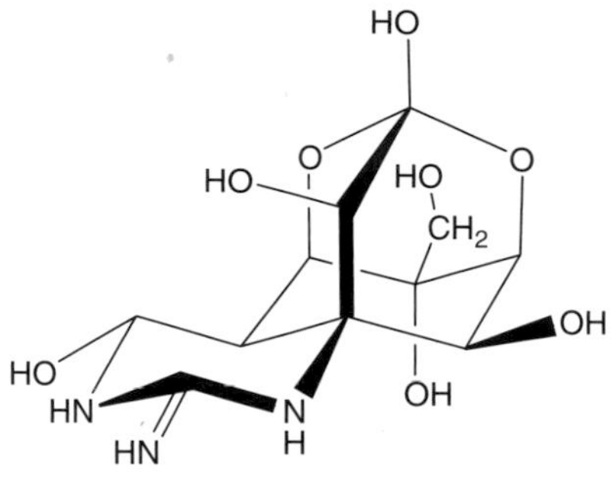

histamine

Figure 11.9. The tetrodotoxin responsible for the toxicity of *fugu*.

tetrodotoxin

affect fish caught by amateur and sport fishermen if their catches are not properly cooled; as Robert Gravani, a past president of the IFT, says, if you are spending a warm summer day out fishing on your boat, "ice the fish, as well as the beer." Because histamine is thermostable, cooking does not render contaminated Scombridae fish safe. While unpleasant, Scombroid fish poisoning usually is not fatal.

Fugu and Tetrodotoxin

Puffer fish, or *fugu* in Japanese, are fish of the family Tetraodontidae that are primarily found in warm waters of the tropics and subtropics, with a few species occurring in temperate zones. As the hybrid Greek/Latin technical name implies, these fish have four strong front teeth (two upper and two lower) that are used to crush the shells of the mollusks and crustaceans that are their primary food. From these food species the puffer fish become inoculated with bacteria of the genus *Vibrio* that produce a complex organic toxin which, because of its association with the puffer fish, is called tetrodotoxin (Figure 11.9). This toxin concentrates in the skin and internal organs of the puffer fish, making them very dangerous to eat. Tetrodotoxin binds to a site at the opening of the fast voltage-gated sodium channel of muscle cells, preventing them from contracting and leading in severe cases to muscle paralysis and rapid death from respiratory failure. As a food, puffer fish have a mild, almost bland taste, and are expensive, which makes them less popular outside of Japan and Korea. In Japan, chefs are required to undergo extensive training in the careful removal of the viscera of the puffer fish before being licensed to prepare and sell *fugu*. Nonetheless, many cases of tetrodotoxin poisoning are reported annually in Japan, usually with more than a dozen fatalities per year, but this type of poisoning is rare elsewhere. No antidote or antivenin for tetrodotoxin exists, and patients suffering from tetrodotoxin poisoning

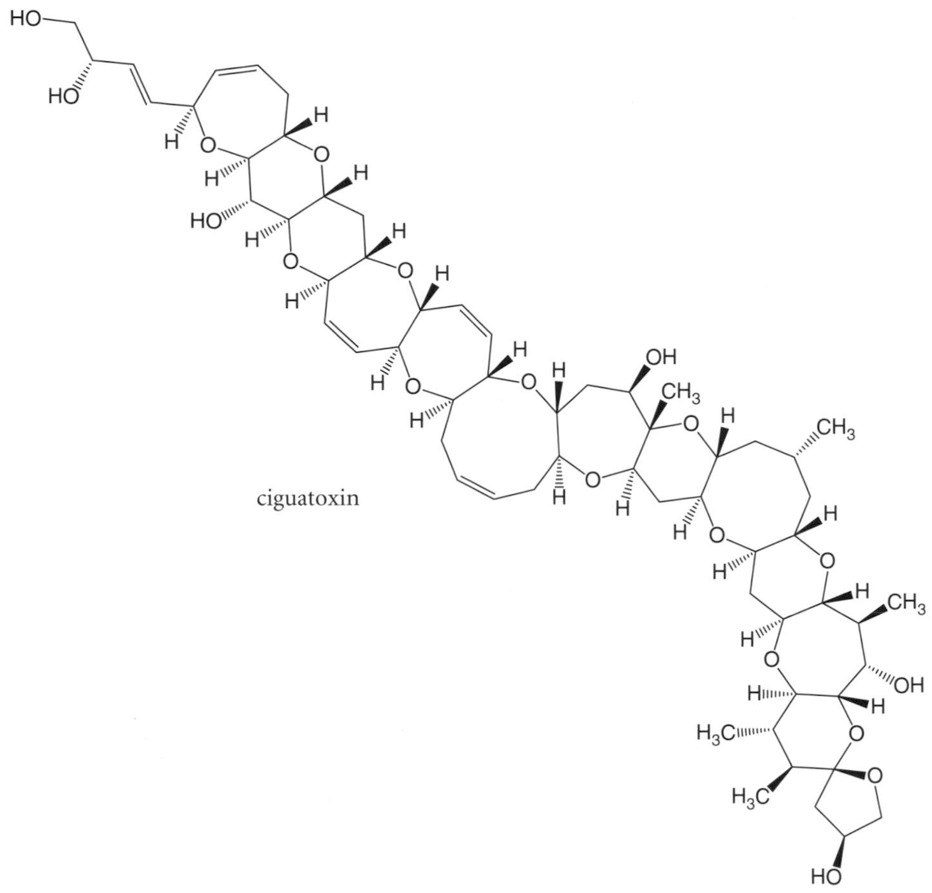

Figure 11.10. The structure of ciguatoxin.

must be kept on total life support until the toxin gradually clears from their system. Because tetrodotoxin does not cross the blood-brain barrier, the patient remains conscious during this period, although paralyzed and unable to speak or otherwise move.

Ciguatoxin

Another illness found in tropical regions that is caused by a fish-borne toxin is called ciguatera. This illness is caused by ingesting fish contaminated with a toxin called ciguatoxin (Figure 11.10), which is produced by the dinoflagellate *Gambierdiscus toxicus* that lives on marine algae. As is the case with methyl mercury and PCBs, this algae and the associated dinoflagellates are consumed by invertebrates, which are in turn eaten by small fish, which are then eaten by larger fish, and so on. Because the ciguatoxin bioaccumulates and is not easily eliminated, it becomes progressively more concentrated as it moves up the food chain, reaching dangerous levels in top-level predators such as barracuda and sharks, and particularly in certain Pacific and Caribbean reef fish.

Ciguatoxin poisoning is rarely fatal, but can cause severe neurological symptoms, including numbness or tingling of the extremities, headaches, hallucinations, unco-ordinated muscle movements (ataxia), as well as muscle pain, and unpleasant gastro-intestinal disturbances. Ciguatoxin is a remarkable polyether molecule consisting of an unusual long series of fused heterocyclic rings of several sizes (6, 7, and 8 atoms), each containing a single oxygen atom, and with a large number of chiral carbon atoms with specific stereochemistries. It is heat stable, so cooking the contaminated fish does not affect the toxicity. Because ciguatoxin is so slowly eliminated from the body, the symptoms of ciguatera can persist for years, although usually the duration is weeks or months.

Brevetoxins and Shellfish Contaminated by Red Tide

Red tides are frequent phenomena in shallow coastal waters, caused by blooms of phytoplankton, including various algae containing pigments such as red-colored carotenoids. These algal blooms are not really directly associated with the tides, and are not necessarily red colored. Some of these phytoplankton blooms are harmless, but some are quite harmful indeed and contain organisms that produce dangerous toxins. In these cases, filter-feeding shellfish living in the shallow waters can become dangerously contaminated with these toxins. The causes of red tides are not always clear; some surely must be due to natural causes because they have been reported for many centuries, including the reports of explorers in relatively undisturbed waters, but many others are apparently due to industrial or agricultural pollution, particu-larly to an overabundance of nutrients from agricultural runoff. Red tides in the Gulf of Mexico often contain large numbers of another dinoflagellate, *Karenia brevis*, which produces a neurotoxin called brevetoxin B that is similar in structure to cigua-toxin (Figure 11.11). Like ciguatoxin, brevetoxin B is a large polyether consisting of a series of fused heterocyclic rings. It is also a neurotoxin that disrupts normal neuro-logical functioning, although poisoning by this toxin is not usually fatal.

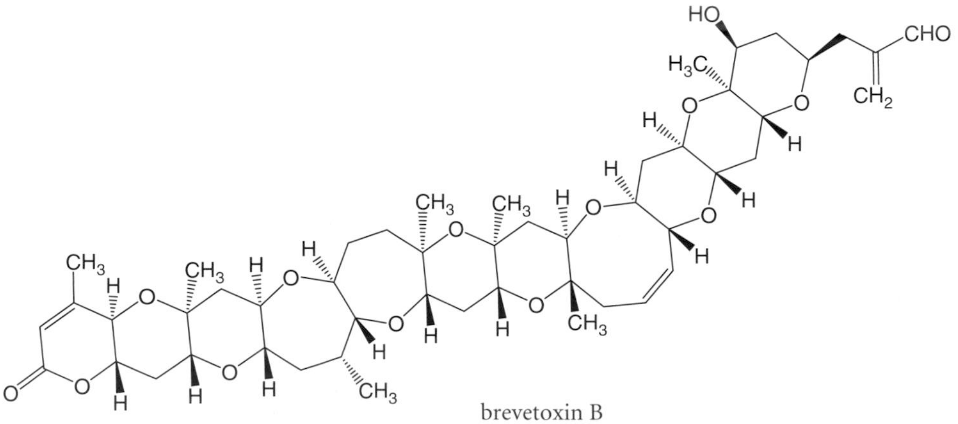

brevetoxin B

Figure 11.11. The structure of brevetoxin, one of the neurotoxins responsible for the poisoning of shellfish by red tides.

Mycotoxins

If the preceding litany of industrial and natural toxins in fishes has scared you away from seafood, let us briefly consider the dangers posed by fungal toxins in some land-based foods, including mold-contaminated grains and the well-known cases of poisonous mushrooms.

Molds

Molds are microscopic fungi that grow in multicellular filamentous structures called hyphae. Interconnected masses of these hyphae can grow into visible colonies; most people are familiar with such growths on foods left for too long in the refrigerator. Molds are often responsible for food spoilage, and their growth is favored by damp conditions. Molds generally subsist by secreting enzymes that hydrolyze polysaccharides and other polymers in their host matrix to produce simple sugars as food. Some also produce toxins that can be a serious source of toxic contamination, particularly in cereal grains. Probably the two most important of these from a food safety standpoint are aflatoxins, produced by various species of *Aspergillus*, and fusarium toxins, produced by molds of the genus *Fusarium*.

Aflatoxin

Aflatoxins are a class of mycotoxins that commonly contaminate cereal grains such as wheat, rice, and corn (maize), as well as many other types of foods, including nuts, seeds, and certain spices. Aflatoxin molecules have a complex structure containing five fused rings (Figure 11.12), three of which are heterocyclic containing oxygen atoms. More than a dozen different types of aflatoxins exist; the most toxic is aflatoxin B_1, shown in Figure 11.12. Aflatoxins are produced by molds of the genus *Aspergillus*, including *Aspergillus flavus* and *Aspergillus parasiticus*, which grow well in warm, moist conditions. *Aspergillus* growth and contamination with aflatoxin are a danger when harvested grains are stored under these conditions, but grains can become contaminated with aflatoxin from *Aspergillus* even while still growing in the field. When aflatoxin-contaminated grain is used as feed for dairy cattle, their milk can become contaminated with metabolic derivatives of the aflatoxins, such as aflatoxin M_1 (Figure 11.12), a carcinogenic metabolic product of aflatoxin B_1.

Figure 11.12. The structures of some examples of aflatoxins from *Aspergillus*.

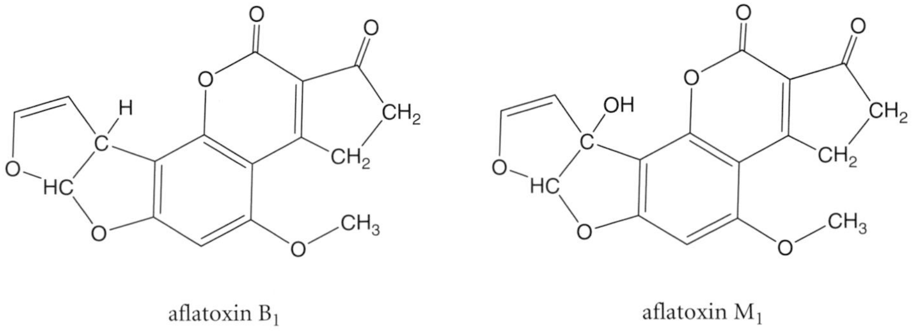

aflatoxin B_1

aflatoxin M_1

Aflatoxin B_1 is one of the most potent carcinogenic substances known. The oral acute LD_{50} dose for aflatoxins is around 7.2 mg/kg in mice, which makes it somewhat less toxic than strychnine. However, aflatoxin poisoning can cause serious liver damage, including cirrhosis, necrosis, or liver cancer in the case of chronic exposure, even though acute poisoning is usually not fatal in humans. Aflatoxin outbreaks tend to be most common in regions with warm, moist climates, such as parts of Africa and Southeastern Asia. The aflatoxins fluoresce with a greenish-gold color when irradiated with ultraviolet ("black") light with a wavelength of 365 nm (Figure 9.4), which forms the basis for one of the tests for contamination of grain with aflatoxins.

Fusarium

Molds of the genus *Fusarium* also can infest cereal grains stored under moist conditions, and these molds produce toxins that can cause potentially fatal alimentary toxic aleukia, ATA, with a distressing series of unpleasant symptoms resembling hemorrhagic fever, as well as possibly leading to a permanent immune system suppression. The principal toxins produced by *Fusarium* are fumonisins, which take their name from the genus, and trichothecenes. The trichothecene toxins include neosolaniol, T-2 toxin, and T-2-tetraol. These toxins, typified by T-2 toxin, shown in Figure 11.13, are thought to be primarily responsible for ATA. T-2, with an oral LD_{50} in rats of 3.8 mg/kg, has a fused polycyclic structure capped by an epoxy ring.

Ochratoxins, such as ochratoxin A, shown in Figure 11.14, which are dihydroisocoumarins produced by *Aspergillus ochraceus*, are another class of common mycotoxins found on grains, particularly corn, dried fruits, and wine. These molecules, which can cause kidney damage, are less acutely toxic than the trichothecenes, but may be carcinogenic under chronic exposure conditions.

Ergot

Ergot refers to a group of fungi of the genus *Claviceps*, particularly *Claviceps purpurea*, that infects various grain crops, especially rye. These fungi produce alkaloid

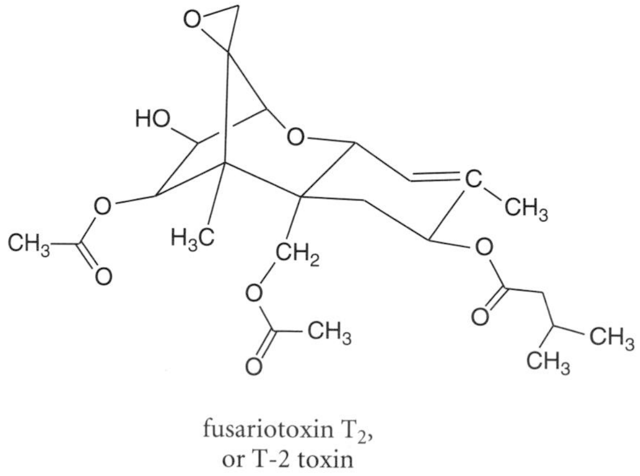

Figure 11.13. The chemical structure of fusariotoxin T2.

fusariotoxin T$_2$,
or T-2 toxin

Figure 11.14. The chemical structure of ochratoxin A from *Aspergillus ochraceus*.

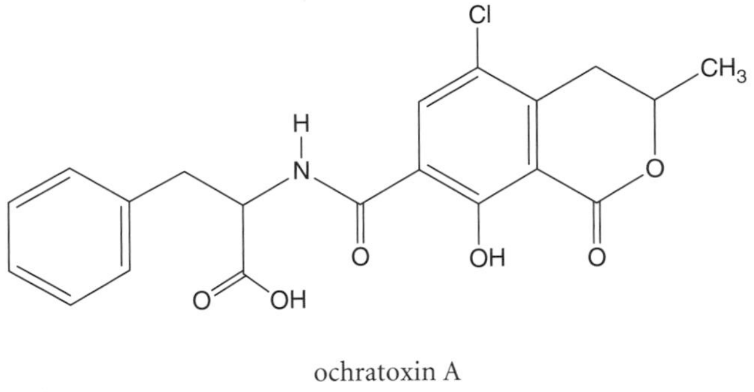

ochratoxin A

Figure 11.15. Molecular diagram of ergotamine.

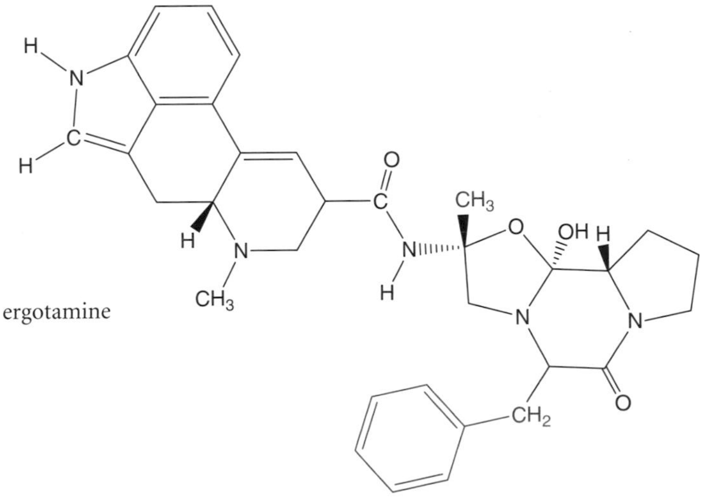

ergotamine

toxins that cause a serious disease called ergotism. Sclerotia from these fungi contain a number of these alkaloid compounds that cause a range of undesirable effects when ingested. The most important of these compounds is ergotamine, shown in Figure 11.15. Several variants of this basic molecular skeleton, including ergocristine, ergostine, and ergosine, have differing substituents at the two positions shown in red in Figure 11.15. Ergotamine is a precursor in the synthesis of lysergic acid diethylamide (LSD), and a number of the ergot alkaloids have hallucinogenic properties at higher concentrations. The effects of ergot poisoning include hallucinations and convulsions, but because these alkaloids are vasoconstrictors, ergotism also often is characterized by the development of gangrene, and the loss of limbs. Ergotism was a serious problem in the past in some northern, rye-growing regions, but has been virtually eliminated through the treatment of rye seeds with fungicides.

Mushrooms
Mushrooms are also fungi, and as is well-known, some of these fungi pose a serious risk of death if eaten. The most deadly of the common mushrooms belong to the genus *Amanita*, and contain a unique peptide toxin called **amanitin**. Several other types of mushrooms are also toxic, and these include species from the genuses

Gyromitra and *Cortinarius*. Identifying the toxic species is not always easy for untrained individuals, and cases of mushroom poisoning occur every year in North America and Europe. In general, amateurs should not gather and eat wild mushrooms, because mistaking a poisonous species for an edible one is a real possibility.

Amanitin

Virtually everyone is aware that some mushrooms are toxic, and that some are so toxic that they can quickly kill an adult human. Unfortunately, most of us do not know how to distinguish the dangerous mushrooms from the harmless (and tasty) varieties. The most deadly mushrooms belong to the genus *Amanita*, including the green "death cap" mushroom (*Amanita phalloides*), commonly found in Central Europe in late summer and autumn, and the white "destroying angel" mushroom (also called deadly agaric), *Amanita verna*, of North America, which is rare in Europe. These two species cause 95% of the fatalities from mushroom poisoning, perhaps because they are easy to confuse with the harmless *Tricholoma equestre* champignon that is popular as a food, and is commonly gathered in the wild by amateurs. Both of these deadly mushrooms contain the same toxin, which belongs to an unusual class of thermostable, bicyclic octapeptides called amatoxins, or amanitins. In these molecules, eight amino acids are linked together in a loop, with a γ-hydroxylated cysteine residue covalently linked at its γ-carbon to the indole ring of a tryptophan residue on the opposite side of the loop, creating a second covalent ring (Figure 11.16).

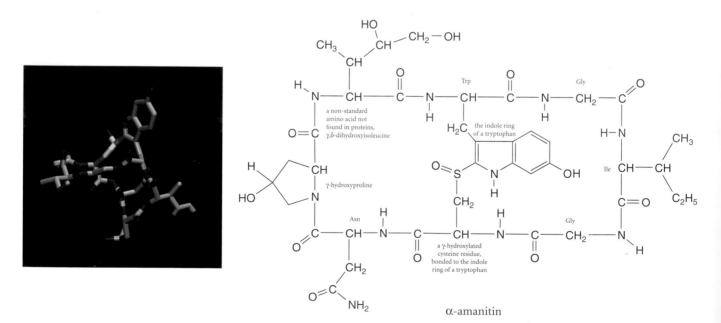

Figure 11.16. On the right is a schematic illustration of the bicyclic structure of the α-amanitin toxin (T. Wieland, *Science* 159:946–952, 1968); on the left is a "licorice" drawing of the three-dimensional structure of this molecule. (D.A. Bushnell, P. Cramer, and R.D. Kornberg, *Proc. Natl. Acad. Sci. USA* 99:1218–1222, 2002.)

One of the residues in α-amanitin, γ,δ-dihydroxyisoleucine, is a nonstandard amino acid not found in proteins, and the indole group of the Trp residue has a nonstandard hydroxyl group on its 6-carbon ring as well.

Amanitin is extremely poisonous. The LD_{50} for humans is less than 0.1 mg/kg, more toxic than cyanide, so that the 7 mg of amanitins that might be found in a single 50 g mushroom would be a lethal dose for an average adult male. In individuals who consume these mushrooms, a latency period of 10 to 24 hours ensues before severe gastrointestinal symptoms begin, which are not caused by the amanitin but rather by other factors in the mushroom. The amanitin causes irreversible and often fatal liver and kidney damage by the time these other unrelated symptoms develop. It functions as an inhibitor of RNA polymerase II, binding to the enzyme in a 1:1 stoichiometry with a binding constant of $10^{-6} M$, essentially shutting down its synthesis of RNA, thus inhibiting the production of proteins and preventing cell metabolism, leading to cell death. A crystal structure of α-amanitin bound to RNA polymerase II has been reported (Figure 11.16; Bushnell et al. 2002). Due to its inhibitory action, amanitin is useful in laboratory studies of RNA synthesis.

Other Mushroom Toxins

Other mushrooms also contain organic toxins. The false morel mushroom *Gyromitra esculenta*, found in conifer forests in Europe and North America, contains the toxin gyromitrin (Figure 11.17), which can be very toxic, even fatal, if the mushrooms are eaten raw. Nevertheless, false morel is popular as a food in Scandinavia, particularly Finland, where the gyromitrin is partially removed by drying and then boiling, followed by rinsing, because it is both volatile and water soluble.

Orellanine, also shown in Figure 11.17, is a toxin found in the webcap mushrooms of the genus *Cortinarius*, including fool's webcap, *Cortinarius orellanus*, and deadly webcap, *Cortinarius rubellus*, found in Europe. It is a bipyridine, and the charged nitrogen atoms of bipyridines are generally toxic regardless of source. They interfere with biological redox reactions through electron capture by the charged nitrogen atoms.

gyromitrin

orellanine

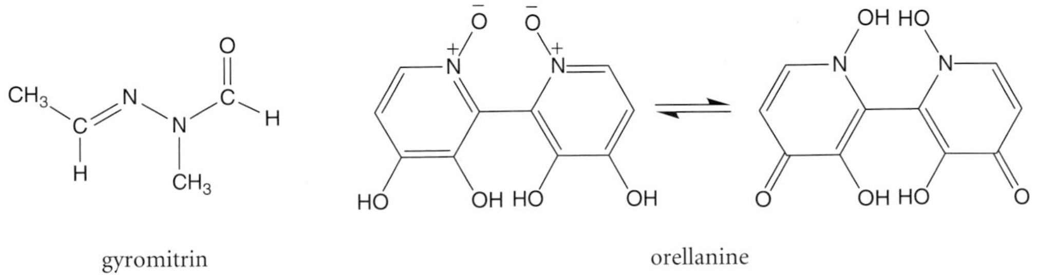

Figure 11.17. The chemical structures of gyromitrin and orellanine.

SUGGESTED READING

Concon, J.M. 1988. *Food Toxicology.* New York: Marcel Dekker.

Dabrowski, W.M., and Z.E. Sikorski. 2005. *Toxins in Food.* Boca Raton: CRC Press.

Mosby, I. 2009. " 'That won-ton soup headache': The Chinese restaurant syndrome, MSG, and the making of American food, 1968–1980." *Social History of Medicine* 22:133–151.

Nicolaou, K.C., and T. Montagnon. 2008. *Molecules That Changed the World.* Weinheim, Germany: Wiley-VCH.

Omaye, S.T. 2004. *Food and Nutritional Toxicology.* Boca Raton: CRC Press.

Scallan, E., P.M. Griffin, F.J. Angulo, R.V. Tauxe, and R.M. Hoekstra. 2011. "Foodborne illness acquired in the United States: Unspecified agents." *Emerging Infectious Diseases* 17:16–22.

Williams, A.N., and K.M. Woessner. 2009. "Monosodium glutamate 'allergy': menace or myth?" *Clinical & Experiemntal Allergy* 39:640–646.

12.
Vitamins

WHAT ARE VITAMINS?

Vitamins are essential organic nutrients that organisms require in small amounts, which they cannot synthesize endogenously and which thus must be obtained from their diet. Generally, this class of molecules is considered to exclude essential amino acids (Chapter 5) and fatty acids (Chapter 6), which are also essential organic nutrients that cannot be synthesized in sufficient quantities. Essential minerals such as iron and zinc are also excluded from this category because they are not organic molecules. Not all species require the same nutrients, so that a vitamin for one species, such as ascorbic acid, which is required by humans and many primates, but not by most other animals, is not a vitamin for these other species. Many of the vitamins are enzymatic co-factors, while two, vitamins A and D, function as hormones. Several exhibit antioxidant activities.

For humans, thirteen different vitamins are necessary: vitamins A, B_1, B_2, B_3, B_5, B_6, B_7, B_9, B_{12}, C, D, E, and K (the B vitamins are also variously written as either B1, B2, etc., or B-1, B-2, etc.; here we will use the subscript notation). Some of these vitamins are water soluble and thus easily excreted from the body; they include all of the B vitamins and vitamin C. Vitamins A, D, E, and K are fat soluble, and thus are stored in fatty tissues and are not easily removed from the body if consumed in excess. Because the water-soluble vitamins are readily excreted, a continuous daily intake is necessary, and taking large doses (as has often been recommended by certain individuals, most notably the late Linus Pauling) simply results in the excess being eliminated. In several cases, related compounds exhibit similar vitamin activity; in such cases, these different forms are referred to as **vitamers**. This does not mean, however, that all of these vitamer forms are equally potent or effective.

Because vitamins are required for the normal functioning of organisms, deficiencies in these nutrients can lead to serious physiological disorders or diseases. Unbalanced diets can cause disorders such as the scurvy suffered by sailors on long ocean voyages before the discovery of the importance of vitamin C in fresh fruits and vegetables, the beriberi caused by a lack of thiamine in some East Asian diets consisting predominantly of milled white rice, or the pellagra common in the American South in the late nineteenth and early twentieth centuries, before the discovery of the role of niacin deficiency in diets consisting largely of maize. Vitamins are abundantly present in fresh foods, however, and healthy individuals normally receive all the vitamins that they need from balanced diets and do not require supplements, with limited exceptions. In fact, evidence exists that taking significant excesses of some vitamins as supplements can be harmful, or even dangerous. As we saw in the previous chapter, any substance can be toxic when consumed at sufficiently high levels. Certain processing or storage operations can degrade the vitamin content of foods, however, and ensuring that cooked or processed foods remain as nutritious as possible requires understanding the reactions that might remove or degrade the vitamins in foods. In the United States, the Nutrition Labeling and Education Act requires that product labels disclose information about only vitamin A and C contents, in terms of the percentage of the Recommended Daily Allowance (RDA) or Recommended Daily Intake (RDI); labeling information about other vitamin contents is voluntary. We will briefly survey the various vitamin types and their basic chemistry, including their stability under typical processing and storage conditions.

FAT-SOLUBLE VITAMINS

Vitamin A

Vitamin A refers to several vitamer variants of the β-ionone-containing carotenoid component retinol, shown in Figure 12.1. As we saw in Chapter 9, **retinol** is essentially one half of the prototypical carotenoid β-carotene, and two retinol molecules can be produced from β-carotene through cleavage by the enzyme carotenoid oxygenase. Vitamin A is necessary for good vision because the aldehyde variant **retinal** produced by oxidation of retinol is the essential light-absorbing molecule in rod cells involved in vision. The 11-*cis* isomer of retinal absorbs light in the visible range and isomerizes to the all *trans* isomer shown in Figure 12.1. In the eye, retinal is bound to various proteins of the G protein-coupled receptor class called opsins, and the interactions

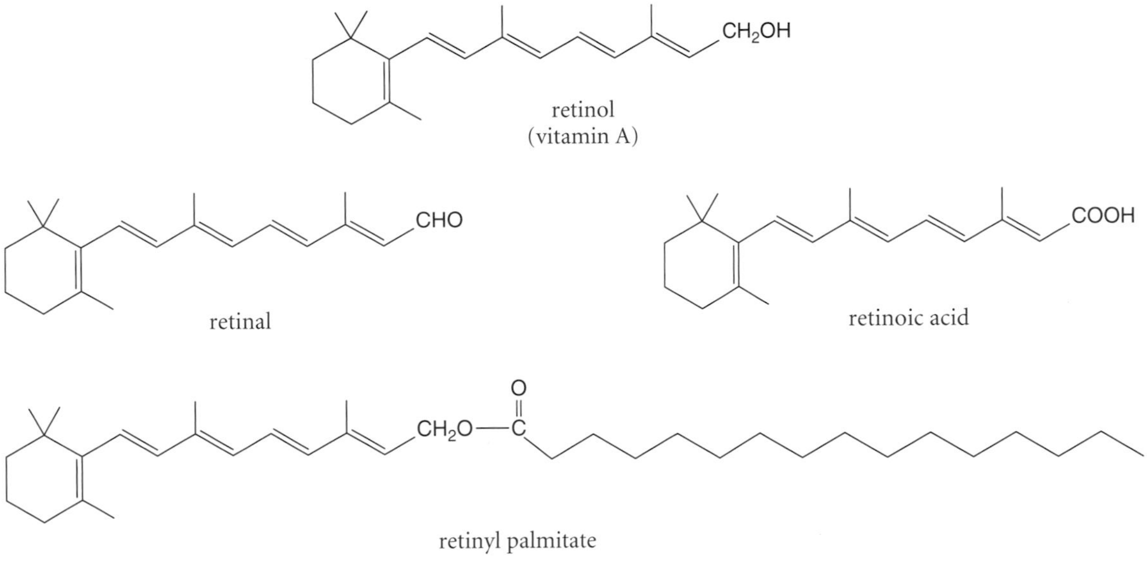

Figure 12.1. Vitamer forms of vitamin A.

of the retinal with the different opsin proteins are slightly different, causing different wavelengths of light to be absorbed in the *cis-trans* isomerization. Deficiencies of vitamin A can affect night vision and can cause a condition called xerophtalmia, or dry eye. Severe deficiencies of vitamin A can lead to much more serious disorders, and even death, as it plays a number of other physiological roles unconnected with vision.

Vitamin A is obtained in the diet from β-carotene-rich foods such as carrots, broccoli, spinach, oranges, peppers, and tomatoes, as well as liver, and to a lesser extent butter, eggs, and cheese. Raw palm oil is also rich in vitamin A. Because vitamin A is fat soluble, the vitamin A content of many fruits and vegetables is less bioavailable, particularly if not consumed with fats. For this reason, some vegetarian diets can counterintuitively produce vitamin A deficiencies. Most carotenoids do not have vitamin A activity; in addition to β-carotene, α-carotene, γ-carotene, and β-*apo*-8′-carotenal also exhibit provitamin A activity.

Vitamin A from foods of animal sources is generally found in the form of an ester linked with palmitic acid, called retinyl palmitate (Figure 12.1). Retinyl palmitate is hydrolyzed to give palmitic acid and the alcohol form of vitamin A, retinol, in the small intestine. Retinol in turn is oxidized in the body to produce retinoic acid.

Vitamin A is susceptible to destruction by oxidation in a manner similar to the oxidation of lipids, and it also functions as an antioxidant. Thus, oxidizing environments can lead to the destruction of vitamin A. Due to its absorption of light, vitamin A can be sensitive to degradation initiated by visible and UV light. It is relatively stable to heat, but it can be destroyed by fragmentation as the result of heating in cooking and thermal processing.

Vitamin A deficiency is common in the developing world and leads to vision impairment in hundreds of thousands of people who do not receive enough fresh vegetables and fruits in their diets. For such individuals, supplementation can be beneficial, although often not practical due to the expense. So-called golden rice, genetically engineered to have high levels of β-carotene, has been proposed as one possible solution to this problem, although there are many difficulties involved with this controversial approach. Conversely, excess vitamin A can be toxic. As a fat-soluble molecule, excess vitamin A is stored in fat tissues rather than excreted, and thus is difficult to eliminate. A wide range of problems can result from vitamin A toxicity, and unnecessary supplementation should be avoided. In developed nations, few people have dietary deficiencies of vitamin A.

Vitamin D

Vitamin D (Figure 12.2) belongs to a class of steroid derivatives called secosteriods (from the Latin "*secare*", to cut), because one of the steroidal rings is "cut," or broken. Two principal vitamer forms of vitamin D, called D_2 and D_3, are shown in Figure 12.2, which together are called **calciferol**. Several minor vitamer forms (called D_1, D_4, and D_5) also exist. Vitamin D_3, also called **cholecalciferol**, is synthesized in the skin from cholesterol using ultraviolet (UV) light. Because cholesterol is found only in animals, vitamin D_3 is obtained in the diet from animal foods such as meat, eggs, and fish. In the United States, milk is fortified with vitamin D_3, and margarine in the United Kingdom is also fortified to levels higher than those found in butter. Vitamin D_2 is produced in plants and fungi from ergosterol, again using UV light, and is obtained in the diet from plant foods. Vitamin D is necessary for proper calcium absorption and bone growth and development. Deficiencies in vitamin D lead to bone abnormalities, including rickets in children and osteomalacia in adults.

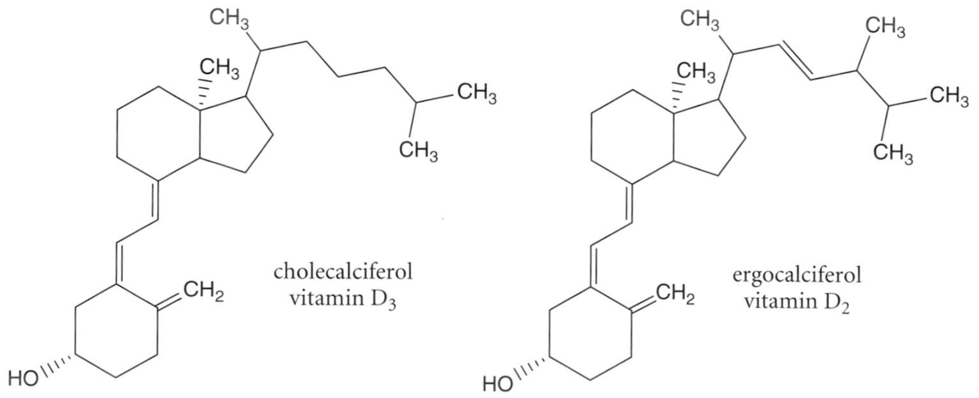

Figure 12.2. The structures of cholecalciferol, or vitamin D_3, and ergocalciferol, or vitamin D_2, the two principal vitamer forms of vitamin D.

Vitamin D_3 is produced in the skin from a precursor, 7-dehydrocholesterol, which is in turn produced enzymatically from cholesterol (Figure 12.3). In the liver, vitamin D_3 is enzymatically hydroxylated at the carbon-25 position in the long hydrocarbon chain attached to the bicyclic moiety to produce 25-hydroxycholecalciferol. This compound is in turn enzymatically converted in the kidneys to 1,25-dihydroxycholecalciferol, or calcitriol, which is the active form in the body. Like many steroid molecules, it functions in the human body as a hormone rather than as an enzyme cofactor. Vitamin D is necessary for proper calcium absorption and bone growth and development. Deficiencies in vitamin D lead to bone abnormalities, including rickets in children and osteomalacia in adults.

Exposure to ultraviolet light in the wavelength range just below 300 nm is necessary for the crucial bond-breaking step in the synthesis, so people who live at high latitudes may not receive enough exposure to light in the winter to produce adequate supplies of choleciferol. In such cases vitamin D supplements might be recommended. Severe vitamin D deficiencies in healthy individuals are rare in the United States, although mild winter deficiencies may lead to minor irritations such as dry, flaky skin.

Vitamin D is relatively stable under processing conditions but is susceptible to degradation when exposed to both heat or light. It has been shown, for example, that the vitamin D in milk can be degraded by exposure to light when packaged in clear glass bottles but not in opaque cardboard containers. It is also subject to degradation by oxidation. Vitamin D is pH stable, however, over the normal range of physiological pHs.

Vitamin E

The term vitamin E collectively refers to various vitamers classified as **tocopherols** or **tocotrienols**. At least eight active forms are known, named α-, β-, γ-, and δ-tocopherol, and α-, β-, γ-, and δ-tocotrienol (see Figure 12.4 and Table 12.1). Not all of these forms have the same biological effect; the prototype of the class is α-tocopherol, which is the most active form. Table 12.1 lists the potencies of the other forms relative to that of α-tocopherol. Dietary sources include various fruits and vegetables, such as leafy greens, cereals, seeds and nuts, and certain sea foods. As with other fat soluble vitamins, supplementation is not usually necessary, and excess intake can be dangerous.

Vitamin E functions in the body as an antioxidant, to prevent the formation of reactive oxygen species when lipids are oxidized, to protect cells and tissues from oxidative damage. It accomplishes this by reacting with lipid radicals formed by lipid peroxidation (Chapter 7) to form tocopherol radicals (Figure 12.5), which can subsequently be regenerated by further reaction with reducing agents such as ascorbate (vitamin C). Deficiencies of vitamin E can lead to various disorders, including anemia, resulting from oxidative damage to red blood cells, neurological dysfunction, and myopathies (muscular diseases).

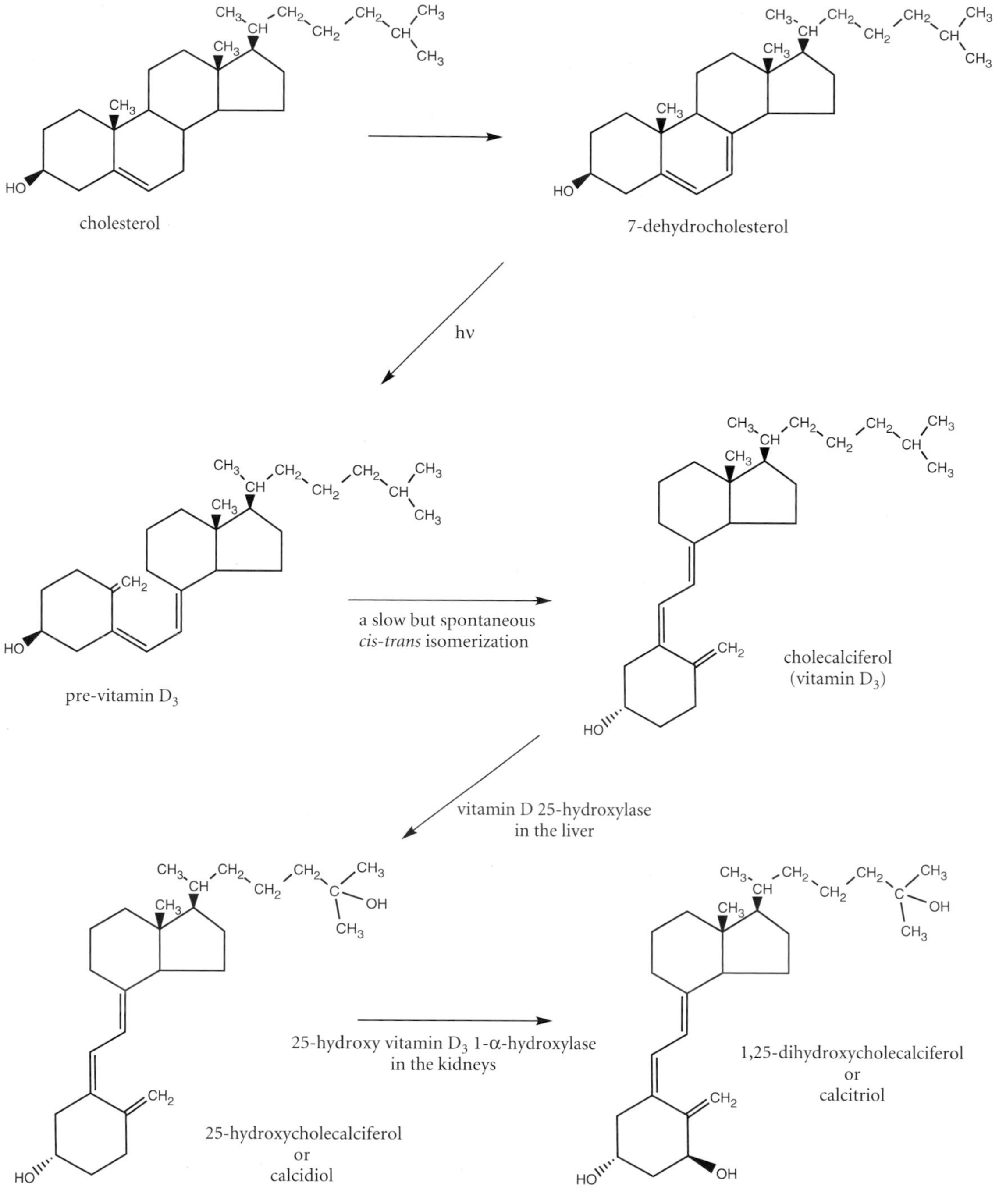

Figure 12.3. The synthetic pathway for the formation of vitamin D_3 in the skin when exposed to UV light, and its subsequent conversion to calcitriol.

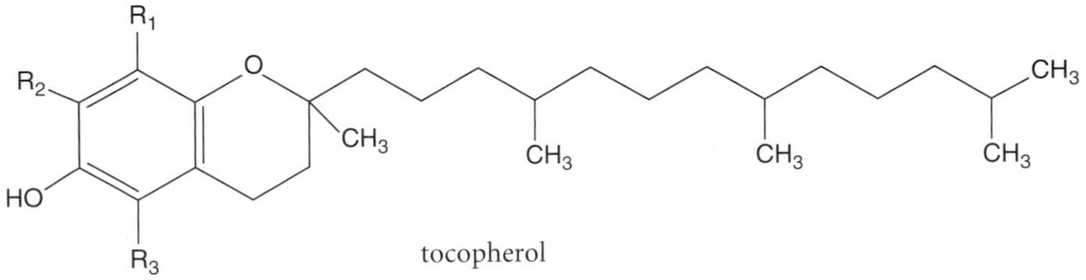

Figure 12.4. The structures of the E vitamins. See Table 12.1 for the definition of the R_n identities for the various tocopherols and tocotrienols.

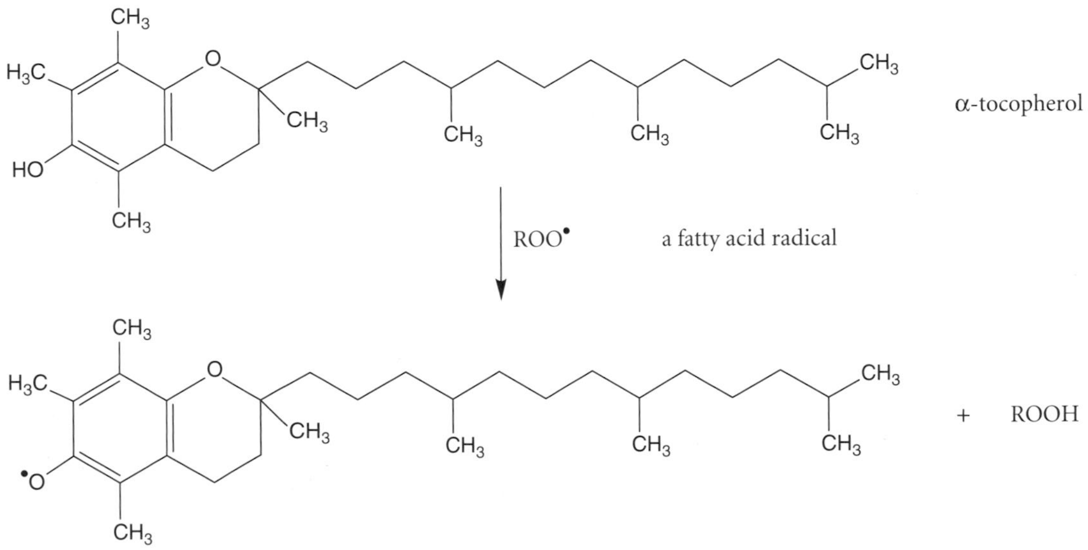

tocopherol

tocotrienol

α-tocopherol

ROO•   a fatty acid radical

+   ROOH

Figure 12.5. The reaction of tocopherol with a fatty acid radical to regenerate the fatty acid.

Table 12.1 Identities of the R_n substituents

Form	R_1	R_2	R_3	Biopotency
α	CH_3	CH_3	CH_3	100
β	CH_3	H	CH_3	10
γ	CH_3	CH_3	H	50
δ	CH_3	H	H	3

Note: Substituents in the various forms of tocopherol and tocotrienol shown in Figure 12.4, along with their biological potencies, relative to the most effective, α-tocopherol.

Vitamin E is generally stable in the absence of oxygen and oxidizing agents, as in stored canned foods; however, it is easily oxidized, particularly in deep fat fried foods. Because of its antioxidant capacity, it is often used as a food additive to prevent oxidation, and in meats to prevent the formation of *N*-nitrosamines (Chapter 5).

Vitamin K

The last major family of fat-soluble vitamins consists of the various forms of vitamin K. Several vitamers of vitamin K occur, with the two most important forms being vitamins K1 and K2, shown in Figure 12.6. All have a naphthoquinone ring, with a methyl group in the 2-position (Figure 12.6) and aliphatic chains attached to the 3-position. Vitamin K1 is obtained in the diet from various plant foods, such as cauliflower, cabbage, and leafy vegetables. Normally vitamin K2 is produced by the action of bacteria in the intestines, and as a result, deficiencies are generally rare in healthy individuals. A synthetic form, menadione, is used to treat certain cases of vitamin K deficiencies, but by prescription only, because excess consumption has been shown to be toxic. Vitamin K is important in the posttranslational carboxylation of certain proteins to give γ-carboxyglutamate residues in proteins involved in blood coagulation as well as other processes. The rare cases of deficiency can thus produce anemia, bleeding, and bruising. Longer-term deficiencies can potentially lead to heart disease and osteoporosis. Because adequate supplies are generally obtained both from the diet and are produced endogenously, the stability of vitamin K in cooking and

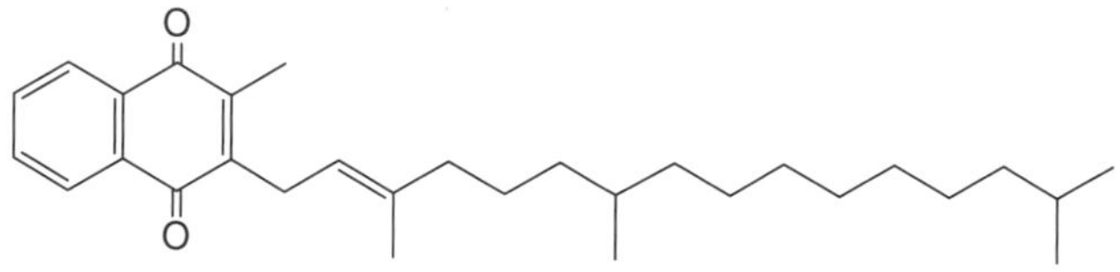

vitamin K1 - phylloquinone

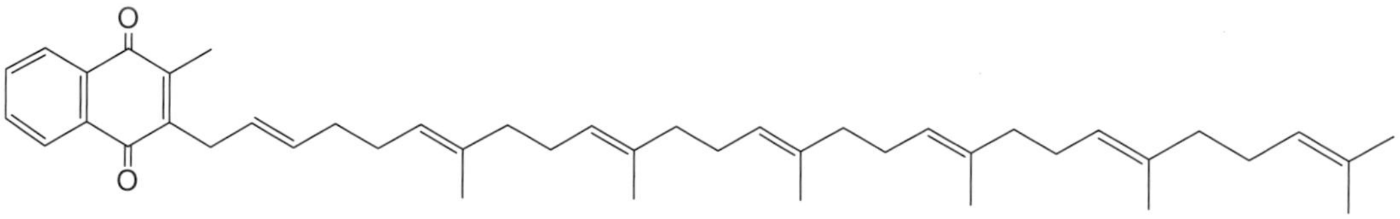

vitamin K2 - menaquinone

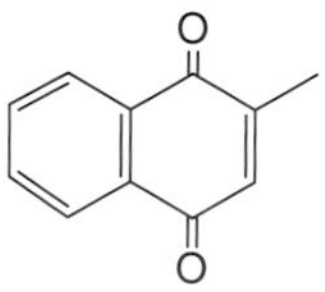

menadione - 2-methyl 1,4-naphthoquinone - synthetic vitamin K

Figure 12.6. The structures of the principal vitamin K vitamers.

processing is generally not an important issue. However, it can be sensitive to pH and oxidation.

WATER-SOLUBLE VITAMINS

B Vitamins

The majority of water soluble vitamins are classified as B vitamins for historical reasons, because they were originally thought to be a single type of molecular species. Eight forms are now recognized, vitamins B_1, B_2, B_3, B_5, B_6, B_7, B_9, B_{12}, again for historical reasons, because several molecules originally thought to be vitamins, such as *myo*-inositol, initially labeled as vitamin B_8, were subsequently shown to be produced endogenously in sufficient quantities. Similarly, the nucleotide adenine was originally thought to be a required nutrient and was labeled as vitamin B_4, but was subsequently found to be produced in adequate levels by the body. Collectively the eight B vitamins are referred to as the **vitamin B complex**. Structurally and chemically, these vitamins are quite dissimilar, and their principal unifying features, apart from historical mislabeling, are that they are water soluble, and are important in cellular metabolism.

Vitamin B_1: Thiamine

Thiamine, or thiamin, and also called vitamin B_1, is a complex molecule containing two heterocyclic rings bridged by a methylene carbon (Figure 12.7) A related vitamer, thiamine pyrophosphate, has a pyrophosphate group linked to the hydroxyl oxygen atom of the five-membered ring. This form is a cofactor for several enzymes involved in vital metabolic reactions.

For animals, thiamine is required in the diet. Cereal grains, particularly whole grains, and some meats are good sources of thiamine, as are eggs and some seeds and vegetables such as cauliflower and asparagus. A dietary deficiency of thiamine can lead to the development of a serious disease of the peripheral nervous system called beriberi, from the repeated Sinhalese word for "weakness." This malady is most common in Eastern and Southeastern Asia, in some diets based predominantly on white rice that is milled for storage stability, thus losing the thiamine normally found in the husk, bran, and germ.

The reaction of thiamine with hydrochloric acid produces thiamine hydrochloride (Figure 12.7). This variant, which crystallizes as a while powder, is used as a food additive to impart meaty flavor. Thiamine stability is influenced by temperature, pH, ionic strength, and the presence of metal ions. Thermal processing is a particular problem. Heating vegetables to just 38 °C (100 °F) can lead to thiamine losses ranging from 20 to over 90% of the total originally present, depending on the temperature and time. It is more thermostable at pHs below 5. Thiamine is relatively stable toward pH under the weakly acid conditions of most foods, but is rapidly degraded at basic pHs. Low water activities promote stability, even at 38 °C. The presence of tannins causes the formation of adduct complexes that inactivate the thiamine. Sulfating

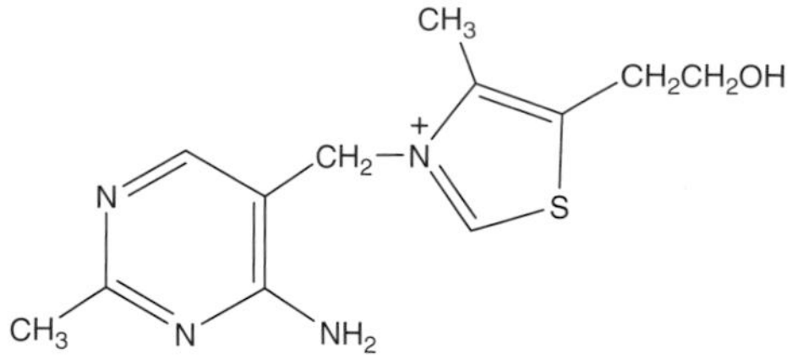

thiamine

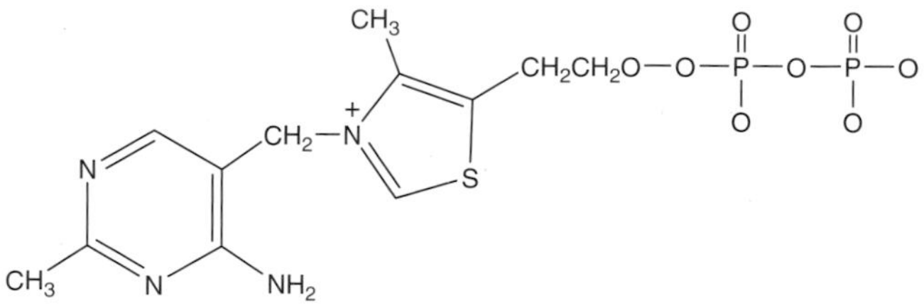

thiamine pyrophosphate

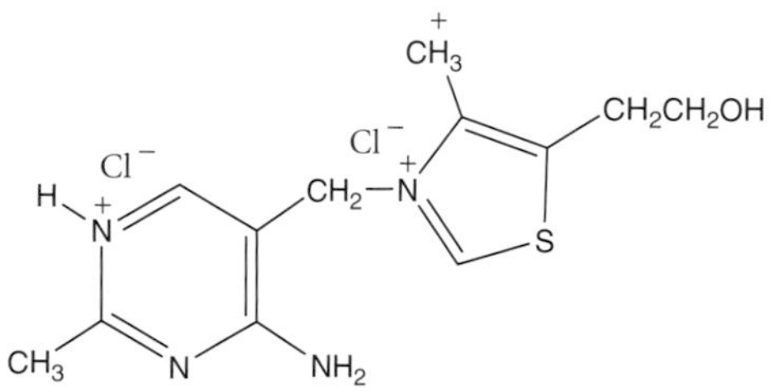

thiamine hydrochloride

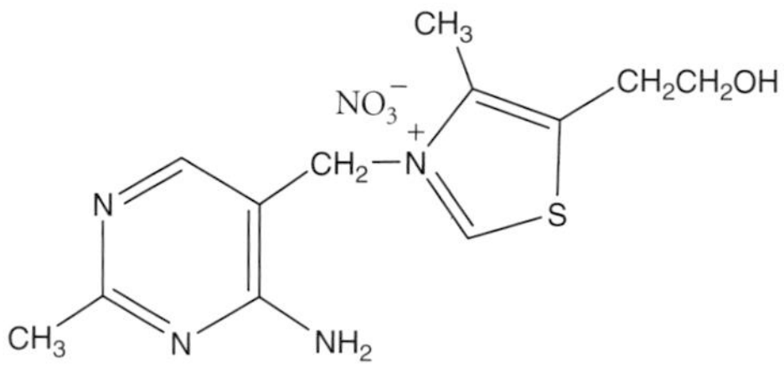

thiamine mononitrate

Figure 12.7. The structure of thiamine vitamers.

agents rapidly degrade thiamine; sulfites used as preservatives will cleave the methylene bridge between the two rings. The thiazole ring of thiamine is oxidized by various acids found in plant foods, including tannic acid, chlorogenic acid, and caffeic acid. Thiamine hydrochloride is more stable than other forms at temperatures above 95 °C, while thiamine mononitrate is more stable than other forms at temperatures below 95 °C. Like many water-soluble vitamins, B_1 is often lost in the water used in cooking.

Vitamin B_2: Riboflavin

Vitamin B_2, or **riboflavin**, is a component of the cofactor for the redox reactions catalyzed by the flavoprotein enzymes. These redox reactions are an essential part of the cellular metabolism of sugars, fats, and amino acids. Riboflavin consists of a derivative of a flavin molecule with a ribose molecule attached at its C5 carbon (Figure 12.8). Due to its extended conjugated double bond system, riboflavin is colored yellow, and, as we saw in Chapter 9 (Table 9.6), it is allowed for use in the United States as a food colorant.

Riboflavin is used in the body to synthesize the flavoprotein cofactors flavin mononucleotide (FMN) and flavin adenine dinucleotide (FAD). Riboflavin is obtained in the diet from foods such as bran, milk, eggs, meats, and cheese. It is present in smaller amounts in cereals, with more in whole grain products. Deficiencies of riboflavin are usually accompanied by deficiencies in other vitamins, but lead to

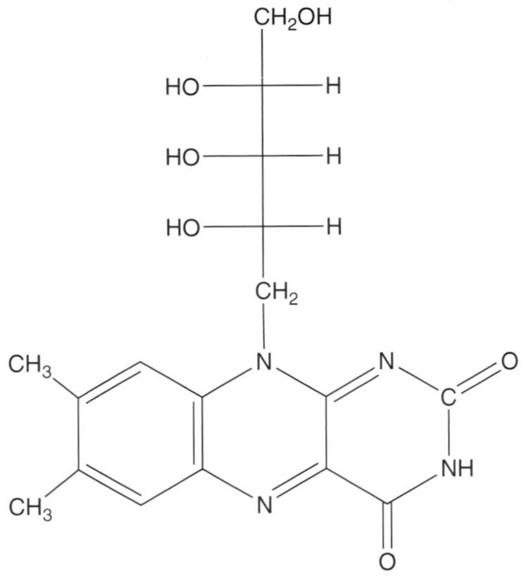

Figure 12.8. The structure of riboflavin, or vitamin B_2.

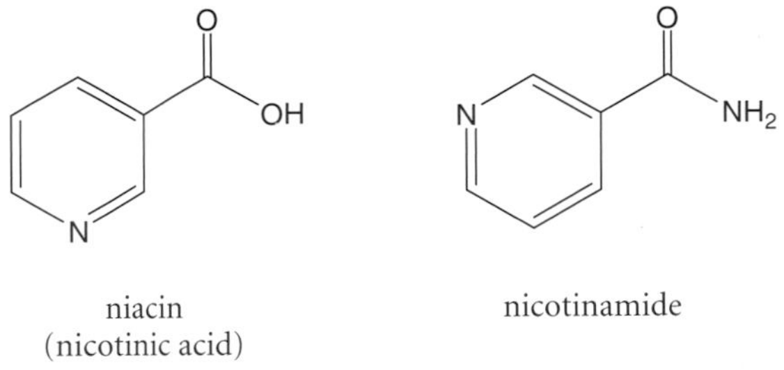

niacin
(nicotinic acid)

nicotinamide

Figure 12.9. The structures of niacin and nicotinamide.

metabolic disruption. Symptoms include skin and tissue dryness, scaling, inflammation, or cracking. Significant amounts of vitamin B_2 are lost during milling, and in the United States, white bread is routinely enriched with riboflavin.

Riboflavin is generally stable during thermal processing and is also stable under acid conditions. Like vitamin B_1, however, it is destroyed by alkaline conditions. It is also sensitive to light. Riboflavin absorbs in the blue, violet, and ultraviolet, producing its yellow-orange color. The excited state produced by this absorption of radiation can lead to the destruction of riboflavin. The products of this breakdown in milk are responsible, for example, for the "sunlight flavor" defect in milk that has been exposed to light. As with the loss of vitamin D from milk, opaque cardboard cartons substantially reduce the loss of vitamin B_2 due to light exposure.

Vitamin B_3: Niacin or Nicotinic Acid

Vitamin B_3, also called **niacin** or nicotinic acid, shown in Figure 12.9, is a simple carboxylic acid derivative (at the 3-position) of pyridine (Figure 4.68). In the body, niacin is converted to nicotinamide, also shown in Figure 12.9, which is then subsequently converted to NAD, and then NADP.

A deficiency of vitamin B_3 can cause or contribute to the severe disease pellagra, the "disease of the three Ds," which is characterized by diarrhea, dementia, and dermatitis, including severe lesions on the neck, arms, and hands. Pellagra is caused by inadequate intakes of both niacin and lysine. Historically it was most common in populations that lived primarily on maize, which, as we saw in Chapter 5, is deficient in lysine, and which is also low in niacin. In addition, the niacin that is present in corn is also generally not readily bioavailable. Pellagra was common among the rural poor in some parts of the American South in the early years of the twentieth century, before its connection to diet was determined by the US Public Health Service. As a disease of poverty, it still occurs commonly in the developing world among those subsisting on corn.

Maize was first domesticated in the New World, probably in the Valley of Mexico, and Native Americans developed a method of treatment for maize that increased the bioavailability of niacin in the corn, and which helped prevent the development of pellagra. This process, called **nixtamalization**, which is derived from a Nahuatl word, the language of the Aztecs, involves soaking and cooking the kernels of corn in an alkaline solution (the name tamales, which are made from maize treated in this fashion, derives from the same root word for corn dough). The alkaline soaking and cooking degrades the pectin and hemicellulose of the plant cell walls and loosens the pericarp, or hull, which is subsequently removed and discarded. The kernels are then washed to remove the base, which has an unpleasant taste. The resulting product is called **nixtamal**. It can be used fresh or dried and subsequently ground into a flour called masa. The preparation of **hominy** in the American South also followed a similar procedure. Southern **grits** are then prepared from dried hominy by grinding. Both procedures were adopted from Native American practices, and all nixtmalization processes improve the bioavailability of free niacin (they of course cannot increase the lysine content of the zein proteins in maize, however, but can make the proteins of maize more bioavailable).

Vitamin B₅: Pantothenic Acid

Vitamin B$_5$, or **pantothenic acid**, is an essential nutrient used in the endogenous synthesis of coenzyme-A, which is necessary for the synthesis and oxidation of fatty acids, and the oxidation of pyruvate in the citric acid cycle. Structurally it consists of the amide formed between pantoic acid and the amino acid variant β-alanine (Figure 12.10). Vitamin B$_5$ is found in many foods, particularly meats and grains, and deficiencies of this vitamin are rare.

Vitamin B$_5$ is fairly stable toward processing and storage conditions. Like nearly all of the B vitamins, pantothenic acid is subject to losses from leaching, and it can be degraded during thermal processing. It is stable at pHs between 5 and 7 and exhibits good stability at low water activities. For purposes of fortification, and in supplements, the calcium salt, calcium panthothenate, is used, due to its greater stability.

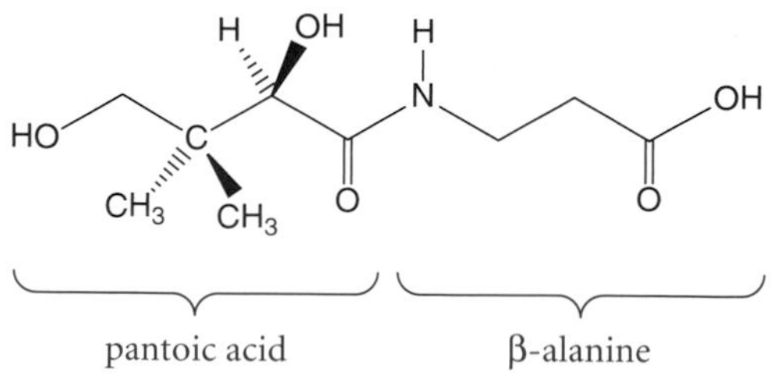

Figure 12.10. The structure of pantothenic acid, or vitamin B$_5$.

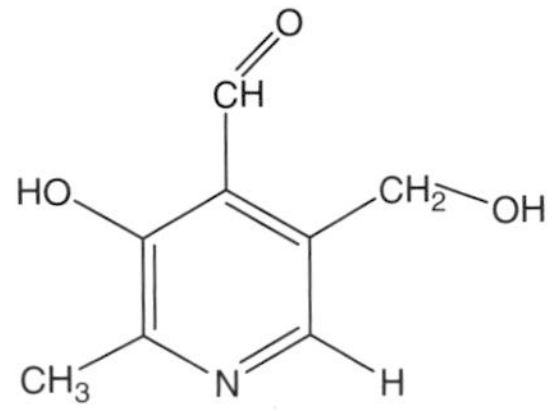

pyridoxal

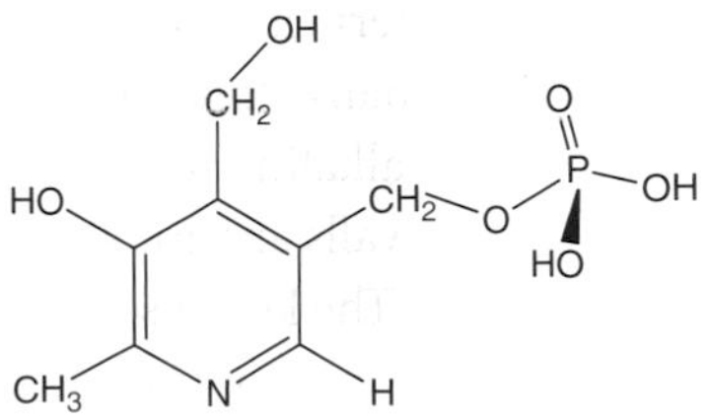

pyridoxine phosphate

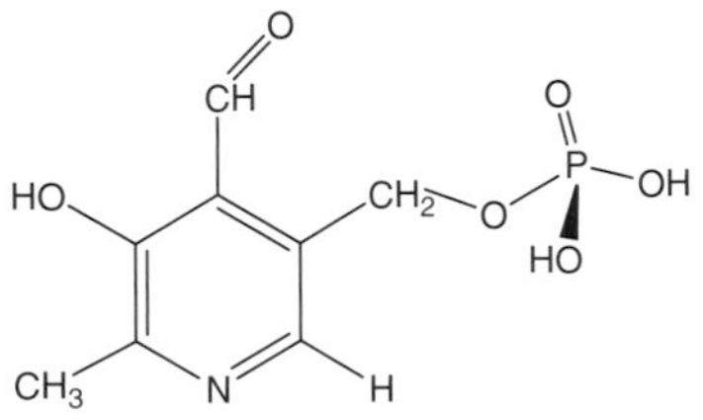

pyridoxal phosphate

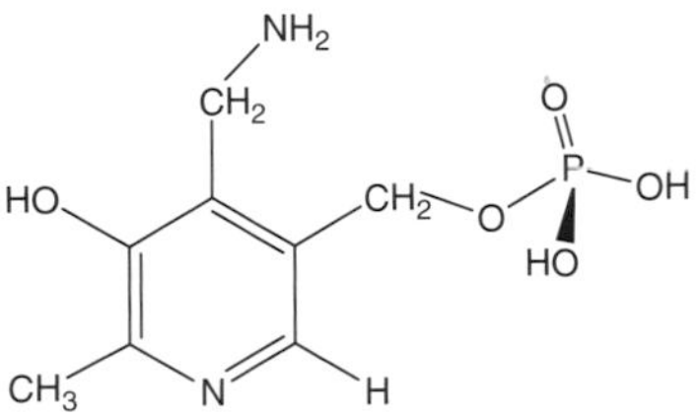

pyridoxamine phosphate

Figure 12.11. The structures of the vitamers of vitamin B_6.

Vitamin B_6

Vitamin B_6 has several basic vitamer forms, all derivatives of pyridine, and based on the **pyridoxal** molecule (Figure 12.11). The principal forms are pyridoxal and pyridoxamine; all three are commonly phosphorylated at the 5′ position, such as in pyridoxal-5′-phosphate, abbreviated as PLP, or sometimes P5P, which is the active form of vitamin B_6. It is a cofactor of amino acid decarboxylase enzymes in metabolic reactions and is important in red blood cell metabolism, and for the production of niacin and neurotransmitters such as serotonin and dopamine from tryptophan and tyrosine, respectively. Deficiencies of vitamin B_6 can thus lead to anemia and depression, dermatitis, and neural damage. Meats and fish are particularly good sources of vitamin B_6, while most fruits and many vegetables have low levels. In vegetarian diets, vitamin B_6 is best obtained from whole grains, particularly wheat germ and gluten (seitan), and some nuts.

Vitamin B_6 is susceptible to degradation from heat during cooking and processing, and during pasteurization of milk. Under such elevated temperature conditions, it can react with the thiol group of the amino acid cysteine to form an inactive thiazolidine derivative. Losses in meats during cooking, particularly at high temperatures or to "well-done" conditions, can approach 50%. As with all of the water-soluble vitamins, losses can also occur due to leaching during boiling or blanching of plant foods. The most thermostable form is pyridoxine, while pyridoxal is the most thermostable active form. Vitamin B_6 exhibits good pH stability, but is inactivated by free radicals, and is also susceptible to degradation by light.

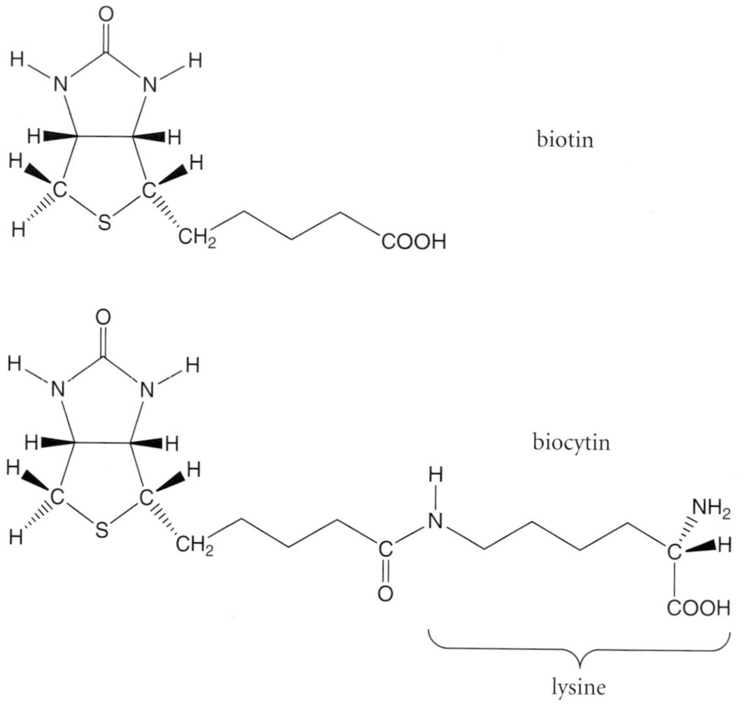

Figure 12.12. The covalent structures of biotin (vitamin B_7) and biocytin, its lysine-bound complex.

Vitamin B_7: Biotin

Vitamin B_7, or **biotin**, is a cofactor for several carboxylase enzymes involved in fatty acid synthesis and the deamination of some amino acids. Biotin is found in many foods, often binding to the lysine of proteins in a complex called biocytin (Figure 12.12). Biotin can also be synthesized by some bacteria in the intestines. Liver, egg yolks, and wheat germ and gluten are particularly rich in biotin. Deficiencies are uncommon but may be a problem for pregnant women. The protein avidin found in egg whites tightly binds biotin and renders it unavailable. Thus, consuming raw egg whites (a potentially hazardous practice if their microbial safety is not guaranteed) can decrease biotin availability. Avidin is denatured by cooking, which releases the bound biotin.

Biotin generally exhibits good processing stability. It is stable toward heat and is not degraded by cooking, and is stable toward light and exposure to oxygen. It can be degraded by extremes of pH but is generally pH stable at physiological values.

Vitamin B_9: Folic Acid

Folic acid, or vitamin B_9 (Figure 12.13), is a coenzyme factor required for the endogenous synthesis of nucleic acids, including DNA and RNA, and thus is of clear importance for cell division and growth. Deficiencies of this vitamin lead to megaloblastic anemia, and possibly serious neural tube birth defects in fetuses in the womb

Figure 12.13. The covalent structure of folic acid (vitamin B$_9$).

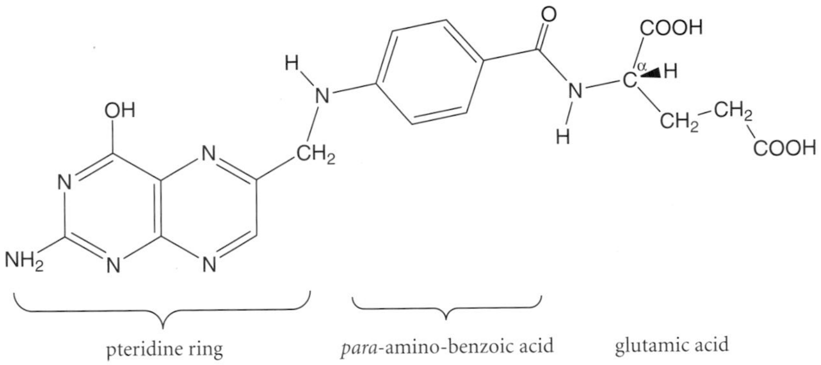

if the mother is deficient in folate. Vitamin B$_9$ is obtained in the diet from many plant foods, including green and leafy vegetables, legumes, and whole wheat, and among meats, liver is particularly rich in folate.

Folic acid is converted in the liver into dihydrofolic acid, which in turn is reduced to tetrahydrofolic (THF) acid by the enzyme dihydrofolate reductase. THF is the biologically active form of the vitamin.

Most forms of folate can be oxidized, so that the fully oxidized folic acid is the most stable form of this nutrient. Folate is destroyed by alkaline conditions and by prolonged heating. It is also destroyed by copper ions.

Because of the clear link between folate deficiency and birth defects resulting from disorders in neural tube development, many of which are extremely severe, most nutritionists suggest that pregnant women receive daily supplements of folate. In the United States, the FDA now mandates that flour and other grain products be enriched with folate supplementation. This practice is controversial, because folic acid supplements can potentially mask deficiencies in vitamin B$_{12}$ (next section; see also the discussion of homocysteine in Chapter 5), while at the same time some workers contend that the fortification levels are inadequate to prevent neural tube defects. Since the fortification regulations went into effect in the United States, however, neural tube defects have declined significantly.

Vitamin B$_{12}$

Vitamin B$_{12}$, or **cobalamin**, is also a coenzyme involved in the metabolism of amino acids and is required for normal growth and development. It is actually produced by bacteria, and neither animals nor plants have the capacity to produce this essential nutrient de novo. A number of sources are available in the diet, including meats, milk, fish, and certain fermented foods. It is not present in plant foods, however. Deficiencies in vitamin B$_{12}$ can lead to both megaloblastic anemia and neurological disorders. Often the anemia is the first and more prominent symptom observed. As we just saw, folate can prevent the development of the anemia, but it cannot prevent the development of the neurological problems. Only very small amounts are required in the diet, so the symptoms of Vitamin B_{12} deficiency require very long time spans, perhaps years, to manifest themselves. Deficiencies of this vitamin are uncommon, and generally occur in strict vegans or in those who suffer from absorption difficulties.

Vitamin B_{12} has the most complex molecular structure of all of the vitamins. In some ways, it resembles myoglobin and chlorophyll, because it consists of a metal atom complexed in the center of a tetrapyrrole ring structure. In vitamin B_{12}, the metal coordinated in the center of this heterocyclic organic ring is cobalt. Unlike the case for these other examples, however, the ring in vitamin B_{12} is not a porphyrin, because two of the pyrrole rings are directly linked, instead of being jointly bonded to an intervening carbon atom (see Figure 12.14). This type of ring is called a corrin ring. Unlike the porphyrin structures, this ring in not completely planar because the rings contain fully saturated, sp^3 hybridized carbon atoms. There are several vitamers of vitamin B_{12} that vary in their substituents on the rings of the corrin frame. Like myoglobin, vitamin B_{12} is colored red, due to the interactions of the cobalt atom with the atoms of the corrin ring. Industrially produced vitamin B_{12} supplements are sometimes in the form of cyanocobalamin, in which the sixth substituent of the cobalt atom is a cyanide group.

Vitamin B_{12} is stable toward heat at normal processing temperatures, and in the pH range from 4 to 6. It is destroyed, however, by exposure to oxygen and oxidizing agents, as well as basic conditions. Because it absorbs light in the visible range, it is susceptible to degradation from exposure to light.

Vitamin C

Vitamin C, or ascorbic acid (Figure 12.15), is a six-carbon sugar acid derivative. The hydroxyl group at C3 is resonance stabilized as a ketone, which makes it weakly acidic (Figure 12.16). It is a cofactor in a number of enzymatic reactions, including the synthesis of collagen. A deficiency of vitamin C thus can lead to a number of problems, most notably and famously, scurvy, which was a serious problem among sailors on long voyages, subsisting on diets lacking in fresh vegetables and fruits, or soldiers or civilians living under siege conditions on similar restricted diets. Humans and the other primates except prosimians are unable to synthesize vitamin C, while most other land animals are able to make sufficient amounts endogenously and thus do not require this nutrient in their diets. Adequate supplies are obtained from most vegetables and fruits in the diet, and in some meats as well, particularly liver. Presumably, when this mutation arose among our distant primate ancestors, living almost exclusively on diets of fresh fruits and vegetable matter, it was not selected against due to the dietary abundance. A normal balanced diet provides all the vitamin C needed to avoid scurvy, but there has been a long history of debate about the putative health benefits from consuming vitamin C supplements, including very large doses. In his later years, the famous American chemist Linus Pauling strongly advocated consuming megadoses of vitamin C, and it has been claimed that vitamin C supplements can help control oxidative stress, inhibit cancers, and prevent or cure the

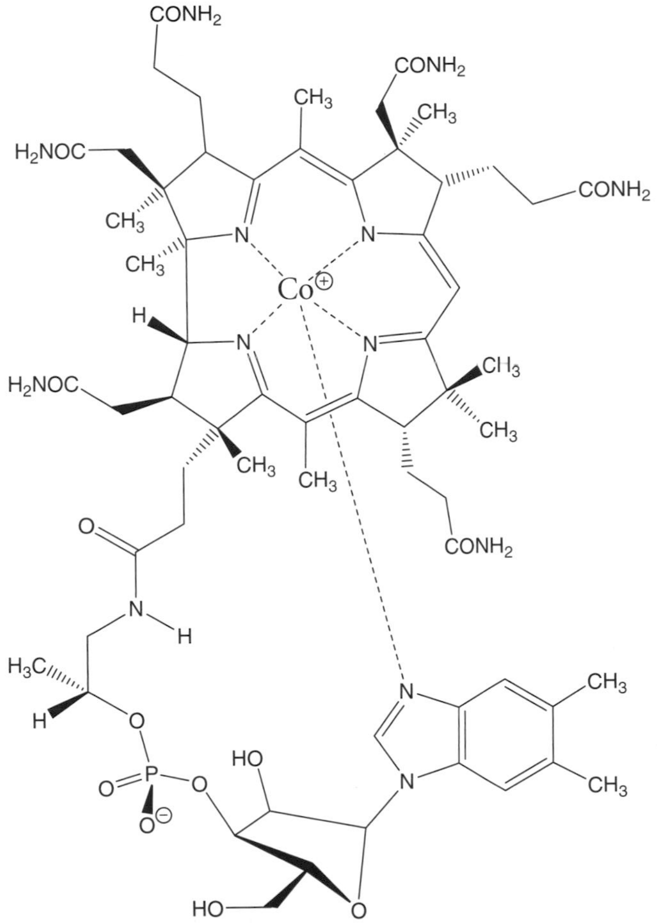

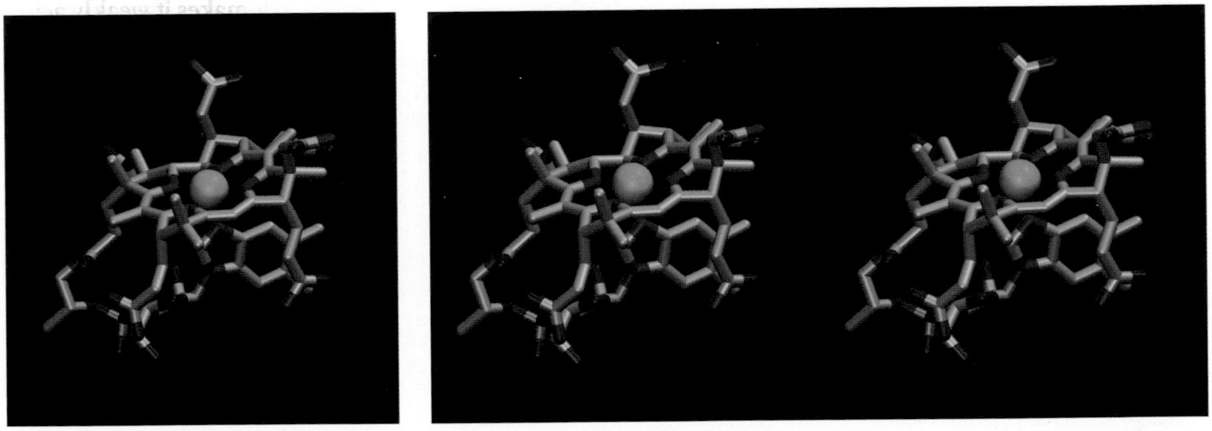

Figure 12.14. Top: the structure of vitamin B_{12}, Bottom: the molecular structure of vitamin B_{12}, as determined from X-ray studies of a complex with RNA. The central cobalt atom is shown in green, and the phosphorous atom in the lower left is shown in gold. Oxygen atoms are red, and nitrogen atoms are blue. The image on the right is a side-by-side stereo image of the same view of the molecule. (D. Sussman, J.C. Nix, and C. Wilson, 2000.)

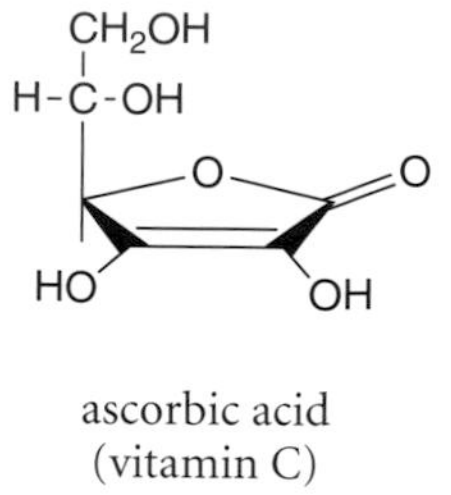
CH₂OH
H-C-OH
O
O
HO
OH
ascorbic acid
(vitamin C)

Figure 12.15. Ascorbic acid.

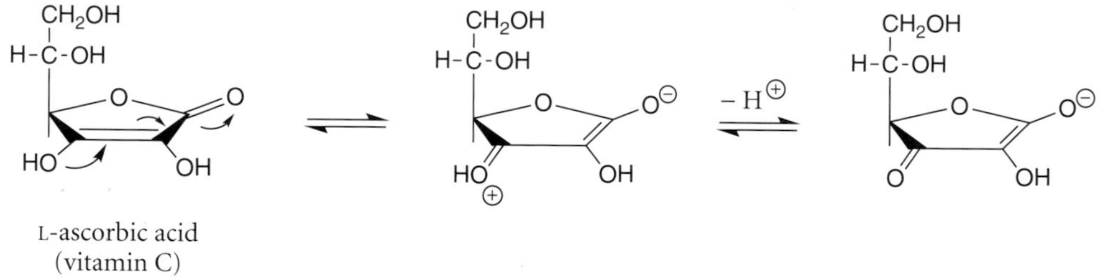
CH₂OH
H-C-OH
HO
OH
L-ascorbic acid
(vitamin C)
– H⊕
O⊖

Figure 12.16. Ascorbic acid acidity.

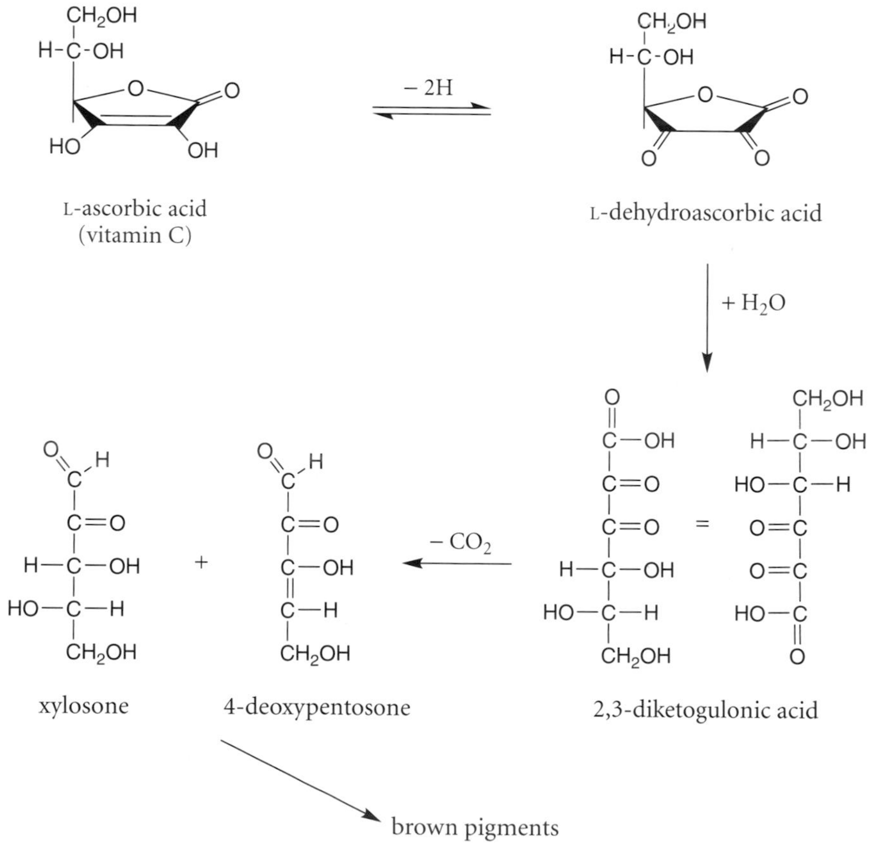
L-ascorbic acid
(vitamin C)
– 2H
L-dehydroascorbic acid
+ H₂O
2,3-diketogulonic acid
– CO₂
xylosone
4-deoxypentosone
brown pigments

Figure 12.17. Aerobic degradation of ascorbic acid.

Figure 12.18. Anaerobic degradation of ascorbic acid.

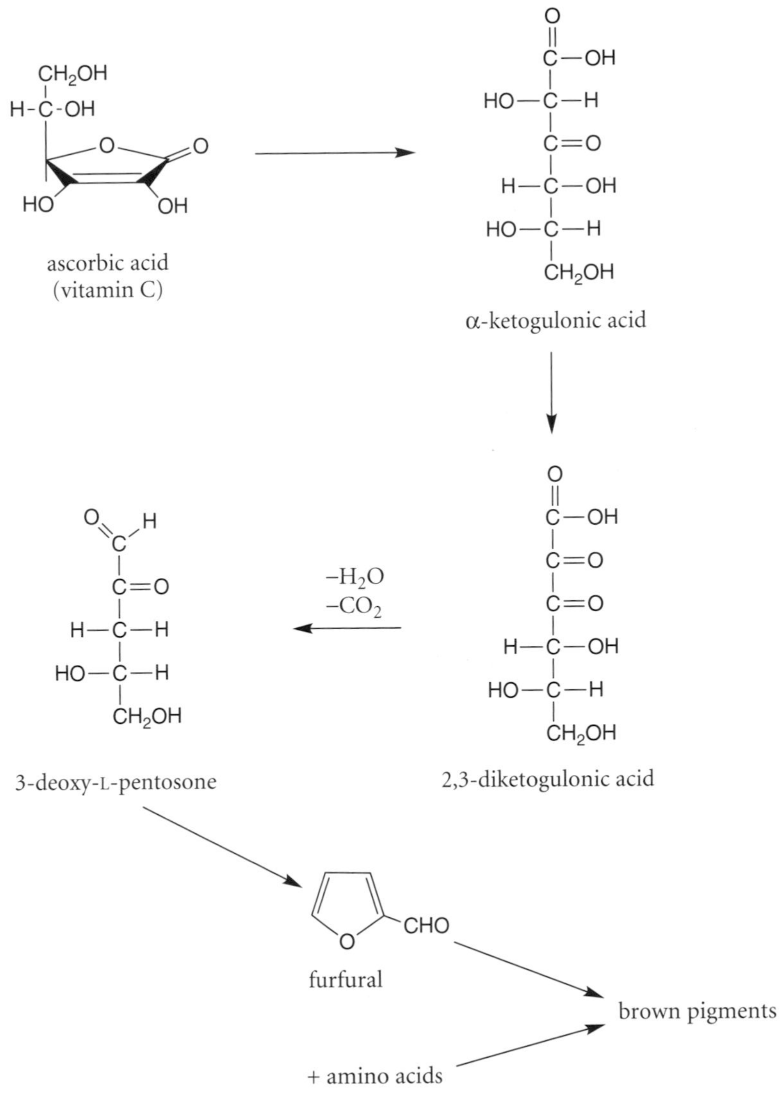

ascorbic acid
(vitamin C)

α-ketogulonic acid

2,3-diketogulonic acid

3-deoxy-L-pentosone

−H$_2$O
−CO$_2$

furfural

+ amino acids

brown pigments

common cold. Little data exists to support this latter claim. As a water soluble vitamin, excess vitamin C is rapidly removed from the blood stream by the kidneys and excreted.

The C5 carbon of ascorbic acid is chiral (Figure 12.15), and the vitamin form of ascorbic acid consists exclusively of the L enantiomer; the D-form has no vitamin activity. As we saw in Chapters 7 and 10, ascorbate also functions as a reducing agent and antioxidant, which of course is not dependent on the stereochemistry of the C5 carbon atom.

Vitamin C is the most susceptible of the vitamins to loss, degradation, or destruction by cooking and processing (Figures 12.17–12.19). It is unstable toward oxygen, heat, light, and exposure to some metals, and care must be taken to store it in the

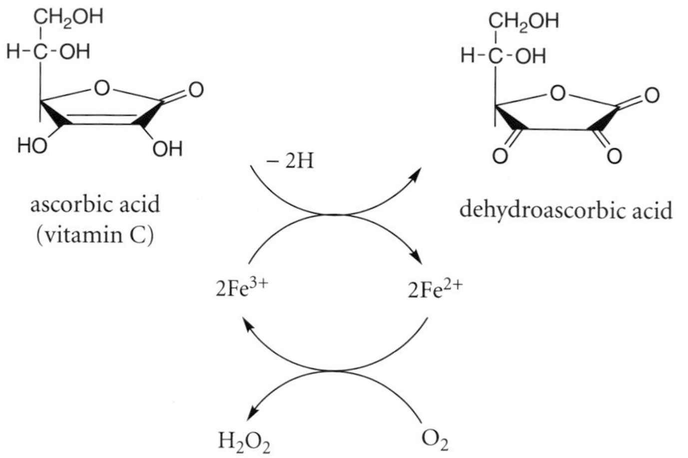

Figure 12.19. Metal catalyzed oxidation of vitamin C (ascorbic acid).

dark and at low temperatures. Ascorbic acid is readily degraded by oxidation when exposed to oxygen, even at moderate temperatures, as shown in Figure 12.17, leading to the formation of brown pigments. This reaction is a fourth type of browning in foods (we have already discussed caramelization and Maillard browning in Chapter 4, and enzymatic browning catalyzed by PPOs in Chapter 6).

SUGGESTED READING

Ball, G.F.M. 2006. *Vitamins in Foods: Analysis, Bioavailability, and Stability.* Boca Raton: CRC Press.

Combs, G.F., Jr. 2012. *The Vitamins,* 4th ed. San Diego: Academic Press.

Kamas, E., and R.S. Harris, eds. 1988. *Nutritional Evaluation of Food Processing,* 3rd ed. New York: Van Nostrand Reinhold.

Ottaway, P.B., ed. 1993. *The Technology of Vitamins in Food.* Glasgow: Blackie.

Stipanuk, M.H., and M.A. Caudill, eds. 2012. *Biochemical, Physiological, and Molecular Aspects of Human Nutrition.* 3rd ed. St. Louis: Elsevier-Saunders.

Bibliography

Achyuthan, K.E., A.M. Achyuthan, P.D. Adams, S.M. Dirk, J.C. Harper, B.A. Simmons, and A.K. Singh. 2010. "Supramolecular self-assembled chaos: Polyphenolic lignan's barrier to cost-effective lignocellulosic biofuels." *Molecules* 15:8641–8688.

Acree, T.E., J. Bernard, and D.G. Cunningham. 1984. *Food Chem.* 14:273–286.

Acree, T.E., E.H. Lavin, R. Nishida, and S. Watanabe. 1990. In *Flavour Science and Technology*, Y. Bessiere and A.F. Thomas, eds., 49–52. Chichester: John Wiley and Sons.

Adachi, M., Y. Takenaka, A.B. Gidamis, B. Mikami, and S. Utsumi. 2001. *J. Mol. Biol.* 305:291–305.

Aguzzi, A., and C. Weissmann. 1997. *Nature* 389:795–798.

Akutagawa, S. 1997. *Topics Catal.* 4:271–274.

Allen, M.S., and M.J. Lacey, 1998. In *Chemistry of Wine Flavor*, ACS Symposium Series 714, A.J. Waterhouse and S.E. Ebeler, eds. Washington: American Chemical Society.

Allevi, P., M. Anastasia, P. Ciuffreda, A. Fiecchi, A. Scala, S. Bingham, M. Muir, and J. Tyman. 1991. *J. Chem Soc. Chem. Comm.* 18:1319–1320.

Anfinsen, C.B. 1973. *Science* 181:223–230.

Anfinsen, C.B., E. Haber, M. Sela, and F.H. White. 1961. *Proc. Natl. Acad. Sci. USA* 47:1309–1314.

Angyal, S.J. 1969. *Angew. Chem. Int. Ed. Engl.* 8:157–166.

Atkins, P.W. 1984. *The Second Law*. New York: Scientific American.

Atkins, P.W. 1997. *Physical Chemistry*, 6th ed. New York: Freeman.

Atkins, P. 2003. *Atkins' Molecules*, 2nd ed. Cambridge: Cambridge University Press.

Baad-Hansen, L., B.E. Cairns, M. Ernberg, and P. Svensson. 2009. *Cephalalgia* 30:68–76.

Bachmanov, A.A., X. Li, D.R. Reed, J.D. Ohmen, S. Li, Z. Chen, M.G. Tordoff, P.J. de Jong, C. Wu, D.B. West, A. Chatterjee, D.A. Ross, and G.K. Beauchamp. 2001. *Chem. Senses* 26:925–933.

Ball, P. 2002. *Nat. Hist.* 111:64–74.

Barger, J.L., T. Kayo, J.M. Vann, E.B. Arias, J.L. Wang, T.A. Hacker, Y. Wang, D. Raederstorff, J.D. Morrow, C. Leeuwenburgh, D.B. Allison, K.W. Saupe, G.D. Cartee, R. Weindruch, and T.A. Prolla. 2008. *PloS ONE* 3(6), e2264, 10 pp.

Barham, P. 2001. *The Science of Cooking*. Berlin: Springer.

Baud, F., E. Pebaypeyroula, C. Cohenaddad, S. Odani, and M.S. Lehmann. 1993. *J. Mol. Biol.* 231:877–887.

Baur, J.A., K.J. Pearson, N.L. Price, H.A. Jamieson, C. Lerin, A. Kaira, V.V. Prabhu, J.S. Allard, G. Lopez-Lluch, K. Lewis, P.J. Pistell, S. Poosala, K.G. Becker, O. Boss, D. Gwinn, M.Y. Wang, S. Ramaswamy, K.W. Fishbein, R.G. Spencer, E.G. Lakatta, D. Le Couteur, R.J. Shaw, P. Navas, P. Puigserver, D.K. Ingram, R. de Cabo, and D.A. Sinclair. 2006. *Nature* 444:337–342.

Becalski, A., B.P.Y. Lau, D. Lewis, and S.W. Seaman. 2003. *J. Ag. Food Chem.* 51:802–808.

Becalski, A., B.P.Y. Lau, D. Lewis, S.W. Seaman, S. Hayward, M. Sahagian, M. Ramesh, and Y. Leclerc. 2004. *J. Ag. Food Chem.* 52:3801–3806.

Beja-Pereira, A., G. Luikart, P.R. England, D.G. Bradley, O.C. Jann, G. Bertorelle, A.T. Chamberlain, T.P. Nunes, S. Metodiev, N. Ferrand, and G. Erhardt. 2003. *Nat. Genet.* 35(4): 311–313.

Belitz, H.-D. and W. Grosch. 1986. *Food Chemistry*. Berlin: Springer-Verlag.

Bennion, B.J., and V. Daggett. 2003. *Proc. Natl. Acad. Sci. USA* 100(9): 5142–5147.

Bergenstråhle, M., J. Wohlert, M.E. Himmel, and J.W. Brady. 2010. *Carbohydr. Res.* 345:2060–2066.

Bhagwal, S., G. Beecher, D. Haytowitz, J. Holden, J. Dwyer, J. Peterson, S. Gebhardt, A.L. Eldridge, S. Agarwal, and D. Balentine. 2003. "Flavonoid composition of tea: comparison of black and green teas." 12th World Congress of Food Science and Technology, Chicago, IL, July 13–16, 2003.

Bhattacharyya, S., A. Borthakur, K. Pradeep, and J.K. Tobacman. 2008. *J. Nutr.* 138(3):469–475.

Blankenfeldt, W., N.H. Thoma, J.S. Wray, M. Gautel, and I. Schlichting. 2006. *Proc. Natl. Acad. Sci. USA* 103:17713–17717.

Blennow, A., S.B. Engelsen, T.H. Nielsen, L. Baunsgaard, and R. Mikkelsen. 2002. *Trends Plant Sci.* 10:445–450.

Borlaug, N.E. 2000. *Plant Physiol.* 124: 487–490.

Borenstein, B., and R.H. Bunnell. 1967. In *Advances in Food Research*, vol. 15, C.O. Chichester et al., eds. New York: Academic Press.

Braccini, I., R.P. Grasso, and S. Pérez. 1999. *Carbohydr. Res.* 317:119–130.

Braccini, I., and S. Pérez. 2001. *Biomacromolecules* 2:1089–1096.

Bradford, M.M. 1976. *Anal. Biochem.* 72:248–254.

Brouillard, R., S. Chassaing, and A. Fougerousse. 2003. *Phytochemistry* 64:1179–1186.

Brown, M.S., and J.L. Goldstein. 2006. *Science* 311:1721–1723.

Brooks, C.L., III, M. Karplus, and B.M. Pettitt. 1988. *Proteins: A Theoretical Perspective of Dynamics, Structure, and Thermodynamics.* Advances in Chemical Physics, vol. LXXI; I. Prigogine and S.A. Rice, eds. New York: Wiley-Interscience.

Brummer, Y., W. Cui, and Q. Wang. 2003. *Food Hydrocolloids* 17:229–236.

Bushnell, D.A., P. Cramer, and R.D. Kornberg. 2002. *Proc. Natl. Acad. Sci. USA* 99:1218–1222.

Calder, P.C. 2012. *J. Nutr.* 142:592S–599S.

Caldwell, J.E., F. Abildgaard, Z. Dzakula, D. Ming, G. Hellekant, and J.L. Markley. 1998. *Nature Structural Biology* 5:427–431.

Cantor, C.R., and P.R. Schimmel. 1980. *Biophysical Chemistry*. San Francisco: Freeman.

Castellan, G.W. 1983. *Physical Chemistry*, 3rd ed. Reading, MA: Addison-Wesley.

Caughey, B., and D.A. Kocisko. 2003. *Nature* 425:673–674.

Cesàro, A. 1986. "Thermodynamics of carbohydrate monomers and polymers in aqueous solution," in *Thermodynamic Data for Biochemistry and Biotechnology*, H.-J. Hinz, ed. Berlin: Springer-Verlag.

Cesàro, A., E. Russo, and V. Crescenzi. 1976. *J. Phys. Chem.* 80:335–339.

Cesàro, A., and F. Sussich. 2001. "Plasticization: The softening of materials," in *Bread Staling*, P. Chinachoti and Y. Vodovotz, eds. Boca Raton, FL: CRC Press.

Cescutti, P., C. Campa, F. Delben, and R. Rizzo. 2002. *Carb. Res.* 337:2505–2511.

Chalutz, E., J.E. DeVay, and E.C. Maxie. 1969. *Plant Physiol.* 44:235–241.

Chandler, D. 2005. "Interfaces and the driving force of hydrophobic assembly." *Nature* 437:640–647.

Chapman, G.M., E.E. Akehurst, and W.B. Wright. 1971. *J. Am. Oil Chem. Soc.* 48:824–830.

Chaudhari N., A.M. Landin, and S.D. Roper. 2000. *Nature Neuroscience* 3:113–119.

Chen, J.C., L.J. Miercke, J. Krucinski, J.R. Starr, G. Saenz, X. Wang, C.A. Spilburg, L.G. Lange, J.L. Ellsworth, and R.M. Stroud. 1998. *Biochemistry* 37:5107–5117.

Cheynier, V., M. Duenas-Paton, E. Salas, C. Maury, J.M. Souquet, P. Sami-Manchado, and H. Fulcrand. 2006. *Am. J. Enol. Vitic.* 57:298–305.

Chin, T.W., M. Loeb, and I.W. Fong. 1995. *Antimicrob. Agents Chemother.* 39(8): 1671–1675.

Chinachoti, P., and Y. Vodovotz, eds. 2001. *Bread Staling.* Boca Raton, FL: CRC Press.

Cohen, F.E., and S.B. Prusiner. 1998 *Annu. Rev. Biochem.* 67:793–819.

Collings, V.B. 1974. *Percept. Psychophys.* 16:169–174.

Coultate, T. 2009. *Food: The Chemistry of Its Components,* 5th ed. Cambridge: RSC Publishing.

Crozier, A., M.E.J. Lean, M.S. McDonald, and C. Black. 1997. *J Agric Food Chem.* 45:590–595.

Creighton, T.E. 1993. *Proteins*, 2nd ed. New York: Freeman.

Damodaran, S., K.L. Parkin, and O.R. Fennema, eds. 2007. *Fennema's Food Chemistry*, 4th ed. New York: Dekker.

Davè, V., and S.P. McCarthy. 1977. *Environ. Polym. Degrad.* 5:237–243.

Davies, H.T. 1957. *Proc. 2nd Int. Congr. Surface Activity* 1:426.

Deleault, N.R., R.W. Lucassen, and S. Supattapone. 2003. *Nature* 425:717–720.

deMan, J.D. 1999. *Principles of Food Chemistry.* Gaithersburg, MD: Aspen Publishers.

DeMarco, M.L., J. Silveira, B. Caughey, and V. Daggett. 2006. *Biochemistry* 45:15573–15582.

Diamond, R. 1974. *J. Mol. Biol.* 82:371–391.

Dickinson, E., and G. Stainsby. 1982. *Colloids in Foods.* London: Applied Science Publishers.

Donati, I., S. Holtan, Y.A. Morch, M. Borgogna, M. Dentini, and G. Skjak-Braek. 2005. *Biomacromolecules* 6:1031–1040.

Donovan, J.W., and K.D. Ross. 1973. *Biochemistry* 12:512–517.

Eisenberg, D., and W. Kauzmann. 1969. *The Structure and Properties of Water.* New York: Oxford University Press.

Elliot, S.S., N.L. Keim, J.S. Stern, K. Teff, and P.J. Havel. 2002. *Am. J. Clin. Nutr.* 76:911–922.

Fennema, O. 1996. *Food Chemistry*, 3rd ed. New York: Dekker.

Ferrell, H.M., et al. 2004. *J. Dairy Sci.* 87:1641–1674.

Floriano, W.B., N. Vaidehi, W.A. Goddard III, M.S. Singer, and G.M. Shepherd. 2000. *Proc. Natl. Acad. Sci. USA* 97:10712–10716.

Food and Nutrition Board, Institute of Medicine, National Academy of Sciences. 1996. *Food Chemicals Codex*, 4th ed. Washington: National Academy Press.

Fox, P.F., ed. 1982. *Advanced Dairy Chemistry*, vol. 2: *Lipids*. London: Chapman & Hall.

Friberg, S., ed. 1976. *Food Emulsions*. New York: Marcel Dekker.

Galitsky, N., et al. 2001. *Acta Crystallogr. Sect. D* 57:101–1109.

Ganem, B. 2002. *Chem. Eng. News* 80:6–9.

Gibson, K.D., and H.A. Scheraga. 1966. *Biopolymers* 4:709–712.

Gidley, M.J., P.J. Lillford, D.W. Rowlands, P. Lang, M. Dentini, V. Crescenzi, M. Edwards, C. Fanutti, and J.S.G. Reid. 1991. *Carbohydr. Res.* 214:299–314.

Gilman, V. 2003. *Chem. Eng. News* 81(34):34.

Gledhill, J.R., M.G. Montgomery, A.G.W. Leslie, and J.E. Walker. 2007. *Proc. Natl. Acad. Sci. USA* 104:13632–13637.

Gray, M.C., A.O. Converse, and C.E. Wyman. 2003. *Appl. Biochem. & Biotech.* 105–108:179–193.

Griffin, W.C. 1949. *J. Soc. Cosmet. Chem.* 1:311–xxx.

Griffin, W.C. 1954. *J. Soc. Cosmet. Chem.* 5:249–256.

Groves, M.R., V. Dhanaraj, M. Badasso, P. Nugent, J.E. Pitts, D.J. Hoover, T.L. Blundell. 1998. *Protein Eng.* 11:833–840.

Gunst, F., and M. Verzele. 1978. *J. Inst. Brew.* 84:291–292.

Hager, T. 2008. *The Alchemy of Air*. New York: Harmony Books.

Hall, S.E., W.B. Floriano, N. Vaidehi, and W.A. Goddard III. 2004. *Chem. Senses* 29:595–616.

Hanson, M.A., and R.C. Stevens. 2000. *Nature Structural Biology* 7:687.

Hanson, M.A., T.K. Oost, C. Sukopan, D.H. Rich, and R.C. Stevens 2000. *J. Am. Chem. Soc.* 122:11268–11269.

Hatano, K., M. Kojima, M. Tanokura, and K. Takahashi. 1996. *Biochemistry* 35:5379–5384.

Hiemenz, P.C. 1986. *Principles of Colloid and Surface Chemistry*, 2nd ed. New York: Dekker.

Hofmann, T. 2009. *Ann. N.Y. Acad. Sci.* 1170:116–125.

Holland, J.W., H.C. Deeth, and P.F. Alewood. 2005. *Proteomics* 5:990–1002.

Horne, D.S. 2006. *Curr. Opin. Colloid Interface Sci.* 11:148–153.

Huang, D.J., B.X. Ou, and R.L. Prior. 2005. *J. Agric. Food Chem.* 53:1841–1856.

Huxtable, R.J. 2001. *Reflections: Science in the Cultural Context* 1(3):141–144.

Imberty, A., and S. Peréz. 1988. *Biopolymers* 27:1205–1221.

Israelachvili, J.N. 1992. *Intermolecular and Surface Forces*, 2nd ed. London: Academic Press.

Jeffrey, G.A., and W. Saenger. 1997. *An Introduction to Hydrogen Bonding*. Oxford: Oxford University Press.

Jie, Z., L. Bang-Yao, X. Ming-Jie, L. Hai-Wei, Z. Zu-Kang, W. Ting-Song, and S.A.S. Craig. 2000. *Am. J. Clinical Nutr.* 72:1503–1509.

Kader, A.A. 1985. *HortSci.* 20:54–57.

Kamphuis, I.G., K.H. Kalk, M.B.A. Swarte, and J. Drenth. 1984. *J.Mol.Biol.* 179:233–256.

Karathanos, V.T., I. Mourtzinos, K. Yannakopoulou, and N.K. Andrikopoulos. 2007. *Food Chem.* 101:652–658.

Kato, M., K. Mizuno, A. Crozier, T. Fujimura, and H. Ashihara. 2000. *Nature* 406:956–957.

Katsuraya, K., K. Okuyama, K. Hatanaka, R. Oshima, T. Sato, and K. Matsuzaki. 2003. *Carb. Polymers* 53:183–189.

Keys, A. 1980a. *Seven Countries: A Multivariate Analysis of Death and Coronary Heart Disease*. Cambridge, MA: Harvard University Press.

Keys, A. 1980b. *Lancet* 2:603–606.

Kim, D.-O., and C. Y. Lee. 2004. *Crit. Rev. Food Sci. Nutr.* 44:253–273.

Kim, N.-H., T. Imai, M. Wada, and J. Sugiyama. 2006. *Biomacromolecules* 7:274–280.

Kim, S.H., A. de Vos, and C. Ogata. 1988. *Trends Biochem. Sci.* 13:13.

Kitagawa, M., Y. Kusakabe, H. Miura, Y. Ninomiya, and A. Hino.. 2001. *Biochem. Biophys. Res. Commun.* 283:236–242.

Kloer, D.P., S. Ruch, S. Al-Babill, P. Beyer, and G.E. Schulz. 2005. *Science* 308:267–269.

Ko, T.P., J.D. Ng and A. McPherson. 1993. *Plant Physiol.* 101:729.

Kosaka, K., and T. Yokoi. 2003. *Biol. Pharm. Bull.* 26:1620–1622.

Kovalevsky, A.Y., A.K. Katz, H.L. Carrell, L. Hanson, M. Mustyakimov, S.Z. Fisher, L. Coates, B.P. Schoenborn, G.J. Bunick, J.P. Glusker, and P. Langan. 2008. *Biochemistry* 47:7595–7597.

Kunieda, H., and K. Ohyama. 1990. *J. Colloid Interface Sci.* 136:432–439.

Kurihara, K., and L.M. Beidler. 1968. *Science* 161:1241–1243.

Kurokawa, H., B. Mikami, and M. Hirose. 1995. *J. Mol. Biol.* 254:196–207.

Kurtén, B. 1986. *How to Deep Freeze a Mammoth*. New York: Columbia University Press.

Kwok, R.H.M. 1968. *N. Engl. J. Med.* 278(14):796.

Labuza T.P. 1971. *Proc. 3rd Intl. Cong. Food Sci. Technol.*, 1970, 618. Washington, DC.

Langan, P., Y. Nishiyama, and H. Chanzy. 1999. *J. Am. Chem. Soc.* 121:9940–9946.

Langan, P., Y. Nishiyama, and H. Chanzy. 2001. *Biomacromolecules* 2:410–416.

Lawless, H.T. and H. Heymann. 1998. *Sensory Evaluation of Food: Principles and Practices*. New York: Chapman and Hall.

Ledl, F., and E. Schleicher. 1990. *Angew. Chem. Int. Ed. Engl.* 29:565–594.

Le, A., L.D. Barton, J.T. Sanders, and Q. Zhang. 2011. *J. Proteome Res.* 10:692–704.

Le Denmat, M., A. Anton, and G. Gandemar.1999. *J. Food Sci.* 64:194–197.

Lee, C.Y., J.A. McCammon, and P.J. Rossky. 1984. *J. Chem. Phys.* 80:4448–4455.

Lee, D. 2007. *Nature's Palette: The Science of Plant Color*. Chicago: University of Chicago Press.

Lee, J.-J., J.-N. Kong, H.-D. Do, D.-H. Jo, and K.-H. Kong. 2010. *Bull. Korean Chem. Soc.* 31:3830–3833.

Lee, K.W., H.J. Lee, and C.Y. Lee. 2002. *J. Nutr.* 132:785.

Lehmann, K., K. Schweimer, G. Reese, S. Randow, M. Suhr, W.M. Becker, S. Vieths, and P. Rosch. 2006. *Biochem. J.* 395:463–472.

Li, L., Y. Fang, R. Vreeker, I. Appelqvist, and E. Mendes. 2007. *Biomacromolecules* 8:464–468.

Li, X., M. Inoue, D.R. Reed, T. Huque, R.B. Puchalski, M.G. Tordoff, Y. Ninomiya, G.K. Beauchamp, and A.A. Bachmanov. 2001. *Mamm. Genome* 12:13–16.

Li, X., L. Staszewski, H. Xu, K. Durick, M. Zoller, and E. Adler. 2002. *Proc. Natl. Acad. Sci. USA* 99:4692–4696.

Liang, C., C. Liang, C.S. Ewig, T.R. Stouch, A.T. Hagler. 1994. *J. Am. Chem. Soc.* 116:3904–3911.

Licker, J.L., T.E. Acree, and T. Henick-Kling. 1998. In *Chemistry of Wine Flavor*, ACS Symposium Series 714, A.J. Waterhouse and S.E. Ebeler, eds. Washington: American Chemical Society.

Lilley, T.H., H. Linsdell, and A. Maestre. 1992. *J. Chem. Soc. Faraday Trans.* 88:2865–2870.

Lim, W.K., J. Rösgen, S.W. Englander. 2009. *Proc. Natl. Acad. Sci. USA* 106(8):2595–2600.

Lineback, D. 2012. "Acrylamide in foods: A review of the science and future considerations." *Annual Review of Food Science and Technology* 3:15–35.

Liu, R.-H. 2004. *J. Nutr.* 134:3479S–3485S.

Manu-Tawiah, W., L.L. Ammann, J.G. Sebranek, and R.A. Molins. 1991. *Food Tech.* 45:94–102.

Marenduzzo, D., K. Finan, and P.R. Cook. 2006. *J. Cell Biol.* 175:681–686.

Masuda, T., Y. Ueno, and N. Kitabatake. 2001. *J. Agric. Food Chem.* 49:4937–4941.

Mata, V.G., P.B.Gomes, and A.E. Rodrigues. 2005. *AIChE J.* 51:2834–2852.

Matthews, J.F., M. Bergenstråhle, G.T. Beckham, M.E. Himmel, M.R. Nimlos, J.W. Brady, and M.F. Crowley. 2011. *J. Phys. Chem. B* 115:2155–2166.

Matthews, J.F., G.T. Beckham, M. Bergenstråhle-Wohlert, J.W. Brady, M.E. Himmel, and M. Crowley. 2012. *Journal of Chemical Theory and Computation* 8:735–748.

Maurer, A., and D. Fengel. 1992. *Holz als Roh- und Werkstoff* 50:493.

McCammon, J.A. and S.C. Harvey. 1987. *Dynamics of Proteins and Nucleic Acids* Cambridge: Cambridge University Press.

McCann, D., A. Barrett, A. Cooper, D. Crumpler, L. Dalen, K. Grimshaw, E. Kitchin, K. Lok, L. Porteous, E. Prince, E. Sonuga-Barke, J.O. Warner, J. Stevenson. 2007. *Lancet* 370:1560–1567.

McCoy, M. 2010. *Chem. Eng. News* 88(35):15–16.

McGee, H. 2004. *On Food and Cooking: The Science and Lore of the Kitchen*, 2nd ed. New York: Scribner.

McGee, H.J., S.R. Long, and W.R. Briggs. 1984. *Nature* 308:667–668.

McMahon, D.J., and B.S. Oommen. 2008. *J. Dairy Sci.* 91:1709–1721.

McMurry, J., and R.C. Fay. 1995. *Chemistry.* Englewood Cliffs, NJ: Prentice Hall.

Mead, S., M.P.H. Stumpf, J. Whitfield, J.A. Beck, M. Poulter, T. Campbell, J.B. Uphill, D. Goldstein, M. Alpers, E.M.C. Fisher, and J. Collinge. 2003. *Science* 300:640–643.

Mercier, J.C., G. Brignon, and B. Ribadeau-Dumas. 1973. *Eur. J. Biochem.* 35:222–235.

Miean, K.H., and S. Mohamed. 2001. *J. Agric. Food Chem.* 49:3106–3012.

Millane, R.P., and T.L. Hendrixson, *Carb. Polymers* 25:245–251.

Miller, K.B., W.J. Hurst, M.J. Payne, D.A. Stuart, J. Apgar, D.S. Sweigart, and B. Ou. 2008. *J. Agric. Food Chem.* 56:8527–8533.

Momany, F.A., D.J. Sessa, J.W. Lawton, G.W. Selling, S.A.H. Hamaker, and J.L. Willett. 2006. *J. Agric. Food Chem.* 54:543–547.

Monsivais, P., M.M. Perrigue, and A. Drewnowski. 2007. *Am. J. Clin. Nutr.* 86:116–123.

Moore, W.J. 1972. *Physical Chemistry*, 4th ed. Englewood Cliffs, NJ: Prentice-Hall.

Moradian, A., and S.A. Benner. 1992. *J. Am. Chem. Soc.* 114:6980–6987.

Mosby, I. 2009. *Social History of Medicine* 22:133–151.

Mulder, H., and P. Walstra. 1974. *The Milk Fat Globule.* Farnham Royal, Bucks, England: Commonwealth Agricultural Bureaux.

Murray, S., N.J. Gooderham, A.R. Boobis, and D.S. Davies. 1988. *Carcinogenesis* 9:321–325.

Narten, A.H., M.D. Danford, and H.A. Levy. 1967. *Discuss. Faraday Soc.* 43:97–107.

Navarro-Perán, E., J. Cabezas-Herrera, F. García-Cánovas, M.C. Durrant, R.N.F. Thorneley, and J.N. Rodríguez-López. 2005. *Cancer Res.* 65:2059–2064.

Nelson G., J. Chandrashekar, M.A. Hoon, et al. 2002. *Nature* 416:199–202.

Nicolaou, K.C., and T. Montagnon. 2008. *Molecules That Changed the World.* Weinheim, Germany: Wiley-VCH.

Oates, C.G. 2001. "Bread microstructure," in *Bread Staling*, P. Chinachoti and Y. Vodovotz, eds. Boca Raton, FL: CRC Press.

Offengenden, M., M.A. Fentabil, and J. Wu. 2011. *Glycoconj. J.* 28:113–123.

Ohta, K., T. Masuda, N. Ide, and N. Kitabatake. 2008. *FEBS Journal* 276:3644–3652.

Olsen, L.R., A. Dessen, D. Gupta, S. Sabesan, J.C. Sacchettini, and C.F. Brewer. 1997. *Biochemistry* 36:15073.

Pantuck, A.J., J.T. Leppert, N. Zomorodian, W. Aronson, J. Hong, R.J. Barnard, N. Seeram, H. Liker, H.J. Wang, R. Elashoff, D. Heber, M. Aviram, L. Ignarro, and A. Belldegrun 2006. *Clin. Cancer Res.* 2:4018–4026.

Papiz, M.Z., L. Sawyer, E.E. Eliopoulis, A.C.T. North, J.B.C. Findlay, R. Sivaprasadarao, T.A. Jones, M.E. Newcomer, and P.J. Kraulis. 1986. *Nature* 324:383–385.

Photchanachal, S., A. Mehta, and N. Kitabatake. 2002. *Biosci. Biotechnol. Biochem.* 66:1635–1640.

Pike, A.C., K. Brew, and K.R. Acharya. 1996. *Structure* 4:691–703.

Pugliese, L., A. Coda, M. Malcovati and M. Bolognesi. 1993. *J. Mol. Biol.* 231:698–710.

Putnam, J.J., and J.E. Allshouse. 1999. *Food Consumption, Prices, and Expenditures.* Washington, DC: USDA, Economic Research Service.

Qin, X., L. Ren, X. Yang, F. Bai, L. Wang, P. Geng, G. Bai, and Y. Shen. 2011. *J. Struct. Biol.* 174:196–202

Raharjo, S., J.N. Sofos, and G.R. Schmidt. 1993. *J. Food Sci.* 58:921–932.

Raschke, T.M., and M. Levitt. 2005. *Proc. Natl. Acad. Sci. USA* 102:6777–6782.

Raschke, T.M., J. Tsai, and M. Levitt. 2001. *Proc. Natl. Acad. Sci. USA*, 98(11): 5965–5969.

Regenstein, J.M., and C.E. Regenstein. 1984. *Food Protein Chemistry: An Introduction for Food Scientists*. Orlando, FL: Academic Press.

Reineccius, G. 2006. *Flavor Chemistry and Technology*, 2nd ed. Boca Raton, FL: CRC Taylor & Francis.

Riek, R., S. Hornemann, G. Wider, M. Billeter, R. Glockshuber, and K. Wüthrich. 1996. *Nature* 382:180–182.

Riek, R., S. Hornemann, G. Wider, R. Glockshuber, and K. Wüthrich. 1997. *FEBS Lett.* 413:282–288.

Rizvi, S.S.H., and A.L. Benado. 1984a. *Food Tech.* 38:83–92.

Rizvi, S.S.H., and A.L. Benado. 1984b. *Drying Tech.* 2:471–502.

Rodríguez, A., C. Nerín, and R. Batlle. 2008. *J. Agric. Food Chem.* 56:6364–6369.

Robyt, J.F. 1998. *Essentials of Carbohydrate Chemistry*. New York: Springer.

Rosen, J., and K.-E. Hellenas. 2002. *Analyst* 127:880–882.

Rovner, S.L. 2006. *Chem. Eng. News* 84:30–32.

Rudd, P.M., M.R. Wormald, D.R. Wing, S.B. Prusiner, and R.A. Dwek. 2001. *Biochemistry* 40:3759–3766.

Saldana, T.M., O. Basso, R. Darden, and D.P. Sandler. 2007. *Epidemiology* 18(4):501–506.

Schaumburg, H.H., R. Byck, R. Gerstl, and J.H. Mashman. 1969. *Science* 163:826–828.

Schenk, H., and R. Peschar. 2004. *Rad. Phys. Chem.* 71:829–835.

Schiraldi, A. 2010. *Food Biophys.* 5:177–185.

Schwartzberg, H.G., and R.W. Hartel, eds. 1992. *Physical Chemistry of Foods*. New York: Dekker.

Seeram, N.P., W.J. Aronson, Y. Zhang, S.M. Henning, A. Moro, R.P. Lee, N. Sartippour, D.M. Harris, M. Rettig, M.A. Suchard, A.J. Pantuck, A. Belldegrun, and D. Heber. 2007. *J. Ag. Food Chem.* 55:7732–7737.

Seljasen, R., H. Hoftun, and G.B. Bengtsson. 2001. *J. Sci. Food Agric.* 81:54–61.

Shallenberger, R.S., and T.E. Acree. 1969. *J. Ag. Food Chem.* 17:701–703.

Shallenberger, R.S., T.E. Acree, and C.-Y. Lee. 1969. *Nature* 221:555–556.

Shimizu-Ibuka, A., Y. Morita, T. Terada, T. Asakura, K. Nakajima, S. Iwata, T. Misaka, H. Sorimachi, S. Arai, and K. Abe. 2006. *J. Mol. Biol.* 359:148–158.

Somoza, J.R., F. Jiang, L. Tong, C.H. Kang, J.M. Cho, and S.H. Kim, 1993. *J. Mol. Biol.* 234:390–404.

Song, B.K., W.T. Winter, and F.R. Taravel. 1989. *Macromolecules.* 22:2641–2644.

Song, H.K., and S.W. Suh. 1998. *J. Mol. Biol.* 275:347.

Specter, M. 2009. *Denialism: How Irrational Thinking Hinders Scientific Progress, Harms the Planet, and Threatens Our Lives*. New York: Penguin Press.

Stadelmann, W.J., and O.J. Cotterill, eds. 1997. *Egg Science*. Westport, CT: AVI Publishing.

Stein, P.E., A.G.W. Leslie, J.T. Finch, and R.W. Carrell 1991. *J. Mol. Biol.* 221:941.

Stick, R.V. 2001. *Carbohydrates: The Sweet Molecules of Life*. San Diego, CA: Academic Press.

Stillinger, F.H. 1980. "Water revisited." *Science* 209:451–457.

Stoddart, J.F. 1971. *Stereochemistry of Carbohydrates*. New York: Wiley-Interscience.

Subramanian, N., P. Venkatesh, S. Ganguli, and V.P. Sinkar. 1999. *J. Agric. Food Chem.* 47:2571–2578.

Sun, H.-J., M.-L. Cui, B. Ma, and H. Ezura. 2006. *FEBS Lett.* 580:620–626.

Sun, H.-J., H. Kataoka, M. Yano, H. Ezura. 2007. *Plant Biotech. J.* 5:768–777.

Sussman, D., J.C. Nix, and C. Wilson. 2000. *Nature Struct. Biol.* 7:53–57.

Svanberg, L., L. Ahrné, N. Lorén, and E. Windhab. 2011. *J. Food Eng.* 104:70–80.

Takahashi, Y., T. Kumano, and S. Nishikawa. 2004. *Macromolecules* 37:6827–6832.

Tareke, E., P. Rydberg, P. Karlsson, S. Eriksson, and M. Tornqvist. 2000. *Chem. Res. Toxicol.* 13:517–522.

Temussi, P. 2006. *J. Mol. Recognit.* 19:188–199.

This, H. 2006. *Molecular Gastronomy*. Translated by M.B. Debevoise. New York: Columbia University Press.

Thompson, J.R., and L.J. Banaszak. 2002. *Biochemistry* 41:9398–9409.

Tobacman, J.K. 2001. *Environ. Health Perspect.* 109(10): 983–994.

Vaeck, S.V. 1960. *Manuf. Confect.* 40:35–46.

Vega, C., and R. Mercadé-Prieto. 2011. *Food Biophysics* 6:152–159.

Voet, D., and J.G. Voet. 1995. *Biochemistry*, 2nd ed. New York: Wiley & Sons.

Walstra, P. 2003. *Physical Chemistry of Foods.* New York: Marcel Dekker.

Walstra, P., and R. Jenness. 1984. *Dairy Chemistry and Physics.* New York: Wiley-Interscience.

Wang, H., G. Cao, and R.L. Prior. 1997. *J. Agric. Food Chem.* 45:304–309.

Wenkert, E., A. Fuchs, and J.D. McChesney. 1965. *J. Org. Chem.* 30:2931–2934.

Werner, M.H., and D.E. Wemmer. 1992. *Biochemistry* 31:999.

Wieser, H. 2007. *Food Microbiol.* 24:115–119.

Williams, A.N., and K.M. Woessner 2009. *Clinical & Experimental Allergy* 39:640–646.

Willie, R.L., and E.S. Lutton. 1966. *J. Am. Oil Chem. Soc.* 43:491–496.

Wu, Y., W. Cui, N.A.M. Eskin, and H.D. Goff. 2009. *Food Res. Int.* 42:1141–1146.

Wurtman, R.J., J.J. Wurtman, M.M. Regan, J.M. McDermott 2003. *Am. J. Clin. Nutr.* 77:128–132.

Wüst, M., and A. Mosandl. 1999. *Eur. Food. Res. Technol.* 209:3–11.

Yang, F., and G.N. Phillips. 1996. *J. Mol. Biol.* 256:762–774.

Yang, F., G.P. Lim, A.N. Begum, O.J. Ubeda, M.R. Simmons, S.S. Ambegaokar, P.P. Chen, R. Kayed, C.G. Glabe, S.A. Frautschy, and G.M. Cole. 2005. *J. Biol. Chem.* 280:5892–5901.

Yang, W.H., M.A. Drouin, M. Herbert, Y. Mao, and J. Karsh. 1997. *J. Allergy Clin. Immunol.* 99:757–762.

Yoshioka, M., Y. Shimomura, and M. Suzuki. 1994. *J. Nutr.* 124:539–547.

Yoon, S.-Y., J.-N. Kong, D.-H. Jo, and K.-H. Kong. 2011. *Food Chem.* 129:1327–1330.

Zanchi, D., C. Poulain, P. Konarev, C. Tribet, and D.I. Svergun. 2008. *J. Phys.: Condens. Matter* 20(49):494224.

Zimmerman, S.S., M.S. Pottle, G. Némethy, and H.A. Scheraga. 1977. *Macromolecules* 10:1–9.

Index

Page numbers in italics indicate a figure or illustration; a page number followed by "t" indicates a table